Handbook of Research on Modeling, Analysis, and Control of Complex Systems

Ahmad Taher Azar
Faculty of Computers and Artificial Intelligence, Benha University, Benha, Egypt & College of Computer and Information Sciences, Prince Sultan University, Riyadh, Saudi Arabia

Nashwa Ahmad Kamal
Faculty of Engineering, Cairo University, Giza, Egypt

A volume in the Advances in Systems Analysis, Software Engineering, and High Performance Computing (ASASEHPC) Book Series

IGI Global
PUBLISHER of TIMELY KNOWLEDGE

Published in the United States of America by
IGI Global
Engineering Science Reference (an imprint of IGI Global)
701 E. Chocolate Avenue
Hershey PA, USA 17033
Tel: 717-533-8845
Fax: 717-533-8661
E-mail: cust@igi-global.com
Web site: http://www.igi-global.com

Library of Congress Cataloging-in-Publication Data

Names: Azar, Ahmad Taher, editor. | Kamal, Nashwa Ahmad, 1978- editor.
Title: Handbook of research on modeling, analysis, and control of complex systems / Ahmad Taher Azar and Nashwa Ahmad Kamal, editors.
Description: Hershey, PA : Engineering Science Reference, an imprint of IGI Global, [2021] | Includes bibliographical references and index. | Summary: "This book is mainly focused on the recent achievements and applications in the field of control and analysis for complex systems with a special emphasis on how to solve various control design and/or observer design problems for nonlinear systems, interconnected systems, and singular systems and involves modeling, non-ideal systems and applications, synchronization, and control for nonlinear systems, such as mechanical, electrical, electromechanical, mechatronic, and very complex systems"-- Provided by publisher.
Identifiers: LCCN 2020018652 (print) | LCCN 2020018653 (ebook) | ISBN 9781799857884 (hardcover) | ISBN 9781799857907 (ebook)
Subjects: LCSH: Automatic control--Simulation methods.
Classification: LCC TJ213 .M53745 2021 (print) | LCC TJ213 (ebook) | DDC 629.8--dc23
LC record available at https://lccn.loc.gov/2020018652
LC ebook record available at https://lccn.loc.gov/2020018653

This book is published in the IGI Global book series Advances in Systems Analysis, Software Engineering, and High Performance Computing (ASASEHPC) (ISSN: 2327-3453; eISSN: 2327-3461)

British Cataloguing in Publication Data
A Cataloguing in Publication record for this book is available from the British Library.

Advances in Systems Analysis, Software Engineering, and High Performance Computing (ASASEHPC) Book Series

Vijayan Sugumaran
Oakland University, USA

ISSN:2327-3453
EISSN:2327-3461

Mission

The theory and practice of computing applications and distributed systems has emerged as one of the key areas of research driving innovations in business, engineering, and science. The fields of software engineering, systems analysis, and high performance computing offer a wide range of applications and solutions in solving computational problems for any modern organization.

The **Advances in Systems Analysis, Software Engineering, and High Performance Computing (ASASEHPC) Book Series** brings together research in the areas of distributed computing, systems and software engineering, high performance computing, and service science. This collection of publications is useful for academics, researchers, and practitioners seeking the latest practices and knowledge in this field.

Coverage

- Parallel Architectures
- Distributed Cloud Computing
- Computer Networking
- Engineering Environments
- Network Management
- Storage Systems
- Metadata and Semantic Web
- Virtual Data Systems
- Computer Graphics
- Performance Modelling

IGI Global is currently accepting manuscripts for publication within this series. To submit a proposal for a volume in this series, please contact our Acquisition Editors at Acquisitions@igi-global.com or visit: http://www.igi-global.com/publish/.

The Advances in Systems Analysis, Software Engineering, and High Performance Computing (ASASEHPC) Book Series (ISSN 2327-3453) is published by IGI Global, 701 E. Chocolate Avenue, Hershey, PA 17033-1240, USA, www.igi-global.com. This series is composed of titles available for purchase individually; each title is edited to be contextually exclusive from any other title within the series. For pricing and ordering information please visit http://www.igi-global.com/book-series/advances-systems-analysis-software-engineering/73689. Postmaster: Send all address changes to above address.

Titles in this Series

For a list of additional titles in this series, please visit: www.igi-global.com/book-series

Impacts and Challenges of Cloud Business Intelligence
Shadi Aljawarneh (Jordan University of Science and Technology, Jordan) and Manisha Malhotra (Chandigarh University, India)
Business Science Reference • © 2021 • 300pp • H/C (ISBN: 9781799850403) • US $195.00

Artificial Intelligence Paradigms for Smart Cyber-Physical Systems
Ashish Kumar Luhach (The PNG University of Technology, Papua New Guinea) and Atilla Elçi (Hasan Kalyoncu University, Turkey)
Engineering Science Reference • © 2021 • 392pp • H/C (ISBN: 9781799851011) • US $225.00

Advancements in Model-Driven Architecture in Software Engineering
Yassine Rhazali (Moulay Ismail University of Meknes, Morocco)
Engineering Science Reference • © 2021 • 287pp • H/C (ISBN: 9781799836612) • US $215.00

Cloud-Based Big Data Analytics in Vehicular Ad-Hoc Networks
Ram Shringar Rao (Ambedkar Institute of Advanced Communication Technologies and Research, India) Nanhay Singh (Ambedkar Institute of Advanced Communication Technologies and Research, India) Omprakash Kaiwartya (School of Science and Technology, Nottingham Trent University, UK) and Sanjoy Das (Indira Gandhi National Tribal University, India)
Engineering Science Reference • © 2021 • 312pp • H/C (ISBN: 9781799827641) • US $245.00

Formal and Adaptive Methods for Automation of Parallel Programs Construction Emerging Research and Opportunities
Anatoliy Doroshenko (Institute of Software Systems, Ukraine) and Olena Yatsenko (Institute of Software Systems, Ukraine)
Engineering Science Reference • © 2021 • 279pp • H/C (ISBN: 9781522593843) • US $195.00

Agile Scrum Implementation and Its Long-Term Impact on Organizations
Kenneth R. Walsh (University of New Orleans, USA) Sathiadev Mahesh (University of New Orleans, USA) and Cherie C. Trumbach (University of New Orleans, USA)
Engineering Science Reference • © 2021 • 268pp • H/C (ISBN: 9781799848851) • US $215.00

Balancing Agile and Disciplined Engineering and Management Approaches for IT Services and Software Products
Manuel Mora (Universidad Autónoma de Aguascalientes, Mexico) Jorge Marx Gómez (University of Oldenburg, Germany) Rory V. O'Connor (Dublin City University, Ireland) and Alena Buchalcevová (University of Economics, Prague, Czech Republic)

IGI Global
PUBLISHER of TIMELY KNOWLEDGE

701 East Chocolate Avenue, Hershey, PA 17033, USA
Tel: 717-533-8845 x100 • Fax: 717-533-8661
E-Mail: cust@igi-global.com • www.igi-global.com

List of Contributors

Table of Contents

Detailed Table of Contents

to their structure, dynamics, and chaos. These three dimensions constitute the dimensionality of systems, which can be used to define and categorize all types of systems. A system dynamics model to quantify the progress and interactions among the United Nation's Sustainable Development Goals (SDG) is introduced. The benefits and limitations of a system dynamics modeling approach in this context are then discussed.

Software startups are increasingly under high pressure to deliver successful products to survive and thrive in the modern highly competitive technology market. Larger organizations with deep pockets can replicate the same business ideas used by startups with relative ease. So how does the average startup stand a chance at succeeding at this seemingly David vs. Goliath contest? This chapter looks at the available literature and identifies factors that can affect the success of software development startups. Using causal loop constructs from the field of system dynamics, the interactions among the various identified factors are visualized to reveal the dynamics of the system. The result is as a three-dimensional view of success factors in form of time, capital, and (product) differentiation. The chapter also explores the cornerstones in the context of the product and business dimensions of software startups. This research acts as groundwork for analyzing the workings of software startups and sets the stage for a more holistic study upon which further research can be carried out.

This chapter proposes a framework for integrating lean and systems thinking tools to explore the dynamics of lean implementations in manufacturing. The value of this integrated approach is in supplementing the operational level principles of lean with the strategic outlook of systems thinking to mitigate adverse impacts of operational complexity on system performance. In particular, the focus of the framework is to enable the sustainment of lean gains in the long-term, a major challenge in manufacturing settings. The application of the framework is illustrated with a case study of a lean implementation for reducing work-in-process (WIP) at a clothing manufacturer dealing with a number of operational complexities such as demand uncertainty, a high product mix and inefficient processes. The case study highlights the usefulness of system dynamics modeling in revealing counterintuitive system behaviors that could compromise the success of the lean initiative. The simulation results demonstrate the application of the framework for sustaining lean implementations in practice.

In every nation, it is important that the higher education system fulfils the community expectation for perceived quality. One aspect of improving quality is to adopt a better funding strategy for the facilities that are provided by a university. This chapter focuses on the funding strategy for Egyptian universities. The chapter provides the background to the problem distinguishing between private and public universities thenceforward describes the process, employing the causal diagrams, to prioritise the six most important facility services. The budgeting for these services are individually modelled (using both the qualitative and quantitative system dynamics) and then incorporated into a combined stock and flow model. The models allow a facility service administrator to optimise spending in a particular area and also to rank the spending between the six facilities. The models for private and public universities have the same structure but differ with the data. Results of the analysis are discussed in detail.

Higher technical education is one of the indicators of national growth. Higher technical education of the country is responsible for its global progress in technology and standard of living as technical advancement helps to make the country economically progressive as the availability of quality technical staff attracts global entrepreneurs who, in turn, increase the revenue of the nation and therefore the higher technical education system is the backbone of the country's economic progress. In the present research paper, India's higher technical education system is being studied, which is currently undergoing uneven expansion due to the implementation of the various policies responsible for this plethora of expansion. The main objectives of this study are to improve the quality of higher technical education by developing a dynamic system model by incorporating some of the important parameters of higher technical education system stakeholders to test various long-term policies that can improve the quality of higher technical education.

When different stakeholders work together, they can implement a strategy to improve the performance of the electricity industry. This chapter shows how the cooperation and reputation influence on the electricity industry. A simulation model was developed to illustrate how related variables such as altruism, reputation, and strategic alliances contribute to promote the renewable energy industry. To do this, the chapter illustrates a case study for the electricity market in Colombia. The results show a novel perspective that contributes to marketers and engineers in the analysis of the relationship between cooperation and reputation in electricity firms.

Hooman Abdollahi, School of Business and Economics, The Arctic University of Norway (UiT), Tromsø, Norway
Amsalu Tadele Alamneh, System Dynamics Group, Department of Geography, University of Bergen, Norway
Nazanin Sharif, Department of Mechanical Engineering, Université Laval, Québec, Canada

Theater is referred to as a cultural product which plays a major part in the cultural paradigm of civilized societies. Public theater is a type of performing arts in which the production receives financial support from the government. Therefore, the public theater production follows a complex trend controlled by a particular socio-economic system. The aims of this study are to identify interactions among financial and economic factors in the Iranian public theater and to find the possible threats and suggest effective policies in order to improve the position of the Iranian theater. For this purpose, the authors use system dynamics methodology to build a model that can explain or mimic the system. They present their analysis in accordance with principles of microeconomics. The analysis is conducted in three acts, namely production, distribution, and consumption. Findings reveal that the position of the Iranian public theater is undesired due to accumulated workforce, lack of distribution networks, and underconsumption. Finally, policies are suggested to overcome the shortcomings.

Sanjay Bhushan, Department of Management, Faculty of Social Sciences, Dayalbagh Educational Institute, India
Saurabh Mani, Department of Applied Economics, Faculty of Commerce, Dayalbagh Educational Institute, India

The shift in focus from hazards to underlying vulnerabilities has provided disaster managers with a richer understanding of the factors that erode the coping capacities of communities and social systems. This chapter presents ideas such as a globally shared digital platform developed as a functional web-portal branded—A.A.D.I.GYAN—augmenting action for disaster management through indigenous knowledge-gyan. Humanitarian supply chain (HSC), on the other hand, plays a central role to any developmental program meant for sustainable capacity building in the eco-communities. The chapter also highlights that HSC and community-based IK resource management can collectively promote more resilient communities promoting social, economic, and environmental equity and ethical imperatives. Towards this end, system dynamics modeling and simulation can be invoked to capture and simulate the causal dynamics of the inter linkages prevailing within the system and suggest some critical intervention strategies for policymakers.

Assessing the impact of existing and emerging mobile technologies on individuals and societies is a complex task. Various models of the telecommunication industry have been developed and differ from country to country. The study analyses the Kenyan mobile industry's governance, complexity, and resilience. It investigates the emerging notion of "resilience" as a perspective for understanding how societies can cope with, and develop from, disturbances and change. A resilience approach to governance issues shows a great deal of promise to enable a more refined understanding of the dynamics of rapid, interlinked, and multiscale change. Simulation is done of the Kenyan mobile industry. The simulation result proves that the regulator Communications Authority of Kenya (CAK) is resilient amid the continuous change in the telecom industry. This study also uses the model to analyze mobile industry policy and concludes with policy suggestions to improve the expected mobile industry services and governance for developing countries.

In this chapter, relation between IFS and IVFS are studied and reviewed along with some existing distance measures on IFSs. Then, some distance measures for IVFSs are derived from IFSs and some properties on distance measures for IVFSs are ascertained. In the end, an attempt has been made to perform pre-screening test for common diseases of patients in two scenarios under this setting. In the first scenario, for medical prescreening IFSs data are taken and converted into IVFSs while IVFSs data are taken into consideration for the same in the second scenario.

Wireless sensor networks (WSNs) may be described as a self-configured wireless networks that can be used to track physical objects or monitor environmental features, such as temperature or motion. The sensed data is then passed across the network to the main location or sink node, where the data can be processed and analyzed. Sensor nodes in WSN are fundamentally resource-constrained: they have restricted processing power, computing, space, and transmission bandwidth. Object tracking is considered as one of the major applications. However, many of the recent articles focused on object localization. In this chapter, the authors suggest an effective approach for tracking objects in WSNs. The aim is to achieve both minimal energy consumption in reporting activity and balanced energy consumption across the WSN lifetime extension of sensor nodes. Furthermore, data reliability is considered in our model. The

The integration in the isolated areas and rural sectors is a better solution for producing the electrical energy needed for applications such as pumping systems. The rural water demand for crop irrigation and domestic water supplies is increasing. For this, one of the most conceived solutions is the photovoltaic water pumping technology which has the advantage of being sustainable and respectful of the environment to supply water to rural areas. This chapter describes a robust control of a standalone photovoltaic water pumping system using induction motor drive coupled with a centrifugal hydraulic pump. The induction motor is controlled by algorithm called an adaptive nonlinear control uses a combination of the adaptive observer for rotor flux and nonlinear control technique. The variables to be controlled are the rotor speed and the rotor flux norm required to implement the nonlinear control algorithm is estimated by adaptive flux observer. Simulations are carried out in order to show the effectiveness of the drive and the robustness to parameters variations.

Atom search optimization algorithm (ASOA) has recently been explored to develop a novel algorithm for distributed optimization and control. This chapter proposes the ASOA-based design of maximum power point tracking controllers (MPPTCs) for controlling the boost converter voltage to harvest the maximum power and enhance the damping of oscillations in the output power of the photovoltaic power plants. The proposed ASOA-based MPPTCs are PI and fractional-order PI controllers. ASOA is utilized to search for optimal controller parameters by minimizing a candidate time-domain based objective function. The performance of the proposed ASOA-based MPPTCs has been compared to the MPPTCs optimized by grey wolf optimizer (GWO) to demonstrate the superior efficiency of the ASOA-based MPPTCs. Simulation results emphasis on the better performance of the proposed MPPTCs compared to MPPTCs and GWO-based PI- FOPI controllers over a wide range of operating conditions.

This chapter discusses modeling and analysis methods for fault detection and diagnosis of stator inter-turn short circuit in three-phase induction machines. dq frame was used to model the induction motor for both health and fault cases to facilitate recognition of motor current and simulate motor environment. Fault diagnosis system was designed with adaptive neuro-fuzzy inference system (ANFIS) to provide an efficient online diagnostic tool. ANFIS diagnostic tool was trained with simulated data that generated

Preface

Many real systems can be modeled using dynamic equations. With the increasing demand for high system performance, more complex systems must be tackled, as dynamic systems used for modeling reality are more and more complicated. The complexity is primarily due to non-linearity, insecurity, delay of time, the system's singularity, robust connections, stochastic processes, etc. Both scientific and control theory advances have offered the opportunity to examine complex dynamic systems. The analysis and control of complex systems are becoming more critical and have given renewed impetus to developing new techniques and skills in complex control systems.

In many fields of science, complex systems are prevalent and can be found every day and everywhere. Examples of the above include financial markets, motorway networks, telecommunications networks, world and country ecosystems, social networks, immunological systems, live humans, electrical and mechanical systems, and energy systems. Complex systems often consist of many interconnected and interactive entities with much wealthier global dynamics than can be deduced by individual entities' characteristics and behavior. In numerous natural, social, engineering, and mathematical sciences, complex systems are studied. The main aim is to increase our knowledge of general principles that apply to various scenarios. As simple as possible, models must be. The study of complex systems does not make it fascinating to study complicated models but study the complexity of unsuspected numerical simulation results. As a multidisciplinary discipline, researchers from many different backgrounds published in many scientific journals attract the study of complex systems. The literature is rapidly growing, and trying to achieve all sorts of comprehensiveness would be a hopeless task.

On the other hand, control engineering is concerned with the understanding and control of machines, processes, and industrial automation systems to provide useful economic products for society. Control engineering is based on the foundations of feedback theory and linear systems analysis. The subject is multidisciplinary since it deals with dynamic systems drawn from electrical, electronics, chemical, mechanical, aerospace, and instrumentation engineering. Control and automation, in its broadest sense, plays a fundamental role in process industries. Automatic control systems involve mathematics more than it is usual in other engineering disciplines. Even though the subject has a strong mathematical foundation, emphasis throughout the text is not on mathematical rigor or formal derivation (unless they contribute to understanding the concept). Instead, it is on the application methods associated with the analysis and design of the feedback system.

This book is intended to demonstrate how complex systems' models are constructed to provide the mathematical methods necessary to research the systems' dynamics. It aims to include several illustrative examples to indicate that the actions of apparently completely different techniques can be interpreted using modeling methods. Also, it presents a comprehensive analysis and design of various types of control systems such as intelligent control, robust control, chaos control, and adaptive control. The prerequisite for the reader is some elementary knowledge of differential equations, vector-matrix analysis, and mechanics.

The main advantage of this book is that it is multidisciplinary. It will attract many researchers working in system dynamics, optimization, control design, automation, simulations, modeling, and related fields. With broad coverage of the contents essential for analyzing and designing various modeling analysis and control problems, this book can also be used as a handy desktop reference with recent applications.

ORGANIZATION OF THE BOOK

This well-structured book consists of 23 full chapters. They are organized into two sections.

Section 1: Modeling and Optimization of Complex Systems
Section 2: Control Design and Nonlinear Dynamics

Book Features

- The book presents the dissemination of advanced research in the area of modeling, optimization, control design and nonlinear dynamics
- The book chapters are lucidly illustrated with numerical examples and simulations.
- The book chapters discuss details of advanced applications and future research areas.
- The book chapters give a good literature survey with a long list of references.

The chapters deal with the recent research problems in system dynamics, optimization, automation, control engineering, intelligent control, chaotic systems, and applications.

Summary of the Book Chapters

Section 1 contains Chapters 1-14 focusing on the area modeling and optimization of complex systems.

Chapter 1 discusses systems concerning their degree of structural and dynamic complexity and should be read in order. Together they discuss the science of simple, complex, coupled, complex, and chaotic systems in the context of sustainable development. This chapter focuses on the structural component of systems by modeling them as static cross-impact networks; frames sustainable development – referred to as anthropocentric systems – as an interconnected group of sectors, such as the economy, using a nexus approach. This chapter avoids dynamic complexity, saved for the next chapter while maintaining a system perspective. Finally, this chapter introduces networks that can complement the nexus and systems approach and how to transcode systems through a matrix.

Chapter 2 introduces a third dimension, chaos, or randomness, and that all systems can be defined according to their structure, dynamics, and chaos. Such a framework, which is called the dimensionality of systems, can define and categorize types of systems. This chapter proposes a system dynamic modeling and cross-impact analysis to look at the SDGs' spatial dimension. A case study is used to detail how various systems approaches can be used to understand different system dimensions but are related and complementary.

Chapter 3 focuses on the available literature and identifies factors that can affect software development startups' success. Software startups are increasingly under high pressure to deliver successful products to survive and thrive in the modern, highly competitive technology market. Larger organizations with deep pockets can replicate the same business ideas used by startups with relative ease. Using Causal Loop constructs from the field of System Dynamics, the interactions among the various identified factors are visualized to reveal the system's dynamics. The result is a three-dimensional view of success factors in the form of Time, Capital, and (Product) Differentiation. The chapter also explores the cornerstones in the context of the product and business dimensions of software startups. This research acts as a groundwork for analyzing software startups' workings and sets the stage for a more holistic study, upon which further analysis can be carried out.

Chapter 4 proposes a framework for integrating lean and systems thinking tools to explore the dynamics of lean implementations in manufacturing. This integrated approach's value is in supplementing the operational level principles of lean with the strategic outlook of systems thinking to mitigate adverse impacts of operational complexity on system performance. In particular, the framework's focus is to enable the sustainment of lean gains in the long-term, a major challenge in manufacturing settings. The framework's application is illustrated with a case study of a lean implementation for reducing work-in-process (WIP) at a clothing manufacturer dealing with a number of operational complexities such as demand uncertainty, a high product mix, and inefficient processes. The case study highlights the usefulness of system dynamics modeling in revealing counterintuitive system behaviors that could compromise the lean initiative's success. The simulation results demonstrate the application of the framework for sustaining lean implementations in practice.

Chapter 5 focuses on this strategy for Egyptian universities. In every nation, the higher education system must fulfill the community expectation for perceived quality. One aspect of improving quality is by adopting a better funding strategy for the facilities provided by a university. The chapter provides the background to the problem distinguishing between private and public universities. Then, it describes the process (using causal diagrams, interviews, questionnaires, and regression analysis) to prioritize the six most essential facility services. The budgeting for these services is individually modeled (using both the qualitative and quantitative System Dynamics) and then incorporated into a combined stock and flow model. The models allow a facility service administrator to optimize spending in a particular area and prioritize the expenditure between the six facilities. The models for private and public universities have the same structure but differ from the data. The results of the analysis are discussed in detail.

In Chapter 6, India's higher technical education system is being studied, which is currently undergoing uneven expansion due to the implementation of the various policies responsible for this plethora of development. This study's main objectives are to improve higher technical education quality by developing a dynamic system model to incorporate some of the essential parameters of higher technical education system stakeholders. Testing various long-term policies can improve the quality of higher technical education so that it can achieve the highest level of excellence that can be met.

Higher Technical Education is one of the indicators of national growth. Higher technical education of the country is responsible for its global progress in technology and standard of living as technological advancement helps to make the country economically progressive as the availability of quality technical staff attracts global entrepreneurs who, in turn, increase the revenue of the nation. Therefore the higher technical education system is the backbone of the country's economic progress. The technical education system is of paramount importance in generating the specialized human resources needed to build a strong nation. In India, demand for engineers has increased over the last few decades. This has resulted in an exponential quantitative expansion of technical institutions without a qualitative standardization of people of different backgrounds, which has been motivated by huge profits from these technical institutions' establishments. Despite the ongoing economic liberalization process, more and more foreign universities are also trying to find a foothold in the Indian educational space by entering joint ventures with Indian universities. However, in the current scenario, technical education suffers from poor quality outputs from major engineering graduates. Due to inadequate planning and sudden expansion of technical education, a chaotic scenario has been created. Qualitative standards are severely hindered and result in unworkable, technically qualified human resources growth. A significant increase in the number of seats across technical institutions, a poor student-faculty ratio, and a lack of industrial institution partnerships, a lack of trained and qualified teachers was seen as the main reasons for the depletion of the technical education system's quality. Defining quality in education is a major challenge as it deals with the most sensitive creation on earth of human beings. The quality of education is a vague and controversial concept. At the broadest level, the quality of education is a vague and controversial concept. At the broadest level, the quality of education can be seen as a set of elements that make up the input, process, and output of the education system and provide services that fully satisfy both internal and external strategic groups by meeting their explicit and implicit expectations. If higher education is seen as a system, any quality management program must evaluate inputs, processes, and outputs. The expansion of engineering education in countries with a fast-growing economy has increased rapidly as it directly impacts the country's economy. This is especially true in the case of India. But maintaining a balance between supply and demand is always a challenge. Unless engineering institutes produce engineers, who are employable on the market or entrepreneurs, most engineering institutes will be agents to increase the unemployed engineers' population.

Chapter 7 addresses the importance of cooperation in the electricity industry, which could affect its growth. If different stakeholders work together, they can implement a strategy to improve the electricity industry's reputation. To illustrate how to identify and model-related variables such as altruism, reputation, and strategic alliances to evaluate the electric utility industry's cooperation, this study used simulation with the system dynamics (SD) methodology. To do this, the chapter illustrates a case study for the electricity market in Colombia. The results show a novel perspective that contributes to marketers and engineers in analyzing the relationship between cooperation and reputation in electricity firms.

Chapter 8 aims to identify interactions among financial and economic factors in the Iranian public theater and find possible threats and suggest effective policies to improve the Iranian theater position. For this purpose, the authors use the System Dynamics methodology to build a model that can explain or mimic the system. The authors present an analysis following the principles of microeconomics. The study is conducted in three acts, namely production, distribution, and consumption. Findings reveal that the Iranian public theater position is undesired due to the accumulated workforce, lack of distribution network, and underconsumption. Finally, policies are suggested to overcome the shortcomings.

Chapter 9 presents a globally shared digital platform developed as a functional web-portal branded – A.A.D.I.GYAN – acronymed- Augmenting Action For Disaster Management through Indigenous Knowledge-Gyan (www.aadigyanproject.in, Dayalbagh Educational Institute, INDIA. Humanitarian Supply Chain (HSC), on the other hand, plays a central role in any developmental program meant for sustainable capacity building in the eco-communities. The paper also highlights that HSC and Community-based IK Resource management can collectively promote more resilient communities promoting social, economic, and environmental equity and ethical imperatives. Towards this end, System Dynamics modeling and simulation can be invoked to capture and simulate the inter-linkages' causal dynamics within the system and suggest some critical intervention strategies for Policymakers.

Chapter 10 analyses the Kenyan mobile industry's governance, complexity, and resilience. Assessing the impact of existing and emerging mobile technologies on individuals and societies is a complicated task. Various models of the telecommunication industry have been developed and differ from country to country. This chapter investigates the emerging notion of "resilience" as a perspective for understanding how societies can cope with and develop from disturbances and change. A resilience approach to governance issues shows a great deal of promise to enable a more refined understanding of rapid, interlinked, and multiscale change dynamics. Simulation is done in the Kenyan Mobile Industry. The simulation result proves that the regulator Communications Authority of Kenya (CAK) is resilient amid the continuous change in the telecom industry. This study also uses the model to analyze mobile industry policy and concludes with policy suggestions to improve mobile industry services and governance for developing countries.

Chapter 11 shows the relation between the intuitionistic fuzzy set (IFS) and interval-valued fuzzy set (IVFS) studied and reviewed some existing distance measures on IFSs. Some distance measures for IVFSs are then derived from IFSs and ascertain some properties on distance measures for IVFSs. In the end, an attempt has been made to perform the prescreening test for common diseases of patients in two scenarios under this setting. In the first scenario, medical prescreening IFSs data are taken and converted into IVFSs and perform medical prescreening while IVFSs data are considered for the same in the second scenario.

In general, the disease is characterized by some directly noticeable symptoms which persuade the patient to approach a medical expert. A set of clinical examinations are undertaken to identify the occurrence of a disease. In medical diagnosis, lots of variables influence the decision-making process and, consequently, differentiate the opinions of the medical experts. Due to so many factors to analyze for the diagnosis of a patient's disease makes the expert's job difficult. The uncertain and complex nature of medical documentation and information makes medical diagnosis the most demanding and exciting research area for applications of fuzzy set theory (FST) in the medical diagnosis process. In FST, assigning membership degree between 0 and 1 to each element of the discourse is too restrictive, and assigning an interval-valued is more realistic.

Consequently, an interval-valued fuzzy set (IVFS) comes into the picture. However, both IFS and IVFS don't incorporate a non-membership degree and hesitant degree. Assimilation of membership, non-membership degree, and a hesitant degree in the same framework can better communicate and process uncertainty. Therefore, the intuitionistic fuzzy set (IFS) is widely employed in real-world problems and one of the most demanding research fields nowadays.

Chapter 12 suggests a practical approach for tracking objects in Wireless Sensor Networks (WSNs). Wireless Sensor Networks (WSNs) may be described as a self-configured wireless network that can track physical objects or monitor environmental features, such as temperature or motion. The sensed data is then passed across the network to the main location or sink node, where the data can be processed and analyzed. Sensor nodes in WSN are fundamentally resource-constrained: they have restricted processing power, computing, space, and transmission bandwidth. Object tracking is considered as one of the major applications. However, many of the recent articles focused on object localization; this chapter aims to achieve minimal energy consumption in reporting activity and balanced energy consumption across the WSN lifetime extension of sensor nodes.

Furthermore, data reliability is considered in the proposed model. The chapter starts by formulating the multi-object tracking problem using 0/1 Integer Linear programming. Also, the authors adopted the Swarm Intelligence technique to solve the optimization problem.

Chapter 13 uses Evolutionary Grey wolf (GW) search optimization as a feature selection technique to improve classifier efficiency. It is also envisaged that software error detection would consider the error's nature when repairing it for remedial action instead of simply finding it either faulty or non-defective. Software bugs (or malfunctions) pose a serious threat to software developers. Many known and unknown bugs can be vulnerable to computer systems, requiring new approaches, analysis, and techniques for efficient bug identification and repair of new unseen programs later. This chapter uses Evolutionary Grey wolf (GW) search optimization as a feature selection technique to improve classifier efficiency. It is also envisaged that software error detection would consider the error's nature when repairing it for remedial action instead of simply finding it either faulty or non-defective. To address this problem, we use bug severity multi-class classification to build an efficient and robust prediction model using a multilayer perceptron (MLP), logistic regression (LR), and random forest (RF) for bug severity classification. Both tests are performed on two software error datasets, namely: Ant 1.7 and Tomcat.

Chapter 14 deals with the problem of energy storage inside an electric vehicle. The primary source of energy is based on a wireless system. This recharge tool regroups inside several components like the storage system, consisting of an ensemble of batteries and a primary power source, a unique electronic converter based on the buck-boost principle, and a coil receiver placed undo the vehicle. One or more than coil transmitters are placed on the road where the vehicle is on from the other side. Modeling all of these components and expressing their mathematical models seems interesting for defining the possible control method that can guarantee a high autonomy when the vehicle is moving. So, taking into account if the care is driving or stopped, this recharge system is studied to verify the effectiveness of this recharge system and show the relationship between the vehicle situation and the quantity of received power. Using the platform Matlab/Simulink, the results were shown and discussed.

Section 2 contains Chapters 15-23 focusing on the area control design and nonlinear dynamics.

Chapter 15 addresses the dynamic analysis and two different control strategies for synchronizing the new Colpitts oscillator topology presented to uncertainties and external disturbances. Precisely, the resulting diagrams reveal bifurcation spirals and chaos when for a specific value of the system parameters. This striking phenomenon in the system is controlled based on the appropriate method. The first (control) is the Sliding Mode Control (SMC) method. Some essential aspects of the design and implementation are considered to reach an appropriate controller for the applications. The second presents an adaptive robust tracking control strategy based on a modified polynomial observer that tends to follow exponentially the chaotic Colpitts circuits brought back to the Chua oscillator's topology (master system) with disturbances. Some requirements are derived for the exponential and asymptotic stability of adaptive robust control systems at a finite time. The proof of stability and convergence (hence, that synchronization takes place) is derived through Lyapunov's stability method. To highlight our contribution, simulation results are shown to compare the conventional polynomial observer's proposed method.

Chapter 16 proposes the synchronization and anti-synchronization of coupled unidirectional and bidirectional chaotic systems by terminal sliding mode control strategy. Unidirectional synchronization consists of setting up a chaotic drive system and a chaotic response system to synchronize the response system variables in a finite time. The unidirectional anti-synchronization consists of keeping the chaotic systems' drive and response variables in the opposite direction. In contrast, the bi-directional anti-synchronization consists of the anti-synchronization of the coupled chaotic system in both directions. Terminal sliding mode control techniques are implemented for these purposes, considering that this strategy provides faster time-to-reach while maintaining the sliding phase. Compared to other control strategies, synchronization and anti-synchronization of unidirectional and bidirectional chaotic systems are achieved while maintaining the error variable at the origin, synchronizing the case, and maintaining the error variable's desired value to achieve anti-synchronization for the unidirectional and bidirectional chaotic system. There are three chaotic systems considered for analysis and experimental purposes in this study. The first is a Lorenz, chaotic system. The second is a Rossler system for synchronization and anti-synchronization of unidirectional systems. Also, the Ikeda system is used to analyze and experiment with bi-directional anti-synchronization. The theoretical results of this study are validated with three numerical experiments to test the performance of synchronizers and anti-synchronizers. Finally, this study's respective discussions and conclusions are presented to demonstrate the theoretical and experimental analysis results obtained.

Chapter 17 considers various soft-computing techniques for control of self-balancing two-wheel mobile robot (TWMR). Initially, a mathematical model of the system was developed using Newton second law. After that, a Simulink of the proposed system was developed in Matlab Simulink environment. Two different controllers, namely fuzzy logic controller and ANFIS controller, were used to control the proposed system. Finally, a real-time model of TWMR was designed, which was controlled using the Arduino Uno microcontroller, and its results were used for the training of ANFIS controller.

Chapter 18 describes a robust control of a standalone photovoltaic water pumping system using an induction motor drive coupled with a centrifugal hydraulic pump. The integration in the isolated areas and rural sectors is a better solution for producing the electrical energy needed for applications such as pumping systems. The rural water demand for crop irrigation and domestic water supplies is increasing. For this one of the most conceived solutions is the photovoltaic water pumping technology, which can be sustainable and respectful of the environment to supply water to rural areas.

The PV panel comprises two stages of power conversion. The first stage extracts the maximum power from a solar PV array by controlling a dc–dc boost converter's duty ratio. The dc bus output voltage control fed the induction motor via the voltage source inverter. The induction motor is controlled by an adaptive nonlinear control algorithm using a combination of the adaptive observer for rotor flux and nonlinear control technique. The variables to be controlled are the rotor speed, and the adaptive flux observer estimates the rotor flux norm required to implement the nonlinear control algorithm. The parameters identified adaptively are stator and rotor resistance, which vary with motor temperature using the adaptive rotor flux observer. Simulations are carried out to show the effectiveness of the drive and the robustness to parameter variations.

Chapter 19 proposes the Atom Search Optimization Algorithm (ASOA)-based design of maximum power point tracking controllers (MPPTCs) for controlling the boost converter voltage to harvest the maximum power and enhance the damping of oscillations in the output power of the photovoltaic power plants. The proposed ASOA-based MPPTCs are PI and fractional-order PI controllers (FOPI). ASOA is utilized to search for optimal controller parameters by minimizing a candidate time-domain based objective function. The performance of the proposed ASOA-based MPPTCs has been compared to the MPPTCs optimized by Grey Wolf Optimizer (GWO), to demonstrate the superior efficiency of the ASOA-based MPPTCs. Simulation results emphasize the better performance of the proposed MPPTCs than MPPTCs and GWO-based PI- FOPI controllers over a wide range of operating conditions.

Chapter 20 discusses modeling and analysis methods for fault detection and diagnosis of stator inter-turn short circuit in three-phase induction machines. A dq frame was used to model the induction motor for both health and fault cases to facilitate recognition of motor current and simulate motor environment. Fault diagnosis system was designed with Adaptive Neuro-Fuzzy Inference System (ANFIS) to provide an efficient online diagnostic tool. ANFIS diagnostic tool was trained with simulated data that generated by induction motor healthy and faulty models. Approached tool is verified online with a motor under different loading conditions. It determines the fault severity values using the Motor Current signature Analysis (MCSA). Developed tool performance is investigated with a case study of 2 HP three-phase induction motor using Matlab/Simulink® software.

Chapter 21 discusses modeling and control of a hybrid electric vehicle. Transport systems are becoming more electrified and the major country in the world program uses electric scooters, electric bicycles, electric trains, electric busses and electric vehicles to transport them. Traditional energy resources are still declining rapidly, and the world is frightened by the transport sector's future. Several international restrictions and laws therefore limit the use of this type of energy in the transport sector. This is done by encouraging public transport and making high taxes on top-of-the-line energy cars. Numerous installations exist inside the new versions relating to electrified vehicles, and several related forms appear. The robustness and efficiency of these transport system designs are primarily associated with the internal electrical motor used and the battery's capacity.

On the other hand, energy management is a serious factor that needs to be optimized to ensure overall efficiency and profitability. With these essential criteria, this paper presents a detailed study of the key internal equipment to facilitate and improve the transport system's global energy efficiency. A unique investigation is being carried out within this work and an attempt is made to define a new specific electric motor in the propulsion system. This type of machine is described by the machine's internal design, by the implementation of the corresponding mathematical model and the corresponding electronic inverter. The related control method is also described and performed. As this electrical machine can be used for a hybrid electric vehicle, the results shown are the robustness and efficiency of this electric motor. After implementing the Matlab Simulink platform of the overall Hybrid electric vehicle blocks needed, these results were obtained, such as the framework vehicle model, the battery model, the electric machine model, and others. On the other hand, the necessary control blocks for charging the battery or controlling the vehicle's speed are carried out, implemented, and tested.

Chapter 22 discusses new chaotic discrete-time two-dimensional maps derived from simple modifications to the well-known Hénon, Lozi, Sine-sine, and Tinkerbell maps. Chaotic behavior is a term that is attributed to dynamical systems whose solutions are highly sensitive to initial conditions. This means that small perturbations in the initial conditions can lead to completely different trajectories in the solution space. These chaotic dynamical systems arise in various natural or artificial methods in biology, circuits, engineering, computer science, and more.

Numerical simulations are carried out for different parameter values and initial conditions. It is shown that the mappings either diverge to infinity or converge to attractors of many different shapes. The application to random bit generation is then considered using a collection of the proposed maps by applying a simple rule. The resulting bit generator successfully passes all statistical tests performed.

In Chapter 23, a new model of tumor growth is dynamically investigated. The model is presented in three ordinary differential equations, which describe the avascular, vascular, and metastasis tumor growth, respectively. For the investigation of system's dynamics and especially of the effect of the population of the immune cells in system's behavior, some of the most well-known tools from nonlinear theory, such as the phase portrait, the Poincaré map, the bifurcation diagram, the Kaplan-Yorke dimension and the Lyapunov exponents have been used. Interesting phenomena related to chaos theory, such as a period-doubling route to chaos, crisis phenomena, and antimonotonicity have been revealed for the first time in this model.

AUDIENCE

The book is primarily meant for researchers from academia and industry working in the research areas such as system dynamics, Control Engineering, Automation, nonlinear dynamics, Computer Science, and Information Technology. The book can also be used at the graduate or advanced undergraduate level as a textbook or major reference for mathematical modeling, computational science, numerical simulation, nonlinear dynamical systems and chaos, control systems, applied artificial intelligence, and many others.

Ahmad Taher Azar
Faculty of Computers and Artificial Intelligence, Benha University, Benha, Egypt & College of Computer and Information Sciences, Prince Sultan University, Riyadh, Saudi Arabia

Nashwa Ahmad Kamal
Faculty of Engineering, Cairo University, Giza, Egypt

ACKNOWLEDGMENT

As the editors, we hope that the chapters in this book will stimulate further research in modeling, analyzing, and controlling complex systems and utilizing them in real-world applications. We hope that this book, covering so many different aspects, will be valuable for all readers. We want to thank also the reviewers for their diligence in reviewing the chapters. Special thanks go to IGI-Global publisher, especially the book Editorial team.

Section 1
Modeling and Optimization of Complex Systems

Chapter 1
Modeling the Sustainable Development Nexus as a Complicated-Coupled System:
Cross-Impact Network Analysis

David Zelinka
https://orcid.org/0000-0003-2457-5813
Civitas Systems, USA

Bassel Daher
Energy Institute and the Institute for Science, Technology, and Public Policy, Texas A&M University, USA

ABSTRACT

This chapter discusses the universal concept of a system through the lens of system thinking and science and applies it to the concept of sustainable development. Because system science is inherently multi-disciplinary, examples and concepts are borrowed from many disciplines. Any system can be explained through multiple dimensions or components of which this chapter details 'structure' by modeling systems as static cross-impact networks; it frames sustainable development—referred to as anthropocentric systems—as an interconnected group of sectors, like an economy, using a nexus approach. As nations work toward implementing the complex and interconnected United Nation's Sustainable Development Goals (SDGs), this chapter offers methods to identify the type of interactions and quantify the strengths among them. This approach supports the process of priority setting and highlights areas in particular needed for policy coordination across different sectors.

DOI: 10.4018/978-1-7998-5788-4.ch001

SYSTEMS SCIENCE: AN INTRODUCTION

Systems Thinking

Systems science is a meta-science in that it applies universally to all fields (Mobus & Kalton, 2015). As such, it is also a mindset and a perspective – a philosophy of sorts – that one holds. It is not just an approach, a tool, or a methodology; systems science is the science of thinking, and more specifically, how to strategize for problems. The application of systems science to thinking is systems thinking. Therefore, systems thinking is the application of a universal understanding and is directly based on scientific principles. Systems thinking provides a conceptually easy way to introduce systems science and more complex topics. Before the science of systems can be discussed, it seems only appropriate that thinking with systems be explored. Systems thinking is not particularly that difficult or involved, and even just being open-minded is a part of systems thinking.

Systems thinking is a mindset so that one's perception of the world accounts for its inherent complexity. It opposes reductionism and linear thinking – the outdated worldview dominating scientific reasoning over the last 500 years – which is no longer sufficient to solve contemporary complex, ill-defined, and messy problems, namely rapid population growth, conflict, environmental degradation, and climate change among others. From a letter published in the New York Times (1946), Albert Einstein wrote that "a new type of thinking is essential if [humanity] is to survive and move toward higher levels." The letter was meant to address the challenges of the newly invented atomic weapons, but the quote fits any complex issue today. Systems thinking is that new type of thinking that can provide the overall approach for genuinely understanding and effectively strategizing for complex global challenges. A detailed description of systems thinking is left up to the myriad of excellent publications (see Braun, 2002; Capra & Luisi, 2016; Meadows, 2008, among many others).

Because systems thinking is a metascience, it shows up in virtually all areas such as the following: logistics and supply chain management (Barorikar, 2020); sustainable development & education (Zelinka & Amadei, 2017); construction projects (Bajracharya, 2014); socioeconomic development (Baporikar, 2016); biology, ecology, and the environment (Capra & Luisi, 2016; Ford, 2010); information & communications technology (Elharakany, Moscardini, Khalifa, & Elghany, 2018); startups (Shanbhag & Pardede, 2019); and many more.

As defined by Richmond (1994), systems thinking "is the art and science of making reliable inferences about behavior by developing an increasingly deep understanding of underlying structure" (p. 6). Another definition proposed by Sterman (2006) considers systems thinking as a new mindset with both depth and breadth and "an iterative learning process in which we replace a reductionist, narrow, short-term, static view of the world with a holistic, broad, long-term, dynamic view, reinventing our policies and institutions accordingly" (p. 509). Systems thinking and systems tools such as cross-impact analysis (CIA) and network analysis (NA), which are considered in this chapter, enable decision-makers to explore the cross-impact and interconnections across the SDGs more rigorously and analytically than just using intuition.

A systems or integrated approach requires a certain level of awareness and decision-making maturity. System thinking is learned, and anyone can become a systems thinker by adjusting their perspectives and adopting new habits (Table 1), according to the Waters Foundation (Benson & Marlin, 2017). These habits can also be understood as thinking strategies (visual, listening and speaking, and kinesthetic) that a decision-maker might want to follow to address complex problems. In the following table, habits in bold are particularly salient for this chapter, while italicized habits are essential for the next chapter; all are important for systems thinking and science. They *cannot* be thought of as individuals as they are fundamentally interdependent of one another.

Table 1. Habits of system thinkers and the SDGs (from Benson and Marlin, 2017)

Habit	Description
Big Picture	A systems thinker "steps back" to examine the dynamics of a system and the interrelationships among its parts. They see the forest rather than the details of any individual tree.
Change with time	Dynamic systems are made up of interdependent elements, the values of which change over time.
System's structure	Focusing on the structure of the system facilitates an understanding of the outcomes of the system (i.e., structure determines behavior).
Interdependencies	A systems thinker knows that the cause-effect relationships within dynamic systems are circular rather than linear. Complex cause and effect relationships include balancing feedback, in which the system is trying to reach and maintain a goal, e.g., the heating system in a house. There also may be reinforcing feedback, such that the more you start with, the more it increases over time, e.g., population.
Connections	A systems thinker intentionally makes connections in order to understand the relationships in systems better. A systems thinker creates meaning by considering how new information connects to previous knowledge by adding, modifying, transferring, and synthesizing the information into a more in-depth understanding.
Change perspectives	To understand how a dynamic system actually works, a systems thinker looks at the system from a variety of different angles and perspectives, perhaps in collaboration with others.
Assumptions	A systems thinker will rigorously examine assumptions in order to gain insight into a system. Insight put into action can lead to improved performance. It is crucial to understand how and why assumptions are made, beliefs are developed, and actions are taken based on perceived data to avoid incorrect assumptions.
Fully considers issues	A systems thinker is patient. S/he will take time to understand the system's structure and its behaviors before recommending and implementing a course of action. A systems thinker also understands that succumbing to the urge for a quick solution can create more problems in the long term.
Mental models	In any given situation, an individual perceives and interprets what is happening, thus creating a picture, or mental models, which are comprised of assumptions, beliefs, and values that people hold, sometimes for a lifetime. A systems thinker is aware of how these mental models influence perspectives and, ultimately, any actions that are to be taken.
Leverage	Based on an understanding of the structure, interdependencies, and feedback within a system, a systems thinker implements the leverage action that seems most likely to produce desirable outcomes.
Consequences	Before acting to change a dynamic system, a system thinker weighs the possible short-term, long-term, and unintended outcomes of the action.
Accumulations	Systems are made up of many elements, including accumulations, i.e., amounts that can increase and decrease over time, and their rates of change.
Time Delays	A systems thinker recognizes that when an action is taken within a complex, dynamic system, the outcome of the action may not be seen for some time.
Successive approximations	By trying a solution and then assessing (monitoring and evaluating) the results, understanding of the issue will increase. Over time, each cycle or successive approximation, of checking results and changing actions if needed, will move the system closer to the desired goal.

Dimensions

Complex systems are composed of numerous concepts, topics, and ideas each of which has been the subject of many papers (Braun, 2002; Meadows, 1999), dissertations (Zelinka, 2019), books (Capra & Luisi, 2016; Mandl, 2019; Meadows, 2008; Mobus & Kalton, 2015), and life's works (Jay Forrester,

Donella Meadows, and John Sterman, among others). The point of this chapter is not to discuss them all in detail but to present them in a way that makes them easier to understand in the context of sustainable development. To that end, complex systems are presented on a continuum from simple to highly complex. Certain concepts like chaos, for example, can exist anywhere on the continuum, while others like determinism are focused on simple systems.

All systems can be defined along two dimensions: space and time. A system's spatial component is its structure, the various pieces, and, fundamentally, how those pieces connect to form the whole. The temporal component focuses on how those components and interactions cause the system to change (i.e., system dynamics). Complex and chaotic systems fall under the umbrella of these dynamical systems and will be discussed and modeled in the next chapter, on which the structural component is a foundation. All systems have structure and dynamics and the degree to which is referred to as structural and dynamic complexity, respectively.

This chapter is focused on the **structural dimension** of complex systems through the lens of simple, complicated, and complicated-coupled systems by viewing these systems with a mostly static representation. The next chapter focuses on the **dynamic dimension** through the lens of complex-coupled systems by viewing them with a dynamic perspective. Each chapter introduces modeling tools and applications for each dimension, spatial and temporal.

Defining a System

Economies, the human body, insect colonies, cities, a bridge, a pendulum, a car, a train, a heard of animals, governing a city, providing electricity, thinking, growing food, orbital mechanics, sets of equations, are all systems. *Nearly* everything is a system, but what is not a system? A system is not a completely isolated asteroid in space not interacting with anything, although it is part of a system – the solar system – as a single element, especially when considered over long timescales and deep time. Because virtually anything is a system, the systems can apply to virtually everything, making it a truly universal approach.

There are many ways to define a system – they differ by their specific application and discipline – but in general, a **system** is a universal and transdisciplinary term that refers to anything with at least two interacting components. A less technical and applicable definition is the way through which something is organized and structured – like a set of rules. In the context of society, social norms, value systems, and laws are the rules that quite literally govern society (i.e., social system). The international financial system is composed of organizations and countries that interact according to international law, rules, and trade. For an engineering example, a car is a system of thousands of pieces interacting chemically, thermally, electronically, and mechanically to get its occupant(s) to their destination safely. The laws of physics are the rules that control the movement of a pendulum, the effect of gravity, air density, and other parameters being the variables of the system that interact.

A system has some kind of physical or abstract **boundary**. Everything inside the boundary (**endogenous**) constitutes the whole system, and everything outside of it is **exogenous** of the system. At the national-level, society is a system that is organized by social norms (generated endogenously over time from within society itself) as well as laws (generated exogenously by an external government). In this example, geography is not within the boundary of the system.

When it interacts with its outside environment or other systems, it is an open system. If the system does not interact with anything beyond its boundary, the system is closed. It can also be partially open, the degree to which is the system's openness. A pot with a lid on it can gain energy from the burner below and lose energy to the atmosphere, both through conduction, but it cannot gain or lose mass because of the physical barrier of the pot and lid. Governments, economies, and societies are bounded geopolitically by international borders or national boundaries. Free trade is a term to describe the openness of the international economic system concerning the global movement of goods. Migration applies to the flows of people across national borders. Borders can also be abstract, like between racial groups living in the same area a part of a website behind a password protected barrier. A boundary is essentially arbitrary and subjective based on the observer's perspective, context, and scale.

All systems have **goals or functions**; the various components interact in such a way so that they strive to reach a desired state or goal – either actively and purposely or passively and organically or – and then maintain the system's function and capacity at that desired state. System capacity is similar to a goal, but it has some nuance; it is the maximum ability of a system to effectively carry out its function, like the environment's capacity to sustain a certain number of a population while not degrading the quality of the environment. The goal of an economy is to supply people with the goods and services they need and desire. Metrics like GDP per capita and Gini coefficient (economic inequality) measure the effectiveness of an economy to do that. The function of the climate system is to move heat and water around the planet and can be directly measured with humidity, temperature, and wind speed. The carrying capacity of an area is the maximum population that a land area can hold while still sustaining that population without harming the environment. From the perspective of people and animals living on that land, its function is to provide them with the resources and land to survive. The function of a road network is to get people to where they need to go safely, and the speed limit is the maximum safe speed a car *should* travel that minimizes danger to maximize someone's chance of getting to their destination.

A system is a **process** through which some abstract or tangible thing is received as an input, and then the system converts (processes) those inputs into an output after a nonzero amount of time. A process is how the system attempts to meet its goal. More sophisticated systems can learn and adapt to changing conditions and shocks to more effectively reach some desired state. A human requires food, water, and air as material resources and outputs work (in the simplest sense), as well as waste and heat as byproducts. This metabolism is the overall process by which the body keeps a person alive, its function. A manufacturing process takes some raw materials or some simple good and turns it into something useful or more advanced (Graedel & Allenby, 2015; Mobus & Kalton, 2015). A process is how the system carries out its function to reach its goal. It is fundamentally dynamic and describes the internal mechanisms of the system over time. These system dynamics are the focus of the next chapter.

Interconnectedness

Systems are described by their level or **degree of complexity**. In general, as the number of elements in a system grows, it will become increasingly complex. Although this is not necessarily always the case, it is still highly correlated. More specifically, a system's complexity is, in reality, governed by the structure and number of interconnections among elements, not directly on the number of elements themselves. Two systems can have the same number of elements but be at different levels of complexity if the pattern of interconnections differs. Those interactions also increase at a faster rate than the individual pieces leading to emergent properties and behavior; when the number of system elements double, the number of interactions will more than double. For example, this can be likened to neurons being the elements and the interconnections among them representing the brain's complexity.

There are many ways to view and measure complexity, most of which are out of the scope of this work (see Fath, 2017; Mitchell, 2011; Mobus & Kalton, 2015) but they can be described as on a spectrum from simple to chaotic (Grabowski & Strzałka, 2008; Rickles, Hawe, & Shiell, 2007) or simple causality to dis-ordered complexity (Fath, 2017). Regardless of the terminology, the degree of interconnectivity among components leads to degrees of complexity in the overall system. This structural aspect of complexity is referred to as structural complexity.

Simple systems are easy to understand and have a small number of elements that interact in a well-defined and deterministic way and are devoid of chaos and unknowns. In classical mechanics, a swinging pendulum is an example of a simple system; thermodynamically, a pot of water being heated is another. Simple systems need not be small or, counterintuitively, that simple. Orbits are simple systems; the positions of planets and asteroids in a solar system can be predicted hundreds and thousands of years in advance to near-perfect accuracy. Similarly, the exact time and location of the eclipse can be predicted to the second, hundreds of years in the future, which is determined by a few equations and parameters that, if known, determine the entire system of planetary motion (i.e., celestial mechanics). We have known these laws for a long time; they are Kepler's Laws of Planetary Motion. A pot of heating water is governed by the Second Law of Thermodynamics and Newton's Law of Cooling. Most simple problems can technically be solved with a pencil and paper.

Complicated systems are an extension of simple systems with many pieces (too many to count). Their various pieces are still well-defined and understood but take more effort to do so. The main difference between simple and complicated systems stems from the number of parts (on the order of hundreds to tens of thousands). Buildings, cars, airplanes, and engineering projects epitomize complicated systems. No matter the problem, there technically, mathematically, or otherwise exists a solution or an optimum state, just as the case with simple systems. In another way: with enough brute force, someone can figure out a complicated problem, if they work hard enough. Solving complicated problems almost always requires computational assistance.

Hierarchy

Coupled systems are another type of system very closely related to complicated systems. The coupling comes from the interconnectedness among the subsystems or layers (i.e., hierarchy). The more two subsystems within a more extensive system interconnect, the more coupled they are, and systems with few interactions are loosely coupled. Most complicated systems are also coupled, and the defining characteristic of a coupled system is the existence of hierarchy and scale, which are not necessary for complicated systems. Hierarchy organizes and structures a system based on different levels and groupings of similar elements.

Coupled systems are built on and from other systems (i.e., subsystems) that can be simple or other complicated systems, which tend to be more numerous and less complex at lower levels. In the same way that interconnectivity relates elements of the systems, hierarchy relates parts of a coupled system. It comes in two general variations: vertical and horizontal, terms borrowed from organizational management. Relations can occur vertically among different levels or might occur horizontally within the same layer. All the examples of complicated systems listed above are also coupled. Buildings (albeit with a wide variety from a shack to a skyscraper) have clear subsystems. There is a heating, ventilation, air-condition, electricity, communications subsystems that work together to maintain the operations of the building. Larger (complicated) buildings generally require more coupling among their systems to operate correctly. Skyscrapers are too big to cool or heat without an HVAC system, but a shack might just open a window for the same function.

Parts of a coupled system can be viewed as sectors in an economy that work together to produce goods and deliver services. Economies exemplify hierarchy in that there are specific levels with noticeable functions. Each layer is a subsystem type layered on top of less complicated layers; the quantity of these layered subsystems indicates the scale of a **vertical hierarchy**. Let us continue with the national economy example. There are many ways to structure an economy, but they all fundamentally adhere to a typical structure of five overarching sectors (de Grandbois, 2016), although some models use three or four:

Primary Sector: Extraction of raw materials from the environment or through recycling for use in all other economic sectors; fishing, agriculture, mining, and fossil fuel extraction.

Secondary Sector: Manufacturing takes raw materials and processes them (see the previous section) into something more useful and of higher quality; food production, factories, workshops, breweries, artisans, and other manufacturing industries of any size and output.

Tertiary Sector: The service sector is the commercial retail face of manufacturing, restaurants, IT services, banking, and transportation services.

Quaternary Sector: The knowledge sector focusing on human capital to innovate and advance the economy overall | academia, consulting, think tanks, education, hi-tech, entrepreneurship.

Quinary Sector: Top-level decisions in the country are made in government, commerce, and industry that manage, guide, or otherwise influence the other sectors of the economy; policymakers in the national government, executives in business and industry, academic experts.

In the above list, each sector represents a different level. The primary sector is at the bottom extracting and sending resources to the secondary sector, which then manufactures those resources into something more useful of higher quality to be sold by the tertiary commercial retail sector. The top two layers directly control and require the bottom three. Vertical hierarchy tends to be where supply chains operate.

Within the five main sectors, many sublevels might constitute a national economy. All subsectors of an economy fall into one of those five main sectors; the North American Industry Classification System divides those five main sectors of the US economy into 20 sectors, 99 subsectors, 311 industry groups, 709 industries, 1057 subindustries, and the smallest operating entity (similar to an enterprise or business) totaled over 17 million (US Office of Management and Budget, 2017). There is an evident vertical hierarchical structure for an economy. Higher-level sectors are more substantial with fewer pieces, while lower-level industries are smaller with more elements and businesses.

Some industries even cross levels like aerospace, defense, and hi-tech. These industries are relatively complex and, thus, require more advanced understanding and techniques to manufacture the associated goods. Without direct involvement from industries from all main sectors, those three subsectors will not function. Because economies are highly complex, they have many interrelated industries, businesses, and components in general. As such, national economies are composed of numerous layers and industries.

There can be fluidity and overlap among the layers. As a small business grows to become a larger one, it can vertically integrate to include many, or sometimes all, levels of its supply chain from the extraction to the manufacturing of the products (upstream inputs) to the retailing and engineering of its products (downstream outputs). A company might also horizontally integrate by expanding into similar industries within the same level. A company that manufactures furniture can decide to start making jackets, for example; this is precisely what Apple and Tesla have done as they have grown (i.e., increasing hierarchy and structural complexity of their organizations) by vertically controlling more levels of the supply chain and expanding into other markets and industries with their products and services. Most large technology companies exemplify horizontal and vertical hierarchy or integration within an economy.

Essentially, a hierarchy can fall along a spectrum of possible levels from the micro-scale to the macro-scale. Macro-scale (low-resolution) focuses on a few large and important areas or categories but holistically looks at the entire system, as opposed to the micro-scale (high-resolution), which focuses on many individual elements but does not interact with different parts of the system. There can be any amount of levels between – and fractals mathematically have infinite – but even the most complex systems usually only have a dozen or so.

Elements, parts, and subsystems can exist at more than one level or move among them over time. Although a municipality is considered a geopolitical level below counties, provinces, districts, and their equivalents, there are cities larger than countries (any megacity) and cities that are countries (e.g., Singapore and Vatican). Some cities might grow large enough to become two levels at the same time, like the City and County of Denver, for example. Many capital cities and other large and prominent cities do this.

Thus far, only vertical hierarchy has been described; on the other hand, **horizontal hierarchy** simply refers to the number of parts and interconnection within each level – not between them. The more pieces within each level, the more complicated that specific level becomes; it exhibits increasing levels of horizontal hierarchy. Continuing with the example of an economy, these individual pieces can be thought of as individual companies, corporations, or specific subindustries within a sector. All entities in the primary sector that work in agriculture compete and interact with each other.

Recall earlier the concept of boundary and national borders. The five sectors can be expanded to include trade with other national economies, which makes an economy open. Trade is an explicit economic activity that happens across borders. Virtually all economies are open to some extent, even North Korea, of which the vast majority of all trade is with China. Trade is part horizontal and vertical. It connects parts of the supply chain (vertical) of a product through other countries, but it also might horizontally meet the demand of a product for consumers (horizontal) when a specific industry cannot compete or supply enough product.

This economic structure has a clear hierarchy with more people at the bottom, performing more straightforward and less demanding jobs and fewer people at the top, performing complex tasks. As an economy develops (literally, advanced economies), the distribution among the sectors changes; it advances from extraction and agriculture to a knowledge economy – from a focus on simpler processes to more complex ones.

THE SUSTAINABLE DEVELOPMENT NEXUS

Complex systems are composed of innumerable components interacting at many scales yielding a collective behavior. Complexity is beyond complicated in that the outcome is not well-defined or deterministic, and the effect of dynamics (changes with time) leads to difficult to model – or impossible – behavior. Complex systems are highly sensitive to initial conditions, and a small change in some variables can result in a significant change, one that is often unexpected.

Complex problems include rapid population growth, natural resource management, corruption, conflict, and climate change, among many others. Many complex problems fall under the umbrella of sustainable development. For that reason, sustainable development, specifically the Sustainable Development Goals (SDGs), will be the case study for this and the next chapter as they are salient issues that face all of humankind. Furthermore, the SDGs are a global framework that provides a way to organize the complexity of sustainable development into a well-structured coupled system.

Sustainable development and the SDGs are intimately tied to human social interactions at varying scales. For that reason, it is referred to as an *anthropocentric system*, i.e., a system whose main actors are directly and strongly influenced by complex human socio-economic-political interactions but are themselves not humans (Zelinka & Amadei, 2019a; Daher et al., 2018). Anthropocentric systems are always complex because they involve ill-defined context with large groups of people interacting with many other complex systems. People are not always rational (they do not think in systems), and their interactions within the sustainable development systems often create complex problems, such as those mentioned in the previous paragraph.

Although these systems are complex, sustainable development can be modeled as an economy of interacting sectors, and the SDGs provide a straightforward way to do that. This method simplifies complexity so that it can be viewed and modeled as a complicated-coupled system through a nexus approach, while the next chapter looks at them as a complex-coupled system by including system dynamics.

The Sustainable Development Goals

In 2015, the United Nations introduced the Sustainable Development Goals (SDGs) as a new 15-year long road map for worldwide sustainable development (United Nations, 2016). The 17 SDGs listed in Table 2 are structured hierarchically with 169 targets and 230 indicators (not shown). As described in the resolution adopted by the General Assembly on 25 September 2015, the SDG framework aims to cultivate and expand humanity's desire to "do good" while also organizing its ability to do so. The SDGs "...seek to build on the [previous] Millennium Development Goals (MDGs) and complete what they did not achieve" (United Nations General Assembly, 2015, p.1). In launching the SDGs, the General Assembly of the United Nations "recognize[d] that eradicating poverty in all its forms and dimensions (including extreme poverty) is the greatest global challenge and an indispensable requirement for sustainable development" (United Nations General Assembly, 2015, p. 1). To that end, the SDGs represent "a plan of action for people, planet, and prosperity," which, in addition to peace and partnership, define the five "P's" of the mission of the SDGs. To accomplish that mission, there needs to be a "balance [among] the three dimensions of sustainable development: the economic, [sociopolitical], and environmental" (United Nations General Assembly, 2015, p. 3). It must be acknowledged that fulfilling the mission of sustainable development is indeed a daunting task, a truly formidable undertaking, unparalleled in human history.

From the inception and ultimately, the adoption of the SDGs, there has been an increasing interest in understanding and quantifying their interactions (Costanza et al., 2016). Multiple papers and reports have been produced by various organizations to identify important linkages among the SDGs, how the SDGs affect each other (Coopman, et al., 2016; ICSU & ISSC, 2015; Nilsson et al., 2017; Nilsson et al., 2013; Vladimirova & Le Blanc, 2016), and how to measure qualitatively or semi-quantitatively those interactions (ICSU & ISSC, 2015; Nilsson, et al., 2016; Nilsson et al., 2017; Sustainable Development Solutions Network, Schmidt-Traubet, al., 2015). Other publications have looked at the interactions of the SDGs with other frameworks and international agreements such as the Paris Climate Agreement, most notably (Von Stechow et al., 2016).

Even though it is recognized that the SDGs are interconnected, the authors are aware of only a few publications that analyze the effects of the SDG interactions in a systemic or integrated manner (Allen,

Table 2. The 17 Sustainable Development Goals (adapted from Global Goals, 2016)

Goal Number	Title (Sector)	Description
SDG 01	Poverty	End poverty in all its forms everywhere
SDG 02	Food Security	End hunger, achieve food security and improved nutrition and promote sustainable agriculture
SDG 03	Health	Ensure healthy lives and promote well-being for all at all ages
SDG 04	Education	Ensure inclusive and equitable quality education and promote lifelong learning opportunities for all
SDG 05	Gender equality	Achieve gender equality and empower all women and girls
SDG 06	Water and Sanitation	Ensure availability and sustainable management of water and sanitation for all
SDG 07	Energy	Ensure access to affordable, reliable, sustainable, and modern energy for all
SDG 08	Economy	Promote sustained, inclusive, and sustainable economic growth, full and productive employment, and decent work for all
SDG 09	Infrastructure	Build resilient infrastructure, promote inclusive and sustainable industrialization, and foster innovation
SDG 10	Inequality	Reduce inequality within and among countries
SDG 11	Cities	Make cities and human settlements inclusive, safe, resilient, and sustainable
SDG 12	Consumption	Ensure sustainable consumption and production patterns
SDG 13	Climate	Take urgent action to combat climate change and its impacts
SDG 14	Ocean	Conserve and sustainably use the oceans, seas, and marine resources for sustainable development
SDG 15	Land	Protect, restore, and promote sustainable use of terrestrial ecosystems, sustainably manage forests, combat desertification, and halt and reverse land degradation and halt biodiversity loss
SDG 16	Governance	Promote peaceful and inclusive societies for sustainable development, provide access to justice for all and build effective, accountable, and inclusive institutions at all levels
SDG 17	Partnerships	Strengthen the means of implementation and revitalize the Global Partnership for Sustainable Development

Metternicht, & Wiedmann, 2019; Barbier & Burgess, 2017; Collste, Pedercini, & Cornell, 2017; Zelinka & Amadei, 2019a, 2019b). However, some mathematical tools already exist in the literature that could be used for that purpose, such as: (i) equations to calculate the indirect impacts (the effect due to the interconnections) from direct observations of a mathematical system (Bollen, 1987); (ii) the economic input-output and Leontief models (Moffatt & Hanley, 2001; Sachs et al., 2016; Todorov & Marinova, 2011); (iii) the MICMAC structural analysis (Arcade et al., 1999; Glenn, 2004); and (iv) the cross-impact analysis (Weimer-Jehle, 2006, 2008; Zelinka & Amadei, 2017). These publications are great starting points, as they focus on the science of linkages, but fall short of quantifying the impact of these interactions. They serve as the scientific foundation (ICSU & ISSC, 2015) on which we build a methodology to understand the long-term trajectory and success of the SDGs with a systems perspective.

While addressing and implementing the SDGs, accounting for their effects on one another is crucial to their success. Because of their complex formulation, some goals are bound to affect others negatively or positively, whereas others may not have any direct effect, while all SDGs will affect all others through some indirect path. Furthermore, the interactions are context- and scale-specific; each country, city, or entity attempting to adhere to the SDG framework has unique characteristics and situations. For example, rural Kenya will not have the same conditions as an island-nation; a city will have different issues than a country; a water-poor country will struggle with problems that a water-rich community will not; political factors might constrain or further enable certain outcomes for the SDGs. Likewise, a country with strong environmental regulations might not have such large negative impacts on the environment resulting from energy-use and economic activity. The factors leading to the success or failure of implementing the SDGs are innumerable.

Current guidelines for implementing the SDGs do not account for the interconnectivity of the SDGs and most focus on only one, or a group of them. Adopting a systems thinking mindset will enable development practitioners to understand how prioritizing certain aspects of sustainable development could influence all SDGs. For example, as discussed later in this chapter, attempting to directly reduce poverty (SDG 01) is ultimately self-defeating, as poverty is dependent on virtually all other goals and areas of sustainable development (energy, sustainable consumption, reducing inequalities, environmental sustainability, good governance, et cetera.) not the other way around. By directly addressing poverty, the other goals are neglected and suffer as a result. Because all the other goals influence poverty and poverty overall negatively affects the other goals, focusing on poverty is, counterintuitively, detrimental to its alleviation and is highly counteracting to the eventual goal of poverty eradication.

A systems thinking approach to addressing the SDGs is an appropriate means to account for the complexity and uncertainty at play in sustainable development at the country or other scales of interest. It has the potential to provide an alternative to the deterministic decision-making process that has been used in development around the world over the past 50 years. It can be used to and complements the habits of systems thinkers from Table 1:

- better understand how meeting the goals and their targets depend on context along with the capacity, vulnerability, and structure at the local, city, country, or global levels
- explore how certain structural variables or factors influence that dynamic, including feedback mechanisms, by carrying out various parametric and sensitivity studies
- explain existing patterns of human development at different spatial and temporal scales
- interpret emergent properties of human development such as poverty/wealth, peace/conflict, health, resilience, and sustainability that may involve one or several goals;
- examine various perspectives of human development interventions and consider possible (intended and unintended) consequences of decision making and policies and their implications
- identify and explore leverage or tipping points, those critical places to intervene at different scales that will most benefit human development
- predict how countries or regions may respond to various constraints and disturbances and/or strategies of capacity development
- monitor and evaluate the performance of development interventions and decide on how to adjust as development projects unfold, thus leading to more successful projects in the long term; and
- develop approximate but good enough sustainable development solutions that are more flexible and adaptive to change than the traditional deterministic, rigid, and ultimately inappropriate long-term solutions

Coupled Systems and the Nexus Approach

Complicated-coupled systems lend themselves over to being understood through the lens of a **nexus**, a connected group, or a center of focus like a hub (Merriam-Webster, 2020). The next chapter uses the nexus as a hub explicitly as a modeling approach. Essentially, a nexus refers to the complex interplay among various components and subsystems within a complex system. The nexus approach is the way to view systems as interconnected subsystems or networks and has been applied to community development (Amadei, 2015; Amadei & Zelinka, 2018; Zelinka, 2019), large cities (Wang, Cao, & Chen, 2017), national economies (Hendrickson et al., 2006), and global sustainable development (Armenia, Pompei, Castaño Barreto, Atzori, & Fonseca, 2019; Liu et al., 2018) among many others.

The food-energy-water (FEW) nexus is one of the most well-known and typical applications of the nexus approach. See Amadei & Zelinka (2018), Mohtar and Daher (2012, 2016), (Daher et al. 2017, 2018) for a detailed explanation of the FEW nexus and its approach to modeling. Food, energy, and water (FEW) resource systems are tightly interconnected. They represent critical components in the overall discourse on human and sustainable development, and the nexus approach has proven itself a highly useful way to frame and model sustainable development as a complex-coupled system. These interconnected resource systems are also regulated, managed, and consumed by different governmental, business, and societal players, with differing sets of goals, value systems, and decision-making powers, across scales (Figure 1).

Figure 1. Interconnected water, energy, food systems governed by interacting governmental, business, and societal actors (Daher et al., 2018)

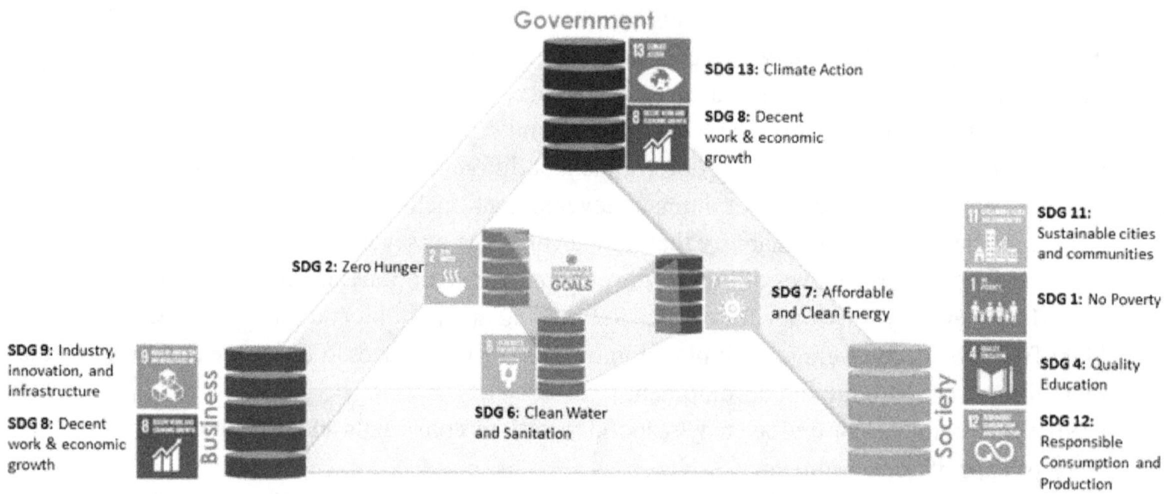

The SDGs connect to most of these components, thus allowing the FEW research community the opportunity to leverage the momentum towards reaching these goals by 2030. As countries work toward achieving the 17 goals, critical attention needs to be given to the level of interconnectedness and trade-offs that might exist between them.

It should be noted that the FEW nexus is not unique as it is one of the many interlinkages that are at play in human and sustainable development. To the three sectors of the FEW nexus, one could add health, soils, land, climate, transport, communication, state security, human rights, labor, trade, and so on. For instance, land is sometimes associated with the FEW nexus, since land is vital to produce food, energy, and water supply as well as other ecosystem services (Hurni et al., 2015; Müller, Janetschek, & Weigelt, 2015; Ringler, Bhaduri, & Lawford, 2013; Weigelt, Janetschek, Müller, & Töpfer, 2015). The acronym WELF is used instead of FEW in the literature (and herein) when land is also considered along with water, energy, and food. Figure 2 shows a tetrahedron representation of the WELF nexus, where each node corresponds to one sector of the nexus. A fifth node (defined as X) connected to the other nodes could be added at the centroid of the tetrahedron. It may correspond to climate or population.

Figure 2. The WELF nexus represented as a tetrahedron; the node X can represent any other sector like climate or population

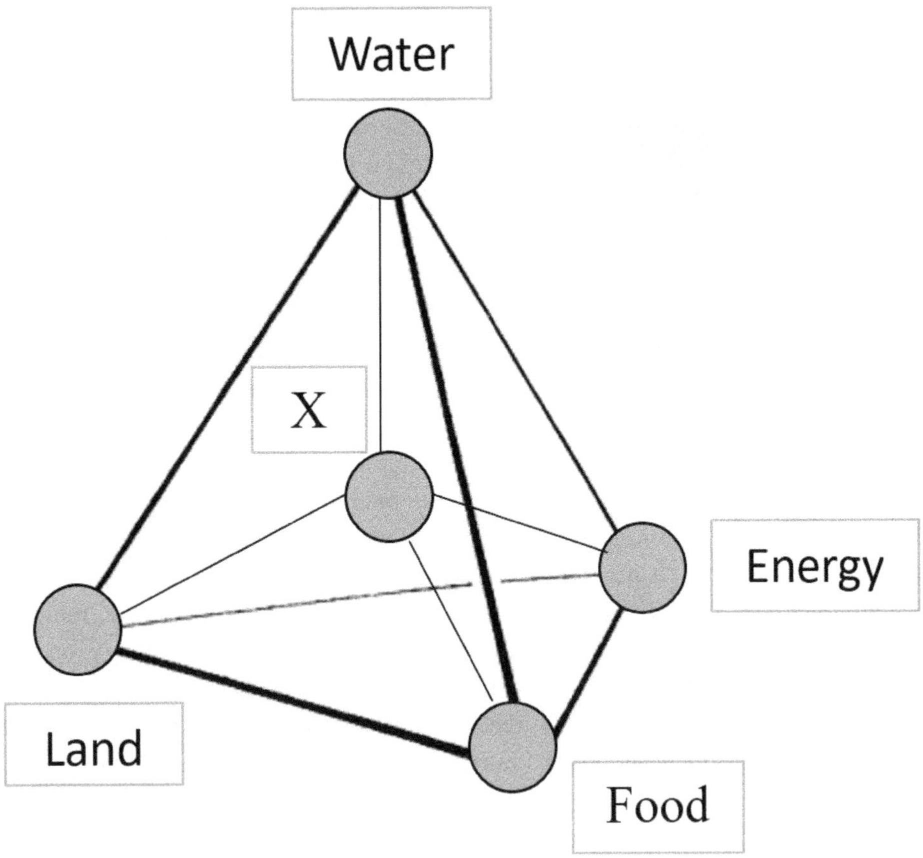

Cities are excellent examples of complex-coupled systems. They have sectors of food, water, health, education, political, and many more systems that interact with humans and the environment to make sure its inhabitants live at a certain level of development. The city's economy is the entity that emerges when all sectors are involved. The same goes for a national and the global economy.

The SDGs are a globally recognized framework that already categorizes and structures sustainable development into 17 sectors (one for each SDG), so it is an excellent application for the nexus approach. There is no need to identify further and define the SDGs, and they can be analyzed as is. The following section introduces cross-impact analysis and network analysis and applies them to all 17 SDGs as a proxy for sustainable development.

CROSS-IMPACT NETWORK ANALYSIS

Systems Versus Networks

One principle of a system is that they can be represented as a network of relations (Mobus & Kalton, 2015), which implies that systems and networks can be depicted using the same tools; this principle underlies the tools in this and the next chapter. A system is how two or more components interact, and a network is a group or set of related and discrete (mutually exclusive) objects (nodes) that are interrelated (Mobus & Kalton, 2015; Newman, 2018) by connections (edges) between them. The definition of a network is very similar to a system, so describing how they differ might yield more clarity. The difference is in what exactly is being connected and to what; the keyword is "related" in the network definition.

In a system, the components that interact are not necessarily similar. Systems focus on the interaction of different components with different key variables. For example, the water-energy-land-food nexus is a system of four interacting subsystems, each with a different key variable. In a water system, it is the flow of water, with energy, it is the flow of electricity, and so on. The system looks at how, why, and what causes the interactions among those four components. Using the building example from earlier, the electrical system interacts with the HVAC system by supplying power to it. The electrical system focuses on electricity, while the HVAC system focuses on the movement of air and heat, but both are part of the same building system.

A network focuses on the flow or movement of one unit of something: a migration network is the movement of people among places (cities, usually); the internet (short, for interconnected network) is the flow of information among computers; water flows among buildings, treatment plants, and water pumping stations (each of which is an example of a node) through a network of pipes (edges).

Let us explain this with a city, which generally sees net positive migration (immigration), while rural areas see net negative migration (emigration) A city might be only one node in a migration network where people either immigrate to or emigrate from, but it is probably the entire entity of interest when viewed as a system, the urban system. A network approach cannot identify why people migrate, but it will tell you how they migrate and critical pieces of information regarding each node (city) or of the network overall. Network analysis can determine which city is vital in the overall migration network, but if someone wants to understand why analyzing the city as a system is needed. How do the various components (water network, policy, energy, health, education, and so on) interact to supply the necessary services to encourage people to move to the city? What is it about rural areas that cause people to want to leave? Depending on the context, scale, and questions asked, a systems approach might be better, or a network approach might be better, but in most cases, it is a mix of both, which is the approach outlined herein.

Mathematically, interactions in systems almost always require unit conversions, whereas networks do not because everything in a network is the same unit. When water is used for power generation, it requires that water and energy networks interact, i.e., how many units of water are required to generate how many units of energy. This unit conversion happens at nodes where networks overlap in the overall systems. At the power plant, water is used for thermoelectric cooling; it takes about 1000 L of water to generate a megawatt-hour (MWh) of electricity for fossil fuel plants in Europe (Larsen & Drews, 2019).

This section treats complicated-coupled systems as a nexus of interconnected networks. The internet functions in the same way; in fact, the word 'internet' is an amalgam of the words 'interconnected' and 'network' (Newman, 2018) and, therefore, is a nexus. Before a system can be analyzed as a network, it must first be put into a matrix format – specifically, a cross-impact matrix – as a way to transcode a system to a network through a soft-systems approach.

The cross-impact network matrix (CINM) approach outlined herein is based on a semi-quantitative unitless strength to depict interactions. As such, it can be treated as a network or a system, and mathematically there is no difference, which allows for a system to be transcoded into a network and vice versa. Even if a fully quantitative data-driven approach is employed, the system and networks are valid, although it might add complications.

Cross-Impact Matrix

Cross-impact analysis (CIA), also called double-causality analysis, is a mathematical approach used in Futures Research studies. It originated in the mid- to late-1960s (Gordon, 2014; Gordon & Hayward, 1968) and was initially developed to analyze weakly (soft) structured systems for which theory-based computational (hard) models do not work due to the systems complexity, uncertainty, and disciplinary heterogeneity. Since its inception, cross-impact analysis has been a general soft-system method for assessing the "interrelations [among] the most important influential factors in a system by experts who evaluate [subjectively] pairs of these factors" (Weimer-Jehle, 2006, p. 336). The different methods of cross-impact analysis differ on how the interrelations are formulated (probabilistically or deterministically) and whether a qualitative, quantitative, or mixed approach is used to describe causalities (Serdar Asan & Asan, 2007).

The cross-impact analysis is particularly appropriate to tackle large-scale human societal issues such as those addressed by the SDGs. Its approach is flexible and lends itself to expert discourse, which is important for effectively identifying the fundamental causes of a problem as opposed to using a mere intuitive and educated guess that risks addressing only the symptoms. The cross-impact analysis is a method to understand complex-coupled systems with relatively little effort in a concise and informative way. Due to its flexibility and versatility, it is apt for use in comparing multiple scenarios and finding out the most appropriate ones based on analytical tools rather than, ideally, guesses, bias, and partisan influence. A cross-impact analysis is used in this chapter and the next to analyze the interactions of the SDGs.

Let us consider a system consisting of n interacting variables. Cross-impact analysis of that system is usually presented by a square (n x n) matrix with zeroes along with its diagonal and n^2-n off-diagonal terms. The off-diagonal terms define how each variable (row) directly influences or impacts the other variables and how each variable (column) depends on or is sensitive to the other components. It should be noted that the matrix is not symmetric. The strengths of the influence between two variables can be described qualitatively using qualifiers such as high, medium, or low, or scored semi-quantitatively over an appropriate scale, say, between 0 for no influence up to 3 (or larger) for a strong influence. The selected scale is specific to the system being analyzed. The scale used herein can be seen below (Table 3).

Figure 3 shows an example of a cross-impact matrix for the N = 17 SDGs, where semi-quantitative scores have been assigned to describe the strength between interacting SDGs, two at a time. Using the

Table 3. Semi-quantitative scale of sustainable development goals' interactions (adapted from Nilsson et al., 2017)

Strength, Name	Explanation	Examples
+3, Indivisible	Inextricably linked to the achievement of another goal. Accomplishing this goal will accomplish another goal by default.	Ending all forms of discrimination against women and girls is indivisible from ensuring women's full and effective participation.
+2, Reinforcing	Aids the achievement of another goal. This goal will by default achieve a significant portion of another goal, with only a relatively small amount of work required to further accomplish both.	Providing access to electricity reinforces water-pumping and irrigation systems. Strengthening the capacity to adapt to climate-related hazards reduces losses caused by disasters.
+1, Enabling	Creates conditions that further another goal's progress. This goal could be a tool to accomplish another. There is positive overlap.	Providing electricity access in rural homes enables education, because it makes it possible to do homework at night with electric lighting.
0, Neutral	No significant positive or negative interaction. The interconnections between two goals are purely indirect, if they exist at all. The positive and negative impacts are equal.	Ensuring education for all does not interact significantly with infrastructure development or conservation of ocean ecosystems.
-1, Constraining	Limits options on another goal. This makes another goal noticeably more difficult to accomplish.	Water requirements for thermoelectric power generation reduce available water for WASH. A climate change can constrain the options for energy access.
-2, Counteracting	Clashes with another goal. Unless direct actions and efforts are taken by people then this goal will make it significantly more difficult for another to be achieved.	When a population consumes more, economic growth can counteract waste reduction and climate mitigation and degrade the environment.
-3, Canceling	Makes it impossible to reach another goal. Nothing can be done to have these two goals coexist.	Fully ensuring public transparency and democratic accountability cannot be combined with national-security goals.

scale proposed by Nilsson (2017) in Table 3, the scores were assumed to range from -3 (most positive dependence) to +3 (most positive influence). Negative values depict an SDG that either constrains, counteracts, or cancels another goal, broadly defined as limiting; the opposite is true for positive values, which either enable, reinforce, or are indivisible to another goal and are generally defined as enhancing. Each of the three limiting and enhancing titles refers to the relative magnitude of those interactions. Interactions are 0 when there are no discernable effects and are called neutral. Values were assigned for each one of the 272 off-diagonal terms of the double-causality matrix of Figure 1, although -3 never appeared in our analysis.

Several remarks can be made about the CIM. First, the values represent the direct impact of one SDG onto another. The point of the cross-impact analysis is to map the direct effects between variables, which, when analyzed further, could help to determine indirect effects among variables. As an example, let us consider two of the SDGs, the effect of the ocean (SDG 14) and gender equality (SDG 05). These two SDGs have no direct influence per se on each other, but an indirect influence is possible. For instance, cleaner oceans might improve water and sanitation that eventually improve health and education, eventually improving human rights and gender equality.

A second remark about Figure 3 relates to the selection of the interaction scores. Metric and indicators must be developed to determine what represents a positive or negative score. Nilsson et al. (2017) developed, for instance, such metrics for SDG 02 (Food Security), SDG 03 (Health), SDG 07 (Energy), and SDG 14 (Oceans). The scores in Figure 3 were based on some of these metrics and other publica-

Figure 3. Cross-impact matrix of the Sustainable Development Goals

Goal	Goal 01	Goal 02	Goal 03	Goal 04	Goal 05	Goal 06	Goal 07	Goal 08	Goal 09	Goal 10	Goal 11	Goal 12	Goal 13	Goal 14	Goal 15	Goal 16	Goal 17	Total	Description
Goal 01	0	-1	-1	-1	-1	-1	-1	-1	1	3	-1	-1	1	0	0	3	2	1	*Poverty*
Goal 02	3	0	3	2	2	0	-1	1	0	3	-1	-1	-1	-1	3	0	0	12	*Food Security*
Goal 03	3	1	0	2	2	1	-1	2	-1	3	1	0	0	1	0	0	0	14	*Health*
Goal 04	2	2	2	0	2	2	1	2	2	3	2	2	0	2	1	2	2	29	*Education*
Goal 05	3	3	2	2	0	-1	-1	3	0	3	2	-1	0	0	2	3	1	21	*Gender Equality*
Goal 06	3	3	3	2	2	0	1	3	1	3	3	1	1	2	3	0	0	31	*WASH*
Goal 07	3	1	3	2	0	2	0	3	3	3	3	-1	-1	-2	-2	0	2	19	*Energy*
Goal 08	3	2	2	3	3	2	2	0	2	-1	2	3	-2	-2	-2	0	2	19	*Economy*
Goal 09	2	2	2	1	0	2	3	3	0	-1	3	-1	-1	0	-2	2	2	17	*Infrastructure*
Goal 10	3	2	2	0	3	-2	-2	3	0	0	-2	-2	-1	-1	-2	0	1	2	*Inequality*
Goal 11	1	1	2	1	3	3	2	1	3	1	0	2	-1	0	1	0	1	21	*Cities*
Goal 12	2	2	1	2	1	1	2	3	2	3	2	0	-1	-1	-1	0	0	18	*Consumption*
Goal 13	2	1	-1	0	1	-1	-2	-1	-1	2	-2	-1	0	3	2	0	0	2	*Climate Change*
Goal 14	-1	1	1	0	0	1	1	-1	1	-1	1	1	0	0	2	0	0	6	*Ocean*
Goal 15	2	3	2	0	0	2	-1	-1	-1	0	0	-1	3	3	0	0	0	11	*Land*
Goal 16	3	3	3	3	3	3	3	3	3	3	3	2	2	2	2	0	3	44	*Governance*
Goal 17	3	2	2	2	2	2	2	3	2	3	2	2	3	3	3	3	0	39	*Partnerships*
Total	37	28	28	21	23	16	8	26	17	30	18	4	2	9	10	13	16	306	

tions (Coopman et al., 2016; Sustainable Development Solutions Network et al., 2015; Vladimirova & Le Blanc, 2016). They were assumed to reproduce the interaction dynamics among the SDGs representing an *average-sized country of average population and at an average level of human development*. The country does not exist, but our goal was to generate a cross-impact analysis of the SDGs for a hypothetical average country. This should be understood as a proof-of-concept that could be used for specific countries.

Finally, it should be noted that the scores in Figure 3 depend on the context in and scale at which the cross-impact analysis of the SDGs is conducted. The proof-of-concept could be extended to analyze the interactions of the SDGs at the global, regional, or community level. It could also be applied to any culture, socioeconomic group, or groups of multiple communities.

Priority Analysis

The total influence (impact) of each SDG on the others and the total dependence (sensitivity) of each SDG on the others were calculated by summing the row scores and column scores (Figure 3). The dependence and influence values were plotted on a single dependence (x) vs. influence (y) graph (Figure 4). The first diagonal shown on Figure 4 divides the goals that are more dependent (above the line) from those that are more influential (below the line). The graph can then help to separate the variables (goals) into several groups. According to Arcade et al. (1999), they include:

Influential variables (goals) with low dependence and high influence; SDG 06 (WASH), SDG 16 (Governance) and SDG 17 (Partnerships)

Excluded variables (goals) with low influence and dependence: SDG 13 (Climate Change), SDG 14 (Oceans), and SDG 15 (Land)

Dependent variables (goals) with high dependence and low influence: SDG 01 (Poverty), SDG 02 (Food Security), SDG 03 (Health), and SDG 10 (Inequality)

Regulating variables (goals) that lie near the center of the graph with similar dependence and influence: SDG 04 (Education), SDG 05 (Gender Equality), SDG 08 (Economy), SDG 09 (Infrastructure), and SDG 11 (Cities)

Because of their high influence and low dependence, influential goals are goals with high leverage; lower is the dependence, higher is the leverage. As their dependence increases, these goals become more challenging to address. On the other side of the dependence vs. influence spectrum, are the dependent variables that are the outputs of all the variable interactions. They are "strong indicators of the 'health' of the entire system of variables" (Ritchie-Dunham & Rabbino, 2001) and are sensitive to changes in the other variables. In Figure 4, both poverty reduction (SDG 01) and inequality reduction (SDG 10) are strongly dependent variables and are critical indicators of the overall state of sustainable development at the country-level. Other dependent variables such as health (SDG 03) and food security (SDG 02) are also indicators of the health of the system but have more influence on the other SDGs than poverty reduction and inequality reduction.

Goals that are clustered anywhere near the diagonal line on the graph have similar influence and dependence values meaning that these goals are not significant one way or another. A focus on these goals is much more tactical than focusing on influential goals. They might not improve sustainable development overall, but they could improve a specific goal. For instance, if a policymaker wants to specifically spur energy and economic development, then allocating resources to infrastructure (Goal 09) makes sense.

Using the total influence and dependence scores from Figure 3, three indices were calculated for each SDG and are listed in Table 4. The first index, the net influence (NI), was calculated as the difference between influence and dependence. The NI shows the absolute power of a goal. The larger the NI, the more "power" that goal has when influencing the other goals. The second index is the *influence (I) ratio.* It focuses more on the efficiency of a goal. Since the net influence of a goal can be positive or negative and IR needs to be positive, the following algorithm was used to calculate IR for each goal SDG_x, where N is the number of SDGs:

Finally, the third index defined as the *priority index* (PI) is a proportionally weighted average of the influence ratio (IR) and net influence (NI) relative to the smallest and largest values for each set of goals:

$$PI = A\left(\frac{IR_x - \min IR}{\max IR - \min IR}\right) + B\left(\frac{NI_x - \min NI}{\max NI - \min NI}\right)$$

$$A + B = 1$$

A: weight of influence ratio
B: weight of net influence
min: minimum value in set of all goals
max: maximum value in set of all goals

Figure 4. Influence and dependence graph

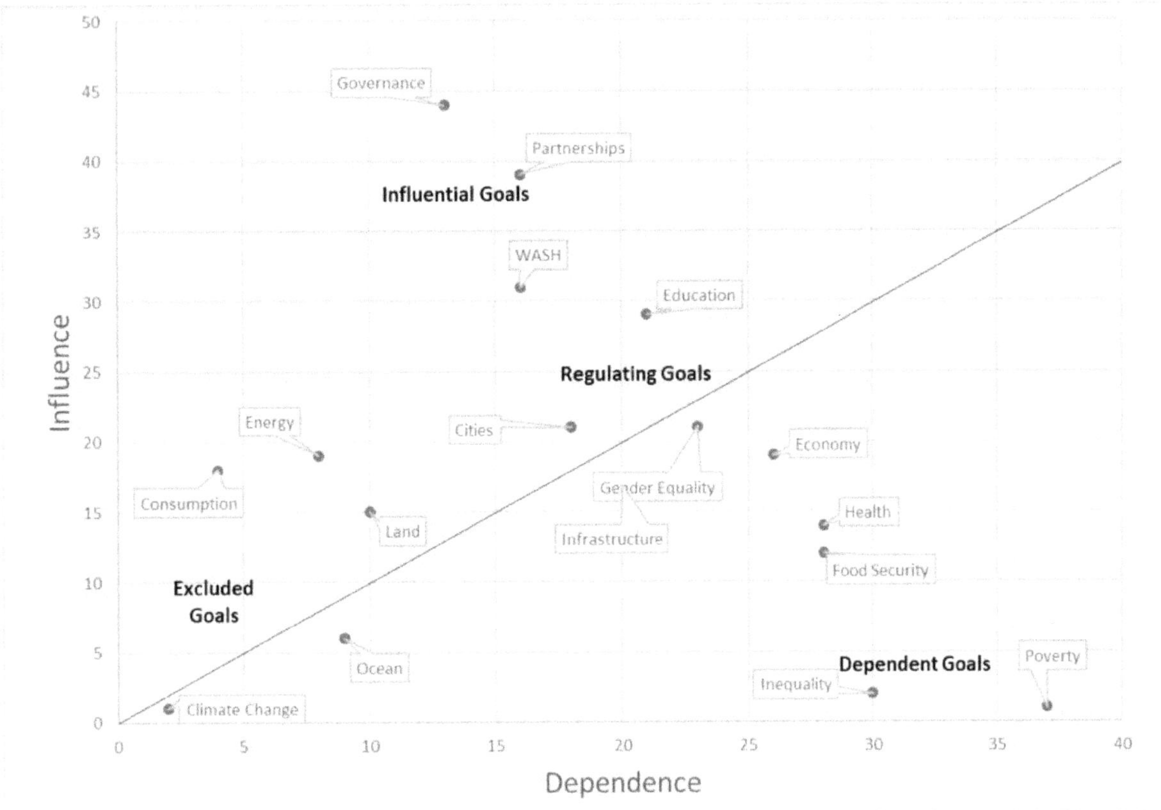

Box 1.

	I>0	**I=0**	**I<0**
D<0	$IR = \frac{I_x}{D_x}$	*IR*=0	
D=0	*IR*=3(*N* – 1)		
D>0	$IR = \frac{I_x}{I_x + \lvert D_x \rvert}$	$IR = \frac{1}{3(N-1)}$	$IR = \frac{D_x}{I_x}$

This index ranges between 0 and 1 with small values indicating low priority goals and high values representing high priority goals. The priority index values listed in Table 4 assumed equal weights (A = B = 0.5) for the influence ratio and the net influence. It should be noted that the two weights A and B do not always have to be equal; however, their sum must always be equal to 1. There might be circumstances, however, for which the weights A and B are not the same. For instance, if policymakers want to focus on generating as much impact as possible, then they would focus solely on the net influence and select B = 1 (A = 0). Conversely, A = 1 (B = 0) would correspond to the most efficient allocation of resources.

Dependence is to input as influence is to output. The influence ratio illustrates the efficacy of a goal. A large ratio implies that *relatively* few resources will yield a *relatively* significant benefit in terms of aggregate impacts. The higher the IR, the more it helps to accomplish the other goals. The net influence NI tells the likely size of that *absolute* impact. Although similar, RI and NI tell two different pieces of information. Just because RI is large does not necessarily indicate it will have a significant impact. Consumption (Goal 12) is an example: it has the largest ratio but is fourth for NI. Goal 12 will produce the highest amount of good, or utility, relative to its cost, but it will not generate the most good in absolute terms. It will be cheaper to allocate resources with the highest relative pay-out, but if the goal is to generate the most significant overall impact, then Goal 16 or 17 would be better.

Using the IR and NI together should aid policymakers in identifying what is best for their situation, which is why the priority index was introduced in the first place. Poorer countries might elect to focus on the influence ratio since resources are scarce, but richer countries might want to focus on a higher net influence to see positive change faster. In short, the main differences are relative vs. absolute and efficiency vs. cost for the influence ratio and net influence, respectively. Regardless, an influence ratio below one, and a net influence below zero should be avoided. Only when calculable and noticeable progress is made on the more attractive goals should the dependent goals be addressed. Both the influence ratio and net influence were used to produce a priority ranking, as shown in Table 4 of the SDGs, according to the equation above.

Table 4. Priority order of the SDGs

Description	Goal	IR	NI	PI
Governance	Goal 16	3.38	31	0.88
Consumption	Goal 12	4.50	14	0.87
Partnerships	Goal 17	2.44	23	0.71
Energy	Goal 07	2.38	11	0.61
WASH	Goal 06	1.94	15	0.60
Education	Goal 04	1.38	8	0.48
Cities	Goal 11	1.17	3	0.42
Land	Goal 15	1.10	1	0.40
Infrastructure	Goal 09	1.00	0	0.38
Climate Change	Goal 13	1.00	0	0.38
Gender Equality	Goal 05	0.91	-2	0.36
Ocean	Goal 14	0.67	-3	0.32
Economy	Goal 08	0.73	-7	0.30
Health	Goal 03	0.50	-14	0.22
Food Security	Goal 02	0.43	-16	0.20
Inequality	Goal 10	0.07	-28	0.07
Poverty	Goal 01	0.03	-36	0.01

The priority analysis depicted in this section is only one method (not *the* method) to analyze a cross-impact matrix. Because it is semi-quantitative, any scale can be used, and one that avoids negative numbers would simplify the process. For example, the algorithm in Table 3 can be removed entirely in lieu of just one equation that is applicable in all cases. The scale can be expanded to add more precision; the current scale has four states (none, low, medium, and high), but a scale can range from zero to ten (negative and positive). The scale is entirely dependent on the context and the goal of the analysis, i.e., the scale is relative.

Network Analysis

The cross-impact analysis helps to describe the influence and dependence of multiple interacting variables – the SDGs in this case. The interaction can also be represented as a network diagram or graph consisting of multiple nodes (vertices or actors) representing each SDG and links (edges or ties). In directed graphs, arrowheads pointing inward to or outward from some nodes can be added to map how the SDGs influence or depend on each other. When the strength of and the connectivity between nodes are represented by symbols of varied sizes such as lines with different thicknesses or colors, the network is defined as a valued network (Borgatti et al., 2002) or weighted network (Newman, 2010).

Several remarks can be made about using network analysis in analyzing the SDG network. First, networks are usually classified in the literature into different groups, such as social networks, technological networks, biological networks, and information networks (Newman, 2010). The network shown in Figure 5 does not fall into any of these categories, however. Although not technically a social network, the SDG network is intimately tied to human social interactions at varying scales. For that reason, we refer to it as an *anthropocentric network*, the network version of an anthropocentric system.

Network analysis complements cross-impact analysis as it provides a mathematical approach to understanding patterns of interactions among the SDGs. The term cross-impact network analysis is proposed to describe the synergist aspect of these two forms of analysis.

A network approach to the SDGs has not received much attention until recently. Le Blanc (2015) looked at the SDGs as part of a network (before their release), where the interconnectivity was assumed to take place at the target-level. The analysis was carried out considering the relationships that exist between the 17 goals and their 169 targets. In the network's literature, this type of analysis is referred to as a two-mode network analysis. The approach selected herein is slightly different as it considers how the 17 goals interact with each other, which can be described as a one-mode network analysis. It starts with the same (17 x 17) matrix considered in the cross-impact analysis (e.g., Figure 3). The matrix, also called the adjacency matrix in network analysis, was input into the network analysis software UCINET, which is available through *Analytic Technologies* based in Lexington, KY (Borgatti et al., 2002). Figure 5 shows a directed, valued network representation of Figure 3, where each node size depends on the value of the priority index (PI) calculated in the cross-impact analysis. The thickness of the links in Figure 5 represents the strengths of the SDG interactions.

Figure 5 is more than just a graphical representation of how the SDGs are interconnected. As discussed by Borgatti et al. (2013) for social networks, a graph such as Figure 4 can be used to identify the strengths, weaknesses, and patterns of interaction; identify potential attractors; make predictions; assess the network resilience; map assets and forms of vulnerabilities; conduct simulations; and plan interventions that leverage or strengthen existing network connections. Borgatti et al. (2013) reviews some of the measures and properties than can be inferred from directed graphs such as Figure 4 at the

Figure 5. Network diagram map of the SDGs; nodes are sized based on the value of their priority index; the thickness of each line depicts the interaction strength

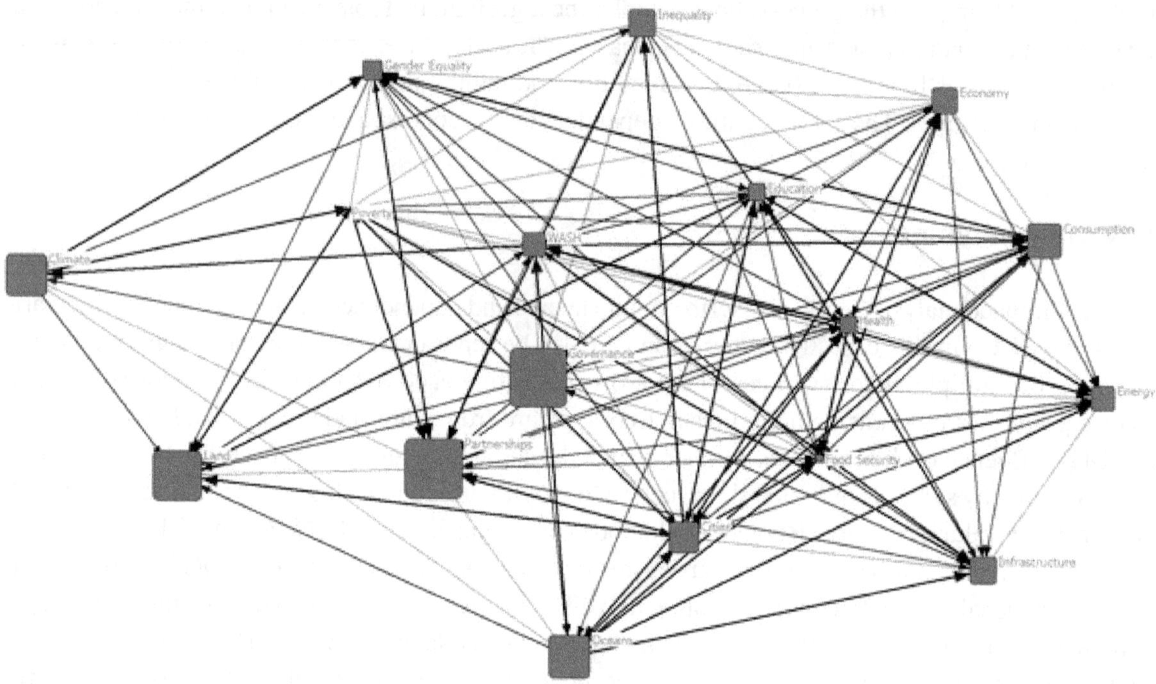

network-level (cohesion, shape), the node-level (different forms of centrality, betweenness), and the link-level (cohesion. equivalence).

Centrality represents a family of concepts that help indicate the significance of the contribution that a node makes on the network as a whole. Centrality can be understood as the extent to which a specific actor in a network is important. An actor with high centrality could be considered a hub or connected to many other actors. It translates into having a large influence on the cross-impact analysis. Among the many measures of centrality, the simplest one is the **degree centrality,** which is the number of actors each other actor connects (Borgatti et al., 2013). Since the network of the 17 SDGs is directed and valued, the centrality of each node of the network (each SDG) needs to be broken down into *outdegree centrality* and **in-degree centrality** (Borgatti et al., 2013). The outdegree centrality (sum of values of outgoing links) and in-degree centrality (sum of values of ingoing links) of the SDGs are respectively equal to the row sums (total influence) and column sums (total dependence) of the adjacency matrix in Figure 3.

Figure 5 is one possible representation of the SDG network, among many others. As discussed further by (Borgatti et al., 2013), the nodes in Figure 5 do not have to be placed at random. Other graph layouts can be generated by using the centrality values as a priority. Likewise, multidimensional scaling (MDS) can be used to graphically represent the nodes so that the "distances between the points [nodes] correspond in a predetermined way to the proximities among objects in the data" (Borgatti et al., 2013). Essentially, the closeness of the distance between two nodes could be used to represent how similar or dissimilar they are to one another.

DISCUSSION

Combining cross-impact analysis and network analysis has many advantages when analyzing the SDGs. More specifically, the approach enables decision-makers to: (i) attain a better qualitative and quantitative understanding of the way the SDGs interact; (ii) detect emerging patterns resulting from those interactions; (iii) use context-specific information about direct impacts to identify indirect effects; and (iv) identify leverage points hidden within the SDGs. CINA presented above can be repeated at the SDG target-level. This could help in understanding better how the SDGs interact with each other. The same could also be done at the indicator level.

Cross impact analysis also has limitations. As noted by Checkland and Poulter (2006), the limitation of a soft systems approach is that it uses an action-oriented learning process for learning to address problem areas and not the content of the problems. The second limitation of cross-impact network analysis is that it is static; it represents a snapshot of a system in time and cannot handle dynamic (time-dependent) issues.

A third limitation is that CINA is entirely based on human input, ideally, expert judgment. The cross-impact matrix, which is at the fundamental core of cross-impact network analysis, requires judging experts to estimate how variables interact with each other, the degree of their interactions, and the possible results of their impacts (Weimer-Jehle, 2006). Essentially, the quality of the analysis depends on the accuracy and expertise of the people undertaking that analysis. The experts filling out the cross-impact matrix are "expected to possess insights which rather should be the results of an analysis (Weimer-Jehle, 2006, p.337)." This is the classic *Catch-22*, where input depends on output, rendering the analysis ineffective through *circle-logic*.

Fourth, like the cross-impact analysis, network analysis should only be used to strategically understand the interactions across a network. It pays less attention to the *nature* and *dynamic* of those interactions, which can be handled better both qualitatively and quantitatively by hard-system tools such as system dynamics. Network analysis provides information indicative of the network as a whole, but errors and problems could arise as the analysis becomes more precise, especially when decisions are made about intervening at the SDG-level.

Moreover, lastly, for the numerical example considered in Figure 1 and represented with the influence-dependence graph (Figure 3) and the network diagram (Figure 4), the centrality of the SDGs is very high, which we expect to hold in most scenarios. Most SDGs have some form of direct connection (one degree of separation) with the other SDGs (Le Blanc, 2015; Nilsson, 2017; Nilsson et al., 2017). Those that do not are indirectly connected with at most only two degrees separating them from any other SDG. This indicates that the SDG network is tight, and tighter networks are more impacted by fewer effects than looser networks.

Case Study

San Antonio, Texas, USA, is home to a growing and rapidly urbanizing population, has significant agricultural activity surrounding the city and lies over the Eagle Ford shale play - with its growing production of oil and natural gas due to hydraulic fracturing technology. San Antonio region represents a resource hotspot whose stakeholders across sectors compete for its limited water, land, and financial resources. To effectively address future resource challenges and ensure the sustainable urbanization of the city, it is essential that stakeholders are aware of the extent of interdependence between their decisions. Figure 6 provides a visual representation of the San Antonio hotspot. The green, red, and blue dots represent groundwater wells for agriculture, oil & gas production, and municipal uses. These groundwater wells for the growing municipal, agricultural, and energy sectors, are competing over the same water aquifers. Addressing the challenge posed by the hotspot requires holistic, localized, transdisciplinary, multi-stakeholder approaches. Plans for strategic water reserves exit but carry a high price tag. Understanding the interconnections (nexus) between these resource systems can help reduce the competing demands, and lower both the costs and the risks associated. Achieving such an understanding demands more than knowledge of sectoral geophysical realities: it requires cross-sectoral communication and cooperation.

Figure 6. FEW hotspot and social network in San Antonio, Texas (Daher et al., 2019)

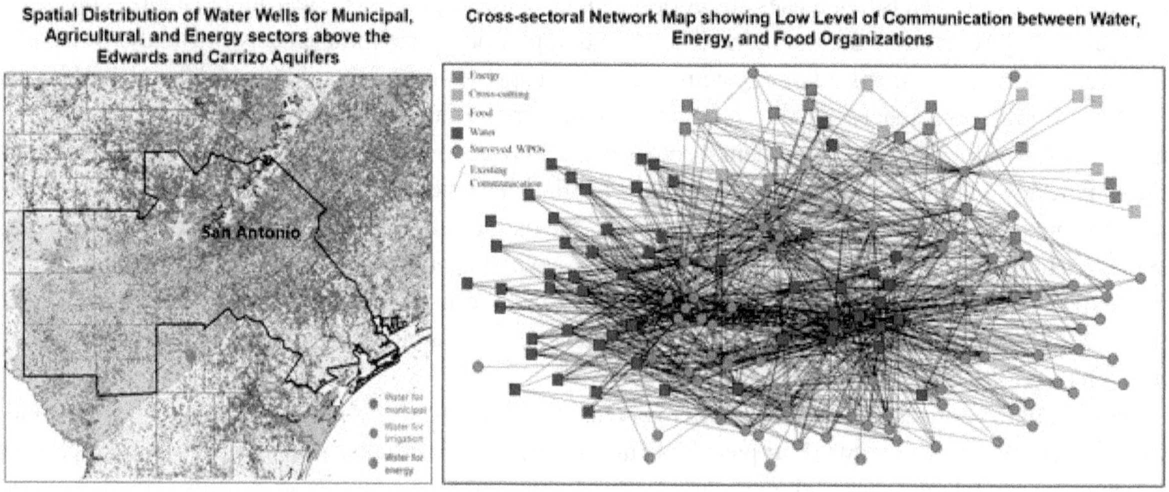

Social Network Analysis (SNA) is used to provide an understanding of the relations between stakeholders (Scott, 2000; Wasserman and Faust, 1994). Daher et al. (2019) sent a survey to public water officials San Antonio asking them about their frequency of communication with other water, energy, and food stakeholders in the region. The survey results showed modest levels of communication between stakeholders. There was a higher frequency of communication among people within the water sector, but much lower with energy and food. Different barriers result in this low level of communication, including *financial* (who will pay for the time and effort involved in pursuing increased communication?), *difference in language* (units, abbreviations, syntax and context of problems and solutions *), and institutional barriers (*the lack of mechanisms for increased communication and integrative planning*). C*omplementing the analysis of the completion between the physical resource systems, with that of stakeholder interactions connected to those systems, allows for identifying a broader set of barriers and solutions which could be explored.

CONCLUSION

By organizing a complicated system into categories, called sectors, they can also be considered a complicated-coupled system and instead viewed as an interconnected nexus. This approach, broadly defined as the nexus approach, opens up a system to be analyzed and modeled through a whole host of tools and techniques. This nexus approach is versatile and is applicable in looking at economies, cities, communities, and many areas across sustainable development and anthropocentric systems.

Cross-impact network analysis is a powerful tool that simplifies the process to prioritize sectoral importance by relying on an analytical method rather than pure educated guesses and intuitions. It illuminates hidden properties and emergent patterns that can be used to identify leverage points where policy can be applied most effectively to design and implement sustainable development strategies.

CINA is but one way that the SDGs can be viewed and examined as an integrated system or network, but other hard-system approaches are required to extract a deeper understanding, such as system dynamics (SD), which focuses on interactions affecting the system through time. A cross-impact matrix is mathematically ingrained as is a structural nexus in that approach, but it is iterated hundreds of times to model the long-term effects of interactions instead of the static analysis at one point in time as is the case from CINA.

REFERENCES

Allen, C., Metternicht, G., & Wiedmann, T. (2019). Prioritising SDG targets: Assessing baselines, gaps and interlinkages. *Sustainability Science*, *14*(2), 421–438. doi:10.100711625-018-0596-8

Amadei, B. (2015). *A Systems Approach to Modeling Community Development Projects* (M. Noori, Ed.). Momentum Press Engineering.

Amadei, B., & Zelinka, D. (2018). *Integrated Modeling of the Water-Energy-Land-Food Nexus: Importance of Context, Scale, Boundary, and Community Capacity*. In *24th International Sustainability Development Research Society Conference*. Messina, Italy.

Arcade, J., Godet, M., Meunier, F., & Roubelat, F. (1999). *Structural analysis with the MICMAC method & Actor's strategy with MACTOR method.* Futures Research Methodology, American Council for the United Nations University: The Millennium Project.

Armenia, S., Pompei, A., Castaño Barreto, A. C., Atzori, A. S., & Fonseca, J. M. (2019). The Rural-Urban Food Systems' Links with the Agenda 2030: From FAO Guidelines on Food Supply and Distribution Systems to a Dairy Sector Application in the Area of Bogota. *Systems*, *7*(3), 45. doi:10.3390ystems7030045

Bajracharya, A. (2014). A Qualitative Systems Thinking Approach in Understanding the Implementation of Innovation on Construction Projects. *International Journal of System Dynamics Applications*, *3*(4), 87–108. doi:10.4018/ijsda.2014100105

Baporikar, N. (2016). Infrastructure Development as a Catalyst for Social-Economic Advancement. *International Journal of System Dynamics Applications*, *5*(4), 101–113. doi:10.4018/IJSDA.2016100106

Barbier, E. B. & Burgess, J. C. (2017). The sustainable development goals and the systems approach to sustainability. *Economics, 11*.

Barorikar, N. (2020). Logistics Effectiveness Through Systems Thinking. *International Journal of System Dynamics Applications*, *9*(2), 16. doi:10.4018/IJSDA.2020040104

Benson, T., & Marlin, S. (2017). *Systems Thinking Overview*. Retrieved from http://watersfoundation.org/systems-thinking/overview/

Bollen, K. A. (1987). Total, direct, and indirect effects in structural equation models. *Sociological Methodology*, *17*, 37–69. doi:10.2307/271028

Borgatti, S. P., Everett, M. G., & Freeman, L. (2002). *UCINET 6 for Windows: Software for Social Network Analysis*. Analytic Technologies.

Borgatti, S. P., Everett, M. G., & Johnson, J. C. (2013). *Analyzing Social Networks*. SAGE Publications.

Braun, W. (2002). The System Archetypes. *System*, *2002*, 27. https://www.albany.edu/faculty/gpr/PAD724/724WebArticles/sys_archetypes.pdf

Capra, F., & Luisi, P. L. (2016). *The Systems View of Life: A Unifying Vision*. Cambridge University Press.

Checkland, P., & Poulter, J. (2006). *Learning for action: a short definitive account of soft systems methodology and its use for practitioner, teachers, and students*. Wiley.

Collste, D., Pedercini, M., & Cornell, S. E. (2017). Policy coherence to achieve the SDGs: Using integrated simulation models to assess effective policies. *Sustainability Science*, *12*(6), 921–931. doi:10.100711625-017-0457-x PMID:30147764

Coopman, A., Osborn, D., Ullah, F., Auckland, E., & Long, G. (2016). *Seeing the whole: Implementing the SDGs in an Integrated and Coherent Way*. Retrieved from https://sf.stakeholderforum.org/index.php/our-publications-sp-1224407103/reports-in-our-publications/625-seeing-the-whole-implementing-the-sdgs-in-the-an-integrated-and-coherent-way

Costanza, R., Daly, L., Fioramonti, L., Giovannini, E., Kubiszewski, I., Fogh, L., … Wilkinson, R. (2016). The UN Sustainable Development Goals and the Dynamics of Well-being. *Solutions*, *7*(1), 20–22. Retrieved from https://www.thesolutionsjournal.com/article/the-un-sustainable-development-goals-and-the-dynamics- of-well-being/%0AThe

Daher, B., Hannibal, B., Portney, K. E., & Mohtar, R. H. (2019). Towards creating an environment of cooperation between water, energy, and food stakeholders in San Antonio. *The Science of the Total Environment*, *651*, 2913–2926. doi:10.1016/j.scitotenv.2018.09.395 PMID:30463143

Daher, B., Mohtar, R. H., Pistikopoulos, E. N., Portney, K. E., Kaiser, R., & Saad, W. (2018). Developing socio-techno-econo-political (STEP) solutions for addressing resource nexus hotspots. *Sustainability*, *2018*(10), 512. doi:10.3390u10020512

de Grandbois, Y. (2016). The Story of Service Science. In *Service Science and the Information Professional* (pp. 27–46). Chandos Publishing. doi:10.1016/B978-1-84334-649-4.00002-0

Easterly, W. (2007). *The White Man's Burden: Why the West's Efforts to Aid the Rest Have Done So Much Ill and So Little Good*. Oxford University Press.

Elharakany, R. A., Moscardini, A., Khalifa, N. M., & Elghany, M. M. A. (2018). Modelling the Effect on Quality of Information and Communications Technology (ICT) facilities in Higher Education: Case Study—Egyptian Universities. *International Journal of System Dynamics Applications*, *7*(3), 1–30. doi:10.4018/IJSDA.2018070101

Fath, B. D. (2017). Systems ecology, energy networks, and a path to sustainability. *International Journal of Design & Nature and Ecodynamics*, *12*(1), 1–15. doi:10.2495/DNE-V12-N1-1-15

Ford, A. (2010). *Modeling the environment* (2nd ed.). Island Press.

Glenn, J. C. (2004). *Introduction to the Futures Research Methods Series*. Futures Research Methodology - Version 2.0. Retrieved from http://mp.cim3.net/file/project/mp-sofi-sd/reference/01-Introduction.PDF

Global Goals. (2016). i. Retrieved from http://www.globalgoals.org

Gordon, T. J. (2014). Cross-Impact Analysis. In *Introduction to the futures methods research series. Futures research methodology, V3.0*. The Millennium Project.

Gordon, T. J., & Hayward, H. (1968). Initial experiments with the cross impact matrix method of forecasting. *Futures*, *1*(2), 100–116. doi:10.1016/S0016-3287(68)80003-5

Grabowski, F., & Strzałka, D. (2008). Simple, complicated and complex systems - The brief introduction. *2008 Conference on Human System Interaction, HSI 2008*, 570–573. 10.1109/HSI.2008.4581503

Graedel, T. E., & Allenby, B. R. (2010). *Industrial Ecology and Sustainability Engineering*. Pearson.

Hendrickson, C., Matthews, H. S., Matthews, D. H., Joshi, S., Maclean, H. L., Lave, L. B., … Bergerson, J. A. (2006). Environmental Life Cycle Assessment of Goods and Services: An Input-Output Approach. Taylor and Francis.

Hurni, H., Giger, M., Liniger, H., Mekdaschi Studer, R., Messerli, P., Portner, B., Schwilch, G., Wolfgramm, B., & Breu, T. (2015). Soils, agriculture and food security: The interplay between ecosystem functioning and human well-being. *Current Opinion in Environmental Sustainability*, *15*, 25–34. doi:10.1016/j.cosust.2015.07.009

ICSU & ISSC. (2015). *Review of Targets for the Sustainable Development Goals: The Science Perspective*. Paris, France: International Council for Science. Retrieved from http://www.icsu.org/publications/reports-and-reviews/review-of-targets-for-the-sustainable-development-goals-the-science-perspective-2015/SDG-Report.pdf

Larsen, M. A. D., & Drews, M. (2019). Water use in electricity generation for water-energy nexus analyses: The European case. *The Science of the Total Environment*, *651*, 2044–2058. doi:10.1016/j.scitotenv.2018.10.045 PMID:30321726

Le Blanc, D. (2015). Towards Integration at Last? The Sustainable Development Goals as a Network of Targets. *Sustainable Development*, *23*(3), 176–187. doi:10.1002d.1582

Liu, J., Hull, V., Godfray, H. C. J., Tilman, D., Gleick, P., Hoff, H., Pahl-Wostl, C., Xu, Z., Chung, M. G., Sun, J., & Li, S. (2018). Nexus approaches to global sustainable development. *Nature Sustainability*, *1*(9), 466–476. doi:10.103841893-018-0135-8

Mandl, C. E. (2019). *Managing Complexity in Social Systems*. Springer., doi:10.1007/978-3-030-01645-6

Meadows, D. (1999). *Leverage Points: Places to Intervene in a System*. The Sustainability Institute. Retrieved from http://donellameadows.org/archives/leverage-points-places-to-intervene-in-a-system/

Meadows, D. (2008). *Thinking in Systems: A Primer* (D. Wright, Ed.). Chelsea Green Publishing.

Merriam-Webster. (2020). *Nexus*. Author.

Mitchell, M. (2011). *Complexity: A Guided Tour*. Oxford University Press.

Mobus, G. E., & Kalton, M. C. (2015). *Principles of Systems Science*. Springer. doi:10.1007/978-1-4939-1920-8

Moffatt, I., & Hanley, N. (2001). Modeling sustainable development: Systems dynamic and input-ouput approaches. *Environmental Modelling & Software*, *16*(6), 545–557. doi:10.1016/S1364-8152(01)00024-X

Mohtar, R. H., & Daher, B. (2012). Water, Energy, and Food: The Ultimate Nexus. In *Encyclopedia of Agricultural, Food, and Biological Engineering* (2nd ed., pp. 1–15). Taylor & Francis.

Mohtar, R. H., & Daher, B. (2016). Water-Energy-Food Nexus Framework for facilitating multi-stakeholder dialogue. *Water International*, *41*(5), 655–661. doi:10.1080/02508060.2016.1149759

Moyo, D. (2009). Dead Aid: Why Aid Is Not Working and How There Is Another Way For Africa. New York: Academic Press.

Müller, A., Janetschek, H., & Weigelt, J. (2015). Towards a governance heuristic for sustainable development. *Current Opinion in Environmental Sustainability*, *15*, 49–56. doi:10.1016/j.cosust.2015.08.007

New York Times. (1946, May 25). Atomic Education Urged by Einstein: Scientist in Plea for $200,000 to Promote New Type of Essential Thinking. *New York Times*, p. 11. Retrieved from https://timesmachine.nytimes.com/timesmachine/1946/05/25/100998236.html?pageNumber=11

Newman, M. E. J. (2018). *Networks: An Introduction* (2nd ed.). Oxford University Press. doi:10.1093/oso/9780198805090.001.0001

Nilsson, M. (2017). *Important interactions among the Sustainable Development Goals*. High-Level Political Forum 2017 (No. 2017-06). Retrieved from www.sei-international.org

Nilsson, M., Griggs, D., & Visback, M. (2016). Map the interactions between Sustainable Development Goals. *Nature*, *534*(15), 320–322. doi:10.1038/534320a PMID:27306173

Nilsson, M., Griggs, D., Visbeck, M., Ringler, C., Mcollum, D., Mollier, L., … Unger, S. (2017). A Guide to SDG Interactions. In Science to Implementation. Paris, France: International Council for Science. doi:10.24948/2017.01

Nilsson, M., Lucas, P., & Yoshida, T. (2013). Towards an integrated framework for SDGs: Ultimate and enabling goals for the Case of energy. *Sustainability*, *5*(10), 4124–4151. doi:10.3390u5104124

Richmond, B. (1994). Systems thinking/system dynamics: let's get on with it. In *1994 International Systems Dynamics Conference*, (vol. 10, pp. 135–157). 10.1002dr.4260100204

Rickles, D., Hawe, P., & Shiell, A. (2007). A simple guide to chaos and complexity. *Journal of Epidemiology and Community Health*, *61*(11), 933–937. doi:10.1136/jech.2006.054254 PMID:17933949

Ringler, C., Bhaduri, A., & Lawford, R. (2013). The nexus across water, energy, land and food (WELF): Potential for improved resource use efficiency? *Current Opinion in Environmental Sustainability*, *5*(6), 617–624. doi:10.1016/j.cosust.2013.11.002

Ritchie-Dunham, J. L., & Rabbino, H. T. (2001). *Managing from Clarity: Identifying, Aligning, and Leveraging Strategic Resources*. John Wiley.

Sachs, J., Schmidt-Traub, G., Kroll, C., Durand-Delacre, D., & Teksoz, K. (2016). *SDG Index & Dashboards - Global Report*. Bertelsmann Stiftung and Sustainable Development Solutions Network.

Schmidt-Traub, G., Karoubi, E. de la M., & Espey, J. (2015). *Indicators and a Monitoring Framework for the Sustainable Development Goals: Launching a Data Revolution*. Sustainable Development Solutions Network.

Scott, J. (2000). *Social Network Analysis: A Handbook*. SAGE Publications.

Serdar Asan, S., & Asan, U. (2007). Qualitative cross-impact analysis with time consideration. *Technological Forecasting and Social Change*, *74*(5), 627–644. doi:10.1016/j.techfore.2006.05.011

Shanbhag, N., & Pardede, E. (2019). The Dynamics of Product Development in Software Startups: The Case for System Dynamics. *International Journal of System Dynamics Applications*, *8*(2), 51–77. doi:10.4018/IJSDA.2019040104

Sterman, J. D. (2006). Learning from evidence in a complex world. *American Journal of Public Health*, *96*(3), 505–514. doi:10.2105/AJPH.2005.066043 PMID:16449579

Thwink.org. (2014). *The Three Pillars of Sustainability*. Retrieved from http://www.thwink.org/sustain/glossary/Sustainability.htm

Todorov, V., & Marinova, D. (2011). Modelling sustainability. *Mathematics and Computers in Simulation*, *81*(7), 1397–1408. doi:10.1016/j.matcom.2010.05.022

United Nations. (2016). *Sustainable Development Goals Report*. Author.

United Nations General Assembly. (2015). Transforming our world: the 2030 Agenda for Sustainable Development, Pub. L. No. 70/1, 16301 General Assembly Seventieth Session 35. New York.

US Office of Management and Budget. (2017). *North American Industry Classification System*. NAICS.

Vladimirova, K., & Le Blanc, D. (2016). Exploring Links Between Education and Sustainable Development Goals Through the Lens of UN Flagship Reports. *Sustainable Development*, *24*(4), 254–271. doi:10.1002d.1626

Von Stechow, C., Minx, J. C., Riahi, K., Jewell, J., McCollum, D. L., Callaghan, M. W., Bertram, C., Luderer, G., & Baiocchi, G. (2016). 2°C and SDGs: United they stand, divided they fall? *Environmental Research Letters*, *11*(3), 1–15. doi:10.1088/1748-9326/11/3/034022

Wang, S., Cao, T., & Chen, B. (2017). Urban energy–water nexus based on modified input–output analysis. *Applied Energy*, *196*(February), 208–217. doi:10.1016/j.apenergy.2017.02.011

Wasserman, S., & Faust, F. (1994). *Social Network Analysis: Methods and Applications*. Cambridge University Press. doi:10.1017/CBO9780511815478

Weigelt, J., Janetschek, H., Müller, A., & Töpfer, K. (2015). Editorial overview: Environmental change issues: Soils in the nexus. *Current Opinion in Environmental Sustainability*, *15*, v–viii. doi:10.1016/j.cosust.2015.09.006

Weimer-Jehle, W. (2006). Cross-impact balances: A system-theoretical approach to cross-impact analysis. *Technological Forecasting and Social Change*, *73*(4), 334–361. doi:10.1016/j.techfore.2005.06.005

Weimer-Jehle, W. (2008). Cross-impact balances: Applying pair interaction systems and multi-value Kauffman nets to multidisciplinary systems analysis. *Physica A*, *387*(14), 3689–3700. doi:10.1016/j.physa.2008.02.006

Zelinka, D. (2019). *Systems Approaches to Transdisciplinary Sustainable Development Issues (No. 13810159)*. Proquest. ProQuest.

Zelinka, D., & Amadei, B. (2017). A Methodology to Model the Integrated Nature of the Sustainable Development Goals: Importance for Engineering Education. *American Society for Engineering Education Conference*.

Zelinka, D., & Amadei, B. (2019a). A Systems Approach for Modeling Interactions Among the Sustainable Development Goals Part 1: Cross-Impact Network Analysis. *International Journal of System Dynamics Applications*, *8*(1), 41–59. doi:10.4018/IJSDA.2019010103

Zelinka, D., & Amadei, B. (2019b). A Systems Approach for Modeling Interactions Among the Sustainable Development Goals Part 2: System Dynamics. *International Journal of System Dynamics Applications*, *8*(1), 41–59. doi:10.4018/IJSDA.2019010103

Chapter 2
Modeling the Sustainable Development Nexus as a Complex-Coupled System:
System Dynamics Modeling

David Zelinka
https://orcid.org/0000-0003-2457-5813
Civitas Systems, USA

Bassel Daher
Energy Institute and the Institute for Science, Technology, and Public Policy, Texas A&M University, USA

ABSTRACT

This chapter begins with defining complex systems, presents an overview of the applied science of dynamical systems by focusing on the main components of complexity and chaos, and introduces the concept of dimensionality of systems. Systems have structural and temporal (dynamic) components – they exist in space and time. This chapter focuses on the time dimension, called temporality. The authors classify a third dimension, chaos (randomness), and illustrate that all systems can be defined according to their structure, dynamics, and chaos. These three dimensions constitute the dimensionality of systems, which can be used to define and categorize all types of systems. A system dynamics model to quantify the progress and interactions among the United Nation's Sustainable Development Goals (SDG) is introduced. The benefits and limitations of a system dynamics modeling approach in this context are then discussed.

INTRODUCTION

There are many types of systems (e.g., simple, complicated, complex, and so on), but they all have a structural and temporal (dynamic) dimension – they exist in space and time (Krakauer, 2019). It is recommended – but not necessary – that the reader looks at the first chapter for a detailed discussion of

DOI: 10.4018/978-1-7998-5788-4.ch002

the spatial (structure) component. This chapter will cover the dimension of time (temporality) using the overarching concept of complexity, as well as system dynamics (SD) modeling.

The analysis herein is based on the rationale that sustainable development issues form an anthropocentric network and coupled system consisting of many structural and spatial interactions among its subsystems (i.e., sectors). This holds true for the dynamic aspect of systems through time, so the Sustainable Development Goals (SDGs) will function as a case study in the second part of this chapter.

Although many authors have acknowledged the interactions among the SDGs in the sustainable development literature little has been done to model how the SDGs influence or depend on one another through time, and their long-term system dynamics have been relatively unexplored (see Allenet et al., 2017; Coopmanet et al., 2016; Le Blanc, 2015; Nilsson et al., 2016; Nilsson, 2017; Nilsson et al., 2017; Nilsson et al., 2013; UN Water, 2016; United Nations Economic and Social Council, 2015; United Nations General Assembly, 2015; Vladimirova & Le Blanc, 2016; Weitz et al., 2014; Daher et al., 2018; Stephan et al., 2018). Some notable works do exist including the SDGs and their targets as a network (Allen, Metternicht, & Wiedmann, 2019; Zhou & Moinuddin, 2017); the SDGs as various interacting sectors (Collste, Pedercini, & Cornell, 2017); and the SDGs as a dynamic system (Collste, Pedercini, & Cornell, 2017; Millennium Institute, n.d.; Zelinka & Amadei, 2019a). It is the goal of this chapter to show that the dynamics of the SDGs can be modeled in a way that is not too complex to understand while still producing informative results. Following this introduction section, the rest of the chapter is organized as follows:

Section 2 defines a complex system and presents an overview of the applied science of dynamical systems by focusing on the main components of complexity and chaos and introducing the concept of dimensionality of a system

Section 3 describes the limitations of more static approaches (see the previous chapter) to modeling systems and how system dynamics can be used to complement them, and it provides a brief overview of the components of causal-loop diagrams, stock-and-flow diagrams, and system dynamics modeling

Section 4 introduces the lower-level of the system dynamics model from the context of system archetypes based on the logistic diffusion model for the SDGs

Section 5 finishes building the system dynamics model by connecting the many structures in the lower-level together using a nested cross-impact matrix based on the nexus approach

Section 6 discusses the benefits and limitations of system dynamics modeling

Section 7 provides a brief overview of potential future work and then concludes

SYSTEMS SCIENCE: DYNAMICAL SYSTEMS

Interconnectivity and hierarchy describe a system's spatial and structural dimension, but systems and particularly complex ones, also have a time dimension called **temporality**. Systems for which time has an impact on their state and function are called dynamical systems as their temporality is strong. More accurately, though, the change in time is mathematically governed by a well-specified rule or set of rules, like an equation or algorithm (Feldman, 2019). This well-defined rule also makes the system deterministic by definition.

The effect of time is characterized by **iterative** cycles (recurring sequences) and **oscillations** (periodic movement between two bounded thresholds) and is inferred through the concept of feedback and causal loops in which the system can influence its future state based on 'memories' of its past, like learning. By learning, systems can adapt to changing conditions more effectively. Just as the interconnections connect the structure of the system (vertically or horizontally), the feedback loops and cycles connect a system with itself through time. The day after December 31st will always be January 1st; when each year's calendar cycle ends, it starts over.

Additionally, **delays** (time between a cause and effect) and **timescales** constitute the second component of the temporal dimension. A delay is a localized effect on an individual part of the system. A timescale mimics the structural element of scale (i.e., short, medium, long, and deep time) of the entire system itself.

Intuitively larger systems – that can scale temporally, spatially, or both – have a much higher likelihood to have more significant cycles and frequent feedback loops, which in turn increase the opportunity that any two cycles will interact and affect each other at any timescale. The global climate exemplifies a dynamical system with many interacting cycles happening at various timescales that can vary with:

- the four seasons (technically, weather) changing due to orbital position every year; they can also move later due to the feedback from economic activity via the emission of greenhouse gases
- solar minima and maxima occurring with eleven-year periods alter incoming solar radiation
- global climate patterns such as the Pacific Decadal Oscillation, the El Niño-Southern Oscillation, and the Pacific / North American Pattern are recurring regional climate patterns that affect the global climate; they can affect each other, especially when minima and maxima coincide (North Carolina Climate Office, 2019)
- the Milankovitch cycles, which are changes in the shape of Earth's tilt (obliquity), wobble (precession), and orbit about the sun (eccentricity), that vary on the order of tens to hundreds of thousands of years affect ice ages most noticeably and the changing position of the North Star every 26,000 years (Buis, 2020)
- economic activity according to human activity that has cycles (recessions and booms, bear and bull markets, and more) every year to every decade depending many factors and events

The list above is not exhaustive. They are examples of different cycles that happen at different intervals and scales whose effects interact with each other in seemingly chaotic ways, which can and usually do cause unforeseeable outcomes (i.e., unintended consequences). Cycles, timescales, and the impact of time are as ubiquitous as systems themselves since time is a fundamental dimension of a system.

Feedback and cycles describe one component of the temporality, and delays and timescales (i.e., short-, medium-, and long-term) describe the other part. Together they are what constitute dynamic hierarchy, a counterpart to interconnectivity and structural hierarchy, which itself also has two components – vertical and horizontal hierarchy.

Complex Systems

It might be best to think of dynamical systems as not just a type of system but a characteristic of them. A dynamical system is a system with a significant dynamic component relative to its structural one, which is the inverse of a complicated system. The structural and temporal components together make up a system's **complexity**. Large cities are more complex than small towns, and planning for a 100-

year timeframe requires considerable strategizing than merely planning for next year. Semantics aside, structure and dynamics are the two fundamental and interacting dimensions of a system.

Although all systems are dynamic because all change requires the passage of time as an enabling environment, the implication is that a dynamical system is one in which time has a significant influence on the structural interactions and function of the system. A pendulum moves through time as it swings, but its location can always be determined because the system is simple. If time were to accelerate, the pendulum would speed up, but otherwise, the system would behave in the same predictable way.

A main discerning – albeit abstract – characteristic that divides a complicated system from a **complex system** is that a complex system cannot be fully comprehended; there are just too many parts interacting and doing too many things. A complex system has high-levels of structural and temporal complexity; a complicated system has only high levels of structural complexity.

No matter how much effort is put in, there does not – necessarily – exist a perfect answer or optimum solution; there exists a continuum of possibilities or a multitude of desired states. For that reason, complex problems usually should not try to meet a goal, but rather minimize a cost or maximize a benefit function (a significant aspect of machine learning), which combines all the various trade-offs held within the system. For example, in economics, most decisions attempt to maximize utility between two or more options. What is the opportunity cost of choosing one product over another? Even between two options, the decision will incur some cost (i.e., an opportunity cost of picking one option over another). Sometimes, the best decision is simply the less bad one. These systems have no single solution, which is a reliable indicator of high complexity.

Sometimes entire new fields must be created to aid people in addressing complex problems. The economic field of game theory exists for that reason (think Prisoner's Dilemma); it is the "study of the ways in which interacting choices of economic agents produce outcomes with respect to the preferences (or utilities) of those agents (Hayes, 2019)." The tools to understand complexity exists, but one must often look across the disciplinary divides to find them. This chapter stresses multidisciplinary in all of its forms and applications.

All the interacting parts of a complex system cause behavior that is ultimately unpredictable, leading to unintended consequences. This unpredictability, aka noise and randomness, is why complex systems require statistical distributions. Outcomes are understood through a range of possibilities and average states. For example, the chance of precipitation in Boulder, Colorado, for March 21st is 17%. Go out there on March 21st, and odds are it will not be 17%; it could be any value. No matter how much effort is put forth, it is impossible to accurately say what the precipitation is going to be without directly measuring it at a specific time and location. Predicting the weather (as well as climate) is the epitome of a complex problem. There is no perfect answer, just one that is good enough that can report probability and precision. That precision, i.e., deviation, allows us to account for randomness (chaos, discussed soon) in complex systems.

It is a complex challenge to achieve food security in a drier world with growing frequency shocks and rapidly changing and unpredictable climatic conditions. The food system is interconnected with a large number of other natural and human-designed systems. These systems do not exist in a vacuum. They are managed, regulated, and consumed by different actors that vary in their goals, preferences, and value systems. There exists no optimum solution for addressing such a challenge; instead, there is a need to consider a continuum of possibilities or desired states which would come as a result of scenario planning. While hard to predict, patterns could be recognized and contextualized while dealing with complex systems.

Complex-coupled systems (CCS) are very similar to the complicated-coupled discussed in the previous chapter, but they have higher levels of structural and dynamic complexity. They are composed of many vertical and hierarchical levels and short-, medium-, long-term, and sometimes deep time factors must be considered. Economies can be thought of as either a complicated-coupled system or complex-coupled system depending on the precision of the analysis or the size and level of economic development. If a national economy is a complicated-coupled system, the global economy can be considered a complex-coupled system. Another definition could divide them by *structural coupling* and *dynamic coupling*: how is the system coupled in space and in time?

Because all elements of a system are connected, none can be ignored, which is the concept of **holism**. It is generally a philosophical term but takes on technical significance in systems science. All dimensions of a system need to be understood together from various perspectives and not individually. Reducing a system and analyzing it only by its parts (subsystem-level) destroys crucial insight and perspectives that can only be seen at the system-level. Even though, mathematically, all the pieces of the puzzle are there, the meaning of the picture is lost or incomplete because their relationships are missing. Holism leads directly to the concept of **nonlinearity**. When systems behave in unexpected ways, they are said to be nonlinear. Proportional and linear are the same, mathematically. In the most simplistic sense, if a system is composed of two variables, x & y, and x directly affects y but not the other way around, then if x were to double, so would y. If x were halved, y would be halved. A linear system can be represented by $mx+b$ where the parameter that determines the linearity and proportionality of the systems is the slope, m. In a nonlinear system, this is not the case; some other equation or relationship determines the system such as a polynomial (ax^2+bx+c) power relationship (kx^a) or exponential (ab^x), to name a few. What does that actually mean; Rickles et al. (2007) describes it perfectly:

*Complex and chaotic systems are both examples of nonlinear dynamical systems. A linear system is characterized by the satisfaction of the superposition principle. The superposition principle says that if A and B are both solutions for some system (ways in which the system could evolve), then so is their sum A + B—this implies that a linear system can be decomposed into its parts and each part solved separately to construct the full solution. For nonlinear systems, this is not possible because of the appearance of nonlinear terms, functions of the variables such as sin(x), x^3 and x*y. In this sense then, the whole here is more than the sum of its parts. Given this, a nonlinear system is one for which inputs are not proportional to outputs: a small (large) change in some variable or family of variables will not necessarily result in a small (large) change in the system. (Rickles et al., 2007)*

Because of this, complex systems often exhibit unintended consequences or 'strange' behavior in general. It is technically possible to be able to know everything about a complicated system like a car – how it works, diagnose it, everything – because people built it. However, it is impossible to perfectly predict or understand a complex system in its totality and without error. Every ornithologist in the world can tell someone everything there is to know about a specific species but not why a flock of them flies in the exact paths and takes on the forms they do. This is an example of self-organization and emergent behavior, unique to complex systems (Grabowski & Strzałka, 2008).

Emergence is the effect or behavior that stems from interaction among components. Think of two strangers as two parts of a linear system. The system has two elements, but nothing is happening because the two people are not talking to each other. The system is not producing anything beyond what intrinsically exists within it. The system is linear, so it equals the sum of its parts: 1 + 1 = 2. If the two people start to talk to each other and form a relationship, this interaction now counts for something adding value to the system. A relationship between two people is more complex than the two people in isolation. This is why in systems 1+1=3: two people plus one relationship equals three.

Another example is the human system. A human is more than the resources they take in; a person is what is done with those resources. A human body can power a brain which is capable of advanced thought and emotion, producing art, holding conversations, and thinking critically. What emerges from the human system is consciousness, arguably more significant than the aggregate atoms that make up a human body.

Chaos and the Dimensionality of Systems

Modeling chaos is somewhat out of the scope of this chapter, but it conceptually merits a discussion (see Feldman, 2019 & Mitchell, 2011 for a detailed description). Chaos is beyond structure and temporality but is ingrained in both. Simple systems can act just as chaotic as complex ones can. Chaotic systems tend to be more dynamical but can exist within structurally complicated systems, as well, since all systems, by definition, are affected by time. Chaos is noise and randomness, which is defined here as something which does not have a repeating (cyclic) pattern – it is aperiodic (Feldman, 2019). Because chaos can exist in all systems, it is also a dimension along with structure and temporality.

Chaotic systems can be deterministic and governed by a simple equation; for example, the logistic equation can exhibit chaotic behavior, yet it is a simple equation with only a couple of variables. Chaos is counterintuitive: it is generated from well-determined and entirely predictable equations, but it produces random and aperiodic behavior. The equations that generate chaos do so only under certain conditions (Mitchell, 2011). The state of dynamical systems at any point in time is affected by the initial conditions, but chaotic systems take this to an extreme. Beyond just the initial state, there are often keystone variables that, if changed even slightly, will cause massive effects that propagate throughout the rest of the system. This conditional hypersensitivity is also known as the *Butterfly Effect* or the *Ripple Effect*. The degree of chaos in a system is its *ergodic hierarchy* (Feldman, 2019), which is a component of chaos, just like hierarchy is a component that exists in structure and temporality. It can be likened to the size of chaos in a system. This form of chaos just described is **deterministic chaos** because it is based on well-defined mathematical equations like the logistic equation.

The second type of chaos (randomness) is **stochastic chaos** or random chance. It occurs from a non-deterministic outcome like a coinflip (Feldman, 2019), the power output of a hydroelectric dam, or the global climate system. Statistical probability is often employed in stochasticly chaotic systems. This chapter will not discuss stochastic chaos further and instead focus on determinism because it is mathematically compatible with system dynamics. That being said, stochastic elements can usually be included with random generator functions according to specified statistical distributions and probabilities.

Figure 1 lays out the various system types relative to one other and according to their three dimensions (structure, temporality, and chaos). Together, the three can be referred to as the **dimensionality of systems**, which is a framework for classifying and defining any system. The dimensionality of systems is, of course, a simplification, but it can help to define what a system is and how to represent the various dimensions of a system with each other. It outlines a multidimensional space in which a system is located.

Chaos is randomness and is, therefore, generally best described using probability and distributions. Structure and dynamics describe the system's average state, while chaos describes its variance and deviation at any given time from that mean. As the system increases in chaos (entropy), its state moves further

Figure 1. Dimensionality of Systems

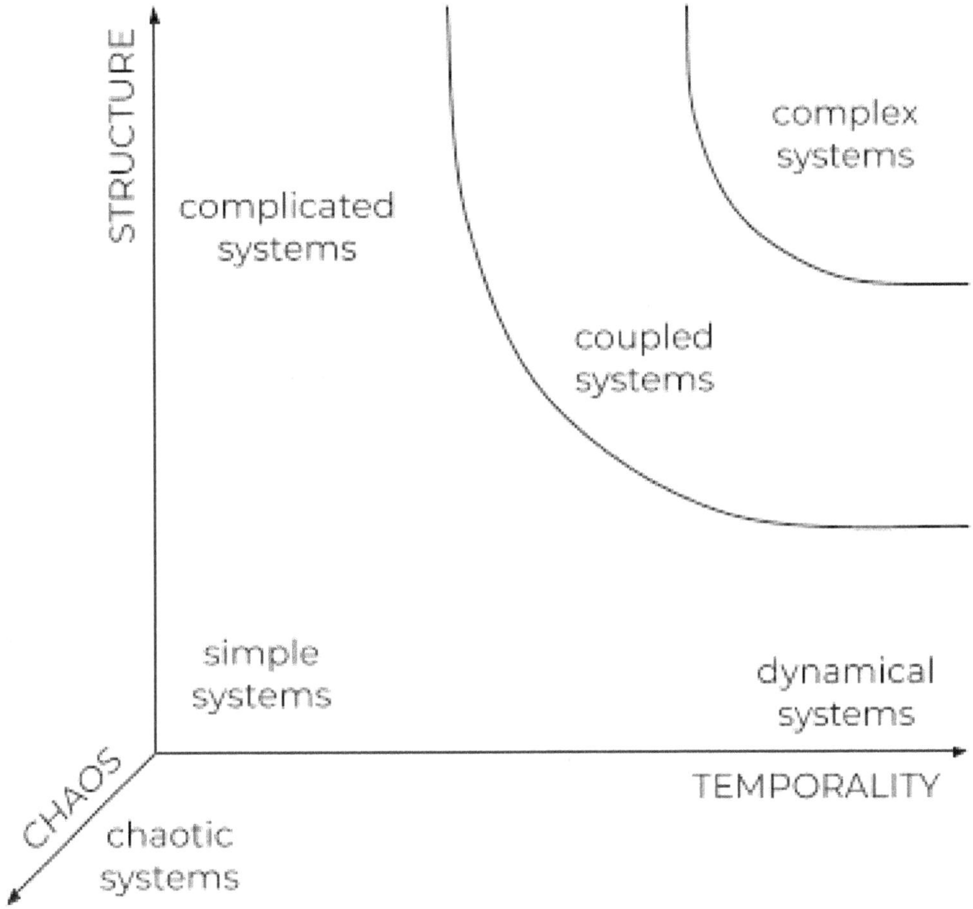

from the desired state (out of the page) of a highly ordered structure and temporality. A system whose state has perfect order and no chaos would be on the plane created by the structure axis and temporality axis.

The size of a system's relative temporal and structural component defines the type of system. Complicated systems are structurally complex with low temporality, and dynamical systems are dynamically complex but have relatively few interconnections and structural complexity. Simple, coupled, and complex systems have roughly similar relative ratios of structural and dynamic complexity, but at vastly different levels. Chaos is more common – or, rather, noticeable – in complex and dynamical systems but can exist anywhere. It can be likened to a hidden dimension buried in a system, much like the complex numbers represent a hidden dimension attached – and often unnoticed hence colloquially imaginary – to real numbers. Even if a system's structure and temporality can be wholly understood, chaos can lead to unforeseen outcomes. Chaos exists alongside the other two dimensions, but its effect is not always apparent, or instead, the conditions are not being met for chaos. Chaos is a dimension or state that might or might not exist in a system, but it is technically always possible. All numbers can have a complex component, even though it might just be zero.

FROM A STATIC STRUCTURAL ANALYSIS TO A DYNAMIC MODELING APPROACH

Limitations of the Cross-Impact Network Analysis

The cross-impact analysis (see the previous chapter for a detailed description) is often used to holistically understand and conceptualize a system. When more advanced understanding is needed and the systems becomes more dynamically complex other methods are required, such as system dynamics and agent-based modeling, although agent-based modeling is out of the scope of this chapter.

Two main limitations of cross-impact analysis are discussed – the possibility of human-induced errors and its static nature. The latter of which is addressed explicitly by the system dynamics model detailed herein. A cross-impact analysis is entirely based on human input, and ideally, expert judgment. The cross-impact matrix, which is the fundamental core of cross-impact analysis, requires "the estimation of at least one of the following quantities by the judging experts: conditional probabilities or joint probabilities of event pairs, or the marginal probability of events. To be able to do these estimations properly, the experts not only need to know which interrelations exist in a system, but they also have to recognize" (Weimer-Jehle, 2006, p. 337) the possible results of these impacts. Primarily, the quality of the analysis depends on the accuracy and expertise of the person or people undertaking that analysis. The experts filling out the matrix are "expected to possess insights which rather should be the results of an analysis (Weimer-Jehle, 2006, p. 337)." This is the classic *Catch-22*, where input depends on output, seemingly rendering the analysis ineffective through *circle-logic*.

Although this issue exists in system dynamics (SD) as well, to a lesser extent, SD allows for what-if and scenario analyses to model ranges in values, while CIA can only have one value, which represents only one possible state. System dynamics can automatically and quickly run models countless times to see the long-term effect of varying a number along a continuum of possibilities. Doing so generates a visual representation of the various states the system can be, usually in the forms of confidence intervals on behavior-over-time charts. System dynamics, discussed in detail in the next section, was designed explicitly for modeling dynamical systems, but CIA in the previous chapter was built to analyze data that is not complex, i.e., simple and complicated and at a high-level. That being said, CIA can be supplemented with other approaches like SD to do so. CIA is a way to simplify systems to provide a strategic overview, but in doing so, it might oversimplify and be too reductionist, hence the need for expert judgment and input.

Cross-impact analysis should not be used to provide specific or detailed information about a system to begin with. So, as long as people understand that limitation, it is a relatively insignificant issue to the analysis itself. CIA is a soft-systems tool. Hence, specific numbers, equations, and quantitative data are not always necessary. A cross-impact analysis is to be used for grasping the strategic view of the system, and not to describe systems in detail. This brings us back to the problem of relying on the judgment of experts as imperfect humans (or just humans). The structure and purpose of cross-impact analysis provide a possible mechanism to aid in this. Experts have a deep understanding of specific elements within a system, and when many experts come together, any deficiencies are hopefully smoothed out, and inaccurate or overly biased information holds less weight when many experts are involved. It is much like smoothing algorithms used for filling data gaps. In short, a cross-impact analysis relies on academic consensus. If most experts determine one outcome, then the minority view should not significantly skew the overall outcome, leaving the analysis valid.

This limitation is also a strength. Expert consensus allows cross-impact analysis to be built with relative ease depending on the desired detail of the output. In many cases, a qualitative approach is sufficient, but a semi-quantitative or quantitative approach is needed. Technically, a qualitative approach does not need external data beyond the expertise of those that fill it out, which makes it a relatively simple and straightforward approach compared to others, but it also tells less about the system. A qualitative approach is best for many situations if, for instance, the system merely needs to be described to the stakeholders or policymakers. A semi-quantitative or a static method (as used in the previous chapter) is usually sufficient enough for most analyses of complex systems – at least, at the top-level to generate a general understanding – as it requires little primary data from the system itself. That can lead to biases from the experts skewing the results subconsciously or otherwise.

Quantitative methods, specifically numerical modeling, require a great deal of empirical data but lead to the highest accuracy and are usually reserved for more detail modeling approaches like system dynamics modeling. As the amount of data increases, so does the possibility and propagation of errors. Regardless of the method employed, surprising levels of insight can be gleaned from any complex system, beyond what could be learned without a systems approach. In short, pick the method most appropriate while weighing the benefits and problems of each, keeping in mind a hybrid-approach might be best.

The second limitation of the cross-impact analysis is that it represents a snapshot of a system at a specific time and cannot handle dynamic issues. Mathematically from calculus, a cross-impact matrix is the derivative of the system. When dynamics are considered, it is integrating the system over time. There are many situations, however, when a static approach will suffice: when a system changes very slowly with time relative to human time scales or when the duration over which the system operates is short. The first could be geological time scales or long-term macroeconomic impacts. When a system changes with time, a new cross-impact analysis must be carried out at each time interval; as time affects the system overall, the magnitude and direction of the interaction will also change. When that interval is long, or few intervals are desired, this is not a problem, but the effort quickly proves counterproductive to the analysis when many intervals are needed.

Since cross-impact analysis pays less attention to the *nature* and *dynamic* of interactions in a system, it needs to be supplemented with hard-systems tools if the dynamic needs to be addressed, which is usually the case with complex systems. This point is addressed in this chapter and focuses on how to use system dynamics (SD) (hard-systems approach) to model the interactions among the SDGs based on the soft-systems approach presented in the previous chapter. (If you have not yet read the previous chapter, the author recommends skimming the relevant sections.) The idea is that once a cross-impact analysis has reached its limits, it can be supplemented by the system dynamics analysis, which automates much of the work in creating the analysis, generating background equations, and calculating the interactions.

This chapter presents below a supplementary *hard*-systems approach to analyze the interactions among the SDGs, which helps address the limitations mentioned above. The approach uses system dynamics, which, as the name suggests, accounts for how the components of systems and their linkages change with time. System dynamics modeling frees the users' time – once the model is built – so that they can perform complex, comparative, and informative analyses on which decision-makers should rely to craft policy. Furthermore, it takes into consideration the components of a system in addition to following a systemic process of inquiry. Unlike cross-impact analysis that can be carried out using simple tables in Excel, system dynamics analysis requires using more sophisticated software packages. The system dynamics (SD) models presented in this paper use the STELLA (Systems Thinking Experiential Learning Laboratory with Animation) Professional software v1.8.2 by *isee systems, inc.*

It should be noted that both cross-impact network analysis and system dynamics are not mutually exclusive; they may appear to be dissimilar but, in reality, complement each other. It is recommended, for instance, first to carry out cross-impact analysis and network analysis to map how the SDGs interact, i.e., identify its structure. Doing so forces decision-makers to conceptualize the strength and centrality of each component within the SDG network, as well as the strength of the linkages among the goals. The results of that analysis are critical to developing robust and meaningful SD models. In system thinking lingo, they help develop so-called *mental models* of how the SDGs interact and outline hypotheses of their interaction dynamics.

An earlier version of the model discussed herein was presented initially at the *United Nations' Multi-Stakeholder Forum on Science, Technology and Innovation for the SDGs* held in New York on 14-16 May 2017, which focused on six of the 17 SDGs, namely, SDG 01 (Poverty Eradication), SDG 02 (Food Security), SDG 03 (Health), SDG 05 (Gender Equality), SDG 09 (Infrastructure), and SDG 14 (Oceans). The model presented herein can be extended further to include all SDGs and other more extensive applications, as well as more detailed and complex modeling approaches using computer programming languages like R, Python, and Julia.

System Dynamics

Systems dynamics is a relatively new branch of systems science (specifically, dynamical systems) that originated with the work of Dr. Jay Forrester at the Massachusetts Institute of Technology in the 1950s and 1960s around the same time as the cross-impact analysis was developed. It represents a milestone in the overall evolution in the application of systems thinking and the development of tools to address complex issues in a wide range of disciplines such as engineering, business, economics, health, planning, management, to name a few. A simple way to see system dynamics is through "how systems change over time" (Ford, 2010). The System Dynamics Society (2016) defines system dynamics as a computer-aided approach to policy analysis and design. It applies to dynamic problems arising in complex social, managerial, economic, or ecological systems (non-inclusively) – literally any dynamic systems characterized by interdependence, mutual interaction, feedback, and causality.

Since its inception, system dynamics have accrued popularity in various STEM and social science fields. It can be used for a wide range of complex systems and applications (Azar, 2012), for example, in the areas of business and innovation (Shafiei & Hakaki, 2018); mobile telecommunication networks in Kenya (Omamo et al., 2018); corruption (Zelinka & Amadei, 2018b); e-waste (Nikabadi & Hajihoseinali, 2018); photovoltaic systems (Yatimi & Aroudam, 2018); even Shakespeare (Haslett, 2012) and comic books (Wyburn & Roach, 2013) and many more – the possibilities are endless. Landmark books, among many others that have promoted the applications of system dynamics include those by Sterman (2000) on business dynamics, Ford (2010) on modeling environmental/ecological processes, Amadei (2015) involving the application of systems thinking and SD to community development and the water-energy-land-food (WELF) nexus.

In general, systems dynamics has unique characteristics that warrant its use in modeling the interaction of the SDGs.

- It is a method that can be used to study how systems continuously change over time due to possible changes in and relationships among components and changes in the overall direction of systems allowing for both qualitative and quantitative modeling.

- It requires precise formulation, mapping, and an iterative approach to model the issue(s) at stake.
- SD models are defined by closed boundaries (causally closed models) where endogenous components, those originating from within, predominantly dictate the behavior of the systems. Models are designed to be self-contained in terms of cause-and-effect inside their boundaries so that no external (exogenous) influence needs to be explicitly considered. In other words, SD models are designed to contain the components that "are important to explain [their] dynamic behavior" (Richardson & Andersen, 2010; Vennix, 1996) of systems, including their internal rules.
- Nonlinearities in the system are included in the form of first-order differential equations.
- Information feedback mechanisms in the system can be included in the form of interconnected closed loops and circular causality, allowing for reinforcing and balancing trends; this can help in explaining the counterintuitive forms of behavior of some systems.
- The method emphasizes that the structure of systems (i.e., their components, mutual interactions, and environmental interactions) affects their continuous behavior. Combining feedback loops, various dynamic patterns can be simulated and used to model different behavioral patterns of system changes such as growth, decay, overshoot, oscillations, equilibrium, randomness, and chaos. As the structure of a system changes, so does its behavior.
- More emphasis is placed on the structure of a system (its aggregated nature) than on trying to figure out the details of individual components.

Causal-Loops

A fundamental characteristic of SD modeling is that it captures the feedback mechanisms inherent to complex systems using two types of cause-and-effect circular causations: reinforcing and balancing feedback loops. *Reinforcing (R)* loops are used to model self-reinforcing, feedback processes that have the potential to cause permanent growth or eventual collapse. Reinforcing loops are prevented from perpetual growth or decline with *balancing (B)* loops, which create self-correcting processes leading to stability and equilibrium. In addition to these two primary components, delays may be included to model the more complex behaviors regarding time (dynamics) in linking causes and effects or any adjustment processes.

As described in the system dynamics literature, various combinations of loops and delays can be created to model the behavior patterns of complex systems and unique repetitive patterns called archetypes. The models can be represented in an object-oriented form of causal-loop diagrams or stock-and-flow diagrams. *Causal loop-diagrams* show how elements of a feedback mechanism interact through a causal chain and how a particular element of a system might affect itself through time. They are useful in mapping, inferring, and visualizing what contributes to growth, decline, delay, or stability, and are mostly used at the *strategy-level*. They show trends and connections and causal feedback mechanisms in a system. They are *not* used to conduct numerical simulations of systems.

On the other hand, *stock-and-flow diagrams*, as discussed below, consist of combinations of several building blocks. They help visualize flow, accumulation, delay, and dissipation processes (*qualitative*). They also allow for numerical (*quantitative*) simulations and parametric or sensitivity studies and can, therefore, be used at the *operation-level*.

In general, both causal-loop and stock-and-flow diagrams are useful tools for: (i) depicting how parts of a system interact and create patterns of behavior; (ii) communicating the dynamics of systems with others; (iii) designing and planning interventions to address issues faced by the system. A causal-loop diagram is not shown in this chapter as it generally falls under the realm of systems thinking. Stock-and-flow diagrams directly inform the structure of the system dynamics model.

Stock-and-Flow Diagrams

All stock-and-flow diagrams that are used in system dynamics software packages such as STELLA consist of several of the components shown in Figure 2 (Richmond, 1994). *Stocks*, represented by rectangular boxes, are net accumulations of something at one point in time. Possible SDG-related accumulations include currency for GDP in economic systems, people as population, CO_2 as an indicator for climate change, and volume of water per person for a village, town, city, or country. Stocks are *state variables*; they define the current state of a system.

Flows, represented in the form of pipelines (with a faucet controlling the flow), refer to processes that cause *change over time* of information or materials (number of births per year, inflation rate, the flow of a river, cash flow, carbon emission, carbon sequestration, rate of cutting or planting trees, and so on). Flow (flux or rate), in turn, results in changes (dynamic behavior) in the stock accumulations within the entire system. Flows are *control variables*; they create change in the state of a system. Flow can be a *uniflow* or a *biflow* – a uniflow flows one direction, while a biflow flows in both. Flows can be connected to stocks and convertors through connectors (visualized by thing arrows). Flows could be attached on both ends, one end, or neither end to a stock. When the end a flow is unconnected, a *cloud* is shown, which represents an infinite source or sink outside the boundaries of the model.

Figure 2. System dynamics modeling structures

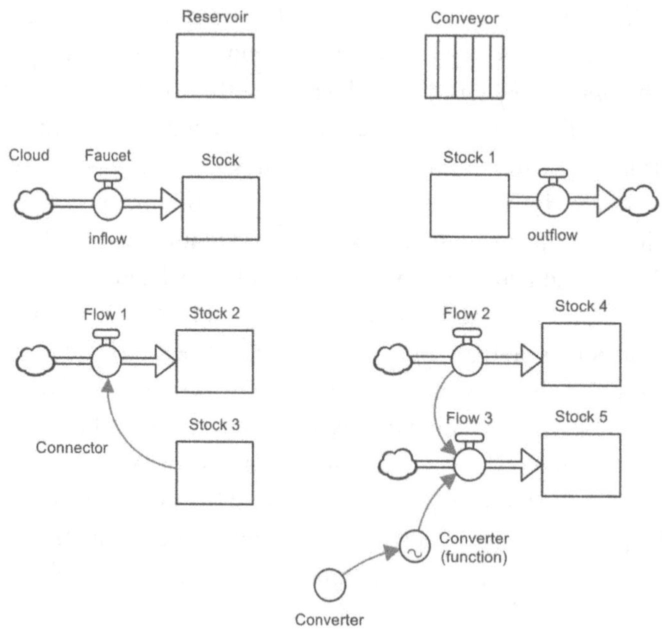

Convertors, represented by just a name or a circle-and-name, convert or transform information from a stock-and-flow path to another path. For example, if the population growth (a stock) requires a related growth in food production (another stock), then a converter can be used to link the two. For each living person, a certain quantity of food is required, and a converter will be included to carry out this calculation. Converters can be described in a functional form that depends on time or a connected variable. In this case, a "~" appears inside the converter.

Connectors, represented by arrows, connect the various components and variables and transmit information throughout the model. Multiple variables can be connected to a single variable or vice versa. Connections include stock-to-flow, flow-to-flow, or between converters. The tail of the arrow represents the cause (dependent variable), and the head represents the effect (independent variable).

Figure 3 shows an example of a simple stock-and-flow model that uses the primary elements described above to represent the dynamics of exponential population change (growth or decay). In this basic example, the population is aggregated into one total population and does not account for various demographics. There are two reinforcing feedback loops depicted in Figure 3. The more people there are, the more people can be born, which increases the overall population causing more people to be born; the same happens with deaths.

Figure 3. Simple population model

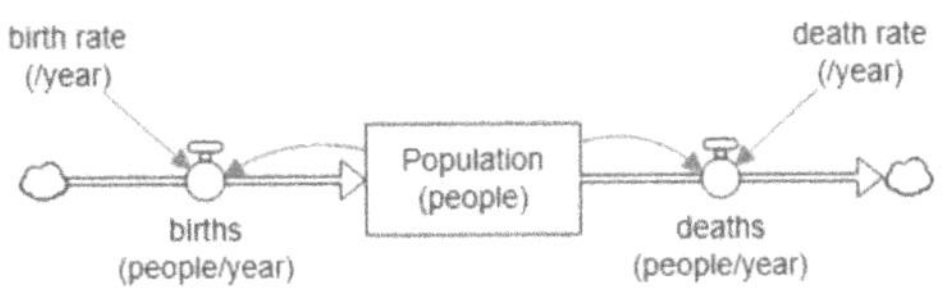

The last chapter discussed a system's structure; a stock-and-flow is the visual representation of that structure. The structure of the population system depicted in Figure 3 is characterized by one stock (population), two uniflows (births and deaths), and two exogenous variables (i.e., the parameters, birth rate, and death rate). It can be considered a simple system, but one that exhibits non-linear behavior in the form of exponential growth decay. Visually, the behavior is determined (the same context as the concept of determinism) by the structure, but each variable has a value or an equation that mathematically determines the behavior of the system. The birth and death rates control the dynamics of the system and connect the temporal dimension with the structure of the system with arrows leading to births and deaths; the units are per time. If the birth rate or death rate increases, the behavior of the system will change. It will speed up or slow down and can behave erratically and chaotically under certain circumstances; these circumstances require the birth and death rates to be changed to some specific number or range of numbers, which is a discussion beyond this chapter's scope and unlikely to be a problem in this system.

Each dimension in the representation of the population system can be clearly defined and seen. It has low structural and dynamic complexity, meaning that it can be placed where it says *simple system* in Figure 1 on the plane of *STRUCTURE* and *TEMPORALITY*. If the conditions required for chaos were met, this system would move off the page into the dimension of *CHAOS* but would still be at the same relative location on the *STRUCTURE* and *TEMPORALITY* axes, but with the axes just shifted into a new dimension.

System Dynamics Modeling

Modeling a problem with system dynamics necessitates following a road map consisting of several interconnected steps. Figure 4 shows a road map proposed by (Ford, 2010) consisting of eight steps: problem familiarization (step 1); problem definition (step 2); model formulation by constructing causal loop diagrams (step 3) and then stock-and-flow diagrams (step 4); parameter estimation (step 5); simulation to explain the problem being addressed (step 6); and simulation analysis consisting of a sensitivity analysis (step 7) and a policy analysis (step 8).

Figure 4. Road map for system dynamics modeling

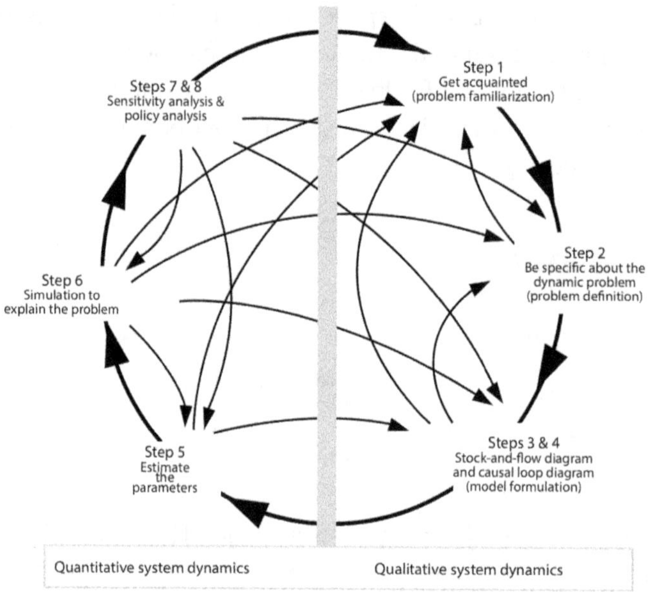

Figure 4 shows that SD modeling follows a cyclical and iterative process. The first two steps suggested by Ford (2010) are about defining the problem being addressed and developing an initial *mental model*, a so-called *reference mode of behavior,* and proposing dynamic hypotheses for the problem. Modelers (system dynamicists) need to pay special attention to these two steps as they define: what is being modeled; why it is modeled; who is involved in the modeling; the where and when of the modeling (i.e., the context and scale, respectively, physical and temporal); and how the modeling is undertaken.

Figure 3 also shows that the SD modeling process is divided into qualitative and quantitative modeling. Steps 1-4 can be interpreted as the qualitative and conceptual components of SD modeling. The other activities emphasize the quantitative dimension of that modeling. Whether qualitative or quantitative modeling is used depends mostly on the purpose of the system analysis, the availability of data and information about the system components, and the participating audience (Wolstenholme, 1999).

CHANGE-OF-STATE ARCHETYPE

The SDGs are not static and progress forward, ideally improving, through time. In this section, the authors use system dynamics to model their progress and interactions with a logistic, S-shaped growth behavior. This type of growth appears throughout multiple fields of science. It is used to model epidemics (susceptible-infected-recovered epidemiological model, also called SIR), population growth under carrying capacity (Limits to Growth archetype model), and market penetration of products (Bass diffusion model) (Sterman, 2000).

S-shaped growth can also be described when a group is acquiring something that they did not have before. As state A turns into state B, state B shows an S-shaped pattern of growth. For epidemics that is a potentially susceptible population becoming infected with an illness; for innovation in markets that is a population adopting a product or service; for population dynamics that is population growth under the constraints of the system's ecological carrying capacity.

The same logistic growth can be applied to each SDG since the SDGs can be seen as going from non-being adopted to being adopted or implemented. Growth is initially slow after an SDG is first implemented because it takes time (delay) to inform relevant agencies, propose projects, et cetera. Eventually, work towards addressing the SDG will reach its fastest rate (inflection point), at which point it will start to slow, which results from constraints such as "the last-mile problem," an issue especially pervasive in sustainable development (Balcik et al., 2008; Chambers, 1983; Minten et al., 2013). It refers to the difficulty in reaching those last few people at the end of the supply line due to their geographical remoteness, for instance.

The change-of-state archetype is the name given to a family of archetypes related to the transdisciplinary concept of diffusion and system behavior of goal-seeking. It is specifically a *dynamic* system archetype because it was specifically built for use in system dynamics modeling.

The Logistic Diffusion Model

Of the three S-shaped growth models mentioned above, the logistic diffusion model (LDM) was selected to model the dynamics of the SDGs and their interaction. Modifications to the original LDM were made to capture that dynamic.

In the fields of marketing and management science, the innovation diffusion model is used to represent the adopting rate of new products and technologies in a market by people who, in turn, are affected by internal and external forces. Internal forces represent word-of-mouth (positive feedback) from people that have already adopted the product and the difficulty in a new product being adopted in the absence of people that have already done so (negative feedback). Initially, the IDM was designed to model market penetration and has been used for mobile phones (Swinerd & McNaught, 2014) and many products, including refrigerators, record players, TVs, air conditioners, and many more (Kumar et al., 2015). External forces represent external influences on the current adopters, such as advertisements from the media. Figure 5 shows a generic LDM stock-and-flow diagram consisting of a uni-directional flow from *potential_adopters* to *current_adopters*. The *adopting_rate* depends on *transfer_rate* and external factors.

In general, the dynamic shown in Figure 5 yields a positively skewed normal distribution (bell-shaped) for the adopting rate and a logistic S-shaped growth curve for the current adopters. Once a technology or product is adopted, a person cannot un-adopt it unless a new, superior, and similar-functioning product is released. If that happens, the person migrates to the new technology; they do not technically un-adopt it.

Figure 5. General Innovation Diffusion Model

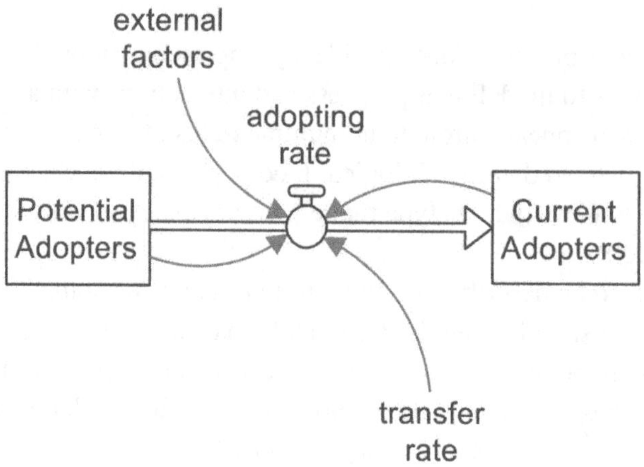

For this reason, the LDM is uni-directional. Internal factors are illustrated by the arrow (uniflow) from potential adopters to current adopters. External factors are whatever the user's specific context requires. The adopting rate may represent, for instance, market saturation and word-of-mouth. The transfer rate is a user-entered value (like population growth rate) that describes how a product or service is adopted by the population.

The generic LDM model depicted in Figure 5 was modified to generate the diagram in Figure 6, which now allows for flow in both directions (from potential adopters to adopters and from adopters to potential adopters). The terms "potential adopters" and "adopters" used in LDM no longer apply and are replaced by *State_A_(Insecure)* and *State_B_(Secure)*, respectively, to be able to depict nonlinearity and feedback more accurately. The bi-flow model of Figure 6 more accurately depicts what is happening in sustainable development. As an example, when addressing poverty, people can escape poverty, but they can also fall back into poverty any number of times. It is also similar to the confidence model used by Featherston and Doolan (2013), where a person could gain confidence in something (board of directors in their case) or lose confidence in something. Instead of confidence, we use a *secure_state*, or desired state, which for SDG 01 is "no poverty."

Figure 6. Insecure-secure diffusion model

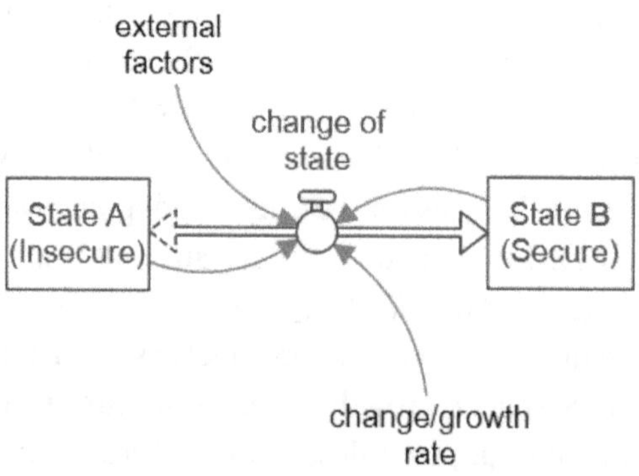

Similarly, losing confidence is akin to being impoverished, which is the *insecure_state*, or undesired state. Change of state is determined by the *Insecure_State* and *Secure_State* (poor versus non-poor population), the effect of the other SDGs (explained later), and the effect of outside influences. The first two are endogenous to the system (i.e., the system effects itself), while the latter two are exogenous to the system.

It might be helpful to think of what is happening in Figure 6 as analogous to a seesaw. Instead of two populations, consider two closed chambers filled with a finite liquid. The *change_of_state* is like a pump that moves the liquid in a pipe between the two chambers; the speed at which this liquid moves between chambers is controlled by the relative amount of liquid in each chamber and the pipe size and friction referred to as internal factors represented by arrows from the stocks to the *change_of_state* biflow. If there is more liquid in one chamber, that side is "heavier" and causes more liquid to flow from the other chamber to that chamber. This phenomenon is known as a positive feedback loop. The external factors might enable or balance this feedback loop. If the effect of the external factors is larger than the push and pull of the internal factors, then the flow will happen in the opposite direction.

Instead of a liquid, it is a population going between an insecure state and a secure state due to various internal and external factors. Because we are presenting a generic methodology, the names for each stock – a current *Insecure_State* and the desired *Secure_State* – are placeholders representing whatever the user's situation requires. We chose those names to signify a transition from one state to another. For each one of the SDGs considered in this chapter (the six mentioned earlier), insecure and secure states are defined in Table 1. When certain conditions are met, the state of the SDG begins to change from insecure to secure, signifying progress towards the implementation of that specific SDG.

The change-of-state method requires a binomial variable (one for each state): an SDG is accomplished when it passes a certain threshold (Table 1) or it is unsuccessful if not. Until the progress of an SDG passes a predefined threshold, it has not been accomplished and mathematically is zero. On completion, it becomes 1. The definitions are left purposely broad because they are dependent on the user's specific case. There are other versions of the archetype that add more chambers and alter the structure, but they are all based on the movement of populations between two states. To keep this chapter at a reasonable length, they were omitted (see Zelinka & Amadei, 2018a to learn more about them).

Table 1. SDG-specific secure and insecure state qualitative definitions

SDG	Insecure State	Secure State
SDG 01 Poverty	Below international poverty line (USD$2/day)	Above the international poverty line (USD$2/day)
SDG 02 Food Security	Lacks access to the proper quantity of calories/nutrients	Has access to the proper quantity of calories/nutrients
SDG 03 Health	General poor health	General good health
SDG 09 Gender Equality	General gender inequality	General gender equal
SDG 11 Infrastructure	Lacks access to basic infrastructure services	Access to basic infrastructure services
SDG 14 Oceans	More than a certain percentage of fishers are overexploited, and pollution is problematic.	Most fisheries are not overexploited, and pollution is minimal

A SYSTEM DYNAMICS MODEL OF THE SDGS: PUTTING IT ALL TOGETHER

SD models are often best explained in a story-form. We will start with an operational-level explanation of poverty, as an example for all individual SDGs, by building on the insecure-secure diffusion model introduced in the previous section and illustrated in Figure 6. Then, we discuss the methodology selected to account for the interactions between SDGs using a simple matrix inversion technique.

In system dynamics modeling, it is assumed that the system's structure determines its behavior, and since all SDGs are assumed to behave the same way (from unsecured to secured states), all SDG modules can have the same structure. Hence, for clarity's sake, we will only depict the modular structure for SDG 01 (poverty); the only difference between the SDG 01 module and the others is the specific input data and names of the module variables. To generalize the methodology to all 17 SDGs: (i) repeat the SDG 01 module an additional 16 times; (ii) connect all modules as explained later; (iii) rename the module variables for each SDG. The reader can follow along with the interactive STELLA model to use the interface: https://exchange.iseesystems.com/public/dave-zelinka/sdgs-interactions-v2.

Modeling the Progress of an Individual SDG

Entry Points and Caps

The generic insecure-secure diffusion model of Figure 6 was repurposed for SDG 01, as shown in Figure 7. There are two types of variables to keep in mind: variables that depend on external data (e.g., expected years for completion, poverty; poverty delay; poverty interconnection factor) and variables that convert data (e.g., poverty change rate, economic mobility, and impoverished and secure population). In the user-interface section below, only variables that depend on external data will be presented to avoid any confusion and for simplicity's sake.

The variables that require external data are simply user inputs and extrinsic factors to the model; they are not fundamental structures, just variables. They are where the model is "capped," they act as entry points into the model, and they are where further complexity can be added. If the user wants to include more factors (e.g., economic, cultural, technological, and so forth.), this is the location where the model accepts other inputs as desired by the user or dictated by the situation.

Variables that receive the input of other data are converters that the user does not change as they contain the underlying equations that govern the model. Changing them could cause errors or erratic behavior. These are not to be modified unless the users have a clear understanding of what they are doing with SD modeling.

In Figure 7, the variable *poverty effect* is introduced to calculate the nominal effect of SDG 01 (Poverty) on all the other SDGs. These appear once in each SDG module and should remain unchanged. It is merely colored red to be easily identifiable and outline its importance. It is the variable that is used as the input to calculate the actual interactions later.

Incorporating Population Growth

Figure 7 has two stocks, one for a population of *Impoverished* people and another one for those that have escaped or were never in poverty, which is categorized as *Secure*. Because the SDGs are expected to progress over a long enough time-horizon when population change becomes significant, Figure 8 was

Figure 7. Insecure-secure diffusion model depicting SDG 01 Poverty

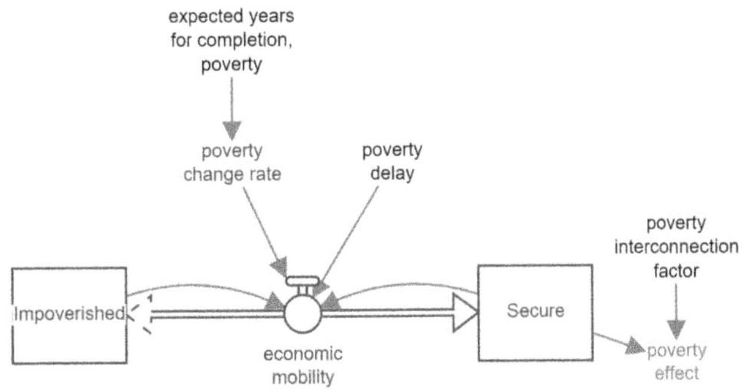

created by combining the population growth model (Figure 3 with the insecure-secure diffusion model for SDG 01 Poverty (Figure 7). Changes, for better or worse, in one's economic status, that is, the flow between the impoverished and secure states for poverty is described by the term *economic mobility*. When a population improves its economic situation, it is said to be upwardly-mobile; when its economic situation worsens, it is said to be downwardly mobile.

The stock-and-flow structure of Figure 8 allows for population changes in the impoverished and secure populations; different socio-economic groups and other population demographics might have different birth and death rates from one another. If population change is not essential to the user, birth and death rates can be left at 0. The birth and death rates could be influenced by numerous factors, but our model leaves that up to the user to decide, so we "cap" the model here and assume constant values.

Figure 8. Insecure-secure diffusion model with population change for SDG 01

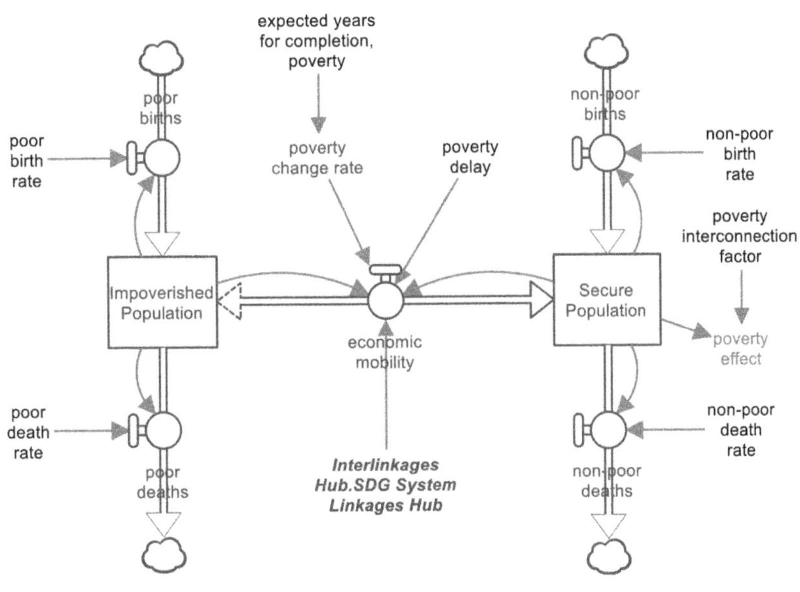

To illustrate different population growth rates for groups of people, we look again at poverty: low-income people and those living in developing countries tend to have higher birth rates compared to higher-income people (Aassve et al., 2005; Schoumaker, 2004) and those living in developed countries (Ahlburg, 1996; Bank, 2017; United Nations Development Programme, 2017). High population growth rates would increase the size of the poor population (*Impoverished*) faster than the non-poor population (*Secure*), causing a feedback loop that exacerbates poverty. To combat this, the number of people escaping poverty needs to be higher than the number of people being born into it.

As a cause, effect, or both, poverty is linked with low food security, poor health, poor gender equality, poor infrastructure, poor oceans' conditions, and to some extent, with all the SDGs. Population changes for impoverished people could, therefore, be a significant factor in designing a policy that addresses poverty and contribute to the successful implementation of the SDGs.

Realistically, there are not just impoverished and non-poor people, they fall on a spectrum commonly referred to as an income distribution, usually categorized by low, medium, and high levels of development and at varying income levels (in income per year, $USD/year for the United States). For simplicity's sake, the model of Figure 8 considers two aggregated populations. The same model could be extended to include multiple socio-economic groups. Future work will aim to address this.

Time Delay and lag Effects

Delays do exist in the real world and should be included in any modeling. When an order for some number of units is placed, the factory cannot immediately begin manufacturing; it takes time to ramp-up production. This lag-time exists in sustainable development as well: when the decision to implement a specific policy is made, it takes time for the proper institutions to be notified, funding to be sourced, and implementation plans and strategies to be devised. Likewise, when an NGO, governmental agency, or some other development entity decides to address poverty, it is not immediately reduced; there is always preparation time before the policy is fully implemented.

This principle was included in our system dynamics model by using a delay function, which is another factor that has the potential to affect economic mobility. Each SDG has a time delay similar to the poverty delay; examples include health delay, food security delay, and more. Delay is defined as the time it takes for change to go from zero to fully "ramped-up." This delay was added using the DELAY1 function available in the STELLA software and can be written as follows:

DELAY1(SDG X Change Rate,(SDG X Delay)/12,0)

This function delays the rate at which flow between two states (SDG X change rate) by a certain number of months (SDG X delay), which is converted to years by dividing by 12. The zero in the third position of the DELAY1 function signifies that there is no rate of change at the simulation's start. In other words, it takes time to sufficiently initiate the SDGs implementation strategy to see progress, which starts at 0.

Migration Between States

Just as each SDG has specific names for the starting and end states (e.g., impoverished and secure for SDG 01), the rate of change between those states has unique names. Change between impoverished and secure states for poverty is economic mobility: for food security, it could be *access rate to food*; for governance, it could be *change in governance effectiveness*; and for partnerships, it could be *change in partnerships*; this is user-defined.

In Figure 8 (SDG 01 Poverty), economic mobility (rate of change between the impoverished and secure state) is controlled by five variables *(poverty change rate, poverty delay, Impoverished population, Secure population,* and *Interlinkages Hub.SDG System Linkages Hub)*, depicted by the five arrows heading to the biflow. The *Interlinkages Hub.SDG System Linkages Hub* variable will be discussed in the following part, but in short, it is the coefficient, calculated endogenously, that accounts for the effects of the other SDGs on a specific SDG. It is included in the economic mobility flow rate as follows:

```
economic_mobility =
DELAY1(poverty_change_rate, poverty_delay/12, 0)*
Impoverished_Population *
Secure_Population *
(1+Interlinkages_Hub.SDG_System_Linkages_Hub[SDG_01_No_Poverty])
```

This equation differs from one SDG to the next by the names assigned to the variables: poverty change rate becomes food security change rate, health change rate, et cetera. Other than that, the mathematical structure remains unchanged for all SDGs.

Interactions Among the SDGs

Figure 9 shows the overall system dynamics model that links six of the 17 SDGs. Each module is in this figure has the same structure as that shown in Figure 8 for SDG 01 (poverty). The arrows emanating from each SDG module to the center *Interlinkages Hub* are described in the next section (*Effect of one SDG on all* SDGs), while the arrows radiating out from the center hub are described in the following

Figure 9. Top-Level View Of The SDG Interconnections

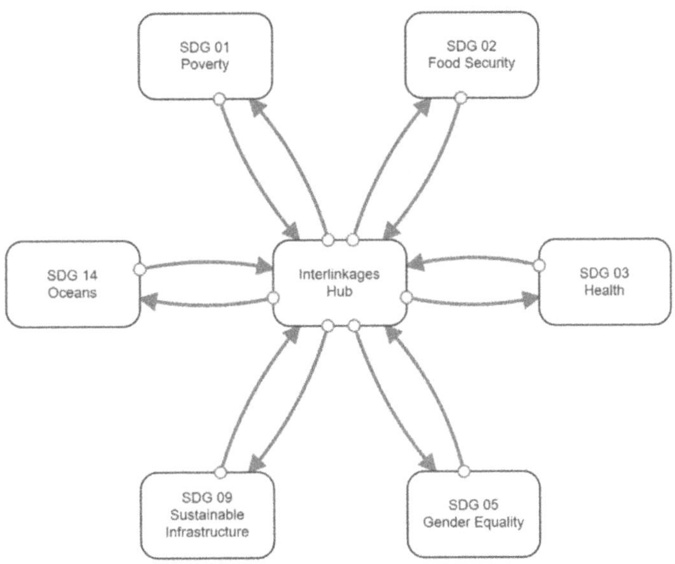

section (*Effect of all SDGs on one SDG*). These two sets of arrows are used to model how the SDGs interact with one another by way of the center *Interlinkages Hub*. Removing the center hub and arrows would be functionally the same as thinking in "silo's."

Effect of one SDG on all SDGs

The effect of SDG 01 (poverty) on the other SDGs is modeled using the *poverty interconnection factor* in Figure 8. This is an n-by-1 array, where n is the number of SDGs linked to the SDG being considered; the array would be 6-by-1 in the present example. The numbers in this array fall between -1 (complete dependence) and 1 (complete influence). A value of 0 signifies a net-zero or no interaction; -1 equates to a maximally-negative interaction; 1 implies a maximally-positive interaction. Table 2 lists the input values used in our STELLA model. Poverty's effect on, say, the SDG 14 (Ocean) is -0.1, which means 10% of poverty's full effect *(Secure)* is removed from the effect of the oceans. Poverty slows the progress of SDG 14 because it is a negative interaction, but the overall effect of all the other SDGs might make the effect positive. These values are multiplied by the scalar value of the *Secure* state stock population. This interconnection value, which exists for all SDGs, is pulled out of the poverty module and sent to the center *Interlinkages Hub*.

Table 2. Poverty interconnection factor

SDG	Interconnection Strength
SDG 01 Poverty	0
SDG 02 Food Security	-0.4
SDG 03 Health	-0.6
SDG 09 Gender Equality	-0.2
SDG 11 Infrastructure	0.2
SDG 14 Oceans	-0.1

Each SDG module has an equivalent interconnection factor with values unique to each SDG. Combining all six interconnection factor values generates a cross-impact matrix, as discussed in our companion paper. The *column* vector depicted in Table 2 is the same as a *row* vector in the cross-impact matrix, but it is transposed to be depicted as vertical. Aside from that, the critical difference between the two is that this cross-impact matrix is quantitative, and the one described in our previous paper is semi-quantitative.

Effect of all SDGs on one SDG

Each SDG sends its interaction effect (Figure 8) to the *Interlinkages Hub*, where a transposition function occurs as shown in the following equation:

```
SDG_System_Linkages_Hub[SDG_01_No_Poverty] =
SDG_01_Poverty.poverty_effect[SDG_01_No_Poverty]+
SDG_02_Food_Security.food_security_effect[SDG_01_No_Poverty]+
```

```
SDG_03_Health.health_effect[SDG_01_No_Poverty]+
SDG_05_Gender_Equality.gender_equality_effect[SDG_01_No_Poverty]+
SDG_09_Sustainable_Infrastructure.infrastructure_effect[SDG_01_No_Poverty]+
SDG_14_Oceans.oceans_effect[SDG_01_No_Poverty]
```

This equation takes the effect of each SDG on poverty (poverty effect in Figure 8), extracts that specific part of the array regarding poverty, and sums those effects into a single value creating the *SDG System Linkages Hub[SDG 01 No Poverty]* variable. The first part of the arrayed variable name refers to the specific variable, while the part in brackets refers to the part of the array, in this case, poverty. For food security, the part in the brackets becomes *SDG 02 Food Security*, while everything else remains unaltered.

We can now explain in its entirety all components of the *economic mobility* equation mentioned above. Taking the poverty part from the interconnection equation above, *SDG System Linkages Hub [SDG 01 No Poverty]*, and applying it to the *economic mobility* equation, the effect of interconnections is a coefficient that either increases or decreases the rate of change between two states (i.e., from *Impoverished* to *Secure* in the case for poverty). This is similar to interest rate, where the rate is added to one and then multiplied by the principle for a given interval of time.

A virtually identical approach can be used for each of the SDGs considered, and at this point, a single iteration of our poverty story is complete. Each iteration represents one year (model's default value), but that could be any user-defined timescale (day, week, decade, et cetera.). Our model is set to a duration of 85 years, from 2015 to 2100, implying 85 iterations. That number will climb to 1020 iterations (85 x 12) if the analysis uses a monthly time-interval. Those iterations are a manifestation of dynamic complexity.

DISCUSSION

System dynamics models provide many benefits over the cross-impact analysis (discussed in Part 1) when modeling the behavior associated with the progress of SDGs and their interconnection over time. Its fundamental limitation, however, is that constructing system dynamics models is time-consuming. That being said, system dynamics is versatile as long as the models are carefully built, appropriate variables are selected, and data are available about these variables. SD models are best suited for more complex and ill-defined situations, such as those associated with the SDGs. System dynamics require that detailed analyses be carried out first about the problems being addressed. Cross-impact analysis and network analysis discussed in our companion paper are good examples of such analyses, as they provide a general overview of the SDGs and their interactions. For this reason, our two papers should be read together and considered as two complementary steps (soft and hard approaches) in modeling the complexity of the SDGs systemically.

System dynamics models can represent more complex situations than static approaches like the cross-impact analysis. A cross-impact analysis is mostly spreadsheet-based, which makes complex interactions among variables challenging to visualize and build. On the other hand, system dynamics models utilize versatile components on a blank canvas allowing for relatively quickly created models to depict complex interactions. Furthermore, it allows for more "creativity" and can handle more detailed analyses of more complicated systems than cross-impact analysis. Finally, system dynamics modeling by its very nature looks at change over time, which is not possible with cross-impact analysis.

The system dynamics model presented herein is limited to six of the 17 SDGs. Its structure is general enough that it can be extended to all SDGs, although the interface proposed in this paper is simple enough that it can be used by development practitioners and decision-makers to explore various scenarios for each SDG and their interactions. Finally, the interface's general nature can be used to model the real-world complexity of ill-defined systems in which several subsystems interact in an intra- and inter-connected manner.

CONCLUSION AND FUTURE RESEARCH AVENUES

With the versatility of SD modeling, future work will apply it to applications outside the SDGs. Early work towards that already shows promise for many complex issues, including urbanization, migration, and community development. Furthermore, it has been especially useful in modeling the water-energy-land-food (WELF) nexus and for other uses of the nexus approach, where multiple relatively discrete sectors interact with one another. These sectors can be technical, environmental, political, social, economic, or more (as with the SDGs), and are useful at any scale from the individual community to countries or international. Future work will expand on the S-shaped archetype in section 3 to include other variants and further complexity.

The three components that make up the dimensionality of systems will be further developed. The idea that structure, time, and chaos are three interacting but explicit aspects that all systems share is relatively new but is influenced by a long history of systems and complexity science research and philosophy. The three dimensions will be further refined but currently holds promise.

Future research directions could more significantly incorporate the social, physical, technical, and applied (engineering) sciences as part of the systems dynamic modeling process. Different applications, including planning for circular economies (from the field of industrial ecology) of the future and governance (political science and economics) of the water-energy-food nexus, could be further explored. Developing multiple case studies around these applications within different social, economic, political, technological, and physical resource contexts would lead to the emergence of patterns that could further be studied and eventually leading to a better understanding of the complexity of the systems in hand. Computer programmers, developers, data scientists, and machine learning engineers not just have a pivotal role to play in that endeavor, but without increased input from those fields, understanding the hypercomplex and chaotic systems that are prevalent in sustainable development will increasingly difficult and slow.

System dynamic modeling excels at depicting feedback, cycles, and coupled systems in general, however, as the number of sectors and parts increases, the time and effort required to make changes to the model quickly become counterproductive at best. Graphical representations such as most system dynamics modeling software become particularly cluttered before they become too cumbersome and difficult to edit and use. At this point, it is best to enter the realms of data science, programming, and machine learning. 17 SDGs already represented too many sectors to model, but what happens when there are hundreds, thousands, or more parts? Python and Julia are two programming languages well-purposed for such undertakings, especially Julia, for when the structural and dynamic complexity becomes massive.

For future avenues of research, machine learning can be coupled with system dynamics. Neural networks and deep learning can be leveraged to add another dimension of understanding of systems modeling, just as system dynamics did with temporality and cross-impact analysis did with the structural components. Dimensionality reduction, for example, can reduce the cross-impact matrix into one line of code as opposed to a two-dimensional matrix simplify without eroding the system's structural complexity. Recurrent neural networks (RNNs) could be used for very complex dynamical systems, where each layer in the network represents a small change in time (*dt*) away from the next layer just as each iteration in a system dynamics model is one *dt* away from the next. The model described in this chapter was bound between 2015 and 2030 and equated 1020 total iterations. Adding one iteration for the initial condition, 1021 total layers would exist in the neural network. Much work needs to be done but represents a fruitful potential direction to include further complexity and unforeseeable capabilities to modeling. System dynamics requires massive amounts of data and computing power, and that is especially the realm of computer and data scientists.

Methods from other disciplines should be included with systems dynamics and systems science in general. One such possibility is to combine with qualitative methods such as grounded theory, which is a systematic approach to structure purposive text to identify emergent patterns inductively by approaching the coding with a particular coding system (Charmz, 2014; Saldana, 2016). Again, RNNs can be used but to analyze text to focus on more social and political complex systems. RNNs are the natural next step after system dynamics modeling as a tool to model complex systems and perhaps chaos.

System dynamics can also be a tool to bridge disciplines to use them in novel ways. Peace/diplomacy engineering – applying engineering approaches with social understanding for peacemaking and peacekeeping goals and for sociopolitical systems in general – provides an excellent example of a hard-systems tool that can be used for social purposes (Amadei, 2019). System dynamics provides an approach for qualitative and quantitative backgrounds to work together and reliably shared insights and can foster transdisciplinary teams to work on these complex problems effectively.

As nations work toward achieving the SDGs, the complexity within each goal and the interconnections among them must be understood and reflected in their policies and investment strategies. System dynamics modeling offers a promising tool for developing such understanding. For that endeavor to be productive, different enabling environments need to be fostered. No single discipline can address the current and anticipated complex challenges single-handedly in their disciplinary silos. This will require investing in and developing new research at the interface of existing disciplines – possibly creating new ones. Addressing these complex challenges will also often require a high level of cooperation among various governmental, industrial, and societal stakeholders. An environment that catalyzes such collaboration will represent critical infrastructure toward improving our resilience as humankind addresses complex challenges.

REFERENCES

Aassve, A., Engelhardt, H., Francavilla, F., Kedir, A., Kim, J., Mealli, F., ... Prskawetz, A. (2005). *Poverty and Fertility in Less Developed Countries: A Comparative Analysis* (ISER Working Paper Series No. 2005–13). ISER Working Paper 2005-13. Retrieved from http://hdl.handle.net/10419/92124

Ahlburg, D. A. (1996). Population Growth and Poverty. In *The Impact of Population Growth on Well-being in Developing Countries* (pp. 219–258). Springer., doi:10.1007/978-3-662-03239-8_7

Allen, C., Metternicht, G., & Wiedmann, T. (2017). An Iterative Framework for National Scenario Modelling for the Sustainable Development Goals (SDGs). *Sustainable Development, 25*(5), 372–385. doi:10.1002d.1662

Allen, C., Metternicht, G., & Wiedmann, T. (2019). Prioritising SDG targets: Assessing baselines, gaps and interlinkages. *Sustainability Science, 14*(2), 421–438. doi:10.100711625-018-0596-8

Amadei, B. (2015). *A Systems Approach to Modeling Community Development Projects* (M. Noori, Ed.). Momentum Press Engineering.

Amadei, B. (2019). Engineering for peace and diplomacy. *Sustainability (Switzerland), 11*(20). Advance online publication. doi:10.3390u11205646

Azar, A. T. (2012). System dynamics as a useful technique for complex systems. *International Journal of Industrial and Systems Engineering, 10*(4), 377–410. doi:10.1504/IJISE.2012.046298

Balcik, B., Beamon, B., & Smilowitz, K. (2008). Last Mile Distribution in Humanitarian Relief. *Journal of Intelligent Transport Systems, 12*(2), 51–63. doi:10.1080/15472450802023329

Bank, W. (2017). *GNI per capita, PPP*. Retrieved July 12, 2017, from https://data.worldbank.org/indicator/NY.GNP.PCAP.PP.CD

Buis, A. (2020). *Milankovitch (Orbital) Cycles and Their Role in Earth's Climate – Climate Change: Vital Signs of the Planet*. Retrieved May 4, 2020, from https://climate.nasa.gov/news/2948/milankovitch-orbital-cycles-and-their-role-in-earths-climate/

Chambers, R. (1983). *Rural Development: Putting the Last First*. Longman Inc.

Charmz, K. (2014). *Constructing Grounded Theory* (J. Seaman, Ed.; 2nd ed.). SAGE Publications.

Checkland, P., & Poulter, J. (2006). *Learning for action: a short definitive account of soft systems methodology and its use for practitioner, teachers, and students*. Wiley. Retrieved from https://library.wur.nl/WebQuery/clc/1928849

Collste, D., Pedercini, M., & Cornell, S. E. (2017). Policy coherence to achieve the SDGs: Using integrated simulation models to assess effective policies. *Sustainability Science, 12*(6), 921–931. doi:10.100711625-017-0457-x PMID:30147764

Coopman, A., Osborn, D., Ullah, F., Auckland, E., & Long, G. (2016). *Seeing the whole: Implementing the SDGs in an Integrated and Coherent Way*. Academic Press.

Daher, B., Mohtar, R. H., Pistikopoulos, E. N., Portney, K. E., Kaiser, R., & Saad, W. (2018). Developing socio-techno-econo-political (STEP) solutions for addressing resource nexus hotspots. *Sustainability, 2018*(10), 512. doi:10.3390u10020512

Featherston, C. R., & Doolan, M. (2013). Using System Dynamics to Inform Scenario Planning : A Case Study. In *31 st International Conference of the System Dynamics Society*. System Dynamics Society. Retrieved from https://www.systemdynamics.org/conferences/2013/proceed/papers/P1193.pdf

Feldman, D. P. (2019). *Chaos and Dynamical Systems*. Princeton University Press.

Ford, A. (2010). *Modeling the Environment* (2nd ed.). Island Press.

Gordon, T. J. (2014). Cross-Impact Analysis. In *Introduction to the futures methods research series. Futures research methodology, V3.0*. The Millennium Project.

Gordon, T. J., & Hayward, H. (1968). Initial experiments with the cross impact matrix method of forecasting. *Futures*, *1*(2), 100–116. doi:10.1016/S0016-3287(68)80003-5

Grabowski, F., & Strzałka, D. (2008). Simple, complicated and complex systems - The brief introduction. *2008 Conference on Human System Interaction, HSI 2008*, 570–573. 10.1109/HSI.2008.4581503

Haslett, T. (2012). Simulating Hamlet: Questions, Cautions, and Critique. *International Journal of System Dynamics Applications*, *1*(4), 77–87. doi:10.4018/ijsda.2012100104

Hayes, A. (2019). *Game Theory Definition*. Retrieved May 4, 2020, from https://www.investopedia.com/terms/g/gametheory.asp

Krakauer, D. C. (Ed.). (2019). *World Hidden in Plain Sight: the Evolving Idea of Complexity at the Sante Fe Institute*. SFI Press.

Kumar, A., Shankar, R., & Momaya, K. (2015). The Bass Diffusion Model does not explain diffusion. *Proceedings of the 33rd International Conference of the System Dynamics Society*. Retrieved from https://www.systemdynamics.org/conferences/2015/papers/P1233.pdf

Le Blanc, D. (2015). Towards Integration at Last? The Sustainable Development Goals as a Network of Targets. *Sustainable Development*, *23*(3), 176–187. doi:10.1002d.1582

Millennium Institute. (n.d.). *iSDG - Integrated Sustainable Development Goals Model*. Retrieved May 6, 2020, from https://www.millennium-institute.org/isdg

Minten, B., Koru, B., & Stifel, D. (2013). The last mile(s) in modern input distribution: Pricing, profitability, and adoption. *Agricultural Economics*, *44*(6), 629–646. doi:10.1111/agec.12078

Mitchell, M. (2011). *Complexity: A Guided Tour*. Oxford University Press.

Nikabadi, M. S., & Hajihoseinali, A. (2018). E-Waste Recycling System in Closed Loop Supply Chain. *International Journal of System Dynamics Applications*, *7*(2), 55–80. doi:10.4018/IJSDA.2018040104

Nilsson, M. (2017). *Important interactions among the Sustainable Development Goals under review at the High-Level Political Forum 2017* (No. 2017–6). Retrieved from www.sei-international.org

Nilsson, M., Griggs, D., & Visback, M. (2016). Map the interactions between Sustainable Development Goals. *Nature*, *534*(15), 320–322. doi:10.1038/534320a PMID:27306173

Nilsson, M., Griggs, D., Visbeck, M., Ringler, C., Mcollum, D., & Mollier, L. … Unger, S. (2017). A Guide to SDG Interactions: From Science to Implementation. Paris, France: International Council for Science. doi:10.24948/2017.01

Nilsson, M., Lucas, P., & Yoshida, T. (2013). Towards an integrated framework for sdgs: Ultimate and enabling goals for the Case of energy. *Sustainability (Switzerland)*, *5*(10), 4124–4151. doi:10.3390u5104124

North Carolina Climate Office. (2019). *Global Patterns*. Retrieved May 5, 2020, from https://climate.ncsu.edu/climate/patterns

Omamo, A. O., Rodriguez, A. J., & Muliaro, J. W. (2018). A Systems Dynamics Model for Mobile Industry Governance in the Context of the Kenyan Vision 2030. *International Journal of System Dynamics Applications*, *7*(2), 81–100. doi:10.4018/IJSDA.2018040105

Richardson, G. P., & Andersen, D. F. (2010). Systems Thinking, Mapping, and Modeling in Group Decision and Negotiation. In C. Eden & D. N. Kilgour (Eds.), *Handbook for Group Decision and Negotiation* (pp. 313–324). Springer. doi:10.1007/978-90-481-9097-3_19

Richmond, B. (1994). Systems thinking/system dynamics: Let's get on with it. *System Dynamics Review*, *10*(2/3), 135–157. doi:10.1002dr.4260100204

Rickles, D., Hawe, P., & Shiell, A. (2007). A simple guide to chaos and complexity. *Journal of Epidemiology and Community Health*, *61*(11), 933–937. doi:10.1136/jech.2006.054254 PMID:17933949

Saldana, J. (2016). *The Coding Manual for Qualitative Researchers* (J. Seaman, Ed.; 3rd ed.). SAGE Publications.

Schoumaker, B. (2004). Poverty and fertility in sub-Saharan Africa: evidence from 25 countries. In Intergovernmental Panel on Climate Change (Ed.), Population Association of America Meeting (pp. 1–30). Boston: Cambridge University Press. 10.1017/CBO9781107415324.004

Shafiei, M., & Hakaki, A. (2018). A Dynamic Model of Effective Factors on Open Innovation in Manufacturing Small and Medium Sized Companies. *International Journal of System Dynamics Applications*, *7*(1), 1–26. doi:10.4018/IJSDA.2018010101

Stephan, R. M., Mohtar, R. H., Daher, B., Irujo, A. E., Hillers, A., Ganter, J. C., Karlberg, L., Martin, L., Nairizi, S., Rodriguez, D. J., & Sarni, W. (2018). Water–energy–food nexus: A platform for implementing the Sustainable Development Goals. *Water International*, *43*(3), 472–479. Advance online publication. doi:10.1080/02508060.2018.1446581

Sterman, J. D. (2000). *Business Dynamics: Systems Thinking and Modeling for a Complex World*. McGraw-Hill Higher Education.

Swinerd, C., & McNaught, K. R. (2014). Simulating the diffusion of technological innovation with an integrated hybrid agent-based system dynamics model. *Journal of Simulation*, *8*(3), 231–240. doi:10.1057/jos.2014.2

System Dynamics Society. (2016). *Introduction to System Dynamics*. Retrieved July 4, 2017, from https://www.systemdynamics.org/what-is-s/

United Nations Development Programme. (2017). *Human Development Index*. Retrieved July 12, 2017, from http://hdr.undp.org/en/indicators/137506

United Nations Economic and Social Council. (2015). *A Nexus Approach For The SDGs*. Retrieved from https://www.un.org/ecosoc/sites/www.un.org.ecosoc/files/files/en/2016doc/interlinkages-sdgs.pdf

United Nations General Assembly. (2015). *Transforming our world: the 2030 Agenda for Sustainable Development* (Vol. 16301). United Nations General Assembly. doi:10.100713398-014-0173-7.2

Vennix, J. A. M. (1996). *Group Model Building: Facilitating Team Learning Using System Dynamics. Quality Progress* (Vol. 30). Wiley. Retrieved from https://www.amazon.com/Group-Model-Building-Facilitating-Learning/dp/0471953555

Vladimirova, K., & Le Blanc, D. (2016). Exploring Links Between Education and Sustainable Development Goals Through the Lens of UN Flagship Reports. *Sustainable Development*, *24*(4), 254–271. doi:10.1002d.1626

UN Water. (2016). *Integrated Monitoring Guide for SDG 6: Targets and global indicators*. UN.

Weimer-Jehle, W. (2006). Cross-impact balances: A system-theoretical approach to cross-impact analysis. *Technological Forecasting and Social Change*, *73*(4), 334–361. doi:10.1016/j.techfore.2005.06.005

Weitz, N., Nilsson, M., & Davis, M. (2014). A Nexus Approach to the Post-2015 Agenda: Formulating Integrated Water, Energy, and Food SDGs. *SAIS Review (Paul H. Nitze School of Advanced International Studies)*, *34*(2), 37–50. doi:10.1353ais.2014.0022

Wolstenholme, E. F. (1999). Qualitative vs Quantitative Modelling: The Evolving Balance. *The Journal of the Operational Research Society*, *50*(4), 422–428. doi:10.1057/palgrave.jors.2600700

Wyburn, J., & Roach, P. A. (2013). A System Dynamics Model of the American Collectable Comic Book Market. *International Journal of System Dynamics Applications*, *2*(1), 37–58. doi:10.4018/ijsda.2013010103

Yatimi, H., & Aroudam, E. (2018). Standalone Photovoltaic System with Maximum Power Point Tracking: Modeling and Simulation. *International Journal of System Dynamics Applications*, *7*(3), 94–111. doi:10.4018/IJSDA.2018070105

Zelinka, D., & Amadei, B. (2018a). A new insecure-secure diffusion archetype to model complex anthropocentric systems. *36th International Conference of the System Dynamics Society*.

Zelinka, D., & Amadei, B. (2018b). A Systems Approach to Modeling Corruption: Bridging the Disciplinary Divide Between the Social and Technical Sciences. In *24th International Sustainability Development Research Society Conference*. Messina, Italy: International Sustainability Development Research Society.

Zelinka, D., & Amadei, B. (2019). A Systems Approach for Modeling Interactions Among the Sustainable Development Goals Part 1: Cross-Impact Network Analysis. *International Journal of System Dynamics Applications*, *8*(1), 41–59. doi:10.4018/IJSDA.2019010103

Zhou, X., & Moinuddin, M. (2017). *Sustainable Development Goals Interlinkages and Network Analysis: A practical tool for SDG integration and policy coherence. The Institute for Global Environmental Strategies*. IGES. doi:10.1128/MCB.05460-11

Chapter 3
A Triple Cornerstone Framework for Software Startups:
A Systems Thinking-Based Analysis

Narendranath Shanbhag
La Trobe University, Australia

Eric Pardede
La Trobe University, Australia

ABSTRACT

Software startups are increasingly under high pressure to deliver successful products to survive and thrive in the modern highly competitive technology market. Larger organizations with deep pockets can replicate the same business ideas used by startups with relative ease. So how does the average startup stand a chance at succeeding at this seemingly David vs. Goliath contest? This chapter looks at the available literature and identifies factors that can affect the success of software development startups. Using causal loop constructs from the field of system dynamics, the interactions among the various identified factors are visualized to reveal the dynamics of the system. The result is as a three-dimensional view of success factors in form of time, capital, and (product) differentiation. The chapter also explores the cornerstones in the context of the product and business dimensions of software startups. This research acts as groundwork for analyzing the workings of software startups and sets the stage for a more holistic study upon which further research can be carried out.

1. INTRODUCTION

In recent times, software startups have disrupted the status quo and taken the market by storm with new, innovative products and services, which attempt to simplify everyday tasks in both business and personal spaces. From a product or service perspective, most software startups face stringent resource

DOI: 10.4018/978-1-7998-5788-4.ch003

constraints (Fayad et al., 2000) such as time (Paternoster et al., 2014), human resources and finances (Song et al., 2010). Additionally, startups need to be flexible and steadily evolve to stay relevant in the modern rapidly changing technology landscape (Duchesneau & Gartner, 1990). To this end, there is evidence to support the adoption of adaptable process frameworks like lean startup methodology among newer startups (Ries, 2011). There is also a good amount of interest in research directed towards startups, particularly in software development sectors. However, the software industry has some of the highest failure rates when compared to most industries. Many cases of failure can be seen in particular in early stages of the startup journey (Akkaya, 2020). Therefore, an in-depth study of the factors that can lead to better insight on factors that influence success of a product development undertaking in an aspiring software startup and would merit further study (Davis & Zweig, 2005).

Startups and the processes such as product development within it, can be considered as an example of a Socio-Economic or Socio-Technical system (Groen et al., 2008) and the complexities involved in such systems cannot be studied easily by most tools and techniques which view the problem at hand from a linear or deterministic viewpoint. According to reviewed literature, such Socio-Economic or Socio-Technical systems fall under the category of multi-loop non-linear feedback systems (Forrester, 1971). Forrester goes on to state that, as a consequence of the human evolutionary process, humans have not evolved to be good at understanding the dynamic behaviour of such complex systems. To add to this, a study found that policies and solution which have their base in linear viewpoints tend to produce solutions of a temporary nature, along with a higher number of escalating problems (Holling, 2001).

However, simply studying the various factors by isolating each one from the entire system might not suffice to provide enough insight. Studying each factor in relation to other factors present in the system can provide much better insight. As an example of this in the context of Socio-Technical systems, System Dynamics had been used to evaluate software development cycle time for effectiveness (Tvedt & Gollofello, 1995). In this study, rather than studying each factor for effectiveness in isolation, the study considered the various factors that would collectively affect the entire system. For instance, if a new software tool was introduced to a company with the intention of reducing cycle time of software development, this might reduce the amount of time the professional (using the software) needs to be involved with that task. However, the time gained by quickly accomplishing this task using the software tool might be spent as slack time by that professional. Therefore, there might be no real benefit or change to the cycle time, which was the original intention of introducing the software tool in the first place. This illustrates that a coordinated study of the various factors should be done in unison and in advance, rather than in isolation done later. By introducing an intervention such as a new software into a system and then modelling how the various elements of the system will be affected can help understand the positive and negative implications better. This is especially relevant in the case of software startups since failing to understand such implications can lead to an increase in cost, effort needed in product development and subsequently, time to market.

This work aims to group success factors for software startups, which are determined after exploring the literature and deducing various elements which affect success in software product development from the perspective of a software startup. This is done while keeping in mind aspects of dynamic behaviour which effects such Socio-Economic/Socio-Technical systems. While there are a variety of factors that lead to the success of a software startup, this would be too broad an area to study considering startups can offer a wide range of services, products, support for products/services as part of their venture. For the purposes of this study, most aspects of software startup success will be considered from a product development perspective.

Cornerstones and goals for each cornerstone are defined from each grouping, which will be used later on during systems modelling. The chapter also explores the interaction among the various identified factors, so as to gain an understanding of how these factors and cornerstones influence one another and the system as a whole. Additionally, the chapter explores the cornerstones in the context of the product and business dimensions of software startups, discussed by the exploration of two case studies. In the same context, the impact of using metrics as part of product and business development initiatives is explored. As a part of this chapter, a review of literature is done by studying over 50 research articles and a total of 14 factors are listed. To understand the impact these factors have on the system, the causal loop construct is applied to these factors.

The sections of this research are structured as follows: in the next section, related work is discussed based on the review of past literature. Research methodology used as part of this work is defined in section three. Section four presents the proposal of the triple cornerstone framework for startup success. Section five introduces the area of System Dynamics and the relevant constructs, from the perspective of this research. Section six presents a causal loop representation of the proposed triple cornerstone model, with a visual depiction of the causal interconnections within the system. Section seven presents the discussion of the cornerstones in context of the product and business model development undertakings in the software startup space with two relevant case studies of unsuccessful products. Section seven also highlights the importance of metrics in the same context, along with contributions of this research and possible directions for future work. Finally, section eight presents the conclusion of the research along with a summary of the previously discussed future work.

2. RELATED WORK

Many research works have explored the area of software development in the context of larger organizations but little to no structured investigation has been undertaken in the startup space (Curcio et al., 2018; Paternoster et al., 2014). This applies to the startup space in general and the software startup space, in particular. Startups are not the same as established organizations and have different strengths and weaknesses. For instance, Blank (2018) points out that startups are in search of a business model which can work and then scale whereas established organization have already found their business model and instead focus on operations. Also startups are more resource constrained (Davis & Zweig, 2005) and need to be concerned with optimal use of their resources, while established organizations might have relatively easier access to resources. Towards this end, the adoption of lean and agile based methodologies has been picking up pace in recent years. Klein (2013) introduces the idea of using a design driven framework for lean startups, which focuses on usage of agile thinking. The traditional process of product development using agile methodologies usually involves the user requirements being gathered, the product then being developed and then user feedback considered. Products are subsequently modified based on the feedback, shown to the end user and once again user feedback is taken. Klein (2013) introduces a simpler but more efficient technique based on agile principles. She points out that by creating a minimum viable product and collecting relevant metrics early on during idea validation, a good product can be built while keeping the cost low. Klein also emphasises the importance of user research, mentioning the high likelihood of heavy rework on the product after it is designed & developed, if the user research is not done previously. Although Klein's approach is geared towards efficiency, the focus is the design driven development along with lowering the overall cost of development and not specifically on the time to market. In such a case, the benefit of faster development time appears to be a by-product.

Hokkanen et al. (2016) delve deeper into the concept of devising a framework to assist designing of products in software startups, while keeping in mind the various factors and constraints faced by software startups. The framework, which is referred to as MVUX, talks about four main elements which include Approachability, Attractiveness, Selling the idea and Professionalism (Hokkanen et al., 2016). The strength of the MVUX framework lies in its base of centring the framework on "Selling the idea", since it is one of the primary end goals of any software startup to get feedback from the end user, who is in a position to appreciate the idea of the product. The remaining elements of attractiveness, approachability and professionalism then contribute by working to pique the user's interest in the product and initiate usage of the product. Furthermore, thinking in factors rather than relying on the expertise of evaluator(s) has been considered more prudent, as seen in instances of usability evaluation (Lin et al., 1997). Although the MVUX approach considers the usage of lightweight tools to assist design decisions for software startups, it does not take into account the capital spent on various tasks involved or the game changing benefits enjoyed by a software startup with early time to market. Keeping in mind these are constraints which software startups are faced with are non-trivial, it is important to take these into consideration.

Even though various frameworks, processes and methods have been proposed to deal with the aspect of design and development, the impact of the adoption of the method is not always studied. For instance, according to Eason, the process of designing the product is still largely techno-centric (Eason, 2001). This implies that the development teams run the show and drive the product design process. The consequences could be that the process results in unforeseen outcomes. Eason goes on to state that the trending predictions over these results are over simplified. This leaves room for systems-based methodologies like System Dynamics to fill the gap and provide a wider and well-rounded perspective on this subject.

It can be observed that over the years, linear models of studying innovation have been abandoned in favour of network based and more interactive models (Rothwell, 1994; Tidd & Bessant, 2018).

3. RESEARCH METHODOLOGY

As a tool to understand systems from the perspective of synthesis rather than analysis, System Dynamics serves as an excellent modelling technique to study the non-linearity in complex systems.

To gather the variables for the systems model, over 50 articles are studies and 14 factors are shortlisted and investigated using causal loop diagrams. Soft variables such as *user experience* and *quality* will be included as a part of the study. The study focuses on listing important factors that play a role in the success of a software startup, based on an analysis of literature and then categorizing them based on the nature of their influence. For instance, decision time and development time would come under the category of Time. The Cornerstones are then defined for each of the categories, since System Dynamics models are created to solve a problem and not just to model a system's working (J. Sterman & Sterman, 2000). The outline of the research methodology used is show in Figure 1.

Along with the categorization and the subsequent development of cornerstones, a causal loop diagram is drawn out to visualize the interactions of the different factors identified. The intention behind using the causal loop construct is to visualize the categorization or grouping of the factors and observe the nature of the influence (positive or negative). Using the causal loop structure to visualize the relationships can result in a much clearer understanding of the dynamics within the different elements of the

Figure 1. Overview of the research methodology

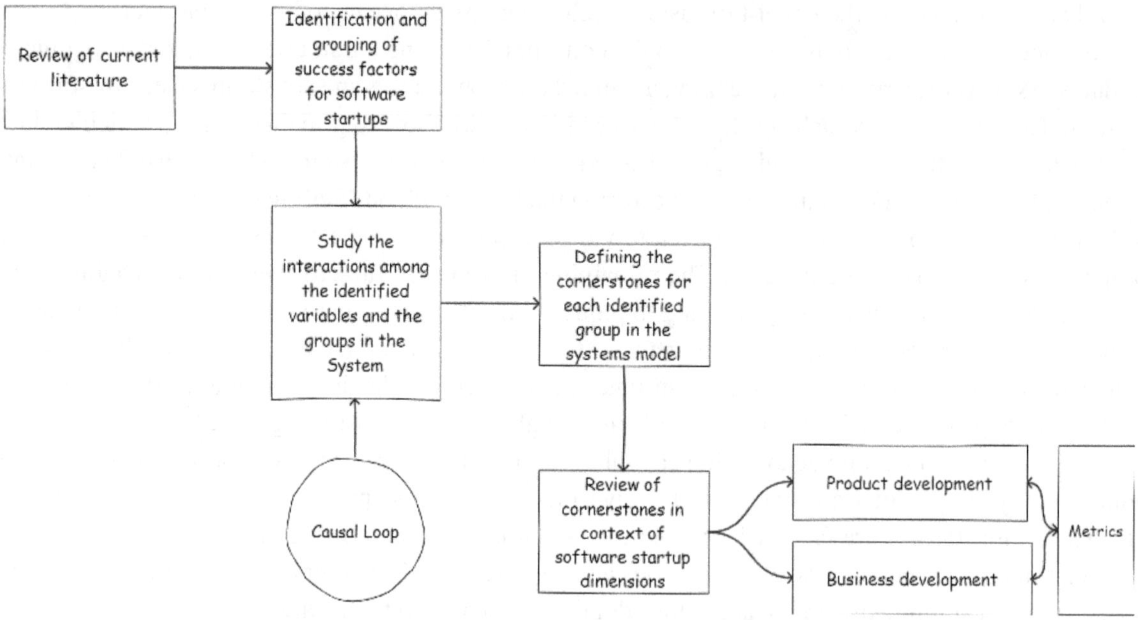

system and of the system as a whole. The cornerstones are then reviewed from the perspective of the various dimensions of software startups such as products and business development. The review takes into account the role of metrics in product and business development of software startups, given the importance of metrics in tracking attributes such as performance and guiding the software startup in the best direction for success.

4. THE TRIPLE CORNERSTONE FRAMEWORK – A PROPOSAL

The reviewed literature resulted in multiple factors which impact the process of product and business development initiatives in software startups. An overview of these factors revealed patterns for possibly groupings of factors which will be discussed more in detail when exploring the causal loop view of the system. Three such groupings were identified and chosen to represent dimensional viewpoints. *Time* dimension represents the time related factors, *Capital* represents resources and related factors for the software startup and *Product (differentiation)* represents factors such as quality and user experience of the offering. The factors uncovered during literature review impact one another but can ultimately be categorized under one or more of the three dimensions. Therefore, to simplify details of the knowledge gained from this study, it is divided into key areas. Furthermore, the grouping takes into account the complexity of the interconnecting factors, which provides a basis for the causal loop view of the system.

4.1 Cornerstone Grouping #1: Capital and Related Factors

4.1.1 Cost of Development

Conventional thinking might lead one to believe that without a large amount of capital, it is not possible to form a software startup. However, Gelderen et al. (2005) provide evidence that in the case of startups in general, getting started with a small amount of capital is easier than a larger amount, since acquiring a smaller amount of capital is easier than a larger amount. Reducing cost for the product development in startups was identified as a critical factor for success (Schwaber & Beedle, 2001). Lowering cost increases the chances of a product's success since more capital is left over to spend on other aspects of a product success like testing for higher quality or marketing. Therefore, the key to a successful outcome might lie in the optimal usage of capital, which is allocated to a development initiative.

4.1.2 Patents and Intellectual Property (IP)

Software startups holding patents and other intellectual property in general are found to have an edge when receiving investments from potential investors (Bessen, 2011; de Wilton, 2011; Dove & Sethumadhavan, 2012) and strategically patenting product assets can prove to be valuable (Noel & Schankerman, 2013). Additionally, software patents are also found to create incentives to invest in product innovation (Mann & Sager, 2007; Noel & Schankerman, 2013). Although there are growing concerns that patenting might result in a drag on innovation, Noel et al. (2013) argues that investing in Research and Development to procure patents will have a positive spill over towards the technology side. However, the costs involved with filing for a patent can usually be quite high (de Wilton, 2011; Graham & Sichelman, 2008). The decision to invest in IP protections must be made while weighing the costs to the benefits they can provide. This can have a substantial impact on the amount of capital a startup can invest in intellectual property, depending on the type of protection the software startup would like to apply for.

4.2 Cornerstone Grouping #2: Time and Related Factors

4.2.1 Time to Market

From the perspective of managing any software product development process as a project, aspects of quality and time (along with the previously noted aspect of cost) can be identified as key factors for success (Schwaber & Beedle, 2001). The aim is to reduce time and cost needed to develop the product, while attempting to increase the quality. Even in the early days of software, when software was viewed as 'software packages', the benefits of faster time to market was acknowledged. Quicker time to market could demonstrate product-market fit earlier on in the development phase, possibly resulting in significantly increased market share, increased customer loyalty and reduced overall cost (Carmel, 1994; Carmel & Sawyer, 1998; Giardino et al., 2014). Time to market can be defined as the time measured from the conception of the software startup to the time when the first product is sold to customers (Hellmann & Puri, 2000).

Another advantage of keeping time in focus, is that the organization could control the choices (both tactical and strategic) during the time of launch of the product. However, Song et al. (2010) emphasize the need for the timing to be right rather than having a faster time to market. This perspective, however, may not be applicable to the case of a software startup since Song et al. (2010) mention that the luxury of checking for the right timing is one that mostly larger organizations can afford, rather than most startups.

Various elements can affect time to market in organizations. Hellman and Puri (2000) state that although not necessarily causal in nature, the presence of venture capital correlates to quicker time to market, and this correlation is particularly evident in companies which innovate. This is understandable since such organization would invest the gained capital to create innovative products and get it to the market sooner to capitalize on the relatively lesser competition during those stages.

4.2.2 Product Decision, Development and Delivery Time

Quicker delivery times for the software product (Fayad et al., 2000; Misra et al., 2009) and shorter decision time with regard to product development is also seen to improve the overall time during product development (Misra et al., 2009). Although decision time is considered in an organizational context in the stated study, it can be argued that shorter decision times has a direct impact on the outcome of the product development process. This is because lesser time spent on making product related decisions will lead to more time available to complete other product related tasks. In cases when the product is developed using agile methodologies, the decision, development and delivery time is relatively shorter, iterative in nature and found to deliver a product with good performance (Poppendieck & Cusumano, 2012).

4.3 Cornerstone Grouping #3: Product Differentiation and Related Factors

4.3.1 Product Differentiation and Innovation

One of the factors which has been seen to positively impact the success of a new product development initiative is Product differentiation (Song et al., 2010). Cozzie et al. (2006) argue that a focus on IP protection could lead to a greater emphasis on quality, but this in turn could negatively impact product differentiation. Therefore, the ideal scenario would be to seek a balance between differentiating the product and seeking IP protection. In addition, attempting to differentiate products from the competition would involve additional investments of capital. The capital used in this case can be used to educate the customer about the product and in advertising to increase awareness of the products differentiating characteristics (Boadway & Tremblay, 2003). Innovation can be a source of competitive advantage for startups (Lichtenthaler, 2020), therefore helping differentiate the startups offering and allowing better product visibility. As a differentiating trait, Innovation has also been known to have a strong impact on the product's success and subsequently the startup's success (Cooper & Kleinschmidt, 1987b; Parry & Song, 1994).

4.3.2 User Experience Design

Terms like User Experience (UX) design are often considered elusive and vague (Hassenzahl & Tractinsky, 2006). However, thought leaders in area of marketing, have pointed out that a powerful differentiator which can offer many competitive advantages is the factor of design (Kotler et al., 2015). Indeed, there are many cases where UX is acknowledged as a key product differentiator (Chapman & Plewes, 2014; May, 2012). This is particularly evident when the landscape of the modern-day competitive technology market is observed. In cases where the experience (UX) in using an application is poor and dissatisfying, it is often found difficult to learn as well. This is especially seen to be the case in enterprise software (Finstad et al., 2009).

4.3.3 Quality

Quality was identified as one of the factors for product success (Schwaber & Beedle, 2001). However, Carmel (1994) mentions in the context of time-to-completion in software startups, that it is usually a matter of balancing the desired level of quality with the amount of time and cost available to spend on it. In case of products developed using the agile methodology, quality is baked in as one of the seven principles of the methodology (Poppendieck & Cusumano, 2012). Despite this, there is still need to further investigate the role of quality in the context of software product development in startups (Klotins et al., 2015).

4.3.4 Technical Skills

Development teams require appropriate technical skills in order to develop the product, innovate and differentiate the product in the market (Song et al., 2010). Alongside the skills, other important elements are experience and the ability for coordination as part of working as a team (Crowne, 2002).

4.4 Model Summary

Multiple factors influencing the success of software startups are explored as a part of this section. A brief overview of the key factors is represented in Figure 2. It is worth noting that certain factors overlap across key areas, but this demonstrates the inherent complexities within such systems and role played by the factors in the outcome of the product's future. This overlap of the various factors within the dimensions is shown with the help of a Venn diagram in Figure 3.

Figure 2. The triple cornerstone framework with the factors illustrated

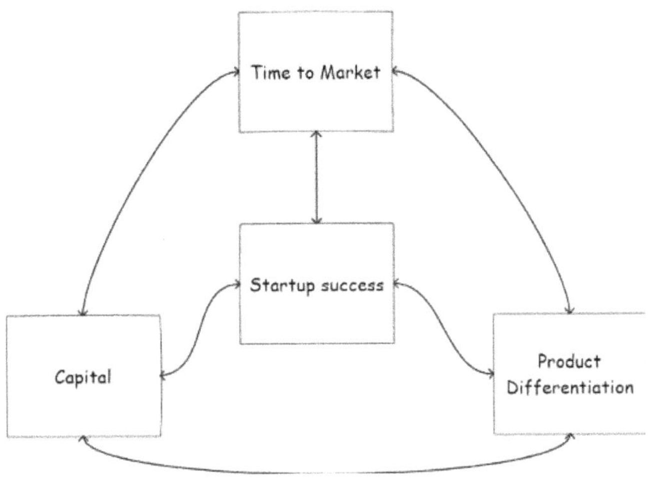

Figure 3. A Venn diagram showing the success factors overlap

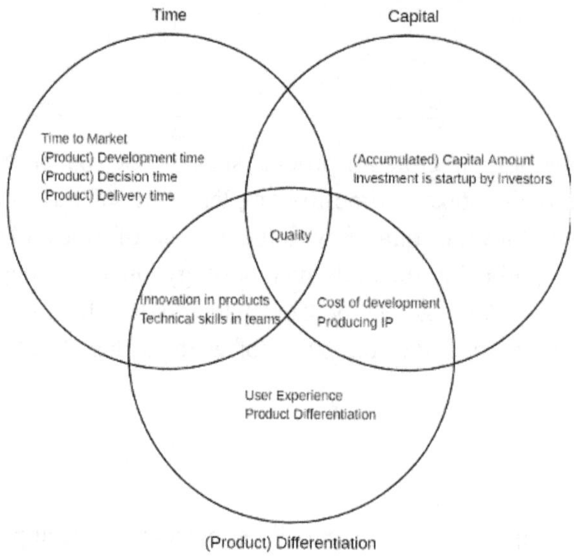

4.5 The Cornerstones – Goals and Ideal Trends

Each factor identified in the previous section has been grouped into one of three areas, based on the nature of influence it has on the system. Establishing ideal cornerstone trends will make the framework actionable and guide modelling efforts as part of any future work in this space. To this end, cornerstones were identified for each of the groupings. The criteria for defining a cornerstone was twofold. Firstly, it had to have a direct influence on the success of a software startup. Secondly, the other identified factors in the grouping had to have a significant influence on the cornerstone. Studying all the factors in the three groupings, the following were determined to be the ideal candidates for the cornerstones.

The first cornerstone defined is to increase the amount of capital that the software startup has. The purpose is for the software startup to be able to invest in resources as and when required throughout the product development process. As startups are typically resource constrained (Fayad et al., 2000), access to greater accumulated capital can help provide better access to resources. The intended trend for this cornerstone is to increase the accumulations in capital amount over time.

Time to market is the second cornerstone for the next grouping. The end goal of time to market is getting the product out to the market sooner than the competition, and in turn get the product in the hands of the customers as quickly as possible (Carmel, 1994; Giardino et al., 2014). The intended trend for this cornerstone is for time to market to decrease over time.

The final cornerstone deals with the aspect of product differentiation. The benefits of product differentiation are well documented (Song et al., 2010). Given the limited initial resources affecting most startups, the best chance a software startup has of succeeding is if they have a differentiating factor which makes their product offering standout from that of the competition. The intended trend for this cornerstone is for product differentiation to increase over time, so that the product offering remains unique when compared with the competition. A summarised view of all the cornerstones along with the goals and intended trends can be seen in Figure 4.

Figure 4. A summarised view of the cornerstones for the systems model.

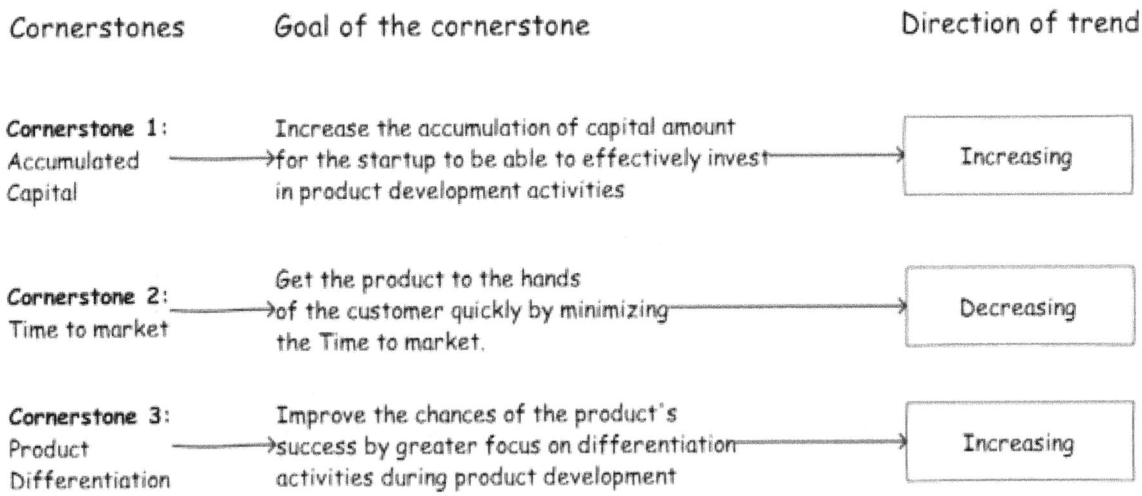

5. SYSTEM DYNAMICS – AN INTRODUCTION

To understand most complex scenarios, which are observable in socio-economic and socio-technical systems, there is a need to use better and more efficient tools which enable the observer to understand the underlying dynamics at work. These tools must also help associate the potential effect of each action on the system and lead us in the direction of possible causes of observed phenomenon within the bounds of that system. System Dynamics provides us tools to achieve this kind of understanding (Sánchez, 2014). Two of the more prominent ones being the causal loop diagrams and simulation models (Ford & Sterman, 1998; J. D. Sterman, 2001). The application of System Dynamics techniques like causal loop diagrams or simulation modelling in the field of software development and related fields has been done before (Hilmola et al., 2003; Rahmandad & Weiss, 2009). Other areas in information technology where System Dynamics have been applied includes data sciences (Casado, 2005; Pruyt, 2017), information systems research (Georgantzas & Katsamakas, 2008), decision support systems (Kljajic et al., 2012; Qudrat-Ullah, 2009), software project staffing (Abdel-Hamid, 1989), systems engineering (Galli, 2019), logistics and application infrastructure (Auxilia et al., 2020; Baporikar, 2020) and market dynamics in the technological sector (Morecroft, 1986). System Dynamics techniques have also been used to study the impact of technology related interventions on populations as a whole (Omamo et al., 2018, 2020) and design frameworks for public policy making (Nasution et al., 2018).

However, an important question many are faced with when considering the usage of System Dynamics is, what complexities and challenges are faced by software startups to warrant the usage of system tools for a detailed study? Apart from having almost no operating history and limited resources to work with, software startups face the challenges of having multiple influences and pressures from various entities within the system. These can include investors, partners, customers and competitors (Sutton, 2000), depending on the circumstance. Sutton (2000) also states the challenge of understanding the dynamic nature of technologies and markets, emphasising the need to work with newer technologies to enter unexplored markets. Besides this fact, software projects in themselves are considered a form of complex adaptive systems (Nguyen-Duc et al., 2015). In addition, there is also a growing need to understand the relationship between product differentiators such as quality, user experience and the challenges in integrating it with the software development process (Kashfi et al., 2016), owing to a rapid increase in competition in most markets. The field of user experience is in itself considered to be complex and subjective, with interrelated elements which are hard to separate (Kashfi et al., 2016). It has been acknowledged that one of the factors affecting usability is the dynamics based on user and system interaction (Goodwin, 1987).

Working with such a varied network of influences can potentially introduce unforeseen complexities. A software startup comprises of a strong interconnected working unit of active components rather than a lose collection of isolated parts. In such systems, the whole is greater and uniquely functional than the sum of its parts and this classifies it to be a complex system. This facet, when combined with the ever-changing nature of the technology market, it is safe to state that a tool like Systems Dynamics is appropriate to visualize and model this problem space. It's usage can provide a more comprehensive understanding of the underlying structure of the process and the overall system.

The nature of feedback and time delays in Socio-Economic and Socio-Technical systems can be realized using Causal loop diagrams (CLD). CLDs are System Dynamics constructs which represent the structure of systems and provide a visual depiction of the nature of relations within that system. As an example, in a linear or event driven view of a system, an issue is encountered in the current circumstance, which leads to the need to take a purpose driven action. On completion of this action an outcome is achieved, which might meet the goal or purpose of action (the overview of this view is depicted in Figure 5).

Figure 5. Example of an event driven view of a system.

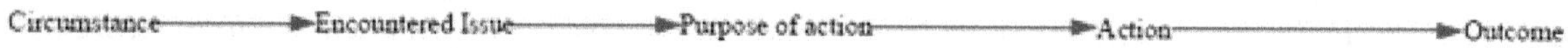

However, in reality, most systems are a lot more complex than this event driven view. The causal view comes closer to representing this complexity. If the same system is observed from a causal viewpoint, an issue which is encountered in the current circumstance leads to the need to take a purpose driven action. However, the action taken will affect the entire ecosystem or the environment in which the issue is taking place. This in turn will guide the subsequent action (also known as feedback) and the original purpose of the action. The action can also have unintended consequences which will also affect the ecosystem or the environment in which the issue is taking place. The action will also affect both the encountered issue and the circumstance leading to that issue as well. Therefore, the causal view sees the system as an interconnected confluence of events which affect each other and the system as whole. A depiction of this is shown in Figure 6.

The feedback occurring in complex systems can be represented in CLDs using feedback loops and these loops can be either positive or negative. A feedback loop can depict reinforcing or balancing patterns. Reinforcing loops represent a compounding pattern, which reinforces itself in a single direction. As an example, employee rewards and recognition can result in better employee performance. In turn, better employee performance can result in more rewards, thereby the cycle of events reinforcing itself. A balancing loop on the other hand represents a goal seeking behaviour. As an example, a thermostat detects the temperature of the area to check if it is at the desired level. If the temperature is lesser, the heat is turned on and a check is performed again after a time interval. This kind of a loop/system balances itself. Irrespective of whether these loops can be found within the system, it is useful to list and map out the nature of influence (positive or negative) which each identified factor have on other factors within the system.

Figure 6. Example of a causal view of a system.

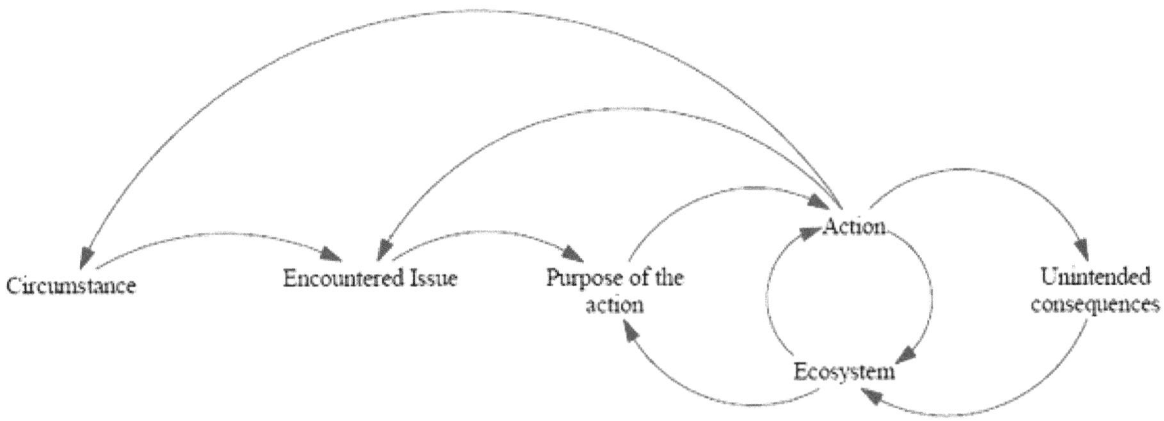

6. THE SYSTEMS MODEL - A CAUSAL VIEW OF THE TRIPLE CORNERSTONE FRAMEWORK USING SYSTEM DYNAMICS

The goal of the systems model is to apply System Dynamics constructs to highlight the factors affecting success in product development in software startups. After a detailed literature study and identification of factors influencing success among software startups, the influence each of the factors might have on one another and the collectively on the system is modelled and visualized (As shown in Figure 7).

6.1 A Causal Analysis View of the System

The CLD is created using the causal loop constructs discussed in the section 5 of this work. Starting from the top, the diagram visually outlines the time dimension of the model. Time to market is critical to ensure the success of any software product development undertaking and quicker time to market has a positive influence on chances of success (Carmel & Sawyer, 1998; Giardino et al., 2014). Decision time is another important factor which impacts the product development time, delivery time and the time to market. Faster decision times could result in reduced product development time (Misra et al., 2009), faster delivery time and quicker time to market (Fayad et al., 2000; Misra et al., 2009). Considering another perspective of the time dimension is that markets tend to value early innovation (Hilmola et al., 2003). To achieve the realisation of planned innovations, Giardino et al. (2014) talks about the idea of releasing prototypes quickly with incremental and iterative development updates. The idea is that faster releases of the product will validate innovative product ideas, which in turn can positively impact the time to market and innovation at the same time.

Figure 7. The Causal loop diagram representing the Systems model

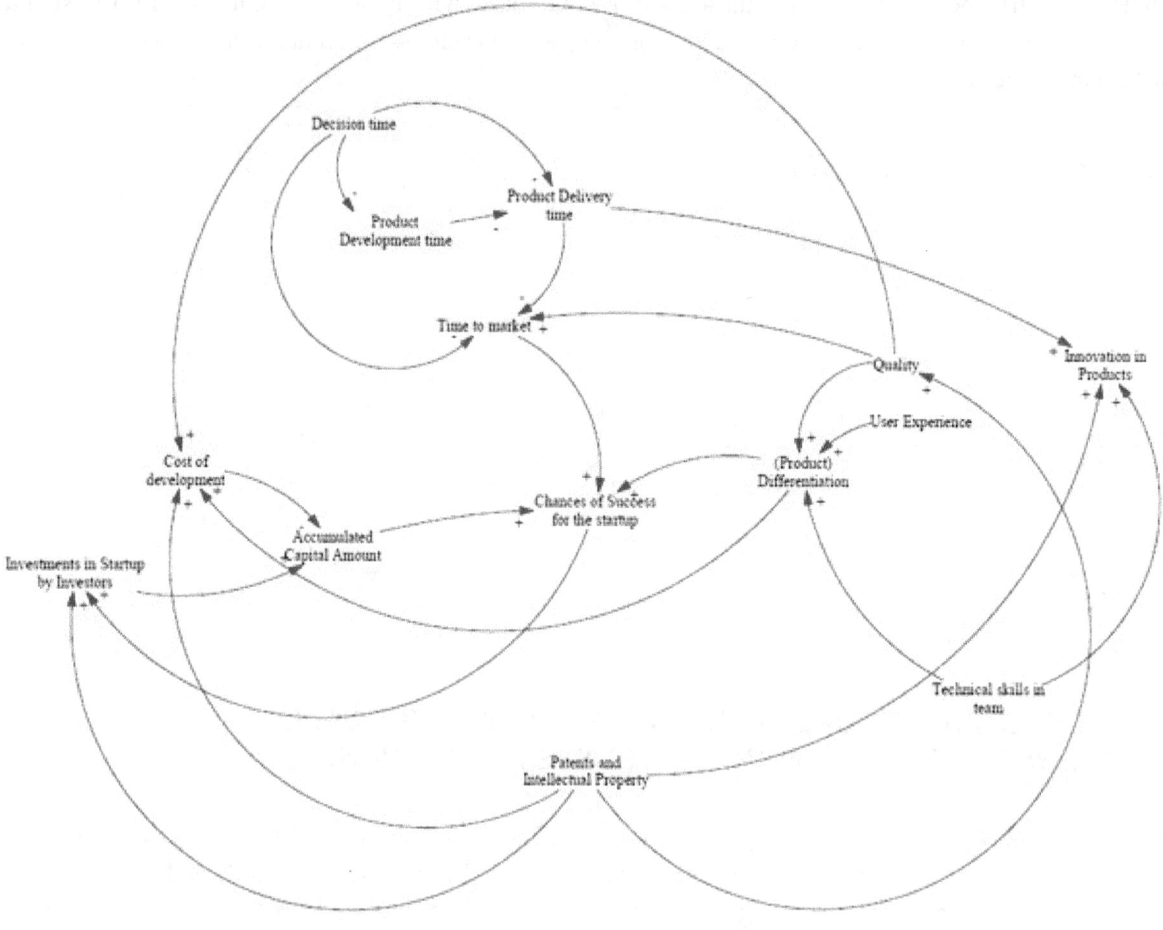

The next consideration is the aspect of startup capital. At some stage, most startups depend on capital to develop the product, so as to give the startup an opportunity at commercial success through monetization of the end product. Startup capital amount can be provided by investors including but not limited to angel investors, venture capitalists and other sources like personal financing or crowd sourcing. The cost of development could be covered by the accumulated capital (Mann & Sager, 2007), thereby depleting some of the startup capital amount. Additional measures to improve quality will further add to the cost of development (Carmel, 1994).

Investors are encouraged to invest in startups which secure intellectual property like patents (Bessen, 2011; Cockburn & MacGarvie, 2009; Mann, 2006), since intellectual properties like patents are typically considered as assets. However, the process of procuring intellectual property is expensive which might deplete some of the capital from the startup (de Wilton, 2011; Graham & Sichelman, 2008), thereby adding to the overall cost of product development. Patents are also found to increase the incentives for innovation among software startups (Mann & Sager, 2007; Noel & Schankerman, 2013), since patents provide a degree of assurance that the investment on innovation will not be infringed upon by competitors. Furthermore, innovation is another aspect that is found to be critical to a product's success (Cooper & Kleinschmidt, 1987b; Parry & Song, 1994).

The final consideration is that of product differentiation. Product differentiation has been found to improve the chances of success in most product development undertakings (Song et al., 2010). This is especially relevant in the case of software startups, since most startups do not have the capital or resources to compete with established players in the market. Established contenders typically have both an established product in the market and in most cases, a loyal customer base to go with it. For new entrants to compete in this space, startups would need to come up with innovative products with a greater degree of product differentiation. Some of the most common differentiators are considered to be quality (Cooper & Kleinschmidt, 1987a; Gatignon & Xuereb, 1997) and user experience (Beauregard & Corriveau, 2007; Pandey & Srivastava, 2014). Gatignon et al. (1997) mentions quality as an enabler of competitive advantage and a means of separating one's own offering from that of other organizations with a competing offering. From a differentiation perspective, Cooper and Kleinschmidt (1987b) go on to highlight the importance for organizations to look for a competitive advantage via product differentiation to achieve success with their product development initiatives. They go on to mention that product differentiation can be achieved by creating higher quality products (Cooper & Kleinschmidt, 1987b). As demonstrated by companies like Apple and Google, a positive user experience can radically differentiate a product offering from that of a competitor's product. However, in order to foster innovation and create differentiation, greater technical skills are necessary (Song et al., 2010).

The main loops for this model can be seen in the dynamics between Time-to-market, Accumulated Capital amount and (Product) differentiation. The objective is to emphasise the factors that serve the greatest contribution to the chances of success for software startups. The other factors eventually contribute to one or more of these three factors. The study of the causal loop diagram provides a better understanding of the overall system (in this case, the startup) and it's functioning.

7. DISCUSSION AND CONTRIBUTIONS

7.1 Cornerstones in Context

Startups in the software space create economic value through the creation of innovative products and development of business models around that product (Tripathi et al., 2019). It can be derived from this that the two major dimensions for software startups to focus on involve product and business development. Most modern software startups also employ the lean/agile philosophy of product and business development (Ries, 2011). A core tenet of this approach is the usage of the *Build-Measure-Learn* cycles as part of their development philosophy. However, 'measure' implies the employment of effective metrics (Kemell et al., 2020). It can therefore be stated that the three dimensions of product development, business development and the employment of metrics are interlinked, in the context of the software startup space (as shown in Figure 8). Since one of the aims of this work explores the cornerstones in the context of software startups, it would be interesting to contrast the role of cornerstones in each of these dimensions. The remainder of this section will explore this role in more detail.

7.1.1 Cornerstones in the Context of Product Development

From the context of the cornerstones covered as part of this work, adopting product development methodologies which inherently promote a "rapid and iterative validated product development" philosophy can

Figure 8. Cornerstones in context of Software startups

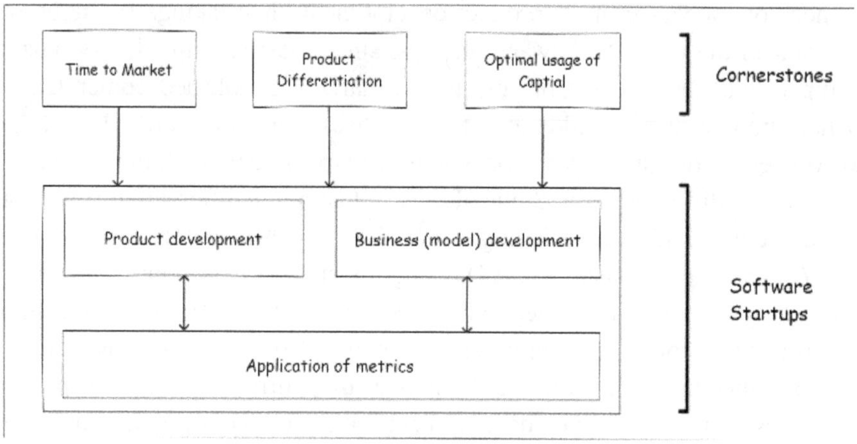

help in the chances of success of any product undertaking. As a consequence of timely validation, such methodologies can save time and capital by ensuring it is invested in building a useful and competitive product offering and making sure it is built right.

Conventional product development approaches in software which follow the outline of the waterfall model have some inherent shortcomings, while applied in the modern product development context. To begin with, the waterfall approach assumes that the requirements for the product development initiative remains unchanged throughout the cycle of development and does not have provisions in place to revisit, revise and manage these requirements (Nurmuliani et al., 2004). This makes the traditional software product development approaches too rigid for use, when observed in context to the nature of modern software products, due to its formal structure. The traditional innovation process also suffers from the same shortcomings (Hannola et al., 2013). In contrast, the more modern agile based approaches start with the assumption that requirements can change in time. The idea is to capture a small set of requirements (established as a hypothesis), create a basic working prototype based on the requirements and then quickly validate if the requirements (or hypothesis) are indeed valid. Changes are made based on the feedback collected as part of the validation process and the next cycle begins with capturing or working through the next set of requirements. An interesting consequence of adopting this method is speed of development. This is because the agile process attempts to quickly develop products in small increments, while making sure that the product being built is indeed catering to the user's requirements. The speed of development falls in line with the cornerstone of reducing the time taken to develop a product which contributes to a reduced time-to-market.

Another noticeable difference in the approach between traditional and agile software product development approaches is the optimal usage of resources, in particular capital. In traditional approaches, owing to the rigid structure of the framework, poor understanding of user requirements or changes in requirements can lead to a significant amount of rework. This can increase the capital expenditure by a significant amount. In contrast to this, since agile frameworks involve frequent validation as part of its iterative approach to software product development, there is little scope for error in understanding of user requirements as any such error would be caught during the validation phase of a given iteration. Agile approaches therefore are better structured to make use of capital and other resources.

While viewing the strengths of the agile methodology, it would appear to be the perfect methodology to adopt for the product development space in software. However, agile approaches themselves present certain shortcomings, in particular for software startups looking for end to end approach to development of innovative products with commercial viability. From the viewpoint of any organization, products are not an end in itself, but rather one part of the business model structure. In the Business model canvas (BMC) methodology for generating business models, this part is represented as a value proposition (Osterwalder & Pigneur, 2010). The agile methodology does inherently not take into consideration the business model aspects, as part of the product development process. Although agile methodologies and other product development methodologies were not created to build and present business models, they could greatly benefit from the early consideration of some aspects of business. Aspects such as preliminary monetization methods for the product, delivery mediums or handling growth, if considered early on, could set the optimal foundation upon which the business model can be built over, eventually. The result can be an end to end process which seamlessly incorporates software product development with business modeling, making it highly applicable for software startups to pick up and run with.

7.1.2 The Cornerstones and Business Modelling

The cornerstones identified as a part of this work can find application in business model development space, as the success factors outlined by the cornerstones can point to the key areas of importance when working on business models. Just as in the case of product development discussed in the previous section, organizations can benefit from a rapid development methodology which aims at quickly and iteratively developing business model hypothesis around a useful but uniquely positioned product offering with minimal expenditure on capital amount.

Traditional business modeling can be rigid as a process and teams working on such business models could take longer times to develop a complete workable business model. If we take into account the modern rapidly changing business landscape, by the time a workable business model is developed using this technique, the value proposition could have become less relevant, or customer segments would have found other solutions to cater to their pain points. The cornerstones emphasize on time to market as being one of the key factors for the success of a product undertaking and by extension to the business.

Business modeling techniques that are similar to the agile product development philosophy, however, are viewed as being more attuned to the needs of the present market conditions. These techniques believe in the principle of "failing quickly" and ensuring that the user learns from that failure and iterate to search for a better business model. One such technique is the Business model canvas (BMC) by Osterwalder and Pigneur (2010). The BMC technique facilitates its users to quickly generate business models by identifying nine key segments of the business modeling space and then allowing the user to rapidly validate the business model. Changes to the business model are made in an iterative manner over time and therefore the business model stays relevant. One of the underlying core principles that can be noted here is the element of speed, wherein the structure of the BMC is developed with the aim of developing business models quickly. This will allow the value proposition defined as part of the BMC to get in the hands of customers as early as possible, therefore reducing the time to market.

The BMC segments such as Cost structure and Revenue streams address Capital related issues directly. Other BMC segments such as Key resources and Key activities also provide insight on expenditures which the business is expected to incur. Studying the customer demographic information in customer segments can provide insight on how the product can be tailored to be better suited to customer needs

(Galli, 2020), at the same time informing the development team on how they can differentiate the product based on other offerings. The differentiation factor can also be provided to the customer as a consequence of working on a unique distribution channel, revenue model or a customer relationship strategy.

The next two sections discuss case studies of products which failed to garner market share and were eventually shelved. The case studies are selected with intention of exploring diverse product types. For instance, the first case study is Google+ by Google, which is an online social networking product, whereas the second case study is Windows phone, which is an operating system product by Microsoft. The common theme among both case studies is the lack of focus on the key factors which culminate in the form of cornerstones, outlined as part of this chapter. This is intended to demonstrate that the success factors presented as part of the cornerstones are independent of software product type.

7.1.3 Case Study 1 - Google+

Google+ was a social networking product announced by Google in mid to late 2011, as a response to the rise of Facebook. Although it received initial praise for its design aesthetics and features, it failed to meet user expectations and saw many initial users abandon the platform (Dieter & Lenze, 2018). From the early onset, a closer look at the data provided evidence of poor user engagement. It showed that active user count was up 3,000 per day which in itself is not a bad indicator for a social network, but when taken in context of the total signups per day, which was 200,000 per day, the problems become evident. This showed that active users were growing 60 times lesser than the registered user base growth (Gonzalez et al., 2013). In fact, much of the initial struggles of Google+ and its lack of meaningful differentiation with the existing competition is well documented (Curran et al., 2012; Dieter & Lenze, 2018).

When announcing its closure, Google announced that in 90 percent of sessions on the network, users spent a mere 5 seconds or less on the network (*Project Strobe*, 2018), while at the same time having 500 million monthly active users (Dieter & Lenze, 2018). Despite the enormous appetite for social networking in the modern age, Google failed to create a unique but useable product experience from the present dominant social networking products (Eadicicco, 2015), such as Facebook's social network, twitter or Instagram. Eadicicco (2015) also cited delayed entry into the social networking market as one of the key reasons for the failure of the social network. Key metrics had revealed early on that Facebook was attracting online browsing time away from Google and that Facebook was on par with Google in collection of valuable user data, which could later be used for the purposes of advertising and/or marketing related monetization efforts (Faktor, 2014). However, this realization had already come too late and Google had to rush to create a hurriedly built product, conceived based on the reaction to the success of the competition rather than driven by data or a focus on the customer problem (Levy, 2011).

It is interesting to note that shortly after the announcement of Google+, Google the parent company behind the social network at the time, attracted $20 billion dollars in capital investment (Schonfeld, 2011). It must be emphasized that this amount added to Google's market capital only after one day of its announcement of the new product undertaking. Although it is worth mentioning that Google significantly invested both human and financial resources into this product (Dieter & Lenze, 2018). The central issue, however, was that the allocation of these resources for the conception of the product was not done based on a focus of the customer's problem or driven by data, but rather by a reaction to current market conditions (Levy, 2011). The capital allocated, therefore, was based purely on a defensive response to being late to the market, rather than planned based on user requirements. To compound this matter, a lot of human and financial capital was expended on this problem in a relatively short 100 day development cycle (Dieter & Lenze, 2018). This highlights the importance of careful consideration on how capital may be best used, when considering product development initiatives in the context of business success.

7.1.4 Case Study 2 – Windows Phone

Windows phone was a classic case of introducing product differentiation for the sake of simply being different rather than being better when compared to the competition.

In 2007, the announcement of the iPhone by Apple triggered a change in the approach towards design and development of smartphones for device manufacturers. The same was the case with smartphone operating systems (OS). The smartphone OS market which was at the time dominated by Symbian OS, Blackberry and Windows Mobile suddenly faced fresh competition from the likes of Android and iOS. Soon, Microsoft found itself late to the smartphone OS party and rushed to create a well-designed smartphone mobile OS in 2008, hoping to learn from the missteps and successes of both Apple's iOS and Google's Android.

The resulting new OS, named Windows Phone 7 arrived in 2010, by which time both iOS and Android had gained significant popularity among the user base and acquired a good amount of developer mindshare (Heisler, 2015). However, Windows Phone 7 was based on an older platform called WinCE and the upgrade to the newer NT based kernel resulted in the release of Windows Phone 8 in 2012, which was the next iteration of their smartphone OS (Cahyani et al., 2017). This meant that Microsoft was late to the smartphone market by 5 years after Apple had introduced the iPhone in 2007. To make matters worse, the Windows Phone platform lagged behind the competition in terms of platform customizability and common functionalities (Sehgal et al., 2019). By this stage, the Apple-Google duopoly ruled the smartphone market with few users or developers looking for a fresh third OS platform to migrate to. This delayed time to market is considered as one of the key factors for the failure of the Windows Phone/ Mobile platform (Herrick, 2018).

While Microsoft created a different interaction experience from their smartphone OS rivals, they neglected one key ingredient of what made a successful smartphone OS platform and that was applications. This was particularly important in the context that the software that ran the devices and the apps developed for them were considered as key differentiators in an increasingly competitive mobile phone market (Giachetti & Marchi, 2017). Giachetti and Marchi (2017) cited lower availability of applications in its stores as one of the key reasons for the failure of Microsoft in the smartphone OS market, which meant it had no real differentiator to separate it from other struggling platforms like Blackberry's OS or Symbian OS.

To add to Microsoft's troubles, the company initially had strict hardware requirements for device manufacturers who made use of the Windows phone platform (Rubino, 2014). Although this allowed Microsoft to better control the experience of the product, it did not incentivize the manufacturers to produce Windows Phone devices. This is because they lacked the level of control over hardware they would have wanted, in addition to not being able to create their own software user interface 'skins' since no one outside Microsoft had access to the source code of the OS (Covert, 2013; Rubino, 2014).

Until much later, Manufacturers were charged a licensing fee for the usage of the Windows Phone which approx. $23-31 per device, depending on the manufacturer (Segan, 2014; Weir, 2012). In comparison to this, Android was completely free and open source under the Android open source project (AOSP) (Weir, 2012). Google also had more relaxed hardware requirements for manufacturers who used the Android OS on their devices (Rubino, 2014), which resulted in many manufacturers working with Google to produce Android devices instead (Weir, 2012).

The exclusive nature of Microsoft's unique partnership arrangements with Nokia further alienated the manufacturing partners. Under this partnership, Microsoft paid 'Platform support payments' to Nokia for using the Windows Phone OS on their devices and got first class Windows Phone platform support (Covert, 2013; Savov, 2012). This resulted in Nokia receiving financial capital from Microsoft, putting them at an advantage when compared to the competition. This ultimately meant that it was substantially more economical for device manufacturers to produce Android devices, while at the same time retain better control over the overall user experience of those devices. The capital amount, which could be saved as a result of going with Android, could be directed towards innovating on newer products. The retention of control over hardware meant that device manufacturers could further differentiate their offerings from the competition. Microsoft eventually performed a write down of its Nokia acquisition for approx. 7.6 billion dollars (Ranger, 2015), cementing the failure of the Windows phone platform. The official announcement of the retirement of the platform was made in October of 2017 (Villas-Boas, 2017).

Based on the analysis of the above-mentioned case studies, it can be understood that Google+ and Windows Phone were instances of two products which failed owing to factors related to poor product differentiation, being late to the market and poor management of capital.

7.1.5 The Importance of Metrics

The importance of determining and using metrics for the success of product and business development undertakings has been well established (Abreu & Carapuça, 1994; Kupiainen et al., 2015). Metrics aid in the success of product and business initiatives by tracking various aspects of the initiative. There are several cases where using metrics can be critical to any commercial undertaking. For instance, tracking metrics such as the *monthly user churn rate* or *daily/monthly active user count* of a product can indicate if the product is indeed useful to the customer and relieving the customer's pain points, as understood by the development team. If the product is shipped to the customer in its current state, it will inevitably impact the success of the product and business development initiative. If the focus on similar important metrics is not placed early on, this will increase the time to market of the final product and can significantly lower the chances for success of that product. The same is the case with creating unique product experiences.

As discussed in the case of Google+ in section 7.1.3 of this work, one of the reasons Google+ primarily failed is simply because it could not create a compelling reason for social networking users to adopt 'yet another social networking platform'. This is because Google+ was created more as a reaction to the competition rather than a product based on current market or user needs. As discussed in the same section, metrics such as proportion of daily new active users as compared to daily new user signups confirmed that although the network was acquiring new users, a very small percentage would remain as active users. Another social network named Tiktok, which took an different approach to social networking has seen significant growth in a relatively short amount of time (Anderson, 2020). Tiktok differentiates its offering by algorithmically (via artificial intelligence) determining its feeds, which are essentially short video content, to be presented to the user rather than only aggregating it from friends and contacts from a registered user's network. This fresh approach to delivering social networking content served as enough of a differentiator to pique the interest of new users. Although introduced recently in 2018, the social network has 800 million monthly active users as of March 2020, with each user interacting with the network over 8 times per day (Sehl, 2020).

To make it easy for product and business development teams to list and apply appropriate metrics, the use of established metric frameworks could be a good option. One such framework is the Goals-Signals-Metrics (GSM) framework (Rodden et al., 2010), which is a product centric framework wherein goals are established based on stakeholder input. Another framework is the Three engines framework by Ries (2011), which caters to tracking growth related initiatives of organizations. A more holistic framework called the MESOPS framework (Metric framework for product development in startups) was proposed by Shanbhag and Pardede (2019), which is geared towards product development initiatives in software startups. Irrespective of the choice of the framework used, utilization of any framework presents its users with guidelines on how to identify useful metrics based on factors such as the current state of progress, performance, revenue and other important metrics depending on the nature of the software startup and other factors.

7.2 Future Work

7.2.1 Stocks, Flows and Simulation of the Model

The causal loop diagram is a mental modelling technique created with the goal of aiding analysis and decision making. However, it is not the only tool in the Systems Dynamics arsenal. One of the major benefits of using System Dynamics techniques is its ability to model and then simulate the working of a system using modelling software. During the process of analysis of any system using System Dynamics techniques, a better understanding of that system can be achieved by quantifying the variables within the causal loop diagram and assigning values to them. Alongside this step, the elements of the CLD can be remapped to create stocks and flows. In such a representation, the models would not only map out the system's structure but rather simulate the system's working by creating a more elaborate version of its mental model. The creation of the stock and flow diagram would therefore enable its user to visualize the trends and use these trends in better decision making.

To achieve a greater degree of accuracy of the model, the stock and flow model can include 'soft' variables. Investor interest and product development effort are good examples of soft variables. There have been many instances where the usage of soft variables have been explored, as a part of systems modelling (Alkadeem et al., 2017; Al-Kadeem et al., 2017; Bajracharya, 2018). Soft variables are harder to quantify. However, this practice is justified since the aim of any System Dynamics model is to show the general trend of the system rather than providing accurate predictions in the form of numbers (Coyle, 2009).

As part of potential future work, the system's stock and flow model can be simulated using software solution such as Vensim. The results of the simulation typically include the trends of the system over time, which are analysed and used in decision-making processes.

7.2.2 A Call for Better Integrated and Holistic Frameworks

As discussed in the contexts section, at present some of the existing frameworks do integrate the critical success factors indirectly. However, these frameworks could benefit from a more direct integration of one or more factors outlined as part of the triple cornerstone approach outlined as part of this work. As an example, the BMC framework discussed in this chapter does not explicitly establish the factors that differentiate the value proposition from that of the competition. In fact, it does not factor in the study of the current competition in the market. The standard BMC structure also does not establish how the time to market of the value proposition can be improved, even though being early to the market can significantly increase the chances for the success of the product and by extension the business.

Another noted issue with frameworks like the BMC is lack of focus on metrics. Without the guiding force of metrics, there is no evidence driven direction for the product development team or organization to follow. The development team or organization will have to rely on the intuition of the management, which does not always lead to success.

Another need for change in current frameworks is that product and business development initiatives of a software startup seem to exist in their own silos. Most frameworks and models in this space cater to one without addressing the importance or details of the other. For instance, the viewpoint considered in this work is from a software product development perspective and does not go in-depth on aspects of strategy, finance, team dynamics, traits of founders or founding members etc. However, it can be argued that one depends on the success of the other, for the overall success of a software startup. The business model built around a product has little chance of success if the product itself is a failure. A metric based, process driven (holistic) end to end frameworks, which address both product and business model development as a single unified effort would likely be useful to software startups. Ideally, the proposal of such a framework would take into account the pivotal role that the factors such as *time to market*, *startup capital* and *product differentiators* play in its success. Software startups can greatly benefit from such a framework, particularly considering the constraints and market conditions that startups operate in.

7.3 Research Contributions

Most research uncovered so far in the domain of software product development, as with many other spheres, has been done so by understanding the area in viewpoints which do not take causal analysis into account (Davis & Zweig, 2005; Ries, 2011; Song et al., 2010). One of the major drawbacks of such perspectives is that it assumes that actions performed when interacting with the system do not create unforeseen side effects or delays in the system, which cannot be anticipated (J. D. Sterman, 1994). Most research work in disciplines like product design or development rarely work in such deterministic patterns. Instead, one can observe causal patterns of behaviour when observing these disciplines (J. D. Sterman, 1994). This study aims to avoid the trap of linear thinking and instead focus on causal thinking by leveraging the potential of Systems Thinking and Systems Dynamics. This is done by building mental models that are likely to be a more accurate representation of the workings of real-world systems and subsequently gaining a better understanding of the working of the said system. In the current case of software development in software startups, this is accomplished by identifying success factors for development and visualizing the interactions within the system using CLDs. Unterkalmsteiner et al. (2016) emphasise the value of studies which explore the dynamics surrounding software startups, which can result in a more holistic understanding of the domain.

Another aspect of this study is the focus on Synthesis based collective assessment rather than Analysis based deconstructive assessment of the entire system. Since it has been observed that most systems work as a collective whole rather than a mere sum of its parts, it should be noted that the method of studying such systems must also take this nature into account. Understanding of the system could therefore be better achieved by understanding each part in its working with the entirety of the system. Perceiving systems in this pattern is known as Synthesis (Barton & Haslett, 2007). Most research techniques instead apply the underlying principle of analysis, which separates each individual aspect of the system from the collective whole and aims to study these aspects in isolation. Using the principle of Synthesis, via System Dynamics, can provide the advantage of observing the systems functioning and understanding the underlying complexity of system as a single functioning unit (Ackoff, 1994).

A lesser emphasised advantage of using a systems based approach would be that it provides superior explanatory abilities and handles deeper contexts, along with being able to better represent newer emerging patterns in the system (Barton & Haslett, 2007). As evidence of this, one can observe the employment of constructs such as feedback, time delays and attribution errors within the sphere of System Dynamics (J. D. Sterman, 2001).

7.3.1 Comparison with Existing Work

Table 1 presents a comparison of the current work's contribution with some of the notable works reviewed as part of the literature study for this body of work. In the first case, the authors use the Gartner's framework to structure their work (Gelderen et al., 2005). When compared with the context of the current study, the area of research focus is the capital amount. Other aspects such as time to market, product differentiation were not a focus of this study. A different study by Carmel (1994), highlights the importance of setting up a core team in order to deliver faster and better differentiated products. This can serve as a reference point for potential investors, when considering investments in software product startups. Meanwhile the central theme of the study by Song et al. (2010) involves product differentiation, with the aim of developing a model to gain competitive advantage for the initial product offering of a startup. The study although does not place emphasis on the other cornerstones noted as part of this work. A more holistic approach is taken in the study by Hellman and Puri (2000), as various aspects of the capital amount and time to market have been discussed in the context of studying product differentiation through the comparative study between Innovator and Imitator firms. However, it does not take a causal view of the system as part of its process, which is ideal to analyse the complex interconnections within the development process.

Another study by Giardino et al. (2015) reviews the main challenges faced by early stage startups and thereby hinder its chances at success. This study utilizes the framework by MacMillan et al. (1987) which defines four dimensions to present challenges in development and learning, which are *product*, *team*, *financial* and *market*. Although the study by Giardino et al. (2015) mentions the area of acquiring capital as one of the top three challenges faced by software startups, it can be argued that the fundamental shortcoming of this dimensional classification is that it was proposed during a time when the software development space was not fully matured and not as competitive as it is today. Hence the dimension of product difference was not considered as relevant, which is considered a success factor today in the software startup space. The same is the case with time to market. The relevance of getting a product out to market earlier than the competition can be considered more relevant in the modern business landscape than it was three decades ago, which is when the initial framework was proposed.

The work done by Tripathi et al. (2019) takes a different approach and greatly emphasis the role of a Minimum viable product (MVP) in software startup success. The emphasis in this work is on how the usage of MVPs can result in many benefits for the startup, which can include improvement of technical skills, shorter development time and consequently time to market. It also emphasises the indirect benefit of greater chances of acquiring interest from potential investors, if the MVP can demonstrate rich user experiences and potential to scale. However, focus in primarily on the MVP and the benefits are viewed as a secondary consequence or a biproduct of using the MVP, since the primary goal is the validation of the value proposition. The work also does not address the role of product differentiation in the success of the software startup.

However, the current body of work attempts to integrate the various considered aspects discussed by reviewed literature. It also extends the study by using a framework based on establishing a set of cornerstones and defining goals for each cornerstone. The outcome is a synthesis-oriented study which views the working of the complex system of the typical software product startup from a causal interconnected viewpoint.

Table 1. Comparative analysis of similar studies

	Identification of Cornerstones and subsequent goals	**[Cornerstone 1] Accumulated capital amount**	**[Cornerstone 2] Time to market**	**[Cornerstone 3] Product Differentiation**
(Gelderen et al., 2005)	Use the Gartner's Framework	✓	-	-
(Carmel, 1994)	-	✓	✓	✓
(Song et al., 2010)	-	-	-	✓
(Hellmann & Puri, 2000)	-	✓	✓	✓
(Giardino et al., 2015)	Use the Macmillan Framework	✓	-	-
(Tripathi et al., 2019)	Some factors identified	partial	partial	-
Current work	✓	✓	✓	✓

8. CONCLUSION

The research findings discussed in this work is exploratory in nature and delves into success factors for product development in a software startup environment. This research provides a view of not only the various factors influencing software startups, but also helps visualize the potential causal effects which the various factors might have on one another. With the application of System Dynamics methodology, the variables which both directly and indirectly impact the chances of success in a new venture can be identified. The findings in the study show that although there are several factors which directly impact the chances of success of a software startup venture, the key factors such as *time to market* and *accumulated capital* are complex factors, which is in turn influenced by many variables and merits further investigation. The reviewed literature also indicates that *product differentiation* has a significant role to play in the success of the product undertaking and in turn the software startup (business).

This work has made use of case studies as part of the discussion section to highlight the importance of the factors in the triple cornerstone framework. As a scope for potential future work, quantitative data can be collected to simulate the working of one or more of the case studies using stock and flow constructs. Another scope for future work can be the creation of a product development framework based on the triple cornerstone model, since such a framework could prove to be useful for software startups as it will be designed with an emphasis on success factors. Similarly, a framework for business model development could also be designed as a continuation of the design for the product development framework. Indeed, a unified product-business model development framework based on the triple cornerstone model could prove to be very useful to software startups looking for a process driven approach to successful operational realisation.

REFERENCES

Abdel-Hamid, T. K. (1989). The dynamics of software project staffing: A system dynamics based simulation approach. *IEEE Transactions on Software Engineering*, *15*(2), 109–119. doi:10.1109/32.21738

Abreu, F. B., & Carapuça, R. (1994). Object-oriented software engineering: Measuring and controlling the development process. *Proceedings of the 4th International Conference on Software Quality*, *186*, 1–8.

Ackoff, R. L. (1994). Systems thinking and thinking systems. *System Dynamics Review*, *10*(2–3), 175–188. doi:10.1002dr.4260100206

Akkaya, M. (2020). Startup Valuation: Theories, Models, and Future. In Valuation Challenges and Solutions in Contemporary Businesses (pp. 137–156). IGI Global.

Al-Kadeem, R., Backar, S., Eldardiry, M., & Haddad, H. (2017). Review on using system dynamics in designing work systems of project organizations: Product development process case study. *International Journal of System Dynamics Applications*, *6*(2), 52–70. doi:10.4018/IJSDA.2017040103

Alkadeem, R., Backar, S., Haddad, H., & Eldardiry, M. (2017). NPD projects simulation model incorporating managerial aspects using system dynamics approach. *International Journal of System Dynamics Applications*, *6*(3), 38–58. doi:10.4018/IJSDA.2017070103

Anderson, K. E. (2020). Getting acquainted with social networks and apps: It is time to talk about TikTok. *Library Hi Tech News*, *37*(4), 7–12. doi:10.1108/LHTN-01-2020-0001

Auxilia, M., Raja, K., & Kannan, K. (2020). Cloud-Based Access Control Framework for Effective Role Provisioning in Business Application. *International Journal of System Dynamics Applications*, *9*(1), 63–80. doi:10.4018/IJSDA.2020010104

Bajracharya, A. (2018). A System Dynamics Approach in Understanding the Mode Choice. In Intelligent Transportation and Planning: Breakthroughs in Research and Practice (pp. 1066–1086). IGI Global. doi:10.4018/978-1-5225-5210-9.ch048

Baporikar, N. (2020). Logistics Effectiveness Through Systems Thinking. *International Journal of System Dynamics Applications*, *9*(2), 64–79. doi:10.4018/IJSDA.2020040104

Barton, J., & Haslett, T. (2007). Analysis, synthesis, systems thinking and the scientific method: Rediscovering the importance of open systems. *Systems Research and Behavioral Science*, *24*(2), 143–155. doi:10.1002res.816

Beauregard, R., & Corriveau, P. (2007). User Experience Quality: A Conceptual Framework for Goal Setting and Measurement. In V. G. Duffy (Ed.), *Digital Human Modeling* (pp. 325–332). Springer Berlin Heidelberg. doi:10.1007/978-3-540-73321-8_38

Bessen, J. E. (2011). *A Generation of Software Patents* (SSRN Scholarly Paper ID 1868979). Social Science Research Network. https://papers.ssrn.com/abstract=1868979

Blank, S., & Euchner, J. (2018). The genesis and future of Lean Startup: An interview with Steve Blank. *Research Technology Management*, *61*(5), 15–21. doi:10.1080/08956308.2018.1495963

Boadway, R., & Tremblay, J.-F. (2003). *Public Economics and Startup Entrepreneurs* (SSRN Scholarly Paper ID 385781). Social Science Research Network. https://papers.ssrn.com/abstract=385781

Cahyani, N. D. W., Martini, B., Choo, K.-K. R., & Al-Azhar, A. M. N. (2017). Forensic data acquisition from cloud-of-things devices: Windows Smartphones as a case study. *Concurrency and Computation*, *29*(14), e3855. doi:10.1002/cpe.3855

Carmel, E. (1994). Time-to-completion in software package startups. *Proceedings of the Twenty-Seventh Hawaii International Conference on System Sciences*, 4, 498–507. 10.1109/HICSS.1994.323468

Carmel, E., & Sawyer, S. (1998). Packaged software development teams: What makes them different? *Information Technology & People*, *11*(1), 7–19. doi:10.1108/09593849810204503

Casado, E. (2005). Expanding Data Mining Power with System Dynamics. Encyclopedia of Information Science and Technology, 1155–1161. doi:10.4018/978-1-59140-553-5.ch204

Chapman, L., & Plewes, S. (2014). A UX Maturity Model: Effective Introduction of UX into Organizations. In A. Marcus (Ed.), *Design, User Experience, and Usability. User Experience Design Practice* (pp. 12–22). Springer International Publishing. doi:10.1007/978-3-319-07638-6_2

Cockburn, I. M., & MacGarvie, M. J. (2009). Patents, Thickets and the Financing of Early-Stage Firms: Evidence from the Software Industry. *Journal of Economics & Management Strategy*, *18*(3), 729–773. doi:10.1111/j.1530-9134.2009.00228.x

Cooper, R. G., & Kleinschmidt, E. J. (1987a). Success factors in product innovation. *Industrial Marketing Management*, *16*(3), 215–223. doi:10.1016/0019-8501(87)90029-0

Cooper, R. G., & Kleinschmidt, E. J. (1987b). New products: What separates winners from losers? *Journal of Product Innovation Management*, *4*(3), 169–184. doi:10.1111/1540-5885.430169

Covert, A. (2013, September 3). *Microsoft and Nokia: A marriage of necessity*. CNNMoney. https://money.cnn.com/2013/09/03/technology/mobile/microsoft-nokia-smartphones/index.html

Coyle, G. (2009). Qualitative and quantitative modelling in system dynamics. *SYSTEM DYNAMICS-Volume*, *II*, 33.

Cozzi, G., & Spinesi, L. (2006). Intellectual appropriability, product differentiation, and growth. *Macroeconomic Dynamics*, *10*(01), 39–55. doi:10.1017/S1365100506050115

Crowne, M. (2002). Why software product startups fail and what to do about it. Evolution of software product development in startup companies. Engineering Management Conference, 2002. IEMC '02. 2002 IEEE International, 1, 338–343. 10.1109/IEMC.2002.1038454

Curcio, K., Navarro, T., Malucelli, A., & Reinehr, S. (2018). Requirements engineering: A systematic mapping study in agile software development. *Journal of Systems and Software, 139*, 32–50. doi:10.1016/j.jss.2018.01.036

Curran, K., Morrison, S., & Mc Cauley, S. (2012). Google+ vs Facebook: The Comparison. *Telkomnika, 10*(2), 379. doi:10.12928/telkomnika.v10i2.814

Davis, A. M., & Zweig, A. S. (2005). The Rise and Fall of a Software Startup. *Journal of Information Technology Case and Application Research, 7*(2), 31–48. doi:10.1080/15228053.2005.10856064

de Wilton, A. (2011). Patent Value: A Business Perspective for Technology Startups. *Technology Innovation Management Review, 1*(3), 5–11. doi:10.22215/timreview/501

Dieter, D. G., & Lenze, J. S. (2018). Google+ Case Study: Create a Social Network or Risk Everything. *Performance Improvement, 57*(10), 26–36. doi:10.1002/pfi.21826

Dove, L., & Sethumadhavan, A. (2012). Breaking Down Barriers: The Interdependence of Research and Design. *Interaction, 19*(6), 76–80. doi:10.1145/2377783.2377799

Duchesneau, D. A., & Gartner, W. B. (1990). A profile of new venture success and failure in an emerging industry. *Journal of Business Venturing, 5*(5), 297–312. doi:10.1016/0883-9026(90)90007-G

Eadicicco, L. (2015, April 27). *Why Google+ failed, according to Google insiders.* Business Insider Australia. https://www.businessinsider.com.au/what-happened-to-google-plus-2015-4

Eason, K. (2001). Changing perspectives on the organizational consequences of information technology. *Behaviour & Information Technology, 20*(5), 323–328. doi:10.1080/01449290110083585

Faktor, S. (2014, May 1). *What Killed Google+ And What Can Save It.* Forbes. https://www.forbes.com/sites/stevefaktor/2014/05/01/what-killed-google-and-what-will-save-it/

Fayad, M. E., Laitinen, M., & Ward, R. P. (2000). Thinking Objectively: Software Engineering in the Small. *Communications of the ACM, 43*(3), 115–118. doi:10.1145/330534.330555

Finstad, K., Xu, W., Kapoor, S., Canakapalli, S., & Gladding, J. (2009). FEATURE: Bridging the Gaps Between Enterprise Software and End Users. *Interaction, 16*(2), 10–14. doi:10.1145/1487632.1487635

Ford, D. N., & Sterman, J. D. (1998). Dynamic modeling of product development processes. *System Dynamics Review, 14*(1), 31–68.

Forrester, J. W. (1971). Counterintuitive behavior of social systems. *Technological Forecasting and Social Change, 3*, 1–22. doi:10.1016/S0040-1625(71)80001-X

Galli, B. J. (2019). Thoughts of Using Economic Decision-Making to Systems Engineering and Systems Thinking: An Exploratory Study. *International Journal of System Dynamics Applications, 8*(3), 1–14.

Galli, B. J. (2020). Economic-Decision-Making in New Product Development: A Review of the Relationship. *International Journal of Applied Management Sciences and Engineering, 7*(1), 1–27.

Gatignon, H., & Xuereb, J.-M. (1997). Strategic Orientation of the Firm and New Product Performance. *JMR, Journal of Marketing Research, 34*(1), 77–90. doi:10.2307/3152066

Georgantzas, N. C., & Katsamakas, E. G. (2008). Information systems research with system dynamics. *System Dynamics Review*, *24*(3), 247–264. doi:10.1002dr.420

Giachetti, C., & Marchi, G. (2017). Successive changes in leadership in the worldwide mobile phone industry: The role of windows of opportunity and firms' competitive action. *Research Policy*, *46*(2), 352–364.

Giardino, C., Bajwa, S. S., Wang, X., & Abrahamsson, P. (2015). Key Challenges in Early-Stage Software Startups. In C. Lassenius, T. Dingsøyr, & M. Paasivaara (Eds.), *Agile Processes in Software Engineering and Extreme Programming* (pp. 52–63). Springer International Publishing.

Giardino, C., Wang, X., & Abrahamsson, P. (2014). Why Early-Stage Software Startups Fail: A Behavioral Framework. In C. Lassenius & K. Smolander (Eds.), *Software Business. Towards Continuous Value Delivery* (pp. 27–41). Springer International Publishing., doi:10.1007/978-3-319-08738-2_3

Gonzalez, R., Cuevas, R., Motamedi, R., Rejaie, R., & Cuevas, A. (2013). Google+ or Google-? Dissecting the evolution of the new OSN in its first year. *Proceedings of the 22nd International Conference on World Wide Web*, 483–494.

Goodwin, N. C. (1987). Functionality and Usability. *Communications of the ACM*, *30*(3), 229–233. doi:10.1145/214748.214758

Graham, S. J. H., & Sichelman, T. (2008). Why Do Start-Ups Patent? *Berkeley Technology Law Journal*, *23*(3), 1063–1097.

Groen, A. J., Wakkee, I. A. M., & De Weerd-Nederhof, P. C. (2008). Managing Tensions in a High-tech Start-up: An Innovation Journey in Social System Perspective. *International Small Business Journal*, *26*(1), 57–81. doi:10.1177/0266242607084659

Hannola, L., Friman, J., & Niemimuukko, J. (2013). Application of agile methods in the innovation process. *International Journal of Business Innovation and Research*, *7*(1), 84–98. doi:10.1504/IJBIR.2013.050557

Hassenzahl, M., & Tractinsky, N. (2006). User experience—A research agenda. *Behaviour & Information Technology*, *25*(2), 91–97. doi:10.1080/01449290500330331

Heisler, Y. (2015, August 3). Microsoft engineer exposes the inside story: Why Windows Phone was doomed from the start. *BGR*. https://bgr.com/2015/08/03/windows-phone-failure-microsoft-engineer/

Hellmann, T., & Puri, M. (2000). The Interaction between Product Market and Financing Strategy: The Role of Venture Capital. *Review of Financial Studies*, *13*(4), 959–984. doi:10.1093/rfs/13.4.959

Herrick, J. (2018, March 30). *We know exactly why Windows Phone failed*. TechnoBuffalo. https://www.technobuffalo.com/windows-phone-failure-reason-terry-myerson

Hilmola, O.-P., Helo, P., & Ojala, L. (2003). The value of product development lead time in software startup. *System Dynamics Review*, *19*(1), 75–82. doi:10.1002dr.255

Hokkanen, L., Kuusinen, K., & Väänänen, K. (2016). Minimum Viable User EXperience: A Framework for Supporting Product Design in Startups. Agile Processes, in Software Engineering, and Extreme Programming, 66–78. doi:10.1007/978-3-319-33515-5_6

Holling, C. S. (2001). Understanding the Complexity of Economic, Ecological, and Social Systems. *Ecosystems (New York, N.Y.)*, *4*(5), 390–405. doi:10.100710021-001-0101-5

Kashfi, P., Nilsson, A., & Feldt, R. (2016). *Integrating UsereXperience Practices into Software Development Processes: Implications of Subjectivity and Emergent Nature of UX.* ArXiv Preprint ArXiv:1605.03783.

Kemell, K.-K., Wang, X., Nguyen-Duc, A., Grendus, J., Tuunanen, T., & Abrahamsson, P. (2020). Startup Metrics That Tech Entrepreneurs Need to Know. In *Fundamentals of Software Startups* (pp. 111–127). Springer. doi:10.1007/978-3-030-35983-6_7

Klein, L. (2013). *UX for Lean Startups: Faster, Smarter User Experience Research and Design.* O'Reilly Media, Inc.

Kljajic, M., Borštnar, M. K., Škraba, A., & Kofjac, D. (2012). System Approach to MIS and DSS and its Modeling within SD. Regional Development: Concepts, Methodologies, Tools, and Applications, 315–335. doi:10.4018/978-1-4666-0882-5.ch208

Klotins, E., Unterkalmsteiner, M., & Gorschek, T. (2015). Software Engineering Knowledge Areas in Startup Companies: A Mapping Study. In J. M. Fernandes, R. J. Machado, & K. Wnuk (Eds.), *Software Business* (pp. 245–257). Springer International Publishing. doi:10.1007/978-3-319-19593-3_22

Kotler, P., Keller, K. L., Manceau, D., & Hémonnet-Goujot, A. (2015). *Marketing management* (Vol. 14). Prentice Hall Englewood Cliffs.

Kupiainen, E., Mäntylä, M. V., & Itkonen, J. (2015). Using metrics in Agile and Lean Software Development–A systematic literature review of industrial studies. *Information and Software Technology*, *62*, 143–163. doi:10.1016/j.infsof.2015.02.005

Levy, S. (2011, June 28). *Inside Google+—How the Search Giant Plans to Go Social | WIRED.* https://www.wired.com/2011/06/inside-google-plus-social/

Lichtenthaler, U. (2020). Agile innovation: The complementarity of design thinking and lean startup. *International Journal of Service Science, Management, Engineering, and Technology*, *11*(1), 157–167. doi:10.4018/IJSSMET.2020010110

Lin, H. X., Choong, Y.-Y., & Salvendy, G. (1997). A proposed index of usability: A method for comparing the relative usability of different software systems. *Behaviour & Information Technology*, *16*(4–5), 267–277. doi:10.1080/014492997119833

MacMillan, I. C., Zemann, L., & Subbanarasimha, P. N. (1987). Criteria distinguishing successful from unsuccessful ventures in the venture screening process. *Journal of Business Venturing*, *2*(2), 123–137. doi:10.1016/0883-9026(87)90003-6

Mann, R. J. (2006). *Do Patents Facilitate Financing in the Software Industry?* Social Science Research Network. https://papers.ssrn.com/abstract=510103

Mann, R. J., & Sager, T. W. (2007). Patents, venture capital, and software start-ups. *Research Policy*, *36*(2), 193–208. doi:10.1016/j.respol.2006.10.002

May, B. (2012). Applying Lean Startup: An Experience Report – Lean amp;amp; Lean UX by a UX Veteran: Lessons Learned in Creating amp;amp; Launching a Complex Consumer App. *Agile Conference (AGILE)*, 2012, 141–147. 10.1109/Agile.2012.18

Misra, S. C., Kumar, V., & Kumar, U. (2009). Identifying some important success factors in adopting agile software development practices. *Journal of Systems and Software*, *82*(11), 1869–1890. doi:10.1016/j.jss.2009.05.052

Morecroft, J. D. W. (1986). The dynamics of a fledgling high-technology growth market: Understanding and managing growth cycles. *System Dynamics Review*, *2*(1), 36–61. doi:10.1002dr.4260020104

Nasution, F. B., Bazin, N. E., Rosalyn, R., & Hasanuddin, H. (2018). Public Policymaking Framework Based on System Dynamics and Big Data. *International Journal of System Dynamics Applications*, *7*(4), 38–53. doi:10.4018/IJSDA.2018100103

Nguyen-Duc, A., Seppänen, P., & Abrahamsson, P. (2015). Hunter-gatherer Cycle: A Conceptual Model of the Evolution of Software Startups. *Proceedings of the 2015 International Conference on Software and System Process*, 199–203. 10.1145/2785592.2795368

Noel, M., & Schankerman, M. (2013). Strategic Patenting and Software Innovation. *The Journal of Industrial Economics*, *61*(3), 481–520. doi:10.1111/joie.12024

Nurmuliani, N., Zowghi, D., & Powell, S. (2004). Analysis of requirements volatility during software development life cycle. *2004 Australian Software Engineering Conference. Proceedings.*, 28–37. 10.1109/ASWEC.2004.1290455

Omamo, A. O., Rodrigues, A. J., & Muliaro, W. J. (2020). A System Dynamics Model of Technology and Society: In the Context of a Developing Nation. *International Journal of System Dynamics Applications*, *9*(2), 42–63. doi:10.4018/IJSDA.2020040103

Omamo, A. O., Rodriguez, A. J., & Muliaro, J. W. (2018). A Systems Dynamics Model for Mobile Industry Governance in the Context of the Kenyan Vision 2030. *International Journal of System Dynamics Applications*, *7*(2), 81–100. doi:10.4018/IJSDA.2018040105

Osterwalder, A., & Pigneur, Y. (2010). *Business model generation: A handbook for visionaries, game changers, and challengers*. John Wiley & Sons.

Pandey, S., & Srivastava, S. (2014). Data Driven Enterprise UX: A Case Study of Enterprise Management Systems. In S. Yamamoto (Ed.), *Human Interface and the Management of Information. Information and Knowledge in Applications and Services* (pp. 205–216). Springer International Publishing., doi:10.1007/978-3-319-07863-2_21

Parry, M. E., & Song, X. M. (1994). Identifying new product successes in China. *Journal of Product Innovation Management*, *11*(1), 15–30. doi:10.1111/1540-5885.1110015

Paternoster, N., Giardino, C., Unterkalmsteiner, M., Gorschek, T., & Abrahamsson, P. (2014). Software development in startup companies: A systematic mapping study. *Information and Software Technology*, *56*(10), 1200–1218. doi:10.1016/j.infsof.2014.04.014

Poppendieck, M., & Cusumano, M. A. (2012). Lean Software Development: A Tutorial. *IEEE Software, 29*(5), 26–32. doi:10.1109/MS.2012.107

Project Strobe: Protecting your data, improving our third-party APIs, and sunsetting consumer Google+. (2018, October 8). Google. https://blog.google/technology/safety-security/project-strobe

Pruyt, E. (2017). Integrating Systems Modelling and Data Science: The Joint Future of Simulation and 'Big Data' Science. Artificial Intelligence: Concepts, Methodologies, Tools, and Applications, 822–840. doi:10.4018/978-1-5225-1759-7.ch033

Qudrat-Ullah, H. (2009). System Dynamics Based Technology for Decision Support. Encyclopedia of Information Science and Technology, Second Edition, 3647–3650. doi:10.4018/978-1-60566-026-4.ch581

Rahmandad, H., & Weiss, D. M. (2009). Dynamics of concurrent software development. *System Dynamics Review, 25*(3), 224–249. doi:10.1002dr.425

Ranger, S. (2015, July 9). *Microsoft and Windows Phone: What went wrong, and where can they go from here?* ZDNet. https://www.zdnet.com/article/microsoft-and-windows-phone-what-went-wrong-and-where-can-they-go-from-here/

Ries, E. (2011). *The Lean Startup: How Today's Entrepreneurs Use Continuous Innovation to Create Radically Successful Businesses*. Crown Publishing Group. https://books.google.com.au/books?id=tvfyz-4JILwC

Rodden, K., Hutchinson, H., & Fu, X. (2010). Measuring the User Experience on a Large Scale: User-centered Metrics for Web Applications. *Proceedings of the SIGCHI Conference on Human Factors in Computing Systems*, 2395–2398. 10.1145/1753326.1753687

Rothwell, R. (1994). Towards the fifth-generation innovation process. *International Marketing Review, 11*(1), 7–31. doi:10.1108/02651339410057491

Rubino, D. (2014, February 23). *Windows Phone is losing some hardware differentiation as Microsoft woos Android manufacturers*. Windows Central. https://www.windowscentral.com/windows-phone-losing-hardware-differentiation

Sánchez, M. A. (2014). Modeling for System´s Understanding. Formal Languages for Computer Simulation: Transdisciplinary Models and Applications, 38–61. doi:10.4018/978-1-4666-4369-7.ch002

Savov, V. (2012, January 26). *Microsoft paid Nokia $250 million to use Windows Phone in Q4 2011*. The Verge. https://www.theverge.com/2012/1/26/2736132/microsoft-nokia-250-million-platform-support-payment

Schonfeld, E. (2011, July 11). *Google+ Added $20 Billion To Google's Market Cap | TechCrunch*. https://techcrunch.com/2011/07/10/google-plus-20-billion-market-cap/

Schwaber, K., & Beedle, M. (2001). *Agile Software Development with Scrum* (1st ed.). Prentice Hall PTR.

Segan, S. (2014, February 27). *OEM: Windows Phone License Fee Plummeting*. PCMag Australia. https://au.pcmag.com/mobile-apps/4192/oem-windows-phone-license-fee-plummeting

Sehgal, K., Jain, A., Nagrath, P., & Kumar, A. (2019). Recent advances in networks and data security survey on various mobile operating systems. *International Conference on Innovative Computing and Communications*, 181–190. 10.1007/978-981-13-2324-9_18

Sehl, K. (2020, March 2). *Everything Brands Need to Know About TikTok in 2020*. Hootsuite Social Media Management. https://blog.hootsuite.com/what-is-tiktok/

Shanbhag, N., & Pardede, E. (2019). A Metrics Framework for Product Development in Software Startups. *Journal of Enterprising Culture*, *27*(03), 283–307. doi:10.1142/S0218495819500110

Song, L. Z., Benedetto, C. D., & Song, M. (2010). Competitive Advantages in the First Product of New Ventures. *IEEE Transactions on Engineering Management*, *57*(1), 88–102. doi:10.1109/TEM.2009.2013836

Sterman, J., & Sterman, J. D. (2000). *Business Dynamics: Systems Thinking and Modeling for a Complex World with CD-ROM* (HAR/CDR edition). McGraw-Hill Education.

Sterman, J. D. (1994). Learning in and about complex systems. *System Dynamics Review*, *10*(2–3), 291–330. doi:10.1002dr.4260100214

Sterman, J. D. (2001). System Dynamics Modeling: Tools for Learning in a Complex World. *California Management Review*, *43*(4), 8–25. doi:10.2307/41166098

Sutton, S. M. (2000). The role of process in software start-up. *IEEE Software*, *17*(4), 33–39. doi:10.1109/52.854066

Tidd, J., & Bessant, J. R. (2018). *Managing innovation: Integrating technological, market and organizational change*. John Wiley & Sons.

Tripathi, N., Oivo, M., Liukkunen, K., & Markkula, J. (2019). Startup ecosystem effect on minimum viable product development in software startups. *Information and Software Technology*, *114*, 77–91. doi:10.1016/j.infsof.2019.06.008

Tvedt, J. D., & Gollofello, J. S. (1995). Evaluating the effectiveness of process improvements on software development cycle time via system dynamics modelling. *Computer Software and Applications Conference, 1995. COMPSAC 95. Proceedings., Nineteenth Annual International*, 318–325. 10.1109/CMPSAC.1995.524796

Unterkalmsteiner, M., Abrahamsson, P., Wang, X., Nguyen-Duc, A., Shah, S., Bajwa, S. S., Baltes, G. H., Conboy, K., Cullina, E., & Dennehy, D. (2016). Software startups–a research agenda. *E-Informatica Software Engineering Journal, 10*(1).

van Gelderen, M., Thurik, R., & Bosma, N. (2005). Success and Risk Factors in the Pre-Startup Phase. *Small Business Economics*, *24*(4), 365–380. doi:10.100711187-004-6994-6

Villas-Boas, A. (2017, October 10). *Microsoft's Windows Phone platform is dead*. Business Insider Australia. https://www.businessinsider.com.au/microsofts-windows-phone-platform-dead-2017-10

Weir, A. (2012, January 19). *ZTE reveals cost of Windows Phone OS licensing*. Neowin. https://www.neowin.net/news/zte-reveals-cost-of-windows-phone-os-licensing

Chapter 4
A Framework for Using Dynamic Modeling to Support Lean Sustainment in a Complex Setting:
Illustrated With a Case Study From Manufacturing

Marc Haddad
https://orcid.org/0000-0002-3010-4033
Lebanese American University, Lebanon

Rami Otayek
Lebanese American University, Lebanon

ABSTRACT

This chapter proposes a framework for integrating lean and systems thinking tools to explore the dynamics of lean implementations in manufacturing. The value of this integrated approach is in supplementing the operational level principles of lean with the strategic outlook of systems thinking to mitigate adverse impacts of operational complexity on system performance. In particular, the focus of the framework is to enable the sustainment of lean gains in the long-term, a major challenge in manufacturing settings. The application of the framework is illustrated with a case study of a lean implementation for reducing work-in-process (WIP) at a clothing manufacturer dealing with a number of operational complexities such as demand uncertainty, a high product mix and inefficient processes. The case study highlights the usefulness of system dynamics modeling in revealing counterintuitive system behaviors that could compromise the success of the lean initiative. The simulation results demonstrate the application of the framework for sustaining lean implementations in practice.

DOI: 10.4018/978-1-7998-5788-4.ch004

INTRODUCTION

Manufacturing systems are inherently complex, meaning they are typically large-scale, interconnected (with feedback connections between different parts of the system or with external sources), and where human agency is strong and conditions are dynamic (sometimes changing rapidly) (Bar-Yam, 1997). The manufacturing environment is also rife with coordination and communication challenges, which are affected by operational complexity and in turn contribute to increasing it, whether due to lack of information (uncertainty), conflicting information (equivocality), or delays and breakdowns in communication. This is why a multitude of process improvement approaches are available and always needed to tackle the countless adverse impacts of complexity in manufacturing. Some are focused on particular outcomes such as total quality management (TQM), while others are meant to be more cross-cutting, such as process reengineering. But perhaps the one that is most encompassing and that has withstood the test of time since its inception in Toyota's production system is the lean approach.

Lean manufacturing is a proven set of tools and techniques for improving an entire production operation from material sourcing to customer delivery. Lean is focused on minimizing wastes in production time, cost, effort and resources in order to maximize overall system performance, such as production throughput and product quality. However, lean is not just a shop floor practice, but also a philosophy of disciplined and continuous improvement, sometimes referred to as "lean thinking" (J. P. Womack & Jones, 2010), which goes beyond the mere tools and techniques for improving a set of processes. Lean thinking is a comprehensive system of beliefs that advocates a change in mindsets about how we perceive work and value beyond our own silos; it redefines value to be always aligned with customer needs; and, it aims to evolve the traditionally disconnected factory towards an integrated enterprise from supplier to customer (Nightingale & Srinivasan, 2011), where workers are empowered and inherently incentivized to create value for the customer. In principle, this is supposed to ensure a sustainable enterprise that can continuously innovate to meet customer needs, while maintaining its own efficiency and effectiveness. However, in practice lean gains are hard to sustain due to emergent behaviors resulting from complexity, which calls for complementary approaches to lean (Haddad & Otayek, 2019).

Systems thinking is such an approach for holistic problem solving in the context of complex systems, with a comprehensive set of principles and tools to that end. The core philosophy of systems thinking is based on exploring the interactions between a problem and all affected parts of the system it's embedded in to find globally optimal solutions instead of a limited local fix (Sterman, 2000). This makes it particularly appropriate for tackling the types of problems encountered in manufacturing, especially those that require a big picture view of the problem before getting to its root cause, or understanding the long-term consequences of proposed solutions. The main qualitative tools of systems thinking are causal loop diagrams (CLD) which enable the elicitation of key variables constituting a system, and the cause-effect relationships between them that govern system behavior. The quantitative arm of systems thinking is system dynamics (SD) modeling which is used to describe the structure of a complex system with mathematical equations in order to simulate its behavior over time (Azar, 2012). These tools have proven useful for tackling 'messy' situations such as recurring problems, or those that have been made worse by quick attempts at fixing them, both of which are common in manufacturing.

Therefore, and in contrast with the reductionist approach to problem solving, both lean thinking and systems thinking have similar underpinning philosophies that advocate for comprehensive and holistic approaches from different perspectives. However, they are not interchangeable approaches but rather complementary. Lean is the vehicle for thoughtful intervention with practical tools, while systems science is the engine of 'big picture' thinking about the intervention, with an eye for potential unintended consequences (Labedz & Gray, 2013). Lean gives systems science a complete toolset to implement change on the ground, while systems thinking gives lean the strategic outlook over the long-term. Therefore, combining the two approaches is a natural and promising step towards better improvement results. For example, mapping out the flow of production material and information using lean's value-stream mapping (VSM) technique while looking out for feedback effects in these flows makes it possible to identify hidden wastes that can build up and become costly over time.

In particular, the combined approach can be useful for helping to sustain lean improvements over the long-term, which is a major challenge facing lean initiatives after an initially successful implementation (Bateman, 2005). Indeed, the use of dynamic modeling to investigate the performance of the manufacturing system during and after lean improvements can reveal undesirable feedback effects over time. Dynamic modeling can also serve as a decision support tool for evaluating different mitigation options to sustain lean gains, and thereby ensure the long-term sustainability of the lean initiative.

Therefore, the use of SD modeling to guide a lean initiative and ensure its long-term sustainment presents promising potential. This chapter proposes a framework combining lean tools and dynamic modeling to guide the implementation of lean techniques in manufacturing. The chapter is structured as follows: The first section presents an overview of the recent literature on the use of system dynamics modeling in lean manufacturing contexts. The subsequent section presents a methodological framework for using dynamic modeling to supplement the traditional value stream mapping (VSM) approach for implementing lean in order to ensure that improvements are sustained in the long-term. This is followed by a case study illustration of the use of the proposed framework in the manufacturing sector, specifically in the apparel industry where complexity is high due to demand variation and high product variety. The first two steps of the framework are used to characterize the current state of production operations using VSM and SD modeling, and the following two steps describe the implementation of lean techniques in the future state with an assessment of their impacts on production efficiency using SD modeling. The results from the simulation and the actual lean implementation are compared and discussed in terms of the usefulness of SD modeling for quantifying and validating lean improvements, as well as the potential for sustaining gains in the long-term. Concluding remarks and proposed directions for future research are presented in the final section.

LITERATURE REVIEW

A new line of research has evolved in recent years based on the use of SD modeling to capture the complexity of manufacturing systems (Georgiadis & Michaloudis, 2012; Poles, 2013) and supply chains (Gafi & Javadian, 2018; Gonçalves et al., 2005), with a smaller subset of studies focusing on lean manufacturing systems, as reviewed in this section. This is the result of increased awareness about the benefits of dynamic modeling and simulation in the study of production systems, in particular as a valuable decision support tool for understanding complex interactions and their impacts in a cost-effective way, something that has long been established in the literature (Baines & Harrison, 1999; Edghill & Towill, 1989).

Of the few studies using the SD approach to explore lean manufacturing systems, the majority have used SD models to assess the impact of one or more lean practices on different measures of performance of production efficiency, but with greatly varying levels of system complexity and model detail. Starting with earlier studies, (Deif, 2010) explored the impact of applying Just-in-Time (JIT) lean policy on inventory levels under stochastic demand, serving as a proof of concept for the benefits of SD modeling to understand complex interactions in a typical production system. (Godinho Filho & Uzsoy, 2011) used SD to assess the impact of different hypothetical continuous improvement programs on production cycle time and lot size. Their results showed that "the relationship between lot sizes, cycle times, and shop floor parameters is complex and nonlinear" and that SD modeling was needed to quantify benefits and determine the best improvement policy. (Deif, 2012) developed an SD model to assess the responsiveness of a lean cell to customer orders under uncertainty of demand and machine availability. The study recommended employing a dynamic capacity policy as a mitigation measure for ramping up production to increase responsiveness in the face of complexity.

A number of the following studies were largely exploratory, proposing elaborate SD models but with basic recommendations or even just validation of commonly known best practices. (Hosseini-nasab et al., 2013) explored the interrelations between an organization's strategies for technology improvement and lean manufacturing (both broadly defined) under increasing "complexity of industry". The authors used SD modeling to argue that introducing new technology in an organization improves its lean manufacturing capabilities irrespective of industry complexity. (Khataie & Bulgak, 2013) developed an SD model of the value-added (VA) and non-value-added (NVA) activities associated with production quality to identify the most beneficial lean manufacturing policy that minimizes production and quality costs. SD modeling in this case was useful for capturing the complex relationships between the many activities and costs directly or indirectly associated with quality. (Pai et al., 2013) built on the earlier work of (Deif, 2010) to explore the complexity of a Just-in-Time (JIT) manufacturing system by testing the outcomes of different inventory policies. The work is a typical example of the majority of earlier studies which focused mostly on performing basic scenario analysis.

Following on the insights from earlier work, Table 1 presents a summary of the most recent studies using system dynamics modeling to tackle various aspects of complexity in lean manufacturing implementations.

The above shows that while the body of literature combining the use of systems thinking with lean thinking in manufacturing remains very small on the whole, there has been a steady increase in the number of such focused studies in recent years. A review of the relevant results from the preceding table highlights the usefulness of SD modeling in tackling various aspects of complexity in lean manufacturing implementations. Indeed, the depth of some of the recent studies is not only in the scope and detail of the developed SD models, but also in the sophistication of purpose in terms of exploring the dynamics of lean interventions with scenario testing. In particular, several studies have recognized the usefulness of SD modeling in testing the impact of one or more production factors on system performance, thereby simulating a lean intervention and its potential benefits. This is generally considered an appropriate use of SD as it can capture system complexity at a macro level without getting lost in the details of the system as with other approaches (for example, discrete-event simulation which typically requires extensive data collection at a more micro-level).

However, when it comes to intervening in complex settings as with implementing lean in manufacturing, the devil is in the operational details where emergent behaviors and counter-intuitive consequences are the norm rather than the exception. This is observed in what is known in systems thinking as "policy resistance" such that operational complexity can counter and sometimes reverse the expected impacts of a strategy or policy, even when the latter is well-founded and validated with modeling and simulation. In manufacturing, human agency and coordination and communication challenges can limit the expected

Table 1. Review of recent studies using SD modeling in lean manufacturing

Reference	Focus	Relevant Results
(Haddad & Otayek, 2019)	Sustaining a lean implementation over the long-term for a traditional apparel manufacturer.	SD modeling can reveal counter-intuitive behaviors during lean implementation in complex manufacturing systems and is thus useful for sustaining lean success.
(Mahmood, 2019)	Evaluating the financial and energy impacts of increasing production capacity for a weaving mill.	SD modeling supports lean by capturing the complex relationships between several systems and subsystems and all material, information, energy and financial flows.
(Segura et al., 2019)	Analyzing the impact of lean strategies on business performance of a print packaging manufacturer.	SD modeling is most useful as a strategic tool that allows capturing an organization's complex value model to focus its lean strategies on the most beneficial targets.
(Shetty et al., 2019)	Investigating the use of Kanban to level production at an automotive parts manufacturer.	In a complex manufacturing environment with high demand fluctuation, SD models allow testing of different production factors simultaneously to control inventory levels.
(Stadnicka & Litwin, 2019)	Extending VSM with additional process data and integrating with SD modeling to facilitate complex analysis of manufacturing systems.	VSM analysis is static, but when combined with SD modeling it increases the potential for identifying and eliminating more wastes in complex manufacturing systems.
(Gupta et al., 2018)	Assessing lean-green performance of a tire manufacturing firm.	SD modeling can represent complex interrelationships between different production wastes, and with worker skill and availability of machines and workers.
(Shaik & Rodrigues, 2018)	Investigating the impact of lean cleanliness and layout optimization on performance of an electronics manufacturer.	SD modeling can capture the combined impacts of different lean improvements applied simultaneously to different parts of a complex manufacturing system.
(Omogbai & Salonitis, 2017)	Assessing impacts of sorting on throughput at a make-to-order print packaging manufacturer.	SD modeling is useful to assess the extent to which a lean improvement should be implemented, as well as its dynamic impact over time.
(Omogbai & Salonitis, 2016)	Exploring hidden interactions between different lean practices and overall improvement benefits.	SD modeling can validate the benefits of lean practices, but a detailed SD model is needed to assess the complex interactions between multiple practices and benefits.
(Ali & Deif, 2016)	Studying the impact of demand uncertainty on the dynamic behavior of a lean production system.	Complexity of a multi-stage manufacturing system with demand variability affects lean performance, and SD modeling can be used for management policy recommendations.
(Godinho Filho & Barco, 2015)	Proposing a framework for choosing among lean improvement programs at an aeronautics manufacturer.	Defining a general framework for applying SD method to assess lean improvement programs should be based on results from actual case studies in the literature.
(Deif & Elmaraghy, 2014)	Investigating the challenges and costs of implementing production leveling in a lean cell.	The benefits of implementing lean policies in manufacturing depend on complex system and market factors, with SD modeling the dynamic assessment tool of choice.

benefits of a lean implementation. This is why a number of the recent studies reviewed above agree that more attention is needed when it comes to dynamic modeling in the lean manufacturing context.

In this chapter, a framework is proposed that taps into the real value of SD modeling beyond simple scenario testing, which is to reveal counter-intuitive behaviors in the actual implementation of lean. For example, SD modeling is most useful in uncovering unexpected change in system response to lean improvements between the short-term and long-term time horizons. This is why it can be a useful tool for investigating the sustainment of a lean intervention by simulating the implementation dynamics both during and post implementation. This can help ensure that the intervention is feasible and sustainable over the long-term.

METHODOLOGICAL FRAMEWORK

The following framework illustrated in Figure 1 proposes a systematic approach for lean sustainment in manufacturing based on the integration of lean thinking and systems thinking tools.

Figure 1. Integrated framework for sustaining a lean manufacturing implementation with systems thinking tools

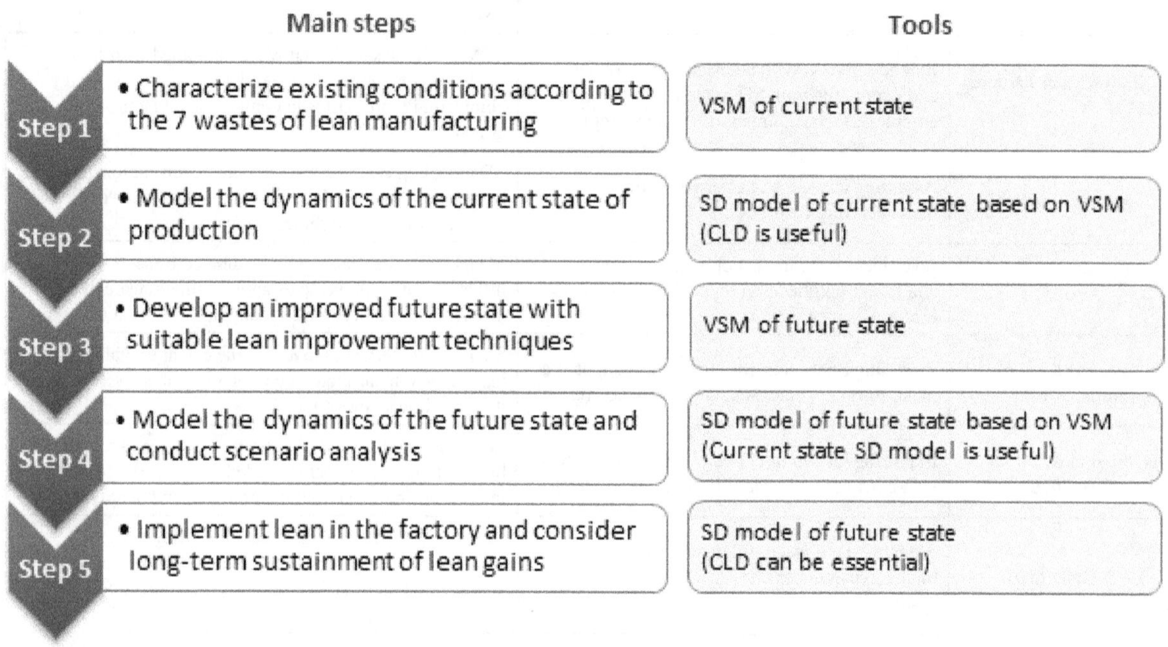

The framework starts with a generic step applicable to any intervention, which is to understand the current state of production, and specifically to identify the wastes in the system using the 7 categories of waste from the lean toolset. This serves as the start of the problem solving process in terms of "learning to see" the components of the problem and visualizing them together using a value-stream map (VSM), another useful tool of lean (J. Womack & Jones, 2011). The VSM depicts all activities involving the flow of material and information for manufacturing a product along with their duration. This includes both the activities which add value and which are non-value-added, meaning the wastes and bottlenecks that must be eliminated or at least reduced in the future (i.e. desired) state. This step requires extensive data collection and observation on the shop floor using a number of techniques including time and motion studies to populate the VSM with the required metrics, interviews with managers and workers to understand the work culture, and surveys to collect other undocumented data (Robson, 2002).

In step 2, a dynamic model of the current state is built using the SD method based on the production data and interactions described in the VSM. In the SD modelling of a production process, inventories are described as stocks of work-in-process (WIP), and work on the product is usually represented by inflows and outflows into different types of WIP stocks. Information transfer is usually represented by feedback links. The model equations capture the dynamics of the production process including the observed inefficiencies, delays and breakdowns. In situations where complexity makes model building difficult, it's usually helpful to make use of CLD's from the systems thinking toolset to further elicit key variables that represent relevant components of the manufacturing system and any feedback relationships between them. This facilitates SD model building and connecting visible problems to their root cause (Maani, 2017). The SD model is then simulated to understand system behavior over time and to identify the most influential causes behind the observed production wastes. This helps inform the building of the future state VSM where the vision is defined for improving the system with appropriate lean techniques.

The third step of the framework is to build the future state VSM according to lean principles for developing an improved vision of the production system (Rother & Shook, 2009). This involves selecting appropriate lean techniques to eliminate the wastes identified in the current state and improve the production flow.

Step 4 relies on the VSM from step 3 to build the future state SD model following the same approach for building the current state SD model, but incorporating the lean improvements in the modeling. In this step, it's essential to capture the details of how each lean technique actually works in improving the production process as this will serve to assess the effectiveness of the proposed lean improvements before their implementation in the factory. As with any simulation exercise, this can save time and resources before costly trial and error attempts are made in practice. Furthermore, making use of scenario analysis in the simulation of the future state model can allow the "tweaking" of some aspects of the lean intervention, such as the sizing of supermarket pull systems or the selection of optimal inventory policies.

Finally, the last step of the framework considers the sustainment of the lean intervention, both during (i.e. short-term) and post implementation (medium to long-term). This is possible when the SD model is detailed enough to capture the complexity of the system, such as by including stocks and flows for all stages of production and their corresponding material and information flows; describing the workings of each lean technique in the model equations; and, representing the human agency in the system (e.g. the response of workers and managers to the lean initiative) with appropriate variables. Such detailed modeling can reveal possible failure points in the lean implementation over time due to emergent behaviors or unintended outcomes. The future state SD model is used here to trace the root causes of problems and to test mitigation measures that can ensure the long-term sustainment of the lean initiative. The use of CLD's can be essential in this step to help troubleshoot the problems faced and to identify appropriate mitigations, before their impacts are assessed in the SD model.

The framework is therefore based on combining lean principles and systems science to improve the effectiveness and sustainment of waste identification and elimination in any complex manufacturing setting. The step-by-step application of the framework is described in the following case study.

CASE STUDY

To illustrate the use of the proposed framework, a case study conducted at an apparel manufacturer of children's clothing is presented step-by-step in this section following the order in the framework. The operation is a medium-sized factory in a developing country, and is run on a family business model using largely traditional and inefficient production techniques. The manufacturer experiences variable demand on a monthly basis and there are seasons of high and low demand, which increases uncertainty and requires high levels of operations planning and coordination. The product mix is also very high due to different clothing model varieties, which further increases operational complexity. As a result, both lean thinking and systems thinking were deemed appropriate in this context. The goal of the lean initiative was to reduce work-in-process (WIP) at all stages of production. The case study is presented in detail in (Haddad & Otayek, 2019, 2018).

STEPS 1 AND 2: MODELING THE CURRENT STATE

The existing conditions of production were characterized through on-site observations and data collection over a total period of 3 months using time and motion studies and interviews with managers and workers. Collected data included machine uptime, VA and NVA process times, and defect counts at each step of the process, and WIP quantities, delay times and transportation times between process steps. Semi-structured interviews were conducted with line workers to understand work methods and the challenges faced across all parts of the process, and unstructured interviews were conducted with management to understand the work culture at the plant. Process records were also reviewed to obtain available data such as actual demand for raw materials, total customer demand, production forecasts and delivery schedules.

The collected data was used to build the current state VSM for one of the manufacturer's main products, as presented in Figure 2. Starting at the top right-hand side of the figure, the customer sends its order (shown by an information arrow) to the manufacturer's production control unit on a quarterly basis based on the customer's own forecast, but actual customer demand almost always fluctuates. This introduces uncertainty, a major aspect of complexity, into the manufacturer's production process. Production control compounds the impacts of this uncertainty by using the customer forecast to request raw materials from foreign suppliers, and sends corresponding production schedules to the shop floor. A monthly schedule is sent to the knitting department where product manufacturing starts, and a daily schedule is sent to the dyeing department where the product is given its final color as per the customer's order. The items are then boarded to be ready for packing the next day, and finally delivered to shipment the day after.

As shown in the timeline of the current state VSM, the total production lead time (LT) is 6 days which includes all VA and NVA times, whereas the actual value-adding process time (PT) is only 8 hours. This is a good illustration of the magnitude of work delays, another main aspect of complexity which leads to unpredictable behaviors on the production line. It also shows that NVA work takes up a major portion of the work, up to 83% of the total work time, reflecting the significant inefficiencies in the production process. The VSM illustrates the most dominant of the seven wastes of lean observed in the factory, namely: push production (indicated with dashed black-and-white arrows between parts of the process); excess inventories of knitted, sewn and assembled WIP before each process (listed under the triangle symbol); long waiting times for dyeing, boarding and packing (shown on the elevated parts of the timeline at the bottom of the VSM); and, low machine uptimes (listed in the data box under each process) with corresponding high scrap rates.

Figure 2. Current state VSM

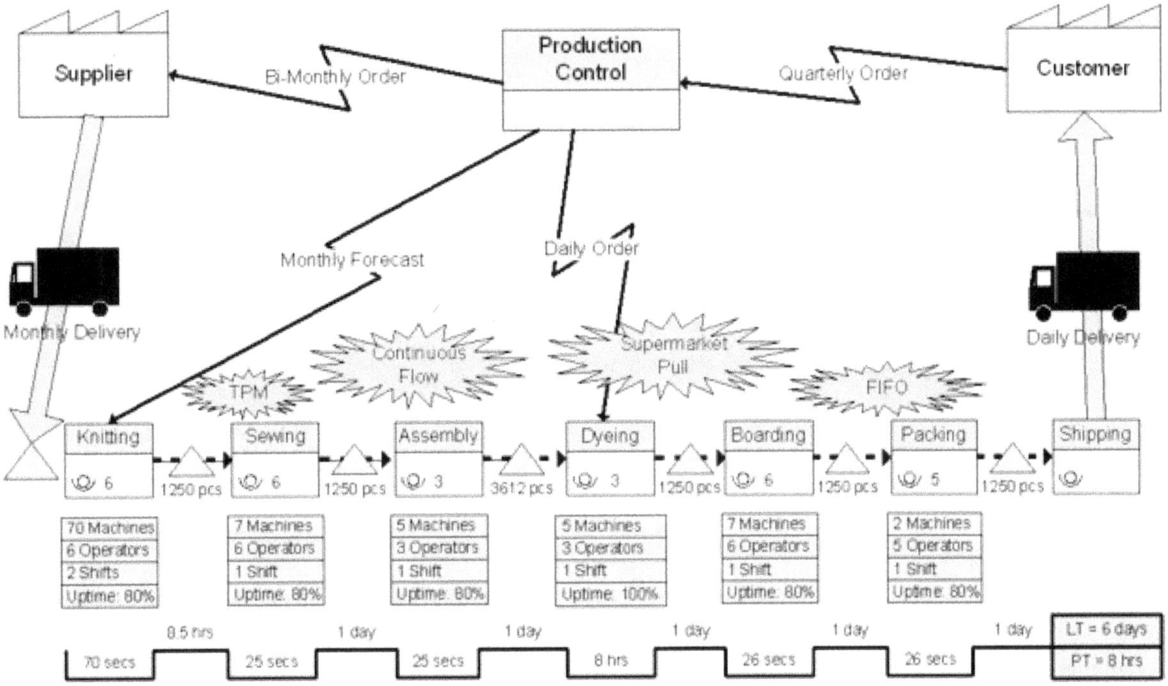

The cause for push production and inventory accumulation is the strategy of overproduction in anticipation of demand (e.g. unnecessary knitting/cutting/dyeing of items above daily customer demand). In addition, management's policy to maintain a buffer stock between departments worth an entire day of WIP (an average of 1,250 pieces) compounded inventory levels. Long waiting times were caused by process bottlenecks, especially at the dyeing stage which had the longest cycle time. A higher inventory (of approximately 3,600 pieces per day) occurs in the dyeing department, as arriving items must be held in waiting until the required machine load capacity is reached for dyeing all pieces of each color together in a batch. Lack of machine maintenance was responsible for high scrap rates and low machine uptimes, especially at the sewing department, leading to product defects (e.g. various imperfections in sewing) and costly rework.

The observed wastes brought about chronic failures in meeting production targets. As a result, four opportunities for improvement were identified and shown on the VSM with Kaizen bursts. The first improvement is the introduction of continuous flow with a pull production strategy using supermarkets. This serves to tackle overproduction, inventory and waiting wastes at the three upstream processes. The second improvement is Total Productive Maintenance (TPM) at the sewing department to increase machine reliability and reduce product defects and scrap; and, a First-In-First-Out (FIFO) inventory control system to mitigate bottlenecking and ensure smooth flow at the downstream processes.

To understand the dynamics of this production system, an SD model was built using the modeling software Vensim DSS version 7.2 (Ventana Systems Inc., 2013), as shown in Figure 3. The model represents the production process as captured in the current state VSM.

Figure 3. SD model of the current-state production process

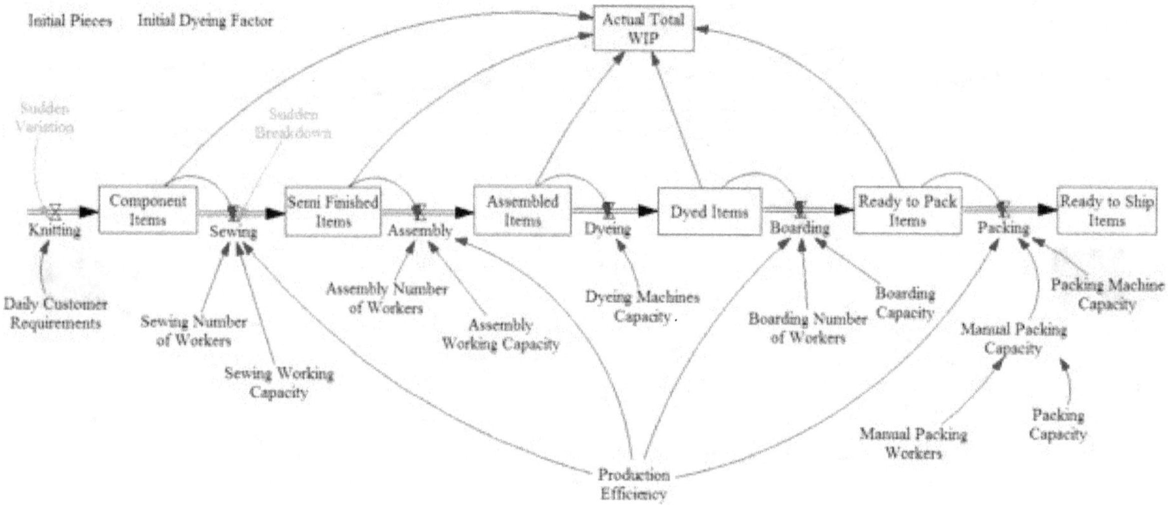

Demand was modeled using a random normal distribution function with an average input equal to the buffer size of 1,250 pieces per day and upper and lower limits (maximum = 1400 pieces, minimum = 1,100 pieces). This was done to capture the stochastic nature of actual customer demand, and a p-value greater than 0.1 was achieved which indicates a good of fit with collected data. WIP inventories at each stage of production are modeled as stocks, with flows representing the actual processing in between stages. Affecting each flow are auxiliary variables representing the number of workers on each machine, the machine's throughput capacity, and an overall production efficiency factor (calculated at 80% in the current state) which reflects machine breakdowns and worker absenteeism as a result of management's overtime policy for dealing with surges in demand.

To capture such disruptions in production operations, a sudden demand surge scenario was included in the model using a pulse increase of 250 pieces occurring early in the modeling timeframe (at day 13), followed by a sudden breakdown scenario of all sewing machines for 24 hours occurring later (at day 30). This causes a significant increase in WIP at all stages of production, as was observed on occasion in the factory. All WIP are accounted for in a single stock (Total WIP) to monitor the total inventory level at the plant throughout the simulation.

The SD model was validated to ensure it's able to reproduce the actual behavior and performance of the production system. In particular, a dimensional consistency test was done to ensure that all units are correct and consistent with each other and that all equations are logically defined and consistent with reality. Sensitivity testing was also used to make sure that small or large parameter changes do not affect the validity of the results (Barlas, 1996).

STEP 3: DEVELOPING THE FUTURE STATE

Based on the identified opportunities for improvement in the current state VSM, it was decided to focus on reducing WIP at each stage of the production process, in addition to eliminating or reducing the other types of waste identified in the current state assessment. This was done using appropriate and complementary lean techniques, namely supermarket pull with Kanban to promote continuous flow and reduce WIP upstream, TPM to improve sewing machine reliability and reduce scrap and defects, 5S to reduce waiting at the dyeing stage, and a FIFO lane to smooth flow at downstream processes. The lean improved state is represented in the future state VSM as shown in Figure 4.

Figure 4. Future state VSM

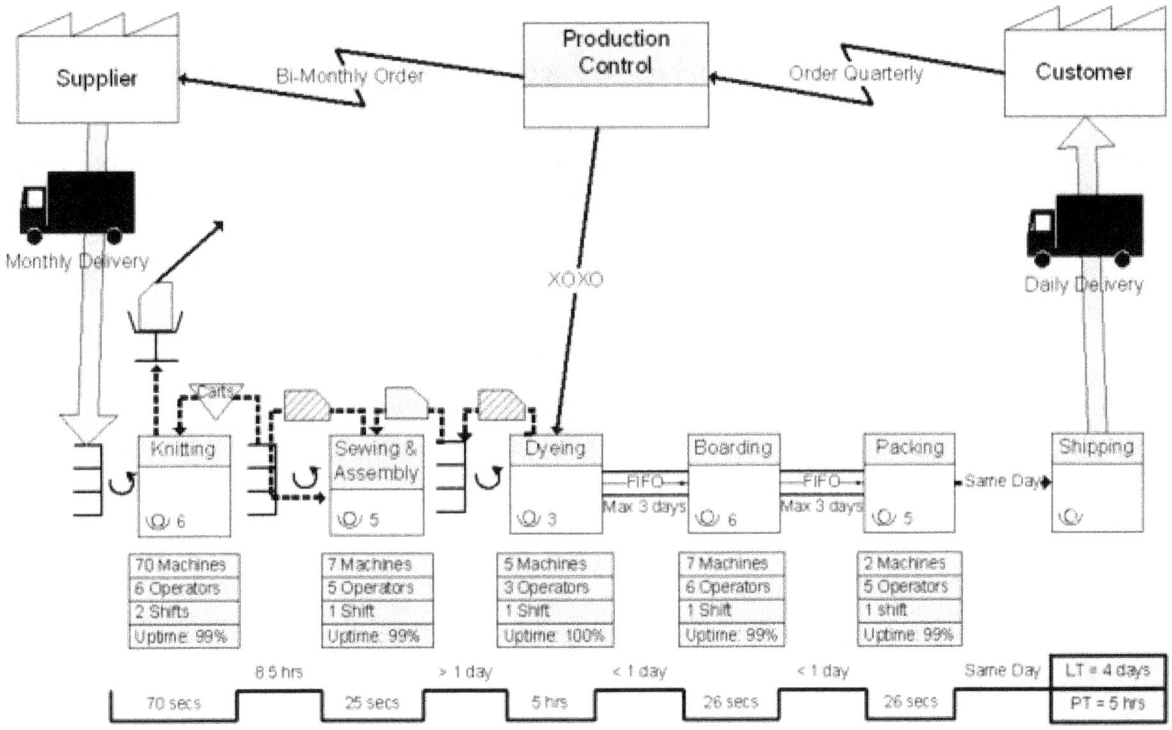

Three supermarkets were located at upstream processes, the first sized to hold 4 containers of yarn from the supplier to satisfy average monthly production needs. The second was sized to hold 4 containers of 300 pieces each to satisfy daily demand within the limits of operator processing capacity. TPM was applied to reduce the high rates of sewing machine breakdowns, which made it possible to sew and assemble at the same time. It thus became possible to join the sewing and assembly processes into one production cell which pulls items from the second supermarket using the Kanban system. Information of what is being pulled is sent back to production control to provide an overview on production progress. At the dyeing process, 5S was applied by organizing work tasks, sorting equipment and setting workspaces in order, which allowed expanding the dyeing process capacity by 1,000 pieces per day. The third supermarket was sized to hold 16 containers of 300 pieces each based on a time study of the improved dyeing process.

In the improved process, production control sends a weekly production schedule (per product type, size, and color) directly to the dyeing process as per a weekly customer forecast (instead of quarterly as in the current state), providing instructions for pulling products from the third supermarket using the Kanban containers. Once the dyeing process pulls the container of assembled items, the sewing/assembly production cell retrieves the empty container and replenishes it, sending a full container back to the third supermarket. The pull system serves to level the production upstream as indicated by the leveling symbol (XOXO) on the VSM. This serves to minimize variation and reduce the adverse impacts of complexity.

Downstream of the dyeing process, a FIFO lane was established with a maximum buffer of 3 days to synchronize downstream processes, ensuring delivery of items from boarding to shipping on a daily basis. The expected lead time in the future state becomes 4 days which amounts to a 33.3% improvement, equivalent to a 32.5% reduction of NVA work time from the current state.

STEP 4: MODELING THE FUTURE STATE AND ANALYZING SCENARIOS

Figure 5 shows the SD model of the lean improved future state. Feedback relationships between every process and its predecessor are added to the current state model to represent the appropriate lean intervention at each production stage. This captures the pull strategy executed through the Kanban system, ensuring that upstream processes are no longer blind to downstream processes and thus avoiding unnecessary WIP.

Figure 5. SD model of the production process post-lean implementation

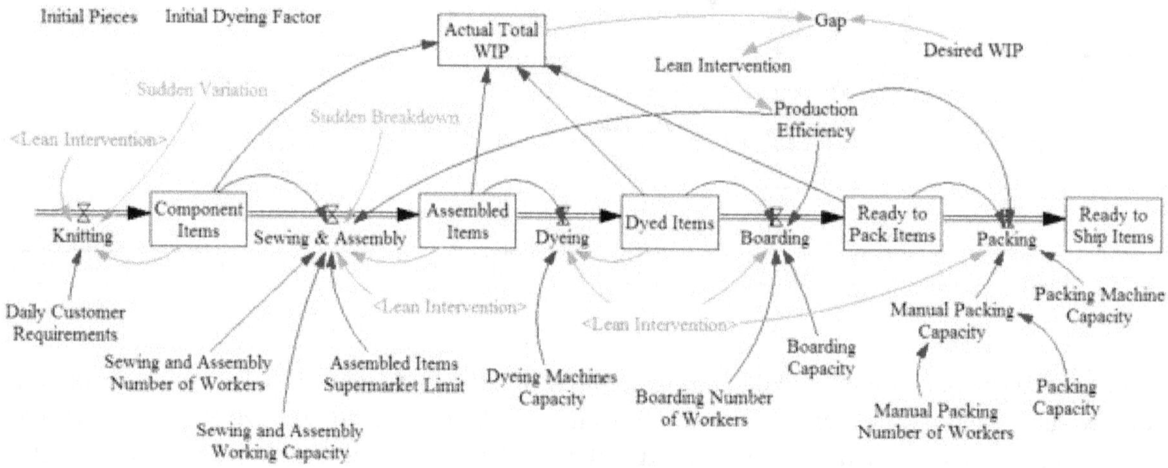

The workings of the model are illustrated by the equation descriptions provided in Table 2. Stocks are simply the difference between their inflows and outflows, therefore only the stock of "Component Items" is shown in the Table for illustration. The stock for Total WIP is the sum of all other stocks. The auxiliary variables are also described in the table. All other model components not listed in the table are constants which take values from the collected data for the production system. The actual equations are provided in (Haddad & Otayek, 2019).

The flow equations illustrate how in the period before lean is applied, each stage of the production process operates blindly based on push from upstream processes, whereas after lean implementation, each downstream stage pulls items from upstream based on actual need. For example, before lean the Knitting flow produces items according to a long-term forecast regardless of inventory levels down-

Table 2. Description of the equations in the future state SD model

Variable Name	Type	Equation Description
Actual Total WIP	Stock	Sum of all stocks (except the stock of Ready to Ship Items).
Component Items	Stock	Difference between inflow to the stock and outflow from the stock.
Knitting	Flow	IF lean is not applied, THEN the daily quantity produced by the knitting department is equal to the forecast of 1,250 pieces plus extra pieces due to sudden customer demand variation. ELSE when lean is applied, knitting is based on pull from the downstream stock such that IF the Component Items stock contains extra items above the buffer quantity of 1,250 pieces, THEN production is reduced by the additional quantity to maintain a steady stock of 1,250 pieces.
Sewing & Assembly	Flow	IF lean is not applied, THEN no items flow out of the stock of Component Items when it holds less than 1,200 pieces to avoid operating at less than maximum capacity per management policy, ELSE the number of items sewn/assembled depends on the number of workers available, machine capacity, machine breakdown rate and overall production efficiency. IF lean is applied, THEN the flow depends on downstream demand such that IF the stock of Assembled Items exceeds the supermarket size, THEN nothing is produced, ELSE the items sewn/assembled equal the production capacity or the Component Items WIP, whichever is smaller.
Dyeing	Flow	IF lean is not applied, THEN no assembled items are dyed before the dyeing machine capacity is reached to avoid operating at less than maximum capacity per management policy, ELSE the number of items dyed is equal to the dyeing machine capacity. IF lean is applied, THEN the dyeing process is activated only when the downstream FIFO lane holds less Dyed Items than its maximum capacity.
Boarding	Flow	IF lean is not applied, THEN no items flow out of the stock of Dyed Items when it holds less than 1,200 pieces to avoid operating at less than maximum capacity per management policy, ELSE the number of items boarded depends on the number of workers available, machine capacity, and overall production efficiency. IF lean is applied, THEN the items boarded equal the production capacity or the Dyed Items WIP, whichever is smaller.
Packing	Flow	IF lean is not applied, THEN no items are packed when the stock of Ready to Pack Items holds less than 1,200 pieces, ELSE the flow is equal to the manual plus automatic packing capacity. IF lean is applied, THEN the items packed are equal to the packing capacity or the Ready to Pack Items WIP, whichever is smaller.
Lean Intervention	Auxiliary	IF the Gap between the Desired and the Actual Total WIP is different than zero, THEN trigger the Lean Intervention.
Gap	Auxiliary	The difference between the Desired and the Actual Total WIP.
Manual Packing Capacity	Auxiliary	The product of Packing Capacity and Manual Packing Workers.
Sudden Variation	Auxiliary	A “Pulse” function of positive magnitude 250 pieces at day 13 to simulate a sudden surge in customer demand.
Sudden Breakdown	Auxiliary	A “Pulse” function of negative magnitude 1,250 pieces at day 30 to simulate a machine breakdown in Sewing & Assembly.
Daily Customer Requirements	Auxiliary	A “Random Uniform” function between a minimum of 1,100 pieces and a maximum of 1,400 pieces, around the average demand of 1,250 pieces.

stream in the Component Items stock, while after lean the Knitting rate is reduced if inventory levels downstream exceed the buffer limit.

The future state SD model was simulated to estimate the impact of the lean intervention in terms of reducing WIP at different stages of production. Several model parameters (variables and constants such as the number of workers and supermarket limits) were tweaked in the simulation process to achieve the highest WIP reduction. Figure 6 illustrates how much the future state is improved over the current state in terms of reduced WIP at the Component Items stage. The lean implementation adjusts the quantity knitted to respect the capacity limit of the stock of Component Items as shown at day 13 when there is a sudden demand surge and at day 30 when there is a machine breakdown in sewing and assembly. Therefore, the impacts of these sudden disruptions are drastically dampened in the future state. This illustrates the usefulness of a detailed SD model in validating the selected lean improvements before actual implementation, which provides a strategic outlook for management about the value of the lean intervention.

Figure 6. WIP of knitted items in the current state vs future state

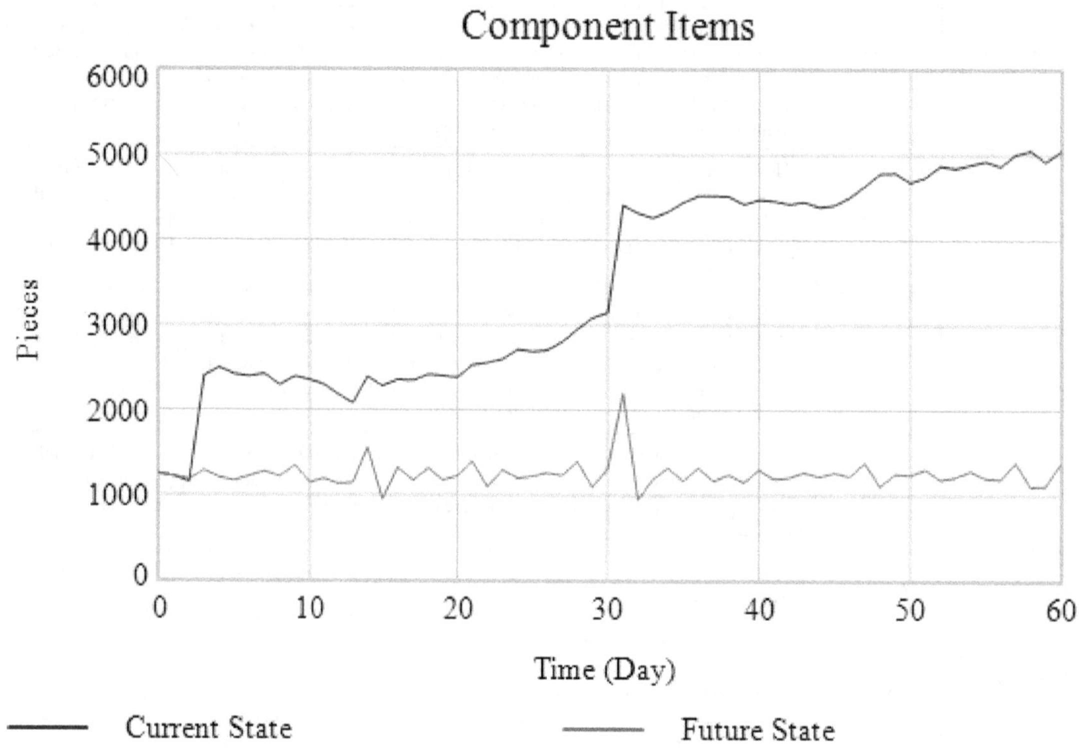

But the results above are not sufficient since they only illustrate improvement in one part of the system, which may not be reflected in other parts of a complex system. This is why it's important to capture all other major components of production so that the SD model can be used as a decision support tool for understanding the holistic impact of the lean initiative on overall system performance. This is done by testing different scenarios of desired total WIP levels (i.e. management targets for the performance of the entire factory) ranging between the minimum possible total WIP of 4,800 pieces per day (reflecting the expanded capacity of the dyeing process after all lean techniques are implemented) and a maximum of 15,000 pieces per day (as in the current state). Figure 7 illustrates such a scenario analysis, but the simulation results can present a deceptive image about the real potential of the lean initiative in the first 40 days due to counter-intuitive behaviors typical in complex systems.

Figure 7. Actual total WIP in the system for different desired WIP policies

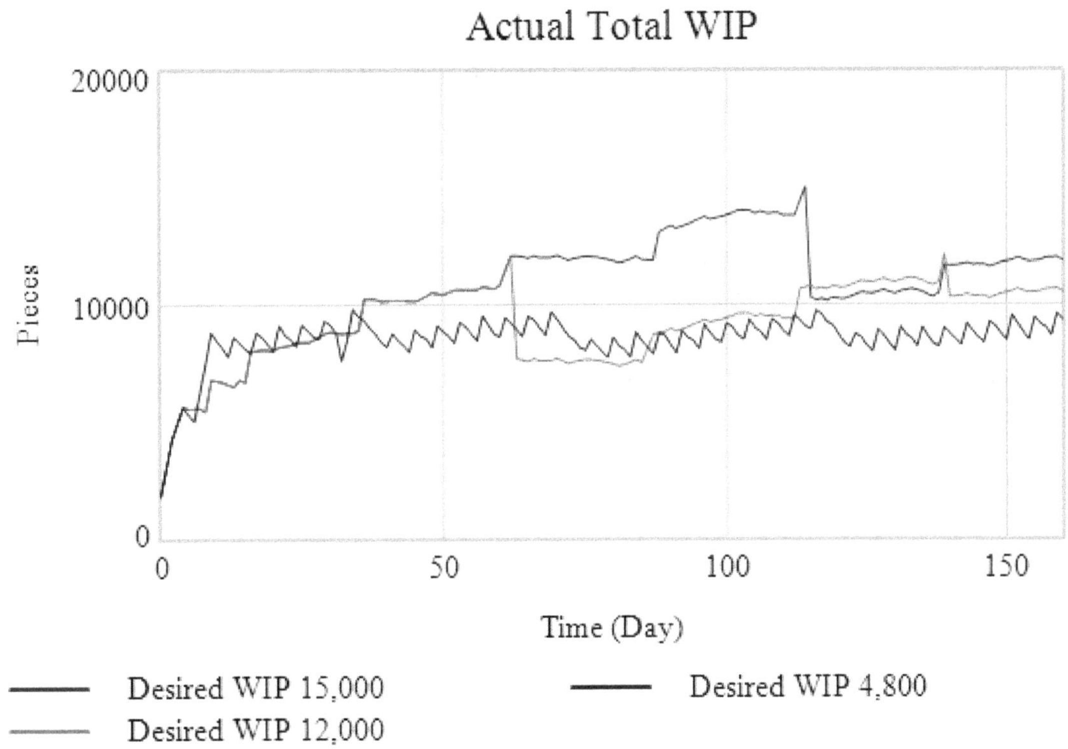

As the trends show, when lean techniques are implemented (i.e. for a Desired WIP policy of 4,800 pieces) actual WIP levels are initially higher than those where lean is not implemented. This is known in systems thinking as a "getting worse before getting better" behavior and is due in this case to the expanded capacity of the dyeing process in the future state (4,800 versus 3,812 in the non-lean state), where now larger WIP levels have to accumulate before the improved capacity is reached. However, total WIP levels are clearly lower and stabilized over time, with only minor fluctuations due to variations in average daily demand. This shows the usefulness of basic scenario analysis to validate the benefits of the lean intervention at the entire system level, saving on potentially costly trial and error efforts on the factory floor.

STEP 5: IMPLEMENTING LEAN AND CONSIDERING SUSTAINMENT

The lean techniques were applied in the factory in accordance with the principles of lean implementation (Feld, 2001). This took approximately 8 weeks to complete. The result was an improvement in productivity and a significant decrease in the total WIP similar to that observed in the model, along with a damping down of the impacts of sudden disruptions. The performance of the implemented lean improvements was monitored in the factory over a period of 3 months in order to ensure the simulated improvement levels are sustained in real world conditions.

In the sustainment phase, questions are usually raised about the risks of reducing buffer stocks below the comfort levels that management and operators have traditionally relied upon when responding to seasonal surges in customer demand. A push back on the lean implementation can be in the form of trying to find a middle ground between the full transformation and an easier more familiar (and thought to be cheaper) alternative. Here SD modeling can be useful for investigating less stringent WIP policies (e.g. 7,000 pieces), which reflects partial efficiency improvements such as at the downstream processes only. In this case, actual total WIP was in fact reduced to lower levels than those in the full lean implementation, as shown in Figure 8, due to more frequent processing downstream. This could have deceptively looked like a better outcome in an actual lean implementation in the factory, potentially compromising the sustainment of the lean initiative.

Figure 8. Actual total WIP for a less stringent WIP policy of 7,000 pieces

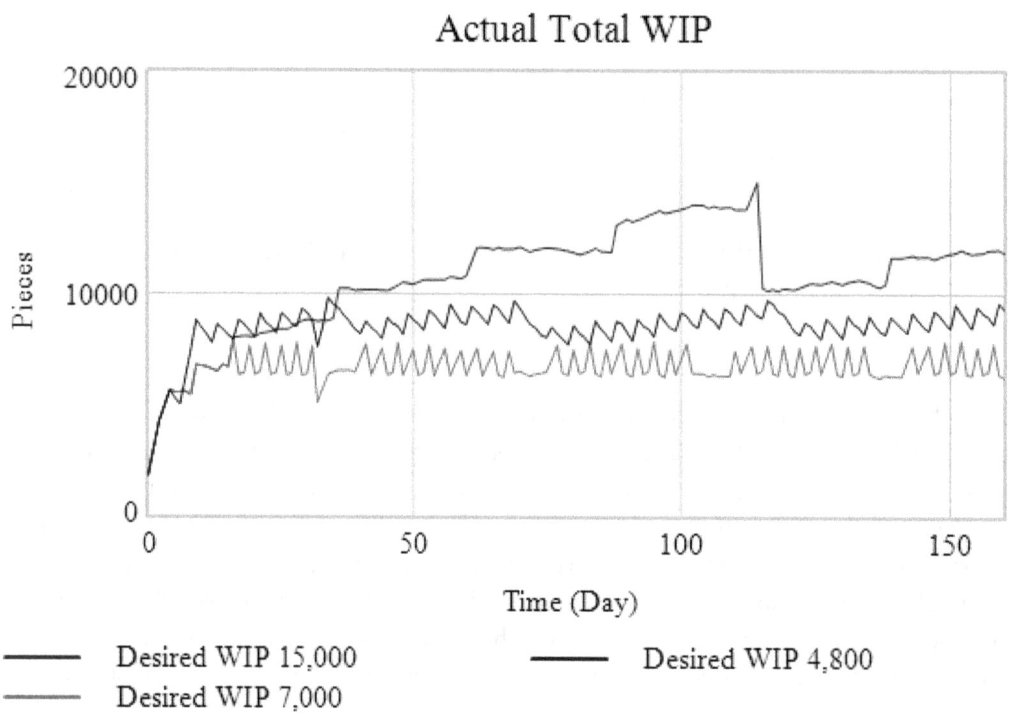

However, since the SD model is able to capture all stages of production, it was found that this would come at the expense of meeting actual customer demand, as shown in Figure 9. This illustrates how SD modeling can play an important role in helping to ensure lean sustainment.

The monitoring of the lean implementation in the factory over the first 3 months showed similar trends to the simulation results, amounting to a net reduction in total WIP of 31% on average and making it possible to reassign labor that was required for production to other areas in need of support. And since the simulation results had shown that improvements were sustained beyond 3 months, no further monitoring was necessary in the plant, showing again that SD modeling can effectively serve to support the long-term sustainment of the lean initiative in practice.

Figure 9. WIP of Ready to Ship Items in the current versus future state scenarios

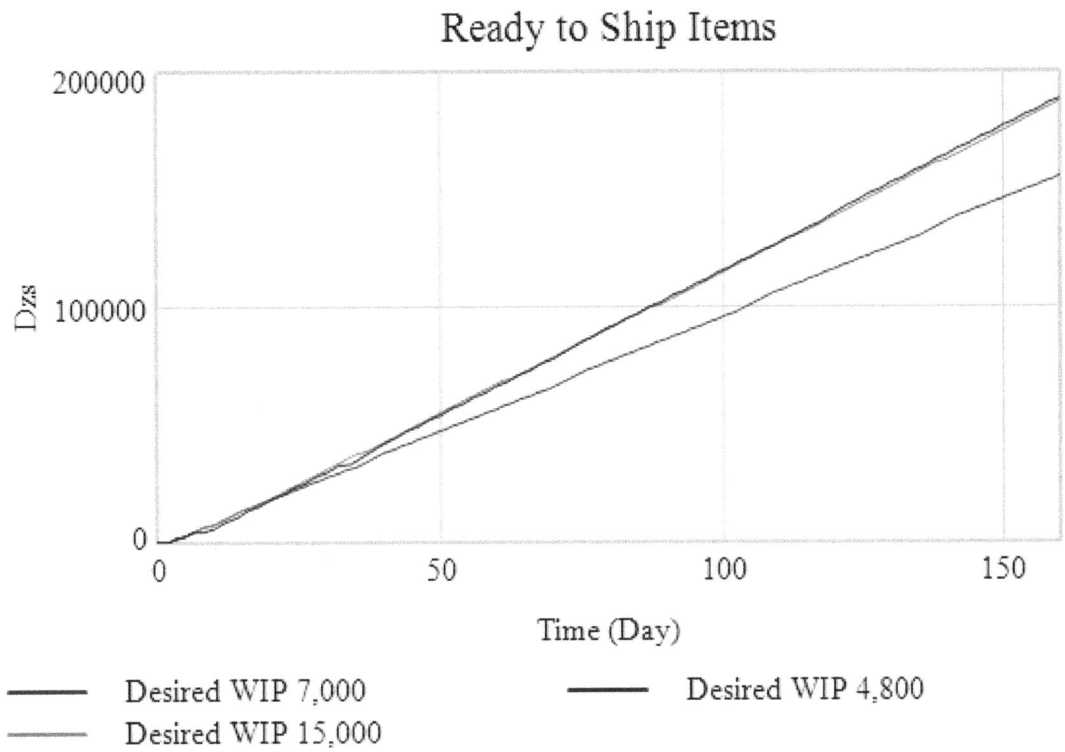

CONCLUSION

This chapter proposed a framework for the use of system dynamics (SD) modeling to explore the dynamics of a lean implementation in a complex manufacturing system, with a particular focus on lean sustainment. The framework consists of using value stream mapping (VSM) as a basis for building SD models of the current and future states of all major stages of a production system in order to capture the complex dynamics of the system. The value of this approach is in assessing the dynamic impact of different lean techniques on system performance, as well as in addressing the major challenge of lean sustainment post implementation.

The application of the framework was illustrated with a case study at an apparel manufacturer with uncertain demand and a high product mix which increase operational complexity. The lean initiative targeted the reduction of WIP in the factory using 5S, Total Productive Maintenance (TPM) and Kanban with supermarket pull at upstream processes, and First-in-First-out (FIFO) processing at downstream processes. Detailed SD models of the current and future states of production were built based on the VSM of each state to capture the main stages of production, the major wastes and the mechanisms of the lean techniques. The SD models were then used to assess the impact of the lean intervention before implementing it on the factory floor. Modeling results showed a net reduction in WIP of 34% on average over the first 3 months, and the actual lean implementation showed a similar decreasing trend and a net average reduction of 31% in WIP over the same period.

However, SD modeling also revealed counter-intuitive behaviors as a result of system complexity, including a "getting worse before getting better" behavior for total WIP levels in the factory after the lean intervention, which is due to the improved process capacity in the future state. The modeling was also useful for understanding the impact of potentially counter-productive policies, such as a partial lean implementation which decreases WIP but at the expense of meeting customer demand. This further illustrates the importance of capturing system complexity in the modeling in order to ensure the long-term commitment of management to the success of the lean implementation. The simulation results showed that the lean implementation was sustained for a simulation period of 6 months, and since the actual improvements in the factory showed a similar trend, only 3 months of monitoring were necessary on the factory floor. This shows the usefulness of SD modeling for reducing costs of implementation and controlling sustainment risks.

In summary, the proposed framework can be used to investigate the implementation of other lean techniques in similar as well as in different contexts, as advocated for by previous studies for understanding the dynamics and long-term sustainment of lean. Note that the use of causal loop diagrams (CLD) was not illustrated in this study for brevity. Separate work by the authors is in process with preliminary results expected to illustrate the usefulness of CLD as a powerful tool for tracing the root causes of unexpected problems during the lean sustainment phase, and for supporting the mitigation of adverse impacts with SD modeling.

REFERENCES

Ali, R. M., & Deif, A. M. (2016). Assessing leanness level with demand dynamics in a multi-stage production system. *Journal of Manufacturing Technology Management*, *27*(5), 614–639. doi:10.1108/JMTM-08-2015-0064

Azar, A. T. (2012). System dynamics as a useful technique for complex systems. *International Journal of Industrial and Systems Engineering*, *10*(4), 377–410. doi:10.1504/IJISE.2012.046298

Baines, T. S., & Harrison, D. K. (1999). An opportunity for system dynamics in manufacturing system modelling. *Production Planning and Control*, *10*(6), 542–552. doi:10.1080/095372899232830

Bar-Yam, Y. (1997). Dynamics of complex systems. Addison-Wesley.

Barlas, Y. (1996). Formal aspects of model validity and validation in system dynamics. *System Dynamics Review*, *12*(3), 183–210. doi:10.1002/(SICI)1099-1727(199623)12:3<183::AID-SDR103>3.0.CO;2-4

Bateman, N. (2005). Sustainability: The elusive element of process improvement. *International Journal of Operations & Production Management*, *25*(3), 261–276. doi:10.1108/01443570510581862

Deif, A. M. (2010). Computer Simulation to Manage Lean Manufacturing Systems. *2nd International Conference on Computer Engineering and Technology (ICCET)*, 677–681. 10.1109/ICCET.2010.5485976

Deif, A. M. (2012). Dynamic analysis of a lean cell under uncertainty. *International Journal of Production Research*, *50*(4), 1127–1139. doi:10.1080/00207543.2011.556154

Deif, A. M., & Elmaraghy, H. (2014). Cost performance dynamics in lean production leveling. *Journal of Manufacturing Systems*, *33*(4), 613–623. doi:10.1016/j.jmsy.2014.05.010

Edghill, J., & Towill, D. (1989). The use of system dynamics in manufacturing systems engineering. *Transactions of the Institute of Measurement and Control, 11*(4), 208–216. doi:10.1177/014233128901100406

Feld, W. M. (2001). *Lean manufacturing: tools, techniques, and how to use them*. CRC Press.

Gafi, E. G., & Javadian, N. (2018). A System Dynamics Model for Studying the Policies of Improvement of Chicken Industry Supply Chain. *International Journal of System Dynamics Applications, 7*(4), 20–37. doi:10.4018/IJSDA.2018100102

Georgiadis, P., & Michaloudis, C. (2012). Real-time production planning and control system for job-shop manufacturing: A system dynamics analysis. *European Journal of Operational Research, 216*(1), 94–104. doi:10.1016/j.ejor.2011.07.022

Godinho Filho, M., & Barco, C. F. (2015). A framework for choosing among different lean-based improvement programs. *International Journal of Advanced Manufacturing Technology, 81*(1–4), 183–197. doi:10.100700170-015-7181-4

Godinho Filho, M., & Uzsoy, R. (2011). The effect of shop floor continuous improvement programs on the lot size-cycle time relationship in a multi-product single-machine environment. *International Journal of Advanced Manufacturing Technology, 52*(5–8), 669–681. doi:10.100700170-010-2770-8

Gonçalves, P., Hines, J., & Sterman, J. (2005). The impact of endogenous demand on push-pull production system. *System Dynamics Review, 21*(3), 187–216. doi:10.1002dr.318

Gupta, V., Narayanamurthy, G., & Acharya, P. (2018). Can lean lead to green? Assessment of radial tyre manufacturing processes using system dynamics modelling. *Computers & Operations Research, 89*, 284–306. doi:10.1016/j.cor.2017.03.015

Haddad, M., & Otayek, R. (2018). Addressing the challenge of lean manufacturing sustainment with system dynamics modeling: a case study on apparel manufacturing in a developing country. *2nd European Conference on Industrial Engineering and Operations Management*, 442–450. http://www.ieomsociety.org/paris2018/papers/89.pdf

Haddad, M., & Otayek, R. (2019). Assessing the Sustainment of a Lean Implementation Using System Dynamics Modeling : A Case Study of Apparel Manufacturing in Lebanon. *International Journal of System Dynamics Applications, 8*(4), 14–29. doi:10.4018/IJSDA.2019100102

Hosseini-nasab, H., Dehghani, M., & Hosseini-nasab, A. (2013). Analysis of Technology Effectiveness of Lean. *Industrial Engineering (American Institute of Industrial Engineers), 2013*, 1–10.

Khataie, A. H., & Bulgak, A. A. (2013). A cost of quality decision support model for lean manufacturing : Activity-based costing application. *International Journal of Quality & Reliability Management, 30*(7), 751–764. doi:10.1108/IJQRM-Jan-2011-0016

Labedz, C., & Gray, J. (2013). Accounting for Lean Implementation in Government Enterprise: Intended and Unintended Consequences. *International Journal of System Dynamics Applications*, 2(1), 14–26. doi:10.4018/ijsda.2013010102

Maani, K. (2017). *Multi-Stakeholder Decision Making for Complex Problems: A Systems Thinking Approach With Cases*. World Scientific. doi:10.1142/9294

Mahmood, A. (2019). Gazing Lean through the lens of System of Systems Dynamics: A case of weaving mill. *Journal of Engineered Fibers and Fabrics*, *14*, 1–16. doi:10.1177/1558925019870705

Nightingale, D., & Srinivasan, J. (2011). *Beyond the lean revolution: achieving successful and sustainable enterprise transformation*. AMACOM.

Omogbai, O., & Salonitis, K. (2016). A lean assessment tool based on systems dynamics. *26th CIRP Design Conference*, *50*, 106–111. 10.1016/j.procir.2016.04.169

Omogbai, O., & Salonitis, K. (2017). The Implementation of 5S Lean Tool Using System Dynamics Approach. *Procedia CIRP*, *60*, 380–385. doi:10.1016/j.procir.2017.01.057

Pai, R. R., Hebbar, S., Kamath, V., & Kamath, G. (2013). Improvement of Process Productivity through Just-in-Time. *Research Journal of Management Sciences*, *2*(12), 1–6.

Poles, R. (2013). System Dynamics modelling of a production and inventory system for remanufacturing to evaluate system improvement strategies. *International Journal of Production Economics*, *144*(1), 189–199. doi:10.1016/j.ijpe.2013.02.003

Robson, C. (2002). Real world research: A Resource for Social Scientists and Practitioner-Researchers (2nd ed.). Blackwell.

Rother, M., & Shook, J. (2009). *Learning to see: value stream mapping to add value and eliminate muda*. Lean Enterprise Institute.

Segura, M. G., Oleghe, O., & Salonitis, K. (2019). Analysis of lean manufacturing strategy using system dynamics modelling of a business model. *International Journal of Lean Six Sigma*. Advance online publication. doi:10.1108/ijlss-05-2017-0042

Shaik, A. S., & Rodrigues, L. L. R. (2018). Plant layout optimisation with implementation of technical cleanliness in an automotive industry: A system dynamics approach. *Int. J. Technology. Policy and Management*, *18*(3), 201–221. doi:10.1504/IJTPM.2018.093850

Shetty, D. K., Abakari, G., Rodrigues, L. L. R., Oommen Mathew, A., & Motlagh, F. G. (2019). To develop Lean Systems in Electronic Automotive Parts Manufacturing Industry: A System Dynamics Modelling Approach. *2019 International Conference on Automation, Computational and Technology Management (ICACTM)*, 277–283. 10.1109/ICACTM.2019.8776803

Stadnicka, D., & Litwin, P. (2019). Value stream mapping and system dynamics integration for manufacturing line modelling and analysis. *International Journal of Production Economics*, *208*, 400–411. doi:10.1016/j.ijpe.2018.12.011

Sterman, J. D. (2000). *Business dynamics: systems thinking and modeling for a complex world*. Irwin/McGraw-Hill.

Ventana Systems Inc. (2013). *Vensim® User Manual*. Ventana System, Inc. http://www.vensim.com

Womack, J., & Jones, D. (2011). *Seeing the Whole Value Stream* (2nd ed.). Lean Enterprise Institute.

Womack, J. P., & Jones, D. T. (2010). *Lean Thinking: Banish Waste and Create Wealth in Your Corporation*. Simon and Schuster.

Chapter 5
The Use of Systems Dynamics in the Analysis of Facilities Management Affecting the Quality of Higher Education in Egypt

Reem A. ElHarakany
College of Management and Technology, Arab Academy for Science and Technology, Egypt

Alfredo Moscardini
Cardiff Metropolitan University, UK

Nermine M. Khalifa
College of Management and Technology, Arab Academy for Science and Technology, Egypt

Marwa M. Abd Elghany
College of Management and Technology, Arab Academy for Science and Technology, Egypt

Mona M. Abd Elghany
College of Management and Technology, Arab Academy for Science and Technology, Egypt

ABSTRACT

In every nation, it is important that the higher education system fulfils the community expectation for perceived quality. One aspect of improving quality is to adopt a better funding strategy for the facilities that are provided by a university. This chapter focuses on the funding strategy for Egyptian universities. The chapter provides the background to the problem distinguishing between private and public universities thenceforward describes the process, employing the causal diagrams, to prioritise the six most important facility services. The budgeting for these services are individually modelled (using both the qualitative and quantitative system dynamics) and then incorporated into a combined stock and flow

DOI: 10.4018/978-1-7998-5788-4.ch005

model. The models allow a facility service administrator to optimise spending in a particular area and also to rank the spending between the six facilities. The models for private and public universities have the same structure but differ with the data. Results of the analysis are discussed in detail.

1 INTRODUCTION

Decision making has always been viewed from a philosophical perspective as a definitely human activity replicating the capability of human beings to practise assessment and decide freely to proceed and take action in a specific selected approach (Pomerol, 2003). Though, that fact has not been consistently acknowledged in science, and other human behaviours, such as language, have long been included in the research of how the brain works; the discovery of Broca's areas in 1861 demonstrates the start of the research of the brain structure. Damasio et al. (1996) argued that decision making is similar to the definition of a human characteristic as a language, and this persuasive argumentation emerged from their decisional biology study.

Decision making and successive actions, predominantly the earlier phases of the process by which individuals focus on their goals and the desired outcomes of their actions, are usually human activities (Pomerol, 2003). For human beings, action is a direct explicit result of the stimuli perceived within the environment, because there occurs an additional layer of look-ahead reasoning between the data acquisition (the stimuli or inputs) and the resultant action. Nevertheless, there are also some instinctive and unconscious responses in human behaviour, but they are closely linked to other biological processes such as closing the windpipe while eating, breathing and certain reflex behaviour, as it takes one's hand out of the flames. Therefore, Action is guided by human thought and this analytical stage offers the freedom to discover what can happen in the big universe (Pomerol, 1997).

Operational research systems, management information systems MIS, and database management systems DMS were built for a long time without any connection to human behaviour as tools that are based on logical models. This led to systems that were more likely to crunch data rather than to relieve the uncertainty inherent in decision-making and decision-makers support. Hence, it rapidly seemed that managers were not going to use MIS systems and some authors were able to criticize the side effects of "MIS- management information systems" (Ackoff, 1967). Rather of addressing the issue of information overload these systems expelled to escalate and led managers to be swamped with data and starving for knowledge. In consequence, systems development is needed, which are capable of raw data translation, represented in facts and figures taken randomly in the heaps of management information reports or in the data warehouse of the organisation, in information given in an ordered collection of relevant data and delivered in a manner that improves its relevance to a specific problem (Pomerol, 2003).

In line with the specialists, DMSS is more technically advanced, offering broader and deeper support for decision-making, as well as more applications. New information and computer technologies are required, the roles of the decision-makers shift, and thus new organizational structures arise to accomplish the changes. Organisations will gain substantial significant benefits from successful DMSS deployments. Researchers and practitioners are stimulated to co-operate their efforts for further enhancements for the pragmatic and theoretical developments of DMSS. Despite technological progress, a key problem still remains how deficiencies in DMSS implementation in organizations can be minimized (Forgionne et al., 2003).

Systems can be used as an idea in numerous scientific fields counting agriculture, cognitive science, earth sciences, ecology and biology, education, evaluation, environmental sciences, human development, management and business, sociology and psychology, sustainability, planning, public health, and other physical sciences. Systems thinking will influence several of the current principles, theories and information (Cabrera et al., 2008). Managers need to stay ahead of rivals in similar business strategies to keep going from niche to market-leader status. In a learning environment, simulations can be used in evaluating the strategies built to avoid violent marketplace (Inelmen, 2006).

Barry Richmond, who is the originator of the word 'systems thinking' in 1987, describes systems thinking as the science and art of building accurate behavioural inferences through gaining an ever deeper understanding of the mechanism underlying. Since then the term has been redefined by a lot of researchers. Hopper & Stave (2008) determined that concerning the different meanings given that covers the huge arena of literature on the subject; common elements appear to encounter the interconnections, the perception of dynamic behaviour, the structure of systems as a resultant of this behaviour, and the concept of seeing structures as wholes instead of parts.

Between the multidisciplinary theories such as: decision theory, control theory, applied computer technology, and simulation technology, System Dynamics (SD) has benefits in solving the complex and dynamic problems of the systems (Sterman, 2000). Therefore, SD is a technique for seeing wholes, understanding trends and interrelationships, and studying how to more effectively and efficiently format those interrelationships (Senge, 1990). SD was applied extensively in macroscopic areas like cultural, social, and environmental and biological systems (Yujing et al., 2015). Several researchers have lately integrated SD into the project management of the product development and used it as a strategic method for understanding the actions of organizations and small and micro small businesses (Mindil et al., 2014). The key benefit of SD theory is that it is easy to see the interrelationship of various system variables in terms of cause and effect, and it is easier to determine the true cause of actions. The other advantage is that structures or parameters, which are needed to change in order to enhance behaviour, can be investigated (Azar, 2012).

Pruyt (2016) believed that in the 'big data' era, there are significant prospects for SD. There are three ways at least in which data science and larger data can perform a role: (i) extracting valuable inputs and knowledge from larger real data; (ii) inferring realistic hypotheses and model structures from larger real data; and (iii) analysing and interpreting massive simulation runs.

Throughout this chapter, system analysis will be followed via the adoption of the approach of the System Dynamics to the development of a modelling method for enhanced management of facilities (ICT), and hence improved perceived quality throughout universities. Two universities, one public university and the other is private university, are considered as case studies and the use of the proposed tool could ultimately be expanded to higher education in Egypt. The proceeding sections begin with relevant research and then implementation of a dynamic system model with a thorough description of causal loops and diagrams of stock and flows. Then a clear illustration of its implementation to both universities, the public and the private, is shown. A conceptual combining these service facilities is subsequently given to come up with a deducing conclusion.

2 RELATED WORK

Work on system dynamics was incorporated into several management areas counting organizational behaviour, operations management, project management, and research on business uncertainties. Abdelsalam (2012) and more recently Al-Kadeem et al. (2017) shed the light the system dynamics use in modelling project organizations' work system design pursuing improved project behaviour.

Knowledge is essential to developing and expanding enterprises. Mindil et al. (2014) provided a conceptual and system dynamics model, which exposes the underpinnings of information management and recognition and thus formulates explicit points of control where changes can be applied. Their research offers a systemic technique of using Information and Communication Technologies (ICTs) as interventions in the institutional foundations of information recognition and management, and modelling them within the complex context of the system. Developed system dynamic model serves as a learning resource for scholars who can make further change and implement it in various scenarios.

Guffey (2015) sought to exploit prevailing comprehensive literature on higher education through the determination of the relationship between the perceived significance of institutional features and the gender, age, and country of origin of cross-border students. As stated by Mazzarol (1998), these considerations include the reputation of an organization for efficiency, profile of the sector, selection of courses, alliances or coalit.

Recent research by Soni (2017) and Soni and Chorasia (2017) investigate the quality of a higher technical institute through combining numerous vital parameters such as the strength of the faculty, faculty satisfaction, student placement, student satisfaction etc. The effect of these factors on the higher technical education quality is investigated by building a dynamic system model for policy planning for obtaining optimal quality. In today's global competitive climate, an open and high-quality system of higher education is essential to economic growth for a country. A sound higher education program promotes and strengthens the economic and social growth cycle for a better future. In a vastly competitive global economy, developing and transition countries face challenges due to the fact that their higher education systems are not sufficiently designed for the development and utilization of available technical manpower. The rapidity with which they adapt to the changing world depends on turning the threats into opportunities.

Little work has been conducted with a particular focus on higher education in Egypt. Many System Dynamics SD practitioners studied few of the higher education management dilemmas in other countries, and their results are outlined next according to the following fields of research.

2.1 External Forces/ Legislation

Diane Green (1994) stressed that the public attention in quality standards has augmented since the mid-1980s as a result of successive British Governments reform for higher education as below:

- Rapid increase in the numbers of student against public limited spending;
- The common quest for the improvement of public services;
- Increased competition within the education 'market';
- And imbalance between quality and performance.

Managers in institutions of higher education therefore have an important role to perform as regards to quality for the sustainability of their effectiveness through:

- Discuss effective methods for the productive use of the resources of the organizations and even generate additional resources;
- Ensuring that the educational institution is compliant with accredited approved university standards;
- Adopt enhanced strategic planning and institutional management systems.

In addition, Mackintosh et al. (1994) made the definition of a quasi-market (i.e. public service competition and contracting when the majority of customers are government agencies or qualified intermediaries instead of end users) and explored the management and decision-making dilemmas in the new paradigm of 'hybrid' organizations - i.e., tax services organizations rely on income returning from the commercial markets. However, they believed that they could be exhibited as functioning on a dual paradigm, guided by quasi-market forces on the tax-supporting side, and via private sector forces somewhere else. Consequently, this would impact the scale and quality of teaching and research outputs, as well as universities' costs, management strategies, and working conditions. Furthermore, they established that joint dependency means that "the management problems of an intensifying fragmented organization can not be solved by simply separating them into two non-commercial and commercial sections".

Gornitzka and Maasen (2000) proposed that the pressures on higher education for "modernisation', 'adaptation', 'diversification', 'marketing' should become more 'entrepreneurial', 'economic', more 'accessible' and more 'effective', more 'service-oriented' and more 'socially important' in enhancing the 'efficiency of its processes and goods', its connection with the labour market, and its institutions, universities and colleges 'governance and management'.

2.2 Corporate Governance

Kennedy and Clare (1999) contrasted between the purely central-pointing "managerialists" and the "collegiumists" who regarded the university as a community of scholars. They argued that the stakeholders of a university can be divided into four distinct groups: employers, government, students and the broader community; where each institution has commitments to promote accessibility to facilities, for serving the academic community, the universal community and the society as a whole.

Watson (2000) considered the essence of management in the university context and denoted Rittel and Webber's seminal paper (1973) to explain the characteristics of policy formulation illustrated in the following:

- The quest for scientific bases to solve social policy issues is inevitable to collapse for the reason of the 'false' nature of the problems;
- Policy issues are certainly not describable;
- There is no objective concept of equity; policies that lead to social problems may certainly not be right or wrong and there is no point in saying 'optimal solutions to social problems' unless there is a significant qualification enforced first.

2.3 Planning, Resourcing & Budgeting

In his extensive chain of researches (Galbraith (1982, 1989, 1998a, 1998b)) and in (Galbraith and Carss, 1989), Peter Galbraith analysed the influence of managerial policy on institutional performance in higher education, showing the gap between policy change and the effects being observed and visible. He has noted other optimistic loops in his papers of the year 1998a & 1998b such as the rise in enrolments needing added funding and staff of academia. However, undesirable correlations like the rise in enrolments leading to a rise in workers and in combination with a rise in the wage bill resulting in a decrease of funding resulting in a lower employment rate for employees in Queensland University. Time delays in years in both loops were considered until the loops closed.

Forrester (1994) defined the causal loop mechanism as a question that is acknowledged, an action that is suggested, an outcome that is predicted but the outcome is not always the case. Symptom, intervention, and response in linear cause-to-effect relationships, that are not isolated, but occur in a nest of interconnecting structures. Mackintosh et al. (1994) found that management accounting information systems are converted into short-term aim financial goals, which typically end up chaotic behaviour. The allocation of operating costs to goods in the higher education sector is subjective and indefinite, resulting in weak allocation of capital to schools and faculties.

Barlas and Diker (1996) explored numerous problems through the creation of an interactive dynamic simulation model, like faculty ratios, low study productivity, student development, and poor teaching quality. They have come up with these main conclusions:

- The research output is positively influenced by the orientation of the graduate versus the undergraduate study;
- Increased enrolments in limited class sizes result in many sections or multiple elective courses to satisfy the students' desired number of class hours.

Correspondingly, the teaching loads for the members of the faculty would rise, which would lead to rises in leave levels and reduction in the prospective pool of faculty supply.

Kennedy and Clare (1999) analysed the potential challenges with traditional approaches to higher education to identify the variables that must be integrated into a SD model to facilitate policy research on issues of resource management. They believed that the key issue with conventional adopted models is kept in their linear and static perspective; ignoring the complex interconnections between the input/output variables and the essence of the process improvement 'transformation'.

Bell et al. (2000) created the Holon Planning and Costing Framework, saying it was intended to "identifying an approved future and finding ways to achieve it despite the constraints of cost". The framework combines a soft methodology (Holon) that answers the problem of what, who, when, and where, in the current state S0 to create a vision of a future state S1 and hard SD methodology capturing 'why' and 'how' questions typed. In an experimental case study they applied the system within SCISM at South Bank University.

Vahdatzad and Mojtahedzadeh (2000) used system dynamics to understand a department's funding allocation mechanism which seeks to expand with only limited resources available. Their approach impeded conflicts between seeking ways to improve the research role without compromising the school's consistency and integrity. They described the University of Yazd's growth in regards of the increase in the numbers of students and the members of faculty yet; university capital did not rise in the same way as university growth. The research purpose was established as the means for further expansion by which the revenues could be increased. There is a bigger difference between the budget assigned and the expected desire. The researchers pointed to the creation of university visions in accordance with the role of stakeholders, contributing to a coherent response. Their study demonstrated the shortcomings of the university's existing research and decision-making progress, and recognised concrete ways of meeting the university's requirements.

2.4 Human Resource Management

Shattock (1997) noted that "the 'massification' effect has come to be more apparent: firstly, in the cost of higher education; secondly, in the uncertainty over the preservation of academic excellence in mass higher education systems; and thirdly, in the structural questions of how institutions should deal with that rise".

Shattock (1999) further emphasized two years later, accompanying the rapid increase in the numbers of student produced a sharp collapse in real and expected unit costs. For example, the student / staff ratio and the necessity for changes in the loads of teaching appeared to worsen considerably, in addition to the incapability for investing in infrastructure of academia like teaching rooms to suit the growth.

Lewis and Altbach (1996) observed that, based on the Carnigie International Academic Profession Survey, there is growing friction between academic faculty staff and administrators. Due to degradation in working conditions, there is also an atmosphere of discontent within the international academic life. Since classes are increasingly becoming bigger, there is demand for educators for further teaching, research funding is diminishing, and wages are not getting up with inflation. It can be shown that the academic profession's morale is starting to be weary. This is besides the occurrence of the universal phenomenon, which puts pressure on budget cuts and centralizes decision-making.

2.5 Teaching Quality

The Information Management and Modelling Research Group (IMMaGe) proposed a primary SD model to examine the quality control problems fronting the South Bank University School of Computing, Information Systems and Mathematics (SCISM), and realised that:

- Concerns like Management, Quality of Staff, Effectiveness of Department, Funding and Research for an upcoming quality control model should be considered;
- And related success measures or metrics for additional SD inquiries should be compiled.

2.6 Teaching Practice

Roberts (1978) studied the performance of the student in the classroom. The SD role in the development of teaching practice was later investigated in several disciplines of academia, in Saeed's (1990) for social sciences, in Saeed's (1993) for economic growth and in Saeed's (1997) for 'Modern Liberal Education'. Saeed stated in his papers that experimental learning in comparison to physical sciences and engineering and a number of the fine arts is hardly enabled in social science teachings. In addition, he tested the application of Kolb's experiential learning model employing the SD as a tool. Also, Carol Frances (2000) addressed the question of influence assessment of emerging technologies in education.

2.7 Microworlds

Barlas and Diker's (1996, 2000) studies intended for creating an interactive dynamic simulation model capable of evaluating a variety of issues related to the academic dimensions of a university management system, such as the rise in student-faculty ratios, low efficiency of research, poor quality of teaching, and testing different policies for its 'Microworld' incapacitation. For a university's three main activities, specifically, teaching, research, and professional projects, a series of performance measures were created. The model was tested using data from the University of Bogaziçi in 1983–1997. Main results from the studies include:

- The output of the research is positively influenced by the evaluation of the graduate with the incorporation of certain interconnected decisions like criteria for instruction-hour, recognition of research and incentives etc.
- Class sizes should remain low to improve teaching efficiency nevertheless, taking into consideration the high student enrolment, numerous sections or multiple electives are the alternatives that are expressed in bigger loads of teaching, hence decreasing the faculty supply for the reason of the increase in faculty leave rates.
- A single independent decision-making process may yield counter-intuitive results; thus collaboration and alignment with other related decisions should be considered.

2.8 Enrolment Demand

During the last two decades, Frances et al. (1994), Frances (1995), and Frances (2000) researched the market for student enrolment. More modelling research was followed by the University of Houston System exploring approaches to generate new demands for enrolment among the Hispanic and African American communities of Houston. Findings are summed up as below:

- SD may aid in the management adjustments or policy having the potential to be successful in achieving preferred outcomes, taking into account the frequent realization of counter-intuitive patterns;
- Furthermore, if incorrect decisions were taken, the reverse of the formulated decisions could not automatically restore the prior state.

The above findings are in consistence with those of studies by Galbraith and studies by Barlas and Diker (1996, 2000), in addition to studies by Kennedy and Clare in 1999.

3 ADOPTION FOR HIGHER EDUCATION IN EGYPT

The research is to investigate the relationship between university resource allocation in facilities management and the perceived quality of education in Egyptian universities. The data for the model was from a qualitative study interviewing in-depth several decision-makers of budget allocation in twenty-three private universities and twenty public universities as shown in the subsequent tables. The interviews were conducted on a specific time schedule, arranged according to the interviewee and interviewer agreement. The researcher initiated a discussion about their adopted policy in budget distribution and the various vital service facilities that should be funded from their perspective to better enhance the student experience in their universities and ultimately the perceived quality of education. These interviews provided he researcher with in-depth data that was flexible, quick and inexpensive and allowed a suitable questionnaire to be constructed.

Figure 1. Initial Causal Diagram (The Authors)

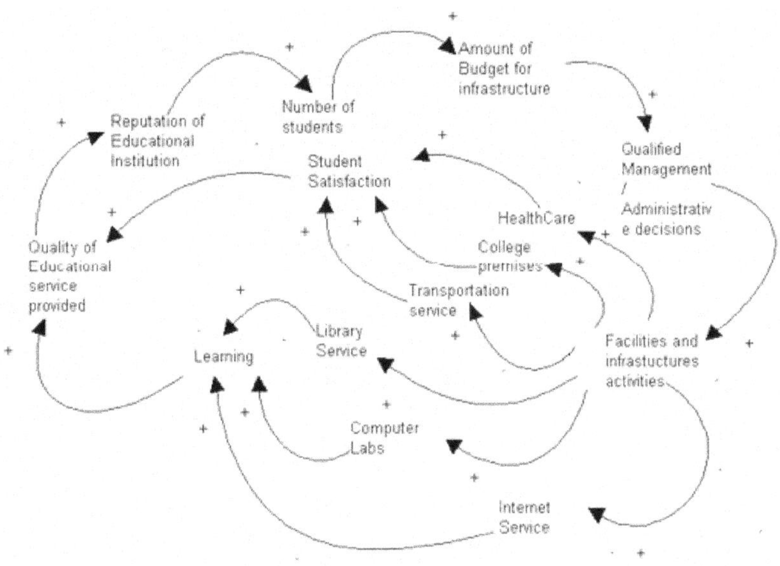

The purpose of the interviews was to identify the dependent variables. This was done by constructing a causal diagram such as shown in figure 1 and discussing it with the interviewees. The feedback enabled a better causal loop to be constructed. This process was repeated five times and the result was the causal model shown in figure 2.

Using the above causal model as a guide, a regression analysis was set up based upon the relationship between the dependent variable, which is the perceived quality of education, and the confounders, which are the thirty questioned service facility administrators

The data gathered was analysed using stratified random sampling using SPSS version 20.

The results showed that the most effective services' facilities that should be assigned the highest prioritization from Top Management when allocating resources for funding budget distribution are: Library Services, Computer Services, Classrooms, Health Care, Sport, and Other premises like transportation, security, and so on. Based on the obtained results, the concluded university service facilities will be used to construct the proposed stimulating modelling tool for Decision-Makers of facility management in the Egyptian Context Higher Education. Each of the six facilities were modelled individually and then a combined model was constructed. This chapter illustrates in detail the process for Library services. The procedure for the other six followed the same process.

Figure 2. Final Causal Diagram (The Authors)

Figure 3 shows the causal loop for the library service budget which has to cover three chief issues; books purchase, more scientific databases addition, and the staff's salaries increase or new members of staff hiring. With a fixed budget, any increase in one area will affect the other two. Thus, one is looking for the optimum proportion that can be spent in these areas.

Figure 4 shows a System Dynamics model developed from the causal model. The individual increase in quality for each area is a "soft variable" and table functions were used. The model shows the table functions which affect the budget required for books, databases and salaries for staff. As stated earlier, the rise in the amount of books available in the library will have a positive impact on library quality. At the other hand, the staff salaries and the library's additional databases are also important for the quality of the library.

Since they all have the same budget, any change in the budget of the books would result in less being obtainable for the databases and the salaries. This model assists in maintaining a balance between all those factors that affect the output. The average staff salaries and the number of new staff, in line with the amount of hypothetically retiring staff, make up the staff payroll budget. The friendliness and efficiency of the library staff is a big factor in the quality of the library, because they are directly contacted with the students, and it is crucial to make sure they are happy and are able to do their job with excellence.

Using table functions, figures 5,6 and 7 show the effect of each factor contributing to the Library service facilities on the perceived quality of education. These graphs were assembled from interviews with relevant staff and the author's own personal experiences with facilities management. The graphs relate to public universities. Similar graphs were produced for the private universities.

Figure 3. Library Service Causal Model (The Authors)

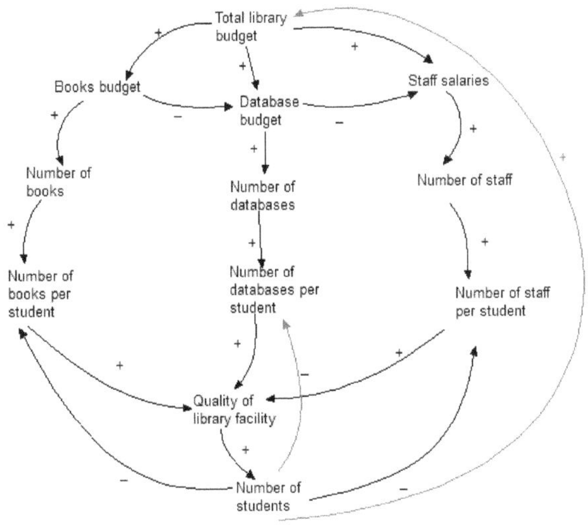

Figure 4. Library Service Stock & Flow Diagram (The Authors)

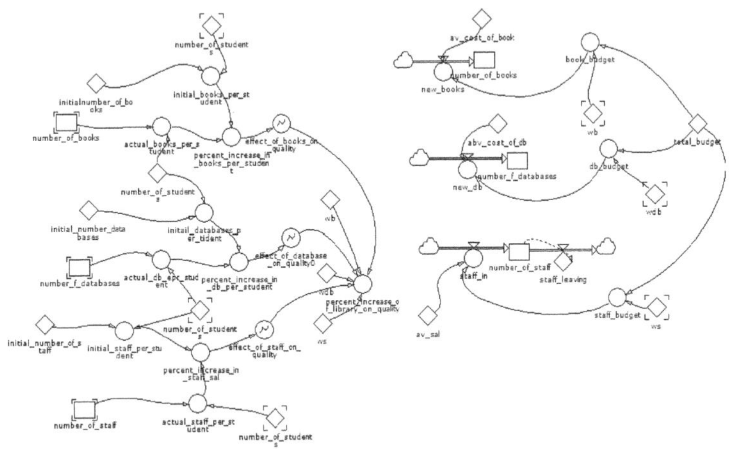

The data for the models (both public and private) is shown in the Appendix, as are the stock flow equations.

Tables 1 and 2 show the output results of the System Dynamics approach that were obtained using POWERSIM.

The estimated poor quality of the library service facility in public universities relates to the low percentage of the budget assigned to the database in comparison to the volume of books given the large percentage allocated to the salaries of staff. However, they do not explicitly collaborate in service facility improvement.

The higher estimated quality in the private universities is because of the availability of the database and the books that the students only require without the consideration of staff salaries.

The derived weights are the ones that are currently in use. The beauty of the model is that these weights can now be varied to find the optimum.

Figure 5. Books Factor Effect on Estimated Quality for Public University (The Authors)

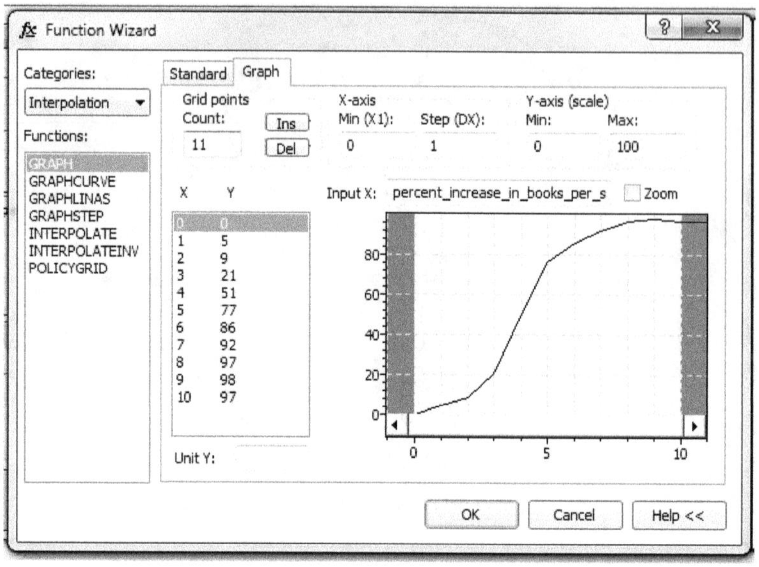

Figure 6. Staff Factor Effect on Estimated Quality for Public University (The Authors)

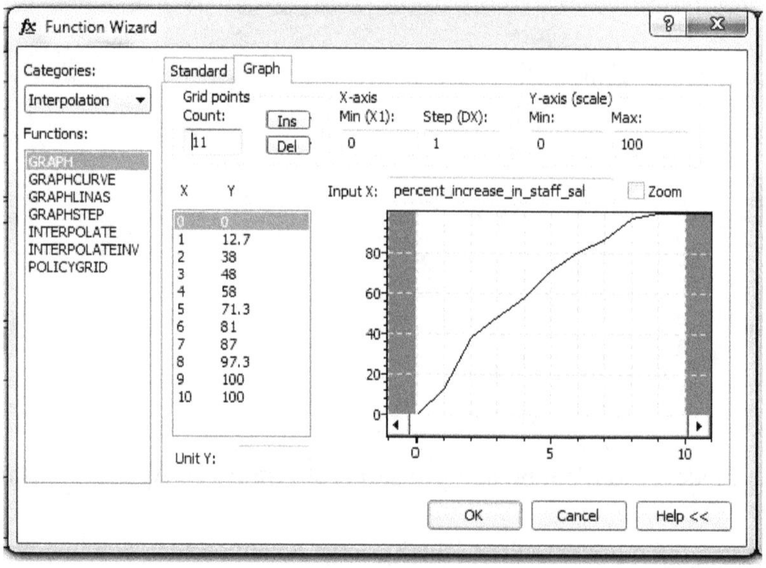

These models were then combined to form a total facilities management model. The stack and flow diagram is too large to show in this chapter, but it consists of the six previous models connected by a series of weights. By changing the weights, one can assign a priority to the facilities.

Figure 7. Database Factor Effect on Estimated Quality for Public University (The Authors)

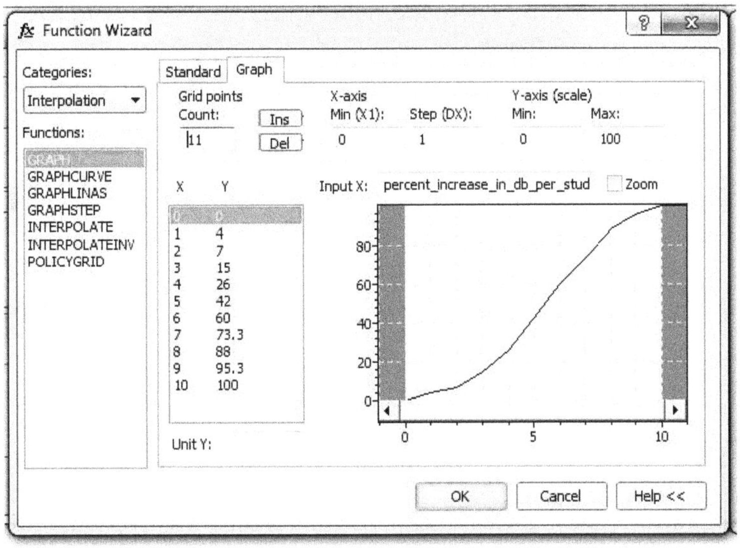

Table 1. The Estimated Quality of the Library Service Facility in Public University (The Authors)

Library Service Facility in Public University			
Service Facility Factors	Quantity of Books	Staff Salaries	Database
Budget Assigned	619800	826400	619800
Derived Weight of Factor	0.3	0.4	0.3
Effect of Factor on Service Facility	19	32	62
Total Estimated Quality	37.1% with 2066000 Egyptian Pounds of Budget		

Table 2. The Estimated Quality of the Library Service Facility in Private University (The Authors)

Library Service Facility in Private University			
Service Facility Factors	Quantity of Books	Staff Salaries	Database
Budget Assigned	1338180	2007270	1115150
Derived Weight of Factor	0.3	0.45	0.25
Effect of Factor on Service Facility	78	78	52
Total Estimated Quality	71.5% with 4460600 Egyptian Pounds of Budget		

4 IMPLEMENTATION OF THE MODEL

The proposed integrated model consists of two levels – it has six service functions which themselves have various budget allocation options. By varying the weights, an optimum allocation for that budget can be achieved.

- The more general problem is when there is a total budget for facilities and the task is how to allocate between the six functions. There are many different ways of proceeding. The procedure followed in this section is as follows:
- Six runs are done where each service function in turn was given a weight of 50% and the others ten per cent. This was to find the service function that had the most significant contribution (all other things being equal).
- The second stage was then to fix this service budget and to try and find the second most significant contributor. The process is then repeated to finally produce a priority list.

This process was done for Alexandria University (the public example) and for the Arab Academy (the private one).

Some results are:

- *Library Facilities*
  This funds books, databases and library staff. An optimum shared allocation for this was found to be a division of 30%, 30% and 40%, which is fixed for the subsequent runs described below.

- *Computer Labs*
  This funds devices, their maintenance, staff salaries (technical) and software. An optimum shared allocation for this was found to be a division of 55%, 50% and 40%, which is fixed for the subsequent runs described below.

- *Classrooms*
  This funds the education resources (including physical classrooms), their maintenance and the salaries of the academic staff. An optimum shared allocation for this was found to be a division of 20%, 30% and 50%, which is fixed for the subsequent runs described below.

- *Sports Services*
  This funds equipment (tools), trainers and the sports fields. An optimum shared allocation for this was found to be a division of 30%, 20% and 50%, which is fixed for the subsequent runs described below.

- *Health Services*
  This covers equipment, medical staff salaries, and medications. An optimum shared allocation for this was found to be a division of 20%, 30% and 40%, which is fixed for the subsequent runs described below.

- *Extra Facilities*
  This is a catch all for what is left which includes premises, salaries and safety. An optimum shared allocation for this was found to be a division of 20%, 30% and 50%, which is fixed for the subsequent runs described below.

The procedure is covered under six scenarios – three for the public and three for the private university.

4.1 First Scenario for Integrated Model (Public)

In table 3, the idea was to see which service is most significant. Accordingly, each service facility was given a weight of 50% and the other ten per cent. The data used in the model is that was used in previous section, which covered the individual facilities.

The six runs show that for Alexandria University, the more significant service is Classrooms.

Table 3. First Scenario for Integrated Model for a Public University (The Authors)

Total Budget 9,469,200 Egyptian pounds						
Six Service Facilities	**Weight Derived**	**Weight Derived**	**Weight Derived**	**Weight Derived**	**Weight Derived**	**Weight Derived**
Library Services	0.5	0.1	0.1	0.1	0.1	0.1
Computer Labs	0.1	0.5	0.1	0.1	0.1	0.1
Classrooms	0.1	0.1	0.5	0.1	0.1	0.1
Sports	0.1	0.1	0.1	0.5	0.1	0.1
Health Care	0.1	0.1	0.1	0.1	0.5	0.1
Extra Facilities	0.1	0.1	0.1	0.1	0.1	0.5
% Increase in Quality	6.1	6.0	6.7	5.5	4.9	5.0

4.2 Second Scenario for Integrated Model for Public University

The classroom facility was then set at 40% and the other five facilities were varied to find the second most significant one. These five runs are shown below in table 4.

It appears that the second most significant factor is the computer provision.

Table 4. Second Scenario for Integrated Model for a Public University (The Authors)

Total Budget 9,469,200 Egyptian pounds					
Six Service Facilities	**Weight Derived**	**Weight Derived**	**Weight Derived**	**Weight Derived**	**Weight Derived**
Library Services	0.05	0.4	0.05	0.05	0.05
Computer Labs	0.4	0.05	0.05	0.05	0.05
Classrooms	0.4	0.4	0.4	0.4	0.4
Sports	0.05	0.05	0.4	0.05	0.05
Health Care	0.05	0.05	0.05	0.4	0.05
Extra Facilities	0.05	0.05	0.05	0.05	0.4
% Increase in Quality	6.5	6.3	5.5	3.0	4.2

4.3 Third Scenario for Integrated Model for Public University

Classrooms and Computer labs were now fixed and four runs were done to find the third most significant which are shown in table 5.

It seems that this is Library services. This process was repeated, to find the fourth, fifth and consequently the sixth considered the most significant factor. The model suggested that the services should have budgets prioritized in the following order: Classrooms, Computer labs, Library services, Sports facilities, extra facilities and Health services. This is an example of how to use the model as a tool for Decision Support for financial administrators at a university.

Table 5. Third Scenario for Integrated Model for a Public University (The Authors)

Total Budget 9,469,200 Egyptian pounds				
Six Service Facilities	**Weight Derived**	**Weight Derived**	**Weight Derived**	**Weight Derived**
Library Services	0.25	0.05	0.05	0.05
Computer Labs	0.3	0.3	0.3	0.3
Classrooms	0.3	0.3	0.3	0.3
Sports	0.05	0.25	0.05	0.05
Health Care	0.05	0.05	0.25	0,05
Extra Facilities	0.05	0.05	0.05	0.25
% increase in Quality	6.5	6.3	5.5	3.0

4.4 Scenarios for a Private University

The next three tables show similar analysis for the Arab Academy, which is an example of a private university. In 2017, their budget was considerably higher than Alexandria University at 15790600 Egyptian pounds. All precise data for 2017 is shown in the appendix. The same six services are examined but the optimum allocations within each individual service are different as shown in suggested in section 7.3 are different. The library is divided in the ratio 35, 25%, 45%; computer labs in the ratio 55%, 50% and 40%; the classrooms are 30%, 50% and 20%; the Sports facilities 30%. 20% and 50%, Health 20%, 40% and 50% and the Extra Facilities 20%, 30% and 50%. Table 6 shows the first stage of the process i.e. to identify the most significant contribution.

In this scenario it can be concluded that the computer labs were deemed most significant.

Table 7 shows the results of five runs to find the second biggest contributor.

The results show that the second is the most significant factor is the sports facility.

Table 8 shows four runs to discover the third most significant factor.

It can be seen that extra facilities (which encompass safety is the third factor). After completing the process, the priority list for the Arab academy as an example of a private university is as follows: computer provision, sports, safety, classrooms, library provision and health services.

Table 6. Finding the Most Significant Contribution for a Private University (The Authors)

Total Budget 15,790,000 Egyptian pounds						
Six Service Facilities	**Weight Derived**	**Weight Derived**	**Weight Derived**	**Weight Derived**	**Weight Derived**	**Weight Derived**
Library Services	0.5	0.1	0.1	0.1	0.1	0.1
Computer Labs	0.1	0.5	0.1	0.1	0.1	0.1
Classrooms	0.1	0.1	0.5	0.1	0.1	0.1
Sports	0.1	0.1	0.1	0.5	0.1	0.1
Health Care	0.1	0.1	0.1	0.1	0.5	0.1
Extra Facilities	0.1	0.1	0.1	0.1	0.1	0.5
% Increase in Quality	6.1	6.8	6.0	5.5	4.9	5.0

Table 7. Finding the Second Most Significant Contribution for a Private University (The Authors)

Total Budget 15,790,000 Egyptian pounds					
Six Service Facilities	**Weight Derived**	**Weight Derived**	**Weight Derived**	**Weight Derived**	**Weight Derived**
Library Services	0.05	0.4	0.05	0.05	0.05
Computer Labs	0.4	0.05	0.05	0.05	0.05
Classrooms	0.4	0.4	0.4	0.4	0.4
Sports	0.05	0.05	0.4	0.05	0.05
Health Care	0.05	0.05	0.05	0.4	0.05
Extra Facilities	0.05	0.05	0.05	0.05	0.4
% Increase in Quality	6.5	5.5	6.3	3.0	4.2

Table 8. Finding the Third Most Significant Contribution for a Private University; Source: (The Authors)

Total Budget 15,790,000 Egyptian pounds				
Six Service Facilities	**Weight Derived**	**Weight Derived**	**Weight Derived**	**Weight Derived**
Library Services	0.25	0.05	0.05	0.05
Computer Labs	0.3	0.3	0.3	0.3
Classrooms	0.3	0.3	0.3	0.3
Sports	0.05	0.25	0.05	0.05
Health Care	0.05	0.05	0.25	0,05
Extra Facilities	0.05	0.05	0.05	0.25
% Increase in Quality	6.5	6.3	5.5	3.0

5 COMMENTS

The applicability of these models is only useful if the models are true- i.e. can be trusted. There are two verification stages:

- Are the models structurally correct?

This was shown to be true by applying the Sterman tests (Sterman, 2000)

- Are the models verifiable?

This is a more difficult question. The universities in Egypt (as in the rest of the world) are ranked according to quality. A list of these rankings was obtained. The quality of AAST and Alexandria University was calculated using the model and compared with the rankings. In both cases there was agreement. For the present, the model is structurally correct and the results of the model do not disagree with reality. In such a situation, one can claim that models are abductively correct using the verification theory of C.S Pierce (Pierce 1897).

The process followed in the scenarios is not the only possible one. It is included to demonstrate how the models can be used to aid decision-making. Possible criticisms are:

a) Since there occur a vast degree of interdependence between variables, the process will not necessarily be accurate.

This is indeed true, but the researchers would argue that it will give a good indication of the priorities. Many more runs could be done to check the accuracy of the prediction. One obstacle is that there could be a minimum budget needed for a facility. This could be due to maintenance i.e. maintaining the provision that is already in place. This can easily be handled by a more thoughtful allocation of the weights.

b) Every university is different. It depends on its immediate (local) competition, its location and its reputation.

This is obviously true but can be accommodated in the models by accurate data and judicious weightings.

The comparison between the private and public universities is interesting. The priorities are summarised in table 9.

The changing priority of classrooms could reflect the fact that this is a problem (due to lack of funding) with the public university whereas the private one has adequate classrooms and budgets are needed only for maintenance. As the private one has a better infrastructure, then this may explain the prominence of the sport facilities.

These are just speculations that arise out of the modelling and the researchers not giving definitive pronouncements. It is mentioned here as it shows the debate that can be generated by using this model as a tool. This is one of the advantages of such research as it allows flexibility in decision-making and enables budget allocations to be investigated without having to actually spend the money

Table 9. Priorities Concluded for Both Public & Private Universities (The Authors)

Alexandria University	Arab Academy
Classrooms	Computer
Computer	Sports
Library	Extra Facilities
Sports	Classrooms
Health Care	Library
Extra Facilities	Health Care

6 CONCLUSION

It is well known that in most higher education institutions, the combination of increasingly growing enrolment and deficit in capital has caused lesser quality perceived. Spending on university ICT facilities through the employment of the system dynamics SD model presented would be precisely useful in enhancing quality perceived in higher education in Egypt.

The principle of system thinking and the implementation of system dynamics methodology offer a valuable context where different performance metrics and measures for change can be assessed. Systems thinking permits for a consistent description of the mechanism, and framework in which causal effects can be extracted and the complexities of the proposed decision-making may be better understood in terms of the overall situation of the institution. And any refinements may occur successively to achieve the wanted result that best fits the context of the institution. For example, more research could be carried out on other facilities in universities such as medical services (healthcare) or other premises such as safety, transport, and so on, by means of the approach of system dynamics SD to enhanced educational quality perceived.

REFERENCES

Abdelsalam, H. A., & El-Tagy, A. O. (2012). A Simulation Model for Managing Marketing Multi-Channel Conflict. *International Journal of System Dynamics Applications*, *1*(4), 59–76. doi:10.4018/ijsda.2012100103

Ackoff, R. L. (1967). Management Misinformation Systems. *Management Science*, *14*(4), B147–B156. doi:10.1287/mnsc.14.4.B147

Al-Kadeem, R., Soheir, B., Eldardiry, M., & Haddad, H. (2017, April-June). Review on Using System Dynamics in Designing Work Systems of Project Organizations: Product Development Process Case Study. *International Journal of System Dynamics Applications*, *6*(2), 52–70. Advance online publication. doi:10.4018/IJSDA.2017040103

Azar, A.T., (2012). System dynamics as a useful technique for complex systems. *International Journal of Industrial and Systems Engineering*, *10*(4), 377-410.

Barlas, Y., & Diker, G. (2000, September). A Dynamic Simulation Game (UNIGAME) for Strategic University Management. *Simulation & Gaming*, *31*(3), 331–358. doi:10.1177/104687810003100302

Barlas, Y., & Diker, G. D. (1996). Decision Support for Strategic University Management: A Dynamic Interactive Game. *Proceedings of the 14th International System Dynamics Conference.*

Bell, G., Cooper, M., Kennedy, M., & Warwick, J. (2000). The Development of the Holon Planning and Costing Framework for Higher Education Management. *Proceedings of the 18th System Dynamics Conference.*

Cabrera, V. E., Hagevoort, R., Solís, D., Kirksey, R., & Diemer, J. A. (2008). The New Mexico dairy industry: An economic engine. *Journal of Dairy Science*, *91*, 2144–2150. doi:10.3168/jds.2007-0791 PMID:18420646

Damasio, A-R. (1994). *Descartes' error*. Putman's Sons.

Damasio, A.-R., Damasio, H., & Christen, Y. (Eds.). (1996). *Neurology of decision making*. Springer. doi:10.1007/978-3-642-79928-0

Forgionne, G. A., Gupta, J. N. D., & Mora, M. (2003). *Decision Making Support Systems: Achievements, Challenges and Opportunities.* Available at https://www.igi-global.com/chapter/decision-making-supportsystems/8080?camid=4v1

Forrester, J. W. (1994). Policies, Decisions, and Information Sources for Modelling. In J. D. W. Morecroft & J. D. Sterman (Eds.), *Modelling for Learning Organisations*. Productivity Press.

Frances, C. (1995). *Using System Dynamics Technology to Improve Planning for Human Resource Development.* Presentation at the Congress of Political Economists, International 6th Annual Conference, Seoul, Korea.

Frances, C. (2000), Using System Dynamics as a Tool for Decision Making in Higher Education Management- U.S. Experience. Academic Press.

Frances, C., Van Alstyne, M., Ashton, A., & Hochstettler, T. (1994). Using System Dynamics Technology to Improve Planning Higher Education: Results in Arizona and Houston, Texas. *Proceedings of the 13th International System Dynamics Conference*, 444-453.

Galbraith, P. L. (1982). *Forecasting Futures for Higher Education in Australia: An Application of Dynamic Modelling* (Ph.D. Thesis). The University of Queensland.

Galbraith, P.L. (1998a). When Strategic Plans are not Enough: Challenges in University Management. *System Dynamics: An International Journal of Policy Modelling*, 55-84.

Galbraith, P. L. (1998b). System Dynamics and University Management. *System Dynamics Review*, *10*(2), 69–84. doi:10.1002/(SICI)1099-1727(199821)14:1<69::AID-SDR139>3.0.CO;2-T

Galbraith, P. L., & Carss, B. W. (1989). Strategies for Institutional Resource Allocation: Insights from a Dynamic Model. *Higher Education Policy*, *2*(2), 31–38. doi:10.1057/hep.1989.23

Galbraith, P. L., & Carss, B. W. (1989). Strategies for Institutional Resource Allocation: Insights from a Dynamic Model. *Higher Education Policy*, *2*(2), 31–36. doi:10.1057/hep.1989.23

Gornitzka, A., & Maasen, P. (2000). The Economy, Higher Education, and European integration: An Introduction. *Higher Education Policy*, *13*(3), 217–223. doi:10.1016/S0952-8733(00)00008-8

Green, D. (1994). *What is Quality in Higher Education*. St. Edmundsbury Press.

Guffey, R. V. (2015). Globalization, Cross Border Education, and Student Migration: Determining Student Institutional. *Choice Factors*. Available at https://www.igi-global.com/chapter/globalization-cross-border-education-andstudent-migration/133730?camid=4v1

Hopper, M., & Stave, K. A. (2008). Assessing the Effectiveness of Systems Thinking Interventions in the Classroom. *The 26th International Conference of the System Dynamics Society*, 1-26.

Inelmen, E. (2006). Philosophy of Education. *The Proceedings of the Twenty-First World Congress of Philosophy*, *4*, 57-64. http://inelmen.boun.edu.tr

Kennedy, M., & Clare, C. (1999). Some Issues in Building System Dynamics Model for Improving the Resource Management Process in Higher Education. *Proceedings of the 17th International System Dynamics Conference*.

Lewis, L. S., & Altbach, P. G. (1996). Faculty Versus Administration: A Universal Problem. *Higher Education Policy*, *9*(3), 255–258. doi:10.1016/S0952-8733(96)00016-5

Mackintosh, M., Jarvis, R., & Henry, E. (1994, November). On Managing Hybrids: Some Dilemmas in Higher Education Management. *Financial Management & Accountability*, *10*(4), 339–353. doi:10.1111/j.1468-0408.1994.tb00156.x

Mazzarol, T. (1998). Critical success factors for international education marketing. *International Journal of Educational Management*, *12*(4), 163–175. doi:10.1108/09513549810220623

Mindil, A., Rodrigues, A., McCormick, D., & Mwangi, R. (2014, January-March). ICT Powered Strategic Flexibility System Dynamic Model: A Pillar for Economic Development in Micro and Small Enterprises. *International Journal of System Dynamics Applications*, *3*(1), 90–110. doi:10.4018/ijsda.2014010105

Pomerol, J.-C. (1997, May 16). Artificial intelligence and human decision making. *European Journal of Operational Research*, *99*(1), 3–25. doi:10.1016/S0377-2217(96)00378-5

Pomerol, J.-C. (2003). *From Human Decision Making to DMSS Architecture*. www.igi-global.com/chapter/human-decision-making-dmssarchitecture/

Pruyt, E. (2016, January-March). Integrating Systems Modelling and Data Science: The Joint Future of Simulation and 'Big Data' Science. *International Journal of System Dynamics Applications*, *5*(1), 1–16. Advance online publication. doi:10.4018/IJSDA.2016010101

Richmond, B. (1994). Systems Dynamics/Systems Thinking: Let's Just Get On With It. *International Systems Dynamics Conference*. 10.1002dr.4260100204

Rittel, H., & Webber, M. L. (1973). *Dilemmas in a General Theory of Planning. Policy Sciences 4*. Elsevier Science.

Roberts, N. H. (Ed.). (1978). System Simulation of Student Performance in the Elementary Classroom, In Management Applications of System Dynamics. Productivity Press.

Saeed, K. (1990). Bringing Practicum to Learning: A System Dynamics Modelling Approach. *Higher Education Policy*, *3*(2), 44–49. doi:10.1057/hep.1990.30

Saeed, K. (1993). Bringing experimental learning to the social sciences: A simulation laboratory on economic development. *System Dynamics Review*, *9*(2), 153–164. doi:10.1002dr.4260090205

Saeed, K. (1997). *System Dynamics as a Technology for New Liberal Education.* Worcester Polytechnic Institute, Report No. 3.

Senge, P. (1990). *The Fifth Discipline*. The Art and Practice of the Learning Organisation, Currency Doubleday.

Shattock, M. (1997, July). The New Managerial Implications of the New Priorities. *Higher Education Management*, *9*(2), 27–34.

Shattock, M. (1999, March). The Impact of the Dearing Report on UK Higher Education. *Higher Education Management*, *11*(1), 7–17.

Soni, S., & Chorasia, B. K. (2017). Policy Planning in Higher Technical Education: A System Dynamic Approach. *International Journal of System Dynamics Applications*, *6*(3), 87–110. doi:10.4018/IJSDA.2017070105

Soni, S., & Chorasia, B. K. (2017, July - September). Policy Planning in Higher Technical Education: A System Dynamic Approach. *International Journal of System Dynamics Applications*, *6*(3), 87–110. Advance online publication. doi:10.4018/IJSDA.2017070105

Sternman, J. D. (2000). *Business dynamics: systems thinking and modeling for a complex world.* McGraw Hill Publications.

Vahdatzad, M. A., & Mojtahedzadeh, M. T. (2000). Some Issues in the Strategic Management in a fast growing academic Institution: The Case of University of Yadz. *Proceedings of the 18th International Conference of the System Dynamics Society.*

Watson, D. (2000, January). Managing in Higher Education: The 'Wicked Issues'. *Higher Education Quarterly*, *54*(1), 5–21. doi:10.1111/1468-2273.00142

Yujing, W., Yongkui, L., & Peidong, G. (2015). Modelling Construction Project Management Based on System Dynamics. *Metallurgical and Mining Industry*, *9*, 1056 – 1060. Available at: https://www.metaljournal.com.ua/assets/Journal/english-edition/MMI_2015_9/0166Yujing-Wang.pdf

APPENDIX

Library Service Data

Type Public University Private University
abv_cost_of_db 10000 10000
av_cost_of_book 130 130
av_sal 3500*12 4000*12
initial_number_databases 50 100
initial_number_of_staff 30 30
initialnumber_of_books 8400 10000
number_of_students 4200 2300
staff_leaving 0.3*number_of_staff
total_budget 2066000 4460600
wb 0.3 0.3
wdb 0.3 0.25
ws 0.4 0.45

Chapter 6
Policy Analysis in Technical Education: A System Dynamic Modelling

Sanjay Soni
Jabalpur Engineering College, India

Basant Kumar Chourasia
Jabalpur Engineering College, India

ABSTRACT

Higher technical education is one of the indicators of national growth. Higher technical education of the country is responsible for its global progress in technology and standard of living as technical advancement helps to make the country economically progressive as the availability of quality technical staff attracts global entrepreneurs who, in turn, increase the revenue of the nation and therefore the higher technical education system is the backbone of the country's economic progress. In the present research paper, India's higher technical education system is being studied, which is currently undergoing uneven expansion due to the implementation of the various policies responsible for this plethora of expansion. The main objectives of this study are to improve the quality of higher technical education by developing a dynamic system model by incorporating some of the important parameters of higher technical education system stakeholders to test various long-term policies that can improve the quality of higher technical education.

INTRODUCTION

The massive Expansion in the engineering Institution across in India during last decades has threatened the standards and quality of technical education system. This irrational growth has resulted in unemployment among engineering graduates and poor academic environment which has lead to closure of sub-standard institutes. In such a situation quality only remains the hope for sustaining in such a competitive

DOI: 10.4018/978-1-7998-5788-4.ch006

environment (Ucharia, 2015).On the other hand growth of global economy has increased opportunities for countries with good levels of Technical education

(Carnoy, 1999). In the global competitive environment of today, an accessible and high quality of higher education system is imperative for a nation's economic progress. A sound higher education system supports and enhances the process of economic and social development for a better future growth of nation. Focus on quality is urgent need for bringing academic excellence and for development of technical manpower for Industry 4.0 requirement for future.

1.1 Overview of Technical Education System in India

The higher education system in India grew rapidly after independence. By 1980, there were 132 universities and 4738 colleges in the country enrolling around five percent of the eligible age group in higher education. Today, in terms of enrollment, India is the third largest higher education system in the world, behind China and the U.S.A. China having the highest enrolment in the world (nearly 23 million) is organized in only about 2,500 institutions. While the average enrolment in a Technical institution in India is about 500-600 students with 6223 technical institutions (source, A.I.C.T.E.).This makes the system of Technical education in India a highly fragmented one that is far more difficult to manage than any other system of higher education in world. However, other than a handful few institutions of national importance providing high quality higher education, the system is failing to produce quality graduates who are much needed in all sectors of society (NAASCOM, 2006). India's technical education has failed to map the future demand for various skills and it has not kept pace with industry's growth (Khemani, 2006).Since higher education in India is an important part of modern Indian society and is intertwined in the political and social systems of the society, this sector is in need to change, development and improvement (Albatch,1993).

1.2 Challenges Faced by Technical Education in India

In India, the entry of the private sector in Technical education is on massive scale and is largely driven by inadequacy in government funding in the education sector, paving way for private investment. This is due to liberal policies in early 90's. With the present growth in GDP it is expected that India's economy will surpass Italy by 2015, France by 2020, Germany by 2025 and Japan by 2035.It has been predicted that India, being driven by Knowledge economy, will become the 3rd largest economy by 2050 (Varghese, 2007). Technical education in India is more privatized than any advanced countries. A large number of Institutes are private and self-financed as prestigious Institutions are not able to accommodate all aspirants (Source, A.I.C.T.E.2012). Most private Institutes concentrate on profit making rather than on imparting quality education to students by limiting the essential requirements needed to fulfill the curriculum, Pay structure variations among faculties makes there retention and availability difficult. The total admission in Engineering institutions went down from 17.53 lakh in 2012-13 to 13.84 lakh in 2018-19, a reduction of more than 20% in 7 years. The placement data on the AICTE website is shocking. al Available data suggests that only an average of 40 per cent of the total number of engineering graduates is getting placed. To propel the nation in 21st century there is the need for planned change in higher education with realistic perception for policy formation as to what is possible and what is not.

1.3 Views on Quality in Technical Education

Quality in Technical Education is defined as, The totality of features and characteristic of a product or service that bear on its ability to satisfy stated or implicit needs (B.S.I., 1978).It is further defined as development of intellectual skills and knowledge that will equip graduates to contribute to society through productive and satisfying engineering careers as innovators, decision makers and leaders in the global economy (Natrajan, 2000). Education quality is a multi-dimensional concept (Eriksen, 1995). It can be viewed as the combination of the quality of input, process, and output of the education system. Many opinions can be observed in the literature about the factors influencing the quality in engineering education. Some of them are teaching process, University – Industry collaboration (Natarajan, 2003), role of management (Gopalan, 2003), student intelligence & interest (Mouly, 2003,), excellence of teachers (Shrivastava, 2003), accreditation standards (Vrat, 2003), e-education (Maji,2003) and proper documentation of activities (Jagdeesh,2001). For some of the academicians, quality is synonymous with the continuous improvement (Bhat, 2000), selected for identifying the barriers to the quality improvement of engineering education in India.

Figure 1. Growth of Enrolments in Technical Education (Source, A.I.C.T.E.)

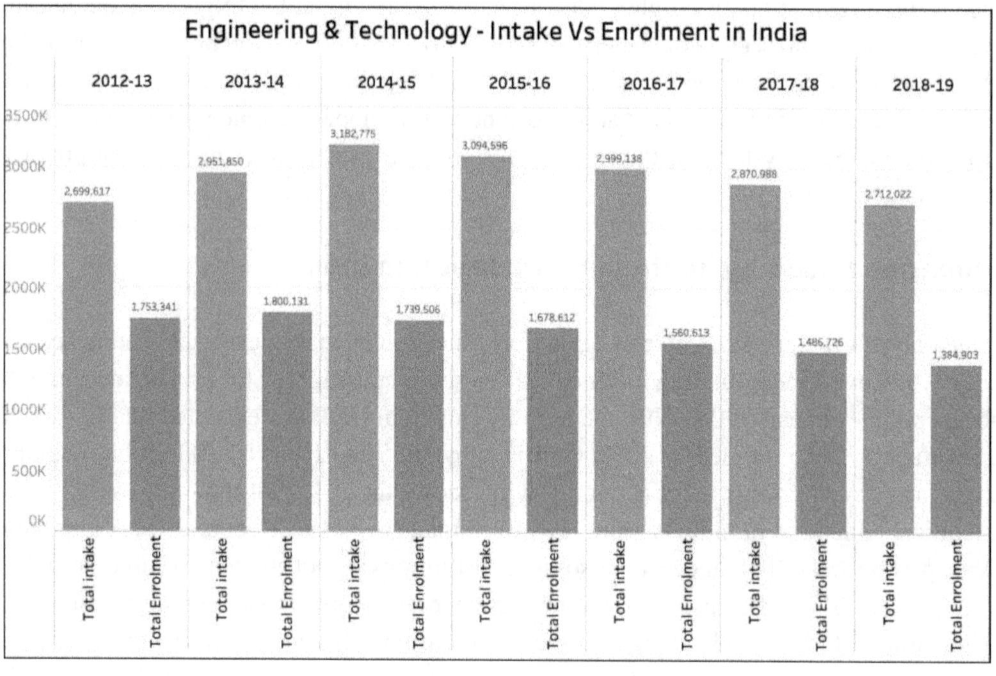

2. LITERATURE REVIEW

The expansion of engineering education in the countries where the economy is under fast growth rate has increased rapidly as it has direct bearing on the economy of the country. This is especially true as far as India is concerned. But maintaining a balance between supply and demand is always a challenging task.

Unless the engineering institutes produce engineers who are employable in the market or to be entrepreneurs, most of the engineering institutes will be agents for enhancing the population of unemployed engineers (Honnutagi, 2012).Technical education system is of paramount importance in generating the technical manpower required for building a strong nation. The demand for engineers has been growing in India during the last few decades. This has resulted in exponential quantitative expansion of technical institutions without qualitative standardization. People from different backgrounds have been motivated by huge profit gains from establishments of these technical institutes. Although with the ongoing process of liberalization and globalization of the economy, more and more foreign universities are also trying to find the foothold in the Indian educational space by entering into joint ventures with Indian universities. But in the present scenario technical education is suffering from poor quality outputs of major engineering graduates. Due to inadequate planning and sudden expansion in technical education chaotic scenario has been created where qualitative standards are badly hampered and are resulting in unemployable technically qualified man power growth. Significant increase in the number of seats across technical institutes, poor student faculty ratio, lack of industry institution partnership, lack of trained and qualified teachers was seen as the main reasons for quality depletion of technical education system (Pant, 2013).

Figure 2. Passed and Placement in Technical Education (Source, A.I.C.T.E.)

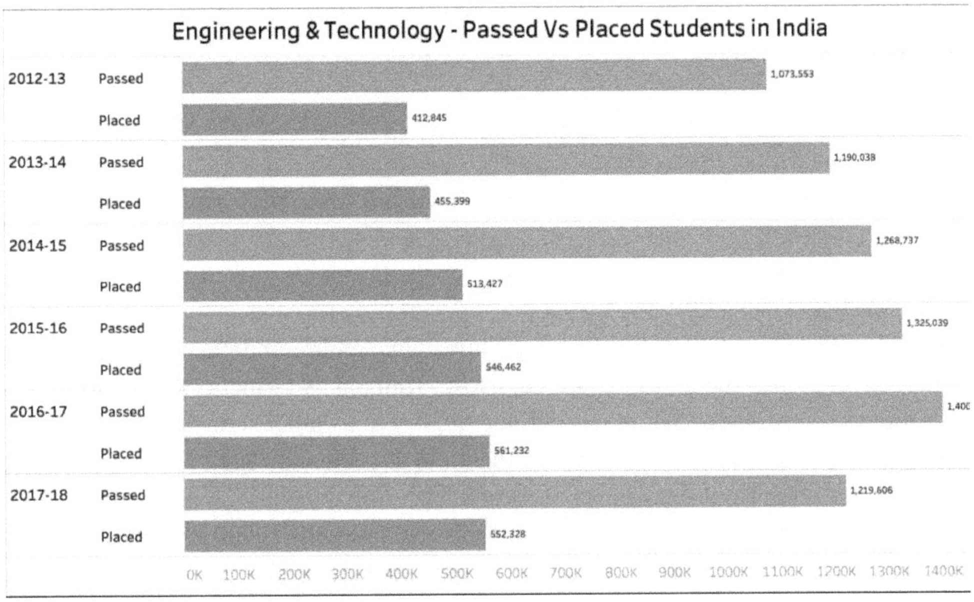

2.1 Quality in Education

Defining quality in education is a massive challenge since it deals with the most sensitive creation on earth the human beings. (Cropley, 2003) suggests that education quality is a rather vague and controversial concept. At the broadest level, education quality is a rather vague and controversial concept. At the broadest level, education quality can be viewed as a set of elements that constitute the input, process and output of the education system, and provide services that completely satisfy both internal and external strategic constituencies by meeting their explicit and implicit expectations. If higher education is viewed as a system, then any quality management program must therefore assess inputs, process and outputs.

2.2 Contextual Perspective

Lindsay (1992) maintains that indeed, the notion of quality in technical education has no agreed technical meaning and its use usually involves a heavy contextual overlay of some political or educational position. For example, reference to the quality of research, student's supervision, assessment, student intake, academic programs, teaching and learning, and program designs are not uncommon. Any attempt to define or attach meaning to the term is largely ignored and one is left to assume that it is high quality that is being referred to as opposed to good or poor quality (Oyo, 2010).

2.3 Stakeholders Perspective

From the stakeholder's perspective, Harvey and Green (1993) highlight the importance and value of considering quality from a variety of stakeholder perspectives. The different HE stakeholders, i.e., government, quality agencies, universities, individual academics, students, employer, parents and society, have the potential to think about quality in different ways. Similarly, five aspects of quality, including: exception, perfection, fitness for purpose, value for money, and transformation, are attributed to stakeholder's perspective. Watty (2003) further elaborates on these aspects as follows:

- Exception: distinctive, embodied in excellence, passing a minimum set of standards.
- Perfection: zero defects, getting things right the first time.
- Fitness for purpose: relates quality to a purpose, defined by the provider.

Quality is a function of competencies in terms of knowledge, skills and attitudes, capacity and competition vis-à-vis needs (Goel and Biswal, 1996).

Efficiency in meeting the set goals, relevance to human and environmental needs and conditions and something more in relation to the pursuit of excellence and human betterment (Rajput and Walia, 1997).

The US National Science Foundation has defined quality in technical education as the development of intellectual skills that will equip graduates to contribute to society through productive and satisfying engineering careers as innovators, decision-makers and leaders in the global economy of the twenty- first century (Natarajan, 2000).As in manufacturing and service industries quality is the hallmark of excellence and effectiveness in engineering education. Every engineering college and polytechnic should define their quality policy and articulate their commitment to achieve quality in all their activities and implement the policies energetically (Natrajan, 2000).Quality assurance is a process-driven approach with specific steps to help define and attain goals. This process considers design, development, production, and service. The most popular tool used to determine quality assurance is the Shewhart cycle, developed by Dr. W. Edwards Deming. This cycle for quality assurance consists of four steps: Plan, Do, check, and Act. These steps are commonly abbreviated as PDCA.An Internal Quality Assurance system under which students, staff and management satisfy themselves that control mechanisms are working to maintain and enhance the quality (Mani and Uma Devi, 2009).The quality assurance strategies that are appropriate for virtual education share common features with other forms of media, but there are also differences. The range and flexibility of information and communications technologies create new opportunities, but also give rise to complexities and challenges for governments, agencies, institutions and faculty (Robin Middlehurst, 2001).

QA handbook (2004) provides 10 tips for quality assurance for digital library such as document your policies, ensure your technical infrastructure is capable of implementing your policies, and ensure your technical infrastructure is capable of implementing your policies, ensure that you have the resources necessary to implement your policies, implement systematic checking procedures to ensure your policies are being implemented, keep audit trails, learn from others, share your experiences, seek "fitness for purpose" not perfections, remember that QA is for you to implement, seek to deploy QA procedures more extensively.

Education committee of Oxford University (2008) suggests major areas of quality assurances or quality enhancement such as admission, induction, and course design, student feedback, student complaints and appeals, statistical information, external input, quality enhancement in learning and teaching, monitoring of teaching, postgraduate research degrees, collaborative provision and placement.

2.4 System Dynamic Modeling

System dynamics is a computer based modeling approach for analyzing and solving complex problems through policy design and analysis. The problems addressed by SD are based on the premise that the structure of a system, that is, the way essential system components are connected, generates its behavior (Sterman,2000). If dynamic behavior arises from feedback within the system, finding effective policy interventions requires understanding system structure. Once a model is built, it can be used to simulate the effect of proposed actions on the problem and the system as a whole. As Forrester (1998) notes, this kind of tool (System Dynamic model) is necessary because, while people are good at observing the local structure of a system, they are not good at predicting how complex, interdependent systems will behave. System Dynamics has been used to handle big data (Pruty, 2016), Model for subsistence farmer food security (Oyo, 2016), Modeling public transportation (Bajracharya, 2016), Societal Agining (Logtens, 2015), Social economical advancement analysis (Baporikar, 2016). Neural Network techniques in electromyography (Azar, 2012), Modeling complex systems (Azar, 2012), Chaotic Maps (Azar, 2017), Control System Design (Azar, 2017).

2.5 System Dynamics Models on Higher Education

2.5.1. Teaching Quality

Kennedy (2002) provides a comprehensive taxonomy of publications describing various system dynamics models on higher education issues. The brief details of investigations of the models and key findings are presented in the following section

2.5.1. 1.Quality, Pedagogical and Socio-Economic Aspects of System Dynamics in Schools

The earliest teaching of system dynamics was almost exclusively to graduate students but relatively early the potential of teaching system dynamics to school children was appreciated. Foster (1972) considers the impact of 'Education in the city' as an extension of the Urban Dynamics programme. As such this included the socio-economic impact of education on a community.Roberts (1976) describes an early

"System Dynamics Curriculum Development Project for Elementary and Secondary Education". As well as examining some practical pedagogical aspects of system dynamics in schools she critically examines Blooms Taxonomy of thinking skills (Anderson, 1997). More recently there has been an increasing interest in assessing the effectiveness of system thinking in the classroom.

2.5.1. 2.Quality, Pedagogical and Socio-Economic Aspects of System Dynamics in Universities

The information management and modeling research group (IMMAGE) have developed an initial system dynamic model to examine quality management issues at London South bank University Kennedy (1998a,1998b). Interview was conducted with academic members of staff to guide the construction of the model. This investigation is considered to be the first part of a long term project.

2.5.1. 3.Key Findings

- The identification of sectors, e.g. Administration, staff Performance, Department Effectiveness, Funding, Research and Funding, Research and funding, needed to be considered for a future quality management model.
- The identification of metrics (or performance indicators) needed to be collected for further system dynamic investigations.

In an ambitious project somewhat reminiscent of Roberts (1976) earlier work in schools Eftekhar & Strong (2005) examine the process of learning in colleges & universities and outline some aspects of the debate among experts in education as to the most effective approaches to influence or reinforce the learning process. This work is an ambitious undertaking that implements a programme that includes the combination of education metrics and computer simulations.

The potential value of system dynamics modeling in addressing higher education quality issues is in its ability to:

- Model feedbacks or interactive views in dynamic systems like higher education.
- Incorporate non- linear relationships inherent in higher educational quality issues.
- Address complexity situations while experimenting their behavior over time.
- Accommodate soft factors such as effectiveness of student's project supervision, staff competence, quality of staff, and quality of research, teaching that underpin higher education quality issues.
- Model time delays that underpin certain policies on quality, e.g. time to recruit new staff, durations of study programmes, staff on training, executing research projects and investment in new student capacity.

2.6. Features of SD Methodology

Six features of SD have been identified by Schwinger (2006).These are further discussed here.

Focus on feedback-driven, mainly internally generated dynamics: The systems modeled are networks of closed loops of information. However, they are not limited to the representation of closed systems, in that

A. Flows can originate from outside the system boundaries.
B. Exogenous factors or systems can be incorporated into any model parameters or special modules.
C. New information can be accommodated via changes to a model.

High degree of operationality: System dynamics relies on formal modeling. This fosters disciplined thinking, assumptions underlying equations and quantifications to be clarified. Feedback loops and delays are visualized and formalized: therewith the causal logic inherent in a model is made more transparent and better discussable than in most other methodologies.

Far- reaching possibilities for the combination of qualitative and quantitative aspects of modeling and simulation. The focus is not on point –precise prediction, but on the generation of insights into the patterns of behavior generated by the system under study.

High level of generality and scale robustness: The representation of dynamical systems in terms of stocks and flows is a generic form which is adequate for an enormous spectrum of potential applications. This spectrum is both broad as to the potential subjects under study, and deep as to the possible degrees of resolution and detail.

Availability of powerful application software: The packages (Stella /I think, Powersim, Venesim and My Strategy) are easy to handle and give access to a high variety of mathematical functions. Part of this applications array offers optimization procedures and validation tools. Also, some support for collaborative modeling and the communication with databases is provided.

Potential synergies: Combination with many other tools and methodologies is possible, both conceptually and technically.Given the diversity of system dynamic strength, an equivalent application area of system dynamic as already highlighted exists. The emerging trend of mixing other methods with system dynamics further confirms the superiority of system dynamic features. Examples of the latter include: dynamic synthesis methodology- involving case study and system dynamic (William, 2002, 2004), soft systems dynamics methodology-involving soft systems methodology and system dynamics, managing from clarity (Rabbino, 2001), group model building, modeling with clients (Vennix, 1999,Anderson et al,2007), action research-system dynamic integration (Scholl, 2004), collaborative systems modeling-GMB integration (Hoppenbrouwers,2008),object role modeling-system dynamic integration (Tulinayo et al, 2008).

2.6.1. Technical Education a Closed Loop System

Any standard industrial activity includes three different stages, Such as the input, the processes and the output, where feedback closes the loop. In this process, feedback gained from the output can be utilized to improve the quality of the process. This model has also been adopted for the quality assessment of education structures (see Figure 2). An educational institution as a system has inputs such as students, infrastructure, financial resources and instructional resources. The processes are Teaching and learning, instruction, evaluation, etc. While the outputs are the graduates and their behavioral, academic and physical qualities. An institution has a well-defined boundary and an environment in which it is set. Another they are interlinked and interdependent in systemic framework.

It can be seen from Figure 3 that without feedback links the above structure it will not be able to capture the dynamic closed loop representations of the technical education system. Therefore for proper assessment of policy measures implemented in technical education system. Accountabilty of feedback interplay are necessary for realistic assessments of dynamics of the system.

Forrester (1989) considers shortcomings in US school education and suggest that system dynamics could from a more satisfactory basis for high school education and proposes a programme of action in order to realize this potential.

2.6.2. Sub-Systems of Technical Education System

A system obviously comprises several sub-systems. The various sub- systems of Technical education system are faculty module, student's module, academics module, and administrative module (Narula, 1992) (see Figure 4).

2.7. Closed Loop Interactions

Technical education system is a complex system in which commotion in one factor impacts other factor which in turn may impact some other factor until they form a structure of closed loops to influence the factor which started the commotion. The interrelationship in the sub-systems can be studied with the help of simple cause and effect diagram called causal loop diagrams of various modules and their impact on quality can be studied. The loop clearly depicts the causal explanations.

Figure 3. Standard Industrial Activity Structure

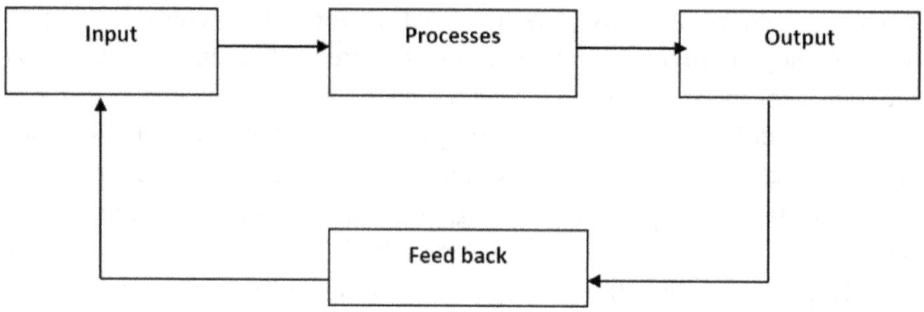

Figure 4. Structure for Quality Assessment in Technical Education System

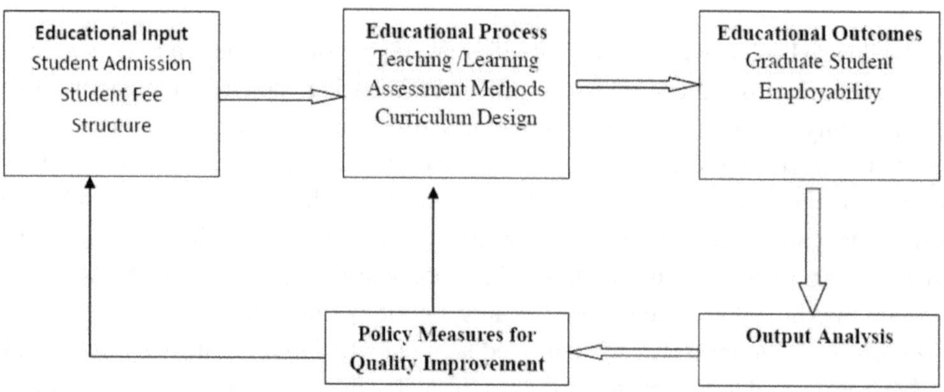

Educational institutions including the national Open University should be ethically and legally stopped from running courses and programmes in which they do not have core faculty. The use of teachers as academic managers and hiring of part time consultants has been helping many to generate resources, but quality is definitely being compromised. In open and distance learning system, the mechanism of quality assurance in India should include recognition and accreditation, institutional and programme assessment, process as well as product assessment (Mishra,2007).

Venkataram and Giridhran has proposed technical education quality assurance and assessment (TEQ-AA) system that uses mobile agents to access the web pages of the technical institutions or colleges and collect relevant quality measurement information. The collected information is analyzed and deducted into the qualitative facts of the institution, standard measurement by using (RARM) Realistic Abductive Reasoning Model (Venkataram & Giridharan, 2007).

Quality assurance can be achieved through accreditation of the institutes, periodically running faculty development program research and institutional development .But there are lots of challenges such as inequity in the number of diploma level institutions vs. degree, absence of vertical mobility for ITI to doctoral levels, Availability of quality and competent faculty, number of PhDs in technical education, outcomes based accreditation criteria, flexible learning, transitional education and distance learning etc. before the institution to go for quality assurance (Prasad,2009).

2.7.1. Reinforcing Loop of Faculty Module (R1)

This loop depicts the feedback of faculty module of higher technical education system which shows that as quality of institute in numbers increases teacher student ratio increases which affects positively sanction posts of faculty in the institute which increases appointment in the institute which again increases the teacher student ratio which again increases the quality of institute in numbers (see Figure 5).

The loop B1 is the balancing loop of faculty module which depicts as quality of institute in numbers increases guest faculty appointment are decreased and as guest faculty appointment increases the regular faculty dropout increases which results in decreases in teacher student ratio which again decreases quality of institute in numbers.

Figure 5. Subsystems of Technical Education System

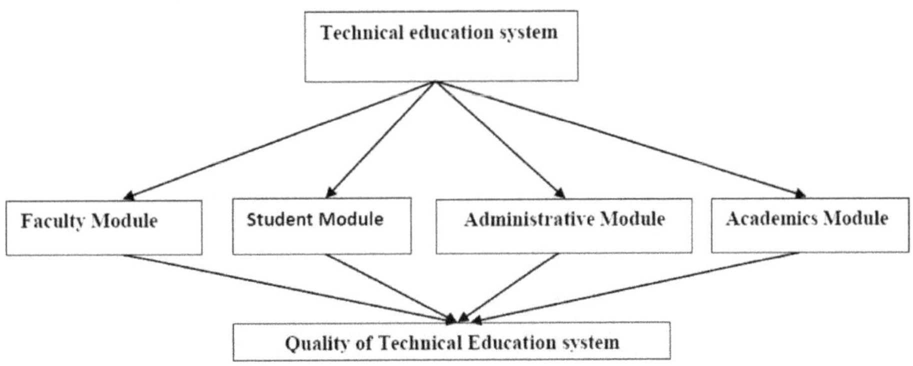

Figure 6. Cuasal loop Diagram

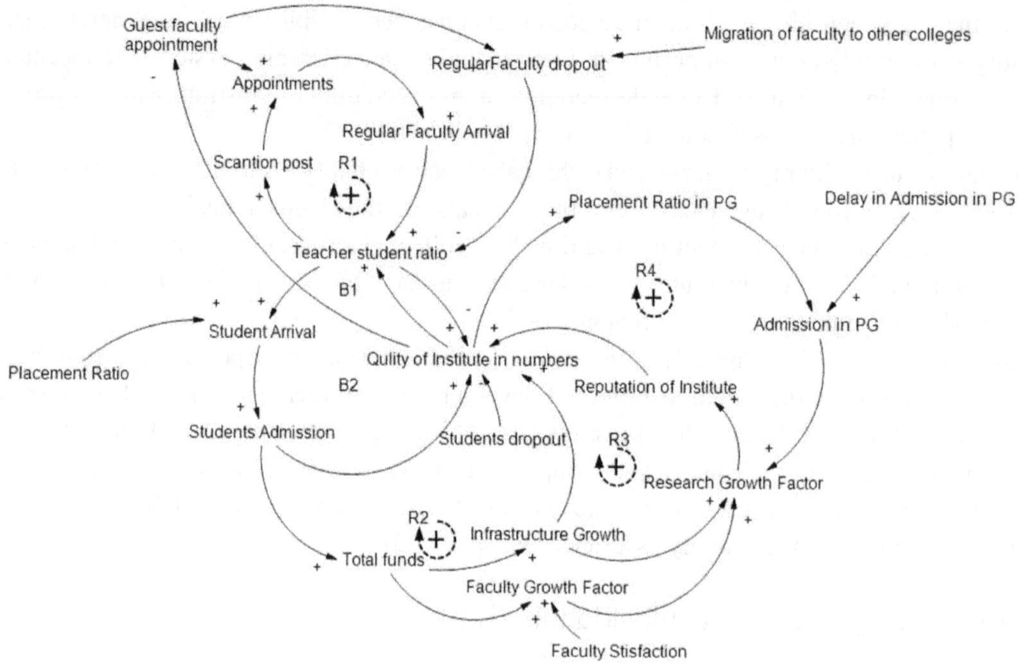

2.7.3. Balancing Loop of Student Module (B2)

This loop shows the feedback of student module of higher technical education system which depicts as quality of institute in numbers increases the teacher student ratio increases which affects the student admission positively which increases student arrival which decreases quality of institute in numbers. Student dropout an exogenous factor decreases the quality of institute in numbers.

2.7.4. Reinforcing Loop of Administrative Module (R2, R3)

Reinforcing loop R2 and R3 captures the feedback of student module as the quality of institute in numbers increases the teacher student ratio increases which increases admissions of students in the institute which increases the funds generated from students which increases infrastructure growth of the institute which increases the quality of institute in numbers (R2) and as total fund increases it increases the faculty growth factor which is influenced by faculty satisfaction which is an extraneous factor which increases the reputation of the institute which increases quality of institute in numbers (R3).

2.7.5. Reinforcing Loop of Academic Module (R4)

Reinforcing loop R4 depicts the feedback of Post graduation module of higher technical education system which indicates that as quality of institute in numbers increases the placement ratio in post graduation increases which increases the admission in post graduation which affects reputation of the institute positively which results in increase in quality of institute in numbers (R4).

2.8. Development of Stock and Flow Diagram

Based on the above causal loop diagram stock and flow diagram is developed for policy planning in higher technical education system. The flow diagram for faculty module is shown in Figure 6 and is discussed briefly

2.9. Faculty Population Module

Total Number of Faculty Working and Joining in an Organization in any given year is Estimated in this module. Three things are considered while estimating total Faculty Population.

1. The Number of Regular Faculty Arrivals every year.
2. The Number of Faculty Departures every year.
3. The Number of drop out Faculty per year.

Faculty Population Modules flow diagram containing variables symbols (Levels, Auxiliaries and Constants) flows and links along with respective equation is described in following discussion of chapter.

Figure 7. Flow diagram of faculty Module

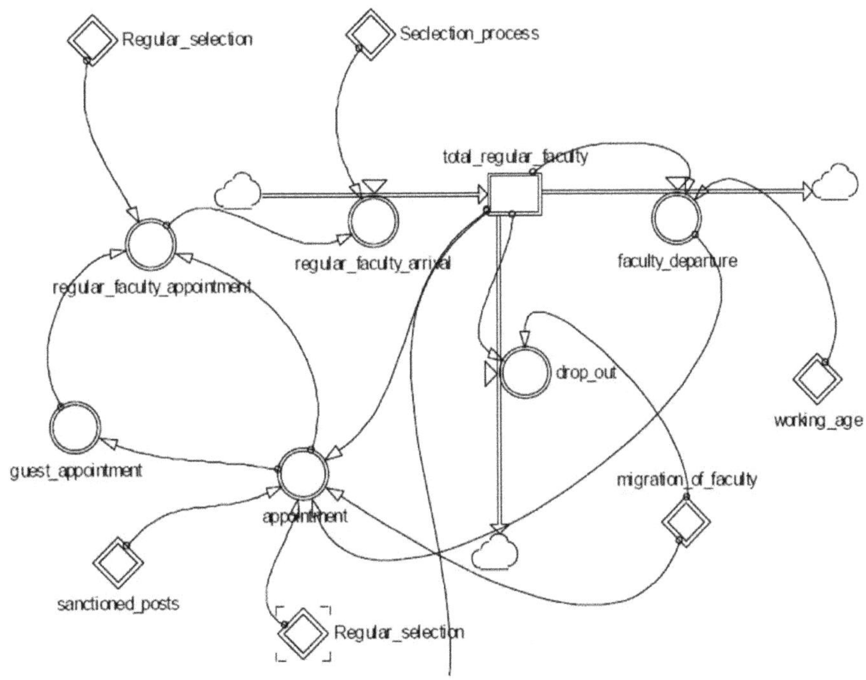

2.10 Total Faculty

Total numbers of faculty working at various positions in an organization in any given year is represented by the level variable total faculty. (Figure 1). It is an array variable which contains information regarding various faculty positions occupied all together in an institute in all the departments, i.e. (Professors, Associate Professors, and Assistant Professors). A level in a model represents accumulations in a system. They are simulated by adding inflows to the initial values and subtracting outflows from them. These computations are made in every time step in the model, so that the value of the level at any point in time equals the value from the previous time step plus any inflow and minus any outflow in the current time step. Each time step corresponds to a period of one year. Inflow to the level is represented by rate variable Faculty arrival. It is influenced by auxiliary variable faculty recruitment every year and also by the level variable to take into account. Outflow is represented by the rate variable faculty departure and drop out faculty. Faculty departure estimates departure of faculty which is retried every year after completion of its full service tenures where drop out faculty are those faculty which are switching to another organization every year.

Level equation is Dimension Total Faculty Initial Total Faculty = 0

Level total-Faculty = +dt* faculty arrival – dt* faculty departure –dt* dropout faculty. Faculty arrival = Number of Faculty joining the institute. (Number of Persons/Year).

Faculty departure = Number of Faculty retired from the Institute. (Number of Persons /Year). Drop-out = Number of Faculty leaving the Institute and Switching over another Organization.

(Number of Persons/ Year).

2.11. Faculty Arrival Rate

It is represented by faculty arrival in the model. It represents arrival of faculty in all the departments of the institute in various sanctioned posts (i.e. Professors, Associate Professors, and Assistant Professors) every year. It is an array variable as it holds more than one variable at the same time that is Professors, Associate Professors, and Assistant Professors. Faculty arrival rate is influenced auxiliary array variable faculty appointment every year and guest faculty appointment every year. Faculty arrival rate is interpreted as:

Faculty arrival (Number of Persons/Year) = Regular Faculty appointment*Selection Process.

2.11.1. Selection Process

It is the process for filling vacant faculty positions in the Institute. It is a review process as to how many seats are to be filled and it is assumed that it is at least once a year. The computation unit is 1/year.

2.11.2. Regular Faculty Appointment

Regular faculty appointment is an array auxiliary variable which computes regular appointments in the Institute on the posts of Professors, Associate Professors and Assistant Professors. It is affected by constant array variable regular selection, auxiliary variables Guest appointment and appointment. It is computed as:

2.11.3. Regular Selection

It is a constant array variable which computes the total regular selection in institute on the post of Assistant Professor, Associate Professor, and Professor in the particular years. Its unit is Number of Persons/year.

2.11.4. Guest Appointment

It is an auxiliary array variable which computes total appointments on period basis in the case when regular appointments are not done for the vacant faculty positions. It is affected by auxiliary variable appointment.

2.11.5. Appointment

It is an auxiliary array variable that computes how many regular appointments are to be made in the institute. It is affected by constant array variable sanctioned posts and regular selection and is computed as:

2.11.6. Sanctioned Posts

It is a constant array variable that computes total sanctioned posts of teaching faculty in the institute as per A.I.C.T.E.

2.11.7. Faculty Departure Rate

It represents flow out from the level variable total faculty. As it is outflow from level variable, it is responsible for reduction of faculty population. Faculty departure rate is an array variable (i.e. at the same time handling more than one variable that is departure of Professors, Associate Professors, and Assistant Professors of the Institute). It is influenced by constant variable service period of faculty and level variable total faculty. Faculty departure is computed as

Faculty departure (Number of Persons/Year) = Total Faculty/service period of faculty

2.11.8. Faculty Dropout Rate

It represents flow out from the level variable total faculty. As it is outflow from level variable, it is responsible for reduction of faculty population. Faculty dropout rate is an array variable (i.e. It is holding three variable at the same time. Assistant Professor, Associate Professor and Professor) which calculates the number of faculties that have left the institute before completing their full tenure of service length. It is affected by constant array variable migration of faculty and level variable total faculty. Faculty dropout rate is calculated as:

Faculty drop out (Number of Persons/year) = Migration of Faculty per year * Total Faculty

Migration of Faculty Per Year: It is a constant array variable which calculates migration of faculty (Professors Associate Professors, Assistant Professors) from various departments of the institute before completing full tenure of service. The units of measurement for migration of faculty are (1/Year). It is the average departure rate per year as percentage of total faculty.

2.12. Description of Stock and Flow Diagram of Student Module

Based on the causal loop the flow diagram for student module is shown in Figure 7 and is discussed briefly as:

2.13. Student Population Module

Total number of student enrolled and joining the institute in any given year is estimated in this module. Three things are considered while estimating total students.

The number of student's arrival every year.
The number of students departure every year.
The drop out students every year.

2.13.1. Total Registered Students

Total number of students registered in the institute in all branches of engineering in any given year is represented by level variable. It is an array variable which contains information regarding registered students of various departments running by the institute. (Civil, Electrical and Mechanical). Levels in a model represent accumulations in the system. Levels are simulated by adding inflows to the initial values and subtracting outflows from them. Levels are the state of the system being modeled. These computations are made in every time step in the model. In flows to the level is represented by rate variable student arrival which is influenced by auxiliary array variable admission per year and also by the level variable total registered student. Outflow to the level are represented by the rate variable student departure and drop out students.

Level equation is:

Dimension Total registered students Initial Total registered students = 0

Level Total registered students = +dt* students arrival – dt* student departure –dt* dropout students. Students arrival = Number of students joining the institute. (Number of persons /Year).

Students departure = Number of students passing out from the Institute after completion of their full tenure desired for graduation (Number of persons /Year).

Drop-out students = Number of students leaving the Institute in the middle of graduation program after registration without completion of their full tenure desired for graduation (Number of persons /Year).

2.13.2. Student Arrival Rate

Student arrival is array rate variable which estimates the total number of enrollments in the institute in all branches in which graduation programs are running. It is an array variable as it calculates number of students arrived in all the branches in which graduation programs is running in the institute i.e. (Civil, Electrical and Mechanical Engineering). Student arrival rate is affected by auxiliary array variable admission per year (Number of persons/year) and array auxiliary variable admission cancellation per year (Number of persons/year). Student's arrival is estimated as:

Students arrival (Number of persons/year) = Admission per year-Admission cancellation per year.

Figure 8. Flow diagram of faculty Module

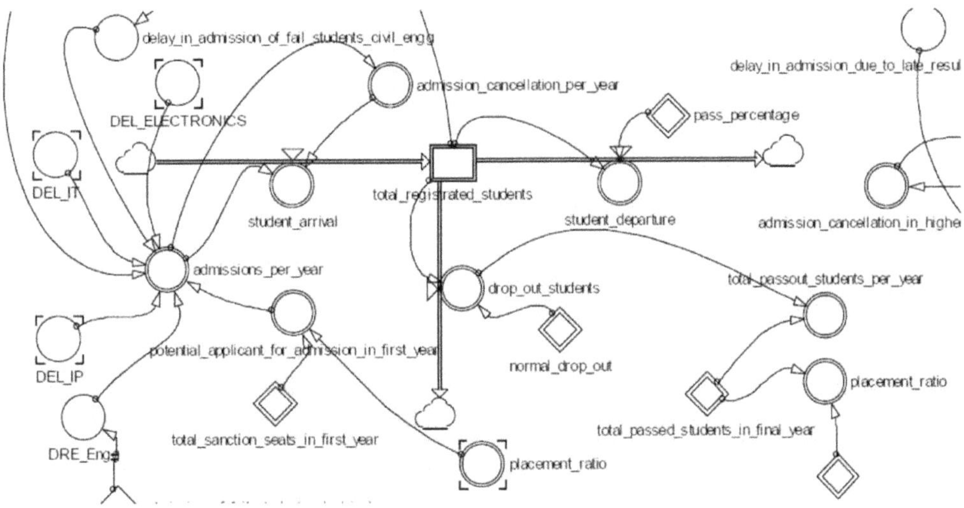

2.13.3. Student Departure Rate

It represents flow out from the level variable total registered students. As it is outflow from level variable, it is responsible for reduction of total registered student's population. Student departure rate is an array variable (i.e. at the same time handling more than one variable that is departure of students from all branches i.e. (Civil, Electrical and Mechanical Engineering) in which graduation programs are offered. It is influenced by constant array variable pass percentage of total registered per year and total registered student registered in the institute out student per year

Student Departure Rate is computed as

Student departure (Number of students/year) = Total registered (Number of students)* pass percentage (1/year).

3. VALIDATION OF MODEL

Validation is the process of establishing confidence in the soundness and usefulness of a model, (Forrester and Sege, 1980) validation begins as the model builder accumulates confidence that a model behaves plausibly and generates problem symptoms or modes of behavior seen in the real system. Validation then extends to include persons not directly involved in constructing the model. Thus validation includes the communication process in which the model builder must communicate the basis for confidence in a model to a target audience. Unless the modeler's confidence in a model can be transferred, the potential of a model to enhance understanding and lead to more effective policies will not be realized. Validation of system dynamics model is complicated by the many relevant audiences, each having its own objectives and criteria for evaluating a model. For a scientist, a model may be considered useful if it generates insight into the structure of real system, makes correct predictions and stimulates meaningful questions for future research. For the public and political leaders a useful model should explain causes of important problems and provide a basis for designing policies that can improve behavior in the future. Model testing is part of a larger validation process. There is no single test which serves to validate a system dynamics model. Rather confidence in a system dynamics model accumulates gradually as the model passes more tests and as new points of corresponds between the model and empirical reality is identified. The validation sequence for the proposed model follows the stepwise progression suggested by Mohapatra (1994).It is known and broadly accepted amongst those who have developed system dynamic model with the issue of validation of system dynamics model, that there is no such thing as absolute validity, only a degree of confidence can be attained, which becomes greater as more and more tests are passed.

3.1. Model Structure Verification Test

Verifying structure means comparing structure of a model directly with structure of the real system that the model represents, therefore these tests were done simultaneously at every stage of model development to detect any structural flaw. Initially the model had to make the estimation of total faculty in the institute from the record of total faculty joining the institute each year and leaving the institute every year and the estimation of total number of students joining the institute from the record of total students joining the institute each year and leaving the institute each year. The stock and flow diagrams comprising of flow representing total faculty appointed each year and total new admissions and total faculty and total registered students as level variables and delay in student's readmissions are able to capture it. This is verified from the results obtained by running the model which is presented in Tables 1 and 2 below.

3.2. Parameter Verification Test

Parameter verification means comparing model parameters to knowledge of the real system to determine if parameter corresponds conceptually and numerically to real life Senge (1980). The model consists of parameters which are building blocks of various modules in the model. Total faculty population at various positions, Guest faculty appointments, Service period of faculty, Total registered students in different streams, Total campus placements, Total registered students in different streams of PG, Total campus placements in PG, Total admissions, Cancellation of admissions, Students attraction factor, Total classes scheduled, Total classes conducted, Total passed students, Total students scoring first division, Students satisfaction, Faculty satisfaction, Institute growth factor. Of all the mentioned parameters the last are easily verifiable in the real system. Parameters values used were numerical data of Jabalpur Engineering College. Excel sheets in the model presents parameters and their values.

3.3. Extreme Conditions Test

The extreme conditions test boots model validity by examining model structure for extreme conditions, one develops confidence in a model ability to behave plausibly for a wide range of conditions and thereby enhances the model usefulness to explore policies that move the system outside of historical ranges of behavior (Forrester and Senge, 1980). The present model on technical education system captures the impact of various parameters on the quality of technical education. The faculty module studies the impact of faculty population on quality of technical education. In the model it is assumed that all the faculty positions are occupied and the regular vacant faculty positions are filled by the guest faculty. Faculty dropout has impact on quality index of technical education. The extreme condition tested in the model is if the situation comes that there is zero number of faculty and zero number of students in the institute than the model results as zero value for quality index of technical education which is similar to the behavior of a real system. Similarly in the present model it is modeled that dropout students and faculty dropout have impact on quality in numbers of technical education. The extreme condition tested in the model is that if there occurs a situation when there is very large number of faculty and students drop out in the institute than the model results as lower value for quality index of technical education which is the similar to the behavior of a real system (see Table 3).

Table 1. Trends of Total Faculty population in the Institute during various Years. 1, 2, and 3 Are Professors, Associate Professors and Assistant Professors

Year	total faculty 1	total faculty 2	total faculty 3
2004	22	21	25
2005	22	21	25
2006	22	21	25
2007	22	21	25
2008	22	21	25
2009	22	21	25
2010	22	21	25
2011	22	21	25
2012	22	21	25
2013	22	21	25
2014	22	21	25
2015	22	21	25
2016	22	21	25
2017	22	21	25
2018	22	21	25
2019	22	21	25
2020	22	21	25
2021	22	21	25
2022	22	21	25
2023	22	21	25
2024	22	21	25
2025	22	21	25

Table 2. Trends of Total Students Registered in the Institute in Various Years 1, 2, and 3 shows Students population from Civil, Electrical and Mechanical Engineering Departments

Year	total Students 1	total Students 2	total Students 3
2004	0	0	0
2005	68	98	97
2006	69	74	118
2007	44	79	91
2008	60	84	105
2009	72	81	95
2010	74	94	108
2011	80	82	80
2012	78	82	95
2013	78	84	89
2014	74	79	90
2015	83	86	95
2016	76	80	90
2017	73	78	88
2018	78	80	93
2019	74	84	91
2020	83	74	90
2021	82	83	94
2022	83	79	95
2023	83	80	71
2024	81	76	87
2025	80	82	82

Table 3. Trend of Quality in Numbers in Extreme Conditions when Faculty Population in the Institute is Zero

year	quality_index(1)	quality_index(2)	quality_index(3)
2,004	0.00	0.00	0.00
2,005	0.00	0.00	0.00
2,006	0.00	0.00	0.00
2,007	0.00	0.00	0.00
2,008	0.00	0.00	0.00
2,009	0.00	0.00	0.00
2,010	0.00	0.00	0.00
2,011	0.00	0.00	0.00
2,012	0.00	0.00	0.00
2,013	0.00	0.00	0.00
2,014	0.00	0.00	0.00
2,015	0.00	0.00	0.00
2,016	0.00	0.00	0.00
2,017	0.00	0.00	0.00
2,018	0.00	0.00	0.00
2,019	0.00	0.00	0.00
2,020	0.00	0.00	0.00
2,021	0.00	0.00	0.00
2,022	0.00	0.00	0.00
2,023	0.00	0.00	0.00
2,024	0.00	0.00	0.00
2,025	0.00	0.00	0.00

4. ANALYSIS OF RESULT FROM THE MODEL

A connive for conducting policy experiments and outcome of the results along with their assay is presented in this chapter. Alternative (What-if) scenarios with suitable assumptions regarding Institute growth factor, changes in faculty satisfaction, changes in faculty dropouts, changes in student satisfaction, changes in reputation of Institute has been accessed. The impact of these changes on quality index of technical education system and impact of institute growth factor and reputation of institute on student's attraction index is presented. What-if scenario has been developed till the year 2025.

4.1. Scenario Specification

Modelling and Simulation is a process of exploration it is not a solution to an existing problem or an answer to a stated question .Most system dynamics models are built with the goal of reaching a new understanding of how a problem arises and then using that understanding to design high leverage for improvement.In this work, the cause of the problem was clearly identified.However the modelling process was not focussed around the assesment of all causes (which are numerous) contributing to the problem but around addressing the influence of one specific cause and its effect on the problem.The following policy design process focuses on producing a set of decisions, which can improve some of the current problems for achieving quality in technical education.

Table 4. Alternative (What-if) Scenarios

S.NO	Scenario Specification	Policy Intervention
1.	Base case scenario	Simulated results (trends) based on historical trends of various parameters of technical education system.
2.	What-if	Faculty Satisfaction is increased to 30%, 50%, 80%.
3.	What-if	Faulty Dropout is increased to 30%, 50%, 80%.

4.2. A General Discussion on the Results

The scenario results discussed here does not focus on future forecasts of quality index of technical education, but possible achievable due to implementations of management measures for quality improvements in quality of technical education. Therefore, while assessing the results, emphasis has been given to trends and percentage improvement achievable rather than to absolute values. What-if scenarios with suitable assumptions regarding possible change in faculty satisfaction, change in institute growth factor, change in reputation of the institute, impact of faculty drop out ratio, change in student satisfaction have been assessed. What if analysis has assisted with the development of policy suggestions for the implementation in the technical institutes for quality improvement in the institutes. Quality index 1 in the results signifies the quality index of the institute when all Assistant professors of the institute are considered, similarly (Quality index2, Qualityindex3) in the results signifies the quality index of the institute when all Associate professors and professors of the institute are considered.

Table 5. Brief description of Scenarios

S.NO	Scenario	Abbreviation	Brief description of Abbreviation
1.	Base case scenario	B	Simulated results (trends) based on historical trends of various parameters of technical education system.
2.	Faculty Satisfaction	FS	Faculty Satisfaction.
3.	Faculty Dropout	FD	Faulty Dropout.

4.3 Base Case Scenario

Table 6. Base case scenario

Scenario	Year	Quality Index1 (Quality index with Total Assistant Professors of the Institute.)	Quality Index2 (Quality index with Total Associate Professors of the Institute.)	Quality Index3 (Quality index with Total Professors of the Institute.)
B	2004	12.10	7.17	12.35
	2005	182.36	182.91	182.89
	2006	170.98	171.30	170.82
	2007	157.78	158.00	157.57
	2008	77.63	77.99	77.53
	2009	133.73	134.13	134.85
	2010	124.32	124.29	124.39
	2011	164.78	164.71	164.73
	2012	165.49	165.40	165.66
	2013	167.28	166.99	166.25
	2014	109.31	109.39	109.03
	2015	111.01	111.09	110.72
	2016	112.71	112.79	112.42
	2017	114.42	114.49	114.12
	2018	116.12	116.19	115.82
	2019	117.82	117.89	117.52
	2020	119.52	119.60	119.21
	2021	121.22	121.30	120.91
	2022	122.92	123.00	122.61
	2023	124.62	124.71	124.31
	2024	126.32	126.41	126.00
	2025	128.02	128.11	127.70

4.4. Faculty Satisfaction is Increased to 30%

Table 7. Faculty Satisfaction is Increased to 30%

Scenario	Year	Quality Index1 (Quality index with Total Assistant Professors of the Institute.)	Quality Index 2 (Quality index with Total Associate Professors of the Institute.)	Quality Index 3 (Quality index with Total Professors of the Institute.)
FS	2004	13.40	8.27	13.65
	2005	183.66	184.21	184.20
	2006	172.28	172.60	172.12
	2007	159.08	159.30	158.87
	2008	78.94	79.29	78.83
	2009	135.03	135.44	136.15
	2010	125.62	125.60	125.69
	2011	166.08	166.02	166.03
	2012	166.79	166.70	166.97
	2013	168.59	168.29	167.55
	2014	110.61	110.69	110.33
	2015	112.32	112.39	112.02
	2016	114.02	114.09	113.73
	2017	115.72	115.79	115.42
	2018	117.42	117.50	117.12
	2019	119.12	119.19	118.82
	2020	120.82	120.90	120.51
	2021	122.52	122.60	122.21
	2022	124.22	124.30	123.92
	2023	125.92	126.01	125.62
	2024	127.62	127.71	127.31
	2025	129.32	129.41	129.01

4.5. Faculty Satisfaction is Increased to 50%

Table 8. Faculty Satisfaction is Increased to 50%

Scenario	Year	Quality Index 1 (Quality index with Total Assistant Professors of the Institute.)	Quality Index 2 (Quality index with Total Associate Professors of the Institute.)	Quality Index 3 (Quality index with Total Professors of the Institute.)
FS	2004	13.60	8.67	13.85
	2005	182.86	184.41	184.40
	2006	172.48	172.80	172.32
	2007	159.28	159.50	159.07
	2008	79.14	79.49	79.03
	2009	135.23	135.64	136.35
	2010	125.82	125.80	125.89
	2011	166.28	166.22	166.23
	2012	166.99	166.90	167.17
	2013	168.79	168.49	167.75
	2014	110.81	110.89	110.53
	2015	112.52	112.59	112.22
	2016	114.22	114.29	113.93
	2017	115.92	115.99	115.62
	2018	117.62	117.70	117.32
	2019	119.32	119.39	119.02
	2020	121.02	121.10	120.71
	2021	122.72	122.80	122.41
	2022	124.42	124.50	124.12
	2023	126.12	126.21	125.82
	2024	127.82	127.91	127.51
	2025	129.52	129..61	129.21

4.6.Faculty Satisfaction is Increased to 80%

Table 9. Faculty Satisfaction is Increased to 80%

Scenario	Year	Quality Index1 (Quality index with Total Assistant Professors of the Institute.)	Quality Index 2 (Quality index with Total Associate Professors of the Institute.)	Quality Index 3 (Quality index with Total Professors of the Institute.)
FS	2004	13.90	8.97	12.15
	2005	184.16	184.71	184.70
	2006	172.78	173.10	172.62
	2007	159.58	159.80	159.37
	2008	79.44	79.79	79.33
	2009	135.53	135.94	136.65
	2010	126.12	126.10	126.19
	2011	166.58	166.52	166.53
	2012	167.29	167.20	167.47
	2013	169.09	168.79	168.05
	2014	111.11	111.19	110.83
	2015	112.82	112.89	112.52
	2016	114.52	114.59	114.23
	2017	116.22	116.29	115.92
	2018	117.92	118.00	117.62
	2019	119.62	119.69	119.32
	2020	121.32	121.40	121.01
	2021	123.02	123.10	122.71
	2022	124.72	124.80	124.42
	2023	126.42	126.51	126.12
	2024	127.12	128.21	127.81
	2025	129.82	129.91	129.51

4.7. Faculty Drop out is Increased to 30%

Table 10.Faculty Drop out is Increased to 30%

Scenario	Year	Quality Index1 (Quality index with Total Assistant Professors of the Institute.)	Quality Index 2 (Quality index with Total Associate Professors of the Institute.)	Quality Index 3 (Quality index with Total Professors of the Institute.)
FD	2004	12.00	7.07	12.24
	2005	182.26	182.80	182.79
	2006	170.87	171.20	170.71
	2007	157.67	157.89	157.47
	2008	77.53	77.88	77.42
	2009	133.63	134.03	134.74
	2010	124.21	124.19	124.29
	2011	164.67	164.61	164.62
	2012	165.38	165.30	165.56
	2013	167.18	166.88	166.14
	2014	109.21	109.28	108.92
	2015	110.91	110.98	110.61
	2016	112.61	112.68	112.32
	2017	114.31	114.38	114.01
	2018	116.01	116.09	115.71
	2019	117.71	117.79	117.41
	2020	119.41	119.49	119.11
	2021	121.11	121.20	120.80
	2022	122.81	122.89	122.51
	2023	124.51	124.60	124.21
	2024	126.22	126.30	125.90
	2025	127.91	128.00	127.60

4.8. Faculty Dropout is Increased to 50%

Table 11. Faculty Dropout is Increased to 50%

Scenario	Year	Quality Index1 (Quality index with Total Assistant Professors of the Institute.)	Quality Index 2 (Quality index with Total Associate Professors of the Institute.)	Quality Index 3 (Quality index with Total Professors of the Institute.)
FS	2004	11.98	7.05	12.23
	2005	182.24	182.78	182.77
	2006	170.86	171.18	170.70
	2007	157.66	157.87	157.45
	2008	77.51	77.86	77.41
	2009	133.61	134.01	134.72
	2010	124.20	124.17	124.27
	2011	164.65	164.59	164.60
	2012	165.36	165.28	165.54
	2013	167.16	166.87	166.12
	2014	109.19	109.26	108.90
	2015	110.89	110.97	110.60
	2016	112.59	112.67	112.30
	2017	114.29	114.37	113.99
	2018	115.99	116.07	115.70
	2019	117.69	117.77	117.40
	2020	119.40	119.48	119.09
	2021	121.10	121.18	120.79
	2022	122.80	122.88	122.49
	2023	124.50	124.59	124.19
	2024	126.20	126.28	125.88
	2025	127.90	127.99	127.58

4.9. Faculty Dropout is Increased to 80%

Table 12. Faculty Dropout is Increased to 80%

Scenario	Year	Quality Index1 (Quality index with Total Assistant Professors of the Institute.)	Quality Index 2 (Quality index with Total Associate Professors of the Institute.)	Quality Index 3 (Quality index with Total Professors of the Institute.)
FS	2004	11.95	7.02	12.20
	2005	182.21	182.76	182.75
	2006	170.83	171.15	170.67
	2007	157.63	157.85	157.42
	2008	77.49	77.84	77.38
	2009	133.58	133.99	134.70
	2010	124.17	124.15	124.24
	2011	164.63	164.57	164.58
	2012	165.34	165.25	165.52
	2013	167.14	166.84	166.10
	2014	109.16	109.24	108.88
	2015	110.87	110.94	110.57
	2016	112.57	112.64	112.28
	2017	114.27	114.34	113.97
	2018	115.97	116.05	115.67
	2019	117.67	117.74	117.37
	2020	119.37	119.45	119.06
	2021	121.07	121.15	120.76
	2022	122.77	122.85	122.47
	2023	124.47	124.56	124.17
	2024	126.17	126.26	125.86
	2025	127.87	127.96	127.56

Figure 9. Trends of Quality Index in Numbers with Base Case Scenario and with Percentage Increase in faculty Satisfaction.

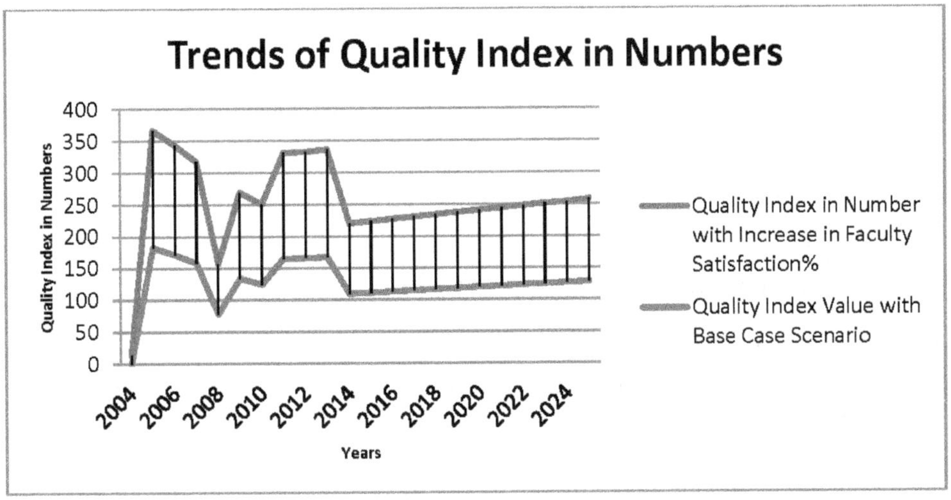

5. GENERAL DISCUSSION ON THE RESULTS OBTAINED FROM THE MODEL

The scenario presented in the table above obtained from various policy interventions in the model are summarized in the table below as:

Table 13. General Discussion on the Results

S.NO	Policy Interventions	Impact on Quality Index	Change in% in value of Quality Index in comparsion with Base case Scenario
1	Faculty Satisfaction is increased to 30%.	Increases.	10.74%.
2	Faculty Satisfaction is increased to 50%.	Increases.	12.39%.
3	Faculty Satisfaction is increased to 80%.	Increases.	14.87%
4	Faulty Dropout is increased to 30%.	Decreases.	0.82%..
5	Faulty Dropout is increased to 50%.	Decreases.	0.99%.
6	Faulty Dropout is increased to 80%.	Decreases.	1.2%.

5.1.Concluding Remarks

With the following discussion and results obtained from the present system dynamic models it is concluded that by various policy interventions, Quality of the institute can be increased and decreased subsequently. The scenario generated from the model clearly indicates the trends of quality index on enhancing particular parameters hence provides trend of behavior of particular parameters with time hence optimistic policy planning can be done for achieving long term goals for the organizations.

REFERENCES

Abdullah, F. (2006). Measuring service quality in higher education: Three instruments compared. *International Journal of Research & Method in Education, 29*(1), 7189. doi:10.1080/01406720500537445

Albach, P. G. (1993). The dilemma of change in Indian higher education. *Higher Education, 26*(1), 3–20. doi:10.1007/BF01575104

Andersen, D. F. (1994). Scripts for group model building. *System Dynamics Review, 13*(2), 102–129.

Andersen, D. F., & Richardson, G. P. (1997). Scripts for group model building. *System Dynamics Review, 13*(2), 107–129. doi:10.1002/(SICI)1099-1727(199722)13:2<107::AID-SDR120>3.0.CO;2-7

Andersen, D. F., Richardson, G. P., & Vennix, J. A. M. (2007). Group model Building: Problem structuring, Policy Simulation and Decision Support. *The Journal of the Operational Research Society, 58*(5), 691–694. doi:10.1057/palgrave.jors.2602339

Arndt, H. (2007). Using System Dynamics based Learning Environments to Enhance System Thinking. *Conference Proceedings, the 2007 International Conference of the System Dynamics Society and 50th Anniversary Celebration.*

Azar, A.T. (2012). System dynamics as a useful technique for complex systems. *International Journal of Industrial and System Engineering, 10*(4), 377-440.

Azar, A. T., & Moysis, L. (2017). New Discrete Time 2D Chotic Maps. *International Journal of System Dynamics Applications, 6*(1), 77–104. doi:10.4018/IJSDA.2017010105

Azar, A. T., & Moysis, L. (2017). Introduction to Control System Design using Matlab. *International Journal of System Dynamics Applications, 6*(3), 130–170. doi:10.4018/IJSDA.2017070107

Azar, A. T., & Valentina, E. B. (2012). Statistical methods and Artificial neural network techniques in electromyography. *International Journal of System Dynamics Applications, 1*(1), 39–47. doi:10.4018/ijsda.2012010103

Bajracharya, A. (2016). Public transportation and Private car A system dynamic approach in understanding the mode choice. *International Journal of System Dynamics Applications, 5*(2), 1–18. doi:10.4018/IJSDA.2016040101

Baporikar, N. (2016). Infrastructure development as a catalyst for social economic advancement. *International Journal of System Dynamics Applications, 5*(4), 31–42. doi:10.4018/IJSDA.2016010103

Barlas, Y. (1989). Multiple tests for validations of system dynamics type of simulation models. *European Journal of Operational Research, 42*(1), 59–87. doi:10.1016/0377-2217(89)90059-3

Barlas, Y. (1996). Formal aspects of model validity and validation in system dynamics. *System Dynamics Review, 12*(3), 183–210. doi:10.1002/(SICI)1099-1727(199623)12:3<183::AID-SDR103>3.0.CO;2-4

Barlas, Y., & Diker, G. (1996a). Decision Support for Strategic University Management: A Dynamic Interactive Game, *Proceedings of the 14th International System Dynamics Conference.*

Barlas, Y., & Diker, V. G. (2000). A Dynamic Simulation Game for Strategic University. *Simulation & Gaming, 31*(3). Advance online publication. doi:10.1177/104687810003100302

Benedict, O. (2016). A System dynamics model for subsistence farmers food security Resilience in sub Saharan Africa. *International Journal of System Dynamics Applications, 5*(1), 17–30. doi:10.4018/IJSDA.2016010102

Carnoy, M. (2000). *Globalization and Educational Reforms in Globalization and Education*. Rowman and Littlefield Publishers.

Cheng, Y. C. (2003). Quality assurance in education: Internal, interface, and future. *Quality Assurance in Education, 11*(4), 202–213. doi:10.1108/09684880310501386

Cropley, D. H. (2003). A case of Compulsory Teaching Accreditation of Engineering Faculty, Guest Editorial. *IEEE Transactions on Education, 46*(4), 24–32. doi:10.1109/TE.2003.819952

Demars, C. E. (2002). *Modeling student Outcomes in a General Education Course with Hierarchial linear models.* Paper presented at 42nd AIR Forum, Ontario, Canada.

Erik, P. (2016). Integrating Systems Modelling and data Science. *International Journal of System Dynamics Applications, 5*(1), 1–16. doi:10.4018/IJSDA.2016010101

Erik, P., & Logtens, T. (2016). Model based Exploration of Societal aging in the Netherlands. *International Journal of System Dynamics Applications, 4*(1), 57–95.

Eriksen, S. D. (1995). TQM and the transformation from elite to a mass system of higher education in the UK. *Quality Assurance in Education, 3*(4), 14–29. doi:10.1108/09684889510146795

Fletcher, M. A., & Zuber-Skerritt, O. (2008). Professional Development through Action Research: Case Examples in South African Higher Education. *Systemic Practice and Action Research, 21*(1), 73–96. doi:10.100711213-007-9083-3

Forrester, J. W. (1994). System Dynamics, System Thinking, and Soft OR. *System Dynamics Review, 10*(2-3), 245–256. doi:10.1002dr.4260100211

Forrester, J. W. (1998). *Designing the Future.* Retrieved April 15, 2016, from http://sysdyn.clexchange.org/people/jay-forrester.html

Forrester, J. W., & Senge, P. M. (1980). Tests for building confidence in system dynamics models. In A. A. Legasto, J. W. Forrester, & J. M. Lyneis (Eds.), *System dynamics* (pp. 209–228). TIMS Studies in the Management Science.

Foster, R. O. (1972). Education in the City D-2144. In MIT System Dynamic Group Literature Collection. System Dynamics Society.

Frances, C. (1995). *Using System Dynamics Technology to Improve Planning For Human Resource Development.* Presentation at the Congress of Political Economists, International 6th Annual Conference, Seoul, Korea.

Frances, C. (2000). *Selected papers presented at an international seminar on Using System Dynamics as a Tool for Decision Making in Higher Education Management held in June 1999 at the Royal Society, London and South Bank University, under the auspices of the Society for Research into Higher Education*. South Bank University Technical Report SBU-CISM-12-00.

Friedman, S., & Raphael, M. (2007). Individual Learning Style and Systems Tool Preferences. *Conference Proceedings, The 2007 International Conference of the System Dynamics Society and 50th Anniversary Celebration.*

Goel, D.R., & Biswal, A. (1998). TQM in Education, Reality, Desirability and Feasibility. *University News, 31*(44), 10-13.

Grandzol, J. R. (2005). *Improving the Faculty Selection Process in Higher Education: A Case for the Analytic Hierarchy Process* (Vol. 6). Association for Institutional Research.

Halbower, M. C. (1993). The First Three Hours D-4230-3. In MIT System Dynamics Group Literature Collection. System Dynamics Society.

Harvey, L., & Green, D. (1993). Defining Quality. *Assessment & Evaluation in Higher Education, 18*(1), 9–34. doi:10.1080/0260293930180102

Hasan, H., & Ilias, A. (2008). Service quality and student satisfaction: A case study at private higher education institutions. *International Business Research, 3*(3), 10–21.

Hill, F. M. (1995). Managing service quality in higher education: The role of the student as primary consumer. *Quality Assurance in Education, 3*(3), 10–21. doi:10.1108/09684889510093497

Ho, W., Dey, K. P., & Higson, H. E. (2006). Multiple criteria decision- making techniques in higher education. *International Journal of Educational Management, 20*(5), 319–337. doi:10.1108/09513540610676403

Ho, W., Higson, H. E., & Dey, P. K. (2007). An integrated multiple criteria decision making approach for resource allocation in higher education. *International Journal of Innovation and Learning, 4*(5), 471–486. doi:10.1504/IJIL.2007.012958

Hopper, M., & Stave, K. (2008). Assessing the Effectiveness of Systems Thinking Interventions in the Classroom. *Conference Proceedings, The 2008 International Conference of the System Dynamics Society.*

Horne, H., & Pierce, A. (1996). *A Practical Guide to Staff Development and Appraisal in Schools.* Kogan Pages.

Jagdeesh, R. (2001). Improvement of quality of higher education in engineering sciences with an emphasis on International aspects. *The Indian Journal of Technical Education, 24*(2), 50-55.

Kennedy, M. (1998a). A Pilot System Dynamics Model to Capture and Monitor Quality Issues in Higher Education Institutions Experiences Gained. *Proceedings of the 16th System Dynamics Conference*, 1-7.

Kennedy, M. (1998b). Some Issues in System Dynamics Model Building to Support Quality Monitoring in Higher Education. *Proceedings of the 16th System Dynamics Conference*, 1-8.

Kennedy, M. (2002). An extended Taxonomy of System Dynamics Models of Higher Education. *Proceedings of the 20th System Dynamics Conference*, 1-19.

Kennedy, M. (2008). A Taxonomy of System Dynamics Pedagogic Techniques. *Conference Proceedings, The 2008 International Conference of the System Dynamics Society.*

Lagrosen, S., Seyyed-Hashemi, R., & Leitner, M. (2004). Examination of the dimensions of quality in higher Education. *Quality Assurance in Education, 12*(2), 61–69. doi:10.1108/09684880410536431

Lindsay, A. W. (1992). Concepts of Quality in Higher Education. *Journal of Tertiary Education Administration, 14*(2), 153–163. doi:10.1080/1036970920140203

Lomas, L. (2004). Embedding quality: The challenges for higher education. *Quality Assurance in Education, 12*(4), 157–165. doi:10.1108/09684880410561604

Maji, S. (2003). Quality Assurance in Technical Education, Possibilities through E-Learning in Distance Education. *The Indian Journal of Technical Education, 26*(2), 19–25.

Maltz, E. N., Murphy, K. E., & Hand, M. L. (2007). Decision support for university enrollment management: Implementation and experience. *Decision Support Systems, 44*(1), 106123. doi:10.1016/j.dss.2007.03.008

Mani, R.S., & Devi, U. (2009). Quality Assurance in Education through Quality Circles. *Global and Indian Context. Online Journal, 7*(4), 1–17.

Morecroft, J. D., & Sterman, J.D. (Eds.). (1992). Modelling for Learning. European Journal of Operations Research, 59.

Mouly, C. M. C. (2003). Quality in Technical Education, A Critical analysis of the Governing Factors. *The Indian Journal of Technical Education, 26*(2), 50–54.

Natrajan, R. (2000). The Role of Accreditation in Promoting Quality Assurance of Technical Education. *International Journal of Engineering Education, 16*(2), 88–96.

Natrajan, R. (2003). The Role of Social and Societal Responsibility as a Core value of the University and Corporate Sectors. *The Indian Journal of Technical Education, 26*(1), 11–21.

Niles, R. (1991). Pre- College Education Project UROP Evaluation D-4229. In MIT System Dynamics Group Literature Collection. System Dynamics Society.

Nodenof, T., Latorcade, P., & Sallaberry, C. (2004). Model based Engineering of Learning for Adaptive Web Based Educational Systems. *Proceedings for the 13th International World Wide Web Conference.* Available from http://Oportal.acm.org.lispac.1sbu.ac.uk/

Owlia, M. S., & Aspinwall, E. M. (1996). Quality in higher education – a survey. Total Quality Management, 7(2).

Owlia, M. S., & Aspinwall, E. M. (1998). A framework for measuring quality in engineering education. Total Quality Management, 7(2).

Oyo, B. (2008). A system Dynamics Tool for Higher Education Funding and Quality Policy Analysis. *Proceedings of System Dynamics Conference.*

Pant, A. S. (2013). *National Workshop on Quality Assurance and Accreditation.* Academic Press.

Potash, P., & Heinbokel, J. (2005). Unleashing the Revolutionary Implications of a System Dynamics Education. *The 23rd International Conference of the System Dynamics Society.*

Prasad, K. (2009). Justification of world class maintence systems using analytic hieracy constant sum method. Journal of Quality in maintence. *Engineering (London), 15*(1), 47–77.

Raisbeck, I. (1994). Royal mail: developing a total quality organization. In D. Green (Ed.), *What is Quality in Higher Education?* SRHE and Open University Press.

Rajput, J.S. & Walia, K. (1997). Quality in Higher Education. *The Journal of Higher Education, 20*(4), 553–567.

Roberts, N. H. (Ed.). (1976). System Dynamics Curriculum Development Project for Elementary and Secondary Education D-2435. In MIT System Dynamics Group Literature Collection. System Dynamics Society.

Roberts, N. H. (Ed.). (1978). System Simulation of Student Performance in the Elementary Classroom. In Management Applications of System Dynamics. Productivity Press.

Runge, D. (1997). A Note on Teaching System Dynamics D-2653. In MIT System Dynamic Group Literature Collection. System Dynamics Society.

Saeed, K. (1990). Bringing Practicum to Learning: A System Dynamics Modelling Approach. *Higher Education Policy, 3*(2), 44–49. doi:10.1057/hep.1990.30

Saeed, K. (1993). Bringing experimental learning to the social science: A simulation laboratory on economic development. *System Dynamics Review, 9*(2), 153–164. doi:10.1002dr.4260090205

Saeed, K. (1996). The Dynamics of Collegial Systems in the Developing Countries. *Higher Education Policy, 9*(2), 75–86. doi:10.1016/S0952-8733(96)90040-9

Saeed, K. (1997). *System Dynamics as a Technology for New Liberal Education*, Worcester Polytechnic Institute, Report No.3.

Saeed, K. (1998). Maintaining professional competence in innovation organizations. *Human Systems Management, 17*, 69–87.

Salazar, P., & Martinez, M. M. (2007). Achievements teaching Systems Thinking and Systems Dynamics to Graduate students through e-learning. *Conference Proceedings, The 2007 International Conference of the System Dynamics Society and 50th Anniversary Celebrations.*

Scholl, H. J. (2004). Action Research and System Dynamics: Can they benefit from each other.Paper presented at the 2004 Hawaiian Internal Conference on System Sciences,Waikoloa,Hawaii, 1-11. 10.1109/HICSS.2004.1265243

Schwaninger, M., & Grosser, S. (2008). System Dynamics as Model- Based Theory Building. *Systems Research and Behavioral Science, 25*(4), 447–465. doi:10.1002res.914

Senge, P. M. (1988). New System Dynamics Learning Tools for Management Education and Training D-3999. In MIT System Dynamics Group Literature Collection. System Dynamics Society.

Senge, P. M. (1990). *The Fifth Discipline*. The Art and Practice of the Learning Organization, Doubleday.

Shaffer, W. A. (1976). An Initial Concept for Organizing System Dynamics Curriculum D-2438. In MIT System Dynamics Group Literature Collection. System Dynamics Society.

Sterman, J.D. (1992). *Teaching Takes Off OR/MS Today*. Academic Press.

Sterman, J.D. (2000). *Business Dynamics. Systems Thinking and Modeling for a complex World*. Irwin McGraw-Hill.

Sterman, J. D. (2002). All models are wrong: Reflections on becoming a system scientist. *System Dynamics Review*, *18*(4), 501531. doi:10.1002dr.261

Stuntz, L., & Lyneis, D. (2007). System Dynamics in K-12 Education: Lessons Learned. *Conference Proceedings, The 2007 International Conference of the System Dynamics Society and 50th Anniversary Celebration*.

Sureshchandar, G. S., & Rajendra, C. (2001). A holistic model for total quality service. *International Journal of Service Industry Management*, *12*(4), 378–412. doi:10.1108/09564230110405299

Sweeney, L., & Sterman, J. (2007). Thinking about systems: student and teacher conceptions of natural and social systems. System Dynamics Review. Accepted July.

Thompson, K., & Reimann, P. (2007). Do school learn more about the environment from a system dynamics model by themselves or with a partner. *Conference Proceedings, The 2007 International Conference of the System Dynamics Society and 50th Anniversary Celebration*.

Tulinayo, P. F., Hoppenbrouwers, S. J. B. A., & Proper, H. A. E. (2008). Integrating System Dynamics with Object- Role Modeling. *IFIP International Federation for Information Processing*.

Umashanker, V., & Dutta, K. (2007). Balanced scorecards in managing higher education institutions: An Indian perspective. *International Journal of Educational Management*, *21*(1), 54–67. doi:10.1108/09513540710716821

Venkataram, P., & Giridharan, A. (2009). Quality Assurance and Assessment in Technical Education System: A web Based Approach. *International Conference on Engineering Education-ICEE 2007*, 1-6.

Vennix, J. A. (1999). Group model building: Tackling messy problems. *System Dynamics Review*, *15*(4), 379–401. doi:10.1002/(SICI)1099-1727(199924)15:4<379::AID-SDR179>3.0.CO;2-E

Vennix, J. A. (2000). Group model-building: Tackling messy problems. *System Dynamics Review*, *15*(4), 379–401. doi:10.1002/(SICI)1099-1727(199924)15:4<379::AID-SDR179>3.0.CO;2-E

Verghese, A. (2007). OD Interventions for Sustainable Excellence in Higher Education. *Conference on Global Competition and Competitiveness of Indian Corporate*, 39-51.

Vrat, P. (2003). Quality Assurance in Technical education, Recent trends and challenges ahead. *The Indian Journal of Technical Education*, *27*(2), 12–14.

Watty, K. (2003). When will Academics Learn about Quality. *Quality in Higher Education*, *10*(1), 17–25.

Williams, D. (2002). An Application of System Dynamics to Requirements Engineering Process Modelling (Unpublished PhD). South Bank University.

Williams, D. (2003). *Challenges of System Dynamics to Deliver Requirements Engineering Projects: faster, better and cheaper*. Paper presented at the 2003 System Dynamics Conference, New York., NY

Williams, R., & Van Dyke, N. (2007). Measuring the international standing of universities with an application to Australian universities. *Higher Education*, *53*(6), 819841. doi:10.100710734-005-7516-4

Chapter 7
Using System Dynamics to Analyse the Cooperation and Reputation in the Electricity Industry

Lina Carvajal-Prieto
Universidad Piloto de Colombia, Colombia

Milton M. Herrera
https://orcid.org/0000-0002-0766-8391
Universidad Militar Nueva Granada, Colombia

ABSTRACT

When different stakeholders work together, they can implement a strategy to improve the performance of the electricity industry. This chapter shows how the cooperation and reputation influence on the electricity industry. A simulation model was developed to illustrate how related variables such as altruism, reputation, and strategic alliances contribute to promote the renewable energy industry. To do this, the chapter illustrates a case study for the electricity market in Colombia. The results show a novel perspective that contributes to marketers and engineers in the analysis of the relationship between cooperation and reputation in electricity firms.

INTRODUCTION

Recently, the renewable energy market has shown significant growth that contributes to the competitiveness of the Latin American countries (Arias-Gaviria et al., 2019; Cárdenas et al., 2017; Gómez et al., 2017; Herrera et al., 2020; Zapata et al., 2019). The rapid growth of the renewable energy market has provoked severe challenges due to the significant financial resources that are required, highlighting the need for marketing of the utility industry in the electricity sector (Castaneda et al., 2017). As it is an industry that needs to work cooperatively, having a reputation increases the added value, and coopera-

DOI: 10.4018/978-1-7998-5788-4.ch007

tion is maintained in the long term. As well-defined Rand & Nowak (2013), it is worth paying the cost of collaboration today to earn the benefits of a good reputation tomorrow.

Customer satisfaction and corporate reputation management can be considered to be promising approaches for energy supply companies (Herrera et al., 2019a). The reputation is one of the most crucial marketing variables because it can change customer perception while impacts the consumption of electricity. Presently, the electricity market liberalisation gives customers the possibility to choose products that best match their preferences (Kishimoto et al., 2017). The client can choose among different options making the industry more competitive, and it requires a structured marketing plan involving cooperative work, reputation and customer value.

One of the fastest-growing products for the electricity industry is rooftop solar panel of greater use in the residential sector (Castaneda et al., 2020; Chen et al., 2014). In the case of Colombia, the potential of solar energy is very high, solar radiation to reach a daily average of 4.5 kWh/m2, which exceed the average world daily of 3.9 kWh/m2 (UPME & BID, 2015). Currently, the Colombian government has taken an important step to support the deployment of renewable energy through law 1715; however, this law has not adequately implemented (Castaneda et al., 2019). Although solar power has great potential in the household sector as well as positive impacts on emission reduction, Colombia has not developed the residential market sufficiently, which not allow rapid response to the challenges of the household market. In response to this concern, there is interested in understanding the effects of different variables related to marketing on the reputation of the utility industry.

Bearing in mind that Colombia has a high potential for solar energy and relevant opportunities because solar radiation throughout the country is mostly uniform during the year (average 4.5 KWh/m2/day) (IDEAM, 2017). It becomes a key product of our industry that requires to be studied and examined. The country needs to diversify its electricity mix by incorporating non-conventional renewable energy sources, but it requires a strategy (Eras et al., 2019; Quiceno et al., 2019). The electricity matrix expansion is clear evidence of the benefits for guarantee the security of supply, which has been experienced in other countries, such as Brazil and Chile (Gómez et al., 2017; Rego & de Oliveira Ribeiro, 2018; Silva et al., 2016).

The development of renewable sources in the electricity industry needs to be reinforced with an accurate organisation and a cooperative working model. Several evolutionary theories can explain why reputation facilitates cooperation (Wu et al., 2016). When the organisation has worked on its reputation, it is easier to get cooperation from the others, and it is a way to get more benefits from different stakeholders like government and other firms (Herrera et al., 2019b). In this sense, it is necessary to go beyond the observation that patterns of social relations matter for cooperation, to study the mechanisms through which social ties may enhance collective outcomes (Baldassarri, 2015).

This chapter, based on a SD model approach, aims at assessing alternative strategies to improve the cooperation and reputation in the industry electricity, to promote the value creation efficiently. This simulation model seeks to understand the dynamic in the electricity sector from consumer behaviour in Colombia to propose an alternative business model based on non-conventional renewable electricity based on cooperation. For this case study, electricity firms must work together to potentiate the electricity utility industry. The chapter updates and improves the findings of a research project recently developed by the authors (Herrera et al., 2019a). Thus, the objective of this chapter is to show a model of cooperation and reputation to assess its effects on the diffusion of the electricity industry in Colombia, particularly renewable energy.

The current socio-technical transition involves modelling and management strategies for understanding the complex dynamics of the energy system (Papachristos, 2019). SD was first developed by Forrester in 1950s to analyse complex behaviour in social sciences and management (Forrester, 1961; Sterman, 2000). Several researchers have highlighted the advantages of using SD modelling for supporting strategic decision-making (Gary et al., 2008; Torres et al., 2017; Warren, 2005). From the literature of SD applications also has identified that this methodology considering factors associated with delays, non-linearity and intangibles (Cosenz & Noto, 2016). In this sense, SD modelling becomes a handy tool when involving complex issues and intangible resources that interacted in the long term. In essence, the SD approach can model and simulate the firm's reputation to design management strategies based on cooperation among actors of the energy system, which contribute to improving the performance throughout the energy supply chain. The main research questions of our chapter are:

Can simulation be of any help to better understanding the relationships between cooperation and reputation? What are the performance drivers that may lead to foster renewable energy diffusion?

This chapter is organised into six sections, including this introduction. First, an analysis of the problem is presented for understanding the relationship between cooperation and reputation and their impacts on value creation in the electricity market. Second, an overview of the studies of cooperation and reputation in the electricity industry are discussed. Third, the chapter describes the simulation and the mathematical model used. Moreover, it considers the validation and assumptions data in the case of study for the electricity industry in Colombia. Fourth, the chapter discusses the obtained results from the simulation. Finally, the conclusions are presented.

BACKGROUND

Previous studies analyse how the seasonality is directly affecting the electricity market due to the variations of hydropower generation in Colombia (Morcillo et al., 2020). Given the considerable disruption happening in the electricity market due to climate variation, utilities need to be transitioning their business modes (Castaneda et al., 2017; Tayal, 2017). In this sense, corporate reputation plays an essential role in transitioning to a novel business model. On the one hand, the corporate reputation is defined as observers' collective judgments of a corporation based on assessments of the financial, social, and environmental impacts attributed to the corporation over time (Barnett et al., 2006). That is, corporate reputation becomes one factor that increases the capability building of firms (Arend & Bromiley, 2009).

On the other hand, the customer value is essential to study the complexity that has a relationship with different marketing concepts such as perception, image, satisfaction, customer service and quality (Oh & Kim, 2017). In this sense, the customer value analysis is equal to the difference between market-perceived quality and market-perceived price (Leroi-Werelds et al., 2014). Understanding the customer value involves assessing the utility of service based on customer's perception, namely what is received and what is given (Leroi-Werelds et al., 2014; Oh & Kim, 2017). This situation comprises a decisive point to reach the goals of the electricity industry. When the customer feels satisfied, there are more possibilities to come back and buy again. The customer value also involves the emotional bond established between a customer and a producer after the customer has used a product or service; therefore, customer value propositions are created to reflect the customer value perceptions (Rintamäki & Kirves,

2017). Such a resulting emotional bond leads the customer to buy repeatedly or, better yet, exclusively from that supplier, to recommend that supplier to friends and family.

The creation of customer value comprises a process where customers evaluate the benefits received and the sacrifices given form using a product/service. Early marketing studies focused on product quality as the primary "get" component and product as the main "give" component (Xu et al., 2015). When a consumer uses a product or service, it knows that has to pay something to receive what needs; in this point is when the value becomes relevant to satisfy a need causing an excellent impression because the customer evaluates the benefits according to his experiences.

The companies also evaluate the benefits that receive when buying a product or service and price to pay to satisfy their requirements. There is a strong relationship between the price and product value concerning the characteristics of the energy service. Thus, a low price is seen as the dominant factor permitting consumers from purchasing energy of clean technologies.

Cooperative Work and Reputation

As renewables displace more expensive technologies, decline both revenues and profits of the incumbent utilities (Castaneda et al., 2017). Firms with several drawbacks associated to reputation have a strong incentive to behave cooperatively and are therefore attractive partners in the social and economic exchange (Kollock et al., 1994; Milinski, 2016; Sylwester & Roberts, 2013). Thus, this situation could be an excellent option to improve processes for the electricity industry in Colombian, when the firms work together, they have the support. They can implement different practices to have more customers. For instance, cooperatives provide an institutional structure that allows its members to 'cooperate' for mutual social or financial benefits (Hopfensitz et al., 2015). In this sense, when they work in a cooperative model, they can have more resources and ideas to invest in the electricity industry.

While the cooperatives have shown a capacity to innovate in the formation of new alliances (Nikabadi & Amir, 2018), there are few innovations in the methods in which advisory services are provided. They do not have enough advisory services to change their processes; they need more people helping them to learn how to organise the method according to the new demands of the market.

People can build up a good reputation by being helpful, or a bad reputation when they tend to be selfish, and reputation may spread through social networks by gossip (Swakman et al., 2016). It is well known that the spread of mouth is one of the most powerful tools in marketing. Thus, it is fundamental to create and keep a good reputation of the firm that contributes to reduce total annual electricity costs as a result of the cooperation (Acuña et al., 2018).

The measurements of reputation can be based on single-item or multi-item approaches (Helm & Klode, 2011). In the first case, there is only one relationship between the general concept of reputation, and in the second one, different variables are affecting the relationship. However, the most important is the contribution of these relationships over the benefits the firm can get, as can observe in Figure 1. This model allows incorporating lower-order factors in several constructs such as altruism, image and strategic alliances.

Measuring reputation is the first step to understand its influence over cooperation to get benefits. Reputation is regarded as a predictor of future behaviour because according to this variable, it is possible to analyse different aspects of the organisation. Reputation systems rely on the feedbacks or ratings provided by the members of the community for a given subject (Bentahar et al., 2012).

Figure 1. Relationship between reputation-cooperation and benefits.
Sources: Own elaboration based on Helm & Klode (2011).

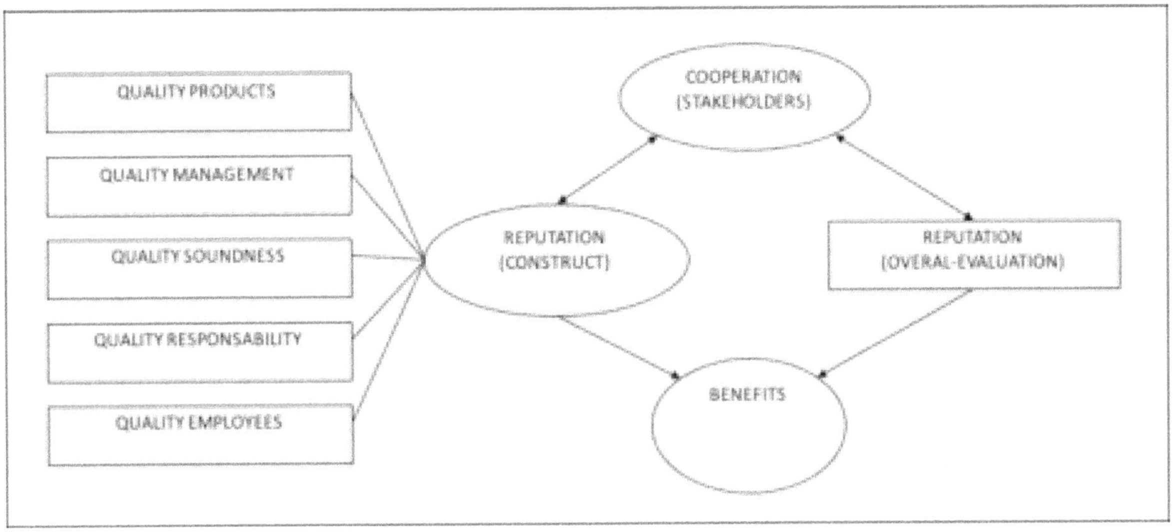

All these concepts have a different relationship with each stakeholder, and customer value is responsible for improving their perception. Customer value takes the perspective of an organisation's customers, considering what they want and believe that they get from buying and using a seller's product. Others studies also think that a company has social responsibilities that go beyond shareholder value creation to encompass other stakeholders such as employees, consumers, and society at large (Salem Khalifa, 2004). This situation is an opportunity for the electric utility industry. In general, companies must understand the importance of creating value their stakeholders. The current businesses model is aimed at creating value. The purpose of a sustainable business is, ðrst, to create value for customers and, second, to extract some of that customer value in the form of proðt, thereby creating value for the ðrm (Cosenz et al., 2019; Kumar & Reinartz, 2016).

Consequently, a simulation process may guide the measurement of the cooperation and reputation to facilitate customer value creation. The simulation could help in the process of prediction and prospective becoming a tool to improve the perceptions and other aspects related to customer satisfaction, as it is mentioned before, reputation and cooperation

MODEL DESCRIPTION

Simulation Model

The rapid growth of the renewable energy market has provoked severe challenges associated with the behavioural patterns of customers that require a systematic analysis for understanding the need for marketing in the electricity sector. Simulation models have been developed and applied to understand both market structures in the electricity industry and to analyse customers' behaviour (Castaneda et al., 2017; Dyner & Franco, 2004; Ford, 1999). Mainly, system dynamics is a methodology of simulation based on feedback structures analysis and time-delays, which define the behaviour of a complex social system

(Ford, 1999; Namujju et al., 2015; Sterman, 2000). The modelling process with system dynamics is an iterative and continuous process of hypothesis formulation, validation and revision of formal model that represent the system in studied (Bala et al., 2017; Mollona, 2008). Scenarios based on simulated results can provide guidelines for policy planning and management, particularly in the electricity industry (Dyner & Larsen, 2001). In this context, Forrester (1961) explains the following steps to simulate a SD model:

a) Describe the problem through a dynamic behaviour hypothesis.
b) Create a simulation structure from stock and flow diagram, which involve a mathematical model based on differential equations.
c) Define the data of the simulation model.
d) Validate and analyse the policy through simulation scenarios.
e) Results.

In essence, the SD methodology contributes to model and simulate the market structure to design management strategies and policy (Davis & Bingham, 2007; Wheat, 2010). The simulation model structure is based on the relationships among the variables and represented by the stock-and-flow diagram, as shown in Table 1 and Figure 2.

Table 1. Essential building elements used in system dynamics.

Elements	Symbol	Description
Stock (level) Flow (rate) Auxiliary (converter)	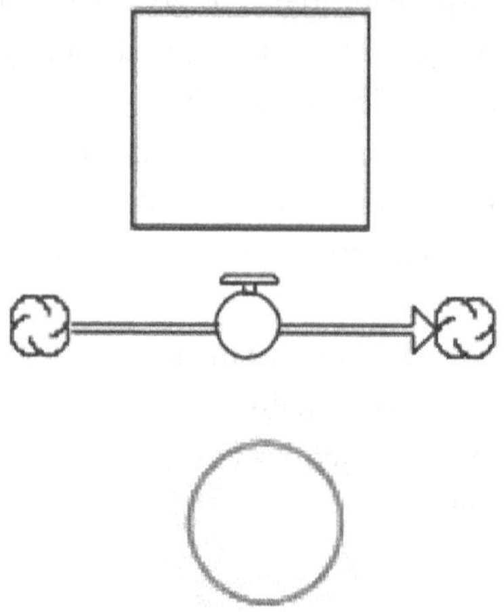	This variable describes the condition or state of the system at any particular time. Alters stock level by an inflow or an outflow. It holds the values for constants, defines external inputs to the model, calculates algebraic relationships and determines graphical functions.

Sources: (Ahmad et al., 2016; Bala et al., 2017).

The decision-making is given by coordinated planning of resources such as humans, financial and technological (Becerra et al., 2016). Previous studies have demonstrated the useful role of simulation for assessing the industrial system (Garwood et al., 2018). For instance, the response capacity of each actor is determined by the adjustment between current capabilities and desired (Repenning & Sterman, 2002). This simulation model shows how the cooperation influence on developing the capacity of renewables and the customer's image (i.e., reputation) affect demand changes as well as capacity margin used by the electricity industry. Figure 2 shows the stock and flow diagram composed by a system of

Figure 2. Stock-and-flow diagram develops to simulate the cooperation and reputation in the electricity industry.

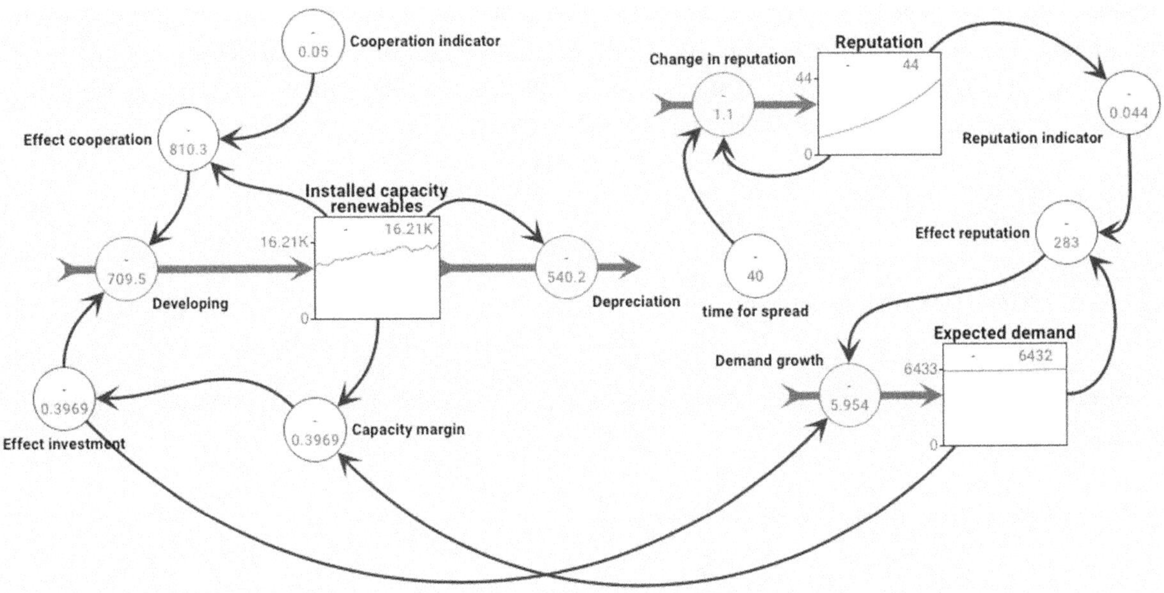

differential equations, which is solved through a simulation structure. For instance, given a stock variable x for any time t can be derived directly from the difference between output flow (*OutF*) and input flow (*InpF*) by the Equation 1.

$$x(t) = \int (InpF - OutF)\, ds\,. \tag{1}$$

Supplementary data associated with this simulation model can be found in the online version at https://sheetless.io/@miltonherrera/behavioural?s=eQChZMz0RjWkjL1pDOynEA.

Table 2. Error analysis of the simulation model

Variable	U^m	U^S	U^C
Installed capacity renewables	0.09	0.4	0.51
Expected demand	0.05	0.48	0.47

Previous studies oriented the mathematical formulation and structure of our model (Dyner, 2000; Ford, 1997; Warren, 2005). The model structure was proposed through the stock and flow diagram for each variable involved. On the one hand, this structure contains three elements. First, the installed capacity renewables that represent the supply of the energy system. Second, the expected demand, including the reputation, and finally, a cooperation indicator. This model represents a basic electricity industry structure that enables the dynamic analysis of the change in reputation and cooperation.

On the other hand, the main equations that oriented the decision-making along the electricity industry are given by cooperation indicator and time for customer reputation to spread. The development of installed capacity renewables is determined by the effects of cooperation (EC). This effect is defined by variation between accumulated capacity (ICr) and cooperation indicator (CIn) as shown in Equation 2. The change in reputation (CR) was calculated, taking into account accumulated reputation (R) and time for the reputation to spread (TFR) as shown in Equation 3.

$$\mathrm{EC} = \mathrm{ICr}^{*}\mathrm{CIn} \quad [\mathrm{Dmnl}]. \tag{2}$$

$$\mathrm{CR} = R/TFR \quad [\mathrm{customers / year}]. \tag{3}$$

Similarly, the accumulated reputation (AR) is estimated using the change in reputation, as illustrated in Equation 4.

$$\mathrm{AR} = \mathrm{AR}(\mathrm{t}-1) + \int CR(t) * dt \quad [\mathrm{customers}]. \tag{4}$$

The stock and flow diagram shows how the development of installed capacity renewables is related to investment for new power (Dyner, 2000; Redondo et al., 2018). Thus, the purchase is determined by the capacity margin (CM)– the relationship between installed capacity (ICr) and expected demand (ED) as shown in Equation 5.

$$\mathrm{CM} = ICr/ED \quad [\mathrm{Dmnl / year}]. \tag{5}$$

Regarding, the behaviour over time of perceived customer value – reputation indicator (RI) allowing understand effects on customer reputation (ER) is calculated in Equation 6.

$$ER = RI * ED \quad [Dmnl]. \tag{6}$$

The degree to which the electricity industry uses renewable sources is characterised by changes in capacity, which accumulated. Thus, the installed capacity renewable (ICr) were estimated as from primary resources (ICr_0) and the difference between the inflow and outflow, namely developing (DVr) and depreciation (DPr) of capacity, as illustrated in Equation 7.

$$\mathrm{ICr}(t) = ICr_0 + \int \left(DVr(t) - \mathrm{DPr}(t) \right) \cdot \mathrm{dt} [\mathrm{Mw / Year}]. \tag{7}$$

The model has a simulation time horizon of 30 years was considered to simulate the mid- to long- term effects of reputation and cooperation. Moreover, the proposed model structure, as well as behaviour, was validated with traditional statistics methods used by the community of system dynamics (Oliva, 2003; Qudrat-Ullah & Seong, 2010; Sterman, 1984). The interpretation of forecast evaluation statistics is an essential step for validation model processes. Amongst the alternatives of statistics available for consideration are Theil's inequality statistic. Theil inequality statistic provides a measure of the error in the predicted values of simulation through the components of the mean (U^m), variance (U^s) and covariance (U^c), as presented in Table 2. Results show that components have a little trend variation; therefore, it is unsystematic. Thus, the fit between model and history is appropriate for behaviour analysis.

Table 3. Simulation scenarios

<table>
<tr><th>Scenarios</th><th>Description</th><th>Main equations</th></tr>
<tr><td>Business as Usual – BAU</td><td>Baseline scenario. This scenario adopts the following parameters: TFR = 40 and CIn = 0.05</td><td rowspan="3">EC=ICr*CIn
CR=R/TFR</td></tr>
<tr><td>Scenario 1</td><td>Cooperation incentive scenario. This scenario assumes the following parameters: TFR = 45 and CIn = 0.06</td></tr>
<tr><td>Scenario 2</td><td>Change of reputation scenario. This scenario assumes the following parameters: TFR = 30 and CIn = 0.04</td></tr>
</table>

Other assumptions in the simulation were considered below:

- An average energy system lifetime of 30 years is assumed in the model (Radomes & Arango, 2015).
- It is determined the growth rate of demand in a value of 3.5% per year (Morcillo et al., 2017).
- This paper assumed the forecast for peak demand in Colombia (UPME, 2016).

RESULTS AND DISCUSSION

In this section, the results of the simulation model for different scenarios are presented. The scenarios are designed to analyse the changes in reputation and cooperation and its effects on expected demand and installed capacity renewables. Moreover, the results show the accumulated reputation and changes of cooperation for each scenario proposed. The analysis of the changes in reputation and cooperation provide essential elements to the evaluation of effects in the long-term for the electricity sector in Colombia. Table 3 exhibits the scenarios that were simulated with the model previously mentioned. The scenarios are the combination of two behaviours associated with changes on cooperation indicator and time for the reputation to spread. Note that the value of TFR and CIn are arbitrary and only indicate an increase or decrease in the change of cooperation and reputation.

Due to the favourable conditions that have Colombia, renewable energy is an alternative for the expansion of the energy matrix (Arias-Gaviria et al., 2019; Jimenez et al., 2016). Figure 3 shows the behaviour of the effect of cooperation for the three scenarios. These results show how an increase in the cooperation indicator through alliances has a positive impact on the installed capacity of renewables (see, Figure 4). In contrast, the lack of cooperation affects the construction of new power. Previous studies have assessed the benefits of collaboration among countries how an essential aspect of improving the diffusion of renewable energy in Latin America (Ochoa et al., 2013; Ochoa & van Ackere, 2014; Redondo et al., 2018).

Figure 3. Simulation results of the effect of cooperation

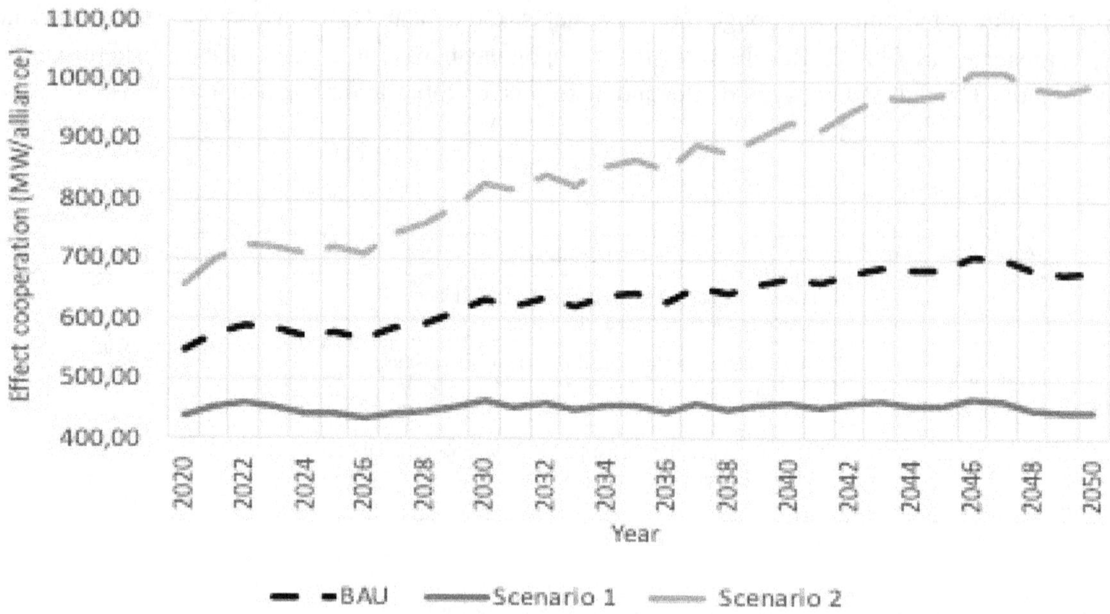

Figure 4. Dynamic of the behaviour of the installed capacity of renewables

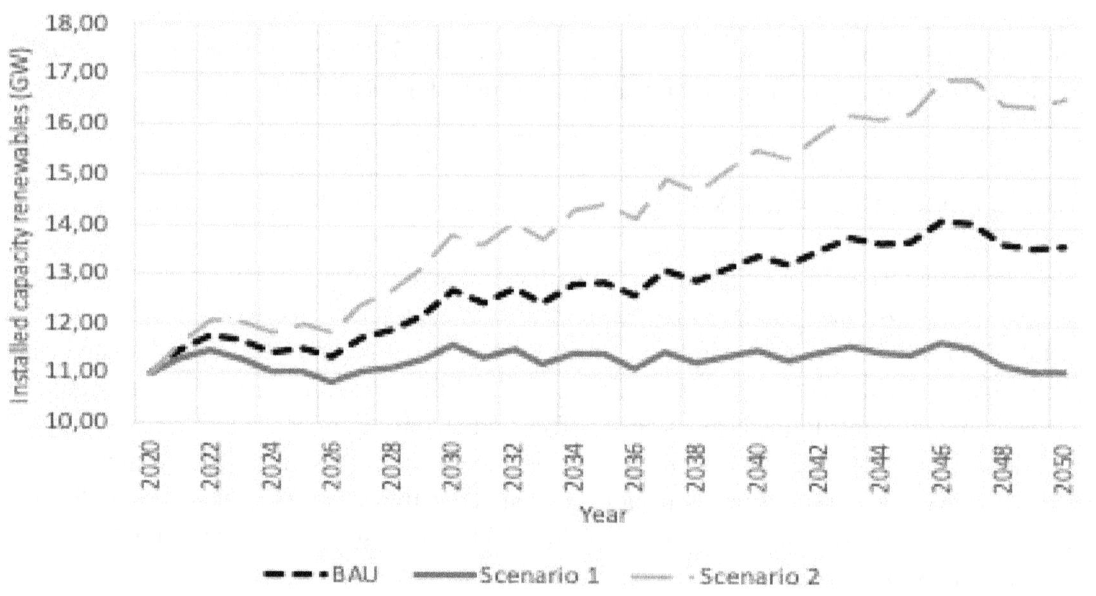

Considering the simulation scenarios, Figure 4 presents the behaviour of the installed capacity of renewable between 2020 and 2050. The accumulation of power exhibits negative impacts due to decrease in terms of alliances or cooperation indicators in the case of scenario 1. While scenarios BAU and 2 suffer an increase of 19% and 30% between 2021 and 2050, scenario 1 reaches an increase of 1% in this same period. This situation involves that the used resources of the energy sector may be related to cooperation of the electricity industry, and therefore profit margin could be positively affected. In this sense, social networks operating between households and energy utility companies may be driving force for developing the benefits of renewables (Bale et al., 2015). Studies have shown that challenges and opportunities of the diffusion of renewable energy may be reached by establishing partnerships and alliances, which contributes to dialogue and motivating cooperative behaviour (Matos & Silvestre, 2013).

Another critical aspect of the analysis is the duration time for the reputation to spread. Steady incremental improvements of the reputation levels in the three scenarios can be observed in Figure 5. However, the simulation shows that scenario 2 has an increase higher than BAU and scenario 1. This situation occurs because when the time for the reputation to spread affects the margin capacity can influence the market values – incentives for the new power of renewable energy. This event means that the reduction of time for the reputation to spread may expand the demand growth, as illustrated in Figure 6. Several studies have shown that installed capacity of clean technologies is directly impacted by the investment (Guo & Guo, 2015; Jimenez et al., 2016), and therefore this help to the electricity prices creating a favourable market environment (Franco et al., 2015; Ibanez-Lopez et al., 2017).

In the process of industrialisation, both economic development and the improvement of people's living standard call for a lot of energy resources consumption (Ge & Fan, 2013). Thus, the deployment of renewable energy industry becomes an opportunity for overcoming drawbacks supply of energy. In this sense, renewable energy diffusion provides new challenges for understanding the decision-making process of consumers (Dyner & Franco, 2004). Figure 6 shows the change in electricity demand between 2020 and 2050 in the case study. Scenario 2 presents an increase of 10% and 25% more than the BAU and scenario 1, respectively. This situation occurs due to the changes of time for the reputation to spread, which affects the expected demand. At the same time, earlier studies analyse rationally bounded energy use and propose policies that may reduce some of the apparent market failures (Dyner & Franco, 2004). Moreover, Ramos et al. (2015) argue that energy certificates and programs can be effective, but only if they are carefully designed and includes a good reputation.

Figure 5. The behaviour of the reputation levels

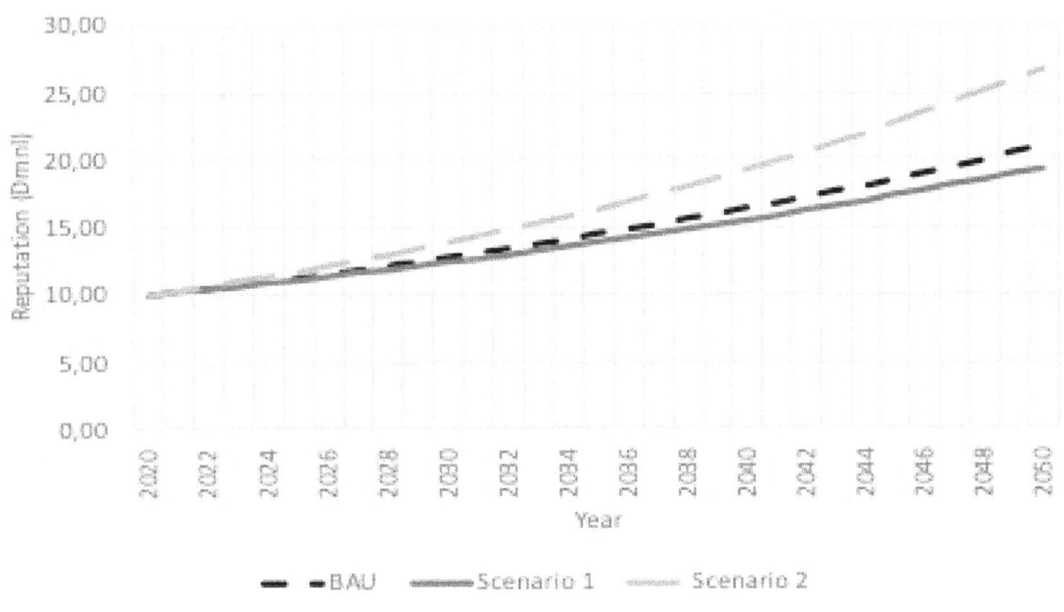

Figure 6. The behaviour of the electricity demand

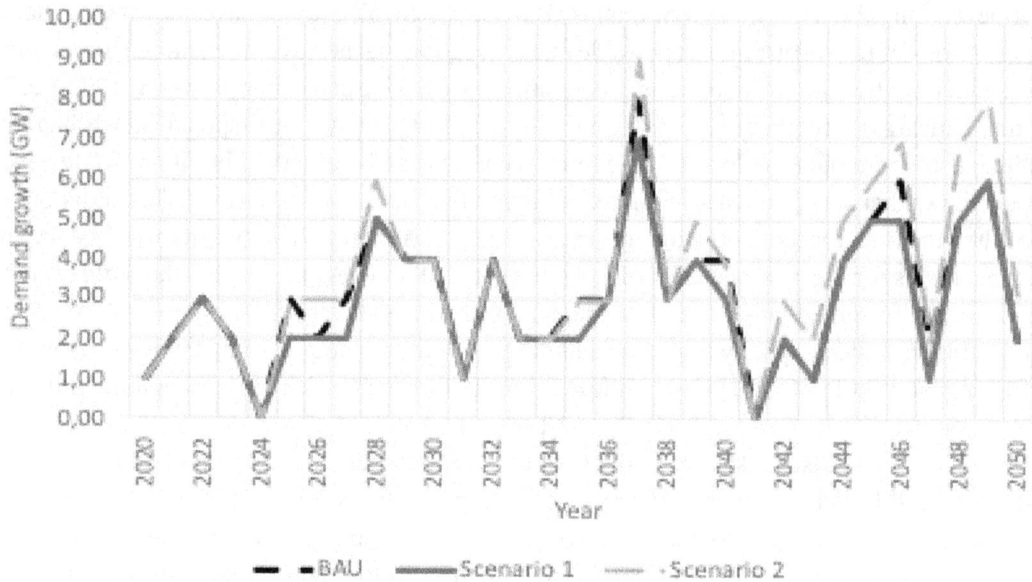

FUTURE RESEARCH DIRECTIONS

Given the challenges and opportunities renewable energy market in Latin America, this chapter considers a simulation modelling approach for better understanding the dynamics of the electricity industry from reputation and cooperation and its effect on the energy sector. This modelling approach is also helpful for assessing alternatives marketing based on non-conventional renewable energy that may foster innovation and performance improvement in SMEs of the electricity sector. It provides a model-based framework for analysing intangible resources of SMEs and offers guidance for designing alternative sustainable policy formulations. Moreover, future research can investigate: Will the expensive electricity generation of conventional renewable energy (e.g., hydropower) decreased reputation and cooperation, affecting the electricity market in Colombia?

CONCLUSION

Given the challenges and opportunities renewable energy market in Latin America (Gan & Li, 2015; Kumar Sahu, 2015), this paper considers a simulation modelling approach for better understanding the dynamics of electricity industry from reputation and cooperation, and its effect on the demand growth and capacity of supply. This chapter developed a simulation model that establishes the dynamic impact of the structure associated with behavioural aspects addressed by earlier studies (Dyner & Franco, 2004). Moreover, the chapter updates and improves the findings of a research project recently developed by the authors (Herrera et al., 2019a). In this sense, the findings contribute to analyse demand-side management incorporating the effects of the reputation on demand, but also the cooperation how an essential aspect of diffusion of renewable energy.

The collaborative behaviour is more driven by a need for trusting relationships and predictability (Christopher & Gattorna, 2005). However, this paper shows that reputation and cooperation should be considered in the decision-making processes. Although the collaboration and coordinated policy are fundamental in the development of the electricity industry (Herrera, 2019b), particularly in renewable energy diffusion, the reputation of customers may contribute to reducing the delays own of the energy system.

Several studies established that non-conventional renewable energy is an alternative to generate sustainable energy for the country, but there are barriers related with policy, innovation and market that do not allow the diffusion of the same (Arias-Gaviria et al., 2019; Zapata et al., 2019). In this context, the modelling of the demand-side management allows us to understand the energy system from the relationships identified between actors and design of policies alternatives, including reputation and cooperation. Moreover, a similar strategy could be used to analyse other marketing policies, using the simulation approach proposed here.

REFERENCES

Acuña, L. G., Ríos, D. R., Arboleda, C. P., & Ponzón, E. G. (2018). Cooperation model in the electricity energy market using bi-level optimisation and Shapley value. *Operations Research Perspectives, 5*, 161–168.

Ahmad, S., Mat Tahar, R., Muhammad-Sukki, F., Munir, A. B., & Abdul Rahim, R. (2016). Application of system dynamics approach in electricity sector modelling: A review. *Renewable & Sustainable Energy Reviews*, *56*, 29–37. doi:10.1016/j.rser.2015.11.034

Arend, R. J., & Bromiley, P. (2009). Assessing the dynamic capabilities view: Spare change, everyone? *Strategic Organization*, *7*(1), 75–90. doi:10.1177/1476127008100132

Arias-Gaviria, J., Carvajal-Quintero, S. X., & Arango-Aramburo, S. (2019). Understanding dynamics and policy for renewable energy diffusion in Colombia. *Renewable Energy*, *139*, 1111–1119. doi:10.1016/j.renene.2019.02.138

Bala, B. K., Arshad, F. M., & Noh, K. M. (2017). *System Dynamics: Modelling and Simulation*. Springer. doi:10.1007/SpringerReference_7284

Baldassarri, D. (2015). Cooperative networks: Altruism, group solidarity, reciprocity, and sanctioning in ugandan producer organisations. *American Journal of Sociology*, *121*(2), 355–395. doi:10.1086/682418 PMID:26594712

Bale, C. S. E., Varga, L., & Foxon, T. J. (2015). Energy and complexity: New ways forward. *Applied Energy*, *138*, 150–159. doi:10.1016/j.apenergy.2014.10.057

Barnett, M. L., Jermier, J. M., & Lafferty, B. A. (2006). Corporate Reputation: The Definitional Landscape. *Corporate Reputation Review*, *9*(1), 26–38. doi:10.1057/palgrave.crr.1550012

Becerra, M., González, E. C., Herrera, M. M., & Romero, O. R. (2016). Collaborative Planning Capacities in Distribution Centers. In L. Zhang, X. Song, & Y. Wu (Eds.), *Theory, Methodology, Tools and Applications for Modeling and Simulation of Complex Systems* (pp. 622–632). Springer Singapore. doi:10.1007/978-981-10-2663-8_64

Bentahar, J., Khosravifar, B., Serhani, M. A., & Alishahi, M. (2012). On the analysis of reputation for agent-based web services. *Expert Systems with Applications*, *39*(16), 12438–12450.

Cárdenas, L., Zapata, M., Franco, C. J., & Dyner, I. (2017). Assessing the combined effect of the diffusion of solar rooftop generation, energy conservation and efficient appliances in households. *Journal of Cleaner Production*, *162*, 491–503. doi:10.1016/j.jclepro.2017.06.068

Castaneda, M., Franco, C. J., & Dyner, I. (2017). Evaluating the effect of technology transformation on the electricity utility industry. *Renewable & Sustainable Energy Reviews*, *80*(65), 341–351. doi:10.1016/j.rser.2017.05.179

Castaneda, M., Jimenez, M., Zapata, S., Franco, C. J., & Dyner, I. (2017). Myths and facts of the utility death spiral. *Energy Policy*, *110*(65), 105–116. doi:10.1016/j.enpol.2017.07.063

Castaneda, M., Zapata, S., Cherni, J., Aristizabal, A. J., & Dyner, I. (2020). The long-term effects of cautious feed-in tariff reductions on photovoltaic generation in the UK residential sector. *Renewable Energy*, *155*, 1432–1443. doi:10.1016/j.renene.2020.04.051

Chen, H. H., Lee, A. H. I., & Chen, S. (2014). Strategic policy to select suitable intermediaries for innovation to promote PV solar energy industry in China. *Applied Energy*, *115*, 429–437. doi:10.1016/j.apenergy.2013.10.034

Christopher, M., & Gattorna, J. (2005). Supply chain cost management and value-based pricing, *34*, 115–121. doi:10.1016/j.indmarman.2004.07.016

Cosenz, F., & Noto, G. (2016). Applying System Dynamics Modelling to Strategic Management: A Literature Review. *Systems Research and Behavioral Science*, *33*(6), 703–741. doi:10.1002res.2386

Cosenz, F., Rodrigues, V. P., & Rosati, F. (2019). Dynamic business modeling for sustainability: Exploring a system dynamics perspective to develop sustainable business models. *Business Strategy and the Environment*, (April), 1–14. doi:10.1002/bse.2395

Davis, J. P., & Bingham, C. B. (2007). Developing Theory Through Simulation Methods. *Academy of Management Review*, *32*(650), 480–499.

Dyner, I. (2000). Energy modelling platforms for policy and strategy support. *The Journal of the Operational Research Society*, *51*(2), 136–144.

Dyner, I., & Larsen, E. R. (2001). From planning to strategy in the electricity industry. *Energy Policy*, *29*(13), 1145–1154. doi:10.1016/S0301-4215(01)00040-4

Dyner, I., & Franco, C. (2004). *Consumer´s Bounded Rationality: The case of competitive Energy Markets*. Medellin, Colombia: John Wiley & Sons Ltda. InterScience.

Eras, J. J. C., Morejón, M. B., Gutiérrez, A. S., García, A. P., Ulloa, M. C., Martínez, F. J. R., & Rueda-Bayona, J. G. (2019). A look to the electricity generation from non-conventional renewable energy sources in Colombia. *International Journal of Energy Economics and Policy*, *9*(1), 15–25. doi:10.32479/ijeep.7108

Ford, A. (1997). System Dynamics and the Electric Power Industry. *System Dynamics Review*, *13*(1), 57–85. doi:10.1002/(SICI)1099-1727(199721)13:1<57:AID-SDR117>3.0.CO;2-B

Ford, A. (1999). Cycles in competitive electricity markets: A simulation study of the western United States. *Energy Policy*, *27*(11), 637–658. doi:10.1016/S0301-4215(99)00050-6

Forrester, J. W. (1961). *Industrial dynamics*. M.I.T. Press.

Franco, C. J., Castaneda, M., & Dyner, I. (2015). Simulating the new British Electricity-Market Reform. *European Journal of Operational Research*, *245*(1), 273–285. doi:10.1016/j.ejor.2015.02.040

Gan, P. Y., & Li, Z. D. (2015). Quantitative study on long term global solar photovoltaic market. *Renewable & Sustainable Energy Reviews*, *46*, 88–99. doi:10.1016/j.rser.2015.02.041

Garwood, T. L., Hughes, B. R., Oates, M. R., O'Connor, D., & Hughes, R. (2018). A review of energy simulation tools for the manufacturing sector. *Renewable and Sustainable Energy Reviews, 81*(August), 895–911. doi:10.1016/j.rser.2017.08.063

Gary, M. S., Kunc, M., Morecroftc, J., & Scott, F. R. (2008). System dynamics and strategy. *System Dynamics Review*, *24*(4), 407–429. doi:10.1002dr.402

Ge, F. L., & Fan, Y. (2013). A system dynamics model of coordinated development of central and provincial economy and oil enterprises. *Energy Policy*, *60*, 41–51. doi:10.1016/j.enpol.2013.05.033

Gómez, C. R., Arango-Aramburo, S., & Larsen, E. R. (2017). Construction of a Chilean energy matrix portraying energy source substitution: A system dynamics approach. *Journal of Cleaner Production*, *162*, 903–913. doi:10.1016/j.jclepro.2017.06.111

Guo, X., & Guo, X. (2015). China's photovoltaic power development under policy incentives: A system dynamics analysis. *Energy*, *93*, 589–598. doi:10.1016/j.energy.2015.09.049

Helm, S., & Klode, C. (2011). Challenges in measuring corporate reputation. In *Reputation management* (pp. 99–110). Springer.

Herrera, M. M., Carvajal-Prieto, L. A., Uriona-Maldonado, M., & Ojeda, F. (2019a). Modeling the Customer Value Generation in the Industry's Supply Chain. *International Journal of System Dynamics Applications*, *8*(4), 1–13. doi:10.4018/ijsda.2019100101

Herrera, M. M., Cosenz, F., & Dyner, I. (2019b). How to support energy policy coordination? Findings from the Brazilian wind industry. *The Electricity Journal*, *32*(8), 106636. doi:10.1016/j.tej.2019.106636

Herrera, M. M., Uriona, M., & Dyner, I. (2020). Dynamics performance of the wind-power supply chain with transmission capacity constraints. *Iranian Journal of Electrical and Computer Engineering*, *10*(2), 1142–1148. doi:10.11591/ijece.v10i1.pp1142-1148

Hopfensitz, A., Miquel-Florensa, J., & Associates. (2015). How forced displacement flows affect public good contributions: The social consequences of conflict in Colombia. Toulouse School of Economics, 1–22.

Ibanez-Lopez, A. S., Martinez-Val, J. M., & Moratilla-Soria, B. Y. (2017). A dynamic simulation model for assessing the overall impact of incentive policies on power system reliability, costs and environment. *Energy Policy, 102*(December), 170–188. doi:10.1016/j.enpol.2016.12.026

IDEAM. (2017). *Promedios mensuales de irradiación global media recibida en superficie para las principales ciudades del país (Wh/m2 por día).* Retrieved from http://atlas.ideam.gov.co/basefiles/Anexo-Promedios-mensuales-de-Irradiacion-Global-Media.pdf

Jimenez, M., Franco, C. J., & Dyner, I. (2016). Diffusion of renewable energy technologies: The need for policy in Colombia. *Energy, 111*, 818–829. doi:10.1016/j.energy.2016.06.051

Kishimoto, J., Goto, M., & Inoue, K. (2017). Do acquisitions by electric utility companies create value? Evidence from deregulated markets. *Energy Policy, 105*(February), 212–224. doi:10.1016/j.enpol.2017.02.032

Kollock, P., Blumstein, P., & Schwartz, P. (1994). The judgment of equity in intimate relationships. *Social Psychology Quarterly*, 340–351.

Kumar, V., & Reinartz, W. (2016). Creating Enduring Customer Value. *Journal of Marketing*, *80*(6), 36–68. doi:10.1509/jm.15.0414

Kumar Sahu, B. (2015). A study on global solar PV energy developments and policies with special focus on the top ten solar PV power producing countries. *Renewable & Sustainable Energy Reviews*, *43*, 621–634. doi:10.1016/j.rser.2014.11.058

Leroi-Werelds, S., Streukens, S., Brady, M. K., & Swinnen, G. (2014). Assessing the value of commonly used methods for measuring customer value: A multi-setting empirical study. *Journal of the Academy of Marketing Science*, *42*(4), 430–451. doi:10.100711747-013-0363-4

Matos, S., & Silvestre, B. S. (2013). Managing stakeholder relations when developing sustainable business models: The case of the Brazilian energy sector. *Journal of Cleaner Production*, *45*, 61–73. doi:10.1016/j.jclepro.2012.04.023

Milinski, M. (2016). Reputation, a universal currency for human social interactions. *Philosophical Transactions of the Royal Society B: Biological Sciences, 371*(1687). doi:10.1098/rstb.2015.0100

Mollona, E. (2008). Strategic Modelling and Business Dynamics. A Feedback Systems Approach. *The Journal of Management and Governance*, (12), 213–217. doi:10.100710997-008-9050-0

Morcillo, J., Angulo, F., & Franco, C. J. (2020). Analysing the Hydroelectricity Variability on Power Markets from a System Dynamics and Dynamic Systems Perspective : Seasonality and. *Energies*, *13*(2381), 1–25. doi:10.3390/en13092381

Morcillo, J., Franco, C., & Angulo, F. (2017). Delays in electricity market models. *Energy Strategy Reviews*, *16*, 24–32. doi:10.1016/j.esr.2017.02.004

Namujju, D. L., Yücel, G., Pruyt, E., & Okou, R. (2015). A Simulation-Based Analysis of Electricity Access in Uganda. *International Journal of System Dynamics Applications*, *4*(1), 1–30.

Nikabadi, M. S., & Amir, H. (2018). A Dynamic Model of Effective Factors on Open Innovation in Manufacturing Small and Medium Sized Companies. *International Journal of System Dynamics Applications*, *7*(1), 1–26.

Ochoa, C., Dyner, I., & Franco, C. J. (2013). Simulating power integration in Latin America to assess challenges, opportunities, and threats. *Energy Policy*, *61*, 267–273. doi:10.1016/j.enpol.2013.07.029

Ochoa, C., & van Ackere, A. (2014). Does size matter? Simulating electricity market coupling between Colombia and Ecuador. *Renewable & Sustainable Energy Reviews*, *50*, 1108–1124. doi:10.1016/j.rser.2015.05.054

Oh, H., & Kim, K. (2017). Customer satisfaction, service quality, and customer value: Years 2000-2015. *International Journal of Contemporary Hospitality Management*, *29*(1), 2–29. doi:10.1108/IJCHM-10-2015-0594

Oliva, R. (2003). Model calibration as a testing strategy for system dynamics models. *European Journal of Operational Research*, *151*(3), 552–568. doi:10.1016/S0377-2217(02)00622-7

Papachristos, G. (2019). System dynamics modelling and simulation for socio-technical transitions research. *Environmental Innovation and Societal Transitions, 31*(September), 248–261. doi:10.1016/j.eist.2018.10.001

Qudrat-Ullah, H., & Seong, B. S. (2010). How to do structural validity of a system dynamics type simulation model: The case of an energy policy model. *Energy Policy*, *38*(5), 2216–2224. doi:10.1016/j.enpol.2009.12.009

Quiceno, G., Álvarez, C., Ávila, R., Fernández, Ó., Franco, C. J., Kunc, M., & Dyner, I. (2019). Scenario analysis for strategy design: A case study of the Colombian electricity industry. *Energy Strategy Reviews, 23*(December), 57–68. doi:10.1016/j.esr.2018.12.009

Radomes, A. A., & Arango, S. (2015). Renewable energy technology diffusion: An analysis of photo-voltaic-system support schemes in Medellín, Colombia. *Journal of Cleaner Production*, *92*, 152–161. doi:10.1016/j.jclepro.2014.12.090

Ramos, A., Gago, A., Labandeira, X., & Linares, P. (2015). The role of information for energy efficiency in the residential sector. *Energy Economics*, *52*, S17–S29.

Rand, D. G., & Nowak, M. A. (2013). Human cooperation. *Trends in Cognitive Sciences*, *17*(8), 413–425. doi:10.1016/j.tics.2013.06.003 PMID:23856025

Redondo, J. M., Olivar, G., Ibarra-Vega, D., & Dyner, I. (2018). Modeling for the regional integration of electricity markets. *Energy for Sustainable Development*, *43*, 100–113. doi:10.1016/j.esd.2017.12.003

Rego, E. E., & de Oliveira Ribeiro, C. (2018). Successful Brazilian experience for promoting wind energy generation. *The Electricity Journal*, *31*(2), 13–17. doi:10.1016/j.tej.2018.02.003

Repenning, N. P., & Sterman, J. D. (2002). Capability traps and self-confirming attribution errors in the dynamics of process improvement. *Administrative Science Quarterly*, *47*(2), 265–295. doi:10.2307/3094806

Rintamäki, T., & Kirves, K. (2017). From perceptions to propositions: Profiling customer value across retail contexts. *Journal of Retailing and Consumer Services*, *37*(October), 159–167. doi:10.1016/j.jretconser.2016.07.016

Salem Khalifa, A. (2004). *Customer value: a review of recent literature and an integrative configuration. Management Decision* (Vol. 42). doi:10.1108/00251740410538497

Silva, A. R., Pimenta, F. M., Assireu, A. T., & Spyrides, M. H. C. (2016). Complementarity of Brazils hydro and offshore wind power. *Renewable & Sustainable Energy Reviews*, *56*, 413–427. doi:10.1016/j.rser.2015.11.045

Sterman, J. D. (1984). Appropriate summary statistics for evaluating the historical fit of system dynamics models. *Dynamica*. Retrieved from https://www.systemdynamics.org/conferences/1983/proceed/plenary/sterm203.pdf

Sterman, J. D. (2000). *Business Dynamics: Systems Thinking and Modeling for a Complex World*. McGraw-Hill.

Swakman, V., Molleman, L., Ule, A., & Egas, M. (2016). Reputation-based cooperation: Empirical evidence for behavioral strategies. *Evolution and Human Behavior*, *37*(3), 230–235.

Sylwester, K., & Roberts, G. (2013). Reputation-based partner choice is an effective alternative to indirect reciprocity in solving social dilemmas. *Evolution and Human Behavior*, *34*(3), 201–206.

Tayal, D. (2017). Leveraging innovation for electricity utilities. *The Electricity Journal*, *30*(3), 23–29. doi:10.1016/j.tej.2017.02.008

Torres, J. P., Kunc, M., & O'Brien, F. (2017). Supporting strategy using system dynamics. *European Journal of Operational Research*, *260*(3), 1081–1094. doi:10.1016/j.ejor.2017.01.018

UPME. (2016). *Boletin estadístico de Minas y energía 2012 – 2016*. Retrieved from http://www1.upme.gov.co/PromocionSector/SeccionesInteres/Documents/Boletines/Boletin_Estadistico_2012_2016.pdf

UPME & BID. (2015). *Integración de las energías renovables no convencionales en Colombia. Ministerio de Minas y Energía*. Retrieved from http://www.upme.gov.co/Estudios/2015/Integracion_Energias_Renovables/INTEGRACION_ENERGIAS_RENOVANLES_WEB.pdf

Warren, K. (2005). Improving strategic management with the fundamental principles of system dynamics. *System Dynamics Review*, *21*(4), 329–350. doi:10.1002dr.325

Wheat, D. (2010). What Can System Dynamics Learn From the Public Policy Implementation Literature? *Systems Research and Behavioral Science*, *27*(3), 425–442. doi:10.1002res.1039 PMID:32327920

Wu, J., Balliet, D., & Van Lange, P. A. M. (2016). Reputation, Gossip, and Human Cooperation. *Social and Personality Psychology Compass*, *10*(6), 350–364. doi:10.1111pc3.12255

Xu, C., Peak, D., & Prybutok, V. (2015). A customer value, satisfaction, and loyalty perspective of mobile application recommendations. *Decision Support Systems*, *79*(September), 171–183. doi:10.1016/j.dss.2015.08.008

Zapata, S., Castaneda, M., Franco, C. J., & Dyner, I. (2019). Clean and secure power supply: A system dynamics based appraisal. *Energy Policy*, *131*(April), 9–21. doi:10.1016/j.enpol.2019.04.028

ADDITIONAL READING

Aslani, A., Akbari, S., & Tabasi, S. (2018). The Robustness of Natural Gas Energy Supply: System Dynamics Modelling. *International Journal of System Dynamics Applications*, *7*(3), 57–71. doi:10.4018/IJSDA.2018070103

Barnabè, F., & Davidsen, P. I. (2020). Exploring the potentials of behavioral system dynamics: Insights from the field. *Journal of Modelling in Management*, *15*(1), 339–364. doi:10.1108/JM2-03-2019-0081

Bianchi, C. (2016). *System Dynamics for Performance Management* (Vol. 49). Springer International Publishing.

Gholizad, A., Ahmadi, L., Hassannayebi, E., Memarpour, M., & Shakibayifar, M. (2017). A System Dynamics Model for the Analysis of the Deregulation in Electricity Market. *International Journal of System Dynamics Applications*, *6*(2), 1–30. doi:10.4018/IJSDA.2017040101

Herrera, M. M., Dyner, I., & Cosenz, F. (2018). Alternative Energy Policy for Mitigating the Asynchrony of the Wind-Power Industry's Supply Chain in Brazil. In H. Qudrat-Ullah (Ed.), *Innovative Solutions for Sustainable Supply Chains* (pp. 199–221). Springer International Publishing.

Herrera, M. M., León, L. S., & Vargas-Ortiz, L. K. (2018). A dynamic analysis of the effects of word-of-mouth on online brand communities. *Suma de Negocios*, *9*(20), 77–85.

Herrera-Ramírez, M. M., Orjuela-Castro, J., Sandoval-Cruz, H., & Martínez-Vargas, M. A. (2017). *Modelado dinámico y estratégico de la cadena agroindustrial de Modelado dinámico y estratégico*. Universidad Piloto de Colombia.

Namujju, D. L., Yücel, G., Pruyt, E., & Okou, R. (2015). A Simulation-Based Analysis of Electricity Access in Uganda. *International Journal of System Dynamics Applications*, *4*(1), 1–30. doi:10.4018/IJSDA.2015010101

Oliva, E. J. D., Prieto, L. A. C. (2015). Organizational identity and its influence on image: A theoretical reflection. Suma de negócios, 6(13), 114-123.

Qudrat-Ullah, H. (2016). *The Physics of Stocks and Flows of Energy Systems Applications in Energy Policy*. Springer London.

Warren, K. (2008). *Strategic Management Dynamics*. John Wiley & Sons.

KEY TERMS AND DEFINITIONS

Behavioral Research: Is the study of behavioral aspect associated to the use of methods of modelling, basically by decision support system.

Cooperation Indicator: Measurement of action or process that assesses the working together to the same end throughout electricity industry.

Electricity Industry: Contemplates the generation, transmission, distribution, and sale of electricity to the commercial and industry sector of a country.

Installed Capacity: The maximum electric power generation capacity available of the system.

Reputation Indicator: Appraisal of an institution based on the economic, social, and environmental impacts attributed to the institution over time.

Stock-and-Flow Diagram: Is a tool used to represent and assess the behaviour of the energy system developed by system dynamics modelling.

System Dynamics Modelling: Is a simulation method particularly applicable for understanding the behaviour of a complex system in the long-term. Moreover, useful when the theoretical focus is nonlinear.

Chapter 8
Production Complexity of the Iranian Performing Arts:
An Economic Analysis in Three Acts

Hooman Abdollahi
https://orcid.org/0000-0001-9733-504X
School of Business and Economics, The Arctic University of Norway (UiT), Tromsø, Norway

Amsalu Tadele Alamneh
System Dynamics Group, Department of Geography, University of Bergen, Norway

Nazanin Sharif
Department of Mechanical Engineering, Université Laval, Québec, Canada

ABSTRACT

Theater is referred to as a cultural product which plays a major part in the cultural paradigm of civilized societies. Public theater is a type of performing arts in which the production receives financial support from the government. Therefore, the public theater production follows a complex trend controlled by a particular socio-economic system. The aims of this study are to identify interactions among financial and economic factors in the Iranian public theater and to find the possible threats and suggest effective policies in order to improve the position of the Iranian theater. For this purpose, the authors use system dynamics methodology to build a model that can explain or mimic the system. They present their analysis in accordance with principles of microeconomics. The analysis is conducted in three acts, namely production, distribution, and consumption. Findings reveal that the position of the Iranian public theater is undesired due to accumulated workforce, lack of distribution networks, and underconsumption. Finally, policies are suggested to overcome the shortcomings.

DOI: 10.4018/978-1-7998-5788-4.ch008

INTRODUCTION

In the modern era, the culture is of a socio-economic formation, thus cultural activities can be considered as a vocation. Durrant (1954) believes four elements constitute the human civilization: economic provision, political organization, moral tradition, and the pursuit of knowledge and the arts. It is imperative that theater, as an important artistic-cultural activity, receives a great attention due to its potentials to civilize societies, nurture constructive thoughts, and create political and societal attitudes. On the other hand, cultural producers need an abundant financial supply because producing cultural products like theater needs a substantial capital. Owing to the fact that economics deals with the studies of production, producers, and their relationships therefore all cultural products, including theater, can be considered as goods and services following a specific trend determined by a particular socio-economic system.

Socio-economic systems investigate the relationships between economics and social/cultural values. This investigation is characterized by interacting patterns which usually compound environmental, behavioral, and dynamic relationships. Therefore, System Dynamics (SD henceforth) approach is an appropriate tool for studying and understanding socio-economic systems. Literature reports that the application of SD has been proved to be functional in modeling cultural and artistic phenomena (see Haslett, 2012; Wyburn and Roach, 2013). The aim of SD is to achieve the standards of controllability and dynamic behavior which are commonplace in various systems (Coyle, 1996). SD is a way of incorporating all available information including written description and personal experience with computer simulation to reach a better understanding of social systems (Forrester, 1976). By means of the result obtained from simulation, a lucid perception of system's future behavior will be achieved so that the policy analysis will be more effective.

The objective of this study is to investigate the relationships between Iran public theater economy and production structure based on a dynamic model including the all economic stages, namely production, distribution, and consumption, in order to achieve practical and theoretical solutions for the dilemmas of the Iranian public theater. The production process is defined as all economic activities that aim to satisfy human needs directly or indirectly. Therefore, in this context production is equivalent to the creation of a cultural product. Before going any further, it is essential to categorize the Iranian theater to gain a comprehensive cognition of the public theater position. Sadeghi (1997) presents a thorough classification of Iranian national theater as follows:

i. Commercial theater: This category encompasses plays performed for entertainment purposes without having a slight tinge of artistic creativity. The aim of this theater is to produce profit for investors.
ii. Academic theater: This class contains theatrical performances produced by students in the Iranian universities and academies using rudimentary facilities. Academic theaters are generally performed sporadically due to lack of available halls. Also, such performances fail to attract public attention due the paucity of desired experience and equipment.
iii. Festival theater: This class is refereed to special theaters produced to be performed in occasional festivals offered by various organizations. Considering the characteristics of the host organization, these theaters have miscellaneous themes but, reportedly, the performance quality is not revered by the experts.
iv. Private theater: This type of theater is a long-standing class with a meandering pattern. In the past, comedy theatrical performances would be performed in private theater houses. However, these theater houses were closed or changed functionality after 1979 revolution. Nowadays, this term is

used for theatrical performances produced by artists who are unwilling to be involved in the tiresome bureaucratic procedures of the public theater. Nonetheless, this class, with its newly formed definition, is at an embryonic stage by the time this chapter is written.

v. Public theater: This type of theater which is replete with knowledge, skill, creativity, and intellectual theme can be considered inspirational and as a pattern for other types of theater. This segment needs financial support as the government funds it through a subsidy system.

The Iranian state supports the public theater via Iran's Dramatic Arts Center, but the vague financial policy has eventuated in a low efficacy and qualified the market place. Hence, this research is being directed to answer these questions: (1) How do financial and economic elements of the Iranian public theater influence one another? (2) What are the threats and dilemmas of Iran public theater and how they can be confronted? Finding answers for these questions forms the motivations behind this study.

The main problem of this study is the severe economic and financial position of the Iranian public theater which led to an accumulated dissatisfaction among artists. Although theater economy has been previously investigated on a factor-by-factor basis using statistical methods without understanding the interactions among them (see Castiglione et al., 2018; Fernández-Blancoet al., 2019; Last and Wetzel, 2010), it is necessary to examine this issue through a systematic approach considering feedback mechanisms contributing to all economic and production stages. The rest of the study proceeds as follows. Background section reviews the literature germane to this study. In the third section the dynamic model is presented. This is followed by model analysis. A few policies are suggested in the fifth section. Finally, the last section concludes the chapter.

BACKGROUND

Cultural economy has been investigated by scholars and some analyses have been applied to conundrums in arts and culture. Economists subsume cultural services under the category of service industries, or better still, tertiary sector of industry in which intangible goods are produced. In other words, this sector gives special services to customers. This area of research has been widely studied by researchers such as Becker (1984), Menger (1999), Lena and Lindeman (2014). Baumol and Bowen (1966) are the first who discuss the implications of general policy on theater economy. Theater economy is a multidisciplinary branch of a wider area called the economy of arts. Zuidervaart (2010) seeks answers for questions like why should government provide funding for theater and other arts? He indicates a vigorous case for government funding based on crucial contributions the arts make to a civil society. According to Feder and Katz-Gerro (2012), it is true that the funding of arts is meant to sustain artistic organizations and to provide more support as the demand for arts in the population grows, however, at the same time, this funding reflects power relations in the society and may have consequences for the preservation of these interactions in a way that parallels cultural policy in authoritarian states. In this case, Fazioli and Filippini (1997) examine a sample of 28 theater in Italy and mention that the subsidy system seems to promote the conditions for its collapse, especially since the public budget needs to confront financial shortage and call for a reform of the state subsidy system which promotes a limited number of high-quality performances. Ginsburgh and Throsby (2006) argue the value of theater and arts and customers' behavior towards it. They also analyze the effect of arts on the economy of developed and developing countries, presenting an overview on the history of arts market.

Theater market cannot be a competitive market because theater products are not considered homogenous and the quantity of the suppliers and producers are not adequate for a competitive market (Abdollahi et al., 2016). The market structure of the public theater is dominated by a few firms and seems highly concentrated; this type of market is known as oligopoly (Heilbrun and Gray, 2001). Research shows theater audience constitutes the industry's primary market, consequently they dictate the nature and design of theater products (Ntangaare, 2001). However, public theater is recognized as a non-profit activity and as Tina Packer states: *The great cathedrals of Europe would never have been built; Van Gogh would never have painted; Michelangelo would never have sculpted if it were just about marketplace demands.*

In a seminal paper, Grisolia and Willis (2012) investigate market segments for theater demand via a latent class model; the findings identify three classes of theatergoers: affluent class, popular class, and intellectual class. The research also provides a clear picture of what theater means for society and suggests that different pricing and marketing strategies should be applied to the intellectual, affluent, and popular classes. By and large, the functioning of art markets is covered by both economists and sociologists of culture (see Breiger, 2005; Beckert and Rossel, 2013; O'Neil, 2008; Velthuis, 2013). Baldin and Bille (2018) investigate the behavioral choice for theater tickets using for the sale system of the royal Danish national theatre. Also, Baldin et l., (2018) propose a multi-objective optimization model to study pricing and seat allocation in Denmark.

Sadeghi (1997) examines Iran public theater position, providing evidence that productivity is low, distribution is inappropriate, and income is inadequate. By comparing the production techniques and economy structure of Iran public theater with France public theater, Jalalipor (2009) finds that the growth of the Iranian theater needs much more funding. Sadeghi (2011) reports the audience crisis is the most important dilemma for the Iranian theater which is rooted in economic troubles compelling the artists to leave theater and participate in movies or TV series productions. Afshar (2013) studies Iran cultural situation and insists the necessity of state subsidy. Mokhtabad (2013) suggests repudiating state's cultural model by replacing a model based on private sector. Afshariasl (2013) uses a survey to assess the various ways of Iran theater development. This research reveals that increasing in theater buildings, advertising, and familiarizing people with theater are chief remedies.

To be in line with the literature developed so far, this research investigates the Iranian public theater as a cultural good comprehensively via modeling the interacting elements between production, distribution, and consumption in order to find the position of this cultural activity and also its potential threats. Despite a large amount of research on theater economy, very few works have studied and analyzed this context quantitatively and empirically. In addition, no research has been found that modeled theater production structure and economy through modern methods, such as system dynamics, but statistical models. Moreover, another contribution of the present paper is to develop the scope of SD application within the boundaries of cultural economics.

RESEARCH METHOD AND MODELLING

In economics context, goods produced not only for private use or satisfying personal needs but also for trading (Abdollahi and Talooni, 2016). Therefore, all cultural goods are produced and distributed within a society for transaction and business purposes. The production trend has three stages consisting of production, distribution, and consumption. In this context, the authors intend to systematically investigate the Iranian public theater with respect to these stages. Of note is that the authors' perspective in this study is based on microeconomics principles. In so doing, a helpful method named SD, created by Professor Jay W. Forrester of Massachusetts Institute of Technology in 1957, is employed.

Applied to problems in a wide variety of disciplines such as economics, the SD is a type of computer modeling method rooted in intellectual origins in management science, control engineering, technology, and business (see Nikabadi and Hakaki, 2018; Moysis et al., 2017; Thakker et al., 2013; Onggo, 2012; Abdelsalam and El-Tagy, 2012). Moreover, the SD has been applied in studying different subjects. For instance, Guma et al. (2018) employ the SD to explore food security indicators for sustainability in order to achieve a lucid understanding of the food security systems. Azhari et al. (2014) applied a case study on construction and demolition waste using the SD as the research methodology. In a seminal paper, Pruyt and Logtens (2015) design a SD model germane to the societal aging in Netherlands and report interesting findings concerning the welfare system. These examples effectively indicate the capacity of the SD in capturing complexities within different systems and efficiently justify the application of this methodology in the present study. In following subsections, the SD method and the proposed model are described.

System Dynamics Definition

SD was originally created as an instrument for helping managers to gain better understanding in controlling corporate systems by building computer simulation models of abstruse behavior within enterprises. SD models are created by identifying and linking the relevant pieces of a system's structure and simulating the behavior generated by that structure. Through an iterative process of structure identification, mapping, and simulation, a model emerges that can explain or mimic a system's problematic behavior and serve as a vehicle for policy design and testing (Radzicki, 2009).

The structure of the SD model consists of stocks, flows, and feedback loops, which are defined as follows:

1. **Stock:** Stocks accumulate or de-cumulate the information that flows into or out of them over time. Stocks can be thought of as bathtubs and are represented by a rectangular in the diagram.
2. **Flow:** Information flows enter and exit a system's stock and create its dynamic behavior. Flows represent the rate at which the stock is changing at any given instant. Flow can be considered as a faucet which fills or drains the stocks. Flows are represented by a faucet in the diagram.
3. **Feedback:** Feedback is the information transmission accumulated in a system's stock. In other words, information moves from a stock return to its flows directly or indirectly. This information movement prompts the system's faucets to close completely, close a bit, open more or stay unchanged. Feedbacks are represented by an arrow in the diagram.

In addition to above elements, in a SD model, auxiliary variables consisting of functions of stocks, constants, or exogenous inputs are necessary. In the SD model, there are two types of feedback: positive loops and negative loops (Azar, 2012). When the return of information reinforces a specific behavior, the loop is said positive. Positive loops engender self-reinforcing behavior and are responsible for the growth or decline of the system. Negative feedback loops generate goal-seeking behavior and are responsible for both stabilizing and oscillating.

In Figure1, an example of stock and flow in an economic context including a stock of employed labor and its inflow of hiring rate and its outflow of separations of labors is presented.

Figure 1. Example of stock and flow diagram

Figure 2 shows the same example with feedback loops. In Figure 2, the transmission of information is presented by arrows, which express the relations among system components. As can be seen in Figure 1, a perceived state of the system current situation is achieved. Comparing the status quo with the system's goal, a gap will emerge that affects the system's behavior. It means if employed labors are more than the goal, there is an increase in labor separation and a decrease in hiring rate. Similarly, if employed labors are less than the goal, there is an increase in hiring rate and a decrease in labor separations.

Figure 2. Example of stock and flow diagram with the presence of feedback loops

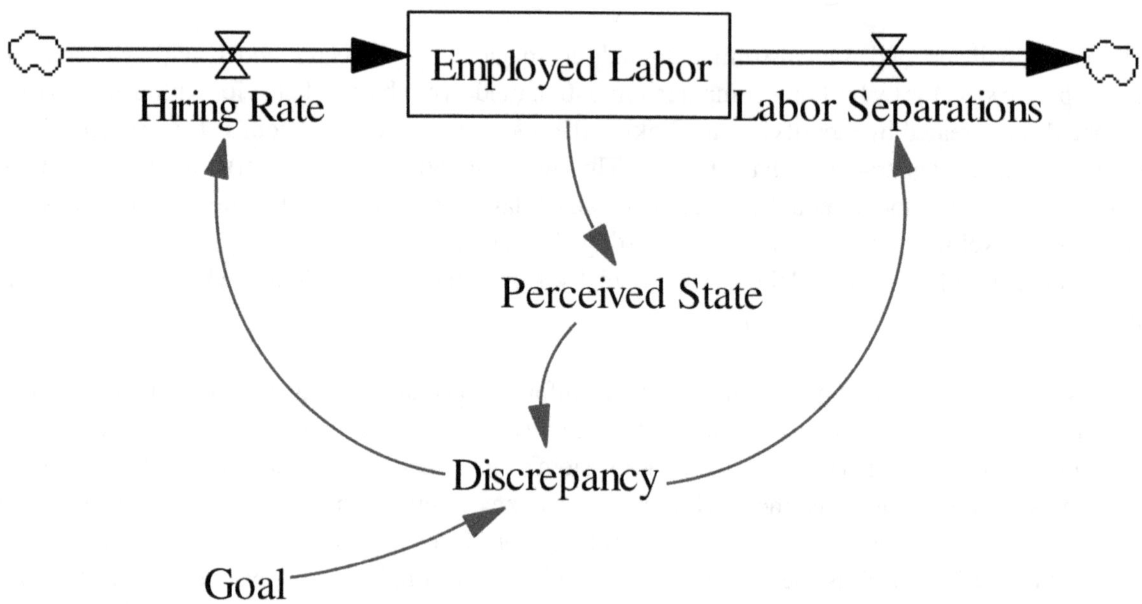

Modeling Process

This section introduces a model to present all components involved in the production structure and economy of the Iranian public theater via the SD method. By spotting relationships among system components, this method predicts the future behavior of the system. Sterman (2000) suggests five steps for system modeling:

Problem Articulation (Boundary Selection)

Problem articulation is the most important step in a modeling process (Abdollahi, 2020). SD model dynamically characterizes the problem, which shows how the problem arose and how it might evolve in the future. Furthermore, to show how the problem emerged and describe its symptoms, time horizon should be considered.

Formulation of Dynamic Hypothesis

Once the problem has been identified and characterized over an appropriate time horizon, dynamic hypothesis must begin to develop to account for the problematic behavior. The hypothesis is dynamic because it must provide an explanation of the dynamic characterizing of the problem in terms of the underlying feedback, stock, and flow structures of the system. It is a hypothesis because it is always provisional, subject to revision or abandonment as modeler learns from the modeling process and from the real world. A dynamic hypothesis is a working theory of how the problem arises.

Formulation of a Simulation Model

After building an initial dynamic hypothesis and a conceptual model, we must test the model. Sometimes modeler can test the dynamic hypothesis directly through data collection or experiments in the real system. Most of the time, modeler's ability to correctly interpret the dynamics of a complex system is extremely poor; therefore, these experiments must be conducted in a virtual world. In so doing, the modeler must move from the conceptual realm of diagrams to a fully specified formal model completed with equations, parameters, and initial conditions.

Testing

In order to gain assurance of model logicality and validity, it must be tested. The model must reproduce historical behavior, also the sensitivity of model behavior and policy recommendations must be assessed in light of the uncertainty in assumptions, both parametric and structural.

Policy Design and Evaluation

By developing confidence in the structure and behavior of the model, it can be used to design and evaluate policies for system improvement.

Model Description and Diagram

In this subsection, the authors present the model for investigating the economic trend and the production structure of the public theater in Iran. Before going any further, it must be noted that the mere and the most prestigious place for the Iranian public theater is Tehran City Theater complex containing five halls. Hence, this complex is the boundary of the proposed model.

Performing a play in Tehran City Theater is only possible for professional and qualified theater groups – e.g. in 2012, there were about 300 professional groups and each group has 23 members on average. On the other hand, 1400 theater graduates enter the market annually; if these graduates cannot be admitted in a professional group, they must constitute amateur or novice groups and gain qualification by performing at least three plays out of the complex's halls. In view of performance traffic in private and academic halls, this process usually lasts about three years.

There are between 270 and 400 demands for performance in the City Theater halls yearly but the productivity of the complex does not allow to meet this demand. Only 8 or 9 plays can be performed in each hall in one year. Restriction of performance is another provision, meaning that when a group staged in one of the complex's halls, there is a two-year prohibition on the next turn. Thus, the demand rejection rate increases and leads to a dissatisfaction among theater artists.

The Iranian government is the sole financial supporter of the public theater. On average, about 1.5 percent of the Iranian public budget is spent on theater and less than 25 percent of that budget is used for production purposes. For instance, the budget per capita for each theater artist was about two hundred thousand Rials (about 1 US dollar) in 2014, which was a pittance. As a result of the high rejection rate and also unsatisfactory financial situation, the skilled, experienced, and well-known theater activists abandoned theater to work in the radio, TV series, cinema, or dubbing. The absence of eminent artists led to the substitution of less-experienced individuals and resulted in a quality decline and dissatisfaction of audience.

State budget is expended for theater production. It means a specific subsidy is paid for each play but it is not enough to defray production costs like inviting experts such as scenic designer, costume designer, lighting designer, thespians, graphic designer, or the rent of rehearsal place, etc. Obviously, a play quality depends upon the skills of the cast; the more professional the cast is, the more expensive the production costs will be. Furthermore, sometimes there is a discrepancy between director's needs and a hall's equipment. In such a case, the group must cover the cost of supplying requirements.

Supply gives rise to consumers' reaction. Each performance in the City Theater, during 30 nights, has around 3000 spectators. These spectators are persons who consider theater as a need in their lives. However, a play gets public attention by using efficacious advertising and prominent performers. Likewise, the budget spent on cultural events of society's individuals (cultural budget), substitute goods, and market segmentation may have a big impact on the number of theater audience.

In addition to subsidy, current wage policy allocates 76 percent of box office to the members of theater group and the rest is used for tax coverage. When sales rate is high, the production is profitable and theatrical groups earn more income. This elevates artists' motivation and, as a result, more theatrical groups ask for performance, whereas the capacity of complex still remains limited. So, the rejection rate rises again. It means, even good salary renders to a dissatisfaction.

In Figure 3, the stock and flow diagram of the Iranian public theater is presented which shows above description as a SD model. State budget, Workforce, Audit, Performing Theater, Archive, Box office, and Income are the stocks of the model. The authors use Vensim PLE software to achieve simulation results. The equations of the model are presented in Table1.

Data

Data sources include literature reviews, reports announced by Public Relation Office of Tehran City Theater complex (2012-2015), the annually reports of Iranian Dramatic Arts Center (2012-2015), Koocheh theater group experts for designing the public theater's model. For gathering data related to cultural budget of Iranian families, the reports of classification of individual consumption by purpose issued by Central Bank of Iran from 2012 to 2015 were studied. Furthermore, in order to estimate the growth in costs, the authors use inflation rate forecast provided by Trading Economics website.

Model Validity

To estimate the model validity, the results of model simulation must be compared with actual data. By using validation process, the authors will be able to confirm model's accuracy and scrutinize whether the modeled system generates expected behavior. In this paper, the researchers apply reproducing past behavior to verify model validity. In this step, stock-flow diagram has been entered in Vensim PLE Software and their formulations have been written.

In Table 2, it is obvious that the deviation between actual and simulated data is insignificant. Thus, the model is able to predict the future behavior of the system.

MODEL ANALYSIS

Act I: Production

In this paper, theater is defined as a cultural good produced for utility creation. To achieve a thorough identification of theater production, the authors categorized it in three segments: workforce or main material, production costs, and production trend. By model simulation, the future situation of the public theater is predicted and practical suggestions for quandaries are presented.

Figure 3. A system dynamics Stock-flow diagram representing the simulated system of Iran's public theater- extracted from Vensim PLE software

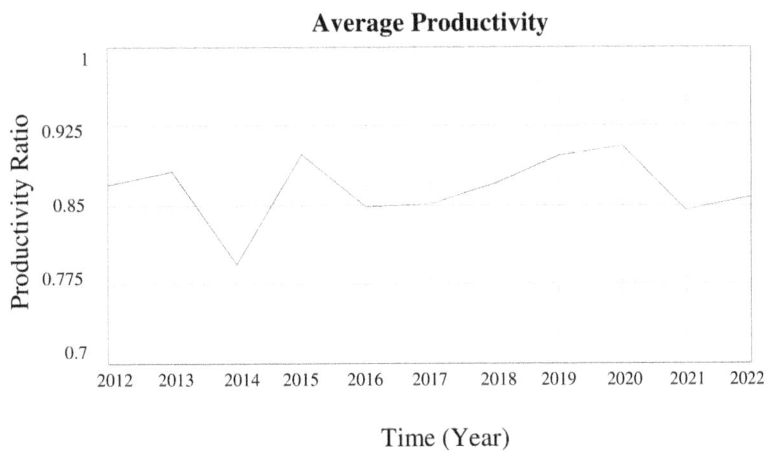

Table 1. The equations of Iran public theater model

Equation	Variable
Addition Rate-Exit Rate	Workforce
Demand Rate-Approbation Rate-Rejection Rate	Control & evaluation
Approbation Rate-Performing Rate	Performing theaters
Õ Performing Rate	Archive
Õ Sales Rate	Box office
Õ Net Income Rate	Income
Augmentation rate-Expenditure Rate	Budget
((Engagement in radio, TV, etc + Job dissatisfaction+ The ratio of Death&Retirement)*workforce)/23	Exit rate
IF THEN ELSE("Budget/Workforce">=1.5e+008, The Effect of Rejection Rate on Dissatisfaction, The Effect of Rejection Rate on Dissatisfaction+0.002)	Job dissatisfaction ratio
Rejection Rate/100000	The effect of rejection rate on dissatisfaction
0.3	Engagement in radio, TV, etc
0.002	The ratio of death & retirement
Graduates = 1400	Addition rate
(2e+009+(2e+009*Time)-STEP(3.3e+009,2)-STEP(9e+008,3)	Augmentation rate
Archive* Subsidy	Expenditure rate
(1e+007)+(1.5e+007*Time)	Subsidy
Income/(total cost*5)	Profitability ratio
Workforce – Qualified groups	Novice groups
DELAY FIXED(Novice Groups, 1, Novice Groups)	Gaining experience
(Qualified Groups-(Archive/5))*(0.5+Profitability Ratio-Job Dissatisfaction Ratio)	Demand for performing
Demand for performing	Demand rate
Control & Evaluation (AUDIT) - Approbation Rate	Rejection rate
Maximum capacity*average Productivity	Approbation rate
RANDOM UNIFORM(0.78, 0.96, 0)	Average productivity
50	Maximum capacity
(5e+006) + (5e+006*Time)	Average cost per play
(1.5e+006) + 3*Time + STEP(1.25e+007, 2)	Advertising cost
Performing Theatres*(Technology Cost + Average Cost per Play + Subside)	Production cost
Advertising Cost + Production Cost	Total cost
IF THEN ELSE((Advertising Effectiveness + Average Demand)*The Effect of Quality on Customers' Demand>=Demand Capacity, Demand Capacity, (Advertising Effectiveness + Average Demand)*The Effect of Quality on Customers' Demand(	Customer's demand
Advertising Cost*0.0001	Advertising effectiveness
3000	Average demand per play
Ave. Demand per Play*Performing Theatres	Average demand
[(0, 0.75), (1, 3.25)], as graph	The effect of quality on customers' demand

continues on following page

Table 1. Continued

Equation	Variable
Production Cost/(Ideal quality level price*Performing Theatres)	The effect of cost on quality
[(2011, 9.8e+007), (2012, 1.1e+008), (2013, 1.21e+008), (2014, 1.31e+008), (2015, 1.4e+008), (2016, 1.49e+008), (2017, 1.69e+008), (2018, 1.8e+008), (2019, 1.99e+008), (2020, 2.19e+008), (2021, 2.37e+008)], as graph	Ideal quality level price
Performing Theatres*1100*30*Performing Period	Demand capacity
IF THEN ELSE (Cultural Budget>=Ticket Price, Customers' Demand, 0.6*Customers' Demand)	Sold tickets
10000 + (700*Time)	Cultural budget
4145 + 1000* (Time)	Ticket price
Sold Tickets*Ticket Price	Sales rate

Source: Authors' computations

Table 2. Comparison between the actual and simulated data of the Iranian public theater (2012-2015)

Year					
2015	2014	2013	2012	Data	
—	1458281440	895432230	537638416	Realized	Sales
—	1492761000	974569000	533797000	Simulated	
—	2.3	8.8	0. 7	Deviation%	
—	236734	190113	129739	Realized	Spectators
—	242923	189421	128871	Simulated	
—	2. 5	0.4	0.7	Deviation%	
40	35	43	43	Realized	Number of performed Theater
44	39	44	43	Simulated	
9	11	2	0	Deviation%	
12000000000	9000000000	5000000000	2400000000	Realized	Budget
10654700000	7989250000	4000000000	2000000000	Simulated	
12	11	25	20	Deviation%	

Source: Research findings

Workforce

In this context, workforce implies individuals involved in public theater production. The main part of this population is directors, thespians, and designers. Approximately 1400 theater graduates add to this workforce yearly. Of note is that, no recruitment opportunity is provided for newly comers and, on the other hand, job security is not available for the experienced artists as well. Lacking job security caused a huge number of resourceful artists to leave theater and less-experienced artists took their place. As a result, the quality of plays does not enhance substantially. Moreover, due to the paucity of funding and standard halls most of these artists cannot take part in production realm. Figure 4 displays the number of groups for Iran's public theater.

By 2022 the workplace is predicted to grow by around 75%. Because of weak planning, there is not a logical relationship between the workforce and production capacity. In other words, the industry will encounter with an accumulated workforce in one side and capacity shortage and deficiency on the other. To solve this problem, Iran public theater needs management reformation and innovation, which increase the production capacity. For instance, organizing the theater groups to perform in plants, mosques, and schools could be practical solutions. However, the production cost is another effective element in the production phase of cultural goods. Next, the production cost of Iran public theater is analyzed and discussed.

Production Cost

Total production cost comprises of fixed and variable costs. The fixed cost does not change with a rise or fall in the amount of goods or services produced. In theater economics, setting up expenses are the fixed cost. On the contrary, the variable cost includes the expenses spent during the performance period like permanent staff's wages and the hall rent-denoted in the model as average cost, technology cost (a cost for hiring a special instrument not supplied in the performance hall in advance), and advertisement cost. In the model equations, these costs are anticipated in accordance with inflation rate forecast.

Figure 4. The simulation result of the national public theater's workforce by Vensim PLE software

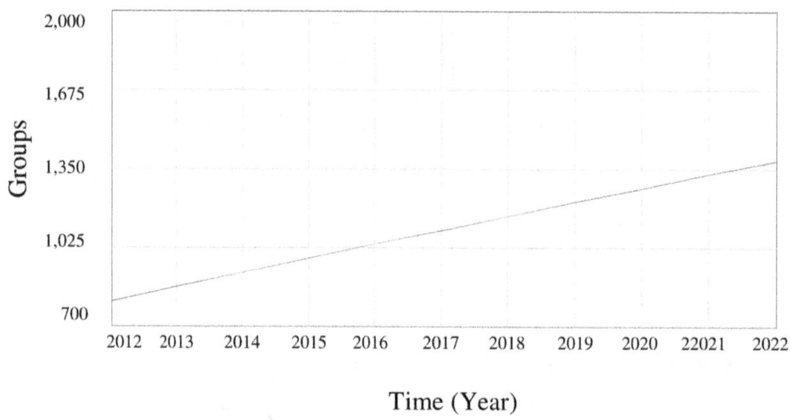

The state plays the leading role in theater production by paying a subsidy for fixed cost however this subsidy has eventuated in a dilemma for the public theater. There is often a delay in subsidy payment due to the lack of lucid instruction. Model's simulation predicts that total production cost will increase three times by 2022, it may be a cumbersome for the state to keep the subsidy up with this growing production cost. The production becomes more efficient if state funding covers variable cost. Although subsidy system makes theater vulnerable, it is essential for institutionalizing theater. Figure 5 illustrates the realized and forecasted total production costs for Iran public theater during the period from 2012 to 2022. The production cost of theater is highly dependent to the production trend. Therefore, following the investigation of production trend of Iran public theater is presented.

Figure 5. The simulation result of national public theater's total cost by Vensim PLE software

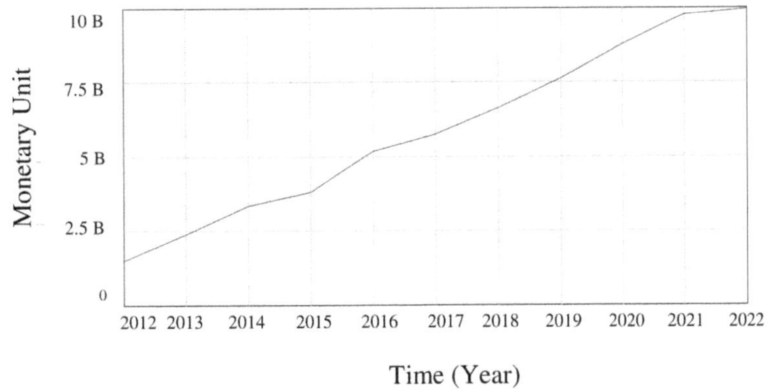

Production Trend

In order for plays to be performed, groups must have City Theater management's agreement. On the other hand, the complex's capacity limitation is a restrictive factor. Over the period from 2012 to 2015, between 270 and 400 performance demands were submitted to the management office of the complex, whilst complex's productivity was not capable of meeting that demand. As a result, about 12% of performance demands were approved and the rejection rate was around 88 percent.

High rejection rate has a negative impact on dissatisfaction of qualified groups resulted in a reduction in the quality level of plays and lowers customers' demand. In model simulation, there is a dramatic increase in rejection rate reaching to 943 plays and also the effect of rejection rate on dissatisfaction rises around three times by 2022. Overall, the public theater will experience a disaster unless its production capacity rises. The simulation result of rejection rate is presented in Figure 6.

Figure 6. The simulation result of national public theater's rejection rate by Vensim PLE software

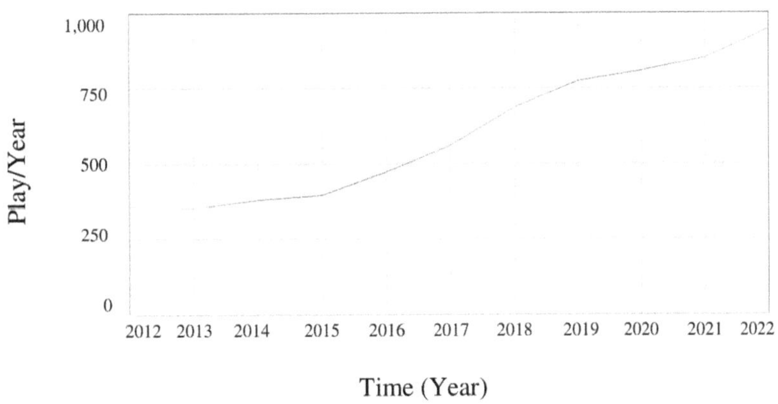

In Figure 3, the performing theater and the archive stocks are seen. The former shows the number of plays performed in the City Theater complex's hall yearly, while the latter accumulates this quantity over the period from 2012 to 2022. According to simulation results presented in Figure 7, it seems that roughly 435 plays would be performed by 2022. This number seems negligible when we make a comparison between Tehran and a city like Paris where about 600 plays are performed daily.

Figure 7 represents the simulation result of the performing theater and the archive stocks. The oldness of complex's halls makes maintenance operation imperative. Therefore, after a steady growing trend in performing theaters, a fall is expected since the halls need rehabilitation and become out of order.

Figure 7. The simulation result of national public theater's archive by Vensim PLE software

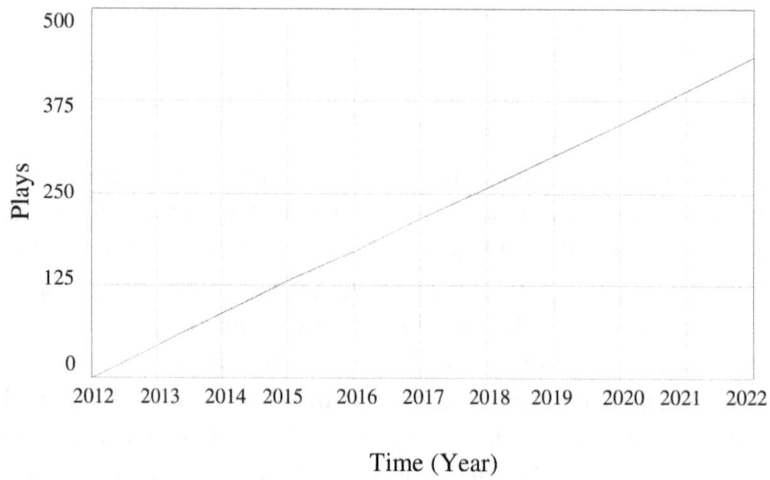

In this subsection the production state of Iran public theater has been modeled and analyzed. According to the findings, the overall condition of this segment is not appropriate. Another economics stage, namely distribution, of this cultural product is investigated in the following subsection.

Act II: Distribution

Distribution means delivering a produced good to consumers. This may be achieved through physical network, financial network, and communication network. The distribution of Iran public theater is not recognized developed enough to benefit from the financial or communication networks and still is based on physical network (theater halls). The geographic distribution is a huge obstacle for theater consumers since City Theater complex is the sole place for public theater, as a consequence the market has been narrowed.

Sales rate has increased between 2013 and 2015, also the result of model simulation demonstrates that this growing trend will continue by 2022. The dramatic increase in theater students and graduates, the presence of movie's superstar in plays, and advertisement are the socio-economic reasons behind this growth. Figure 8 shows actual and simulated sales rate from 2012 to 2022. In the Subsequent subsection, another economics stage, namely consumption stage which includes demand and supply is discussed.

Figure 8. The simulation result of national public theater's sales rate by Vensim PLE software

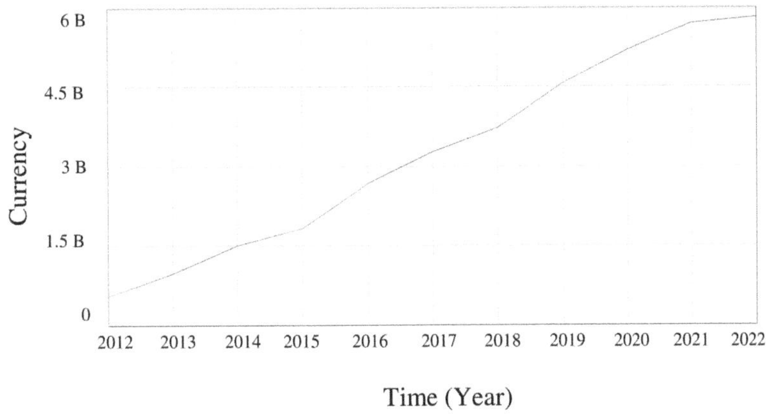

Act III: Consumption

The final purchase and use of goods and services by individuals constitute the consumption. Iran public theater's consumers are divided into two classes. The first group comprises of persons who have a strong tendency to expend time and money on theater. The second class becomes a theatergoer only under certain circumstances. What makes a crisis in producer-consumer relationships is defined as overproduction or underconsumption. According to statistics, the public theater suffers from underconsumption phenomenon. Inadequate income is one of the most important matters for second class theatergoers to ignore theater (afsharïasl, 2013).

Theater, as a cultural product and like other goods, follows the supply and demand principle. Hence, our analysis is presented in the following subsections in accordance with this precept.

Demand

Public affordability is a crucial affair in cultural events because an affluent and growing economy renders a progressive culture. In this case, the amount of budget Iranian families spent on cultural event in one year (family cultural budget) is one of the most important factors on demand quantity. The share of cultural expenditure and Iranian participation in cultural activities is 2.3 percent of family expenditure. Clearly the position of arts and culture is not much high amongst Iranian families, accordingly a dynamic culture is not proved as a vital necessity.

Undoubtedly, advertisement is another instrument for demand stimulation but this factor has a small effect because the public theater seems less-known among people. Advertising will be more effective when individuals are acquainted with theater. Familiarizing youngster with theater in schools may have efficacious consequences for advertisement effectiveness.

Production cost affects the quality of services or goods. Moreover, there is a direct relationship between quality and demand. Demand subsides when production suffers from low quality and this would damage the whole economic system. Fortunately, the demand of the public theater has increased but this is still lower than the expected demand level. There was about 100% increase in consumers' demand between 2012 and 2015 according to the simulation result, Figure 9, this growth will progress and reach to its peak by 2021 with a decline in 2022.

Figure 9. The simulation result of national public theater consumers' demand by Vensim PLE

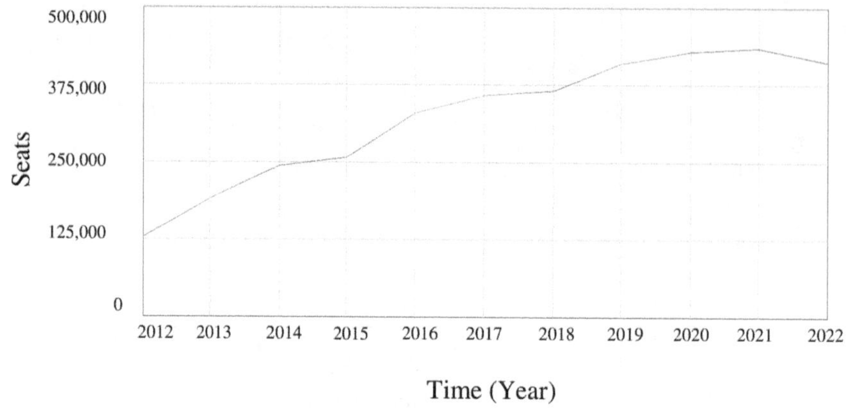

Supply

City Theater complex is popular, in Iran, as the mere place for performing public theatre whose halls' situation is not appropriate for the needs of performing groups due to obsolescence. As a matter of fact, one hall of complex cannot be used for a year on average. Figure 10 shows the productivity of complex and theater performing rate during the period from 2012 to 2022. Sometimes, low level of technical quality, inexperienced performers, and subject unattractiveness led to low sale and lesser than usual audience.

The productivity of City Theater complex is damaging consumption therefore the management of national public theater must show a swift reaction. Some practical solutions are building new and modern complex or halls and diversifying geographic positions of the halls.

POLICY SUGGESTION

Based on the analysis of the existing problems in the Iranian public theater, we can observe that the accumulated workforce cannot be used effectively due to various impactful factors within the theater industry. According to the analysis, the maximum demand is one of the underlying factors, having the capacity to cause a change in the theater market and workforce in the long run. Maximum demand is the greatest number of theatergoers which is proportional to the spectator capacity of Tehran City Theater complex. Number of audiences will increase by designing a long-term plan which includes more advertisement and also making the ticket price more affordable with respect to the Iranian people financial status. As the number of audience increases, the income level for the theatrical group and government (tax) will increase. This reinforces the public theater to cope with the production cost and financial problems.

Another bottleneck of the Iranian public theater is having one and only physical network for the distribution of this cultural product. This fact causes the public theater to lose many potential audiences. To reach a desired level for the audiences, the public theater needs to have enough theater halls. The desired theater is equal to the maximum number of the theater houses where the desired number of spectators, or better still demands, can be met. There are a couple of options to augment the number of theater houses for the Iranian public theater:

- Using schools as a temporary theater hall along with a reduction in the ticket price;
- Using an open theater as far as the weather is clement along with a reduction in the ticket price;
- Using the mosque or hotel halls.

Figure 10a. The simulation results of national public theater's performing rate

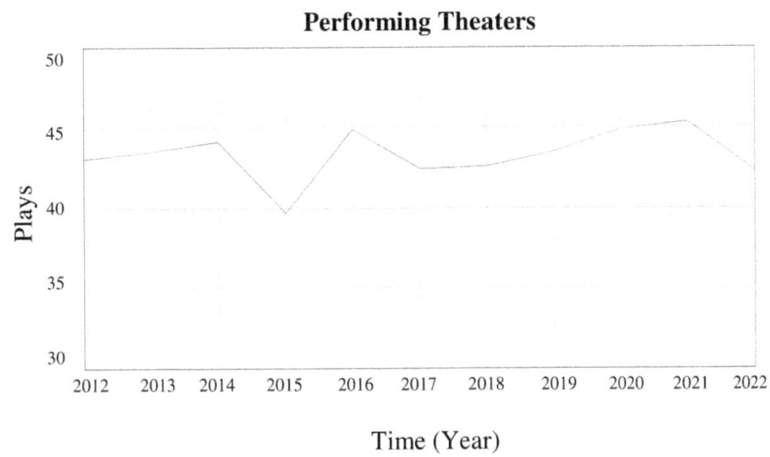

Figure 10b. Average productivity of the City Theater complex by Vensim PLE software

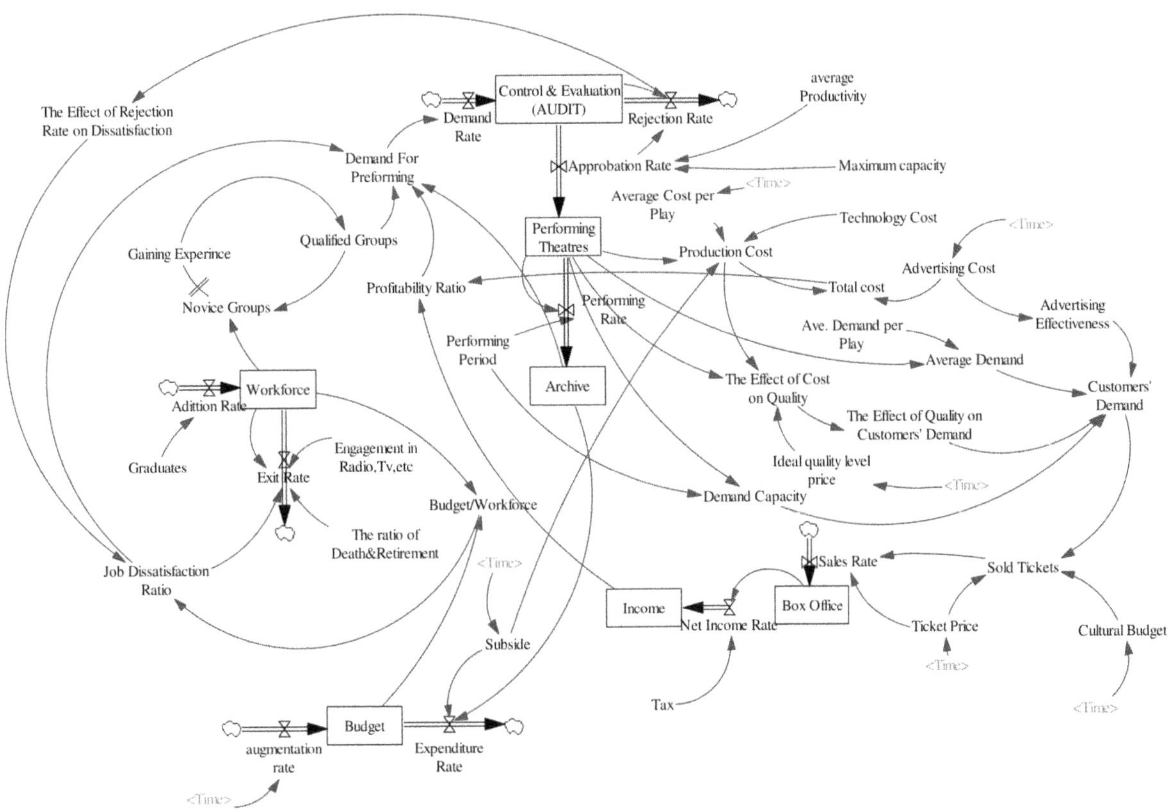

As the number of theater audience increases, the control and evaluation will be more classified and focus only on the quality of plays. The quality of theaters will be classified based on the theater houses and the target group of audiences. That is to say, the quality will be influenced by the audience's group this implies that theaters for children will be different from theaters for adults. Making the theater available for all audiences causes the workforce to have more interest and be more active as an artist rather than changing their profession.

CONCLUSION

In this context Iran public theater, as a cultural product, was analyzed based on economic elements including production, distribution, and consumption. The importance of this research is in identifying the relationships between theater production and theater economics. It means if the paucity of connection remains unchanged, the public theater will experience a decrease in production or even perish in not too distant future.

Using SD method, the interactions of the public theater's factors were modeled and with the help of Vensim PLE software its behavior was forecasted. In view of theater economic approach and current socio-economic condition of Iran society, the public theater's situation is evaluated undesired and some solutions were suggested.

In production section, according to model's simulation results, a growing increase in the workforce occurs whilst the production capacity remains unchanged. Organizing theater groups for performing in different places like plants, mosques and schools could be practical solutions. On the other hand, the public theater suffers from an inappropriate subsidy system that covers the fixed cost instead of variable costs. Furthermore, an increase in performing groups' demand for performance in City Theater complex is observed, but production capacity is restrictive; consequently, rejection rate and its effect on job dissatisfaction will rise and results in a decrease in the quality of performing plays.

Distribution section of Iran public theater is based on physical network, or better still theater halls. This is a huge obstacle for theater consumers because City Theater complex is recognized as the main and mere place for this type of theater, it means the marketplace has been narrowed. According to the result of model simulation, there was a growing trend in sales rate. Dramatic increase in the number of theater students and graduates, the presence of movie's superstar on theater stages, and advertisement are the socio-economic reasons behind this growth.

The public theater encounters with underconsumption phenomenon. Family cultural budget, advertisement, and production costs are the effective factors in consumer's demand. However, an increase in demand needs improvement in quality and people must be acquainted with the public theater. All in all, consumers' demand level was predicted to grow in accordance with simulation results. Due to the paucity of technical facilities, less-qualified performers, and subject unattractiveness of dramas, low revenue and less than usual audience are observed occasionally. Indeed, the low productivity of City Theater complex has a negative effect on consumption hence the management of national theater must show a swift reaction with building new and modern complex or halls and diversifying geographic positions of halls.

Although this paper may be the first study in investigating theater economy via modeling the production trend by SD method, the authors use a parsimonious model to study just a type of the Iranian theater. Other types of theater could be studied in the same way. Also, extending this model to the whole

Iran theater may be a good way to examine a more sophisticated case. Moreover, SD method can be a felicitous instrument for studying cultural economics.

REFERENCES

Abdelsalam, H. M., & El-Tagy, A. O. (2012). A simulation model for managing marketing multi-channel conflict. *International Journal of System Dynamics Applications*, *1*(4), 59–76. doi:10.4018/ijsda.2012100103

Abdollahi, H. (2020). An Adaptive Neuro-Based Fuzzy Inference System (ANFIS) for the Prediction of Option Price: The Case of the Australian Option Market. *International Journal of Applied Metaheuristic Computing*, *11*(2), 99–117. doi:10.4018/IJAMC.2020040105

Abdollahi, H., Ebrahimi, S. B., & Farmani, A. (2016). The inter-relationship between profitability and growth in the emerging Iranian market. *International Journal of Industrial Engineering & Production Research*, *27*(3), 209–217.

Abdollahi, H., & Talooni, M. (2016). Coordinating Contracts for Two-Echelon Supply Chain with Price and Promotional Sensitive Demand. *Modern Applied Science*, *10*(7), 132–140. doi:10.5539/mas.v10n7p132

Afshar, H.R. (2013). An introduction to the economy of theater in Iran. *Journal of Dramatic Art, 18*(2), 13-20. (in Persian)

Afshariasl, I. (2013). Economy in Iran Theater. [in Persian]. *Journal of Namayesh*, *167*, 62–65.

Azar, A. T. (2012). System dynamics as a useful technique for complex systems. *International Journal of Industrial and Systems Engineering*, *10*(4), 377–410. doi:10.1504/IJISE.2012.046298

Azhari, F., Abdi, F., & Abbas Shojaie, A. (2014). Using System Dynamics Method to Manage Construction and Demolition Waste. *International Journal of System Dynamics Applications*, *3*(4), 65–86. doi:10.4018/ijsda.2014100104

Baldin, A., & Bille, T. (2018). Modelling preference heterogeneity for theatre tickets: A discrete choice modelling approach on Royal Danish Theatre booking data. *Applied Economics*, *50*(5), 545–558. doi:10.1080/00036846.2017.1335386

Baldin, A., Bille, T., Ellero, A., & Favaretto, D. (2018). Revenue and attendance simultaneous optimization in performing arts organizations. *Journal of Cultural Economics*, *42*(4), 677–700. doi:10.100710824-018-9323-7

Baumol, W. J., & Bowen, W. G. (1993). *Performing arts-the economic dilemma: a study of problems common to theater, opera, music and dance*. The Twentieth Century Fund.

Becker, H. S. (1984). *Art worlds*. University of California Press.

Beckert, J., & Rössel, J. (2013). The price of art: Uncertainty and reputation in the art field. *European Societies*, *15*(2), 178–195. doi:10.1080/14616696.2013.767923

Breiger, R. L. (2005). Culture and classification in markets: An introduction. *Poetics*, *33*(3-4), 157–162. doi:10.1016/j.poetic.2005.10.003

Castiglione, C., Infante, D., & Zieba, M. (2018). Technical efficiency in the Italian performing arts companies. *Small Business Economics*, *51*(3), 609–638. doi:10.100711187-017-9931-1

Central Bank of Iran. (2014). *The classification of individual consumption by purpose reports (publication No. HBS 1394)*. Central Bank of the Islamic Republic of Iran. (in Persian)

Coyle, R. G. (1996). *System dynamics modelling: a practical approach*. Chapman & Hall. doi:10.1007/978-1-4899-2935-8

Durant, W. (1954). *The story of civilization*. Simon and Schuster.

Fazioli, R., & Filippini, M. (1997). Cost structure and product mix of local public theatres. *Journal of Cultural Economics*, *21*(1), 77–86. doi:10.1023/A:1007341613767

Feder, T., & Katz-Gerro, T. (2012). Who benefits from public funding of the performing arts? Comparing the art provision and the hegemony–distinction approaches. *Poetics*, *40*(4), 359–381. doi:10.1016/j.poetic.2012.05.004

Fernández-Blanco, V., Rodríguez-Álvarez, A., & Wiśniewska, A. (2019). Measuring technical efficiency and marginal costs in the performing arts: The case of the municipal theatres of Warsaw. *Journal of Cultural Economics*, *43*(1), 97–119. doi:10.100710824-018-9330-8

Forrester, J. W., Mass, N. J., & Ryan, C. J. (1976). The system dynamics national model: Understanding socio-economic behavior and policy alternatives. *Technological Forecasting and Social Change*, *9*(1-2), 51–68. doi:10.1016/0040-1625(76)90044-5

Grisolía, J. M., & Willis, K. G. (2012). A latent class model of theatre demand. *Journal of Cultural Economics*, *36*(2), 113–139. doi:10.100710824-012-9158-6

Guma, I. P., Rwashana, A. S., & Oyo, B. (2018). Food Security Indicators for Subsistence Farmers Sustainability: A System Dynamics Approach. *International Journal of System Dynamics Applications*, *7*(1), 45–64. doi:10.4018/IJSDA.2018010103

Haslett, T. (2012). Simulating Hamlet: Questions, Cautions, and Critique. *International Journal of System Dynamics Applications*, *1*(4), 77–87. doi:10.4018/ijsda.2012100104

Heilbrun, J., & Gray, C. M. (2001). *The economics of art and culture*. Cambridge University Press. doi:10.1017/CBO9780511754135

Jalalipor, B. (2009). Theater and a concern named funding: a comparative study between theater in Iran and France. *Journal of Art and Architecture, 79*, 70-97. (in Persian)

Lambert, C. (2012, February). The future of theater in a digital era, is the play still the thing? *Harvard Magazine*. Retrieved from http://www.harvardmag.com/pdf/2012/01-pdfs/0112-34.pdf

Last, A. K., & Wetzel, H. (2010). The efficiency of German public theaters: A stochastic frontier analysis approach. *Journal of Cultural Economics*, *34*(2), 89–110. doi:10.100710824-009-9111-5

Lena, J. C., & Lindemann, D. J. (2014). Who is an artist? New data for an old question. *Poetics*, *43*, 70–85. doi:10.1016/j.poetic.2014.01.001

Menger, P. M. (1999). Artistic labor markets and careers. *Annual Review of Sociology*, *25*(1), 541–574. doi:10.1146/annurev.soc.25.1.541

Mirembe, N. M. (2001). *Theatre and the market in Uganda* (Doctoral dissertation). Available from Makerere university Institutional repository: http://hdl.handle.net/10570/1671

Mokhtabad, S. M. (2013). A cultural economic model for privatizing theater in Iran. *Journal of Art and Architecture, 167*, 55-57. (in Persian)

Moysis, L., Azar, A. T., Kafetzis, I., Tsiaousis, M., & Charalampidis, N. (2017). Introduction to Control Systems Design Using Matlab. *International Journal of System Dynamics Applications*, *6*(3), 130–170. doi:10.4018/IJSDA.2017070107

Nikabadi, M. S., & Hakaki, A. (2018). A Dynamic Model of Effective Factors on Open Innovation in Manufacturing Small and Medium Sized Companies. *International Journal of System Dynamics Applications*, *7*(1), 1–26. doi:10.4018/IJSDA.2018010101

O'Neil, K. M. (2008). Bringing art to market: The diversity of pricing styles in a local art market. *Poetics*, *36*(1), 94–113. doi:10.1016/j.poetic.2007.12.002

Onggo, S. (2012). Adult social care workforce analysis in England: A system dynamics approach. *International Journal of System Dynamics Applications*, *1*(4), 1–20. doi:10.4018/ijsda.2012100101

Pruyt, E., & Logtens, T. (2015). Model-based exploration of societal aging in the Netherlands. *International Journal of System Dynamics Applications*, *4*(1), 57–95. doi:10.4018/IJSDA.2015010103

Radzicki, M. J. (2009). System dynamics and its contribution to economics and economic modeling. In *Complex Systems in Finance and Econometrics* (pp. 727–737). Springer. doi:10.1007/978-1-4419-7701-4_39

Rizzo, I., & Throsby, D. (2006). Cultural heritage: economic analysis and public policy. Handbook of the Economics of Art and Culture, 1, 983-1016.

Sadeghi, G. (1997). The economy of Iran Theater. *Journal of Art, 34*, 53-83. (in Persian)

Sadeghi, G. (2011). This implies the death of theater. *Journal of Art and Architecture, 144*, 42-43. (in Persian)

Sterman, J. D. (2000). *Business dynamics: systems thinking and modeling for a complex world.* Irwin McGraw-Hill.

Thakker, R., Eveleigh, T., Holzer, T., & Sarkani, S. (2013). A System Dynamics Approach to Quantitatively Analyze the Effects of Mobile Broadband Ecosystem's Variables on Demands and Allocation of Wireless Spectrum for the Cellular Industry. *International Journal of System Dynamics Applications*, *2*(3), 73–93. doi:10.4018/ijsda.2013070105

Trading Economics Website. (2015). *Iran Economic forecasts.* http://www.tradingeconomics.com/iran/inflation-cpi/forecast/

Velthuis, O. (2013). *Talking prices: Symbolic meanings of prices on the market for contemporary art*. Princeton University Press.

Wyburn, J., & Roach, P. A. (2013). A system dynamics model of the american collectable comic book market. *International Journal of System Dynamics Applications*, *2*(1), 37–58. doi:10.4018/ijsda.2013010103

Zuidervaart, L. (2010). *Art in public: politics, economics, and a democratic culture*. Cambridge University Press. doi:10.1017/CBO9780511760341

Chapter 9
Indigenous Knowledge Management and Humanitarian Supply Chain for Disaster Mitigation and Sustainable Development in the Eco Communities of India:
Holistic Systems Modeling Approach

Sanjay Bhushan
Department of Management, Faculty of Social Sciences, Dayalbagh Educational Institute, India

Saurabh Mani
Department of Applied Economics, Faculty of Commerce, Dayalbagh Educational Institute, India

ABSTRACT

The shift in focus from hazards to underlying vulnerabilities has provided disaster managers with a richer understanding of the factors that erode the coping capacities of communities and social systems. This chapter presents ideas such as a globally shared digital platform developed as a functional web-portal branded—A.A.D.I.GYAN—augmenting action for disaster management through indigenous knowledge-gyan. Humanitarian supply chain (HSC), on the other hand, plays a central role to any developmental program meant for sustainable capacity building in the eco-communities. The chapter also highlights that HSC and community-based IK resource management can collectively promote more resilient communities promoting social, economic, and environmental equity and ethical imperatives. Towards this end, system dynamics modeling and simulation can be invoked to capture and simulate the causal dynamics of the inter linkages prevailing within the system and suggest some critical intervention strategies for policymakers.

DOI: 10.4018/978-1-7998-5788-4.ch009

INTRODUCTION

As early as in 1994, the United Nations World Conference on Natural Disaster Reduction in Yokohama called for paying more attention than before to traditional knowledge and community based actions. The Kobe earthquake also led to a switch from a technocratic view of natural hazards to a focus on vulnerability. It has also been acknowledged that by appreciating and making use of peoples' knowledge the global community can also promote the principle of equity of knowledge (Pandey, 1998) which in turn being clubbed with formal sciences may result in empowerment, self-determination, security and opportunity for local people. Human ecological perspective thus proves vital in creating the fundamental framework for natural resource management on sustainable basis.

According to the United Nations Environmental Program (UNEP), traditional communities rely on indigenous knowledge to conserve the environment and deal with disasters. These communities, particularly those in hazard-prone areas, have generated a vast body of indigenous knowledge on disaster prevention and mitigation. This knowledge is the sum of facts that are known or learned from experience or acquired through observation and study and handed down from generation to generation (www.unep.org). Further, it is said that the success and the sustainability of environmental management interventions at the community level depend, among a number of factors, on the availability of relevant local culture, knowledge and indigenous practices that can combine with new ideas to generate innovation. Thus, the importance of indigenous knowledge contributes not only to the success of intervention, but more importantly to its sustainability in the longer term. In a particular reference of disaster management, considering the participation and integration of these communities in all disaster-related processes as a necessary means, the importance of indigenous knowledge acquisition can well be appreciated. Even before the advent of high technology based early warning systems, or standard operating procedures (SOPs) for response, numerous local communities worldwide have prepared, operated, acted, and responded to natural disasters using indigenous methods passed on from one generation to the next.

Figure 1. Indigenous Knowledge and Sustainability

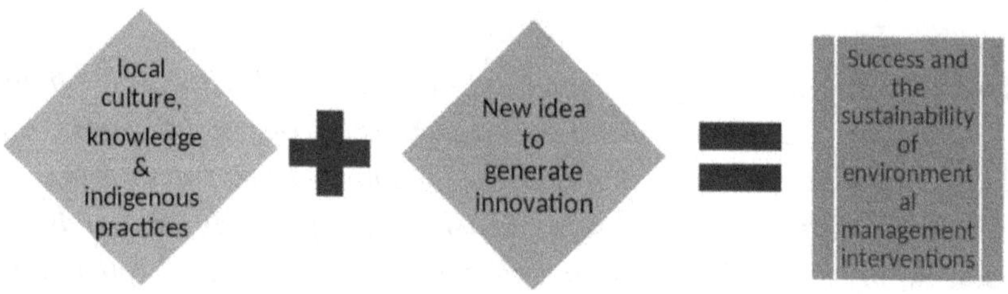

The United Nations already considered indigenous knowledge within Priority-3 of the Hyogo Framework for Action (2005-15), which extensively focused on education and knowledge (WCDR, Hyogo 2005). One of the key activities identified under this priority action focuses on the importance of information management and exchange, and highlights the use of "relevant traditional and indigenous knowledge and cultural heritage" to be shared with and adapted to different target audiences. In order to fulfill this objective, we all need to understand, acknowledge and respect indigenous knowledge as a valuable source of information and as a key contributor to reducing risk in many parts of the world.

Director General of UNESCO (Mayor, 1994) defined traditional knowledge as:

...The indigenous people of the world possess an immense knowledge of their environments, based on centuries of living close to nature. Living in and from the richness and variety of complex ecosystems, they have an understanding of the properties of plants and animals, the functioning of ecosystems and the techniques for using and managing them that is particular and often detailed. Equally, people's knowledge and perceptions of the environment, and their relationships with it, are often important elements of cultural identity.

Most indigenous people have traditional songs, stories, legends, methods and practices as means of transmitting specific human elements of traditional knowledge. In virtually all of these systems, knowledge is transmitted directly from individual to individual. Some of the basic tenets of indigenous knowledge transmission are:

- Practical common sense
- Knowledge of the environment and the relationships between things.
- Holistic
- Rooted in the spiritual health, culture and language
- Dynamic, cumulative and stable

Figure 2. Methods of Transmission

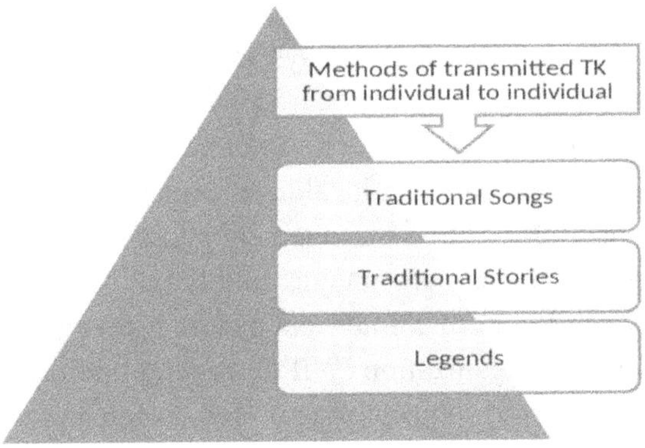

Indigenous and Scientific Knowledge:

In case of indigenous knowledge, some native classification is available only by virtue of its relationship to human activities and feelings. These comparisons sometimes incorrectly lead scientists to trivialize traditional understanding. As we know, science generally excludes the humanistic perspective whereas traditional understanding assumes a holistic view including language, culture, practice, spirituality, mythology, customs and even the social organization of the local communities.

Traditional knowledge incorporates knowledge of ecosystem relationships and a code of ethics governing appropriate use of the environment. This code includes rules and conventions promoting desirable ecosystem relations, human-animal interactions and even social relationships, since the latter continue to be established and reaffirmed through hunting and other activities on the land. Traditional knowledge articulates with nontraditional knowledge to form a rich and distinctive understanding of life and the world environment.

Presently, the communication of traditional knowledge is also hampered by competition from other cultures that capture the imagination of the young generation. They are tempted by technology that teaches them non-indigenous ways and limits the capacity of elders to pass on traditional knowledge to the young. As the elders die, the full richness of tradition is diminished, because some of it has not been passed on and so is lost. It is important therefore to find ways of preserving this heritage of knowledge.

"Each time an elder dies it is as if a library had burned down."

- Amadou Hampate Ba

What is Indigenous Knowledge?

The expressions 'traditional,' 'local' and 'indigenous' knowledge are used in the literature inter-changeably. However, Indigenous knowledge has been implicitly expressed as 'the local knowledge' – knowledge that is unique to a given culture or society. "It is the basis for local-level decision-making in agriculture, health care, food preparation, education, natural-resource management, and a host of other activities in rural eco-communities" (Warren, 1991, as cited in World Bank, 1998). Flavier et al (1995) have also described it as "the information base for a society, which facilitates communication and decision-making. Indigenous information systems are dynamic, and are continually influenced by internal creativity and experimentation as well as by contact with external systems".

Moreover, Indigenous knowledge (IK) is the local knowledge- knowledge that is unique to a given culture or society. IK contrasts with the international knowledge system generated by universities, research institutions and private farms. It is the basis for local level decision-making in agriculture, health care, food preparation, education, natural-resource management and a host of other activities in rural communities (Warren, 1991). Thus, IK refers to the unique, traditional, local knowledge existing within and developed around the specific conditions of women and men indigenous to a particular geographic area.

The World Intellectual Property Organization (WIPO) defines traditional knowledge as "indigenous cultural and intellectual property," indigenous heritage," and customary heritage rights.

Unfortunately, the need to protect the traditional knowledge captured the attention of the international community only recently but the standard setting was left to the national governments and where in the absence of the international standards caused serious negligence for the protection of the traditional knowledge and the benefits of new technology.

Domains of Indigenous Knowledge

This Indigenous Knowledge System (IKS) not only includes the indigenous technological knowledge traits, but also the set of knowledge traits used to maintain a folk life and that is why, IKS is also termed as Peoples' Knowledge. The folk peoples are closely associated with the nature and therefore, the Indigenous Knowledge System is also found to be closely attached with the nature; this close attachment gives IKS the name of Indigenous Ecological / Environmental Knowledge System. So, known by various names, IKS is found closely attached with social, cultural, natural and human capitals within the non-reflective population. It could also be classifies into various systems of gender, and age group. Though being functional in nature, IKS is closely associated with various non-functional symbols of human life.

Additionally, several authors have identified some special features of IK. Mearns, Du Toit and Mukuka (2006) have summarized these as follows:

- IK is local, holistic, and integrative because it is rooted in a particular community and its experiences are situated within broader cultural traditions of the people living in that place.
- IK is essentially functional and is geared to practical response and performance.
- IK is experiential rather than theoretical and is reinforced through continuous experience, trial and error.
- IK is learned through repetition, which aids in its retention and reinforcement.
- IK is constantly changing by way of being produced and reproduced, discovered and lost.
- IK is characteristically shared to a greater degree than other forms of knowledge, although its distribution is socially differentiated, based on gender and age.
- The distribution of IK is always fragmented. It does not exist in its totality in either one place or one individual.
- IK is tacit and cannot easily be codified.
- IK is transmitted orally, or through imitation and demonstration and the process of codification may lead to the loss of some of its properties.

Indigenous and Scientific Knowledge: A Contrast

In case of indigenous knowledge, some native classification is available only by virtue of its relationship to human activities and feelings. These comparisons sometimes incorrectly lead scientists to trivialize traditional understanding. As we know, science generally excludes the humanistic perspective whereas traditional understanding assumes a holistic view including language, culture, practice, spirituality, mythology, customs and even the social organization of the local communities.

Traditional knowledge incorporates knowledge of ecosystem relationships and a code of ethics governing appropriate use of the environment. This code includes rules and conventions promoting desirable ecosystem relations, human-animal interactions and even social relationships, since the latter continue to be established and reaffirmed through hunting and other activities on the land. Traditional knowledge articulates with nontraditional knowledge to form a rich and distinctive understanding of life and the world environment.

Presently, the communication of traditional knowledge is also hampered by competition from other cultures that capture the imagination of the young generation. They are tempted by technology that teaches them non-indigenous ways and limits the capacity of elders to pass on traditional knowledge to the young. As the elders die, the full richness of tradition is diminished, because some of it has not been passed on and so is lost. It is important therefore to find ways of preserving this heritage of knowledge. Differences between traditional knowledge systems and western scientific system as indicated below:

Table 1.

Traditional Knowledge System	Western Scientific system
All parts of the natural world are regarded as animate, all life forms as interdependent.	Human life is generally regarded as superior, with a moral right to control other life forms.
Knowledge is transmitted largely through oral media.	Knowledge is transmitted largely through written media.
Knowledge is developed and acquired through observation and practical experience.	Knowledge is generally learned in a situation, which is remote from its applied context.
Knowledge is holistic, intuitive, qualitative and practical.	Knowledge is essentially reductionist, quantitative, analytical and theoretical.
Knowledge is generated by resource users in a diachronic (long-term) time scale.	Knowledge is generated largely specialist researchers on a synchronistic (short-term) time scale.
The nature and status of particular knowledge is influenced by socio-cultural Factors such as spiritual beliefs, and is communally held.	The nature and status of particular knowledge is influenced by peer review, and is held by individual specialists.
Explanation behind perceived phenomena is often spiritually based on subjective.	Explanation behind perceived phenomena is essentially rational and objective.
Knowledge is used to make suitable decisions under variable conditions.	Knowledge is used to put forward hypothesis and to verify underlying laws and constants.

Disaster Mitigation Management and Sustainable Development

The United Nations also recognized the disastrous impacts of natural hazards on vulnerable communities and, by the year 2000, all countries were encouraged to have comprehensive national assessments of natural hazards and risks integrated into national development plans and to address long-term disaster prevention, preparedness, and community awareness in mitigation plans. In reality, funds dropped – mainly because of the Gulf War (1991) and a series of natural disasters in developed nations, including the Kobe earthquake. The Kobe earthquake demonstrated that developed nations could not prevent disaster and that relief aid was inappropriate.

Figure 3. Causal Linkages to Disaster Risk

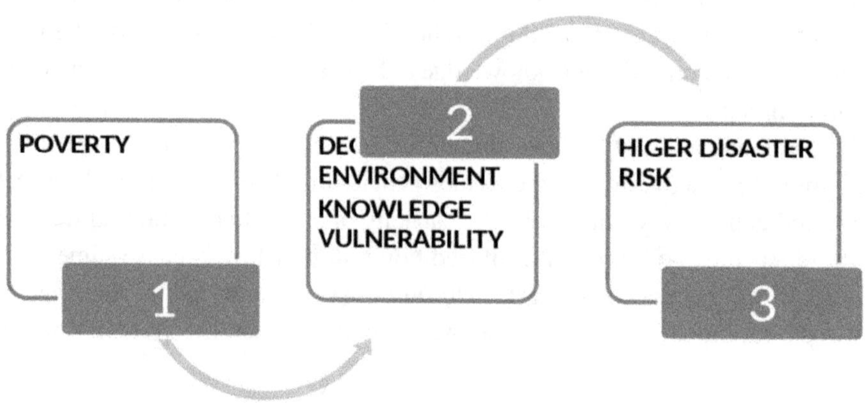

The shift in focus from hazards to underlying vulnerabilities has provided disaster managers with a richer understanding of the factors that erode the coping capacities of communities and social systems. Ecological environment plays a role in many of these factors. There is a strong causal relationship between poverty, a degraded environment, knowledge vulnerabilities and higher disaster risk.

People who live in marginal or environmentally degraded areas and often devoid of rich knowledge heritage, struggle on a daily basis to survive and are unable to cope with any additional stress factors. Limited livelihood alternatives, competition over scarce resources, weak educational-governance structures and lack of access to healthcare can compromise a community's ability to respond to a hazard event. Indigenous environmental management and community based resource management can promote more resilient communities through supporting sustainable livelihoods, conflict prevention and strengthening cooperation for good governance. Moreover, sustainable development is universally considered to be pro-nature, pro-poor, pro-women and pro-job generation that combines social, economic and environmental equity and ethical imperatives for economic situation. A bottom-up approach for eco-restoration calls for holistic management of natural resources in accordance with the live tradition of rural India.

The Hyogo and Sendai Framework for Action (Pre- and Post-2015)

Under Hyogo Framework for Action (HFA) 12 and Sendai Framework, which guide disaster risk reduction activities globally, it is recognized that environmental degradation contributes to disaster risk, and that disasters occur when hazards interact with, among other things, environmental vulnerability. As such, the Framework recognizes both the role of environment as a trigger of disaster risk, and the sensitivity of the environment to the forces of hazards. It equally emphasizes that human societies cannot be dissociated from the environments that they shape and which in turn influence their development and livelihoods. Together they form a comprehensive system with intrinsic levels of vulnerability and inherent coping mechanisms. The less degraded the environmental component of this system, the lower its overall vulnerability and the higher its coping capacity.

Figure 4. Hazard, Vulnerability and Disaster

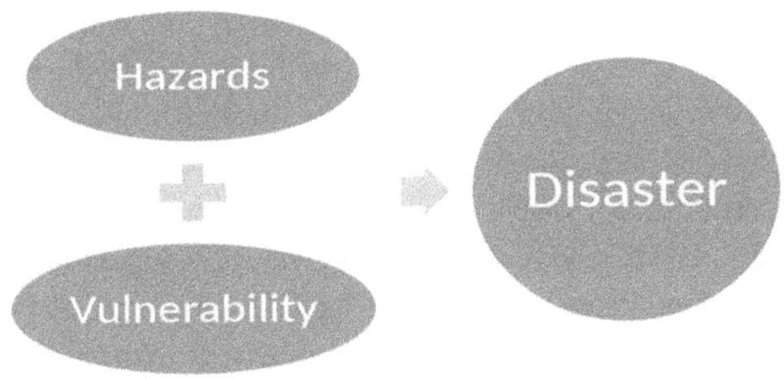

Thus, environmental concerns and opportunities are relevant to the implementation of all disaster risk reduction priorities. According to HFA Priority 3: Use knowledge, innovation and education to build a culture of safety and resilience at all levels, disasters can be substantially reduced if people are well informed and embrace a culture of prevention. It stresses that local communities often maintain vast traditional knowledge on environment and disasters. Still, new technical knowledge of risk reduction, based on ecology and environmental principles, needs to be developed. Knowledge of the environmental dimensions of disaster needs to be made available for use in evidence-based analyses, public awareness, political advocacy, operational decisions and educational curricula. Similarly, Post-2015, the Sendai Framework recognizes that the State has the primary role to reduce disaster risk but that responsibility should be shared with other stakeholders including local government, the private sector and other stakeholders.

Thus, environmental concerns and opportunities are relevant to the implementation of all disaster risk reduction priorities. According to HFA Priority 3: Use knowledge, innovation and education to build a culture of safety and resilience at all levels, disasters can be substantially reduced if people are well informed and embrace a culture of prevention. It stresses that local communities often maintain vast traditional knowledge on environment and disasters. Still, new technical knowledge of risk reduction, based on ecology and environmental principles, needs to be developed. Knowledge of the environmental dimensions of disaster needs to be made available for use in evidence-based analyses, public awareness, political advocacy, operational decisions and educational curricula.

Eco- Community Village and Ecological Dimension for Sustainable Development:

Eco community villages can be termed as being human-scale, full-featured settlements in which human activities are harmlessly integrated into the natural world. It is supportive of healthy human development, and can be successfully continued into the infinite future. Demographically, these can be urban or rural communities of people, who strive to integrate a supportive social environment with a low-impact way of life. To achieve this, the community integrates various aspects of ecological design, permaculture, ecological building, green production, alternative energy, community building practices, and so on. Hence, the motivation for eco community villages is the choice and commitment to reverse the gradual disintegration of supportive social/cultural structures and the upsurge of destructive environmental practices on our planet.

Based on the three basic tenets of eco community village - Social/Community, Ecological, Cultural/Spiritual, indigenous knowledge management gains special significance under the cultural/spiritual dimension. This is with respect to its understanding of the interconnectedness and interdependence of all the elements of life on Earth and the community's place through a scientific perspective in a holistic way. It is prudent to note that this particular feature of an eco community plays a pivotal role in mitigating numerous ecological crises that the global civilization is facing in recent times.

Hence, ecological conditions not only modify the frequency and magnitude of hazard events, but also affect natural barriers that can moderate the impacts of a disaster and protect communities. Moreover, maintaining ecosystems to decrease disaster risk can also contribute to the reduction of greenhouse gases and thereby further minimize the risk of future hazard events. Finally, ecosystems managed to support sustainable livelihoods also help to lessen the social, economic and environmental impacts of disasters on communities. Investments in ecosystems can therefore lead to significant savings, as compared to the cost of a disaster on human livelihoods. The proliferation of new technologies and processes for managing natural resources, including indigenous knowledge of the ecological, social and cultural dimensions of resource management, presents many opportunities for reducing disaster risk.

Figure 5. View of basic tenets of Eco-community village

Disaster Management and Planning

Disaster Management and planning is a key part of government work. Asia tops the list of casualties due to natural disaster. Among various natural hazards, earthquakes, landslides, floods and cyclones are the major disaster adversely affecting very large area and population in the India sub-continent.

These natural disasters are of:

- Geographical origin such as earthquakes, volcanic eruptions, land slides.
- Climatic origin such as drought, flood, cyclone, locust, forest fire.

Rising frequency, amplitude and number of natural disasters and attendant problem coupled with loss of human lives prompted the General Assembly of the United Nations to proclaim 1990s as the International Decade for Natural Disaster Reduction (IDNDR) through a resolution 44/236 in 1989 to focus on all issues related to natural disaster reduction. It is almost impossible to prevent the occurrence of natural disasters and their damages. However it is possible to reduce the impact of disasters by adopting suitable disaster mitigation strategies.

The disaster mitigation works mainly address the following:

- Minimize the potential risks by developing disaster early warning strategies.
- Prepare and implement developmental plans to provide resilience to such disaster.
- Mobilise resources including communication and tele-medicinal services.
- To help in rehabilitation and Post-disaster reduction.

Broadly, the process of Disaster Management involves Six phases: Prevention, Mitigation, Preparedness, Response, Recovery and Rehabilitation. Primarily, any disaster management effort aims to reduce the occurrence of disasters and to reduce the impact of those that cannot be prevented through Mitigation. Mitigation measures include legislation, land-use planning (e.g. the designation of non-essential land like parks to be used as flood zones), and insurance. Mitigation is the most cost-efficient method for reducing the impact of hazards; however it is not always suitable. A precursor activity to the mitigation is the identification of risks. Physical risk assessment refers to the process of identifying and evaluating hazards. The hazards-specific risk (R_h) combines both the probability and the level of impact of a specific hazard. The equation under states that the hazard multiplied by the populations' vulnerability to that hazard produces a risk Catastrophe modeling-

$$R_h = H \times V_n$$

The higher the risk, the more urgent that the hazard specific vulnerabilities are targeted by mitigation and preparedness efforts. However, if there is no vulnerability there will be no risk, e.g. an earthquake occurring in a desert where nobody lives.

Application and Challenges of Traditional Knowledge in Eco-Communities of India

Known by various names, Indigenous Knowledge is closely associated with social, cultural, natural and human capitals within the non-reflective population. It could also be classified into various systems of gender, and age group. Though being functional in nature, IKS is closely associated with various non-functional symbols of human life.

There are many examples that illustrate the resourcefulness of indigenous people in applying traditional knowledge to lessen the impacts of disasters. Some use strips of mangrove forest to absorb the force of tidal surges and tsunamis, others apply genetic diversity in crops to avoid total crop failure, and some communities move between habitats as hazards strike. Their strategies for early warning include observations of conditions and events that are becoming alarming. Out of them, the most common are related to animal behavior and the appearance of the sky. Thus, indigenous people continue to apply traditional knowledge on early warning and strategies for mitigation of the negative consequences of disasters. However, the challenge lies with transferring traditional knowledge between generations. The indigenous knowledge that is passed from generation to generation helps them to minimize risk, cope with natural dangers and learn how to survive apart from creating socio-economic sustenance. Social scientists also observe that these are no longer naïve views of nature. In fact, superstitions, myths, and local know-how are still grounded in people's core beliefs. Scientists who use the latest technology can bridge the gap between the people who stand to benefit from their work. However, they must not underestimate the indigenous capabilities that the people possess and should understand how their work will be received by people who intimately experience and cope with natural disasters and create a sustainable development environment. Thus, it would be desirable to combine modern knowledge and indigenous knowledge in disaster preparedness and mitigation with the goal of building community capacity in a participatory, sustainable and cost-effective manner. The study would aim to encourage policymakers to incorporate a diverse range of indigenous knowledge in the disaster management and sustainable development initiatives of the country.

In the above context, systematic and in-depth studies on eco-communities' indigenous knowledge capacity for disaster risk reduction and management are rare in India. Indigenous knowledge, as such, is not well documented and is in danger of being lost as its custodians pass away. It is crucial to preserve this valuable body of knowledge and integrate it into modern disaster risk reduction efforts. The inclusion of indigenous people into discussions of climate change and risk reduction cannot be ignored. Lack of organized study in this area has led to vulnerability of communities and the nation as a whole in coping with natural disasters. Several studies are available on the value of indigenous knowledge for natural resource management in Africa and Asia-Pacific countries. However, literature on indigenous knowledge for disaster mitigation and preparedness is scattered and very scarce. Systematic and in-depth studies on disaster mitigation in general and indigenous knowledge in particular would be very useful in this regard.

Moreover, people in the disaster prone area have been coping with natural calamities by their own traditional techniques. Often, they face the problems taking different adaptation mechanisms. The coping mechanisms, they adopt, have been originated by their family and community wisdoms, which we can consider as indigenous knowledge. But as intensity of climatic disasters increased, depletion of knowledge among new generation (as the new generations are ignorant about indigenous knowledge) and also occupational migration - this indigenous knowledge base is depleting. Policy documents are lacking to fully address the indigenous knowledge of natural resource dependent people of the coastal area. There are no priority areas to document indigenous knowledge, which are important in the formulation of adaptation and mitigation strategies specific for the location. The policy documents produced by the government do not also consider proper participation of environment dependent occupational groups. Thus, it grossly fails to ensure community level participation in the knowledge capital formation process in order to cope up with natural disasters. Local /Indigenous knowledge is a powerful asset and social capital which is vulnerable to climate change. Policies and programs incorporating such knowledge are more likely to be acceptable and successful in the long run by judiciously prioritizing on community needs, problems, aspirations and local ecology specific.

Hence, special attention would be required to the development and preservation of indigenous knowledge systems through indigenous knowledge documentation and database and its causal relationship to disaster risk reduction and environmental sustainability. A conscious effort would also be made to ensure participation of local inhabitants for long-term acceptance and success of this project. It can be noted that when documented, the information would be available in an easily accessible in terms of format and content form to a wide range of audiences including government decision makers, environmental and disaster management practitioners, medical practitioners, academics, researchers, journalists, indigenous and local communities, as well as international development agents including UNEP and other stakeholders. It would further enable to strengthen capacity-building in indigenous knowledge and help to integrate the knowledge with other knowledge systems and development processes, particularly for environmental conservation and natural disaster management. Towards this goal, it shall also facilitate community engagement and also link teaching; research and practice to better reflect the knowledge creation and knowledge mobilization across the eco communities and academic/research institutions. By this, a vital opportunity to address the issues of community health, community cultures, community practices in sustainable development/natural resources and other related aspects of community knowledge production, application and dissemination may also be facilitated.

Interdisciplinary Relevance

Any holistic work on Indigenous knowledge, Disaster mitigation or Eco village and their mutually adjoining line of investigation would invariably require a facilitating interface between both formal and local methods, as well as local people and formally trained scientists, which shall contribute to co-collect, co-analyse, co-interpreate and co-utilize the data, information and knowledge. In collaborative efforts of such kind, everyone involved may stand to benefit. Both local people as well as external experts need access to the latest scientific developments to see if it can help improve existing conservation knowledge and practices. The policy makers need ready access to the science as well as understanding the difficulties of its application. Indeed, there are numerous examples where local knowledge derived from long-term nature-society interaction has been extremely useful in validating scientific hypotheses and suggesting new research directions (Kimmerer 2002, Robertson and Hull 2001).

Interestingly, Intellectual Property Rights are now also being extended to beyond the conventional domain of mechanical and chemical innovations to include biological and environmental resources. National Biological Diversity Act of India in response to the commitment to the Convention on Biological Diversity and intellectual property rights is also being contemplated to devise operational mechanisms to share benefits of commercial applications of traditional knowledge on biodiversity with local communities. Also useful shall be to ensure a harmonized basket of rules made under the Patent Act, Protected Plant Varieties Act, and the Biological Diversity Act (Utkarsh *et al.* 1999).

Therefore, apart from creating and transmitting scientific knowledge to policy makers and practitioners, through such interdisciplinary and holistic study, scientific community would also be sensitized to collaborate with people to put forth new hypotheses that incorporate aspirations of formal and local systems of knowing and modify their methodologies accordingly. Humanity needs to go beyond disciplinary divide and find a common ground across cultures, faiths and disciplines (Pandey, 2002a). Collective wisdom of humanity for conservation of biodiversity, embodied both in formal science as well as local systems of knowledge, therefore, is the key to pursue our progress towards sustainability.

For the academia, however, this type of community based research projects can become an intrinsic part of learning and teaching. It would facilitate partnerships between eco communities and institutions of higher education so that students and teachers can learn from indigenous knowledge and wisdom, thereby democratizing knowledge production. The proposed study intends to create an *'interface structure'* acting as a communicator, mediator and coordinator of institutional linkages and partnerships with the communities and their indigenous knowledge capital. As an innovative application of data surveys, it is also proposed to ensure eco-community engagement with the assistance of UG and PG students of Management course undergoing internships or course -work. It would provide them much needed opportunity of rural/eco community exposure and sensitization, while also serving the mutually agreed interests of both local communities and the institution.

Research and Development in the subject: Review of International Status

The United Nations Environment Programme (UNEP) recognizes the role of indigenous knowledge in the conservation of natural resources and management of natural disasters. The UN Conference on Environment and Development (UNCED) specifically highlights an urgent need for developing mechanisms to protect the earth's biological diversity through local knowledge. Many documents signed at UNCED reflected the need to conserve the knowledge of the environment that is being lost in communities. Similarly, the World Conference on Science in Budapest (1999) recommended that scientific and traditional knowledge be integrated in interdisciplinary projects dealing with links between culture, environment and development in areas such as the conservation of biological diversity, management of natural resources, understanding of natural hazards and mitigation of their impact.

Globally, the hazard studies show that natural hazards are non-linear and complex events shaped by and resulting from the combination of not only geophysical and meteorological factors but also (and mainly) political, economic, socio-cultural, and psychological (or perceptual) phenomena and factors. Together they make an integral mechanism of communication and exchange of Information, resources and activities between any country's own national or domestic natural system and global or inter-national natural systems (Figure 6). The social dimension of risks and hazards is important because local communities see them through a cultural lens; and this is dependent upon their view of the cosmos and accumulated experience (Linkenbach-Fuchs 2002).

Figure 6. Dynamic Interface between Domestic and Global Natural Systems

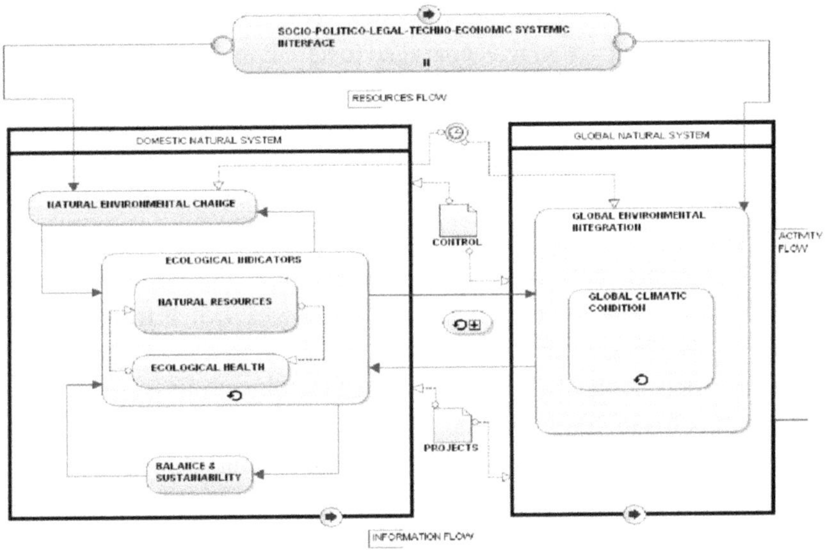

In the 1960s and 1970s, geographers and meteorologists studied the impacts of natural disasters on people mainly from a technical perspective. The dominant approach to natural hazards and disasters focused on hazards as physical events requiring scientific and technical solutions. Natural hazards were understood in the context of simplistic determinism (where physical processes determine human actions) and linear causal relationships from geophysical events to impacts to human responses. People were assumed to live in vulnerable conditions due to a lack of knowledge (Schilderman 2004). In the 1980s and 1990s, researchers in the field of natural hazards and disasters began to criticize the deterministic, ahistorical, and asocial concept of hazards and disasters and its dependence upon the use of choice and decision models (Hewitt 1983; Gardner 2002). Though, as an exception, from the early 1970s on, the Man and the Biosphere (MAB) Program has contributed to work on traditional ecological knowledge, through a series of field studies on local and indigenous peoples and their decision-making and resource-management systems. Lately, the growing focus of research and development in the area of indigenous knowledge has been the need to take the human dimensions of natural hazards into account including local knowledge, practices, and perceptions (Anderson and Woodrow 1989, Johnson et al. 1982). In fact, the studies on hazard perception emerged in the US from Gilbert White's group at Chicago in the

1960s. This work initially focused on wheat farmers' perceptions of and responses to droughts in the Great Plains of North America (Saarinen 1969). However, a lot of this work degenerated into standardized questionnaire surveys and 'official' analyses applied in developing countries (Ken Hewitt 1983).

The Convention on biological diversity (1992) required that every contracting party should respect, preserve and maintain knowledge, innovations and practices of traditional and local communities and promote the wider application with the approval and involvement of the holder of such knowledge, innovations and practices and encourage the equitable sharing of the benefits. As nations implement the Convention on Biological Diversity (CBD) work programs, apply its guidelines, and execute national strategies, its influence on science is likely to grow. CBD-compliant national laws and policies already set priorities for research and affect the way in which scientists can access and use genetic resources (Kate, 2002).

The United Nations also recognized the disastrous impacts of natural hazards on vulnerable communities and, by the year 2000, all countries were encouraged to have comprehensive national assessments of natural hazards and risks integrated into national development plans and to address long-term disaster prevention, preparedness, and community awareness in mitigation plans. In reality, funds dropped – mainly because of the Gulf War (1991) and a series of natural disasters in developed nations, including the Kobe earthquake. The Kobe earthquake demonstrated that developed nations could not prevent disaster and that relief aid was inappropriate. In 1994, the United Nations World Conference on Natural Disaster Reduction in Yokohama called for paying more attention than before to traditional knowledge and community based actions. The Kobe earthquake also led to a switch from a technocratic view of natural hazards to a focus on vulnerability.

It has also been acknowledged that by appreciating and making use of peoples' knowledge we shall also promote the principle of equity of knowledge (Pandey, 1998). Equity of knowledge between local and formal sciences results in empowerment, security and opportunity for local people. If the state and formal institutions incorporate people's knowledge into the resource management decisions, it reduces the social barriers to participation and enhances the capacity of the local people to make choices to solve the problem. Equity of knowledge can also enhance the economic, environmental and social security in its broadest sense. By capitalizing on the collective wisdom of formal and traditional sciences, we shall be able to help people address the problem of global warming as well as to manage the risks they face because of the destruction of the local resources. Equity of knowledge also provides opportunity for local people to participate in the management of local affairs with global implications. It also provides the opportunity for self-determination. Human ecological perspective is vital in crafting the sustainability science for natural resource management.

The UN has specifically, triggered indigenous knowledge (IK) consideration in several publications. For example, Priority 3 of the Hyogo Framework for Action 2005-2015 which focused on education and knowledge, considered IK as a means of building a culture of safety and resilience. It designates one of its key activities to the importance of information management and exchange, and highlights the use of "relevant traditional and indigenous knowledge and cultural heritage" to be shared with and adapted to different target audiences. UNEP has done much work with African and Asia-Pacific local communities who also have well-developed traditional indigenous knowledge systems. The Indigenous knowledge identified in African and Asian countries demonstrate the potential wealth of knowledge available to world from its indigenous peoples which is why such knowledge for assessing risk is fundamental to saving tens of thousands of lives and building resilience in communities.

A Case of Africa and Asia-Pacific Region:

Taking reference of Africa, local communities had well-developed traditional knowledge system for environmental management and coping strategies, making them more resilient to environmental change. This knowledge had, and still has, a high degree of acceptability amongst the majority of populations in which it has been preserved. Specifically, from time immemorial, natural disaster management in Africa has been deeply rooted in local communities which apply indigenous knowledge to master and monitor climate and other natural systems and establish early warning indicators for their own benefit and future generations. These communities can easily identify with this knowledge and it facilitates their understanding of certain modern scientific concepts for environmental management, including disaster prevention, preparedness, response and mitigation. For example, the application of indigenous knowledge for disaster management is prevalent in Swaziland. Floods can be predicted from the height of birds' nests near rivers. Moth numbers can predict drought. The position of the sun and the cry of a specific bird on trees near rivers may predict onset of the rainy season for farming. On the other side, contrastingly, many of the Asia-Pacific communities have been given little attention by other disaster planning mechanisms and have employed their knowledge as a way to help themselves in difficult times. Much of the knowledge embedded in these communities has been dismissed by outsiders as inferior and often ignored as belonging to "backward" and "less-educated" people. Yet many of these communities have developed successful lessons and strategies for managing recurring disasters and surviving extreme events which even high tech instruments are unable to help. All of these communities share a common ability to depend on themselves during disasters and a similar understanding of local threats and how to reduce these risks. So, there are many lessons to be learned from these communities.

Review of National Status in India

In the past, some studies have been done on Bio cultural diversity, climate change and livelihood security of Adivasi community particularly in Arunachal Pradesh. These studies were based on the ccollaborative work with Adi tribes from 14 villages of East and Upper Siang districts of Arunachal Pradesh, and aimed at recording Adi knowledge and experiences related to biocultural resources and their interactions with climate change and livelihood sustainability. The findings revealed that the Adi tribes have a rich knowledge related to biocultural resources, which plays an important role for the tribes in coping with weather anomalies and any abrupt climatic changes in order to sustain their livelihoods. It also emphasizes that the future of the Adi-tribes bio-cultural resources and livelihood sustainability depends very much on their Traditional Environmental Knowledge (TEK) and their active role in research, planning and policy implementation for climate change mitigation and adaptation. In addition to it, certain research papers have reviewed the concepts and associated ecological hypothesis that India has a treasure of indigenous and traditional knowledge of natural resource management and these can be extrapolated to understand the modern concepts of disaster risk management in terms of early warning, preparedness, mitigation, response and relief as well.

Some preliminary researches also discuss this in the context of cases from Rajasthan and Bundelkhand in order to evolve recommendations for science and policies. They argue that revitalization of traditional knowledge should not be considered as strictly implementing more and more sets of simple techniques, but should be viewed as an integrative means to be evaluated in the framework of environmental, productive and cultural conditions of societies. Local people are custodians of traditional systems and are well informed of their resources, what works and does not work.

Very recently, Centre for Equity Studies, New Delhi (India), a think tank prepared a policy note on sub-committee on Community Engagement in Higher Education, which highlights the relevance and necessity of community engagement at national level with the institutions of higher studies and research. This called for community-institutional partnerships promoting research projects which are need based and community oriented leading to policy formulation for sustainable ecological and societal development to achieve the broader goal of national development under the 12th plan. According to this initiative, specific villages and communities can be adopted for providing engagement opportunities to academia from various disciplines and courses to integrate their knowledge with the community based indigenous knowledge. It would then be able to address the challenges of the specific community as well as the broader environmental challenges that the humanity is confronted with. It can thus be argued that systematic integration of cultural heritage and traditional technology, skills and local knowledge systems within the environment and development as effective means of reducing impact of disasters into existing sustainable development goals and policies at international national and local levels are central needs for developing capacities.

Future Prospect and Significance

Environmental conservation and disaster management are critical to the livelihoods of indigenous people who often live in hazard-prone areas. They have built up, through thousands of years of experience and intimate contact with the environment, a vast body of knowledge on hazards and the environment events. This knowledge is a precious resource that continues to contribute to environmental conservation and disaster management in these regions. With the disruption of traditional lifestyles and changing settlement patterns, it is a challenge to maintain the continuity of traditional knowledge through transmission from generation to generation. In this way, Indigenous knowledge should be considered as a complement to scientific knowledge in the development of community based disaster mitigation and risk management plans and programs.

Hence, it can be said that dynamic mechanisms for disaster risk reduction usually demonstrate strong ownership and leadership of knowledge driven risk reduction activities. Not only should indigenous knowledge management be fully integrated into disaster mitigation at the community level, but such programs should also become pillars of efforts to develop broader national knowledge systems at macro level for disaster risk reduction and sustainable community development. An active communion and collaboration is primarily required among various sectors like Political, Social, Economic and Technological (Figure 7). Moreover, it must include environmental and disaster institutions and managers (including Corporate) more closely in the indigenous knowledge acquisition processes that would potentially reduce disaster events through a thorough review and enrichment of their eco-foot print, programs and environmental resources. Despite the knowledge acquisition implications of disaster risk and vulnerability, and the long-term consequences for sustainable development, the role of environmental managers in disaster reduction, response or recovery has always remained ad-hoc. Hence, disaster managers having given little attention to the indigenous knowledge aspects of their work should regularly be engaged in knowledge management programs as active partners.

Figure 7. Systemic Macro-Micro Interfaces in Indigenous Knowledge Management

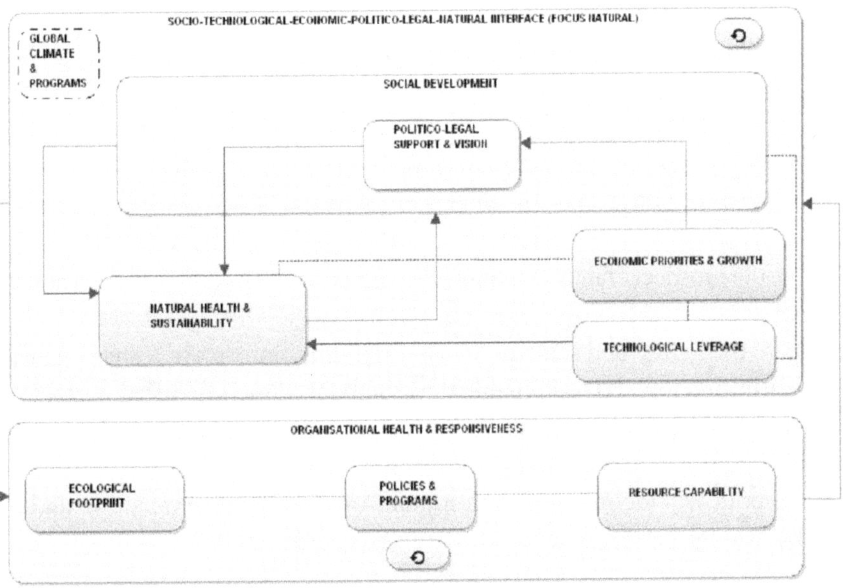

Any such program identifies three essential actions as part of the indigenous knowledge-disaster linking process:

- The disaster mitigation process should include collation of data on indigenous knowledge capital as a fundamental first step in broader environmental project scoping;
- Systematic and Causal-relationship based analysis about knowledge acquisition-natural hazards-vulnerability reduction-capacity building should be undertaken as a central component of ecological assessment and management in areas of risk of natural hazards; and
- A full scale indigenous data compilation and management should be required for certain types of projects in high-risk areas.

Indigenous Knowledge System exists in many developing societies but their future is uncertain. Such knowledge should be recorded and evaluated by people who possess appropriate backgrounds in biology, ecology, resource management, the social sciences, and have the appropriate skills for translating cultural information for the comprehension by other cultures. As discussed above, indigenous and local eco village communities have a stockpile of knowledge about their flora and fauna - their habits, habitats, and seasonal behavior. It is therefore only logical and in consonance with natural justice that they are given a greater say in matters regarding the study, extraction and commercialization of indigenous knowledge. In summary, it is important to note that we need to legitimatize and validate indigenous knowledge for sustainable development on its own terms, and also recognize that it becomes crucial in the development of eco village communities and that the skills and cultures of indigenous people need to be harnessed for the good of all of us. Furthermore, we need to create conditions in which the capacity building in indigenous knowledge may also assist in shaping broader policy framework and encourage benefit sharing for indigenous innovation. In addition, there is a need for thorough research to provide basic direction, particularly in India, that will allow for the success in the indigenous knowledge acquisition process and also transform the mindset of academic and administrative community with regard to indigenous knowledge and other forms of knowledge in general.

Νοω, ανψ μεανινγφυλ εφφορτ ιν τηισ διρεχτιον ωουλδ νατυραλλψ ασκ φορ ανσωερινγ φολλοωινγ θυεστιονσ:

1. What are the macro socio-economic-environmental interfaces in the eco-communities?
2. Which are the facets of indigenous knowledge capacity and its dissemination in the eco community villages?
3. How to explore the prospect of indigenous knowledge capacity building under a holistic 'Stakeholder' framework?
4. How to evaluate the nature and scale of cross-community information sharing and capacity building for disaster mitigation and sustainable development in the identified regions?
5. How to develop causal system model based on the linkages amongst critical decision variables related to indigenous capacity building and value enrichment?
6. How to devise policy guidelines and action plans of large scale acquisition and dissemination of indigenous knowledge capacity of eco communities across India?

Τηεσε μαχρο συβ−σεχτορσ χαν βε αδφοινεδ ανδ ιντεγρατεδ τηουγη ιδεντιφιεδ χαυσαλ λινκαγεσ διστριβυτινγ ϖαρψινγ ποσιτιϖε (ρεινφορχινγ) ανδ νεγατιϖε (βαλανχινγ) ινφλυενχεσ (Φιγυρε 8) ιν ορδερ το ϖισυαλιζε τηε ινχρεασινγ ινστανχεσ οφ νυμερουσ χαυσεσ οφ ϖυλνεραβιλιτιεσ ον τηε ριγητ ηανδ σιδε οφ τηε διαγραμ ωηιχη πρεσεντσ α σεριουσ τηρεατ το τηε εφφιχιενχψ ανδ ρεσιλιενχε οφ τηε εχο−χομμυνιτψ. Τηισ φυρτηερ νεχεσσιτατεσ το ιδεντιφψ τηεσε ϖυλνεραβιλιτιεσ εαρλψ ον ανδ πλαν προαχτιϖελψ το μιτιγατε τηειρ δετριμενταλ χαυσαλ εφφεχτσ. Ιν τηε μοδελ σηοων, ονε σετ οφ χαυσαλ εφφεχτσ ωηιχη ισ προδυχινγ δεσιραβλε εφφεχτ φορ χαπαχιτψ βυιλδινγ χαν βε τερμεδ ασ ϖιρτυουσ χαυσαλιτψ (ον τηε λεφτ σιδε οφ μοδελ), ωηερεασ, τηε οτηερ σετ προδυχινγ υνδεσιραβλε ανδ χουντερ−βαλανχινγ χαυσαλιτψ (ον τηε ριγητ σιδε οφ μοδελ) χαν βε δεσιγνατεδ ασ ϖιχιουσ χαυσαλιτψ.

In the above context, Indigenous knowledge resource management in the identified locations can collectively promote more resilient communities through supporting sustainable livelihoods, disaster prevention and mitigation and strengthening cooperation for overall good governance. Moreover, this kind of knowledge capacity building is universally considered to promote social, economic and environmental equity and ethical imperatives for sustainable community development. The underlying theme of this kind of effort ought to be to nurture an environment where eco-community villages see value and are willing and able to participate in the very activities and purposes of the indigenous knowledge building and promotion activities.

Nevertheless to mention that, people in the disaster prone area have been coping with natural calamities by their own traditional techniques. Often, they face the problems taking different adaptation mechanisms. The coping mechanisms, they adopt, have been originated by their family and community wisdoms, which we can consider as indigenous knowledge. However, as intensity of climatic disasters increased, depletion of knowledge among new generation (as the new generations are ignorant about indigenous knowledge) and occupational migration - this indigenous knowledge base is depleting. Policy documents are lacking to fully address the indigenous knowledge of natural resource dependent people of the coastal area. There are no priority areas to document indigenous knowledge, which are important in the formulation of adaptation and mitigation strategies specific for the location. The policy documents

Figure 8. Integrated Stock-Flow Model of Causal Dynamics in the Eco-Community

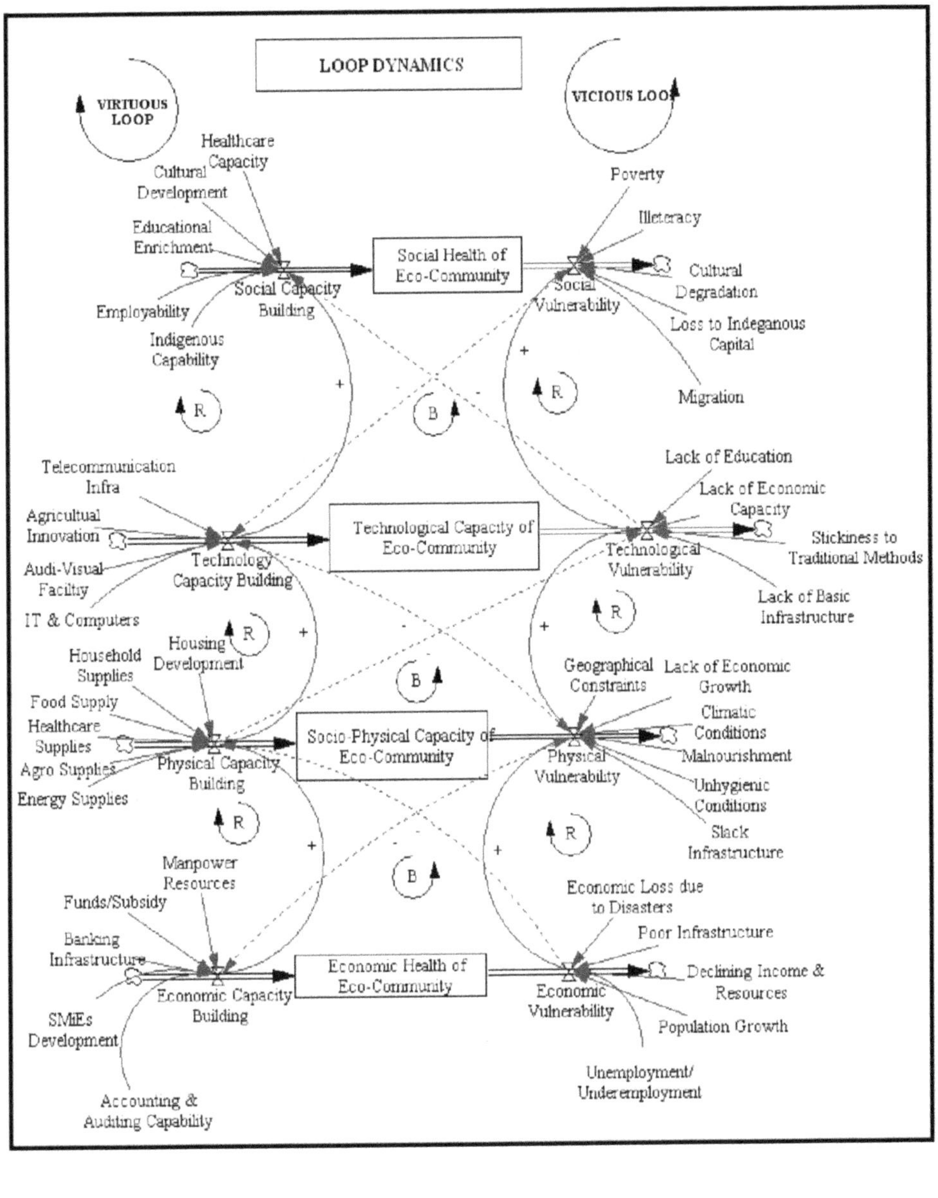

produced by the government do not also consider proper participation of environment dependent occupational groups. Thus, it grossly fails to ensure community level participation in the knowledge capital formation process in order to cope up with natural disasters. Local /Indigenous knowledge is a powerful asset and social capital which is vulnerable to climate change. Policies and programs incorporating such knowledge are more likely to be acceptable and successful in the long run by judiciously prioritizing on community needs, problems, aspirations and local ecology specific.

PROJECT AADIGYAN: A CASE-STUDY IN INDIAN ECO-COMMUNITIES

(A.A.D.I. GYAN- acronym: Augmenting Action for Disaster Management through Indigenous Knowledge, DAYALBAGH EDUCATIONAL INSTITUTE, DAYLABAGH, INDIA)

Sponsored under the aegis of University Grants Commission and Ministry of Human Resource Development of India, the project is about the utility of promoting a global scale and commonly shared digital platform for dissemination and enrichment of Indigenous knowledge and practices across eco-communities of India and abroad based on a fully developed and exclusively functional web-portal branded - A.A.D.I.GYAN.- Augmenting Action For Disaster Management through Indigenous Knowledge (www.aadigyanproject.in, Appendix.1. Screenshots). This portal is a part of digital library initiatives for documenting, preserving and promoting the role of Indigenous knowledge in mitigating natural and man-made disasters, particularly among Eco-communities.

It carries an extensive research database covering various states of India and possibly some other countries and their respective hub-locations (Eco-community/ tribal regions), with a scope to expand at global scale, containing data and information about indigenous activities related to Agriculture, Medicinal plants, Green technology, Healthcare, Education, Environment Management, Socio-Cultural and Spiritual Practices, Employment, Housing etc. It entails a shared web-platform facilitating real-time interactions and exchange of ideas, knowledge and information among all stakeholders -Community, Government, Experts, Researchers, Environmentalists and Academicians. It promises to be a very rich database with an innovative structured classification system for the purpose of systematic arrangement, dissemination and retrieval in order to facilitate further research and distribution of information at global scale. It has a project team constituted of Environmental Scientist, Consultant, Experts and Academicians of high repute. The project has revealed several significant and interesting aspects and applications of traditional knowledge for building long term social, economic and environmental capacities in the eco-communities of India.

Geographical Coverage (Hub-Locations in India) under the Project:

- ✓ Dayalbagh Agra, State of Uttar Pradesh
- ✓ Rajaborari, State of Madhya Pradesh
- ✓ Murar, Distt. Buxar, State of Bihar
- ✓ Udaipur, State of Rajasthan
- ✓ Roorkee, State of Uttarakhand

Indigenous / Tribal Groups studied:

- ✓ M.P.: Baigas, Gond, Bhuia, Birhars, Katkari, Kol
- ✓ Rajasthan: Bhils, Meenas
- ✓ Uttaranchal: Bhotias
- ✓ Bihar: Birhars, ond, Mundas, Oarons
- ✓ U.P.: Khos

ΠΡΟϑΕΧΤ-ΟΒϑΕΧΤΙςΕΣ:

- To analyze the macro socio-economic-environmental interface in the selected eco-communities of India
- To examine the facets of indigenous knowledge capacity and its dissemination in the selected eco community villages of India
- To explore the prospect of indigenous knowledge capacity building under a holistic 'Stakeholder' framework
- To evaluate the nature and scale of cross-community information sharing and capacity building for disaster mitigation and sustainable development in the selected regions
- To develop causal system model based on the linkages amongst critical decision variables related to indigenous capacity building and value enrichment
- To devise policy guidelines and action plans of large scale acquisition and dissemination of indigenous knowledge capacity of eco communities across India

Τηε υνδερλψινγ τηεμε οφ τηισ προφεχτ ισ το νυρτυρε αν ενϖιρονμεντ ωηερε εχο χομμυνιτψ ϖιλλαγεσ σεε ϖαλυε ανδ αρε ωιλλινγ ανδ αβλε το παρτιχιπατε ιν τηε ϖερψ αχτιϖιτιεσ ανδ πυρποσεσ οφ τηε υνδερτακεν ρεσεαρχη.

RESEARCH DESIGN

In the selected eco-village communities of India; apart from doing an extensive review of the relevant literature available on indigenous knowledge, a multiple set of data collection methods have also been used spanning from interactive discussions, open interviews, focus group discussions, meetings, cultural mapping and interactive sessions with rural folk. In particular, personal meetings with people are thought deemed to be important for informing and building community trust with the team of researchers and laying the groundwork for further research. It is thus a promotion of participatory research methodology where research is with communities, and not just 'for' them. This ensures accountability of research process and positive outcomes to a wider community. Attempts have also been made to develop an appropriate checklist of common indigenous knowledge terminology used in environmental resources conservation and natural disaster management. Further, in collecting data, a conscious attempt is made to ensure that there is a clear understanding of the linkages between application and use of indigenous knowledge and cultural and social values, which occupy a pivotal position within the selected communities under study.

Τηε ρεσυλταντ Προφεχτ Δαταβασε ισ χομπιλεδ, αναλψζεδ ανδ υπδατεδ τηρουγη τηε φολλοωινγ διμενσιονσ οφ Τραδιτιοναλ κνοωλεδγε ωιτη ρεσπεχτ το δισαστερ μιτιγατιον:

GREEN TECHNOLOGY:

- Green Technology includes a variety of innovative inventions, ideas, methods and practices with a blend with traditional and modern know-how as possessed by Eco-Indigenous communities and which capitalizes maximum on resources available locally. It develops on the basic premise of preserving and protecting social and ecological balance with that of growing necessities and challenges of modern life.

INDIGENOUS INDUSTRIES AND EMPLOYMENT:

- Optimal utilization of traditional skills and craftsmanship fused harmlessly with economic competitiveness is at the core of promoting traditional industries and employments in eco-communities of India. The thrust is to make local artisans and youth, adequately adept in requisite traits and competencies to explore self-livelihood at native place without migrating to urban locations. Agro and allied industries also flourish as a result of abundance of resources available domestically.

SPIRITUAL AND CULTURAL SUSTAINABILITY:

- Traditional communities have quite resilient cultural, heritage and spiritual systems with strict social and environmental norms helping them sustaining their identity and longevity over ages and changing times. These norms have ensured cohesiveness in the community as well for setting social structure and hierarchy in terms of division of work and roles shared mutually across various gender and age groups.

MEDICINAL PLANTS AND BIO-DIVERSITY:

- The rich reservoir of precious plants and biological resources abundantly distributed and inherited by the Eco-communities in India serves as back-bone for the sustainability of socio-economic systems. The extracts of plants are not only used extensively as medicines in the households but also supplied to the local or outside industries for formulation, manufacturing and marketing of a range of natural drugs and medicines heavily preferred by millions of people across the world.

AGRICULTURE:

- Bio-fertilizers, Crop Cycle-Crop Management, Indigenous Farming practices and tested agricultural know-how are some of the dimensions indicating the richness, diversity and growth of agriculture sector in traditional communities particularly located in disaster-prone areas. Not only providing for the local employment but, it also caters immensely to the secondary and tertiary sectors as resources supply point.

SOCIAL & WELFARE ACTIVITIES:

- Community living with shared social resources creates a resilient framework for survival and growth in any community. Eco-communities are diverse and rich not only in terms of resources, but, mainly through their mechanism to protect the rights and interest of vulnerable social groups such as Old age people, Child and Women. With the aid and support of government agencies and Educational and social organisations, various welfare oriented measures have been taken to uplift the general economic standards and lives of people living in these communities.

EDUCATION:

- Education plays pivotal role in removing score of social ills, evils and habits prevalent in traditional communities of India. Innovative courses and their delivery systems like Distance Education have made it possible to provide world-class education to the local youth at almost no cost and also simultaneously equip them with vocational skills and entrepreneurial traits good enough to sustain an average, peaceful and health community life.

HOUSING AND CONSTRUCTION:

- Indigenous communities have environmental perspective in building houses for their living and constructing facilities for community amenities and functions. Sufficient details are paid attention in order to make houses not only comfortable but also environmentally designed with adequate natural light, ventilation and cooling, heating and structural resilience to withstand disasters like flood, storms and earthquake.

Apart from the above, there are other value-added features and functionality embedded in the project web-portal which makes it a very strong shared platform for interactions and exchange, for example- Seamless navigation through the extensive research database, Posting of Comments, Messages and Reviews, Real-time chatting, E-Mail communication, Search, Gallery (Video and Still photographs), PDF download of detailed documents, References and Subject links. This helps users to get a comprehensive access to the vast repository of traditional knowledge as collected from number of eco-community locations across different states of India, which may greatly enhance their understanding and facilitate sensitization about this grossly untapped knowledge potential of our country

Humanitarian Supply Chain in Eco-Communities

Applying the system dynamics modeling approach towards building long term capacities among the eco-communities of India, key system boundaries and input variables have been identified and depicted in the conceptual model (Figure 9) structure. In this model, humanitarian supply chain links the sources of "supply" (donor institutions or suppliers) to the owners of "demand" (end customers in the identified eco-community). Under this holistic structure, various societies and co-operatives are the primary parties involved under a central administrator agency control. Central Administrator holds main power with the control it has over the entire supply line directly affecting the flows-both Information and Physical, with their decisions. Though, this supply network is huge and complicated with numerous players, it is not that hard to coordinate them along with all the services that are needed to be delivered. Despite the varying objectives and functional differences; collaboration and specialization of the tasks between donor institutions can be improved by clearly defining their goals, scope of activities and mutual interplay.

Now, after defining system boundaries and input variables for modeling, we need to provide a stock-flow structure to them for the intended purpose of simulation and policy analysis. The holistic humanitarian supply chain model can further be segregated into various constituting sub-sectors with respective stock-flow variables and their influence flow as projected below in Figure 3. In addition to the supply line variables identified in the previous figure, the present model also indicates critical vulnerabilities as outflows, which might drench out the inflow-supplies benefits (considered for capacity building) and counter balance the growth of the overall system.

Figure 9. A Schematic Outline of Humanitarian Supply-Chain and Control Framework

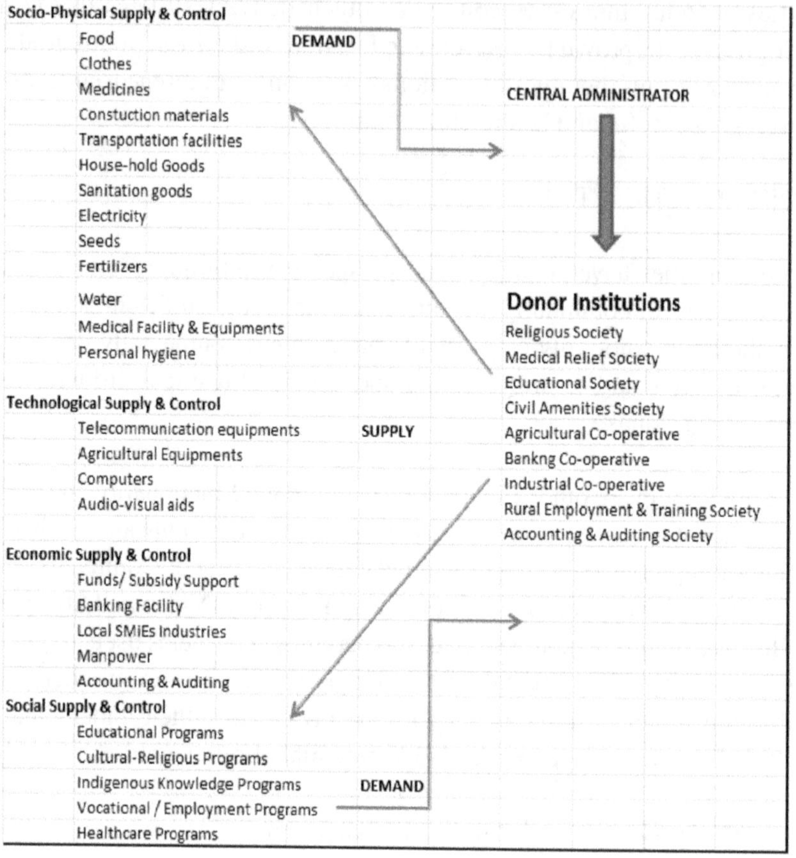

It can be visualized in this model that the increasing instances of numerous causes of vulnerabilities on the right hand side of the humanitarian supply chain model present a serious threat to the efficiency and resilience of the same. This necessitates concept building and systems that can identify these vulnerabilities early in the supply chain planning and react aggressively to their detrimental causal effects. To this end, a dynamic model developed above describes the inherent basic cause-effect relationship between the supply-side and the consumption (loss) side of the humanitarian supply chain for Eco-communities. One set of causal effects which is producing desirable effect for capacity building can be termed as virtuous causality (on the left side of model), whereas, the other set producing undesirable and counter-balancing causality (on the right side of model) can be designated as vicious causality.

Model Development

In this section, an attempt has been made to develop a simulation ready model of humanitarian supply chain to capture certain endogenous and exogenous dynamics of the system under investigation. Though, it was not found feasible to do it for the entire holistic model created in the previous section for the paucity of relevant data. Hence, we have limited our effort to develop a base model for the Economic Sub-sector of HSC supported by the Government assistance program for the Indian tribal communities. It was also considered useful to use a set of additional aiding variables with their quantized values

Figure 10. Humanitarian Supply Chain Capacity Building Vs Vulnerabilities Model

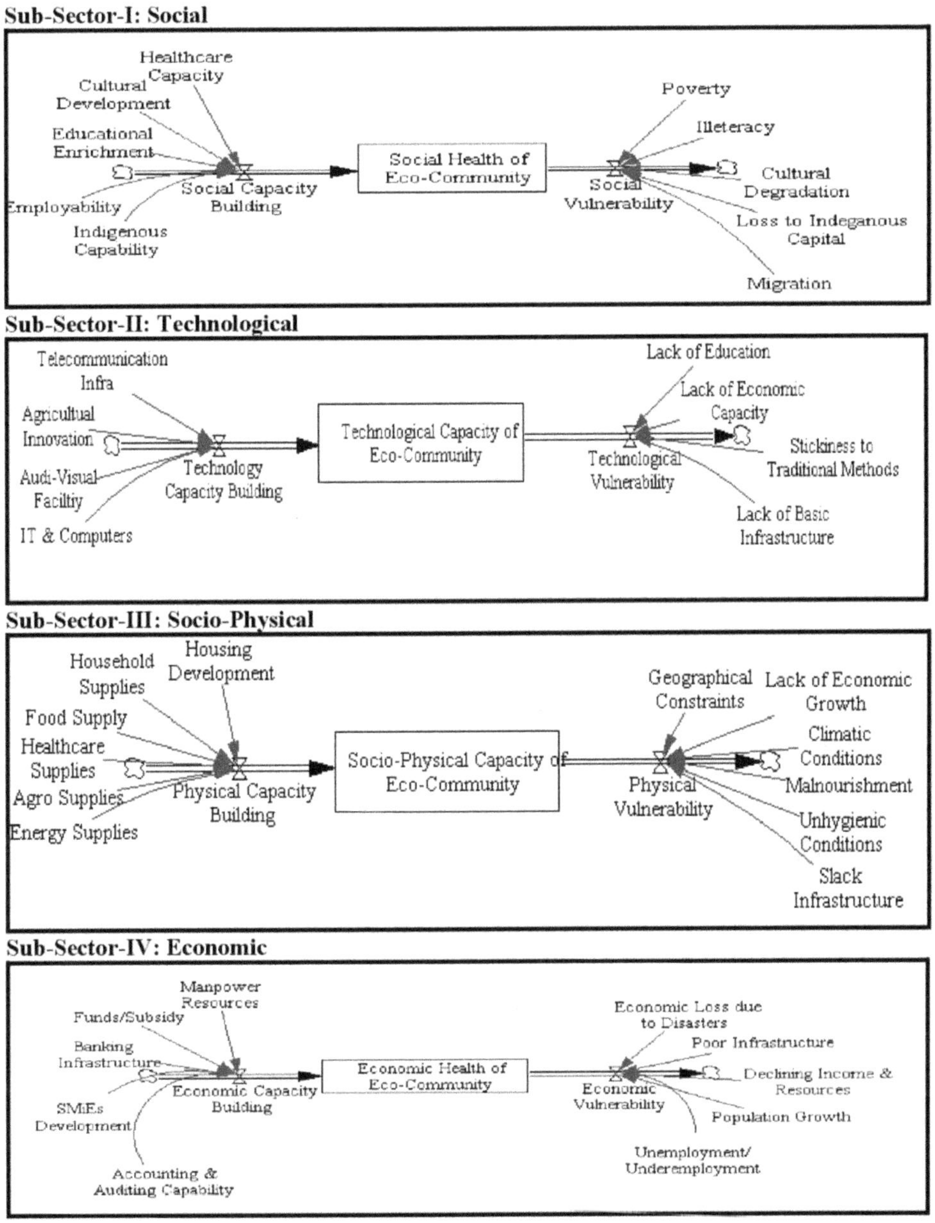

provided by research agencies (Figure 5, Table 1) in order to create a complete supply, consumption and demand scenario.

The key motivations to select tribal regions for our model was that there are thoughtfully crafted developmental programs under India Eco-Development Project (IEP) for Tribal Communities. These programs support the supply of physical infrastructure needs as well as economic needs of the community. Some of the infrastructure activities relate to housing, drinking water, schools, and other amenities whereas on the economic front, activities undertaken are distribution of livestock, providing employment for forestry works, subsidy programs etc. However, the effectiveness of these initiatives were recently

Figure 11. Base Model: Economic Sub-Sector of Humanitarian Supply Line

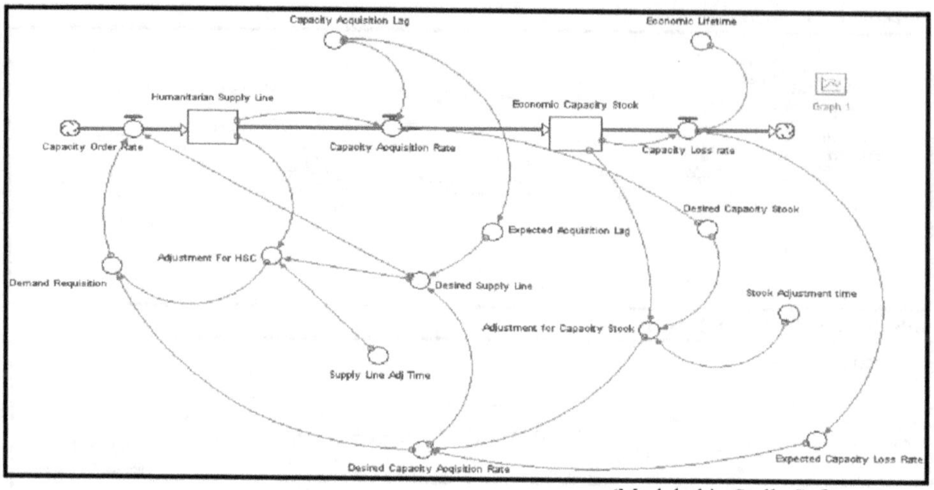

(Modeled in Stella Software)

surveyed to be far from adequate where the project objectives of creating sustainable alternatives and reducing the dependency on the biomass cannot be fulfilled at the present intervention levels. Hence, the current modeling approach aims to explore new insights of Humanitarian supply of various economic capacity building measures with a dynamic perspective and identify crucial intervention decision points under different scenarios.

Sub-Sector-IV: Economic Capacity Building Model for Indian Tribal Community (Base Year 2013-2014):

In the above model, key auxiliary and decision-variables like Capacity acquisition lag, economic lifetime (consumption propensity), capacity stock and supply line adjustment time, have been identified, as they represent the greatest challenge for the humanitarian aid and development assistance under extremely complex economic and social structure of undeveloped eco-communities located across Indian states.

Table 2.

Parameters	Values
Capacity Acquisition (Infrastructure) Lag- India	1.5 Years (McKinsey Report 2009)
Economic Lifetime Value **(HDI-Consumption @ minimum 20% of total per year)***	5 Years (UNDP Report 2011)
Desired Economic Capacity Stock (Budgeted allocation to India Tribal Sub-plan, 2013-2014)	254.98 Billion Rupees (Indian Budget 2013)
Capacity Stock Adjustment Time (Calculated)	2 Years (based on Acquisition lag)
Humanitarian Supply Line Adjustment time (Calculated)	2 Years (based on acquisition lag)

*Inequality-adjusted HDI estimates for backward Indian states facilitate quantification of the potential loss due to inequality with respect to access to education and health.

Table 3.

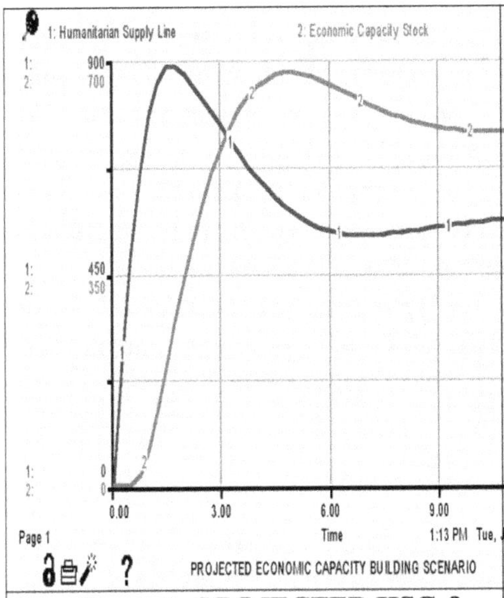	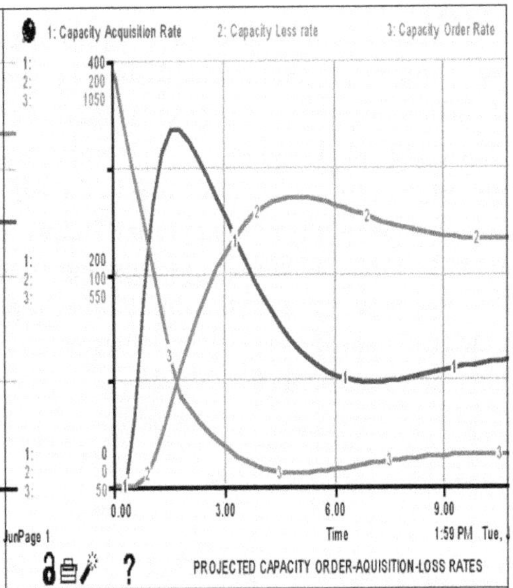
Graph 1: PROJECTED HSC & ECONOMIC CAPACITY	**Graph 2: PROJECTED CAPACITY ACQUISITION-ORDER-LOSS (Utilization) RATES**
Maximum Economic Capacity: Rs. 678 Billion Maximum HSC Supply Line: R 887 Billion **Efficiency: 76%**	
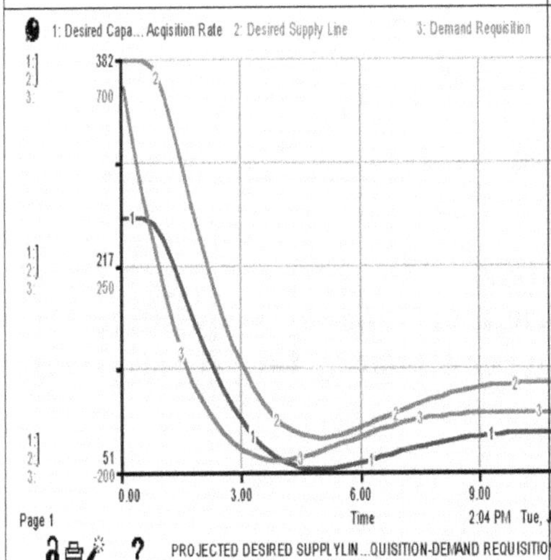	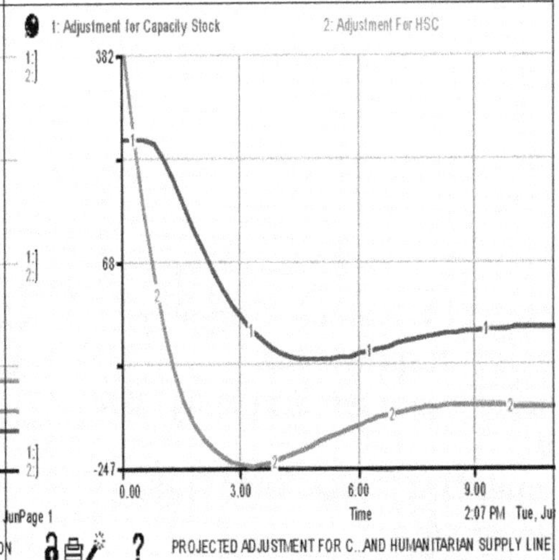
Graph 3: PROJECTED DESIRED SUPPLYLINE-DESIRED ACQUISITION RATE-DEMAND REQUISITION	**Graph 4: PROJECTED ADJUSTMENT CAPACITY STOCK-SUPPLY LINE**

Discussion on Simulation Results: Base-Model

Humanitarian economic assistance in the form of funds, subsidy support, infrastructure development, indigenous industries promotion etc. plays a key part to construct a community's economic capacity, particularly in the aftermath of a disaster. This assistance ensures that the community can develop, instead of sliding back into the vicious trap of vulnerabilities. Seeing the simulation behavior of the base model for economic sub-sector under initial condition, we get to understand that with the present, but, perennially maintained quantum of humanitarian supply of economic inputs to the Indian tribal regions; they can go on building their economic capacity and create a positive surplus by the 12th year (Yr. 2025); as indicated by intersecting point in the simulation Graph 1. Afterwards, even with relatively lower quantum of humanitarian supplies, these communities can remain substantially developed and may also become self-sustained at a later stage.

Table 4.

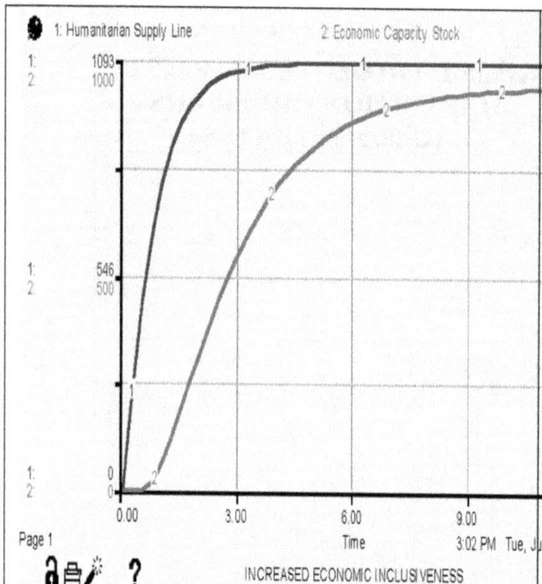

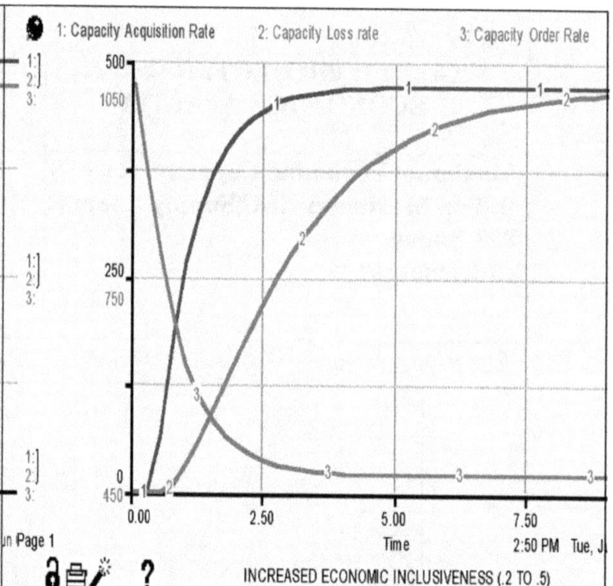

Graph 5: Economic Capacity-Supply Line	Graph 6: Capacity Acquisition-Capacity Order-Capacity Loss (Utilization) Rate
Maximum Economic Capacity: 944 Billion **Maximum HSC Line: 1092 Billion** **Efficiency: 86%**	

*India has an average Country HDI value (using international goalposts) of 0.504 – UNDP 2011

This is further corroborated by seeing the dynamic time-behavior of auxiliary variables like capacity acquisition, order and loss (consumption) rates in simulation Graph 2, desired supply line, acquisition rate and demand requisition patterns in simulation Graph 3 and finally, in the projected values for adjustment in capacity stock and humanitarian supply line depicted in the simulation Graph 4. Under the base simulation, we also analyze the efficiency value of humanitarian supply chain with respect to economic capacity building being maximum 76%, given the fact that the economic capacity is realized with some delay due to lag in acquisition and lower economic life time (consumption) rate.

Analyzing this, here, it should be understood that, an effective and timely humanitarian supply operation and its productive consumption has the capacity to save thousands of lives (Branczik 2004). It is also, however, possible that potential beneficiaries may be located within a zone of disaster or in areas with very poor infrastructure, making it difficult and dangerous for humanitarian agencies to deliver assistance (www.beyondintractability.org). Hence, as a result, one must endeavor to increase the efficiency of humanitarian supply by either increasing the utilization or consumptive capacity of target regions corroborated with decreased capacity acquisition lag or also by decreasing economic stock and supply adjustment times. These scenario effects can be simulated and observed behaviors may provide valuable insight to policy makers for better decision making.

In this backdrop, the following section describes various presumptive scenarios and their respective dynamics indicating the scope of overall performance improvement.

Scenario 1: Increased Consumption of Economic Capacity (Economic Inclusiveness) from Current minimum 20% to India average 50%*:

Table 5.

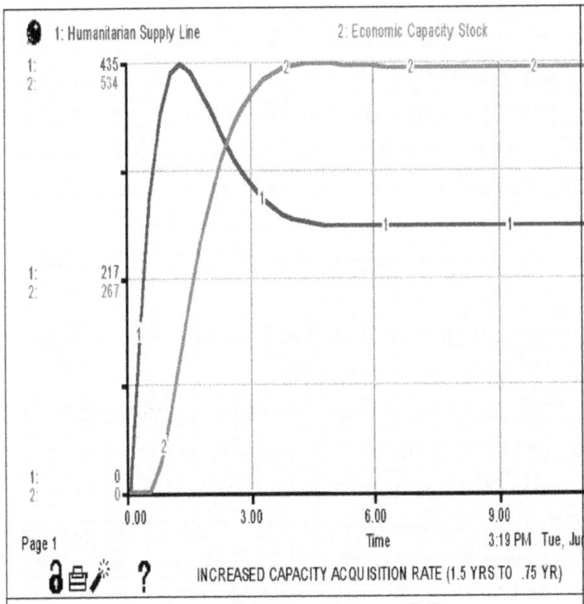	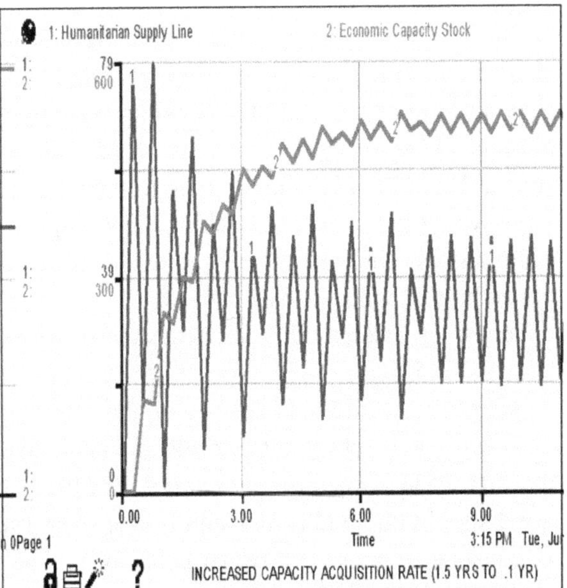
Graph 7: Acquisition Lag: .75 years Maximum Economic Capacity Value: 534 Billion Maximum HSC Supply Value: 434 Billion	**Graph 8: Acquisition Lag: 0.1 Year Maximum Economic Capacity Value: 531 Billion Maximum HSC Supply Value: 79 Billion**

Under this scenario, by increasing the consumptive capacity of tribal regions from current minimum 20% to 50%, we can see the improvement in the maximum efficiency value from earlier 76% to 86%. However, one should accept that this rapid expansion of economic consumption can only be realized by a synchronized stream of consumption of social, physical and technological resources in a holistic sense as modeled and suggested in the section-I of this paper. We can further observe the corroborative behavior pattern of capacity acquisition, order and loss (consumption rates, where, as the acquisition and consumption rates are moving higher and reaching the plateau, there is corresponding declining movement of newer capacity order rate, suggesting that the community is increasingly becoming self-sustained; an idea propounded in the base model simulation of the previous section.

Scenario 2: Decreased Capacity Acquisition Lag Value

As a stand alone measure, the system performance can also be substantially improved by adjusting the capacity acquisition lag value from current base model value of 1.5 years to half of it i.e. .75 years (Graph 7).

One noticeable change is that, once, we decrease the acquisition lag value to the minimum possible value of .1 Year, we find a dramatic reduction in the expected quantum of humanitarian supply from maximum 434 billion in the previous scenario to maximum 79 billion. It goes on to prove that, with faster acquisition rate triggered due to reduction in lag time, comes an ease of supply liquidity putting lesser burden on the donor agencies and government to arrange for the required economic resources. Hence, the created pattern is oscillatory in appearance (Graph 8), indicating the short-termed synchronized lean sequences of smaller quantity supplies and corresponding faster capacity consumption.

If pragmatically attainable, this scenario will effectively fill some of the critical supply-demand gaps in the humanitarian supply system. It will strengthen the coordination and capacity of donor institutions to deliver appropriate, quick, high quality aid, by making resources available in a shorter span of time. It hereby also aims to increase the standing capacity in order to mount immediate responses, and increase the competencies of supporting network agencies within the economic supply sector at both the national and local levels. Not to mention that, it may also facilitate to strengthen humanitarian logistics systems to improve the efficiency of tracking the storage, transport and distribution of relief items.

Similar benefits can also be garnered through an effective adjustment of stock and supply time, as done in the following scenario.

Scenario 3: Decreased Stock Adjustment and Supply-Line Adjustment Time jointly from 2 Years to 1 Year and 0.5 Year:

Under this operational experiment, the effect of stock and supply adjustments at varying rates of 1 to 0.5 years can be observed in terms of change in the patterns of economic capacity stock and humanitarian supply level over time (Graph 9-10). Interestingly, we can also see a somewhat convergence of desired supply line and actual supply line over a long haul, while the value of desired capacity being kept constant (Graph 11). Similar behavior is also observed in case of the desired capacity acquisition rate vis-à-vis the actual capacity acquisition rate (Graph 12). Our thought here is that, this scenario can be attained through the effective coordination between de-centralized agencies responsible for delivery of humanitarian supply services by a central and common command administrator introducing uniform and agile supply chain practices. This scenario can be realized by leveraging a completely IT enabled supply network with the central administrator at its top ensuring standardized supply and service activities delivered to target beneficiary communities.

Table 6.

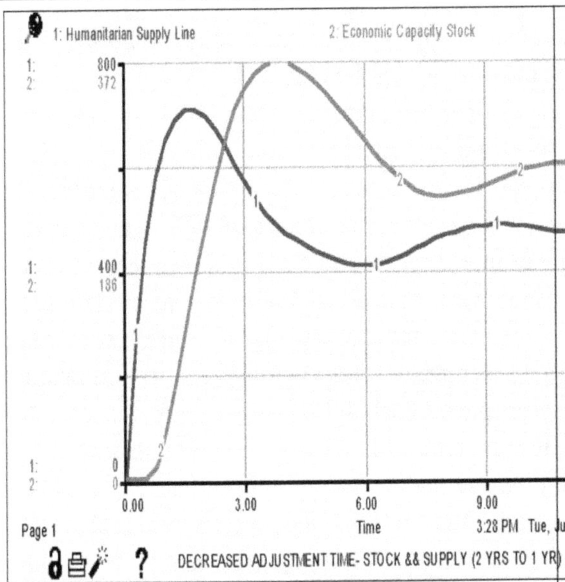

Graph 9: Adjustment Time (Stock & Supply): 1 Year
Maximum Economic Capacity Value: 372 Billion
Maximum HSC Supply Value: 708 Billion

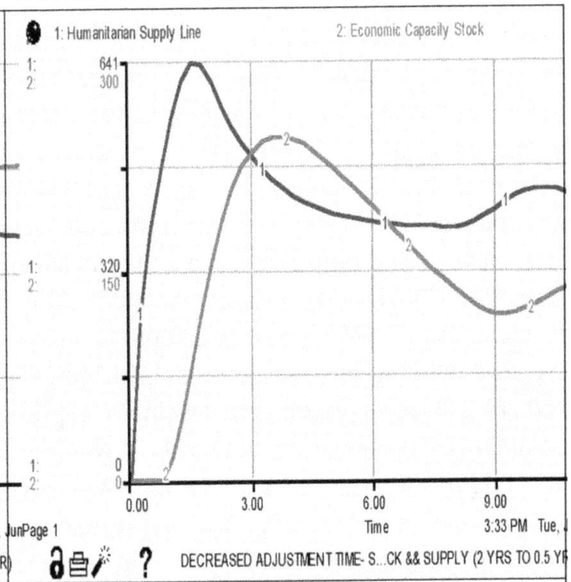

Graph 10: Adjustment Time (Stock & Supply): 0.5 Year
Maximum Economic Capacity Value: 246 Billion
Maximum HSC Supply Value: 640 Billion

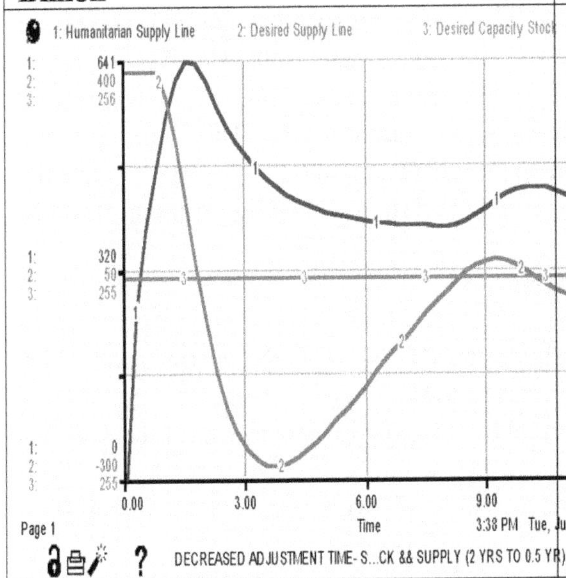

Graph 11: Desired Capacity Stock- Desired Supply Line

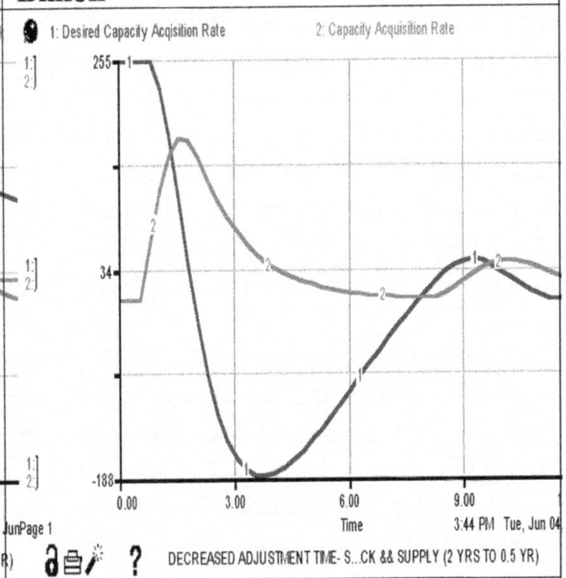

Graph 12: Desired Capacity Acquisition Rate- Actual Capacity Acquisition Rate

CONCLUSION

Based on the discussions under section-I and II above, it can be summed up that special attention should first be given to the development and preservation of indigenous knowledge systems (i.e. indigenous knowledge documentation and database) and its causal relationship to disaster risk reduction and environmental sustainability. A conscious effort should also be made to ensure participation of local inhabitants for long-term acceptance and success of this project. It can be noted that when documented, the information would be available in a convenient form to a wide range of audiences including government decision makers, environmental and disaster management practitioners, medical practitioners, academics, researchers, journalists, indigenous and local communities, as well as international development agents including UNEP, UNSDRR and other stakeholders. This approach would definitely strengthen the capacity building through promoting indigenous knowledge and help to integrate the knowledge with other knowledge-systems and development processes, particularly for environmental conservation and natural disaster management. Further, it would facilitate community engagement and link teaching, research and practices to better reflect the knowledge creation and knowledge mobilization across the eco communities and academic/research institutions as well. This would in turn, offer a vital opportunity to address the issues of community health, community cultures, community practices in sustainable development/natural resources and other related aspects of community knowledge production, application and dissemination.

Aiding to this, the Humanitarian Supply Chain (HSC) can also prove to become quite instrumental to any developmental initiative for capacity building in the backward eco-communities of India, particularly those which are located in the disaster prone regions. Apart from providing immediate relief assistance at the incidence of natural disasters, what is more critical is their long term capacity building with a holistic perspective. This is required to enhance the effectiveness and speed of community response for major humanitarian programs, such as health, food, shelter, water, and sanitation. We can thus say that the success of short-term relief work lies in the long-term capacity building and, is critical to the performance of both current and future humanitarian operations and programs. The scenarios studied in this paper are with respect to those crucial decision environments and their underlying complexities which create an inherent endogenous dynamics perpetuated by various stakeholders giving functional response towards the humanitarian supply chain.

Some key findings of this study can thus be highlighted as-

- Development of a systems model of Humanitarian Supply Chain with the central goal of long-term capacity building in backward eco-communities of India
- Identification of loop-dynamics generated by mutual interplay of various factors at the sub-sector levels
- Creation of a base model of Economic sub-sector of HSC with reference to Indian Tribal communities
- Generation of various scenarios and understanding their behavioural performance over time

Identification of crucial intervention-points by the administrator agency for improving the network output of overall Supply chain.

ACKNOWLEDGMENT

The authors humbly owe their deepest sense of gratefulness to Most Revered Prof. P.S. Satsangi Sahab, Emeritus Chairman, Advisory Committee on Education, Dayalbagh Educational Institutions for His Eternal Blessings, Grace and Guidance at every step of life. The authors also acknowledge the active support provided by their colleagues, project associates and the students of Disaster Management and System Dynamics classes of Dayalbagh Educational Institute in enriching the work immensely.

REFERENCES

Anderson, M. B., & Woodrow, P. J. (1989). *Rising from the Ashes: Development Strategies in Times of Disasters*. Westview.

Anderson, M. B., & Woodrow, P. J. (1991). Reducing vulnerability to drought and famine: Developmental approaches to relief. *Disasters, 15*(1), 43-54.

Anwar, M. A. (2010). Role of information management in the preservation of indigenous knowledge. *Pakistan Journal of Information Management and Libraries, 11*(1).

Azar, A. T. (2012). System Dynamics as A Useful Technique for Complex systems. *Int. J. Industrial and Systems Engineering, 10*(4), 377–410. doi:10.1504/IJISE.2012.046298

Flavier, J. M. (1995). The regional program for the promotion of indigenous knowledge in Asia. Academic Press.

Galli, B. J. (2020). The Application of Systems Engineering to Project Management: A Review of Their Relationship. *International Journal of System Dynamics Applications, 9*(1), 81–106. doi:10.4018/IJSDA.2020010105

Gardner, J. S. (2002). Natural Hazards Risk in the Kullu District, Himachal Pradesh, India. *Geographical Review, 92*(2), 282-306.

Hewitt, K. (Ed.). (1983). *Interpretations of Calamity from the Viewpoint of Human Ecology, The Risks and Hazards Series: 1*. Allen and Unwin.

Johnson, K., Olson, E. A., & Manandhar, S. (1982). Environmental Knowledge and Response to Natural Hazards in Mountainous Nepal. *Mountain Research and Development, 2*(2), 175–188. doi:10.2307/3672962

Kate, K. (2002). Science and the Convention on Biological Diversity. *Science, 295*(5564), 2371–2372. doi:10.1126cience.1070725 PMID:11923511

Kimmerer, R. W. (2002). Weaving traditional ecological knowledge into biological education: A call to action. *Bioscience, 52*(5), 432–438. doi:10.1641/0006-3568(2002)052[0432:WTEKIB]2.0.CO;2

Kohm, K., Boersma, P. D., Meffe, G. K., & Noss, R. (2000). Putting the science into practice and the practice into science. *Conservation Biology, 14*(3), 593–594. doi:10.1046/j.1523-1739.2000.00000-e25.x

Linkenback-Fuchs, A. (2002). *Cultural Strategies for Risk Adjustments in Mountains*. Asia High Summit 2002, Kathmandu, Nepal.

Mearns, M. A., Du Toit, A. S. A., & Mukuka, G. (2006). Conservation of indigenous knowledge at cultural villages: An exploratory study. *Mousaion, 24*(1), 23–50.

Pandey, D. N. (1998). *Ethno-forestry: Local Knowledge for Sustainable Forestry and Livelihood Security*. Himanshu/AFN.

Robertson, D. P., & Hull, R. B. (2001). Beyond biology: Toward a more public ecology for conservation. *Conservation Biology, 15*(4), 970–979. doi:10.1046/j.1523-1739.2001.015004970.x

Saarinen, T. F. (1969). *Perception of the Drought Hazard on the Great Plains*. University of Chicago Press.

Satsangi, A. (2011). Disaster Management. Academic Press.

Schilderman, T. (2004). Adapting traditional shelter for disaster mitigation and reconstruction: experiences with community-based approaches. *Journal of Building Research and Information, 32*(5), 414-426.

Utkarsh, G., & Gadgil, M., & Rao, P.R.S. (1999). Intellectual property rights on biological resources: Benefiting from biodiversity and people's knowledge. *Geographical Review, 92*(2), 282–306.

Warren, D. M. (1991). *Using Indigenous Knowledge in Agricultural Development*. World Bank Discussion Paper No.127. Washington, DC: The World Bank.

Zelinka, D. (2019). Amadei, Bernard., Systems Approach for Modeling Interactions Among the Sustainable Development Goals Part 1: Cross-Impact Network Analysis. *International Journal of System Dynamics Applications, 8*(1).

APPENDIX 1: WWW.AADIGYANPROJECT.IN (WEB-PORTAL SNAPSHOTS)

Figure 13.

Figure 14.

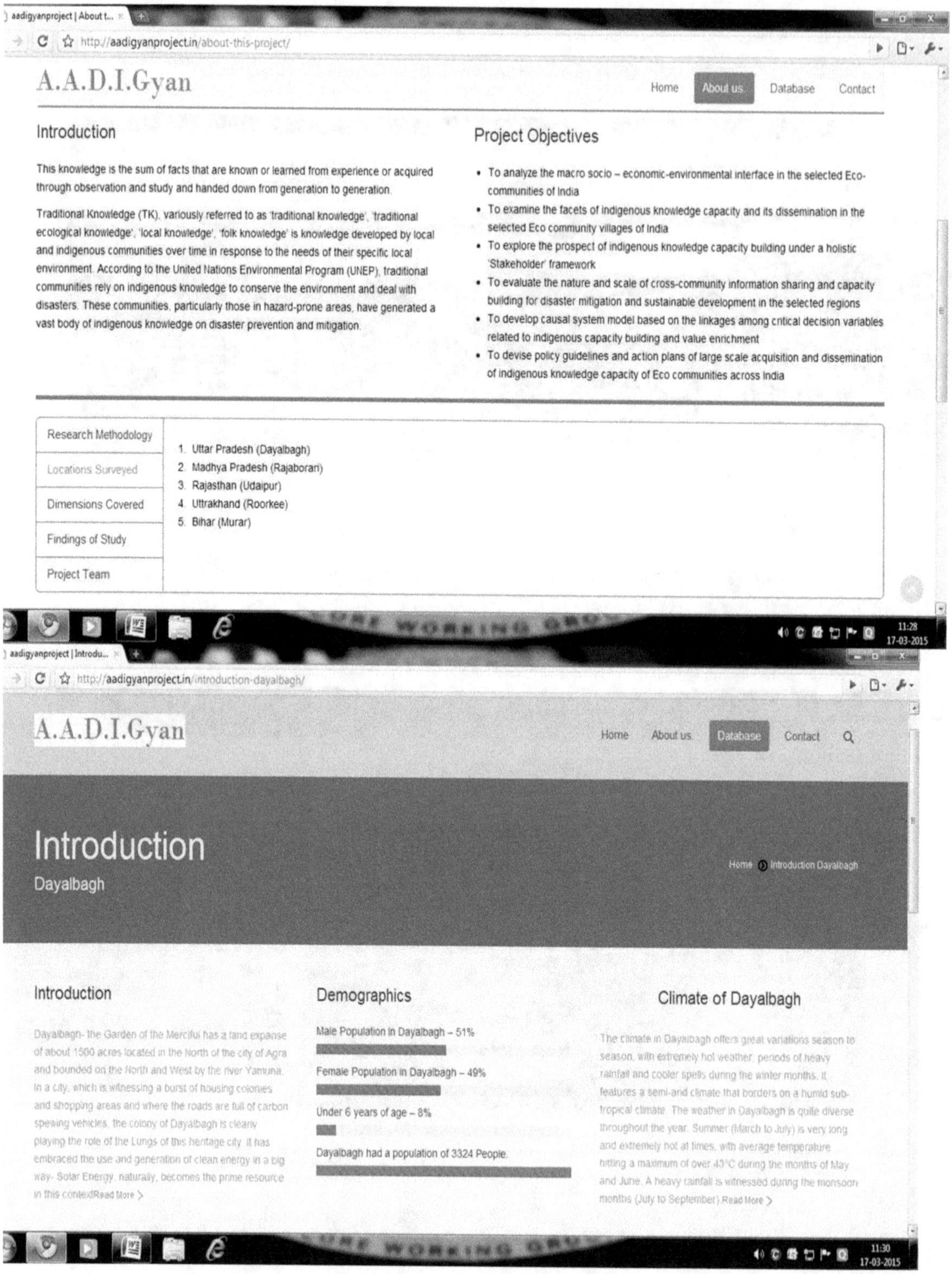
http://aadigyanproject.in/about-this-project/
A.A.D.I.Gyan
Home
About us
Database
Contact
Introduction
This knowledge is the sum of facts that are known or learned from experience or acquired through observation and study and handed down from generation to generation.
Traditional Knowledge (TK), variously referred to as 'traditional knowledge', 'traditional ecological knowledge', 'local knowledge', 'folk knowledge' is knowledge developed by local and indigenous communities over time in response to the needs of their specific local environment. According to the United Nations Environmental Program (UNEP), traditional communities rely on indigenous knowledge to conserve the environment and deal with disasters. These communities, particularly those in hazard-prone areas, have generated a vast body of indigenous knowledge on disaster prevention and mitigation.
Project Objectives
To analyze the macro socio – economic-environmental interface in the selected Eco-communities of India
To examine the facets of indigenous knowledge capacity and its dissemination in the selected Eco community villages of India
To explore the prospect of indigenous knowledge capacity building under a holistic 'Stakeholder' framework
To evaluate the nature and scale of cross-community information sharing and capacity building for disaster mitigation and sustainable development in the selected regions
To develop causal system model based on the linkages among critical decision variables related to indigenous capacity building and value enrichment
To devise policy guidelines and action plans of large scale acquisition and dissemination of indigenous knowledge capacity of Eco communities across India
Research Methodology
Locations Surveyed
Dimensions Covered
Findings of Study
Project Team
1. Uttar Pradesh (Dayalbagh)
2. Madhya Pradesh (Rajaborari)
3. Rajasthan (Udaipur)
4. Uttrakhand (Roorkee)
5. Bihar (Murar)
11:28
17-03-2015
http://aadigyanproject.in/introduction-dayalbagh/
A.A.D.I.Gyan
Home
About us
Database
Contact
Introduction
Dayalbagh
Home
Introduction Dayalbagh
Introduction
Demographics
Male Population in Dayalbagh – 51%
Female Population in Dayalbagh – 49%
Under 6 years of age – 8%
Dayalbagh had a population of 3324 People.
Climate of Dayalbagh
11:30
17-03-2015

Figure 15.

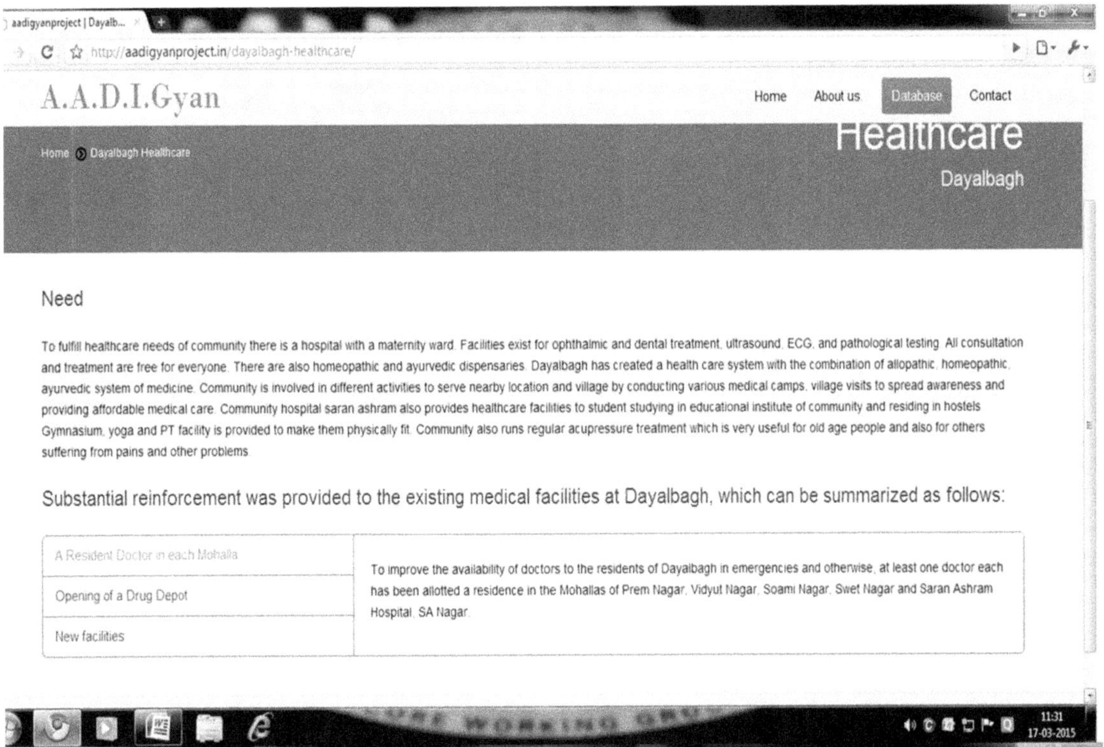
http://aadigyanproject.in/dayalbagh-healthcare/
A.A.D.I.Gyan
Home
About us
Database
Contact
Home Dayalbagh Healthcare
Healthcare
Dayalbagh
Need
To fulfill healthcare needs of community there is a hospital with a maternity ward. Facilities exist for ophthalmic and dental treatment, ultrasound, ECG, and pathological testing. All consultation and treatment are free for everyone. There are also homeopathic and ayurvedic dispensaries. Dayalbagh has created a health care system with the combination of allopathic, homeopathic, ayurvedic system of medicine. Community is involved in different activities to serve nearby location and village by conducting various medical camps, village visits to spread awareness and providing affordable medical care. Community hospital saran ashram also provides healthcare facilities to student studying in educational institute of community and residing in hostels. Gymnasium, yoga and PT facility is provided to make them physically fit. Community also runs regular acupressure treatment which is very useful for old age people and also for others suffering from pains and other problems.
Substantial reinforcement was provided to the existing medical facilities at Dayalbagh, which can be summarized as follows:
A Resident Doctor in each Mohalla
Opening of a Drug Depot
New facilities
To improve the availability of doctors to the residents of Dayalbagh in emergencies and otherwise, at least one doctor each has been allotted a residence in the Mohallas of Prem Nagar, Vidyut Nagar, Soami Nagar, Swet Nagar and Saran Ashram Hospital, SA Nagar.
11:31
17-03-2015

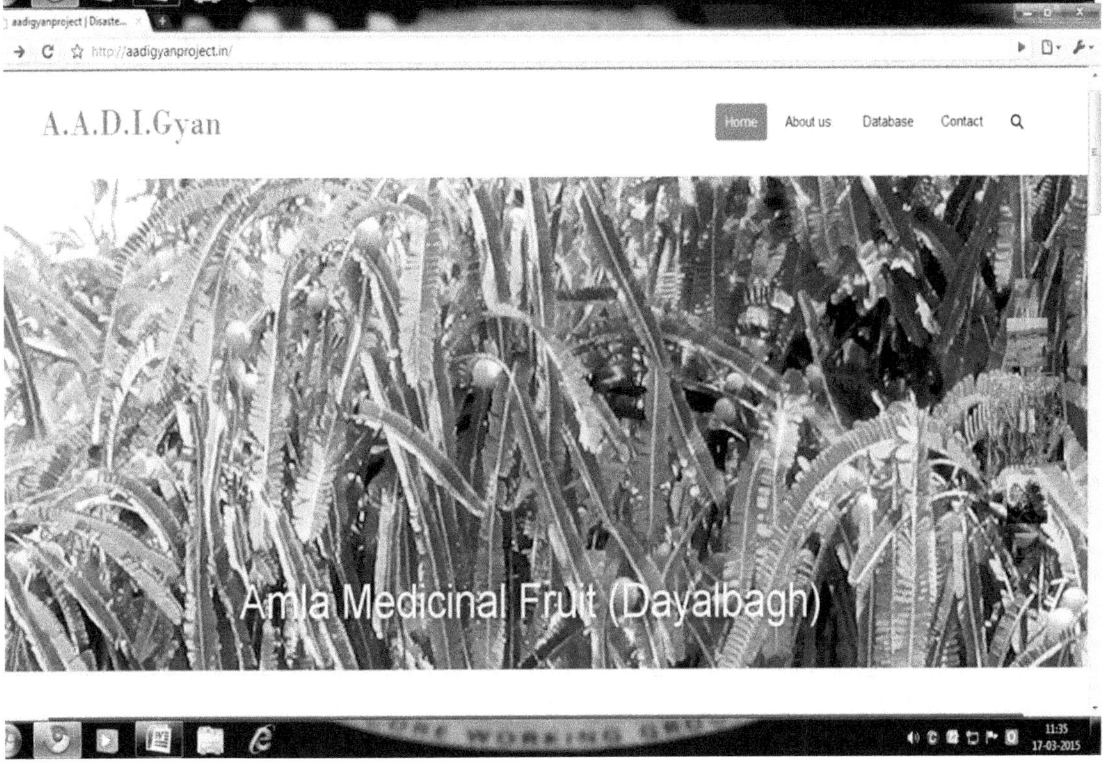
http://aadigyanproject.in/
A.A.D.I.Gyan
Home
About us
Database
Contact
Amla Medicinal Fruit (Dayalbagh)
11:35
17-03-2015

Figure 16.

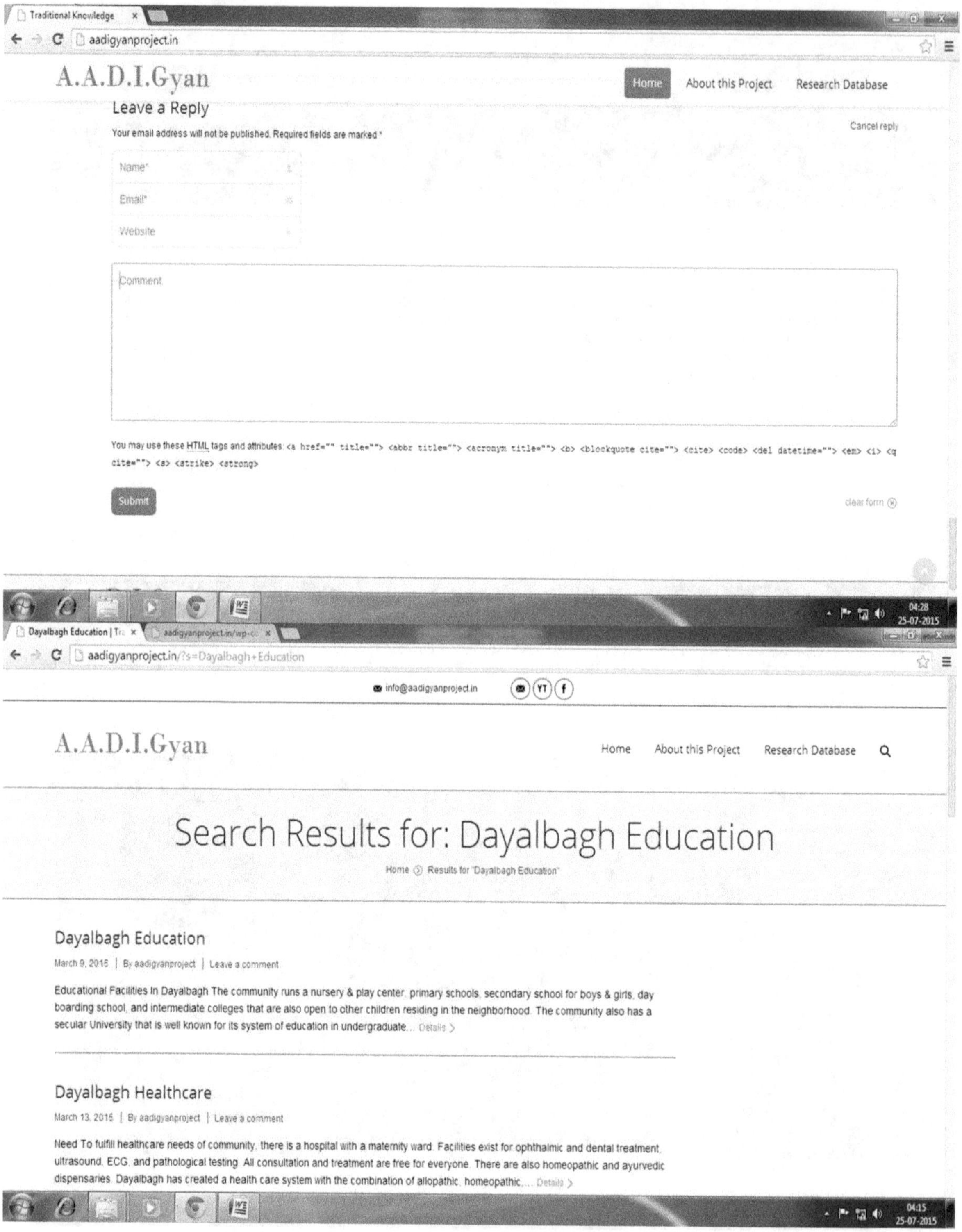
Traditional Knowledge
aadigyanproject.in
A.A.D.I.Gyan
Home
About this Project
Research Database
Leave a Reply
Your email address will not be published. Required fields are marked *
Cancel reply
Name*
Email*
Website
Comment
Submit
clear form
Dayalbagh Education
aadigyanproject.in/?s=Dayalbagh+Education
info@aadigyanproject.in
Search Results for: Dayalbagh Education
Dayalbagh Education
March 9, 2015 | By aadigyanproject | Leave a comment
Educational Facilities In Dayalbagh The community runs a nursery & play center, primary schools, secondary school for boys & girls, day boarding school, and intermediate colleges that are also open to other children residing in the neighborhood. The community also has a secular University that is well known for its system of education in undergraduate... Details
Dayalbagh Healthcare
March 13, 2015 | By aadigyanproject | Leave a comment
Need To fulfill healthcare needs of community, there is a hospital with a maternity ward. Facilities exist for ophthalmic and dental treatment, ultrasound, ECG, and pathological testing. All consultation and treatment are free for everyone. There are also homeopathic and ayurvedic dispensaries. Dayalbagh has created a health care system with the combination of allopathic, homeopathic,... Details
04:28
25-07-2015
04:15
25-07-2015

APPENDIX II: BASE MODEL (ECONOMIC SECTOR)

System Equations:

Economic_Capacity_Stock(t) = Economic_Capacity_Stock(t - dt) + (Capacity_Acquisition_Rate - Capacity__LossRate) * dt
INIT Economic_Capacity_Stock = 0
INFLOWS:
Capacity_Acquisition_Rate = (Humanitarian_Supply_Line/Capacity_Acquisition_Lag)-Desired_Capacity_Stock
OUTFLOWS:
Capacity__LossRate = Economic_Capacity_Stock/Economic_Lifetime
Humanitarian_Supply_Line(t) = Humanitarian_Supply_Line(t - dt) + (Capacity_Order_Rate - Capacity_Acquisition_Rate) * dt
INIT Humanitarian_Supply_Line = 0
INFLOWS:
Capacity_Order_Rate = MAX(0, Demand_Requisition)+Desired_Supply_Line
OUTFLOWS:
Capacity_Acquisition_Rate = (Humanitarian_Supply_Line/Capacity_Acquisition_Lag)-Desired_Capacity_Stock
Adjustment_for_Capacity_Stock = Desired_Capacity_Stock-Economic_Capacity_Stock/Stock_Adjustment_time
Adjustment_For_HSC = Desired_Supply_Line-Humanitarian_Supply_Line/Supply_Line_Adj_Time
Capacity_Acquisition_Lag = 1.5
Demand_Requisition = Desired_Capacity_Acqisition_Rate+Adjustment_For_HSC
Desired_Capacity_Acqisition_Rate = Expected_Capacity_Loss_Rate+Adjustment_for_Capacity_Stock
Desired_Capacity_Stock = 254.98
Desired_Supply_Line = Desired_Capacity_Acqisition_Rate*Expected_Acquisition_Lag
Economic_Lifetime = 5
Expected_Acquisition_Lag = Capacity_Acquisition_Lag
Expected_Capacity_Loss_Rate = Capacity__LossRate
Stock_Adjustment_time = 2
Supply_Line_Adj_Time = 2

Chapter 10
Governance, Complexity, and Resilience of the Kenyan Mobile Industry in the Context of Its Strategic Vision

Amos O. Omamo
Jaramogi Oginga Odinga University of Science and Technology, Kenya

Anthony Joachim Rodrigues
Jaramogi Oginga Odinga University of Science and Technology, Kenya

Joseph W. Muliaro
Jomo Kenyatta University of Science and Technology, Kenya

ABSTRACT

Assessing the impact of existing and emerging mobile technologies on individuals and societies is a complex task. Various models of the telecommunication industry have been developed and differ from country to country. The study analyses the Kenyan mobile industry's governance, complexity, and resilience. It investigates the emerging notion of "resilience" as a perspective for understanding how societies can cope with, and develop from, disturbances and change. A resilience approach to governance issues shows a great deal of promise to enable a more refined understanding of the dynamics of rapid, interlinked, and multiscale change. Simulation is done of the Kenyan mobile industry. The simulation result proves that the regulator Communications Authority of Kenya (CAK) is resilient amid the continuous change in the telecom industry. This study also uses the model to analyze mobile industry policy and concludes with policy suggestions to improve the expected mobile industry services and governance for developing countries.

DOI: 10.4018/978-1-7998-5788-4.ch010

1.0 INTRODUCTION

In Kenya, mobile telecommunication industry has witnessed a tremendous growth over the last few years, as of 31st December 2019 the total number of mobile subscriptions was recorded as 54.5 million up from 46.6 million. Kenya's mobile phone penetration crested the 100 per cent mark as stated by the Communications Authority of Kenya. It joins Morocco, Namibia and Tunisia, the only countries in Africa to achieve the milestone. In the recent times, India has also emerged a fast growing telecom market in the world and witnessed a telecommunication revolution brought about by a collaboration of government, industry, and the scientific community (Bhushnan, 2012). The increase in mobile penetration was "mainly attributed to the fact that most users own more than one SIM card either from the same or different service providers. However, there is a wide variation in mobile diffusion as well as GDP growth across various counties in Kenya, raising questions of socio-economic disparities and how technology diffusion may help in the convergence of growth process among various counties. (Ansah, 2017) elicits the structural mechanism representing the intended effect of structural adjustment policies and the unintended effects observed from the implementation of the structural adjustment policies in many African countries. Empirical studies have found several factors such as per capita income, income inequality, population density, the age profile of the population, competition and regulatory structure to have a positive impact on mobile penetration (Yamakawa, et al., 2013, Chakravarty, 2007). Despite the evolving literature on the development benefits of mobile phones, we still know very little about factors that influence their adoption (Asongu, Nwachukwu & Aziz, 2018). Many studies have investigated determinants of mobile phone penetration in Sub-Saharan African countries using various policy variables. The relation between inequality and mobile penetration has been found to be mixed. In some studies, mobile penetration, was found to be negatively related to income inequality; whereas, it is positively related to inequality in the early stages of diffusion (Roeller & Waverman, 2001, Hytennin & Toivanen, 2011). In another study by (Vokshi et al., 2019), considered that people with access to mobile telephony also have access to Wi-Fi and GPS and that individuals can perform different activities, such as engaging in e-commerce, e-governance, health, and education; paying bills; saving money; and transferring money to other persons. This represented a good foundation for poor persons exit the cycle of deprivation and lead to the development of equal opportunities. In the developing country context, mobile phones serve dual purposes: one, as a consumption good for the rich and two, as a production good for the poor. Case studies from the Africa and Asia have shown the usefulness of mobiles as a production good (Jensen, et al., 2007, Aker et al 2008, Muto, et al, 2008). For this reason, income inequality may influence the spread of mobile penetration in the early stages. Embracing the digital revolution requires good ICT governance. Appearance of the inequality of opportunity might result in losing faith in the markets and might generate preferences for misleading market policies. Thus, understanding the connection between the inequality of opportunity and beliefs is quite important (Brock, 2018). In a study on the determinants of mobile phone ownership, the following were found to be quite important: education, informal work, employment status and type of electricity. To increase mobile phone use, especially among youth, governments should support initiatives involving the development of mobile phone content in local languages (Forenbacher et al., 2019). Although the impact of economic and demographic factors on mobile penetration has been established, there is not much clarity on the relationship between mobile phone penetration, economic growth and the extent to which this leads to convergence of growth process. Systems Dynamics could assist in viewing problems in complex system holistically, while Big Data in e-government is for collecting data and information which are used for developing model of system dynamics and its equations (Nasution et al., 2018). The mobile industry of Kenya is, therefore, an exemplifying case for the phenomenon. The paper uses system dynamics as a framework, firstly analyzes the industry chain, secondly builds a system dynamics model by drawing a flow chart and an archetype analysis, thirdly simulates the model and empirically analyzes the key factors which should be supervised in the industry chain, and analyzes the influence which some policies will bring out. (Such as real mobile identification and unification the service providers' access number)

The outline of the paper includes the introduction in section one; section two covers the mobile industry modelling since it's the main focus of the study; section three covers the simulation to get the different scenarios; section four talks about the policy scenarios and finally section five the conclusion and way forward.

2.0 MOBILE INDUSTRY SYSTEM DYNAMICS MODELING

2.1 Research Boundary and Hypothesis

Since the mobile industry is the main focus of this study, the research boundary only refers to social entities related to the industry, including social influence, telecommunication users, regulator (CAK), service providers, content providers and the government. This paper supposes the system is running under the environment of 4G; there is only one regulator in the market, i.e. the Communications Authority of Kenya. But there are so many service providers and content providers that together form the situation of perfect competition. This hypothesis ensures the system can satisfy the need of evolution. Chaotic behavior is a term that is attributed to dynamical systems whose solutions are highly sensitive to initial conditions (Moysis& Azar, 2017). Disruptive operators are considered to be important for increasing the competitiveness of a telecommunication market comprised of large companies. An analysis of the disruptive impact of the entry of free mobile into the French mobile telecommunication showed that the conditions of entry as set by the regulator played a major role in the disruption. This has had a major impact on provoking the restructuring of companies operating in the telecommunication market, which is actually controlled by regulatory laws. Moreover, low-cost business models were able to be implemented due to complementary assets (Berne, Vialle and Whalley, 2019). The study puts emphasis to understand the problem of corruption under governance while still explaining law and order, social, cultural, economic problems and political instability through a systems approach. This is because corruption is the weakest link for governance and a big problem in Kenya. We achieved this by exploring how the complex social system of corruption develops its stable macro-state. The conceptualization of corruption used in this paper is that it is a social phenomenon, which involves public participation in general; manifested as a social system of corruption that affects all other systems in one way or another.

2.2 Mobile Industry Analysis

To date, there are five major mobile operators in Kenya namely Safaricom PLC, Airtel Networks Kenya Ltd, Finserve Africa Ltd, Telecom Kenya Ltd and Mobile Pay Ltd. According to the mobile subscription and profitability, Safaricom Ltd is in the top position among the five operators CAK (2019), Safaricom PLC was the largest mobile telecom operator in Kenya in 2018 and 2019, with a market share of 64.9 percent in the third quarter of 2019. Airtel Networks Limited accounted for about 24.6 percent in the same period, making it the country's second biggest provider of mobile subscriptions. In Kenya by the end of the third quarter of the 2018/19 financial year, there were a total of 54.5 million subscriptions in the mobile telephony market segment up from 46.6 million posted in the previous quarter. This represented an increase of 114 percent during the period. The mobile service provision process at the industry level is quite complex; very many activities and participants are involved in the production process.

Dynamic behavior in mobile telecommunications markets is often the result of interactions between market actors. As the structure and dynamic complexity of such interactions tend to be difficult to identify, it is challenging for regulators to anticipate the effects of their policy decisions (Casey & Toyli, 2012). The dynamic behavior of the mobile industry is determined by the dynamic complexity of the industry, which involves fewer entities than the detailed complexity of the industry. (Senge, et al, 1994) points out that these two types of complexity exist together in a complex system, and that the dynamic behaviour of the system is determined by the dynamic complexity of the system but not by the detail complexity of the same. While detail complexity is indicated by the number of actors and activities in the system, dynamic complexity is indicated by the number of levels in the feedback control structure of the system. In a mobile industry's production process, the number of system levels is much smaller than the number of actors and activities in the process. Therefore, despite the multiplicity of activities in the mobile service provision process at the industry level in Kenya, the number of activities in the production process, which are responsible for the dynamic behavior of the industry, is rather small.

In this study, the dynamic complexity of the mobile industry was conceptualized in terms of five components in the production process, namely: (i) demand for mobile services in the country's economy; (ii) cost of mobile services (iii) Government services and provision (iv) cost of mobile equipment (v) Quality of services. The five components are fairly distinct occurrences in the mobile service provision process. From the components of dynamic complexity, level variables are conceptualized.

The study developed casual loop diagrams for the mobile phone industry service in Kenya, taking the various stakeholders and factors that would be affected by and affected the system in figure 1, into consideration as outlined.

2.3 Flow Chart Analysis

Flow chart of the whole system consists of a total of 10 key variables, which mainly represent our actual model characteristics, each variable is described in Appendix A and the key leverage loops are shown in Table 2.1. Each variable within subsystems interacts with each other, and there are association relationships between subsystems as shown in Figure 2.

By using the system dynamics methodology, the conceptual framework is converted into a simple system dynamics model by integrating all the sub models. There are 4 sub-models namely Economic, Social, Governance and, Technology. As an attempt to explore the process of technological change affecting the economic growth that the developing country may have some directions more clearly how to design its technology policies; there are three sources of growth (growth scenarios) considered in this study i.e.: population, governance, and innovation. The model is initialized in the full equilibrium. In this equilibrium, the population growth rate is equal to zero. The model is simulated for 25 years, and the changes of the source of growths are introduced into the model in year 5.

Loop Type coded as follows: B indicates a Balancing Loop, R indicates a Reinforcing Loop.
Delay coded as follows: Y indicate the existence of delay, N indicates no delay in the loop.

Figure 1. **Causal loop diagram for the mobile phone Industry service**

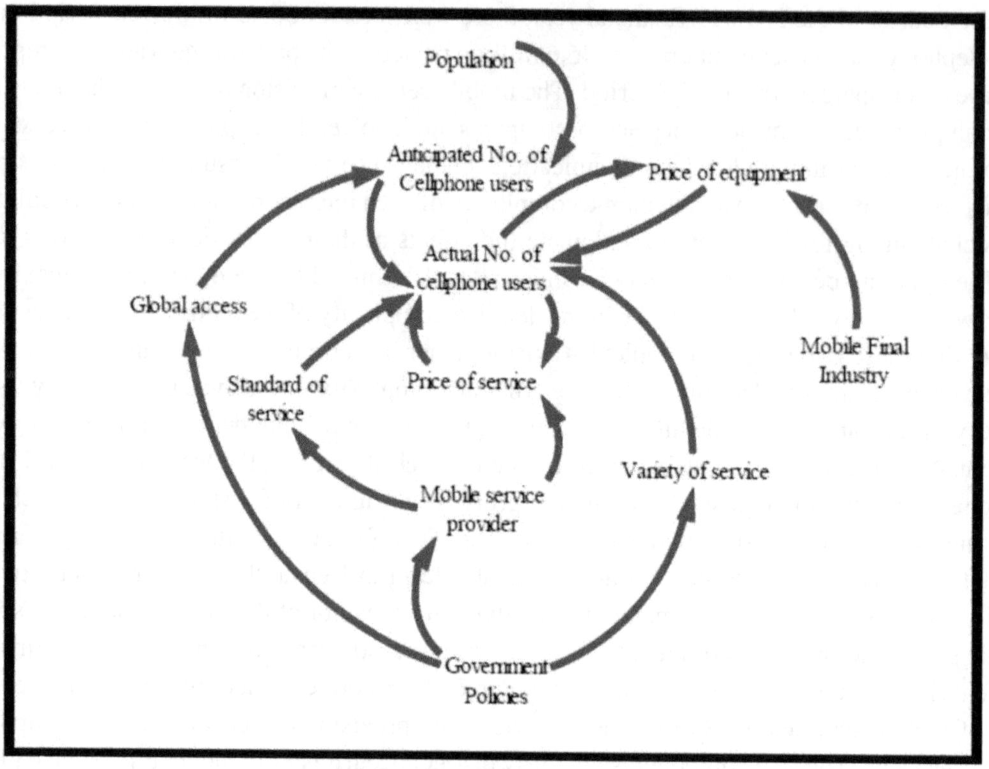

Figure 2. Causal Loop Diagram for Governance, Society, Technology and the economy

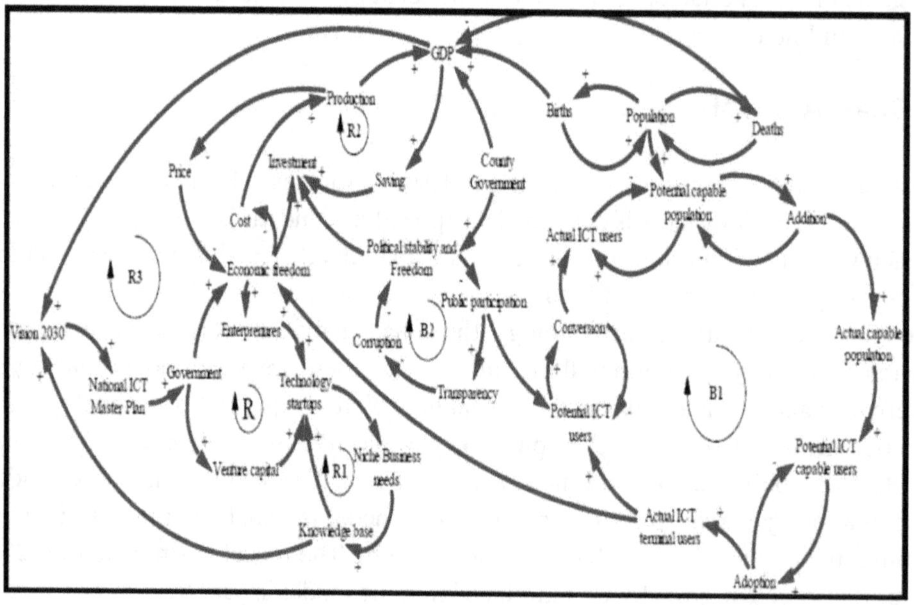

2.3.1 Governance and Mobile Industry Service sub Model

Diversity in governance capacities in successful developing countries means that we cannot necessarily identify simple patterns in the governance capacities of successful countries, but nevertheless, we can identify broad patterns in the functions that successful countries performed, and this can provide useful insights for reform policy in the next tier of developers. The empirical and theoretical issues involved here clearly have critical policy implications for reform efforts in developing countries.

In our sub model there are main feedback loops driving governance as shown in figure 3. On the accountability side investment in the rule of law increases government efficiency and lowers corruption through transparency. This improves Equity and hence more public participation the cost of the technology through mechanisms such as economies of scale in production and learning by doing.

Thus, as the technology diffuses into the market, users' attitude in favor of the technology will increase and it gains legitimacy. As the technology diffuses the venture capitalist increase in size as well as political strength and may influence the regulatory frame work in favor of the technology.

Figure 3. Governance and Mobile Industry service sub model

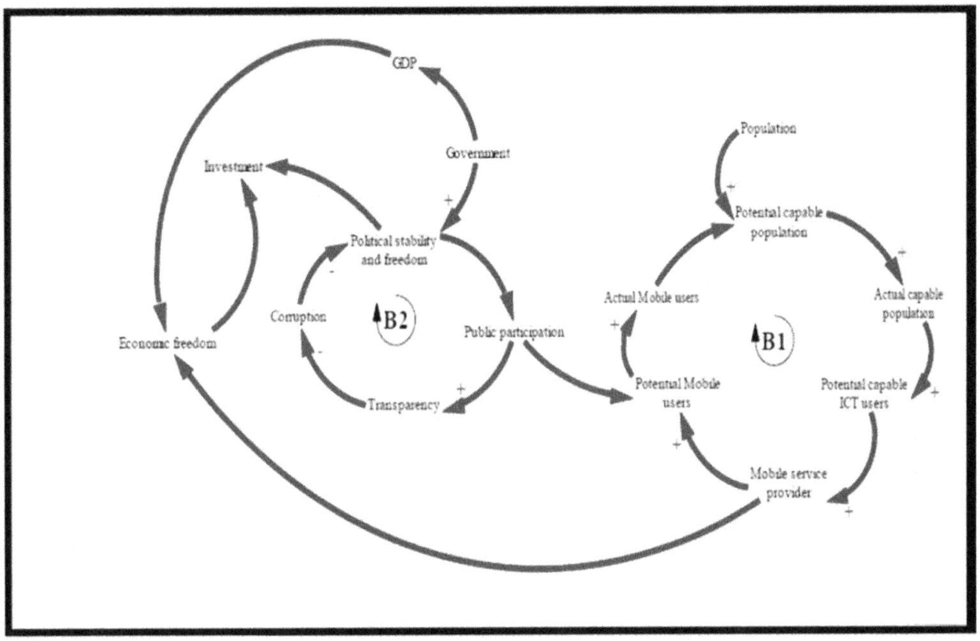

Loop B1- Due to the continuous population increase in Kenya, the number of capable people economically is increasing and some of them translate to potential capable ICT users and due to the high adoption of technology in the country and the emergence of mobile service providers, we get potential mobile users

Loop B2- Political stability and freedom improves governance and hence more public participation from the citizenry or society, this brings about transparency in the way projects are implemented, but due to the problem of rampant corruption, projects and policies are affected

2.3.2 System Dynamic Structural and Mental Model of Mobile Phone Usage and Economic Growth

Qualitative modelling is a starting point in the modelling process. A structured and integrated casual loop diagram follows a mental presentation of the observed system (Munitic, A. and P. Ristov, 2009). In systems dynamics modelling, the generic system archetypes operating in the mobile industry of Kenya from 2005 to 2015 were identified, and the time-variance of mobile output, providers and average mobile output discussed. The paper presents development of a system dynamics model of mobile output, which is based on: (i) mobile service process at industry level in Kenya, and (ii) features of the system archetypes operating in the mobile industry. We further present quantification of the mobile industry as a feedback control system. The system dynamics model of the existing structure of the mobile industry system of Kenya was formulated, conceptualizing the problem of mobile activity variations and growth in terms of the various system variables. The model was particularly for the total mobile output. The mobile industry as a whole was conceptualized as one large corporation, whose annual production is the mobile output captured in national statistics every year. Conceptualization of the model variables was generally based on deductions made in the qualitative data analysis. However, more ideas were added in the formulation of the model structure, in order to replicate a real life mobile industry as closely as practicable. The model variables and their units of measurement were derived from the basic variables. The modeling process took the following steps: (i) review of mobile service provision process in Kenya; (ii) estimation of demand for mobile service facilities in the economy of the country; (iii) articulation of the existing feedback control structure using level rate diagrams and difference equations; (iv) functional definition of the model variables; and (v) simulation runs. After several simulation runs to ensure the soundness of the model mobile services provision, validity of the model was tested using behaviour replication tests, parameter tests and extreme conditions test. The causal loop diagram (CLD) for the mobile phone industry and economic growth is shown in Figure 4.

Figure 4. Causal loop diagram for the mobile phone service and economic growth

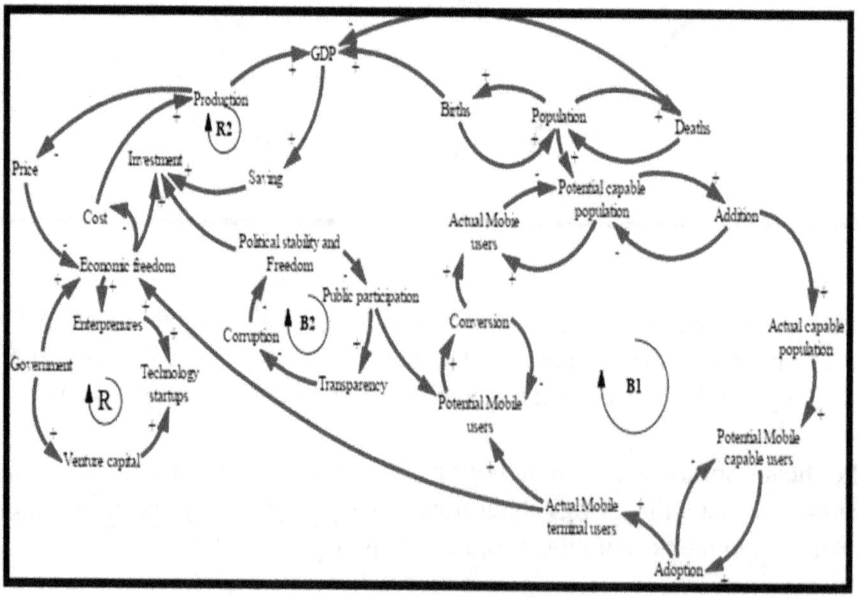

To get an integrated view of the model, we start with the main variable-Number of Mobile Users which represents the subscribers of the basic telecommunication services. Households subscribe to telecom services. An increase in the number of households leads to increase in the number of subscribers and depletes the potential subscribers. This bi-directional relationship forms a loop with negative polarity i.e. negative feedback loop. An increase in economic growth increases the income level of the people and hence the subscriber base for telecommunication services. As the number of subscribers of a telecommunication services increases, the amount of information traffic and frequency of request for connection establishment increase. This causes congestion, which deteriorates network performance. However, if the service provider synchronizes the infrastructural build-up i.e. setting up of switches, access loops and trunks, with the pace of building up of subscriber base, congestion reduces.

Quality of the access loop decides the type of services; a service provider can provide what subscriber has subscribed for. Economically, the access loop cost is one of the major components of total infrastructure cost of telecommunication services. Hence, as the quality of access loop improves, the variety of services being provided increases as well as the cost of services. Increase in cost of providing service enhances the price of subscription of the service, which in turn shows its negative effect on subscriber level based on their price elasticity of demand for services. The price for services gets lowered with increase in industry experience, service providers experience and economy of scale. Variety of services is not only governed by the quality of access loop but by the threat of competition also. If there is no expected competition in the market place, then monopolists do not feel motivated to provide higher quality services even though the access loop is capable of supporting such services. Similarly, even with high threat of competition, the service provider may not be able to provide multiple services immediately because of the low quality of the access loop. Similarly, threat level of competition and variety of services both dictate the time for deployment of services.

The CLD further demonstrates the Kenyan economic sectors interaction. In the model, GDP, saving and investment are also main functions of the model. As to be seen, saving depends on income. Different economic and political factors such as corruption, political rights and civil liberties as effective factors on investment are considered. Simulated behaviour of indicators in comparison with their reference modes were emphasized to test and show if unexpected behaviour and relationships are correct.

Economic systems, being complex adaptive systems, are composed of numerous numbers of components which have the specific internal relationships. The components of this section consist of Gross Domestic Product (GDP), saving, investment, public participation, and inflation as shown. This model evaluates Kenya's economic system and demonstrates how it is possible to integrate a simple static economic model within a dynamic framework using systems dynamics methodology.

2.3.3 Integration of the Kenyan Mobile Industry With the Other Subsectors in Kenya

The composite structure of the model with the interactive feed forward feedback loops is shown in the integrated model sectors Figure 5. The critical subsectors of the Kenyan economy is shown within the four 'sub-sectors' or 'concentration zones' are defined in Table 2.2: They include the Social, Technology, Economic and Governance sectors. These sectors together define the Kenyan strategic Vision. (Zelinka. D.and Amadei.B., 2019) advocates for a systems approach to analyzing the SDGs, categorized as an anthropocentric network. Cross-impact analysis, a semi-qualitative, soft-systems approach is used to explore the connections among the SDGs. This approach can better capture the science underlying the SDGs' interactions in a generic manner compared to existing methods. The suggested approach enables

Figure 5. Sub-sectors of the Kenyan Economy

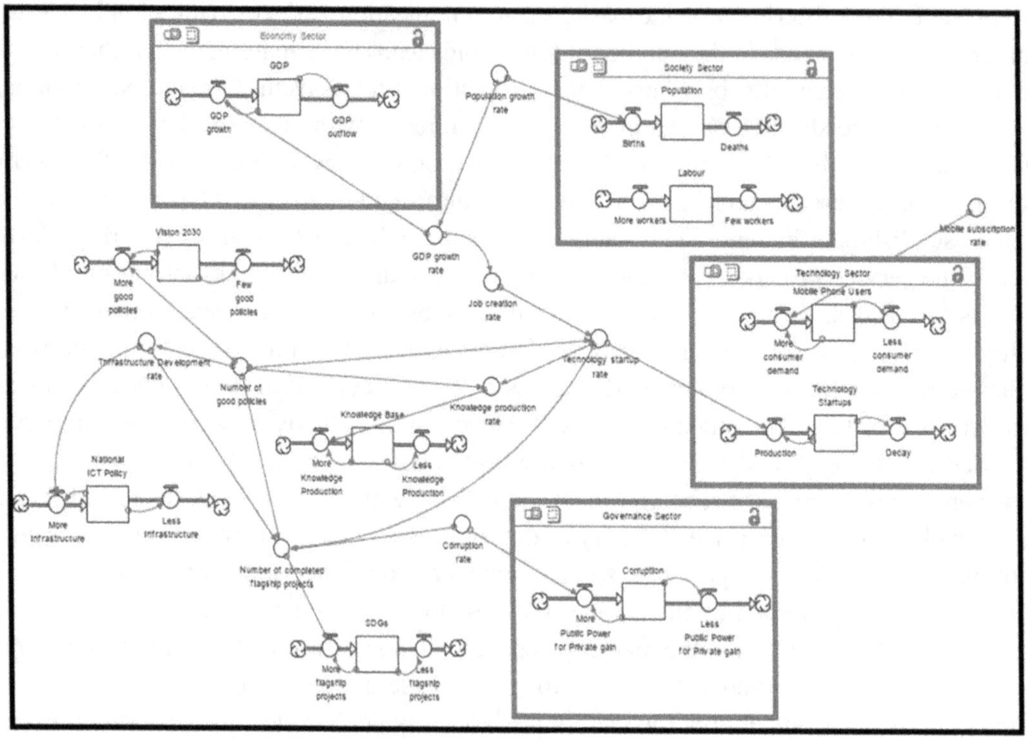

users to: (1) get a better qualitative and quantitative understanding of how the SDGs interact; (2) detect emerging patterns resulting from those interactions; (3) use context-specific information about direct impacts to identify indirect effects; and (4) identify leverage points hidden within the SDGs.

We converted all the above into a simple system dynamics model by integrating all the sub sectors. There are 4 sub-sectors namely Economic, Social, Governance and the Technology Sub-sector. As an attempt to explore the process of technological change affecting the economic growth that the developing country may have some directions more clearly how to design its technology policies; there are three sources of growth (growth scenarios) are considered in this study i.e.: population, governance, and innovation. The model is initialized in the full equilibrium. In this equilibrium the population growth rate is equal to zero. The model is simulated for 25 years, and the changes of the source of growths are introduced into the model in year

Delving deeper in the dynamics of the governance-socio-techno-economic model reveals the causal loop diagram, shown in Figure 5. The causal loop diagram shows the main integrated factors relevant to the research study. The SFD is shown in figure 6

2.4 Archetype Analysis

Archetype is the efficient language for explaining system thinking. Archetype analysis will explain complex phenomena, and resolve the root problem of dividing the work in excess and separately thinking by finding hidden lever result (Jia and Ding, 2002). The qualitative conclusion of archetype analysis will guide the quantitative analysis.

Figure 6. Overall stock and flow diagram for Governance, Society, Technology and the Economy with equations

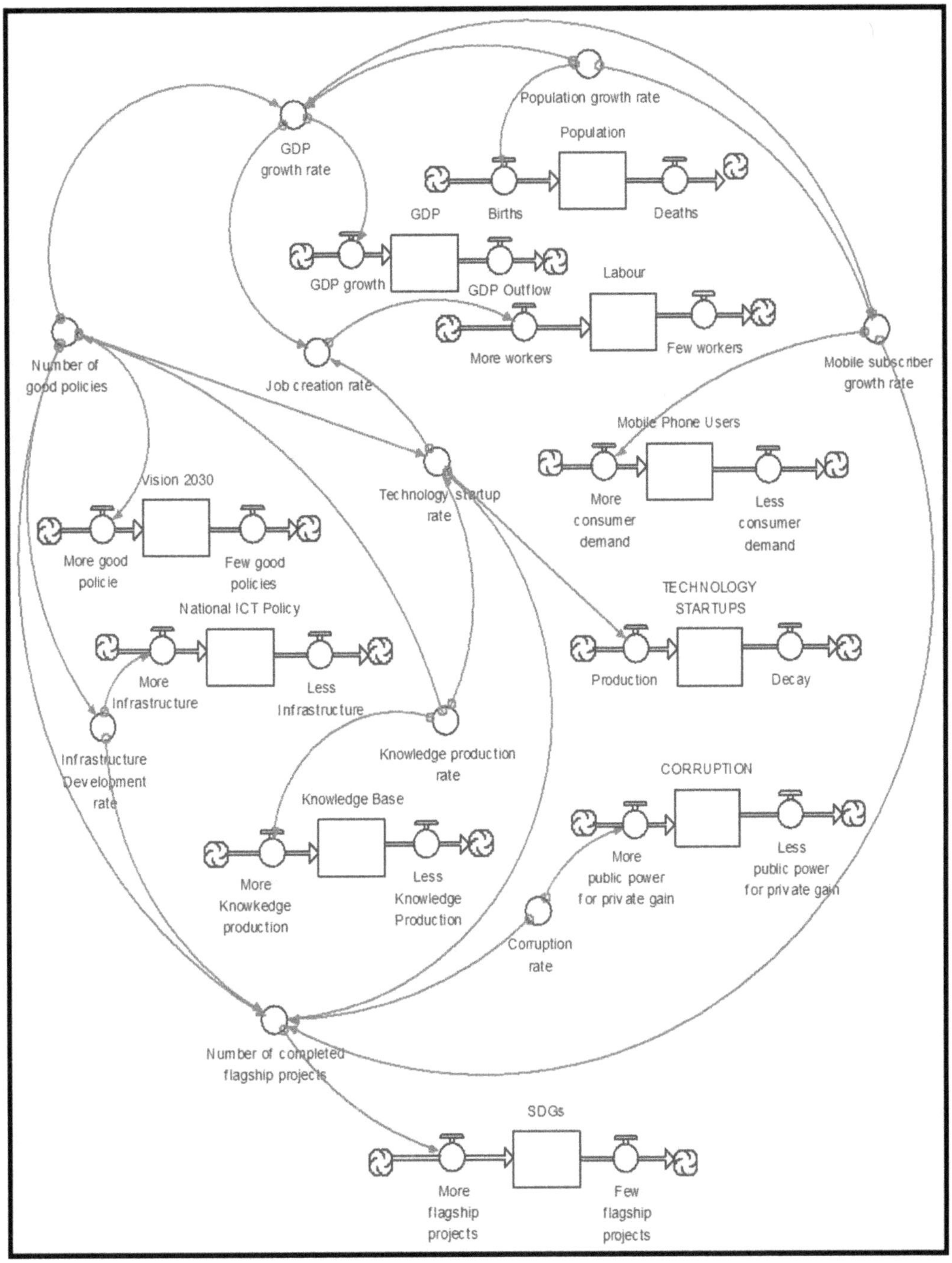

The archetypes eroding goals, limits to growth goals and growth and under investment are very core to the problem of mobile output variations and growth. Table 2.2 gives brief descriptions of the archetypes and examples of situations in which such archetype may apply.

The archetype examples on Table 2.3 include occurrences that are not alien to mobile firms or industries, implying that some insight into the behavior of the mobile industry as a whole (or its parts) can be gained from examining the relevant system archetypes. .

3.0 SIMULATION

Based on the detail analysis and the acquired qualitative conclusions to the system, we input the data into models, and use STELA software for simulation and analysis.

3.1 Source of Data

The parameters of the base level-rate model were set using data collected from publicly available sources. The data compiled show observed and predicted values. These values have been obtained and cross verified through multiple sources, whereas some of them are based on estimates or projections of what they may be in near future, given extraneous information. The consultation of several data sources has proven most useful in providing the information and expertise necessary to build the system model, and in particular to identify underlying feedback feed forward loops. The data was collected by documentary search and interviewing, between 15th January 2014 and 15thDecember 2015.

3.2 Simulation Results and Findings

The results of a simulation run of a typical scenario with a time span of 25 years are shown in Figure 4 and compared in Figure 5.

Figure 7. Mobile Adopters Sector

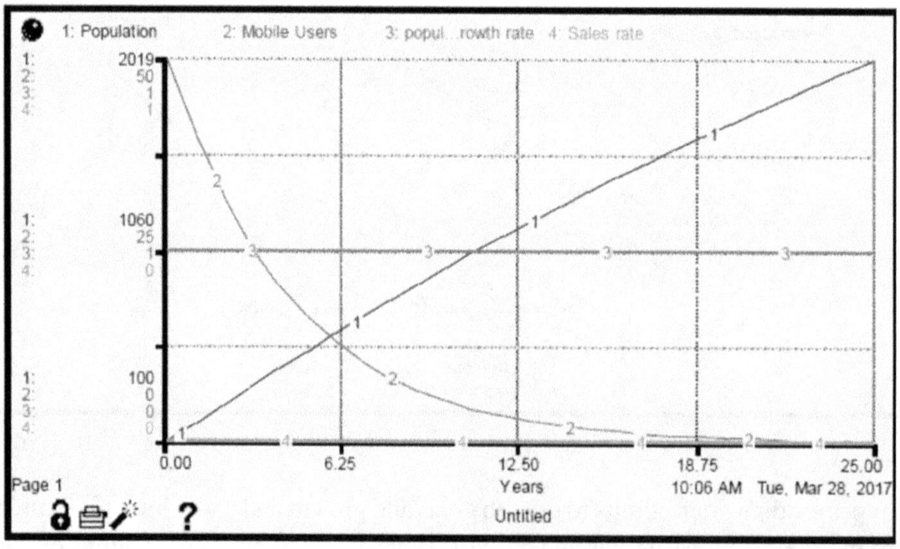

In this module, one can also notice positive feedback loops between Potential number of mobile users, mobile users, and sales rate. However, as one moves along the mobile users path, the respective stock values of potential mobile users diminish as shown in figure 7 indicating a negative feedback loop at work.

Moving forward, we find the proportionate increment in the Mobile market share holding respectively a differential but steeping growth curve of mobile market share initially before flattening out due to stagnation in mobile user's volume. It also creates a negative feedback impact over the quality of services which has an exponential growth slope as it represents accumulated number of subscribers over a period of time. This has also occurred due to a positive causal effect of market feedback to the total number of mobile users at a given point of time.

This market dynamics has also been collaborated with cumulatively increasing market share of leading mobile operator in Kenya having a positive effect on cost of service and equipment. This is shown in figure 8

Figure 8. Society dynamics

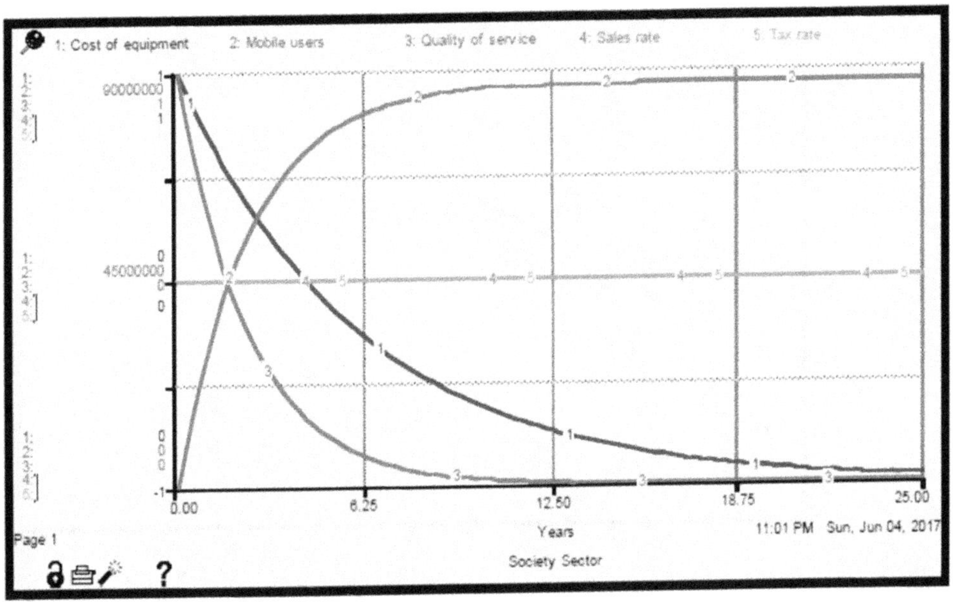

The initial values of stocks in system dynamics models can be determined in a variety of ways. The values are often known or knowable and the analyst needs only to investigate available data sources to find out. In the case of corruption model presented here, we obtained the values for the stock variables from ICRG and WDI for the year 2010.

Table 3.1 presents the interrelations among the stock variables measuring social, economic, political, judicial and cultural factors collectively. It is also evident that high level of corruption deteriorates law and order situation.

As described earlier, if the economy suffers from corruption activities, foreign investment process will imply a bribes payment. Corruption reduces the level of foreign direct investment and, consequently, it will affect the economic growth. An increase in level of corruption increases the complexity of the feedback structure of the economic system since they add new causal relationships connected to those affected by unequal distribution of resources and going into poverty trap. The whole feedback structure can fully explain why the corruption activities influence the socioeconomic conditions in the country as well as the wealth of citizens (Soto-Torres, et al., 2007).

The focus of these simulations is to analyze the impact of different anti-corruption measures on different indicators of interest.

The model assumes that government income distribution improves; the level of corruption in the economy will go down. It seems realistic to assume that if corruption increases then income distribution in the economy deteriorates. As discussed earlier, income inequality has detrimental effect on economic growth. The sub model also assumes that economic openness improves economic growth, as some researchers stated that opening up of economies to international trade is generally viewed as an effective strategy for accelerating economic growth. The following graphs in figure 9 and 10 show the reference modes for levels of corruption and GDP.

Figure 9. Reference Mode for Level of Corruption

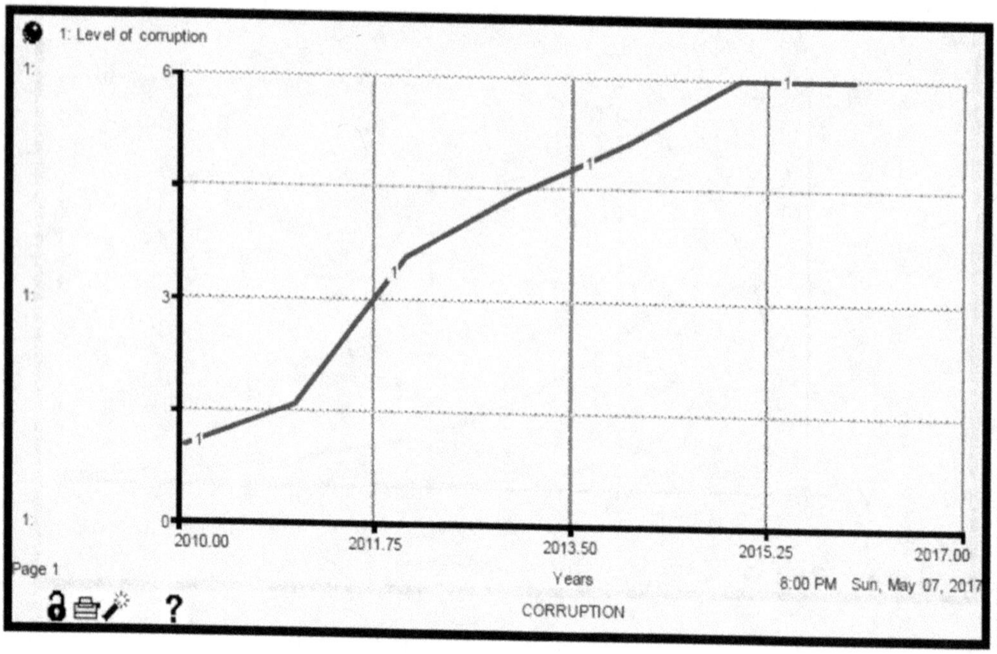

4.0 POLICY ANALYSIS AND CONCLUSION

The dynamic behaviour of corruption model can be studied through simulations by varying the parameters of the model. First the model needs to be placed in equilibrium. At the equilibrium, the model does not generate any dynamic behaviour, i.e. nothing changes over time, and it lies at the equilibrium unless otherwise disturbed. The focus of these simulations is to analyze the impact of different policies on different indicators of interest.

As initialized the model, we simulate the model with reference mode which reflects the status quo. Stocks will approach their highest or lowest value if one or more of the stocks parameter is pushed up or down, although the rapidity of that change is dependent on extent of the push. In other words, when

Figure 10. Reference Mode for Level of GDP

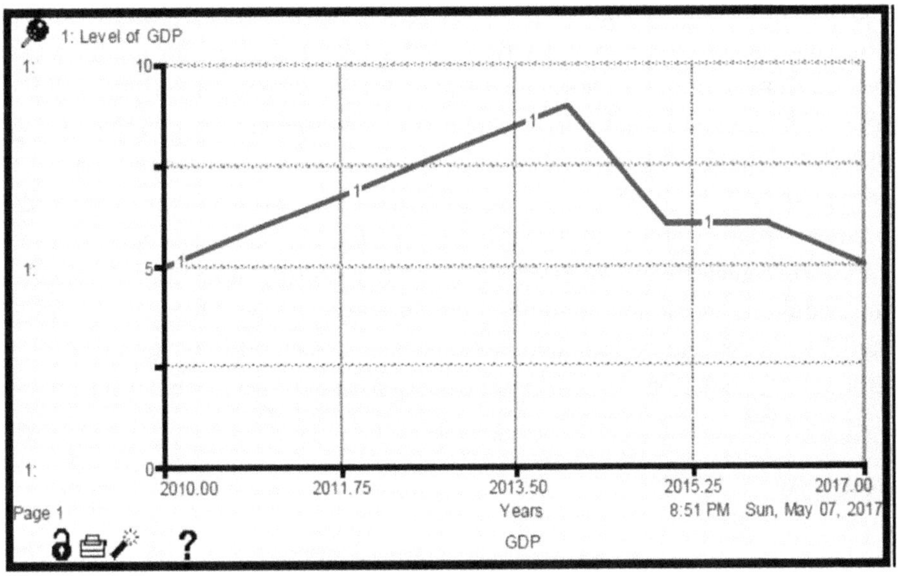

pushed off stocks parameters the model approaches either a very corrupt system or a very clean system. This is the most important stage of the modeling process. At this point, scenario planning is made and the results are drawn from the model. This is an extraordinary guidance to anti-corruption policy for the government, since outcome can base their decision-making process towards focusing on important factors to target corrupt activities. This section highlights the sub model in the following combination of scenarios: The different scenarios focus on changes in variables affecting level of corruption, level of GDP and income distribution. The system dynamics model enables the projection of several different scenarios. As mentioned earlier, the time horizon is selected equal to 25 (2005-2030) years which is a period sufficiently long to show how corruption is affected by change using some important variables like democratic accountability, distribution of income in the country, inflation rate, and organized crime as well as for verifying the necessity to fight against corruption.

Overall, the results show that under the defined circumstances, corruption climbs eventually, with the poor judicial system. Social economic conditions deteriorate as a result of corruption, which worsens income inequality. As income inequality increases, people's attitude against corruption goes up. Along with NGO's involvement, the people are applying pressure to lower the level of corruption through various means (not specified in this model).As a result, the level of corruption is controlled, which in turn improved the effectiveness of the judicial system, which reinforces the reduction in the level of corruption, thus, the betterment of social economic conditions. With such improvements, income inequality is reduced, and we can see that the people's attitude against corruption is also reduced. Given some changes in the 'effect' converters, and a longer simulation period, we may see an increase incorruption again sometime in the future? Since attitudes towards corruption play a critical role in the persistence of corruption. If people's attitude against corruption went high then the level of corruption would be decreased over time. According to (Cameron, et al., 2005), peoples' attitudes towards corruption are shaped by the social, political, legal and economic systems of the countries they live in as well as their everyday experiences of corruption.

A comparison of the scenarios with reference mode suggests that Kenya can achieve different levels of corruption, GDP and income inequality if the different scenarios can be used for policy change.

The study further aimed to develop a generic system dynamics model for mobile phone usage and economic growth in Kenya, for the purpose of understanding influences of various policy interventions on the behaviour of the mobile industry in the country. This research has attempted to propose a new modeling approach for the investigation of mobile phone usage or telecommunications services and economic growth in Kenya. The developed model which incorporates the integrative modular framework of generating causal dynamics prevalent in the telecom sector also successfully acquires the capability of short-mid-long term forecasting. It discards the notion that holistic modeling brings into 'coarseness' in the modeling and therefore cannot be applied for future projections. It has been proved that if the causality of model variables can be founded upon a robust model with the support of conceptual and quantitative information, one can succeed to capture the dynamics of any system in a comprehensive way.

5.0 CONCLUSION

In conclusion, the development of a system dynamics model of mobile services provision output implies that experimental economics and management can be used to learn the behaviour and evolution of a mobile industry. The system dynamics model is a laboratory where various policy suggestions can be experimented with - to gain sufficient insight into the expected response of the mobile industry - before the policies are actually applied to the industry. Such a laboratory should be a great decision tool for policy makers in the mobile industry.

Observations made in this study have three major implications for theory in mobile services provision and economics, and for theory in governance. The observation that there is need for regulation of response of the mobile industry to changes in mobile services provision demand implies that the industry as a whole needs to be managed the way a business corporation is managed. Generally, a mobile industry is not viewed as a 'managed system' the way a corporation is viewed; the industry is therefore managed collectively and simultaneously by all the players in it. This leads to the prisoner's dilemma, whereby bright individual strategies result in collective folly. Efficient mobile services provision management at project (and firm level) should be supported by efficient mobile services provision management at industry level, in order to increase the organizational efficiency of the mobile industry as a whole.

The adoption of systems thinking in mobile services provision economics and management mainly involves change in the way the information available is used in the mobile industry. While need for more comprehensive records of mobile activity is not in dispute, systems thinking is not limited to recorded data. A system's decision criteria depend on data that exists in the system whether the data is recorded or not. The challenge of the mobile industry 'manager' is to imaginatively changes the information links in the system in order to purposefully drive the industry towards clearly defined and mutually beneficial long-term targets.

Poor regulation of mobile activity at industry level and limited market of the mobile industry are the underlying causes of excessive variations and stunted growth of mobile services provision output in Kenya. For that reason, recommendations for addressing this research problem are system ideas that have been tested and found working for similar problems in other complex systems. All the same, the system ideas here are tailored to the specific characteristics of the mobile industry of Kenya, and emphasize what the mobile industry itself can do, without needing the government to shield it from effects of unfavourable political, social or economic conditions in the country.

ICT has a critical role in driving the economic, social and political development of Kenya as espoused in Vision 2030. How the country has steered ICT development over the past four or so years to 2017 and beyond will greatly impact on the achievement of the country's Vision 2030. There is a need to align the National ICT Master Plan to the devolution reality and address key challenges that may hinder the ICT sector from playing its rightful role in national development. The ICT sector increasingly continues to provide essential services for Kenya's socio-economic development. However, sound and responsive ICT sector policy guidelines and regulatory frameworks are required to enhance the socio-economic impact of the sector and to enable it to contribute effectively to the development process.

The study suggests that systems thinking can be effectively used to understand the dynamics of technology and economic growth and especially the mobile industry of Kenya. This research developed a system dynamics model of mobile services provision output in Kenya. It showed and suggested that the system dynamics model developed gives a remarkable framework that could be used to enable ICT to contribute towards achieving national development goals as espoused in Vision 2030. It also finally sets a foundation for further research about system dynamics modelling in the Technology industry worldwide.

REFERENCES

Amos, Rogriguez, & Wafula (2018). A Systems Dynamics Model for Mobile Industry Governance in the Context of the Kenyan Vision 2030. *International Journal of System Dynamics Applications, 70*(1), 65–94.

Ansah, P. J. (2017). Causal Analyses of Public Debt and Structural Adjustment. *International Journal of System Dynamics Applications*, *6*(1), 21–37. doi:10.4018/IJSDA.2017010102

Asongu, Nwachukwu, & Aziz. (2018) Conditional Determinants of Mobile Phones Penetration and Mobile Banking in Sub-Saharan Africa. *Journal of the Knowledge Economy, 9*(1), 81-110.

Berne, M., Vialle, P., & Whalley, J. (2019). An analysis of the disruptive impact of the entry of Free Mobile into the French mobile telecommunications market. *Telecommunications Policy*, *43*(3), 262–277. Advance online publication. doi:10.1016/j.telpol.2018.07.007

Bhushnan. S., (2012). System Dynamics Integrative Modeling and Simulation for Mobile Telephony Innovation Diffusion: A Study of Emerging Indian Telecom Market. *International Journal of System Dynamics Applications, 12*(1), 84-121.

BrockJ. M. (2018) Inequality of Opportunity and Beliefs About Success and Failure, SSRN. doi:10.2139srn.3178010

Casey, R., & Toyli, J. (2012). Mobile voice diffusion and service competition: A system dynamic analysis of regulatory policy. *Telecommunications Policy*, *36*(3), 162–174. doi:10.1016/j.telpol.2011.07.002

Chakravarty, S. (2007). Determinants of mobile penetration in Asia-a panel study. *Indian Journal of Economics and Business.*, *6*(2), 147–165.

Communication Authority of Kenya. (2019). *Second Quarter Sector Statistics Report for the Financial Year 2019/2020.* Author.

Forenbacher, I., Husnjak, S., Cvitić, I., & Jovović, I. (2019). Determinants of mobile phone ownership in Nigeria. *Telecommunications Policy*, *43*(7), 101812. Advance online publication. doi:10.1016/j.telpol.2019.03.001

Jensen, R. T. (2007). The Digital Provide: Information (Technology), Market Performance and Welfare in the South Indian Fisheries Sector. *The Quarterly Journal of Economics*, *122*(3), 879–924. doi:10.1162/qjec.122.3.879

Jia, R., & Ding, R. (2002). *System Dynamics Analysis of Feedback Dynamic Complexity*. Higher Education Press.

Moysis, L., & Azar, A. T. (2017). New Discrete Time 2D Chaotic Maps. *International Journal of System Dynamics Applications*, *6*(1), 77–104. doi:10.4018/IJSDA.2017010105

Nasution, F. B., Bazin, N. E., Rosalyn, R., & Hasanuddin, H. (2018). Public Policymaking Framework Based on System Dynamics and Big Data. *International Journal of System Dynamics Applications*, *7*(4), 38–53. doi:10.4018/IJSDA.2018100103

Roller, L. H., & Waverman, L. (2001). Telecommunications infrastructure and economic growth:A simultaneous approach. *The American Economic Review*, *91*(4), 909–923. doi:10.1257/aer.91.4.909

Roller, L.-H., & Waverman, L. (2001). Telecommunications Infrastructure and Economic Development: A Simultaneous Equation Approach. *The American Economic Review*, *91*(4), 909–923. doi:10.1257/aer.91.4.909

Senge, P. M., Kleiner, A., Robert, C., Ross, R. B., & Smith, B. J. (1994). *The Fifth Discipline Fieldbook: Strategies and Tools for Building a Learning Organization*. Marsh.

Vokshi, M., Youssef, D., & Toçi, V. (2019). *Mobile phone penetration and its impact on inequality in the Western Balkan countries*. *Zagreb International Review of Economics & Business*, *22*(2), 111–131. doi:10.2478/zireb-2019-0023

Yamakawa, P., Gareth, H. R., Jose, M. S., & Nikolai, A. (2013). The diffusion of mobile telephones: An empirical analysis for Peru. *Telecommunications Policy*, *37*(6-7), 594–606. doi:10.1016/j.telpol.2012.12.010

Zelinka, D., & Amadei, B. (2019). Systems Approach for Modeling Interactions Among the Sustainable Development Goals Part 1: Cross-Impact Network Analysis. *International Journal of System Dynamics Applications*, *8*(1), 23–40. doi:10.4018/IJSDA.2019010102

Chapter 11
Medical Pre-Screening of Common Diseases:
An Interval-Valued Fuzzy Set Approach

Palash Dutta
https://orcid.org/0000-0002-1565-4889
Dibrugarh University, India

ABSTRACT

In this chapter, relation between IFS and IVFS are studied and reviewed along with some existing distance measures on IFSs. Then, some distance measures for IVFSs are derived from IFSs and some properties on distance measures for IVFSs are ascertained. In the end, an attempt has been made to perform pre-screening test for common diseases of patients in two scenarios under this setting. In the first scenario, for medical prescreening IFSs data are taken and converted into IVFSs while IVFSs data are taken into consideration for the same in the second scenario.

1. INTRODUCTION

Real world problems such as decision making, operation research, engineering problems, medical decision making or medical pre-screening etc are always tainted with uncertainty. In medical decision making, it is experienced that to prepare a set of symptoms, when medical experts question a patient regarding patient's condition, patient is not to be so confident to describe the conditions and it is also seen that they use the linguistic expression to explain their conditions which are more often vague. Medical expert needs to evaluate a list of possible symptoms for the respective diseases of the patients based on their vague linguistic expressions. However, the relations between symptoms and their corresponding diseases are not often one-to-one. The exhibition of the same disease may not be identical with different patients and even at different disease stages. Moreover, a particular symptom may signify various diseases and again in some situations in a particular patient may disarray the presumed structure of symptoms. However, knowledge base associating the symptom-disease relationship comprises of imprecision, vagueness and uncertainty in medical diagnosis process. To deal with uncertainty Zadeh (1965) developed FST. After

DOI: 10.4018/978-1-7998-5788-4.ch011

that various direct/indirect extensions of fuzzy set have been made and successfully applied in most of the problems of real world situation including medical diagnosis. For example, Alcantud et al., (2019) performed an analysis of survival for lung cancer resections cases with fuzzy and soft set theory in surgical decision making; a process for medical diagnosis through fuzzy decision making was studied in (Elizabeth and Sujatha, 2013). Medical diagnosis decision making based on fuzzy numbers along with compositional rule of inference were discussed in (Yao and Yao, 2001). Divergence measure for intuitionistic fuzzy sets and its application in medical diagnosis was studied in (Maheshwari, and Srivastava, 2016). A study on Bell-shaped fuzzy soft sets and their applications in medical diagnosis were presented in (Dutta and Limboo, 2017). fuzzy decision making for medical diagnosis using arithmetic of generalised parabolic fuzzy numbers was presented by (Dutta and Doley, 2019); Decision Making for Medical Diagnosis Through Credibility Theory was discussed (Dutta and Ali, 2019); medical diagnosis under uncertain environment through bipolar-valued fuzzy sets studied by (Dutta and Doley, 2020);

But in FST to each element of the universe of discourse a degree of membership between 0 and 1 is assigned. However, in some situations it is not always possible for a membership function of the type μ: X→[0,1] to precisely assign one point from [0,1], so it is more realistic to assign interval value. According to Gehrke et al. (1996) many people believe that assigning an exact number to expert's opinion is too restrictive and the assignment of an interval valued is more realistic. In such situations interval valued fuzzy set (IVFS) comes into picture. Sambuc (1975) first presented in his doctoral research (thesis) the concept of IVFS names as ϕ-fuzzy set. After development of IVFSs, different researchers have been studied this issue and applied in different areas. Some authors investigated and suggested some methods for measuring distance between IVFSs. Burillo and Bustine (1996) studied entropy and distances for IVFSs, Grzegorzewski (2004) proposed new distance measure based on Housdroff space which are straight forward generalization of Hamming distance, the Euclidean distance and their normal counter parts. Zeng and Guo (2008) studied the distances of IVFSs and relationship between entropy and similarity measure of IVFS. Park et al., (2008) proposed new distance measures for IVFS by incorporating amplitude of membership concept. Li (2009) discussed on distances of IVFS and some of its properties on metric space.

Some application using IVFSs in medical diagnosis can also be seen in literature. Chetia and Das (2010) extended Sanchez's approach for medical diagnosis using interval valued fuzzy soft sets and exhibited the technique with hypothetical case study. Ahn et al., (2011) presented a fuzzy diagnosis method based on the interval valued interview chart and the interval valued Intuitionistic fuzzy weighted arithmetic average operator and studied the occurrence information symptoms as the weights. Elizabeth and Sujatha (2014) discussed medical diagnosis based on IVFN matrices. Meenakshi and Kaliraja (2011) extended Sanchez's approach for medical diagnosis using interval valued fuzzy matrix, Mariapresenti and Arockiarani (2018) proposed a new interval valued vague weighted arithmetic average operators (IVWAA) are defined to aggregate the vague information from the symptoms. Loganathan & Pushpalatha (2018) introduced geometric mean of an interval valued fuzzy matrix and applied in agriculture field for finding the disease in plants.

On the other hand, an important generalization of FST is the theory of IFS, introduced by Atanassov (1986) ascribing a membership degree and a non-membership degree separately in such a way that sum of the two degrees must not exceed one. Application of IFSs in medical diagnosis is now a day's more interesting area of research among the researchers. De et al., (2001) studied Sanchez's approach for medical diagnosis and extended the concept with the notion of IFS. Szmidt and Kacprzyk (2001) studied medical diagnosis using their proposed distance measures. Own (2009) studied advantages of type-2 fuzzy and switching relation between type-2 fuzzy sets and IFSs defined axiomatically and finally, switching results are applied in medical diagnosis. Samuel and Choi et al., (2012) proposed a fuzzy diagnosis method based on interval valued intuitionistic fuzzy sets, Samuel and Balamurngan (20012a, 2012b, 2013) studied medical diagnosis for IFSs and introduced a new concept of IFS with n parameters and applied in medical diagnosis. Davvaz and Sadrabadi (2016) verified some distance measures on IFSs by applying in medical, Deepa, et al., (2018) presented medical diagnosis using IFS using shortest distance measure, Liu et al., (2018a) proposed cosine similarity measure between hybrid IFSs and applied in medical diagnosis, Liu et al.,(2018b) studied some similarity measures of Neutrosophic sets based on the Euclidean distance and applied in medical diagnosis, Luo & Liang (2018), proposed a novel similarity measure for interval-valued intuitionistic fuzzy sets and its applications, Talukdar & Dutta (2018) studied disease diagnosis using an advanced distance measure for intuitionistic fuzzy sets, Srivastava (2019) carried out medical diagnosis using IFSs, Huang & Guo (2019) presented an enhanced correlation coefficient technique for IFSs and applied in medical diagnosis, Zhou et al., (2020) presented a new divergence measure of Pythagorean fuzzy sets based on belief function and its application in medical diagnosis.

Some application of fuzzy set in medical diagnosis can be found in Azar AT, El-Said SA, Balas VE, Olariu T (2013), Banu PKN, Azar AT, Inbarani HH (2017) while application of neuro-fuzzy technique in medical diagnosis can be seen in Azar (2011), Azar, Hassanien, Kim (2012a), Azar, Hassanien &Kim (2012b), Azar (2013a), Azar (2013b), Azar (2014), Azar & Hassanien (2014). Some other applications of fuzzy sets are seen in Abdel, Elsamahy, Hassan & Bendary (2017), Ismail (2012), Lin & Kuo (2012), Mahmoud, El-Araby, Shehata, Abou Zaid, & El-Samie, (2015), Shereen, Ramadan & Ghali (2016)

Generally a disease is characterized by some directly noticeable symptoms which induce the patient to go to a medical expert. A set of clinical examinations are undertaken to identify the occurrence of a disease. In the realm of medical diagnosis, lots of variables are there which influence the decision making process and consequently, differentiate the opinions of the medical experts. Due to so many factors to analyze for the diagnosis of the disease of a patient makes the medical expert's job difficult. In literature, medical diagnosis problems are solved using purely fuzzy set, IFSs and IVFSs based methods under uncertainty. However, use of distance measures on IVFSs in medical diagnosis have not been seen in literature though more imprecision frequently involve in medical documents. Also no connected study between IFS and IVFS are seen in medical diagnosis to cross confirm the diagnosis results that obtained from IFSs data. These motivate us more to study distance measures on IVFSs in medical diagnosis.

It is obvious that in the diagnosis of a patient always has to dependent on the buoyancy level of a expert which is very limited degree of accuracy. In such situations, IVFS is a more suitable apparatus to embody linguistic opinion of the medical expert as it incorporates the complete degree of confidence of medical expert's opinion in certain closed region (i.e., interval). From the point of view, in this study, a medical data base is assembled in the form of IVFS. In the end, an attempt has been made to perform pre-screening test for common diseases of patients in two scenarios under this setting. In the first scenario, medical pre-screening for IFSs data are taken and converted into IVFSs and accordingly perform medical pre-screening while interval valued IFSs data are taken into consideration for the same in the second scenario.

The major contribution of this paper is to develop a medical diagnostic system based on distance measures on interval valued fuzzy sets (IVFSs) which were basically derived from IFSs. The reason for choosing IVFS is the ability to address imprecision, uncertainty and vagueness properly.

The contributions of our works in this article can be observed in three phases: Establishment of resemblance between IFS and IVFS. Derivation of distance measures on IVFSs and investigations of its some properties, Perform medical diagnosis system via distance measures on IVFSs.

The advantages of our study can be pointed out as follows:

It presents an effort to carry out medical diagnosis by considering expert's medical knowledge in terms of IVFSs. It has more utility because of its ability to include lower and upper membership values of symptoms and diseases as well. It provides assistance to expert to carry out medical diagnosis providing better efficiency to the output.

This present article is organized as follows: Section 1 is the introductory one discussing the need of different types of fuzzy sets along with literature review of different applications of FSs, IVFSs and IFSs. Section 2 discusses the introductory and the related ideas of FST, IVFS and IFS. Section 3 discusses the relation between IFS and IFVS. Section 4 refers to existing distance measures on IFSs. Section 5 describes the derivation of distance measures on IFVSs and some related properties. Section 6 explains a case study to carry out medical diagnosis. Section 7 discusses results and discussion. Finally, in Section 9 conclusions are made.

2. PRELIMINARIES OF TYPE-I/IVFS AND IFS

In this section, some basic definitions and concepts on fuzzy set, IVFSs and IFSs are reviewed.

2.1 Fuzzy Set

(Zadeh, 1965): Fuzzy set is a set in which every element has degree of membership of belonging in it. Mathematically, let X be a universal set. Then the fuzzy subset A of X is defined by its membership function

$\mu A_{:} X \rightarrow [0,1]$

Which assign a real number $\mu A_{(}x)$ in the interval [0, 1], to each element x∈A, where the value of $\mu A(_{x})$ at x *s*hows the grade of membership of x *in* A.

2.2 Interval Valued Fuzzy Set (IVFS)

An interval valued fuzzy set A' defined in the universe of discourse X is represented by

$$A' = \left\{ (x, [\mu_{A'}^{L}(x), \mu_{A'}^{U}(x)] : x \in X) \right\}$$

where $0 \le \mu_{A'}^{L}(x) \le \mu_{A'}^{U}(x) \le 1$ and the membership grade $\overline{\mu}_{A'}(x)$ of elements of x to the interval valued fuzzy set A is represented by an interval

$$[\mu_{A'}^{L}(x), \mu_{A'}^{U}(x)] \left(i.e., \overline{\mu}_{A'}(x) = [\mu_{A'}^{L}(x), \mu_{A'}^{U}(x)] \right).$$

where $\mu_{A'}^{L}(x)$ is called lower membership function and $\mu_{A'}^{U}(x)$ is called upper membership function.

If an interval valued fuzzy set A satisfies the following properties

- A is normal
- A is defined in a closed bounded interval
- A is convex set

Then A is called an interval valued fuzzy number.

The amount $\Delta_{A'}(x) = \mu_{A'}^{U}(x) - \mu_{A'}^{L}$ is called range of uncertainty.

2.3 α-Cut of IVFN

(Hamrawi et al., 2010, Hamrawi, 2011):

Let A' be a continuous and convex IVFN with LMF $\mu_{A'}^{L}$ (or $\underline{A}'$) and UMF $\mu_{A'}^{U}$ (or $\overline{A}$). Let α-cut of LMF ($\underline{A}'$) be ${}^{\alpha}\underline{A}' = [L_{\underline{x_\alpha}}, R_{\underline{x_\alpha}}]$ and of UMF ($\overline{A}'$) be ${}^{\alpha}\overline{A}' = [L_{\overline{x_\alpha}}, R_{\overline{x_\alpha}}]$. Then α - cut of IVFN A' can be evaluated by the following formula:

$$
{}^{\alpha}A' = \begin{cases} ([L_{\underline{x_\alpha}}, R_{\underline{x_\alpha}}],[L_{\overline{x_\alpha}}, R_{\overline{x_\alpha}}]),\ \alpha \le h(\overline{A}')\ \&\ \alpha \le h(\underline{A}'). \\ (\phi,[L_{\overline{x_\alpha}}, R_{\overline{x_\alpha}}]), \alpha \le h(\overline{A}')\ \&\ \alpha > h(\underline{A}'). \\ (\phi,\phi), \alpha > h(\overline{A}') \end{cases}
$$

where $\forall \alpha : L_{\overline{x_\alpha}} \le L_{\underline{x_\alpha}} \le R_{\underline{x_\alpha}} \le R_{\overline{x_\alpha}}, h(\underline{A}')$ is the height of LMF, $h(\overline{A}')$ is the height of UMF and ϕ is an empty set.

2.4 Complement of IVFS

Complement of the fuzzy number A' is defined as

$$
\begin{aligned}
\mu_{A'^c}(x) &= 1 - \overline{\mu_{A'}}(x) \\
&= 1 - [\mu_{A'}{}^{L}(x), \mu_{A'}{}^{U}(x)] \\
&= [1 - \mu_{A'}{}^{U}(x), 1 - \mu_{A'}{}^{L}(x)]
\end{aligned}
$$

Thus, $\mu_{A'^C}{}^{L}(x) = 1 - \mu_{A'}{}^{U}(x)$ and $\mu_{A'^C}{}^{U}(x) = 1 - \mu_{A'}{}^{L}(x)$.

2.5 Intuitionistic Fuzzy Set

A Intuitionistic fuzzy set (Atanassov, 1986) A on a universe of discourse X is of the form

$$A = \{(x, \mu_A(x), v_A(x) : x \in X\},$$

where $\mu A(x) \in [0,1]$ is called the "degree of membership of x in A", $vA(x) \in [0,1]$ is called the "degree of non-membership of x in A", and where $\mu A(x)$ and $vA(x)$ satisfy the following condition:

$$0 \le \mu_A(x) + v_A(x) \le 1.$$

The amount $\pi_A(x) = 1 - (\mu_A(x) + v_A(x))$ is called hesitancy of x which is reflection of lack of commitment or uncertainty associated with the membership or non-membership or both in *A*.

3. ASSOCIATION BETWEEN IFSs AND IVFSs

Extension of Fuzzy set theory introduced by Atanassov (1986) under the name of "intuitionistic fuzzy set (IFS)" has become more popular topic in decision making. Atanassov (1989) established that IVFSs are a version of the IFSs. Basically, IVFSs and IFSs constitute an isomorphism (Dubois et al., (2005)). Some interesting debate between Atanassov group and Dubois group has been seen. Dubois et al., (2005) criticized on terminological difficulties in intuitionistic fuzzy set theory. Atanassov (2005) replied to Dubois et al., (2005). Grzegorzewski and Mrowka (2005) put forwarded some comments on Atanassov (1986) presenting points related to terminology, connections with other mathematical structures and possible interpretations. Tizhoosh (2008) defended Atanassov (1986) and his work (Tizhoosh (2005)) that were criticized by Vlachos and Sergiadis (2007).

Here, we discuss relation between IFSs & IVFSs as well as hesitancy & range of uncertainty.

3.1 Relation between IFS and IVFS

Definition: (Atanassov, 1989)

(a) The map *f* assigns to every IVFS A' an IFS $B = f(A')$ given by

$$\mu_B(x) = \inf M_{A'}(x),\ v_B(x) = 1 - Sup M_{A'}(x).$$

(b) The map *g* assigns to every IFS *B* an IVFS $A' = g(B)$ given by

$$M_{A'}(x) = [\mu_B(x), 1 - v_B(x)]$$

The relation between IVFS and IFS can be shown in the following lemma (Atanassov, 1989).

Lemma: (a) Let A be an IVFS. Then for every $x \in X$,

$$\begin{aligned} M_{g(f(A'))}(x) &= [\mu_{f(A')}(x), 1-\nu_{f(A')}(x)] \\ &= [\inf M_{A'}(x), 1-1+\sup M_{A'}(x)] \\ &= M_{A'}(x) \end{aligned}$$

(b) Let B be an IFS. Then for every $x \in X$,

$$\begin{aligned} \mu_{f(g(B))}(x) &= \inf M_{g(B)}(x) \\ &= \inf[\mu_B(x),\ 1-\nu_B(x)] \\ &= \mu_B(x) \end{aligned}$$

In particular, if A' is a IVFS in universe of discourse X such that $M_{A'}(x) = [\mu_{A'}^L(x), \mu_{A'}^U(x)]$ then it can be converted into IFS B in the same universe X such that

$$\mu_B(x) = \inf M_{A'}(x) = \mu_{A'}^L(x)$$

and

$$\nu_B(x) = 1 - Sup M_{A'}(x) = 1 - \mu_{A'}^U(x)$$

and vice versa.

3.2 Relation Between Hesitancy and Range of Uncertainty

Let B be an IFS in the universe of discourse X then hesitancy of x indicating lack of commitment or uncertainty associated with the membership or non-membership or both in B is

$$\begin{aligned} \pi_B(x) &= 1-(\mu_B(x)+\nu_B(x)) \\ &= 1-(\mu_{A'}^L(x)+1-\mu_{A'}^U(x)) \\ &= \mu_{A'}^U(x)-\mu_{A'}^L(x) \\ &= \Delta_{A'}(x) \end{aligned}$$

That is, hesitancy of lack of commitment of IFS is equivalent to range of uncertainty of IVFS A' and vice-versa.

4. DISTANCE MEASURES ON IFSs

In this segment, the distance measures proposed by Szmidt and Kacprzyk (1997, 2001) as well as Wang and Xing (2005) are reviewed.

For any two IFSs $A=\{(x,\mu_A(x),v_A(x):x\in X\}$, and $B=\{(x,\mu_B(x),v_B(x):x\in X\}$,

(i) The Hamming Distance *d(A,B)*:

$$d(A,B)=\frac{1}{2}\sum_{i=1}^{n}[|\mu_A(x_i)-\mu_B(x_i)|+|v_A(x_i)-v_B(x_i)|+|\pi_A(x_i)-\pi_B(x_i)|] \tag{1}$$

(ii) The normalized Hamming Distance *l(A,B)*:

$$l(A,B)=\frac{1}{2n}\sum_{i=1}^{n}[|\mu_A(x_i)-\mu_B(x_i)|+|v_A(x_i)-v_B(x_i)|+|\pi_A(x_i)-\pi_B(x_i)|] \tag{2}$$

(iii) The Euclidean distance *e(A,B)*:

$$e(A,B)=\sqrt{\frac{1}{2}\sum_{i=1}^{n}[(\mu_A(x_i)-\mu_B(x_i))^2+(v_A(x_i)-v_B(x_i))^2+(\pi_A(x_i)-\pi_B(x_i))^2]} \tag{3}$$

(iv) The normalized Euclidean distance *q(A,B)*:

$$q(A,B)=\sqrt{\frac{1}{2n}\sum_{i=1}^{n}[(\mu_A(x_i)-\mu_B(x_i))^2+(v_A(x_i)-v_B(x_i))^2+(\pi_A(x_i)-\pi_B(x_i))^2]} \tag{4}$$

(v) Wang and Xing (2005) distance measures:
(a)

$$d_{WX1}(A,B)=\frac{1}{n}\sum_{i=1}^{n}\left[\frac{|\mu_A(x_i)-\mu_B(x_i)|+|v_A(x_i)-v_B(x_i)|+|\pi_A(x_i)-\pi_B(x_i)|}{4}+\frac{\max(|\mu_A(x_i)-\mu_B(x_i)|,|v_A(x_i)-v_B(x_i)|,|\pi_A(x_i)-\pi_B(x_i)|)}{2}\right] \tag{5}$$

(b)

$$d_{WX2}(A,B)=\frac{1}{2\sqrt[p]{n}}\sqrt[p]{\sum_{i=1}^{n}(|\mu_A(x_i)-\mu_B(x_i)|+|v_A(x_i)-v_B(x_i)|)^p} \tag{6}$$

5. DISTANCE MEASURES ON IVFSs

In this section, distance measures for IVFSs have been derived from IFS.

Let $A=\{(x,\mu_A(x),v_A(x):x\in X\}$, and $B=\{(x,\mu_B(x),v_B(x):x\in X\}$ be two IFSs and $A'=\left\{(x,[\mu_{A'}^L(x),\mu_{A'}^U(x)]:x\in X)\right\}$ and $B'=\left\{(x,[\mu_{B'}^L(x),\mu_{B'}^U(x)]:x\in X)\right\}$ be their corresponding IVFSs respectively.

(i) The Hamming Distance $d(A',B')$:

$$\begin{aligned} d(A,B)&=\frac{1}{2}\sum_{i=1}^{n}[|\mu_A(x_i)-\mu_B(x_i)|+|v_A(x_i)-v_B(x_i)|+|\pi_A(x_i)-\pi_B(x_i)|]\\ &=\frac{1}{2}\sum_{i=1}^{n}[|\mu_{A'}^L(x_i)-\mu_{B'}^L(x_i)|+|1-\mu_{A'}^U(x_i)-(1-\mu_{B'}^U(x_i))|+|\Delta_A(x_i)-\Delta_B(x_i)|]\\ &=\frac{1}{2}\sum_{i=1}^{n}[|\mu_{A'}^U(x_i)-\mu_{B'}^U(x_i)|+|\mu_{A'}^L(x_i)-\mu_{B'}^L(x_i)|+|\Delta_{A'}(x_i)-\Delta_{B'}(x_i)|]\\ &=d(A',B') \end{aligned} \tag{7}$$

In a similar way, other distance measures can also be derived for IFVSs.

(ii) The normalized Hamming Distance $l(A',B')$:

$$l(A',B')=\frac{1}{2n}\sum_{i=1}^{n}[|\mu_{A'}^U(x_i)-\mu_{B'}^U(x_i)|+|\mu_{A'}^L(x_i)-\mu_{B'}^L(x_i)|+|\Delta_{A'}(x_i)-\Delta_{B'}(x_i)|] \tag{8}$$

(iii) The Euclidean distance $e(A',B')$:

$$e(A',B')=\sqrt{\frac{1}{2}\sum_{i=1}^{n}[(\mu_{A'}^U(x_i)-\mu_{B'}^U(x_i))^2+(\mu_{A'}^L(x_i)-\mu_{B'}^L(x_i))^2+(\Delta_{A'}(x_i)-\Delta_{B'}(x_i))^2]} \tag{9}$$

(iv) The normalized Euclidean distance $q(A',B')$:

$$q(A',B')=\sqrt{\frac{1}{2n}\sum_{i=1}^{n}[(\mu_{A'}^U(x_i)-\mu_{B'}^U(x_i))^2+(\mu_{A'}^L(x_i)-\mu_{B'}^L(x_i))^2+(\Delta_{A'}(x_i)-\Delta_{B'}(x_i))^2]} \tag{10}$$

(v) In a similar way, distance measures can also be derived for IVFSs from Wang and Xing (2005) distance measures between IFSs.

(a)

$$d_{WX1}(A,B)=\frac{1}{n}\sum_{i=1}^{n}\left[\frac{|\mu_A(x_i)-\mu_B(x_i)|+|v_A(x_i)-v_B(x_i)|+|\pi_A(x_i)-\pi_B(x_i)|}{4}+\frac{\max(|\mu_A(x_i)-\mu_B(x_i)|,|v_A(x_i)-v_B(x_i)|,|\pi_A(x_i)-\pi_B(x_i)|)}{2}\right]$$

$$=\frac{1}{n}\sum_{i=1}^{n}\left[\frac{|\mu_{A'}^{L}(x_i)-\mu_{B'}^{L}(x_i)|+|(1-\mu_{A'}^{U}(x_i))-(1-\mu_{B'}^{U}(x_i))|+|\Delta_{A'}(x_i)-\Delta_{B'}(x_i)|}{4}\right.$$
$$\left.+\frac{\max(|\mu_{A'}^{L}(x_i)-\mu_{B'}^{L}(x_i)|,|(1-\mu_{A'}^{U}(x_i))-(1-\mu_{B'}^{U}(x_i))|,|\Delta_{A'}(x_i)-\Delta_{B'}(x_i)|)}{2}\right]$$

$$=\frac{1}{n}\sum_{i=1}^{n}\left[\frac{|\mu_{A'}^{U}(x_i)-\mu_{B'}^{U}(x_i)|+|\mu_{A'}^{L}(x_i)-\mu_{B'}^{L}(x_i)|+|\Delta_{A'}(x_i)-\Delta_{B'}(x_i)|}{4}\right.$$
$$\left.+\frac{\max(|(\mu_{A'}^{U}(x_i))-\mu_{B'}^{U}(x_i))|,|\mu_{A'}^{L}(x_i)-\mu_{B'}^{L}(x_i)|,|\Delta_{A'}(x_i)-\Delta_{B'}(x_i)|)}{2}\right]=d_{WX1}(A',B') \tag{11}$$

(b)

$$\begin{aligned}d_{WX2}(A,B)&=\frac{1}{2\sqrt[p]{n}}\sqrt[p]{\sum_{i=1}^{n}\left(|\mu_A(x_i)-\mu_B(x_i)|+|\nu_A(x_i)-\nu_B(x_i)|\right)^p}\\&=\frac{1}{2\sqrt[p]{n}}\sqrt[p]{\sum_{i=1}^{n}\left(|\mu_{A'}^{U}(x_i)-\mu_{B'}^{U}(x_i)|+|\mu_{A'}^{L}(x_i)-\mu_{B'}^{L}(x_i)|\right)^p}\\&=d_{WX2}(A',B')\end{aligned} \tag{12}$$

Here also some theorem and lemmas on some elementary concept of distance measures between IVFSs have been discussed.

Theorem 1: Let X be a universe of discourse, then the distance measures on IVFSs given by (7)-(12) are metric.

Proof: We give proof only for (8), (11) & (12) and remaining are left for readers.

For any two IVFSs $A'=\{(x,[\mu_{A'}^{L}(x),\mu_{A'}^{U}(x)]:x\in X)\}$ and $B'=\{(x,[\mu_{B'}^{L}(x),\mu_{B'}^{U}(x)]:x\in X)\}$ of the universe of discourse $X=\{x_1,x_2,\ldots,x_n\}$, $l(A',B')\geq 0$ because of absolute value properties.

If $A'=B'$ then $\mu_{A'}^{U}(x)=\mu_{B'}^{U}(x)$, $\mu_{A'}^{L}(x)=\mu_{B'}^{L}(x)$ and $\Delta_{A'}(x)=\Delta_{B'}(x)$ for each $x_i \in X$ and hence $l(A,B)=0$. Conversely, let $l(A,B)=0$ then for each $x_i \in X$, we have, $|\mu_{A'}^{U}(x)-\mu_{B'}^{U}(x)|=0$, $|\mu_{A'}^{L}(x)-\mu_{B'}^{L}(x)|=0$ and $|\Delta_{A'}(x)-\Delta_{B'}(x)|=0$ and so we can conclude that $A=B$.

The symmetry property can be shown as

$$\begin{aligned}l(A',B')&=\frac{1}{2n}\sum_{i=1}^{n}\left[|\mu_{A'}^{U}(x_i)-\mu_{B'}^{U}(x_i)|+|\mu_{A'}^{L}(x_i)-\mu_{B'}^{L}(x_i)|+|\Delta_{A}(x_i)-\Delta_{B}(x_i)|\right]\\&=\frac{1}{2n}\sum_{i=1}^{n}\left[|\mu_{B'}^{U}(x_i)-\mu_{A'}^{U}(x_i)|+|\mu_{B'}^{L}(x_i)-\mu_{A'}^{L}(x_i)|+|\Delta_{B'}(x_i)-\Delta_{A'}(x_i)|\right]\\&=l(B',A')\end{aligned}$$

For triangular inequality we need to show that for any three IVFSs A',B',C' with upper membership functions $\mu_{A'}^U, \mu_{B'}^U, \mu_{C'}^U$ and lower membership functions $\mu_{A'}^L, \mu_{B'}^L, \mu_{C'}^L$ respectively, the following inequality holds:

$$l(A',C') \le l(A',B') + l(B',C')$$

$$l(A',C') = \frac{1}{2n}\sum_{i=1}^{n}\left[|\mu_{A'}^U(x_i) - \mu_{C'}^U(x_i)| + |\mu_{A'}^L(x_i) - \mu_{C'}^L(x_i)| + |\Delta_{A'}(x_i) - \Delta_{C'}(x_i)|\right]$$

$$= \frac{1}{2n}\sum_{i=1}^{n}\Big[|(\mu_{A'}^U(x_i) - \mu_{B'}^U(x_i)) + (\mu_{B'}^U(x_i) - \mu_{C'}^U(x_i))| + |(\mu_{A'}^L(x_i) - \mu_{B'}^L(x_i)) + (\mu_{A'}^L(x_i) - \mu_{B'}^L(x_i))| + |(\Delta_{A'}(x_i) - \Delta_{B'}(x_i)) + (\Delta_{B'}(x_i) - \Delta_{C'}(x_i))|\Big]$$

$$\le \frac{1}{2n}\sum_{i=1}^{n}\left[|\mu_{A'}^U(x_i) - \mu_{B'}^U(x_i)| + |\mu_{A'}^L(x_i) - \mu_{B'}^L(x_i)| + |\Delta_{A'}(x_i) - \Delta_{B'}(x_i)|\right]$$

$$+ \frac{1}{2n}\sum_{i=1}^{n}\left[|\mu_{B'}^U(x_i) - \mu_{C'}^U(x_i)| + |\mu_{B'}^L(x_i) - \mu_{C'}^L(x_i)| + |\Delta_{B'}(x_i) - \Delta_{C'}(x_i)|\right]$$

$$= l(A',B') + l(B',C')$$

Therefore, l satisfies triangular inequality. Hence, l is a metric.

Next, to show $d_{WX1}(A,B)$ is a metric.

The properties $d_{WX1}(A',B') = 0 \Leftrightarrow A' = B'$ and $d_{WX1}(A',B') = d_{WX1}(B',A')$ are straight forward. Before proceed to prove the triangular inequality property of $d_{WX}(A',B')$, we express $d_{WX1}(A',B')$ as a sum of two functions $d^{\#}{}_{WX1}(A',B')$ and $d^{\#\#}{}_{WX1}(A',B')$. That is,

$$d_{WX}(A',B') = d^{\#}{}_{WX1}(A',B') + d^{\#\#}{}_{WX1}(A',B'),$$

where

$$d^{\#}{}_{WX1}(A',B') == \frac{1}{n}\sum_{i=1}^{n}\left[\frac{|\mu_{A'}^U(x_i) - \mu_{B'}^U(x_i)| + |\mu_{A'}^L(x_i) - \mu_{B'}^L(x_i)| + |\Delta_{A'}(x_i) - \Delta_{B'}(x_i)|}{4}\right]$$

and

$$d^{\#\#}{}_{WX1}(A',B') == \frac{1}{n}\sum_{i=1}^{n}\left[\frac{\max(|\,(\mu^{U}_{A'}(x_i)) - \mu^{U}_{B'}(x_i))\,|,|\,\mu^{L}_{A'}(x_i) - \mu^{L}_{B'}(x_i)\,|,|\,\Delta_{A'}(x_i) - \Delta_{B'}(x_i)\,|)}{2}\right].$$

Now, we shall show that $d^{\#}{}_{WX1}(A',B')$ and $d^{\#\#}{}_{WX1}(A',B')$ are metric.

Like $l(A',B')$, easily one can show that $d^{\#}{}_{WX1}(A,B)$ is a metric.

To show $d^{\#}{}_{WX1}(A',B')$ is a metric, because of absolute value property $d^{\#}{}_{WX1}(A,B)=0 \Leftrightarrow A'=B'$ and $d^{\#}{}_{WX1}(A',B') = d^{\#}{}_{WX1}(B',A')$ are straight forward.

The triangular property directly comes from the fact that for any non-negative numbers a_1, a_2, a_3, b_1, b_2, a_3 such that $a_1+a_3 \geq a_3$ and $b_1+b_2 \geq b_3$ implies

$$\max(a_1,b_1) + \max(a_2,b_3) \geq \max(a_3,b_3).$$

i.e., $d^{\#\#}{}_{WX2}(A',C') \leq d^{\#\#}{}_{WX1}(A',B') + d^{\#}{}_{WX1}(B',C')$.

Thus, $d^{\#\#}{}_{WX1}(A',B')$ is a metric.

Next, to show $d_{WX1}(A',B')$ is a metric, we need show the triangular property. As

$$d_{WX1}(A',C') = d^{\#}{}_{WX1}(A',C') + d^{\#\#}{}_{WX1}(A',C')$$

$$\leq (d^{\#}{}_{WX1}(A',B') + d^{\#}{}_{WX1}(B',C')) + (d^{\#\#}{}_{WX1}(A',B') + d^{\#\#}{}_{WX1}(B',C'))\ \text{(by triangular property)}$$

$$= (d^{\#}{}_{WX1}(A,B) + d^{\#\#}{}_{WX1}(A,B)) + (d^{\#}{}_{WX1}(B,C) + d^{\#\#}{}_{WX1}(B,C))$$

$$= d_{WX1}(A,B) + d_{WX1}(B,C)$$

Thus, $d_{WX1}(A',B')$ is a metric.

Finally, to show $d_{WX2}(A',B')$ is a metric.

Clearly, due to absolute property the following two properties can be easily proved.

$$d_{WX2}(A',B') = 0 \Leftrightarrow A' = B' \text{ and } d_{WX2}(A',B') = d_{WX2}(B',A').$$

To prove triangular inequality, for any three IVFSs A',B',C' with upper membership functions $\mu^{U}_{A'}, \mu^{U}_{B'}, \mu^{U}_{C'}$ and lower membership functions $\mu^{L}_{A'}, \mu^{L}_{B'}, \mu^{L}_{C'}$ respectively, the following inequality holds:

$$d_{WX2}(A',C') \leq d_{WX2}(A',B') + d_{WX2}(B',C').$$

$$d_{WX2}(A',C') = \frac{1}{2\sqrt[p]{n}} \sqrt[p]{\sum_{i=1}^{n} \begin{pmatrix} |(\mu_{A'}^{U}(x_i) - \mu_{B'}^{U}(x_i) + (\mu_{B'}^{U}(x_i) - \mu_{C'}^{U}(x_i))| \\ + |(\mu_{A'}^{L}(x_i) - \mu_{B'}^{L}(x_i)) + (\mu_{B'}^{L}(x_i) - \mu_{C'}^{L}(x_i))| \end{pmatrix}^{p}}$$

$$\leq \frac{1}{2\sqrt[p]{n}} \sqrt[p]{\sum_{i=1}^{n} \begin{pmatrix} |(\mu_{A'}^{U}(x_i) - \mu_{B'}^{U}(x_i)| + |(\mu_{B'}^{U}(x_i) - \mu_{C'}^{U}(x_i))| \\ + |(\mu_{A'}^{L}(x_i) - \mu_{B'}^{L}(x_i))| + |(\mu_{B'}^{L}(x_i) - \mu_{C'}^{L}(x_i))| \end{pmatrix}^{p}} \text{ (by absolute value property)}$$

$$\leq \frac{1}{2\sqrt[p]{n}} \sqrt[p]{\sum_{i=1}^{n} \left(|(\mu_{A'}^{U}(x_i) - \mu_{B'}^{U}(x_i)| + |(\mu_{A'}^{L}(x_i) - \mu_{B'}^{L}(x_i))| \right)^{p}} \text{ (by Minkowski Inequality)}$$

$$+ \frac{1}{2\sqrt[p]{n}} \sqrt[p]{\sum_{i=1}^{n} \left(|(\mu_{B'}^{U}(x_i) - \mu_{C'}^{U}(x_i))| + |(\mu_{B'}^{L}(x_i) - \mu_{C'}^{L}(x_i))| \right)^{p}}$$

$$= d_{WX2}(A',B') + d_{WX2}(B',C')$$

Thus, d_{WX2} is a metric.

Theorem 2: For any two IVFSs $A' = \{(x,[\mu_{A'}^{L}(x), \mu_{A'}^{U}(x)] : x \in X\}$, and $B' = \{(x,[\mu_{B'}^{L}(x), \mu_{B'}^{U}(x)] : x \in X\}$, of the universe of discourse $X = \{x_1, x_2, \ldots, x_n\}$, the following inequality holds:

$$0 \leq d(A,B) \leq \frac{3n}{2}, 0 \leq l(A,B) \leq \frac{3}{2}, 0 \leq e(A,B) \leq \sqrt{\frac{3n}{2}}, 0 \leq q(A,B) \leq \sqrt{\frac{3}{2}}, 0 \leq d_{WX1}(A,B) \leq \frac{5}{4}$$

$0 \leq d_{WX2}(A,B) \leq 1$.

Proof: As $\mu_{A'}^{U}(x), \mu_{A'}^{L}(x), \mu_{B'}^{U}(x), \mu_{B'}^{L}(x), \Delta_{A'}(x)$ and $\Delta_{B'}(x) \in [0,1]$. Thus, we have

$$\left|\mu_{A'}^{U}(x) - \mu_{B'}^{U}(x)\right| \leq 1, \left|\mu_{A'}^{L}(x) - \mu_{B'}^{L}(x)\right| \leq 1 \text{ and } \left|\Delta_{A'}(x) - \Delta_{B'}(x)\right| \leq 1$$

for each $x_i \in X$ and hence, if $A=B$ then only all distance measure will be zero.

$$\therefore 0 \leq d(A',B') \leq \frac{1}{2}\sum_{i=1}^{n} 3 = \frac{3n}{2}, 0 \leq l(A',B') = \frac{1}{2n}\sum_{i=1}^{n} 3 = \frac{1}{2n} 3n = \frac{3}{2}, 0 \leq e(A',B') = \sqrt{\frac{1}{2}\sum_{i=1}^{n} 3} = \sqrt{\frac{3n}{2}},$$

$$0 \leq q(A',B') = \sqrt{\frac{1}{2n}\sum_{i=1}^{n} 3} = \sqrt{\frac{1}{2n} 3n} = \sqrt{\frac{3}{2}}, 0 \leq d_{WX1}(A',B') = \frac{1}{n}\sum_{i=1}^{n}\left[\frac{3}{4} + \frac{1}{2}\right] = \frac{1}{n}\frac{5n}{4} = \frac{5}{4}$$

and

$$0 \le d_{WX2}(A',B') = \frac{1}{2\sqrt[p]{n}}\sqrt[p]{\sum_{i=1}^{n} 2^p} = \frac{1}{2\sqrt[p]{n}}\sqrt[p]{n2^p} = 1.$$

Lemma 1: For any two IVFSs $A' = \{(x,[\mu_{A'}^L(x),\mu_{A'}^U(x)]: x \in X\}$, and $B' = \{(x,[\mu_{B'}^L(x),\mu_{B'}^U(x)]: x \in X\}$, of the universe of discourse $X = \{x_1, x_2, \ldots, x_n\}$, the following inequality holds:

$$d_{WX2}(A',B') \le q(A',B').$$

Lemma 2: For any two IVFSs $A' = \{(x,[\mu_{A'}^L(x),\mu_{A'}^U(x)]: x \in X\}$, and $B' = \{(x,[\mu_{B'}^L(x),\mu_{B'}^U(x)]: x \in X\}$, of the universe of discourse $X = \{x_1, x_2, \ldots, x_n\}$, the following inequality holds:

$$d_{WX2}(A',B') \le d_{WX1}(A',B').$$

Lemma 3: For any two IVFSs $A' = \{(x,[\mu_{A'}^L(x),\mu_{A'}^U(x)]: x \in X\}$, and $B' = \{(x,[\mu_{B'}^L(x),\mu_{B'}^U(x)]: x \in X\}$, of the universe of discourse $X = \{x_1, x_2, \ldots, x_n\}$, the following inequality holds:

$$q(A',B') \le d_{WX1}(A',B').$$

Lemma 4: For any two IVFSs $A' = \{(x,[\mu_{A'}^L(x),\mu_{A'}^U(x)]: x \in X\}$, and $B' = \{(x,[\mu_{B'}^L(x),\mu_{B'}^U(x)]: x \in X\}$, of the universe of discourse $X = \{x_1, x_2, \ldots, x_n\}$, the following inequality holds:

$$d_{WX2}(A',B') \le l(A',B').$$

Lemma 5: For any two IVFSs $A' = \{(x,[\mu_{A'}^L(x),\mu_{A'}^U(x)]: x \in X\}$, and $B' = \{(x,[\mu_{B'}^L(x),\mu_{B'}^U(x)]: x \in X\}$, of the universe of discourse $X = \{x_1, x_2, \ldots, x_n\}$, the following inequality holds:

$$q(A',B') \le l(A',B').$$

Lemma 6: For any two IVFSs $A' = \{(x,[\mu_{A'}^L(x),\mu_{A'}^U(x)]: x \in X\}$, and $B' = \{(x,[\mu_{B'}^L(x),\mu_{B'}^U(x)]: x \in X\}$, of the universe of discourse $X = \{x_1, x_2, \ldots, x_n\}$, the following inequality holds:

$d_{WX1}(A', B') \le l(A', B')$.

Lemma 7: For any two IVFSs $A' = \{(x, [\mu_{A'}^{L}(x), \mu_{A'}^{U}(x)] : x \in X\}$, and $B' = \{(x, [\mu_{B'}^{L}(x), \mu_{B'}^{U}(x)] : x \in X\}$, of the universe of discourse $X = \{x_1, x_2, \ldots, x_n\}$, the following inequality holds:

$e(A', B') \le d(A', B')$.

Lemma 8: For any two IVFSs $A' = \{(x, [\mu_{A'}^{L}(x), \mu_{A'}^{U}(x)] : x \in X\}$, and $B' = \{(x, [\mu_{B'}^{L}(x), \mu_{B'}^{U}(x)] : x \in X\}$, of the universe of discourse $X = \{x_1, x_2, \ldots, x_n\}$, the following inequalities hold for $n \ge 2$:

$d_{WX2}(A', B') \le q(A', B') \le d_{WX1}(A', B') \le l(A', B') \le e(A', B') \le d(A', B')$.

Proofs of the corollaries are straight forward and left for readers.

6. MEDICAL PRE-SCREENING TEST

One of the most prolific and appealing areas of applications of FST is medicine. In real life situations, due to the uncertain nature of medical documents and uncertain information gathered for decision making entails the exercise of FST. After the conception of FST, a variety of expansions have been observed and effectively applied in medical decision making. Sanchez (1976, 1979) initiated the diagnosis models involving fuzzy matrices representing the medical knowledge between symptoms and diseases. Subsequently, different researchers have studied medical decision making using fuzzy set, IFS, IVFS, Rough set, soft set etc.

In this section, an algorithm for medical pre-screening test is presented first. For this, suppose *S, D* and *P* are the sets of symptoms, diseases and patients, respectively. Like Sanchez's notation of "Medical knowledge" as interval valued fuzzy relation *R* from *S* to *D* and similarly, another fuzzy relation *Q* from *P* to *S* are formulated.

The route is abridged below:

1. Identification of symptoms and formulation of medical knowledge based on interval valued fuzzy relation *R* from *S* to *D* and another interval valued fuzzy relation *Q* from *P* to *S*.
2. Computation of the distances between patients and diseases considering the composition of S-D relation matrix and P-S relation matrix as the derived distance measures.
3. Shortest distance between patient and disease will indicate that the patient is likely to have the disease.

6.1 Case Study

In this section, medical diagnosis has been studied using the distance measures between IVFSs originally derived from IFSs. In this case study two scenarios are taken into consideration.

6. 1.1 Scenario-1

In this scenario, medical pre-screening is performed based on the data set presented by Davvaz and Sadrabadi (2016) which are originally available in IFSs form and those have been converted into IVFSs using association between IFS and IVFS (section-3).

Let $P = \{P_1, P_2, P_3, P_4\}$ be the set of patients admitted in a Medical Hospital. The symptoms set is $S = \{S_1, S2, S_3, S_4, S_5\}$ and corresponding set of diseases is D= $\{D_1, D_2, D_3, D_4, D_5\}$. Based on the knowledge base of doctor/physician on valued fuzzy relation Q (P → S) and R $(S \rightarrow D)$, we need to diagnose which patient is suffering from which exactly disease. Then a hypothetical data of interval valued fuzzy relation Q (P → S) is given as in Table 1. The interval valued fuzzy relation R $(S \rightarrow D)$ is given as in (hypothetical) Table 2. Step-I: Consider that the patient-symptom relation Q has been determine and given as in table 1. Again, consider that the symptom-disease relation R has been formulated and given as in Table 2.

Table 1. Patient-symptom relation Q

Q	S_1	S_2	S_3	S_4	S_5
P_1	(0.8,0.9)	(0.6,0.9)	(0.2,0.2)	(0.6,0.9)	(0.1,0.4)
P_2	(0.0,0.2)	(0.4,0.6)	(0.6,0.9)	(0.1,0.3)	(0.1,0.2)
P_3	(0.8,0.9)	(0.8,0.9)	(0.0,0.4)	(0.2,0.3)	(0.0,0.5)
P_4	(0.6,0.9)	(0.5,0.6)	(0.3,0.6)	(0.7,0.8)	(0.3,0.6)

Table 2. Symptom-disease relation R

R	D_1	D_2	D_3	D_4	D_5
S_1	(0.4,1)	(0.7,1)	(0.3,0.7)	(0.1,0.3)	(0.1,0.2)
S_2	(0.3,0.5)	(0.2,0.4)	(0.6,0.9)	(0.2,0.6)	(0.0,0.2)
S_3	(0.1,0.3)	(0.0,0.1)	(0.2,0.3)	(0.8,1)	(0.2,0.2)
S_4	(0.4,0.7)	(0.7,1)	(0.2,0.4)	(0.2,0.3)	(0.2,0.2)
S_5	(0.1,0.3)	(0.1,0.2)	(0.1,0.1)	(0.2,0.3)	(0.8,0.9)

Step-II: Patient-disease relationship are calculated using normalized Hamming distance (8) and normalized Euclidean distance (10) which are given in table (3) and (4) respectively.

The distance measures (10) & (11) on IVFSs are also applied in medical diagnosis and results are obtained as (Table 5 and Table 6-15).

Step-III: As shortest distance between patient and disease indicates that the patient is likely to have the disease. Therefore, from all the above Tables, it can be concluded that the minimum value in the first raw indicates P_1 suffers from D_2, in the second raw P_2 suffers from D_4, in the third raw P_3 is suffering from D_3 and in the last raw P_4 is suffering from D_1.

Table 3. Patients-diseases relation

	D_1	D_2	D_3	D_4	D_5
P_1	0.28	0.24	0.28	0.54	0.56
P_2	0.40	0.50	0.31	0.14	0.42
P_3	0.38	0.44	0.32	0.50	0.55
P_4	0.28	0.30	0.38	0.44	0.54

Table 4. Patients-diseases relation

	D_1	D_2	D_3	D_4	D_5
P_1	0.29	0.25	0.32	0.53	0.58
P_2	0.43	0.56	0.33	0.14	0.46
P_3	0.36	0.41	0.32	0.52	0.57
P_4	0.25	0.29	0.35	0.43	0.50

Table 5. Using formula (10)

	D_1	D_2	D_3	D_4	D_5
P_1	0.21	0.20	0.24	0.49	0.53
P_2	0.36	0.48	0.29	0.12	0.38
P_3	0.28	0.36	0.23	0.43	0.49
P_4	0.23	0.26	0.33	0.40	0.49

Table 6. Using formula (11) for p=1

	D_1	D_2	D_3	D_4	D_5
P_1	0.19	0.18	0.28	0.46	0.5
P_2	0.33	0.46	0.26	0.1	0.36
P_3	0.25	0.32	0.18	0.4	0.46
P_4	0.2	0.23	0.29	0.37	0.45

Table 7. Using formula (11), for p=2

	D_1	D_2	D_3	D_4	D_5
P_1	0.21	0.22	0.26	0.51	0.56
P_2	0.4	0.55	0.31	0.1	0.44
P_3	0.28	0.38	0.2	0.48	0.54
P_4	0.2	0.25	0.3	0.4	0.46

Table 8. Using formula (11), for p=3

	D_1	D_2	D_3	D_4	D_5
P_1	0.24	0.27	0.3	0.54	0.58
P_2	0.44	0.59	0.34	0.11	0.48
P_3	0.3	0.43	0.23	0.52	0.58
P_4	0.21	0.28	0.32	0.42	0.48

Table 9. Using formula (11), for p=4

	D_1	D_2	D_3	D_4	D_5
P_1	0.25	0.3	0.32	0.56	0.6
P_2	0.46	0.62	0.37	0.11	0.51
P_3	0.32	0.46	0.25	0.55	0.61
P_4	0.21	0.29	0.33	0.44	0.49

Table 10. Using formula (11), for p=5

	D_1	D_2	D_3	D_4	D_5
P_1	0.26	0.32	0.34	0.57	0.6
P_2	0.48	0.63	0.38	0.11	0.54
P_3	0.33	0.48	0.26	0.57	0.62
P_4	0.21	0.3	0.34	0.45	0.5

Table 11. Using formula (11), for p=6

	D_1	D_2	D_3	D_4	D_5
P_1	0.28	0.35	0.36	0.58	0.6
P_2	0.49	0.64	0.4	0.12	0.55
P_3	0.35	0.49	0.28	0.58	0.63
P_4	0.22	0.32	0.36	0.49	0.51

Table 12. Using formula (11), for p=7

	D_1	D_2	D_3	D_4	D_5
P_1	0.28	0.35	0.36	0.6	0.61
P_2	0.51	0.65	0.41	0.12	0.58
P_3	0.36	0.5	0.28	0.59	0.65
P_4	0.22	0.32	0.36	0.47	0.51

Table 13. Using formula (11), for p=8

	D_1	D_2	D_3	D_4	D_5
P_1	0.28	0.37	0.37	0.61	0.62
P_2	0.51	0.66	0.41	0.12	0.58
P_3	0.37	0.51	0.29	0.6	0.65
P_4	0.22	0.33	0.37	0.47	0.52

Table 14. Using formula (11), for p=9

	D_1	D_2	D_3	D_4	D_5
P_1	0.29	0.37	0.38	0.61	0.62
P_2	0.52	0.67	0.42	0.12	0.59
P_3	0.37	0.52	0.29	0.61	0.66
P_4	0.22	0.33	0.38	0.48	0.52

Table 15. Using formula (11), for p=10

	D_1	D_2	D_3	D_4	D_5
P_1	0.30	**0.38**	0.39	0.62	0.63
P_2	0.53	0.67	0.43	**0.13**	0.6
P_3	0.38	0.53	**0.30**	0.62	0.67
P_4	**0.23**	0.34	0.38	0.49	0.53

Table 16. Patient-symptom relation Q

Q	S_1	S_2	S_3
P_1	(0.3,0.5)	(0.8,1)	(0.4,0.7)
P_2	(0.7,1)	(0.3,0.6)	(0.2,0.4)
P_3	(0.5,0.7)	(0.1,0.5)	(0.7,0.9)

6.1.2 Scenario-2

In this scenario, pre-screening of plans is performed based on the data set of IVFSs adopted from Loganathan & Pushpalatha (2018).

Suppose in an agriculture park there are three plants Mango (P_1), Citrus (P_2), Guava (P_3) and it is observed that leaves of the trees are pretentious. After study it is found that the symptoms are black or dark coloured circular spots with shot holes, powdery growth, and pinkish brown pustules on the leaves. Let the possible diseases relating to the above symptoms be anthracnose, powdery mildew, rust. Suppose the set $S=\{S_1, S_2, S_3\}$ as universal set, where S_1, S_2 and S_3 represent the symptoms black or dark coloured circular spots on leaves with shot holes, powdery growth on the leaves, pinkish brown pustules on the leaves, respectively and the set $D = \{D_1, D_2, D_3\}$ where D_1, D_2 and D_3 represent the parameters Anthracnose, powdery mildew and rust, respectively. The P-S IVFS relation and S-D IVFS relations are presented in Table-16 and Table-17, respectively.

The P-D relation is evaluated applying the distance measure (7) and presented in the Table-18.

From the Table-18, it is clear that Mango (P_1) is affected by the disease powdery rust ($D3$), Citrus (P_2) is affected by the disease powdery mildew ($D2$), and Guava (P_3) is affected by the disease Anthracnose ($D1$).

Table 17. Symptom-disease relation R

R	D_1	D_2	D_3
S_1	(0.1,0.5)	(0.3,0.8)	(0.2,0.4)
S_2	(0.2,0.5)	(0.7,1)	(0.6,0.9)
S_3	(0.5,0.7)	(0.2,0.3)	(0.5,1)

Table 18. Patient-Disease relation

	D_1	D_2	D_3
P_1	0.7	0.5	**0.45**
P_2	0.95	**0.75**	1.3
P_3	**0.55**	1.25	0.9

7. RESULT AND DISCUSSION

Due to vague, imprecision information frequently involve in medical documents uncertainty more often occurs in medical diagnosis or pre screening test. It is important to include all the uncertainties in the system and try to incorporate all the information into analysis. IVFSs are used to represent uncertainty and medical diagnosis has been carried out under interval valued fuzzy environment. Medical pre-screening has been performed through two scenarios. In scenario-1, medical pre-screening is performed based on the data set presented by Davvaz and Sadrabadi (2016) which are originally available in IFSs form and those have been converted into IVFSs using association between IFS and IVFS First we determined symptoms and formulate medical knowledge based on interval valued fuzzy relations. Then, patient-disease relation was obtained via distance measures. From the Tables 3-15, it can be observed that P_1 suffers from D_2; P_2 suffers from D_4; P_3 is suffering from D_3 and P_4 is suffering from D_1 which also corroborate the same facts presented by Davvaz and Sadrabadi (2016). In scenario-2, pre-screening of plans is performed based on the data set of IVFSs adopted from Loganathan & Pushpalatha (2018), it is found that Mango (P_1) is affected by the disease powdery rust (*D3*), Citrus (P_2) is affected by the disease powdery mildew (*D2*), and Guava (P_3) is affected by the disease Anthracnose (*D1*).

8. CONCLUSION

In this chapter, IVFS is studied along with a few distance measures on IVFSs. The perception of distance measures was originally available in the form of IFSs. Here, distance measures on IVFSs are derived from IFSs to deal with uncertainty is being first observed in medical pre-screening. Afterwards, various properties of distance measures on IVFSs are presented. Later, an appropriate algorithm has been formulated for medical pre-screening and accordingly medical pre-screenings have been carried out successfully under two scenarios. In the first scenario, medical pre-screening IFSs data are taken and converted into IVFSs and later performed medical pre-screening while in the second scenario IVFSs data are taken into consideration for the same. It is observed that this approach makes it possible to introduce weights of all symptoms and consequently patient can be diagnosed directly. Lots of possibilities are there to apply similar type of study in different areas such as multi-criteria/group decision making, system of linear equations, fuzzy relation equations, Risk assessment, engineering problems optimization problems etc.

REFERENCES

Abdel, M. S. E. D. A., Elsamahy, M., Hassan, M. A. M., & Bendary, F. M. (2017). Enhancement of Turbo-Generators Phase Backup Protection Using Adaptive Neuro Fuzzy Inference System. *International Journal of System Dynamics Applications*, *6*(1), 58–76. doi:10.4018/IJSDA.2017010104

Ahn, J. Y., Han, K. S., Oh, S. Y., & Lee, C. D. (2011). An application of interval-valued intuitionistic fuzzy sets for medical diagnosis of headache. *International Journal of Innovative Computing, Information, & Control*, *7*, 2755–2762.

Alcantud, J. C. R., Varela, G., Santos-Buitrago, B., Santos-Garcia, G., & Jimenez, M. F. (2019). Analysis of survival for lung cancer resections cases with fuzzy and soft set theory in surgical decision making. *PLoS One*, *14*(6), e0218283. doi:10.1371/journal.pone.0218283 PMID:31216304

Atanassov, K. (1986). Intuitionistic fuzzy sets. *Fuzzy Sets and Systems*, *20*(1), 87–96. doi:10.1016/S0165-0114(86)80034-3

Atanassov K., (1999). *Intuitionistic Fuzzy Sets: Theory and Applications*. Wyrzburg: Physica-Verlag.

Atanassov, K. (2005). Answer to D. Dubois, S. Gottwald, P. Hajek, J. Kacprzyk and H. Prade's paper "Terminological difficulties in fuzzy set theory—The case of Intuitionistic Fuzzy Sets." *Fuzzy Sets and Systems*, *156*(3), 496–499. doi:10.1016/j.fss.2005.06.003

Atanassov, K., & Gargov, G. (1989). Interval valued intuitionistic fuzzy sets. *Fuzzy Sets and Systems*, *31*(3), 343–349. doi:10.1016/0165-0114(89)90205-4

Azar, A. T. (2011). Neuro-Fuzzy System for Cardiac Signals Classification. International Journal of Modelling. *Identification and Control*, *13*(1/2), 108–116. doi:10.1504/IJMIC.2011.040495

Azar, A. T. (2013a). Neuro-Fuzzy Applications in Dialysis Systems. In A. T. Azar (Ed.), *Biofeedback Systems and Soft Computing Techniques of Dialysis* (Vol. 405, pp. 1223–1274)., doi:10.1007/978-3-642-27558-6_10

Azar, A. T. (2013b). Multi-Adaptive Neural-Fuzzy System as a Novel Predictor of in-vivo Blood Side Dialyzer Urea Clearance. *International Journal of Computer Applications in Technology*, *46*(2), 77–90. doi:10.1504/IJCAT.2013.052290

Azar, A. T. (2014). Neuro-fuzzy feature selection approach based on linguistic hedges for medical diagnosis. *International Journal of Modelling Identification and Control*, *22*(3), 195–206. doi:10.1504/IJMIC.2014.065338

Azar, A. T., El-Said, S. A., Balas, V. E., & Olariu, T. (2013) Linguistic Hedges Fuzzy Feature Selection For Erythemato-Squamous Diseases. In Soft Computing Applications, Advances in Intelligent Systems and Computing (AISC). Springer-Verlag. DOI: doi:10.1007/978-3-642-33941-7_43

Azar, A. T., El-Said, S. A., & Hassanien, A. E. (2013). Fuzzy and Hard Clustering Analysis For Thyroid Disease. *Computer Methods and Programs in Biomedicine*, *111*(1), 1–16. doi:10.1016/j.cmpb.2013.01.002 PMID:23357404

Azar, A.T., & Hassanien, A.E. (2014). Dimensionality Reduction of Medical Big Data Using Neural-Fuzzy Classifier. *Soft Computing*, *19*(4), 1115-1127. DOI doi:10.100700500-014-1327-4

Azar, A. T., Hassanien, A. E., & Kim, T. H. (2012) Expert System Based On Neural-Fuzzy Rules for Thyroid Diseases Diagnosis. *International Conference on Bio-Science and Bio-Technology (BSBT 2012)*. 10.1007/978-3-642-35521-9_13

Banu, P. K. N., Azar, A. T., & Inbarani, H. H. (2017). Fuzzy firefly clustering for tumor and cancer analysis. *International Journal of Modelling Identification and Control*, *27*(2), 92–103. doi:10.1504/IJMIC.2017.082941

Bustince, H., & Burillo, P. (1996). Vague sets are intuitionistic fuzzy sets. *Fuzzy Sets and Systems*, *79*(3), 403–405. doi:10.1016/0165-0114(95)00154-9

Chen, T. Y. (2007). A note on distances between intuitionistic fuzzy sets and/or interval-valued fuzzy sets based on the Hausdorff metric. *Fuzzy Sets and Systems*, *158*(22), 2523–2525. doi:10.1016/j.fss.2007.04.024

Chetia, B., & Das, P. K. (2010). An application of Interval Valued Fuzzy Soft Set in Medical Diagnosis. *Int. J. Contempt. Math. Science.*, *5*(38), 1887–1894.

Choi, H. M., Mun, G. S., & Ahn, J. Y. (2012). A medical diagnosis based on interval valued fuzzy set. *Biomedical Engineering: Applications, Basis and Communications*, *24*(4), 349.

Davvaz, B., & Sadrabadi, E. H. (2016). An application of intuitionistic fuzzy sets in medicine. *International Journal of Biomathematics*, *9*(3), 16500371–15. doi:10.1142/S1793524516500376

De, S. K., Biswas, R., & Roy, A. R. (2001). An application of intuitionistic fuzzy sets in medical diagnosis. *Fuzzy Sets and Systems*, *117*(2), 209–213. doi:10.1016/S0165-0114(98)00235-8

Deepa, G., Praba, B., Manimaran, A., Chandrasekaran, V. M., & Rajakumar, K. (2018). Medical diagnosis using intuitionistic fuzzy set in terms shortest distance measure. *Research Journal of Pharmacy and Technology*, *11*(3), 949–952. doi:10.5958/0974-360X.2018.00176.2

Dubois, S., Gottwald, S., Hajek, P., Kacprzyk, J., & Prade, H. (2005). Terminological difficulties in fuzzy set theory—The case of "Intuitionistic Fuzzy Sets. *Fuzzy Sets and Systems*, *156*(3), 485–491. doi:10.1016/j.fss.2005.06.001

Dutta, P., & Ali, T. (2019, March). Decision Making for Medical Diagnosis Through Credibility Theory. In *International Conference on Information Technology and Applied Mathematics* (pp. 713-724). Springer.

Dutta, P., & Doley, D. (2019). Fuzzy decision making for medical diagnosis using arithmetic of generalised parabolic fuzzy numbers. *Granular Computing*, 1-12.

Dutta, P., & Doley, D. (2020). Medical diagnosis under uncertain environment through bipolar-valued fuzzy sets. In *Computer vision and machine intelligence in medical image analysis* (pp. 127–135). Springer. doi:10.1007/978-981-13-8798-2_13

Dutta, P., & Limboo, B. (2017). Bell-shaped fuzzy soft sets and their application in medical diagnosis. *Fuzzy Information and Engineering*, *9*(1), 67–91. doi:10.1016/j.fiae.2017.03.004

Elizabeth, S., & Sujatha, L. (2013). Application of Fuzzy Membership Matrix in Medical Diagnosis and Decision Making. *Applied Mathematical Sciences*, *7*(127), 6297–6307. doi:10.12988/ams.2013.38485

Elizabeth, S., & Sujatha, L. (2013). Application of fuzzy membership matrix in medical diagnosis and decision making. *Applied Mathematical Sciences*, *7*(127), 6297–6307. doi:10.12988/ams.2013.38485

Elizabeth, S., & Sujatha, L. (2014). Medical Diagnosis Based on Interval Valued Fuzzy Number Matrices. *Ann. of Pure Appl. Math.*, *7*, 91–96.

Gehrke, C., Walker, C., & Walker, E. (1996). Some Comments on Interval-Valued Fuzzy Sets. *International Journal of Intelligent Systems*, *11*(10), 751–759. doi:10.1002/(SICI)1098-111X(199610)11:10<751::AID-INT3>3.0.CO;2-Y

Grzegorzewski, P. (2004). Distances between intuitionistic fuzzy sets and/or interval-valued fuzzy sets based on the Hausdorff metric. *Fuzzy Sets and Systems*, *148*(2), 319–328. doi:10.1016/j.fss.2003.08.005

Grzegorzewski, P., & Mrówka, E. (2005). Some notes on (Atanassov's) intuitionistic fuzzy sets. *Fuzzy Sets and Systems*, *156*(3), 492–495. doi:10.1016/j.fss.2005.06.002

Hamrawi, H. (2011). *Type-2 fuzzy α-cuts* (Ph. D. Thesis). De Montfort University.

Hamrawi, H., Coupland, S., & John, R. (2010). A novel α-cut representation for type-2 fuzzy sets in fuzzy systems. *IEEE International Conference*.

Huang, H. L., & Guo, Y. (2019). An Improved Correlation Coefficient of Intuitionistic Fuzzy Sets. *Journal of Intelligent Systems*, *28*(2), 231–243. doi:10.1515/jisys-2017-0094

Ismail, M. M. (2012). Applications of ANFIS and fuzzy algorithms for improvement of the DTC performance for the three phase saturated model of induction motor. *International Journal of System Dynamics Applications*, *1*(3), 54–83. doi:10.4018/ijsda.2012070102

Kacprzyk, J. (1997). *Multistage Fuzzy Control*. Wiley.

Li, C. (2009). Distances between interval-valued fuzzy sets. *Proc. Society Annl. Conf. (NAFIPS2009)*. DOI:10.1109/NAFIPS.2009.5156438

Lin, T. C., & Kuo, C. H. (2012). Adaptive Hybrid Intelligent Tracking Control for Uncertain Fractional Order Chaotic Systems. *International Journal of System Dynamics Applications*, *1*(1), 1–16. doi:10.4018/ijsda.2012010101

Liu, D., Chen, X., & Peng, D. (2018a). Cosine similarity measure between hybrid intuitionistic fuzzy sets and its application in medical diagnosis. *Computational and Mathematical Methods in Medicine*, *2018*, 2018. doi:10.1155/2018/3146873 PMID:30416537

Liu, D., Liu, G., & Liu, Z. (2018b). Some Similarity Measures of Neutrosophic Sets Based on the Euclidean Distance and Their Application in Medical Diagnosis. *Computational and Mathematical Methods in Medicine*. doi:10.1155/2018/7325938 PMID:30622626

Loganathan, C., & Pushpalatha, V. (2018). An Application of Interval Valued Fuzzy Matrices in Agricultural Field. *Int. J. Math. And Appl.*, *6*(3), 387–397.

Luo, M., & Liang, J. (2018). A Novel Similarity measure for interval-valued intuitionistic fuzzy sets and its applications. *Symmetry*, *10*(10), 441. doi:10.3390ym10100441

Maheshwari, S., & Srivastava, A. (2016). Study on divergence measures for intuitionistic fuzzy sets and its application in medical diagnosis. *Journal of Applied Analysis and Computation*, *6*(3), 772–789. doi:10.11948/2016050

Mahmoud, M. I., El-Araby, S. M., Shehata, S. A., AbouZaid, R. M. F., & El-Samie, F. A. (2015). Design and Implementation of a Fast General Purpose Fuzzy Processor. In Research Methods: Concepts, Methodologies, Tools, and Applications (pp. 543-558). IGI Global. doi:10.4018/978-1-4666-7456-1.ch025

Mariapresenti, L., & Arockiarani, I. (2018). A New Approach Of Interval Valued Vague Set In Plants. *International Journal of Pure and Applied Mathematics*, *120*(7), 325–333.

Meenakshi, A. R., & Kaliraja, M. (2011). An application of interval valued fuzzy matrices in medical diagnosis. *International Journal of Mathematical Analysis.*, *5*(36), 1791–1802.

Own, M. (2009). Switching between type-2 fuzzy sets and intuitionistic fuzzy sets: An application in medical diagnosis. *Applied Intelligence*, *31*(1), 283–291. doi:10.100710489-008-0126-y

Papakostas, A., Hatzimichailidis, A. G., & Kaburlasos, V. G. (2013). Distance and similarity measures between intuitionistic fuzzy sets: A comparative analysis from a pattern recognition point of view. *Pattern Recognition Letters*, *34*(14), 1609–1622. doi:10.1016/j.patrec.2013.05.015

Park, H., Lim, K. M., Park, J. S., & Kwun, Y. C. (2008). Distances between interval-valued intuitionistic fuzzy sets. *Journal of Physics: Conference Series*, *96*, 012089. doi:10.1088/1742-6596/96/1/012089

Sambuc, R. (1975). *Fonctions f-floues aplication 'l' aide au diagnostic en pathologie thyroidienne* (Ph.D. Thesis). University of Marseille.

Samuel, E., & Balamurugan, M. (2012a). Intuitionistic fuzzy set with rank correlation technique in medical diagnosis, In *Proceedings of the International Conference on Mathematics in Engineering & Business Management*. Stella Maris College.

Samuel, E., & Balamurugan, M. (2012b). Intuitionistic fuzzy set in medical diagnosis using ranking function. *Surveys in Mathematics and Mathematical Science*, *2*(1), 23–34.

Samuel, E., & Balamurugan, M. (2013). IFS with N-Parameters In Medical Diagnosis. *International Journal of Pure and Applied Mathematics*, *84*(3), 185–192. doi:10.12732/ijpam.v84i3.5

Sanchez, E. (1976). Resolution of Composite Fuzzy Relation Equations. *Information and Control*, *30*(1), 38–48. doi:10.1016/S0019-9958(76)90446-0

Sanchez, E. (1979). Medical Diagnosis and Composite Fuzzy Relations. In Advances in Fuzzy Set Theory and Applications. Amsterdam: North-Holland.

Shereen, A., Ramadan, R. A., & Ghali, N. I. (2016). Classification of EEG Signals for Motor Imagery based on Mutual Information and Adaptive Neuro Fuzzy Inference System. *International Journal of System Dynamics Applications*, *5*(4), 64–82. doi:10.4018/IJSDA.2016100104

Srivastava, A. (2019). Medical diagnosis using intuitionistic fuzzy sets. In. AIP Conference Proceedings: Vol. 2061. *No. 1* (p. 020029). AIP Publishing LLC.

Szmidt, E. (2014). *Distances and Similarities in Intuitionistic Fuzzy Sets*. Springer. doi:10.1007/978-3-319-01640-5

Szmidt, E., & Janusz, K. (2011). Intuitionistic fuzzy sets two and three term representations in the context of a hausdorff distance. *ACTA Univ. Matthiae BellI Ser. Math*, *19*, 53–62.

Szmidt, E., & Kacprzyk, J. (1997). On measuring distances between intuitionistic fuzzy sets. *Notes on IFS*, *3*(4), 1–13.

Szmidt, E., & Kacprzyk, J. (2000). Distances between intuitionistic fuzzy sets. *Fuzzy Sets and Systems*, *114*(3), 505–518. doi:10.1016/S0165-0114(98)00244-9

Szmidt, E., & Kacprzyk, J. (2001). Intuitionistic fuzzy sets in some medical applications. *Lecture Notes in Computer Science*, *2206*, 148–151. doi:10.1007/3-540-45493-4_19

Szmidt, E., & Kacprzyk, J. (2006). Distances between intuitionistic fuzzy sets: straightforward approaches may not work. *3rd International IEEE Conference Intelligent Systems IS06*, 716–721. 10.1109/IS.2006.348507

Szmidt, E., & Kacprzyk, J. (2008). Dilemmas with distances between intuitionistic fuzzy sets: Straightforward approaches may not work. *Studies in Computational Intelligence*, *109*, 415–430. doi:10.1007/978-3-540-77623-9_24

Talukdar, P., & Dutta, P. (2018). Disease diagnosis using an advanced distance measure for intuitionistic fuzzy sets. *Int. Res. J. Microbiol, 7*(2), 29-42.

Tcvetkov, R., Szmidt, E., & Kacprzyk, J. (2009). On Some Issues Related to the Distances between the Atanassov Intuitionistic Fuzzy Sets. *Cybernetics and Information Technologies*, *9*(2), 54–61.

Tizhoosh, H. R. (2005). Image thresholding using type II fuzzy sets. *Pattern Recognition*, *38*(12), 2363–2372. doi:10.1016/j.patcog.2005.02.014

Tizhoosh, H. R. (2008). Interval-valued versus intuitionistic fuzzy sets: Isomorphism versus semantics. *Pattern Recognition*, *41*(5), 1812–1813. doi:10.1016/j.patcog.2007.11.003

Wang, W. Q., & Xin, X. L. (2005). Distance measure between intuitionistic fuzzy sets. *Pattern Recognition Letters*, *26*(13), 2063–2069. doi:10.1016/j.patrec.2005.03.018

Yao, J. F. F., & Yao, J. S. (2001). Fuzzy decision making for medical diagnosis based on fuzzy number and compositional rule of inference. *Fuzzy Sets and Systems*, *120*(2), 351–366. doi:10.1016/S0165-0114(99)00071-8

Zadeh, L. A. (1965). Fuzzy sets. *Information and Control*, *8*(3), 338–356. doi:10.1016/S0019-9958(65)90241-X

Zeng, W. Y., & Guo, P. (2008). Normalized distance, similarity measure, inclusion measure and entropy of interval-valued fuzzy sets and their relationship. *Inform. Sci.*, *178*(5), 1334–1342. doi:10.1016/j.ins.2007.10.007

Zhou, Q., Mo, H., & Deng, Y. (2020). A new divergence measure of pythagorean fuzzy sets based on belief function and its application in medical diagnosis. *Mathematics*, *8*(1), 142. doi:10.3390/math8010142

Chapter 12
Multi-Object Tracking in Wireless Sensor Networks Using Swarm Intelligence

Rabie A. Ramadan
https://orcid.org/0000-0002-0281-9381
College of Computer Science and Engineering, University of Hail, Hail, Egypt

Fatma H. Elfouly
Department of Communication and Computer Engineering, El-Shorouk Academy, El-Shorouk, Egypt

ABSTRACT

Wireless sensor networks (WSNs) may be described as a self-configured wireless networks that can be used to track physical objects or monitor environmental features, such as temperature or motion. The sensed data is then passed across the network to the main location or sink node, where the data can be processed and analyzed. Sensor nodes in WSN are fundamentally resource-constrained: they have restricted processing power, computing, space, and transmission bandwidth. Object tracking is considered as one of the major applications. However, many of the recent articles focused on object localization. In this chapter, the authors suggest an effective approach for tracking objects in WSNs. The aim is to achieve both minimal energy consumption in reporting activity and balanced energy consumption across the WSN lifetime extension of sensor nodes. Furthermore, data reliability is considered in our model. The chapter starts by formulating the multi-object tracking problem using 0/1 Integer Linear programming. In addition, the authors adopted the swarm intelligence technique to solve the optimization problem.

I. INTRODUCTION

Wireless Sensor Network (WSN) is a vast network of low-cost, battery driven, multipurpose, and tiny nodes. Nodes in this network can monitor physical conditions in the real world, such as temperature and humidity, carry out computations, and communicate via wireless media. They notify a base station named sink node of the data that has been gathered (Ilyas & Mahgoub, 2004). The design of WSN faces

DOI: 10.4018/978-1-7998-5788-4.ch012

various challenges where energy efficiency is one of the main challenges. Nonetheless, sensor nodes are subject to resource constraints such as battery power (or energy), bandwidth and storage. The most important resource is energy, because not only the quality of the sensor nodes that is affected by the limited energy, but also of the whole network (Ammari, 2009) . Communication energy consumption has been identified as the main energy source and costs significantly greater than sensing and data processing in WSNs (Pottie & Kaiser, 2000). Thus, energy conservation is considered as the most important performance objective when designing WSN routing protocols (Dahane & Berrached, 2019). Reliability in WSNs is a key aspect, which is always taken into account where multi-hop routing strategies are always used in WSNs and data packets are sent from one node to another (intermediate nodes) before reaching the sink node. Unexpected node loss or unreliable wireless connection (the action of radio links differs unpredictably over time and space) is normal at each hop (Tran-Thanh & Levendovszky, 2009) Baccour et al., 2009); this might lead to packet drop. By contrast, the loss of important information prevents the sensor network from achieving its primary purpose of data transfer (Xu & Wang, 2010). Therefore, routing strategies would emphasize effective transmission. Simultaneously, reducing the number of dropped packets in WSNs is important, and would increase network performance and energy consumption.

In the last decade, swarm-inspired optimization techniques have become more and more common (Shahin et al., 2014). Our research is autonomous and represents the behavior of swarms of collective insects like bees, flocks of birds or fish schools(Shi, 2018). Swarm smart systems are versatile, flexible, adaptable through basic behavior (McCune & Madey, 2014) like the shortest path finding that can solve complicated problems effectively. Ant Colony System (ACS) (El-Fouly & Ramadan, 2020) is one of the most relevant swarm intelligence systems to provide tentative solutions for problems of optimization in a fair period of time.

ACS (Gunes et al., 2002) (El-Fouly & Ramadan, 2020) has been influenced by the food finding behavior of real ants and can be used to find the shortest route in WSNs. This chapter concentrate on ACS which uses two artificial ant agents that is, Forward ant agent which moves from source to destination and find out details about quality of the road. Backward ant agent travel from target to source and gather knowledge regarding pheromone storage. In comparison to other routing strategies (Rocca et al., 2009), the ant colony optimization meta-heuristic suggested in the literature for WSNs is focused only on local knowledge of sensor nodes. That is, no routing tables or other information blocks must be transmitted to neighbors or other network nodes (Gunes et al., 2002) .

The potential implementations of WSNs include object detection, which have been one of the main use of WSNs. The target tracking has multiple real-life purposes such as forest animal control and combat field intrusion detection (Eswari & Vanitha, 2013). However, there are plenty of activities in this field (Vadakkan, & Raghavanpilla, 2015). The target detection method consists of two essential processes. The first is tracking, where sensor nodes are used to identify and control the movement conditions of the mobile object. The second process is monitoring, where nodes observing the target send their findings to the sink node (Chen et al., 2013) . Many object tracking researches concentrate on how to map the location of objects accurately and do not recognize several other criteria such as reliable data reporting (Chen & Liao, 2010) (Liu et al., 2010) (Mahboubi et al., 2012) (Liu et al., 2012) (Chen et al., 2013), nodes energy usage, and nodes energy balance. Therefore, in this chapter, we take these parameters collectively into consideration. We assume that considering these parameters would improve the overall efficiency of the WSNs as well as advance the object tracking process. To do so, our contributions in this chapter concentrate on: 1) formulating the object tracking problem into 0/1 integer programming with previously listed parameters, 2) the energy consumption in reporting activity for WSN lifetime extension, 3) balancing of energy consumption across sensor nodes to retain and balance of residual energy on sensor nodes as well, 4) enhancing data reliability where the sensed data can enter the sink node in a more stable way, and 5) introducing a Swarm Intelligence as a heuristic solution dependent energy reduction and reliability as well as load balancing.

The rest of this chapter is organized as follows: related work is discussed in the next section. Section 3 introduces the problem description and the problem formulation. In addition, the proposed approach is described in section 4. The simulation results are depicted in section 5. Finally, the conclusions are presented in section 6.

II. RELATED WORK

Beni and Wang (Beni & Wang, 1993) coined the word Swarm Intelligence in cellular robotic systems. Nowadays, SI is generally recognized as one branch of Computational Intelligence (CI) (Engelbrecht, 2018) . In thinking about SI, the five SI design concepts must be followed. Such principles have already been established by Milonas (Millonas M., 1993) in 1993 and include the principle of proximity, the principle of quality, the principle of diversity of reaction, the principle of flexibility and the principle of adaptability. The steady development of the research papers on the topic of SI indicates that SI is one of the most successful types of CI. There is no doubt that the SI advantages contribute to this(Olariu & Zomaya, 2013) . One of the most obvious benefits is autonomy. Swarm has no external oversight, but each actor in the swarm manages their actions independently. In this scenario, the actor is a potential answer to a given problem. From this basis, we can deduce the second benefit, self-organization. The intelligence is not based on the single agent, but is evolving in the swarm. Then the problem's solutions (agents) are not identified beforehand, but adjust themselves at operating program time. Self-organization plays a large part of adaptability. The above is evident in evolving settings, where agents react well to the described changes, adjust their behaviour and autonomously adapt. Moreover, because there is no central control, the swarm is stable, because there is no single point of failure. In addition, the swarm allows redundancy, in which two other benefits are hidden. The first is scalability, which means the swarm can consist of a few to thousands of agents and the control architecture remains the same. In addition, because there is no single agent necessary for swarm, the simplicity of the SI advantage is completely fulfilled.

Throughout this section, we concentrate only on the most relevant research to our proposal in this chapter. The section begins by describing the research discussed in(Lin & Lee, 2010) (Orojloo et al., 2012) (Lv, 2012) which are the more applicable research to our proposed solution followed by the variations from our proposal. Our proposal in this chapter will be compared in its performance to the previous work in (Lin & Lee, 2010) (Orojloo et al., 2012) (Lv, 2012) . Finally, explaining some of the related work done in object tracking and SWARM intelligence awareness. Finally, some of the similar research undertaken in object tracking and SWARM intelligence awareness are described. The suggested guideline in (Lin & Lee, 2010), reflects on how to reduce energy use in reporting activity. This proposed object tracking challenge is formulated as 0/1 numerical programming for optimum solution. Then, it suggested a lagrangian relaxation-based (LR-based) to solve the optimization problem. The cumulative communication cost is estimated and reduced by an object tracking tree.

While the proposed LR-based algorithm achieves energy efficient routing, it does not consider the real time reliable data transfer to the sink node. We assume that ignoring such problem could increase the amount of lost packets as well as may cause further energy consumption due to retransmission due to unreliable paths. Therefore, it does not understand how to align residual energy of sensor nodes. It could impact the load balancing on the nodes, as the nodes near to the sink would run out of energy early. In comparison, the network in (Lin & Lee, 2010) is believed to have two kinds of nodes, sensor and communication nodes. Sensor nodes used to detect and track mobile object activity status and communication nodes are used to store the identified object presence information and transfer it to the sink. Each communication node stores the object information sensed jointly by its pair of neighboring sensor nodes. In addition, this is a hard assumption: each communication node must have two neighboring nodes in mobile object mode, which might not always be the case. In fact, using lagrangian relaxation to solve optimization problem can have certain drawbacks, such as solution oscillation, which is a severe and intrinsic drawback. Also, it is sometimes difficult for lagrangian relaxation to find a feasible solution (Guan et al., 2002).

Swarm intelligence based energy balance routing (SEB) (Lv, 2014) (Nguyen et al., 2020) is a routing scheme that uses swarm intelligence to achieve balanced energy consumption across sensor nodes and extend the network lifetime of densely deployed WSNs. SEB should adjust the residual energy of the sensor nodes as much as possible due to their weight. It uses a meta-heuristics method to assign WSNs a suboptimal lifetime. The selection of the next-hop neighbor node when transmitting an event to the sink is determined by residual energy, distance to the sink, and node weight, as well as the pheromone environment responsible for determining the quality of the path. Nevertheless, in SEB, the environmental phenomena are believed to be spread equally. All source nodes, that is to say, have the same number of messages. So, the next hop node 's weight is related to the number of neighbor nodes that might choose to relay their messages. This could be inefficient when non-uniform distribution of the environmental events. Therefore, the route quality is linked to the latency of the message along the whole path. Based on the statement in (Lv, 2014), the path efficiency can be calculated as a function of hop count. Hop count is a basic metric to use and quick to determine, so it doesn't need additional effort to be calculated, in addition to providing a low overhead. However, it does not recognize the dynamic behavior of the wireless link efficiency over time and distance. It can be easily led to the use of low-quality links, which results in poor routes (Xu & Wang, 2010).

(Orojloo et al., 2012) Proposed an energy and path aware Ant Colony Optimization based routing algorithm for wireless sensor network (EPWSN). In EPWSN algorithm, for calculating the probability of selecting the next hop neighbor, the residual energy of node, pheromone value, and location of nodes are considered. In addition, in this algorithm a new pheromone update operator is designed to integrate energy consumption, hop count, and path length into a routing choice which can prolong the lifetime of the whole network. The authors of (Orojloo et al., 2012) see that the use of the presented probability function and new pheromone update operator, can increase the system efficiency, obtain more balanced transmission among sensor nodes, and reduce the length of path and energy consumption of the routing.

The authors of (Lv, 2012) suggested a new routing method for WSNs consisting of stable nodes based on Ant Colony Optimization (ACO) algorithm that explore the network and learn successful routes, utilizing a novel variant of reinforcement learning. The key goal of this ACO algorithm was to preserve full network existence, thus finding the shortest paths from the source nodes to the sink utilizing ACO based swarm intelligence. That is to say, to achieve lower energy consumption, more data packets should be transferred along shorter paths. Furthermore, the energy levels of the nodes are considered as well as the lengths of the paths where, in a routing task, the nodes which have more energy should be selected. This leads to improving the network total lifetime. Finally, a multi-path data transfer is also accomplished to provide reliable network operations where routing becomes more robust by performing acknowledgement-associated data transfers and keeping various paths alive. In other words, the source node resends the packet to different paths in the event of an absent acknowledgment for a data packet. This might increase the energy usage due to the same packet being retransmitted but guarantees the delivery of the packet in case of network problem.

From the research mentioned above (Orojloo et al., 2012) (Lv, 2012), we point out that certain problems are not addressed. The first is the path's reliability before sending the packet along the path to the sink node. As mentioned above, this could increase the amount of lost packets and packet retransmissions which affects the network performance. Secondly, node load which is an influential factor in the energy balance among sensor nodes from our point of view. That's to say, if more sensor nodes may prefer the same node to relay their messages, more resources should be allocated for a node with a heavier load. Therefore, taking residual energy and node load into consideration will manage residual energy effectively as stated in (Lv, 2014).

Our approach firstly formulates the object tracking task as 0/1 integer programming problem. Then, it develops a heuristic algorithm to construct an efficient object tracking in WSNs. We propose a novel protocol based on energy reduction, reliability, and energy balance routing in WSNs for object tracking. The difference from the previous work in (Lin & Lee, 2010) is that all the nodes in our network are considered as sensor nodes only. Sensor node could be used to sense the objects in the environment. At the same time, a sensor node could be used as a relay node where WSNs are usually based on a multi-hop transmission. Additionally, the work in this chapter does not consider the frequency of object movement of each pair of sensor nodes in the calculation of total communication cost. Instead, it considers the frequency of object movement at each sensor node which detects objects in the environment which is more realistic assumption. Furthermore, instead of Lagrangian relaxation, we propose SWARM intelligence which presents feasible and efficient solution (Shi, 2018).

unlike previous works in (Lin & Lee, 2010) (Orojloo et al., 2012) (Lv, 2012) (Lv, 2014), we consider the end-to-end reliability of a multi-hop route based on the Packet Reception Rate (PRR) which is one of the most commonly used reliability metrics (Zhu & Yi, 2012). In this model, the work analyzes the reliability of the whole path from the next hop node to the destination, and then chooses the relay node with the best PRR which results in high reliability instead of dropping packets.

Similar to SEB (Lv, 2014) and unlike the previous work in (Lin & Lee, 2010) (Orojloo et al., 2012) Lv, 2012), our proposed protocol can balance residual energy of the sensor nodes evenly according to their weight. The difference is that the nodes have different number of messages based on our model assumptions. A node's weight is used to express the total number of messages at its neighbor nodes which may choose it to relay their messages. In contrast to SEB, the pheromone value in our model reflects not only the path quality but also the path length. The path length at a fixed transmission range can be calculated by the number of nodes on the path (Gunes et al., 2002). Additionally, we don't use the delay of the message along the whole path to estimate path quality. Instead, the PRR can be used to improve the end-to-end reliability of a multi-hop route. Therefore, the local information in our model refers to each neighbor's residual energy, weight, distance to the sink, and pheromone. Sensor nodes' residual energy, weight, and pheromone change frequently and are exchanged dynamically with their one-hop neighbor nodes if necessary.

III. PROBLEM FORMALATION FOR OPTIMAL SOLUTION

We consider a WSN deployed in a field for the purpose of object tracking. Our objective is to propose a data reporting model for this kind of service, and consider reliable object tracking taking into consideration nodes energy consumption and the energy balancing. The object tracking problem is modeled as a graph based on the nodes location in the monitored environment and their characteristics. The efficient object tracking in WSNs problem can be modeled as a simple undirected graph, *G(V,L)*, where *V* is the set of sensor nodes in the network distributed in a two-dimensional plane and *L* is the set of all links *(i, j)* where, *i, j,є V* Link *(i, j)* exists if and only if $j \in NEB_i$, where NEB_i is the set of neighbors of node *i*. Assuming multi-objects object moving in the environment, they will be detected by some sensor nodes, which are called source nodes. The frequency of object movement at each source node differs according to the number of objects that are within the sensing range of each source node. The information about the presence of the detected objects at each source node should be reported to the sink node. Since WSNs are usually based on a multi-hop transmission, the source nodes send their information to the sink through intermediate sensor nodes which act as a relay nodes. The chosen path from each source node to the sink should be the best path which satisfies some constraints including 1) low communication cost, 2) its reliability greater than or equal target value, and 3) at the same time, sensor nodes on that path should have a maximum residual energy to weight ratio compared with their neighbours to balance energy among sensor nodes.

Based on the previous modeling of the object tracking problem, the problem can be solved optimally. In this section, the problem is mathematically formulated using Integer Linear Programming (ILP); then solved by any of the selected solver (Gunes et al., 2002) . This solution is used to guarantee the optimal solution, if any, to the previously described problem. However, due to the complexity of the problem and its constraints, it is expected and it is well known from the previous experiences in similar problems that no optimal solution could be found in some cases of the problem representation. Therefore, the mathematical formulation is used to solve small-scale problems as well as it is designed to fully understand the problem with its major constraints. In addition, the optimal solution for small-scale problems could be used to measure the quality of any given heuristic that might be used to solve the same problem. In fact, in the next section, the chapter presents a SWARM based optimization solution to the problem. The optimization is to generate decision for the network designer about data reporting operation. This solution is used for large-scale problems. To simplify the description of the problem and its formulation, the notations used to model the problem are given in Table 1. Our model notations.

Due to the use of multi-hop routing technique, the information about the detected objects at each source node should be transmitted as messages to the sink node through intermediate nodes or relay nodes. In order to achieve energy balanced routing, more energy should be reserved for a node with heavier weight. So, we consider a model in which the residual energy to weight ratio is used when choosing the relay node.

Now, let's start with the computation of the weight of a neighbor *j* by equation (1).

$$w_j = \begin{cases} \sum_{i \in NEB_j} Mes_i & if\ dis_j < dis_i \\ 0 & otherwise \end{cases} \quad (1)$$

Table 1. Our model notations

Given Parameters	
Notation	*Description*
S	The set of all sensor nodes that in sensing or sensing-relaying state.
R	The set of all sensor nodes that in relaying state accept sink node.
L	The set of all links, $(I, j) \epsilon L$ and $i \neq j$
C	The set of transmission cost $C_{(i,j)}$ associated with link (i, j).
PRR	The set of packet reception ratio $PRR_{(i,j)}$ associated with link (i, j).
Q	The target end-to end success probability.
wq	Constant value less than or equal 1.
Ds	The set of all messages corresponding to the detected objects at each source node s, $\forall_s \in S$
Dis_i	The distance between node i and sink, $i \epsilon S \cup R$
RE_j	The residual energy of each sensor node $j, j \epsilon NEB_i, NEB_i \epsilon S \cup R$
$se_{(i,j)}$	The energy required to do single hop transmission from i to j, $(i, j) \epsilon L$
Mes_i	The number of messages at node i, $i \epsilon S \cup R$
w_j	The weight of a neighbor $j, j \epsilon NEB_i NEB_i \epsilon S \cup R$
Ewr_j	The residual energy to weight ratio for each neighbor node $j, j \epsilon NEB_i NEB_i \epsilon S \cup R$
NRE_j	The ratio between RE_j and $se_{(i,j)}$ for each neighbor node $j, j \epsilon NEB_i NEB_i \epsilon S \cup R$
Ps	The set of all candidate paths between any pair (s, sink), $\forall_s \in S$
PRR_{Ps}	The set of PRR for all candidate paths between any pair (s, sink), $\forall_s \in S$
NEB_i	The set of neighbors of node i, $i \in S \cup R, NEB_i \in S \cup R$
Indicator **Parameter**	
s_j^p	The indicator function which is 1 if node j is on path p and 0 otherwise.
Decision Variables	
$t_{(i,j)}^{sd}$	1 if the source node s uses the link (i, j) to transmit message d through it to sink node and 0 otherwise, $\forall d \in D_s, \forall_s \in S, and\, (i,j) \in L$
$U_{(i,j)}^{sd}$	1 if the sensor node i uses node j to relay message d of the source node s and 0 otherwise, $\forall d \in D_s, \forall s \in S, i \in S \cup R, j \in NEB_i \in S \cup R$
x_p^{sd}	1 if the source node s select the path p to send message d to sink node and 0 otherwise, $\forall s \in S, p \in P_s, and\, \forall d \in D_s$
z_N	1 if the difference between the residual energy to weight ratio of sensor node j and N is less than zero and 0 otherwise, $\forall j \in NEB_i, N \in NEB_i - \{j\}, and\, NEB_i \in S \cup R$
m_j	1 if the sensor node j has a maximum residual energy to weight ratio compared with other neighbors and 0 otherwise, $\forall j \in NEB_i, and\, NEB_i \in S \cup R$
k_p^{sd}	1 if the selected path p for each source node s and message d has PRR greater than or equal to the target end-to-end success probability and 0 otherwise, $\forall s \in S, \forall d \in D_s, and\, p \in P_s$

Because the objects detected in the monitored environment distribute non-uniformly, node's weight can be defined as the total number of messages at its neighbour nodes which may choose it to relay their messages. Equation (1) means that node *i* may choose node *j* as next hop only when node *j* is closer to the sink than node *i*. This strategy guarantees that messages will get closer towards the sink via relaying and prevents forming a loop.

In addition, the residual energy to weight ratio for each node *j* is defined by equation (2) as follows:

$$Ewr_j = \begin{cases} \left(NRE_j - w_j\right) + 1 & if\ NRE_j \geq w_j \\ NRE_{j/wj} & if\ NRE_j < w_j \\ 0 & if\ w_j = 0 \end{cases} \quad (2)$$

To compute the shortest path from the source node to the sink node, the Dijkstra algorithm has been used. To fit Dijkstra into our formulation, the algorithm is represented mathematically as follows (Makariye, 2017):

The sensor nodes are being processed according to their order. The sensor nodes that are yet to be processed denoted by U, initially $U \in S \cup R$. When a sensor node *i* is processed, the following task is performed:

$$F(j) = \min\left\{F(j), D_{(i,j)} + F(j)\right\}, for\ all\ j \in NEB_i, NEB_i \in U \quad (3)$$

Where *F(j)* denotes the length of the shortest path from node *i* to node *j* which is initially equal to zero for the first processed node. When the sensor node *i* is processed, the *{F(j)}* values of its neighbours that have not yet been processed are updated in accordance with equation (3).

To complete the informal description of the algorithm, it is only necessary to specify the order in which the nodes are processed. The next node to be processed is one whose *F(j)* value is the smallest over all the unprocessed nodes as follows:

$$i = \arg\min\left\{F(j)\right\}, j \in U \quad (4)$$

Recalling that *U* denotes the set of unprocessed nodes, Thus after node *i* is processed it is immediately deleted from *U*. so, *U=U - {i}*

The total communication cost for a graph *G* and object tracking tree *T* is defined as the sum of the individual contributions of all source and relay nodes in *G*:

$$Total\ communication\ \cos t(G,T) = \sum_{s \in S} \sum_{d \in D} \sum_{(i,j) \in L} t^{ds}_{(i,j)} C_{(i,j)} \quad (5)$$

Based on these computations the problem is formulated as follows:

The multi-objective function:

$$Z_{IP} = \min \sum_{s\in S} \sum_{d\in D} \sum_{(i,j)\in L} t^{ds}_{(i,j)\in L} C_{(i,j)} \text{ (IP)}$$

Subject to:

$$\sum_{p\in P_I} x^{sd}_{p} = 1 \qquad \forall s \in S, \forall d \in D_s \tag{6}$$

The constraint defined by equation (6) is a routing constraint which is used to guarantee that any source node s must choose only one path p to the sink node.

$$\sum_{i\in S\cup R} U^{sd}_{(i,j)} = 1 \qquad \forall s \in S, \forall d \in D_s, j \in NEB_i, NEB_i \in S \cup R \tag{7}$$

Another routing constraint is defined by equation (7) which is used to avoid cycles where, the use of any node j as a relay node for the same source node s and message d is equal to 1, except the sink node.

$$\sum_{N\in NEB_i - \{j\}} z_N \left(Ewr_j - Ewr_N\right) < 0 \quad \forall j \in NEB_i, NEB_i, \in S \cup R \tag{8}$$

$$2 - \sum_{N\in NEB_i - \{j\}} z_N = m_j + 1 \qquad \forall j \in NEB_i, NEB_i \in S \cup R \tag{9}$$

$$\sum_{j\in NEB_i - \{j\}} z_N \leq 1 \qquad \forall j \in NEB_i, NEB_i \in S \cup R \tag{10}$$

The constraints defined by equations (8), (9), and (10) are the energy constraints which are used to maintain higher and balance residual energy on nodes. Any node i must choose only one node j from its neighbors which has the highest residual energy to weight ratio compared with other neighbor nodes.

$$\sum_{p\in P_s} k^{sd}_p PRR_p \geq Q \qquad \forall s \in S, \forall d \in D_s, PRR_p \in PRRp_s \tag{11}$$

$$k^{sd}_p + 1 \leq x^{sd}_p + 1 \qquad \forall p \in P_s, s \in S, d \in D_s \tag{12}$$

Equations (11) and (12) define reliability constraints which are used to guarantee that the selected path p for the source node s and message d has a *PRR* greater than or equal the target end-to-end success probability.

Once the path p is selected, and the *PRR* of that path is greater than or equal the target end-to-end success probability. Node j is on the path and has the highest residual energy to weight ratio compared with other neighbour nodes, then the decision variable $U_{(i,j)}^{sd}$ must be enforced equal to 1 as described in the routing constraint defined by equation (13) as follows:

$$\sum_{p\in P_s}\sum_{j\in NEB_i} s_j^p x_p^{sd} m_j k_p^{sd} \leq U_{(i,j)}^{sd} \quad \forall s\in S, \forall d \in D_s, i \in S\cup R, NEB_i \in S \cup R \tag{13}$$

$$\sum_{J\in NEB_i} m_j \leq 1 \quad NEB_i \in S\cup R \tag{14}$$

The constraint defined by equation (14) is a routing constraint which is used to guarantee that any node *i* must choose only one node *j* from its neighbours.

$$\sum_{(i,j)\in L} t_{(i,j)}^{sd} \geq 1 \quad \forall s\in S, \forall d\in D_s \tag{15}$$

The constraint defined by equation (15) is a redundancy constraint.

The decision variable constraints are defined by the following equations:

$$x_p^{sd} = 0\, or\, 1 \quad \forall s\in S, \forall d\in D_s, p\in P_s \tag{16}$$

$$U_{(i,j)}^{sd} = 0\, or\, 1 \quad \forall s\in S, j\in NEB_i, NEB_i \in S\cup R, \forall d\in D_s \tag{17}$$

$$m_j = 0\, or\, 1 \quad \forall j\in NEB_i, NEB_i\in S\cup R \tag{18}$$

$$k_p^{sd} = 0\, or\, 1 \quad \forall p\in P_s, \forall s\in S, \forall d\in D_s \tag{19}$$

$$t_{(i,j)}^{sd} = 0\, or\, 1 \quad \forall s\in S, \forall d\in D_s, (i,j)\in L \tag{20}$$

IV. SWARM BASED SOLUTION

This section describes the second solution approach for the reliable object tracking problem which is Swarm Intelligence techniques. Using swarm intelligence to solve the hard problem has several advantages (Sharvani et al., 2009):

1. Fast route recovery
2. Fault tolerance
3. Scalability and adaptation

Due to the use of multi-hop routing, it is essential that sensor nodes select a neighbour node as the closest hop to the sink. The probability of selecting neighbour *j* at node *i* as the closest hop is given in equation (21).

$$P_r(i,j) = f(x) = \begin{cases} \dfrac{f_m(p_{ij}, n_{ij})}{\sum\limits_{j \in NEB_i} f_n(p_{ij}, n_{ij})} & if\ j \in NEB_i \\ 0 & if\ j \notin NEB_i \end{cases} \tag{21}$$

Where p_{ij} is the pheromone value corresponding to neighbor *j* at node *i*, $0 < p_{ij} \leq 0$ and η_{ij} is the local heuristic value of link *(i,j)* for node *j* (Lv, 2014).

A. The Routing Scheme

At the initial phase of our model, sensor nodes collect the information of their neighbour nodes using local broadcast and estimate their weight using equations (1) and (2) described in section 4. After the initial phase, our model works as follow:

(i) When the source node detects an object, it generates a message and encapsulates into a forward data packet (*FDATA*).
(ii) The *FDATA* is transmitted to the neighbor node which is selected according to equation (31) described in section (D).
(iii) The information of the relay nodes required to estimate pheromone values is appended to the path field in *FDATA*.
(iv) Repeat (ii) and (iii) until *FDATA* reaches the sink.
(v) When the sink node receives *FDATA*, it determines the increasing density of the pheromone value which reflects the relative path quality and length using information contained in the *FDATA* according to equation (29) described in section (B).
(vi) The sink node converts *FDATA* into backward data (*BDATA*) packet and sends it to the source node along the same path in the opposite direction. *BDATA* contains collected path information.
(vii) The relay nodes in the path receive *BDATA* and update their pheromone according to equation (30) described in section (B).
(viii) When the source node receives *BDATA* discards it.

B. Calculation of Local Heuristic Information

In order to maintain higher and balance residual energy on sensor nodes, the combination of residual energy and weight which denoted by n_{ij}^1 can be used as information when choosing a neighbour node as next hop.

$$n_{ij}^{1} = \frac{Ewr_i}{\sum_{j \in NEB_i} Ewr_j} \tag{22}$$

A neighbour node j of node i with greater value of n_{ij}^{1} indicates that it has more energy compared to its weight. So, it will have a much better chance to be selected as next hop.

In order to achieve energy efficient routing, each node i should select a neighbour node j which is closer to the sink as next hop. The distance to the sink from a neighbour node j to the sink node can be used as heuristic information which is denoted by n_{ij}^{2}.

$$n_{ij}^{2} = \frac{1 / D_j}{\sum_{j \in NEB_i} 1 / D_j} \tag{23}$$

A neighbor node j at node i with greater value of n_{ij}^{2} indicates that it is closer to the sink than others. So it will have a much better chance to be selected as next hop.

C. Pheromone Calculation

The *FDATA* is sent from the source node to the sink along a loop-free path p, and it is responsible for estimating the pheromone value which reflects the length of the path, as well as the path quality. The quality of the path p, q_p, is related to the *PRR* and can be expressed as:

$$q_p = PRR_p q_p \tag{25}$$

Since every sensor node has a fixed transmission range, the length of the path p, L_p can be estimated by the number of nodes on the path p as follow:

$$L_p = n_p \tag{26}$$

By estimating every possible path from the same source to the sink, the best path length L_{pbest} is recorded at the sink. The relative quality of path p is defined as the ratio of q_p and wq as follows:

$$Rq_p = \begin{cases} q_p / wq & if\ q_p \leq wq \\ 1 & if\ q_p > wq \end{cases} \tag{27}$$

The relative length of path p is defined as the ratio of L_p and L_{pbest} as follows:

$$Rl_p = n_{p\,best} / n_p \tag{28}$$

The increasing density of pheromone on the path *P* is defined as follows:

$$\Delta p_{ij} = \frac{1/2\left(Rl_p + Rq_p^2\right)}{n_p} \tag{29}$$

The sink node constructs the value of Δp_{ij} which is contained in *BDATA* and sent it back to the source node along the reverse path. When the relay node *i* receives *BDATA*, it updates its pheromone value according to the following rule:

$$p_{ij} = (1-p)p_{ij} + \Delta p_{ij} \tag{30}$$

Where, $p \in (0,1)$ is the evaporation constant that determines the evaporation rate of the pheromone (Lv, 2014).

D. Probability of Neighbour Nodes

Based on the pheromone p_{ij} described in the previous section and the local heuristic information n_{ij}^1 and n_{ij}^2 described in section B, the probability of selecting a neighbor node *j* as a next hop at node *i* is determined as follows:

$$p_r(i,j) = \frac{\left(p_{ij}\right)^a \left(n_{ij}^1\right)^\beta \left(n_{ij}^2\right)^\gamma}{\sum_{j \in NEB_i} \left(p_{ij}\right)^a \left(n_{ij}^1\right)^\beta \left(n_{ij}^2\right)^\gamma} \tag{31}$$

Where α, γ and β are the weight factors that control the pheromone value and the heuristic information parameter respectively.

V. SIMULATION RESULTS

In this section, different experiments are conducted to show the performance of our proposals in this chapter. It is assumed that the sensor network consists of homogeneous sensor nodes, the sink node, and sensor nodes are stationary after being deployed in the field and the sink node is not limited in terms of energy, memory, and computational power. Furthermore, The sink node is placed at (1000,500). All the later experiments are performed with multi-object detection. In addition, it is assumed that the number of the detected objects by each source is from 1 to 10 objects randomly in the case of multi-object detection. The running time of each round during data gathering is 200 Sec. The section starts by describing the simulation setup followed by the different experiments with different settings. Our experiments include the analysis and the performance evaluation of 1) Our method of how to report information about the detected objects from each source node to the sink compared to the previous work stated in (Lv, 2014), 2) Our SWARM approach in terms of network lifetime compared to the previous work in (Lin & Lee, 2010) (Orojloo et al., 2012) (Lv, 2014) (Lv, 2012) for homogenous network with multi-object detection, 3) Our SWARM approach in terms of network obtained reliability compared to the previous work in (Lin & Lee, 2010) (Orojloo et al., 2012) (Lv, 2012) (Lv, 2014) for homogenous network with multi-object detection, and 4) Our SWARM approach in terms of energy balance compared to the previous work in (Lin & Lee, 2010) (Orojloo et al., 2012) (Lv, 2012) (Lv, 2014) for homogenous network with multi-object detection.

A. Simulation Setup

All of the experiments in this section are analyzed using our Matlab 7.5 program as optimization toolbox. Experiments are conducted on a i5-2.3 GHz laptop running Microsoft Windows 7. Nodes are distributed randomly in a square region, see figure 1. In this simulation, we choose a harsh wireless channel Model, which includes shadowing and deep fading affects, as well as the noise. These setting are suitable for both indoor and outdoor environment (Zamalloa & Krishnamachari, 2007). Table 2 Simulation environment parameters shows the used simulation environment parameters.

In the later experiments, we use the combination 132 ($\alpha = 1, \beta = 3, \gamma = 2$), since the evaluation results show this being the best for all experiments.

Table 2. Simulation environment parameters

Parameters	values
Network size	1000×1000
Number of nodes	80
Number of sink nodes	1
Node placement	Random uniform
Packet size	64 byte
Frequency	2400 MHz
Transmission power	-5dBm
Maximum transmission range	223 m
Channel model	Log-normal shadow
Path loss exponent	6
Shadow fading variance	6
Noise power	-145dBm
Reference distance	3 m

Figure 1. Random deployment with different communication ranges

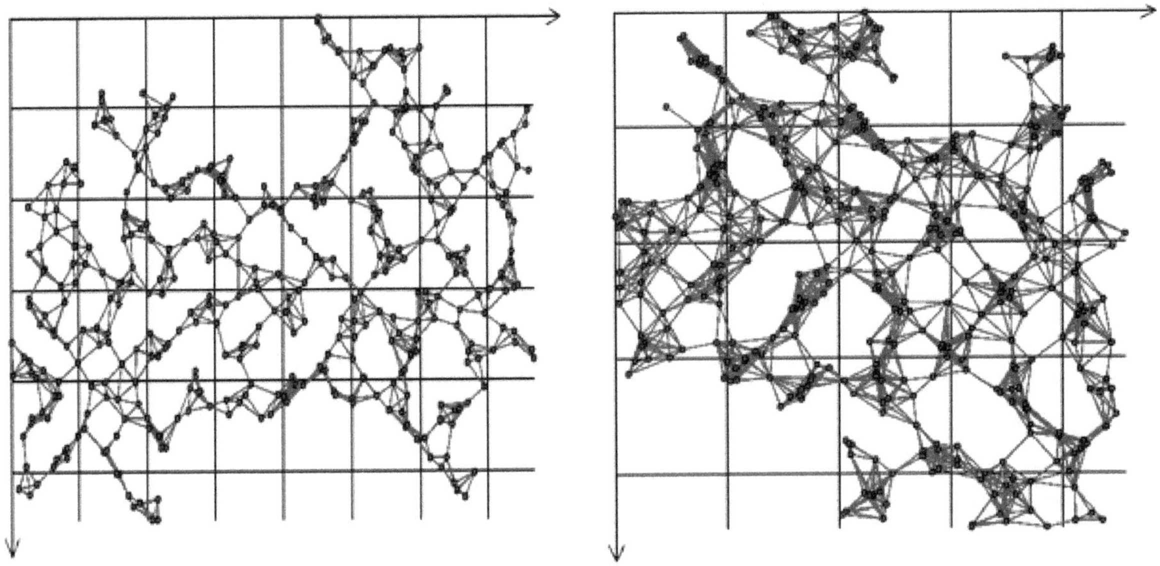

(a) (b)

B. SWARM Approach Performance

In this experiment, we evaluate the performance of our method of how to report information about the detected objects from each source node to the sink compared to the LR-based algorithm (Lin & Lee, 2010) at target reliability equal to 0.9. Assuming that multi-objects are moving in the environment and detected by four sensor nodes S1, S2, S3, and S4 as shown in Fig.1. The scenario of the objects movement is as follows:

1. Five objects move from S1 to S2.
2. Three objects move from S2 to S3.
3. Two objects move from S2 to S4.
4. Two objects move from S3 to S4.

As can be seen in Fig. 2.a, due to the hard assumption in the LR-based algorithm (Lin & Lee, 2010), only information about objects movement from S3 to S4 are reported by S4 to the sink node. Information about other objects movement is not reported to the sink node.

As can be seen in Fig. 2.b, in our approach, information about all detected objects is reported to the sink node. Therefore, the object tracking method in our work is more efficient than the previous work as well as, the reliability and energy balance are considered in our approach. As described above, the object tracking method in LR-based algorithm (Lin & Lee, 2010) is not efficient. Therefore, in all later experiments to compare the performance of our approach with the LR-based algorithm (Lin & Lee, 2010), we consider all the nodes in the network are sensor nodes and each node that detects an object will send its information to the sink node.

Figure 2. Data reporting method for our SWARM approach and the LR-based algorithm (Lin & Lee, 2010).

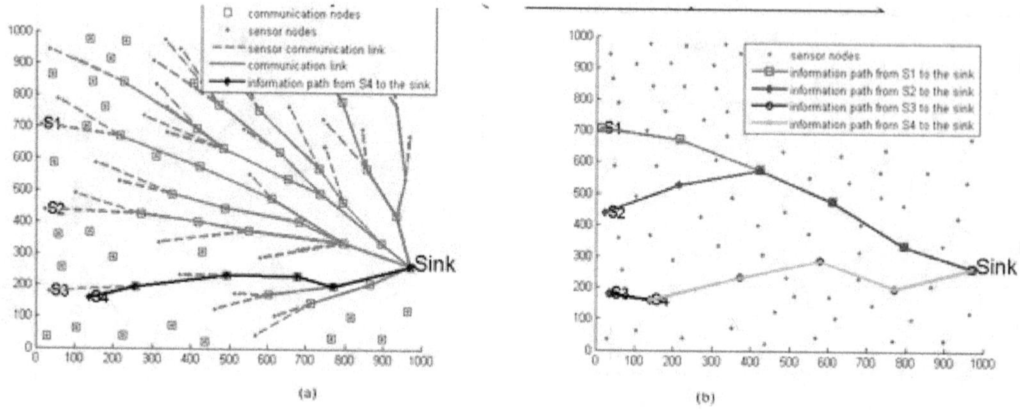

C. Network Lifetime Evaluation

In this set of experiments, we evaluate the performance of our SWARM approach in terms of network lifetime compared to the LR-based algorithm (Lin & Lee, 2010), SEB (Lv, 2014), EPWSN (Orojloo et al., 2012), and ACO proposed in (Lv, 2012) under different number of source nodes ranging from 5 to 15. It is assumed that the initial energy of each sensor node is 48mJ. Fig.2 shows the network lifetime for different number of source node ranging from 5 to 15. The network lifetime decreases with the increment of source nodes. However, Fig.3 shows a useful situation that the network lifetime using our approach is longer than that using SEB (Lv, 2014), EPWSN (Orojloo et al., 2012), and ACO (Lv, 2012). This is because our approach balances energy consumption more efficiently than SEB (Lv, 2014), EPWSN (Orojloo et al., 2012), and ACO proposed in (Lv, 2012) and considers reliability but SEB (Lv, 2014), EPWSN (Orojloo et al., 2012), and ACO proposed in (Lv, 2012) don't. In addition, the LR-based algorithm (Lin & Lee, 2010) considers only the minimization of total communication cost which leads to a much lower network lifetime than our approach. This is because our approach considers nodes' energy balance and reliability but the LR-based algorithm (Lin & Lee, 2010) doesn't. For instance, at 10 source nodes our SWARM achieves about 115%, 48%, 38%, and 102% improvement in network lifetime compared to the previous works in (Lin & Lee, 2010) (Orojloo et al., 2012) (Lv, 2012) (Lv, 2014) respectively.

D. Network Reliability Evaluation

In this experiment, we evaluate the performance of our SWARM approach in terms of network reliability compared to the LR-based algorithm (Lin & Lee, 2010), SEB (Lv, 2014), EPWSN (Orojloo et al., 2012), and ACO proposed in (Lv, 2012) under different number of source nodes ranging from 5 to 15. It's assumed that the initial energy of each sensor node is 48mJ. Fig. 4 shows the obtained reliability under different number of source nodes. As can be seen in the figure, the total obtained reliability using our approach is larger than that using the previous work in (Lin & Lee, 2010) (Orojloo et al., 2012) (Lv, 2012) (Lv, 2014). This is because, our approach consider data transmission in a more reliable way. For instance, at 10 source nodes our SWARM achieves about 58%, 27%, 39%, and 39% improvement in network lifetime compared to the previous work in (Lin & Lee, 2010) (Orojloo et al., 2012) (Lv, 2012) (Lv, 2014) respectively.

Figure 3. The number of source node vs. Network lifetime for homogenous network with multi-object detection.

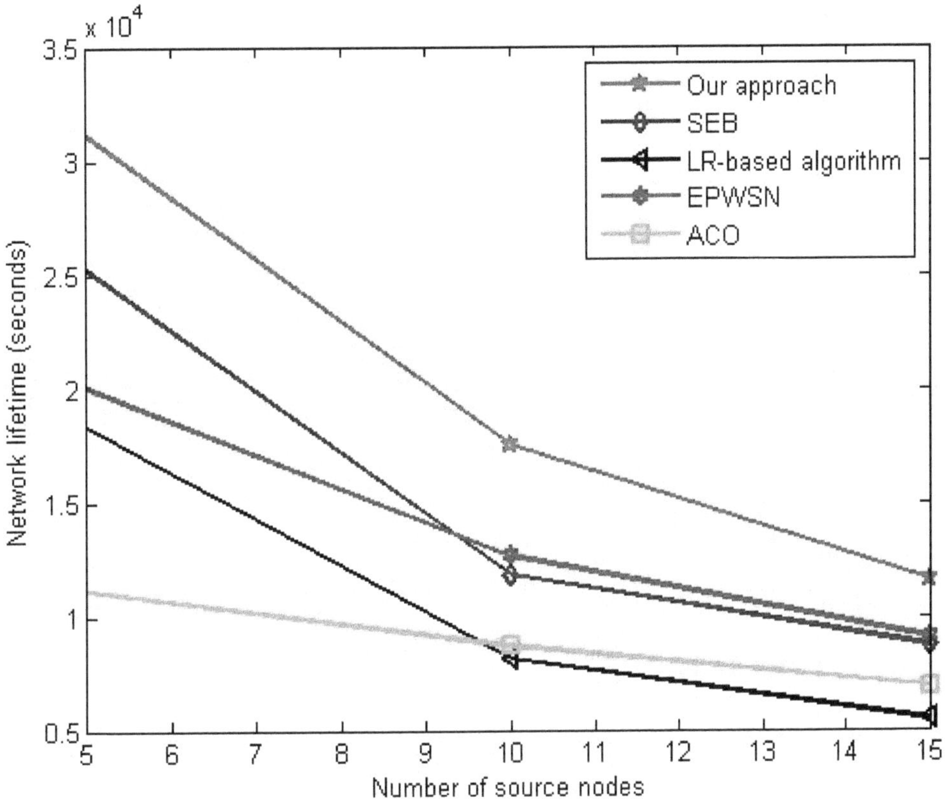

E. Energy Balancing Evaluation

In this experiment, we evaluate the performance of our SWARM approach in terms of energy balance compared to the LR-based algorithm (Lin & Lee, 2010) , SEB (Lv, 2014), EPWSN (Orojloo et al., 2012), and ACO proposed in (Lv, 2012). The minimum residual energy was calculated during running time in order to find the energy balance efficiency of the network. Where, increasing the minimal residual energy of sensor nodes means that the energy balance is achieved. In this set of experiments, it is assumed that 10 sensor nodes can detect the events in the environment and the initial energy of each sensor node is 48mJ. Fig. 5 shows the minimum residual energy verses network lifetime. As shown in Fig.4 the balanced distribution of residual energy using our approach is more efficient than the previous works. After approximately 20% of network lifetime the minimum residual energy for our approach was about 27.82 mJ while were 6.45, 16.6, 24.02,and 9.27 mJ for the previous work in (Lin & Lee, 2010) (Orojloo et al., 2012) (Lv, 2012) (Lv, 2014) , respectively.

Figure 4. Network Reliability vs. No. of source nodes for homogeneous network with multi-object detection.

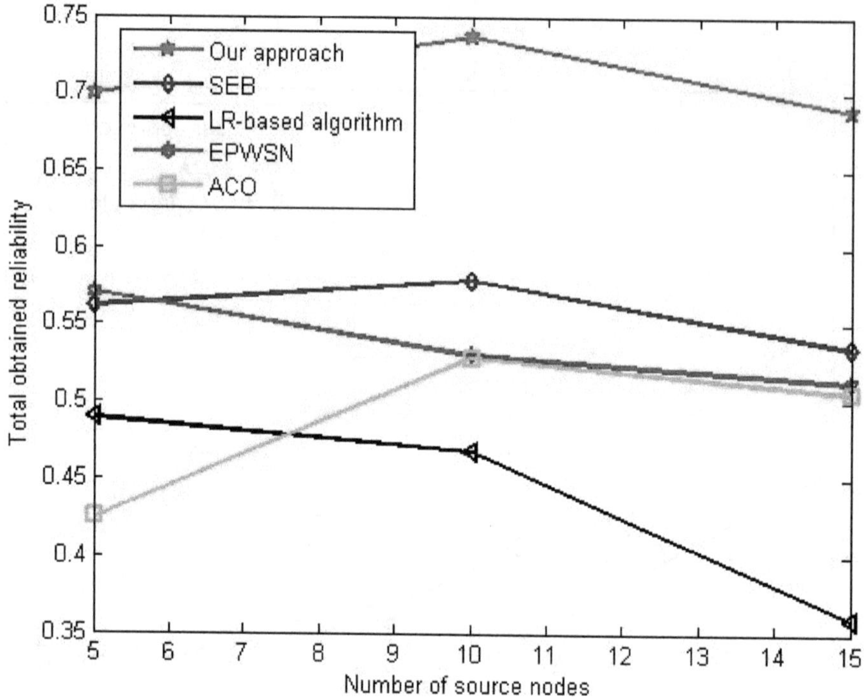

Figure 5. Minimum residual energy vs. Network lifetime for homogeneous network with multi-object detection.

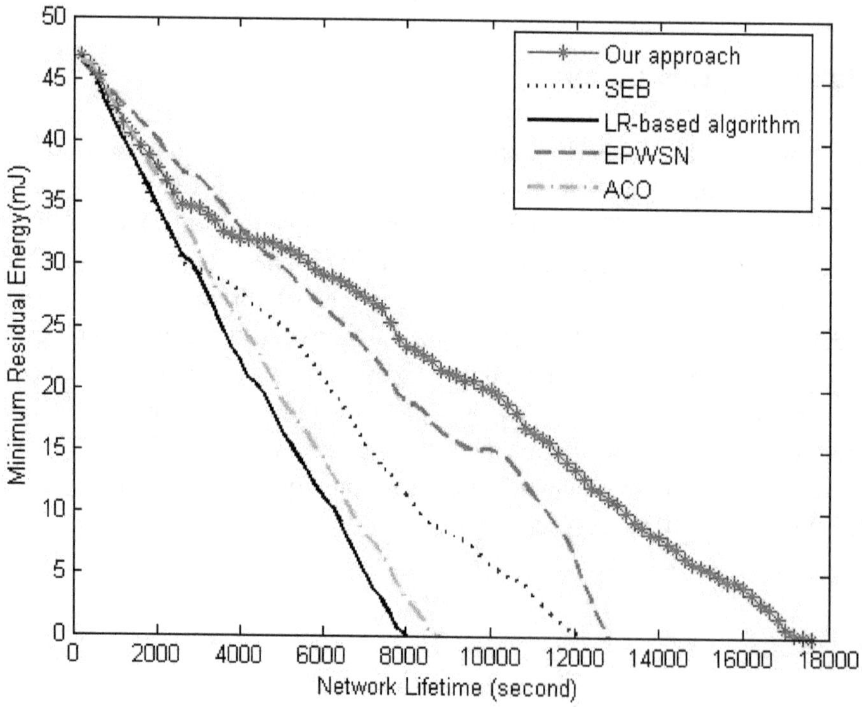

VI. CONCLUSION

An effective method for the detection of objects in WSNs is proposed in this study. In the data reporting phase, the proposed approach not only reduces energy consumption but also balances it between sensor nodes in order to extend the life of the network. Furthermore, the sensed data supplied the reliability needed to the sink. This research formulates the problem as a question of 0/1 integer programming and then proposes SWARM knowledge to address the problem of optimization. Compared to previous works relating to our subject, such as LR-based object tracking algorithm, SEB, EPWSN, and ACO, the output of the proposed method is evaluated and analyzed by simulation. It has been observed that the method of reporting information about the detected objects from each source node to the sink in the proposed approach outperformed the LR-based object tracking algorithm according to the simulation results. Finally, the findings of the simulation revealed that our solution is stable and achieves longer network existence relative to previous works such as the LR-based entity tracking algorithm, SEB, EPWSN and ACO.

REFERENCES

Ammari, H. M. (2009). *Challenges and opportunities of connected K-covered wireless sensor networks: From sensor deployment to data gathering*. Springer Science & Business Media.

Baccour, N., Koubaa, A., Ben Jamaa, M., Youssef, H., Zuniga, M., & Alves, M. (2009). A comparative simulation study of link quality estimators in wireless sensor networks. *2009 IEEE International Symposium on Modeling, Analysis & Simulation of Computer and Telecommunication Systems*. 10.1109/MASCOT.2009.5366798

Beni, G., & Wang, J. (1993). Swarm intelligence in cellular robotic systems. *Robots and Biological Systems: Towards a New Bionics,* 703-712. doi:10.1007/978-3-642-58069-7_38

Chen, C., & Liao, C. (2010). Model-based object tracking in wireless sensor networks. *Wireless Networks, 17*(2), 549–565. doi:10.100711276-010-0296-5

Chen, Y., Lin, Y., & Sun, T. (2013). A prediction scheme for object tracking in grid wireless sensor networks. *2013 Seventh International Conference on Innovative Mobile and Internet Services in Ubiquitous Computing*. 10.1109/IMIS.2013.67

Dahane, A., & Berrached, N. (2019). *Wireless sensor networks: A survey*. Wireless and Sensor Networks., doi:10.1201/9781351190756-1

El-Fouly, F. & Ramadan, R. (2020). Optimal and Meta-Heuristic Algorithms for Object Tracking in Multi-Sink Wireless Sensor Networks. *WAS Science Nature Journal*, (1), 20-38.

Engelbrecht. (2018). *Computational intelligence: An introduction* (3rd ed.). Wiley-Blackwell.

Eswari, T., & Vanitha, V. (2013). A novel rule based intrusion detection framework for wireless sensor networks. *2013 International Conference on Information Communication and Embedded Systems (ICICES)*. 10.1109/ICICES.2013.6508172

Guan, X., Zhai, Q., & Lai, F. (2002). New lagrangian relaxation based algorithm for resource scheduling with homogeneous Subproblems. *Journal of Optimization Theory and Applications, 113*(1), 65–82. doi:10.1023/A:1014805213554

Gunes, M., Sorges, U., & Bouazizi, I. (2002). ARA-the ant-colony based routing algorithm for MANETs. *Proceedings. International Conference on Parallel Processing Workshop*. 10.1109/ICPPW.2002.1039715

Ilyas, M., & Mahgoub, I. (2004). *Handbook of sensor networks: Compact wireless and wired sensing systems*. CRC Press.

Kim, K., Park, S., Park, H., & Ham, Y. H. (2008). Reliable and real-time data dissemination in wireless sensor networks. *MILCOM 2008 - 2008 IEEE Military Communications Conference*. doi:10.1109/MILCOM.2008.4753643

Lin, F. Y., & Lee, C. (2010). An efficient Lagrangean relaxation-based object tracking algorithm in wireless sensor networks. *Sensors (Basel), 10*(9), 8101–8118. doi:10.3390100908101 PMID:22163643

Liu, L., Zhang, X., & Ma, H. (2010). Optimal node selection for target localization in wireless camera sensor networks. *IEEE Transactions on Vehicular Technology, 59*(7), 3562–3576. doi:10.1109/TVT.2009.2031454

Liu, T., Liu, Y., Cui, X., Xu, G., & Qian, D. (2012). MOLTS: Mobile object localization and tracking system based on wireless sensor networks. *2012 IEEE Seventh International Conference on Networking, Architecture, and Storage*. 10.1109/NAS.2012.34

Lv, Y. (2012). Routing in wireless sensor networks using swarm intelligence. *Proceedings of the 10th World Congress on Intelligent Control and Automation*. 10.1109/WCICA.2012.6357964

Lv, Y. (2014). An energy efficient routing based on swarm intelligence for wireless sensor networks. *Journal of Software, 9*(10). Advance online publication. doi:10.4304/jsw.9.10.2758-2763

Mahboubi, H., Momeni, A., Aghdam, A. G., Sayrafian-Pour, K., & Marbukh, V. (2012). An efficient target monitoring scheme with controlled node mobility for sensor networks. *IEEE Transactions on Control Systems Technology, 20*(6), 1522–1532. doi:10.1109/TCST.2011.2167151

Makariye, N. (2017). Towards shortest path computation using Dijkstra algorithm. *2017 International Conference on IoT and Application (ICIOT)*. 10.1109/ICIOTA.2017.8073641

McCune, R. R., & Madey, G. R. (2014). Control of artificial swarms with DDDAS. *Procedia Computer Science, 29*, 1171–1181. doi:10.1016/j.procs.2014.05.105

Millonas, M. (1993). Swarms, Phase Transitions, and Collective Intelligence. Addison Wesley. arXiv:adap-org/9306002

Nguyen, B. H., Xue, B., & Zhang, M. (2020). A survey on swarm intelligence approaches to feature selection in data mining. *Swarm and Evolutionary Computation, 54*, 100663. doi:10.1016/j.swevo.2020.100663

Olariu, S., & Zomaya, A. Y. (2013). *Handbook of Bioinspired algorithms and application* (3rd ed.). Chapman and Hall/CRC.

Orojloo, H., Moghadam, R. A., & Haghighat, A. T. (2012). Energy and path aware ant colony optimization based routing algorithm for wireless sensor networks. *Communications in Computer and Information Science*, *269*, 182–191. doi:10.1007/978-3-642-29219-4_22

Pottie, G. J., & Kaiser, W. J. (2000). Wireless integrated network sensors. *Communications of the ACM*, *43*(5), 51–58. doi:10.1145/332833.332838

Rocca, P., Benedetti, M., Donelli, M., Franceschini, D., & Massa, A. (2009). Evolutionary optimization as applied to inverse scattering problems. *Inverse Problems*, *25*(12), 123003. doi:10.1088/0266-5611/25/12/123003

Shahin, M., Saied, E., Hassan, M. M., & Bendary, F. (2014). Voltage swell mitigation using flexible AC transmission systems based on evolutionary computing methods. *International Journal of System Dynamics Applications*, *3*(3), 73–95. doi:10.4018/ijsda.2014070104

Sharvani, S., Cauvery, C., & Rangaswamy, T. (2009). Different types of swarm intelligence algorithm for routing. *2009 International Conference on Advances in Recent Technologies in Communication and Computing*. 10.1109/ARTCom.2009.157

Shi, Y. (2018). Unified swarm intelligence algorithms. *Critical Developments and Applications of Swarm Intelligence*, 1-26. doi:10.4018/978-1-5225-5134-8.ch001

Vadakkan, J. J., & Raghavanpilla, V. (2015). An energy efficient multi-target tracking in wireless sensor networks based on polygon tracking method. *International Journal of Emerging Trends in Science and Technology*. doi:10.18535/ijetst/v2i8.03

Xu, B., & Wang, H. (2010). A reliability transmission routing metric algorithm for wireless sensor network. *2010 International Conference on E-Health Networking Digital Ecosystems and Technologies (EDT)*. 10.1109/EDT.2010.5496534

Xu, B., & Wang, H. (2010). A reliability transmission routing metric algorithm for wireless sensor network. *2010 International Conference on E-Health Networking Digital Ecosystems and Technologies (EDT)*. 10.1109/EDT.2010.5496534

Zamalloa, M. Z., & Krishnamachari, B. (2007). An analysis of unreliability and asymmetry in low-power wireless links. *ACM Transactions on Sensor Networks*, *3*(2), 7. doi:10.1145/1240226.1240227

Zhu, T., & Yi, P. (2012). Reliable and energy-efficient networking protocol design in wireless sensor networks. *Intelligent Sensor Networks*, 411-426. doi:10.1201/b14300-22

Chapter 13
Hybrid Multi-Objective Grey Wolf Search Optimizer and Machine Learning Approach for Software Bug Prediction

Mrutyunjaya Panda
https://orcid.org/0000-0001-5713-9220
Utkal University, India

Ahmad Taher Azar
https://orcid.org/0000-0002-7869-6373
Faculty of Computers and Artificial Intelligence, Benha University, Benha, Egypt & College of Computer and Information Sciences, Prince Sultan University, Riyadh, Saudi Arabia

ABSTRACT

Software bugs (or malfunctions) pose a serious threat to software developers with many known and unknown bugs that may be vulnerable to computer systems, demanding new methods, analysis, and techniques for efficient bug detection and repair of new unseen programs at a later stage. This chapter uses evolutionary grey wolf (GW) search optimization as a feature selection technique to improve classifier efficiency. It is also envisaged that software error detection would consider the nature of the error when repairing it for remedial action instead of simply finding it either faulty or non-defective. To address this problem, the authors use bug severity multi-class classification to build an efficient and robust prediction model using multilayer perceptron (MLP), logistic regression (LR), and random forest (RF) for bug severity classification. Both tests are performed on two software error datasets, namely Ant 1.7 and Tomcat.

DOI: 10.4018/978-1-7998-5788-4.ch013

1. INTRODUCTION

The defect in the software product appears to be the one where the code developed does not meet the requirements of the end-user. Such a defect in the software product is referred to as a bug in the code that prevents the product from working properly. Looking into today's digital age, a human being totally dependent on software and a little bug (or defect) may have serious consequences in human life (Rana et al., 2015). Software testing and bug fixing is a precondition for software delivery that takes almost 50% to 60% of the total effort required for software development (Mishra & Mishra, 2009). Early identification of such software bugs (or defects) with defective modules saves a huge amount of effort and cost. Since the software product is too large, complex and versatile, checking all test classes after a code change is not a good idea, but software metrics such as code-based matrices and/or software change matrices can be used to test faulty software modules (Choudhary et al., 2018; Malhotra, 2016). While code-based matrices are used to find the size and complexity of the code, there is a change between two versions of the software in the software change matrices. Researchers use either the full or subset of features of these software matrices to build an efficient and accurate model to enhance the quality of the software (Gondra, 2008). It is important to develop a quality software product by accurately detecting software bugs at the early stage of the software development life cycle and then taking action to remove them.

Software development takes place at the following stages while handling software bugs in code (Grishma & Anjali, 2015): a) bug spotting, b) bug assortment, c) bug scanning, d) bug prediction and e) bug removal. The first stage refers to the occurrence of software defects, if any, by means of a code inspection, of such a defect and then a test to detect malfunctions in the software. After this initial identification step, the bug assortment is done for categorization followed by a bug scan to analyze the details of the present bug. Then, in the last two stages, make software bug prediction and removal using some promising methods to improve the quality of software.

It is observed that the prevalent seed of finding bugs in the software product includes: user coolness, ambiguous objectives and puzzling objectives, imperfect specification, lack of assets, lack of communication between group members and poor testing (Venkata et al., 2005; Rajkumar & Alagarsamy, 2013). As per IEEE 104 standard, software fault proneness may be classified as falling within any of the following (Mikyeong & Hong, 2014): (a) bug or defect resulting from the failure of the developer to meet the requirements of the end-user; (b) failure to comply with the previous required task of the software products; or incorrect results are received by the customer for each input entered; (c) error occurs by c Several techniques have been proposed by many researchers to detect software bugs based on their types in the recent past (Jia & Han, 2013).

Feature selection algorithms are dimensionality reduction methods often associated to data mining tasks of classification or clustering (Kaen & Algarni, 2019; Czibula et al., 2016; Hassanien et al., 2015; Xu et al., 2000; Emary et al., 2014a; Banu et al., 2014; Aziz et al., 2012, 2013a, 2020; Anter et al., 2020; Sayed et al., 2019). These methods provide a reduced set of the input features while preserving the relevant discriminatory information. Although different types of wrapper-based feature selection have recently been developed, meta-heuristic optimization algorithms have gained a great deal of attention for their crucial role in detecting optimal feature subsets for efficient classification or prediction in many applications because they are based on simple concepts and easily implementable (Pilla et al., 2019; Gorripotu et al., 2019; Najm et al., 2019; Ammar et al., 2019; Ben Smida et al., 2018; Kumar et al., 2015a; Hassanien et al., 2014a; Zhu & Azar, 2011). Some example are particle swarm optimization (Chengying et al., 2012; Kamal et al., 2020; Sallam et al., 2020; Inbarani et al., 2014a,b, 2015a; Jothi et al., 2013), binary whale optimization (Hussien et al., 2020), chaotic crow search algorithm (Sayed et al., 2019); Jaya optimization (Jothi et al., 2019), firefly algorithm (Jothi et al., 2017), harmony search (Inbarani et al., 2015b), ant-colony optimization (Asad et al., 2014a,b, 2012, 2013), bee-colony optimization (Emary et al., 2014b); binary ant-lion optimization (Emary et al., 2016a), Binary grey wolf optimization (Emary et al., 2016b), salp swarm algorithm (SSA) (Faris et al., 2019) and the evolutionary version of the gravitational search algorithm (GSA) (Taradeh et al., 2019).

It is noted that Grey wolf search-based optimization has not yet been explored by researchers in software bug detection that has motivated us to explore its potential for designing an effective bug detection model. The WSO is simple, easy to use, flexible and scalable. The main objective of this study is to construct a fault prediction model based on software change metrics. The following research questions (RQs) are addressed in this study.

Q1. How Hybrid search based algorithms perform in the prediction of software bugs?
Q2. What is the difference between the performance of the proposed method and the predictive models based on its single classifier counterpart?

Therefore, the main contributions of this work are summarized as follows:

(1) Exploring the use of Gry wolf search-based optimization (WSO) on publicly available two software bug datasets (Ant 1.7 and Tomcat) for best feature selections to improve the performance of the prediction model.
(2) Comparing the performance of the proposed hybrid techniques by combining WSO with three popular machine learning techniques such as MLP, random Forest and Logistic regression with conventional classifiers without feature selection for efficient software bug prediction.

The rest of the chapter is organized as follows. Section 2 presents the related work. Section 3 discusses preliminaries of the used methods. The proposed method is presented in Section 4. Results and analysis are presented in Section 5 followed by a discussion in Section 6. Finally, the chapter is concluded in Section 7 with future scope of work.

2. RELATED WORK

Several research techniques for detecting software bugs include particle swarm optimization, evolutionary computing, genetic algorithms, etc. (Xiaoying, 2011; Chengying et al., 2012). Manjula & Florence (2018) proposed a new hybrid approach combining genetic algorithm and deep neural network. They demonstrated the efficiency of their hybrid approach on PROMISE dataset was due to because of the use of the genetic algorithm as an optimization technique. They compared the results with other existing machine learning approaches, including Naive Bayes, Decision Tree, Support Vector Machine, etc. Machine learning has attracted increasing interest in different fields (Koubaa et al., 2020; Jothi et al., 2020; Inbarani et al., 2020; Mjahed et al., 2020; Ajeil et al., 2020; 63. Elkholy et al., 2020a,b; Mohamed et al., 2020; Ibrahim et al., 2020; Sayed et al., 2020; Soliman et al., 2020; Barakat et al., 2020; Pilla et al., 2020, Aboamer et al., 2019, 2014a,b; Banu et al., 2017; Inbarani et al., 2015c, Anter et al., 2015, 2014; Hassanien et al., 2020, 2019a,b; Hassanien & Azar, 2015; Moftah et al., 2014; ElBedwehy et al.,

2014; Roy et al., 2014; Azar et al., 2013; Giove et al., 2013; Santoro et al., 2013). Machine learning has been used to solve several problems in the wider area of software engineering.

Alsaeedi & Khan (2019) discussed the usefulness of the machine learning techniques to predict the software bugs. They use SMOTE re-sampling techniques to address the data imbalance in the datasets and then used Random forest and its combination with Bagging and Boosting techniques to obtain better accuracy. However, they emphasized to use meta-heuristic feature selection approaches for better feature selection as a future study. Lamba et al. (2019) proposed a machine learning model with performance metrics such as accuracy, R squared and correlation for software bug prediction. They concluded that support vector machine is a best choice with high accuracy and low mean square error.

Jia (2018) proposed a hybrid feature selection approach by using various pre-processing techniques such as Information gain, Pearson correlation coefficient and Chi-square test to select the best features. Finally, Random forest is used to predict the software defects. Adak (2018) proposed a hybrid fuzzy logic-based data mining models for accurate software defect prediction. It uses MANOVA as a statistical tool to find the most significant software metrics from the NASA software quality dataset. The results showed an accuracy of 68.6% and an average specificity of 85.9%. Bashir et al. (2019) presented several feature selections, data balancing and noise filtering techniques along with eight most frequently classification algorithms such as Naive Bayes, Random Forest, MLP, SVM, Decision tree, Logistic regression, K-nearest neighbor (KNN) for accurate software defect prediction. They use three software metrics such as object-oriented, static code and combined metric for their evaluation.

Ruan & Liu (2019) proposed a self-organizing data mining approach to predict the software bugs by finding causal relationships between software metrics. They used ranking and classifications to detect software bugs from the datasets obtained from NASA, PROMISE and SOFTLab repository. They claimed that their proposed self-organizing data mining approach is efficient and a feasible one with optimal F-value and small error. Kaen & Algarni (2019) investigated the use of Chi-square correlation method to select the best possible features. They used support vector machine, naive Bayes and linear discriminant analysis for bug clarification and achieves an increment of 4.15%, 5.12% and 1%, respectively over the baseline methods.

Premalatha & Srikrishna (2018) proposed a cost random forest algorithm with robust similarity clustering techniques to have the most accurate software bug prediction model. Pandey et al. (2019) proposed a hybrid a machine learning algorithm with software metrics for software defect prediction. Moustafa et al. (2018) used ensemble classification methods on the datasets of different sizes and different metrics combinations for software bug prediction and concluded that the ensemble methods outperformed the base classifiers with more accuracy. Rhmann (2018) proposed a hybrid search based algorithms with machine learning techniques for software bug prediction. Panda (2019) introduced a hybrid classification method by combining distribution base balance (DBB) based instance selection and radial basis function (RBF) neural network classifier to obtain the best prediction of software defects in comparison to the existing research algorithms using different software defect datasets namely: NASA MDP dataset, SOFT LAB dataset and JEDIT dataset. The experimental results with post-hoc statistical significance test showed the effectiveness of the proposed approach.

Kumar, et al. (2020) summarized the importance of machine learning techniques in software vulnerability analysis with current and previous studies. They used the ensemble approach with Chidamber and Kemerer metric (CKJM) suite and neural network for source code change proneness prediction by using five different versions of eBay web services.

Alsaeedi & Khan (2019) presented a comparative study using supervised machine learning algorithms such as Support vector machine, Decision Trees and Random Forest based on 10 NASA datasets. They concluded that Random forest was the best performing classifiers among others in detecting software defects. Bishnu & Bhattacharjee (2012) have conducted experiments on four real datasets such as AR3, AR4, AR5 and IRIS dataset. They evaluated the effectiveness of Quad Tree-based K-Means clustering algorithm in predicting software defect modules as compared to the original K-Means algorithm and global K-means clustering algorithm.Gong et al. (2019) proposed a cluster-based Over-sampling with noise filtering (KMFOS) approach with extensive experiments on 24 projects for class imbalance in software defect dataset. They compared KMFOS with some over-sampling approaches such as Synthetic Minority Over-sampling Technique (SMOTE), Adaptive Synthetic Sampling (ADASYN), random over-sampling (ROS), K-means SMOTE etc. Finally, Lingden et al. (2019) proposed a correlation-based feature selection with under-sampling method to address a class imbalance in software defect dataset.

3. PRELIMINARIES

This section describes the various methods that are adopted for experimental analysis in making a robust and efficient software bug detection model.

3.1 Grey Wolf Search Optimization (WSO)

Grey wolf (Canis lupus) belongs to the Canidae family. Grey wolves, being like apex predators, are at the top of the food chain, who prefer to live in a pack. The group size of the wolves is often filled with 5-12 wolf numbers. More specifically, they exclusively follow the dominant social hierarchy. WSO is a modern bio-inspired multi-objective meta-heuristic optimization technique (Tang et al., 2012) where the almost perfect hunting behavior of the wolf colony is accomplished with a simple knowledge of the leading wolf, detecting wolves and fierce wolves. The motivation to use many other swarm intelligence methods instead lies in its ability to produce the best features during the wrapper-based feature selection process. It is easy and effective to take the minimum iteration necessary to find a solution (Yamany, et al., 2014).

The WSO algorithm can completely crisscross and dig the solution space in the progression process by using the moving behavior of the detective wolves. Wolves are the social carnivores that hunt in packs. The wolves are special to each other, but budge in a collective that is elastically coupled. Still, they hunt their devour (or prey) individually. The engendering and gathering behavior of the leading wolf will make the colony of the wolf move towards the good areas where there is every possibility that the devour can be caught. It is because of the beleaguer actions of the detective and vicious wolves that the WSO can not only perform a fine search in good solution domination, but can also create a new solution space among the good solutions. In the end, the worst wolves are replaced by the WSO upgrade process by retaining the good ones. These powerful ones can be used arbitrarily to achieve the diversity of the specification and, in this manner, the WSO algorithm avoids local optima (COLREGS, 2013).

Wolves are search agents in the search space of features that make it possible for them to devour (prey) passively or intuitively on the basis of their intelligence. Each wolf is circumscribed by a visual region with a defined radius for detection. The distance can be determined using the distance from Minkowski. The fitness of the objective function corresponds to the consistency of the current position of the wolf.

Three concurrent devour behaviors are included as follows in the WSO. I Devour intuitively: this action of the wolf represents that devour has the greatest fitness value within its perceptual scope, so that it can change its location for that devour without finding its associates. (ii) Devour passively: in this action, the wolf devours passively as it does not find any nourishment and no other alternative choice than to obey its peers. In this situation, it remains alert to any possible threats and tries to strengthen its current status. (iii) Escape: The wolf senses a threat of certain possibility in this devoured behavior. Then the wolf tries to escape this threat by staying at an escape distance that is more than its visual range. The escape would essentially reduce the population. In this way, the wolf does not collapse to the local optimum. The likelihood of a threat is arbitrarily determined by the user.

The WSO is different from other swarm intelligence optimization algorithms where WSO identifies each wolf as a leader in the search for devour rather than a leader, as in the case of conventional ones. WSO leadership and hunting mechanisms are proposed by Mirjalili et al. (2014). In WSO, wolfs maintain social dominance to preserve their stability and then support each other through the entire hunting cycle. Four forms of Grey Wolf are known to be an intelligent search technique to ensure that the devour (or prey) hunting is successful. Three simple hunting measures are used, such as searching, encircling and attacking during the optimization process.

Four types of grey wolves, such as alpha, beta, delta, and omega, are used to simulate the hierarchy of leadership. In comparison, three main stages of hunting, the search for prey, the encircling of prey, and the attacking of prey are carried out for optimization. As can be seen from Figure 1, the hunting cycle is completely controlled by alpha, with beta and delta being the participants whenever appropriate. The ranks of wolves on the basis of social hierarchy and their duties are shown in Table 1. When simulating mathematically, we consider alpha to be the best solution along with beta and omega, and thus require other search agents (omega) to change their positions by looking at the position of alpha, beta or delta. This update behavior is shown in Figure 2. The WSO pseudo-code is shown in Figure 3 for a better interpretation of the WSO operating principle.

Table 1. Social Hierarchy of Grey Wolves

Hierarchy Level	Rank Name	Roles in the Pack
1	Alpha	Leader (male or female)
		Dictates Decision to the Pack
2	Beta	Adviser to alpha (male or female)
		Maintains discipline in the pack. It is the successor of alpha during any requirement
3	Delta	Aged ones, Watching the territory boundary
		Caretakers of ill and wounded wolves
4	Omega	Obeys other dominant wolves, like a scapegoat.
		They always have to submit to all other dominant wolves. Omega wolves are allowed to eat at last, after eating of all the dominant wolves

Figure 1. The social hierarchy of wolves in GWO (Chantar et al., 2019)

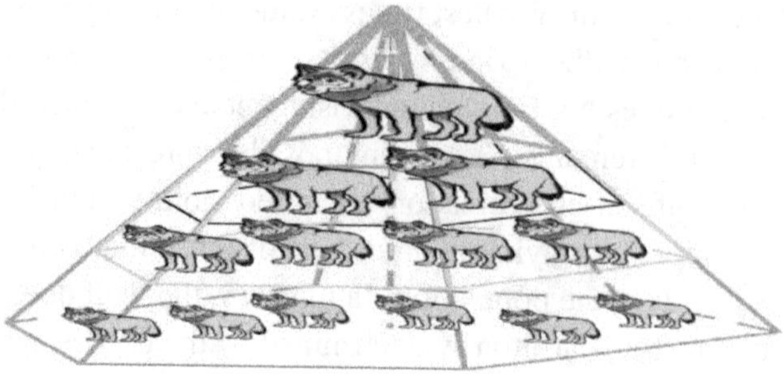

Figure 2. Location update behavior of Omega in 2D search space

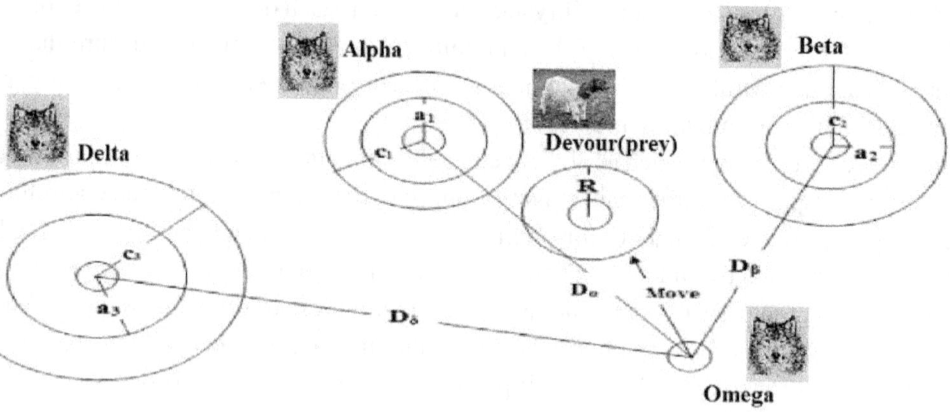

The Mathematical models of Grey wolf pack hunting in WSO are as follows:

$$\vec{D} = \vec{C} \times \vec{X}_p(t) - \vec{X}(t) \tag{1}$$

$$\vec{C} = 2 \times \vec{r}_2 \tag{2}$$

$$\vec{X}(t+1) = \vec{X}_p(t) - \vec{A} \times \vec{D} \tag{3}$$

$$\vec{A} = 2\vec{a} \times \vec{r}_1 - \vec{a} \tag{4}$$

In Equation (1), D is the two-dimensional search space, vector $X_p(t)$ is the position vector of the devour (or prey), and vector $x(t)$ shows the position vector of a Grey wolf. From equations 1 to Equation 4, vector A and C are the coefficient vectors and vector a which is a decision variable decreases linearly from 2 to 0, over the course of iterations. 't' denotes the present iteration. Here, r_1 and r_2 are random vectors with values ranging from 0 to 1. These random vectors enable the Grey wolves to reach any random location by updating the position inside the space and around the devour using equations (2) and (4). As discussed earlier, the location of the omega is updated based on the changes in the locations of alpha, beta and delta in 2-D search space. This is shown in Figure 2. Updating the wolf's positions are performed as follows:

$$\vec{D}_a = \left|\vec{C}_1 \times \vec{X}_a - \vec{X}\right| \tag{5}$$

$$\vec{D}_\beta = \vec{C}_2 \times \vec{X}_\beta - \vec{X} \tag{6}$$

$$\vec{D}_s = \left|\vec{C}_3 \times \vec{X}_s - \vec{X}\right| \tag{7}$$

$$\vec{X}_1 = \vec{X}_a - \vec{A}_1 \times \left(\vec{D}a\right) \tag{8}$$

$$\vec{X}_2 = \vec{X}_\beta - \vec{A}_2 \times \left(\vec{D}\beta\right) \tag{9}$$

$$\vec{X}_3 = \vec{X}_s - \vec{A}_3 \times \left(\vec{D}_s\right) \tag{10}$$

$$X\left(t+1\right) = \frac{\vec{X}_1 + \vec{X}_2 + \vec{X}_3}{3} \tag{11}$$

In this process, the weights of $\vec{a}$ decreases from 2 to zero and the limiting value of $\vec{A}$ lies in between -2a to 2a. The location of each omega wolf is updated using the equation (11).

3.2 Multilayer Perceptron

Multilayer perceptron (MLP) is a most popular artificial neural network architecture that can take any number of inputs, processed through one or more hidden layers with any number of units with activation function (sigmoid function is used in this chapter) (Liu et al., 2020; Ibraheem et al., 2020; Fekik et al., 2018a,b; Lamamra et al., 2017; Hassanien et al., 2014b). The MLP has connections between the input to the first hidden layer, then from first hidden layer to intermediate hidden layers and then final hidden layer to the output layer (Elkholy et al., 2020a). At the output, the desired value and obtained value are compared and the error is computed. The error is backpropagated to input through the hidden layers by updating weights and bias in the network and the process continues till the termination criteria are satisfied. The fast learning of MLP makes it a popular choice in many applications (Elkholy et al., 2020; Kumar et al., 2015b).

Figure 3. Pseudocode of WSO

```
Step 1: Initialize the maximum number of iterations 'L'
Step 2: Initialize the population, xi(i=1,2,..,n)
Step 3: Obtain Objective function f(x), x =(x1,x2,..xd)^T
Step 4: initialize parameters r,s,α, pa, A and C
r = radius of the visual range
s = step size by which a wolf moves at a time
α = velocity factor of wolf
pa = a user-defined threshold [0, -1], to check the possibility of the appearance of an enemy
→A , →C are the coefficient vectors
→a is linearly decreasing from 2 to 0, over the course of iterations
Step 4: WHILE (t<maximum iterations and also for stopping criteria is not met)
step5: FOR i=1: W // in case of each wolf
step6: Devour (or Prey) new food intuitively ();
step7: Generation of the new location ();
step8: check whether the next location suggested by the random number generator
is a new one.
step8: If not, repeat finding a new random location.
Step9:IF(dist(xi, xj) < r and xj is better as f(xi)<f(xj)) then xi moves towards xj // xj is a better
than xi
Step 10: ELSE IF
xi = Prey new food passively ();
Step 11: END IF
Generation of new location ();
IF (rand ()>pa)
xi = xi + rand() + v; wolf escape to a new position.
END IF
END FOR
```

3.3 Random Forest Algorithm

Random forest (RF) is an ensemble of unpruned classification or regression trees, induced from bootstrap samples of the training data, using random feature selection in the tree induction process (Breiman, 2001). Prediction is made by aggregating (majority vote for classification or averaging for regression) the predictions of the ensemble. Random forest generally exhibits a substantial performance improvement over the single tree classifier such as CART and C4.5. It yields generalization error rate that compares favourably to AdaBoost, yet it's more robust to noise. However, similar to most classifiers, random forest can also suffer from the curse of learning from extremely imbalanced training data set. As it is constructed to minimize the overall error rate, it tends to focus more on the prediction accuracy of the majority class, which often results in poor accuracy, for the minority class. In random forests, there is no need for cross-validation or a test set to get an unbiased estimate of the test error. Since each tree is constructed using the bootstrap sample, approximately 1/3rd of the cases are left out of the bootstrap samples and not used in training. These cases are called out of bag (OOB) cases. These OOB cases are used to get a run –time unbiased estimate of the classification error as trees are added to the forest. The error rate of a forest depends on the correlation between any two trees and the strength of each tree in the forest. Increasing the correlation increases the error rate of the forest. The strength of the tree relies on the error rate of the tree. Increasing strength decreases the error rate of the forest. When the forest is growing, random features are selected at random out of all features in the training data (Panda & Patra, 2009).

3.4 Logistic Regression

Logistic regression (LR) can be used also to solve problems of classification (Jothi et al., 2020; Kumar et al., 2014, 2017, 2019; Inbarani et al., 2018; Elshazly et al., 2013a,b; Aziz et al., 2013b). In general, logistic regression classifier can use a linear combination of more than one feature value or explanatory variable as argument of the sigmoid function. The corresponding output of the sigmoid function is a number between 0 and 1. The middle value is considered as a threshold to establish what belongs to class 1 and class 0. In particular, an input producing an outcome greater than 0.5 is considered to belong to class 1. Conversely, if the output is less than 0.5, then the corresponding input is classified as belonging to class 0 (Harrington, 2012).

4. PROPOSED METHOD

4.1 Software bug Datasets

Table 2 provides statistics on the number of instances and distribution of class labels of each dataset used in this study. The nine datasets that we used are accessible and freely available, while initially gathering data from the dataset created by Jureczko et al. (2010) containing 20 types of structural metrics as defined in Table 3, while Palomba et al. (2019) added the new function of the code smell strength by calculating the code smell severity and some of the other features.

Table 2. Description of datasets

System Name	Instances	Buggy Instances	Non- Buggy Instances
ANT 1.7	745	166	579
Tomcat 6.0	858	77	781

Table 3. Summary of structural metrics for the datasets of each software project

Abbrev	Feature	Type
WMC	Weighted Methods per Class	Numerical
DIT	Depth of Inheritance Tree	Numerical
NOC	Number of Children	Numerical
CBO	Coupling between Object classes	Numerical
RFC	Response for a Class	Numerical
LCOM	Lack of Cohesion in Methods	Numerical
CA	Afferent Couplings	Numerical
CE	Efferent Couplings	Numerical
NPM	Number of Public Methods	Numerical
LCOM3	A normalized version of LCOM	Numerical
LOC	Lines of Code	Numerical
DAM	Data Access Metric	Numerical
MOA	Measure Of Aggregation	Numerical
MFA	The measure of Functional Abstraction	Numerical
CAM	Cohesion Among Methods	Numerical
IC	Inheritance Coupling	Numerical
CBM	Coupling Between Methods	Numerical
AMC	Average Method Complexity	Numerical
MAX_CC	Maximum values of methods in the same class	Numerical
AVG_CC	Mean values of methods in the same class	Numerical
Bug	Defect or no-defect	Numerical

4.2 Proposed Model

The proposed software bug classification and evaluation process are presented in Figure 4 and Figure 5.

Each dataset was divided into 10 stratified folds for a 10-fold cross-validation test. Each fold was then held out to form a test set and the other nine folds were combined and randomized (to reduce the ordering effect) to produce the training set. Such stratified cross-validation ensures that there are instances of a defective class in each test set, thus reducing the likelihood of classification uncertainty.

Three different classifiers (MLP, RF and LR) were used using the same training set with all 21 features available in the datasets and then, 10-fold cross-validation testing, as a single classifier as shown in Figure 4. To identify the optimal features, an efficient wolf search feature optimization method was performed for each of the above three classifiers on the training set. The model constructed by each hybrid classifier was used to classify the test set using 10-fold cross-validation as shown in Figure 5.

The predictive performance of all the six classifiers (three baselines as a single classifier and then three hybrid ones) are recorded using four different measures: Time taken to build the model, Accuracy, F-measure and Matthews' Correlation Coefficient (MCC) as shown in Table 4. F-measure gives the harmonic mean of precision and recall that ranges from 0 to 1. In addition, Matthews' Correlation Coefficient (MCC) was chosen because it is relatively easy to understand with a range of-1 to +1 and its reliability in measuring the predictive performance of the classifier (Shepperd et al. 2014). When the classifier is perfect, the value of the MCC is 1. If the classifier is always misclassified, its MCC value is-1. If the MCC is 0, this means that the classifier is nothing but the flipping of a fair coin. A further important feature of the MCC is that it takes all classes of equal importance. In comparison, MCC has the additional advantage that it incorporates all four elements of the confusion matrix, while f-Measure ignores the proportion of true negatives. The tests for each classifier and dataset combination were further evaluated using Violin plots for their reliability in software error detection as defective and non-defective. In comparison, the violin plot for both single and hybrid classifier datasets in terms of classification accuracy is shown in Figure 6, in order to explain the robustness of the proposed hybrid methodology.

Figure 4. Software bug classification process

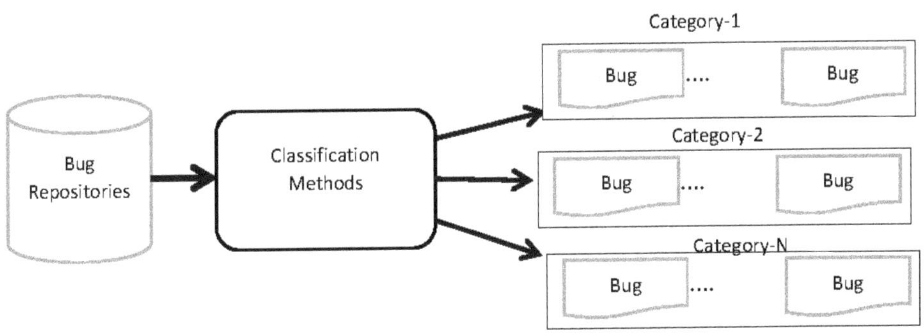

Figure 5. The comparative performance evaluation process

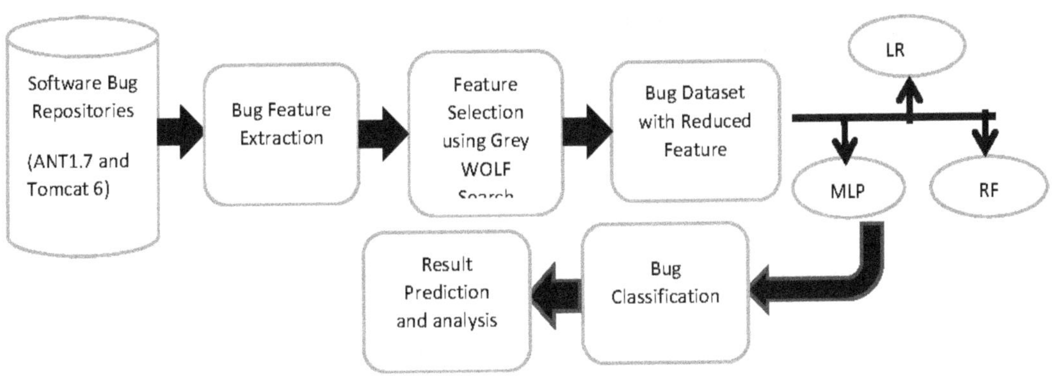

Figure 6 shows violin plots demonstrating the precision of the six different classifiers on both ANT 1.7 and Tomcat datasets in more detail. Violin plots of box plots and jitters in the y-axis reflect the precision of the bug estimation, where the central part of the 'violin' represents the median and the reduction at the ends of the 'violin.' The x-axis displays datasets where a wide 'violin' indicates a high proportion of modules, while a smaller 'violin' reflects a limited number of modules.

5. RESULTS AND ANALYSIS

The ANT 1.7 and Tomcat 6.0 software bug datasets have the class attribute range from 0 to 10, where 0 indicates non-defective and 1 to 10 indicates the severity of the bug. For simplicity, combine a bug value from 1 to 10 as defective and 0 as non-defective when building a binary classifier.

The original ANT 1.7 and Tomcat 6.0 datasets consist of 21 features including the class = bug attribute as shown in Table 3. Since using Gray Wolf Data Search Optimization (WSO), the features have now been reduced to 8 features (wmc, rfc, moa, cam, amc, max cc, avg cc, bug) for ANT 1.7 and 7 features (wmc, rfc, ca, loc, clam, avg cc, bug) for Tomcat 6.0. This can be shown in Table 4.

The performance analysis of all classifiers is shown in Table 4. The result shows that the hybrid approach improves performance in terms of high precision, better MCC and F-value and, most significantly, is faster than single baseline classifiers without a feature selection process. Table 4 also shows that the performance of the classifier depends on the datasets and their reduced characteristics after using WSO, whereas the accuracy of the ANT 1.7 hybrid classifier+WSO ranges from 75.97% to 77.72% and the accuracy of the Tomcat 6.0 ranges from 90.45% to 90.64%. Next, the weighted average MCC and the weighted average F-score are shown in Table 4 for the suitability of the proposed approach.

The weighted average MCC of (0.10, 0.062), (0.31 and 0.245) and (0.29, 0.255) for (ANT 1.7, Tomcat 6.0) using MLP+WSO, LR+WSO and RF+WSO, respectively shows that LR+WSO classifies ANT 1.7 data with highest MCC value of 0.31 while random forest classifier is the best one for Tomcat 6.0 data with highest MCC value of 0.255.

In addition, the weighted average F-score of (0.68, 0.867), (0.71, 0.877) and (0.71, 0.875) for (ANT 1.7, Tomcat 6.0) using MLP+WSO, LR+WSO and RF+WSO, respectively demonstrates that logistic regression-based and random forest-based hybrid approach both equally detect the bugs. Finally, the time taken to build the model for (ANT 1.7, Tomcat 6.0) using MLP+WSO, LR+WSO and RF+WSO is (1.94,0.36), (75.59, 120.25) and (0.59, 0.61) seconds respectively. This time revales that the random forest+WSO is faster in ANT 1.7 data with 0.59 seconds while for Tomcat data, MLP+WSO is faster with 0.36 seconds. However, in comparison to others, logistic regression+WSO takes some more time to build a model.

6. DISCUSSION

Although we focus on the current best practice of using a hybrid classifier approach with feature selection, there are many different ways in which data can be pre-processed and a specific classifier modeled. All of these may have some impact on the predictive performance of the classifier. We use only three popular classifiers for our software bug prediction models combining with the basic wolf search optimization technique to make our approach computationally feasible by agreeing with the Lessmann et al. (2008). To make our experiment easy, we try to use experiments for 10 runs, for which processing time was shorter than running our experiments many times longer than that. We have not addressed the class imbalance problem; wolf search is used as a feature selection technique instead. We used only two datasets with the same features here. This may be true that other datasets can behave differently. We used time taken to develop the model, precision, F-measure and MCC as our success metrics, some using different cutoff points with different indicators with bug prediction (D'Ambros et al., 2012).

Table 4. Experimental analysis

Model	Original features		Reduced features after WSO		Accuracy %		Weighted Average MCC		Weighted Average F-measure		Time is taken to build the model in seconds	
	ANT 1.7	Tomcat	ANT 1.7	Tomcat	ANT 1.7	Tomcat	ANT 1.7	Tomcat	ANT 1.7	Tomcat	ANT 1.7	Tomcat
MLP	21	21	8 (wmc,rfc,moa, cam,amc,max_ cc, avg_cc, bug)	7 (wmc, rfc,ca, loc, clam, avg_cc, bug)	77.72	91.02	0.21	0.098	0.69	0.782	3.94	4.36
LR					75.97	89.04	0.31	0.214	0.71	0.781	122.5	324.01
RF					77.04	90.56	0.25	0.143	0.71	0.869	1.02	1.16
MLP+WSO					77.72	90.55	0.10	0.062	0.68	0.867	1.94	0.36
LR+WSO					77.18	90.45	0.31	0.245	0.71	0.877	75.59	120.25
RF+WSO					77.5	90.64	0.29	0.255	0.71	0.875	0.59	0.61

Figure 6. Violin plot for accuracy of software bug datasets

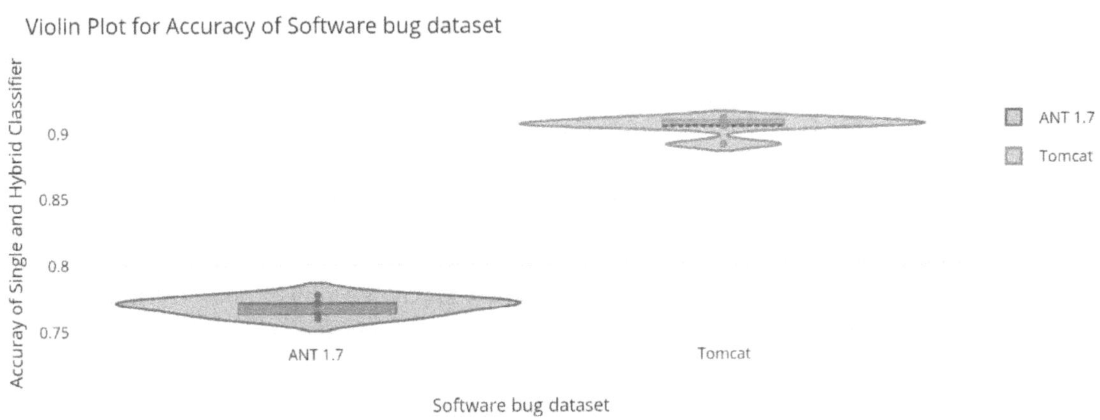

7. CONCLUSIONS AND FUTURE DIRECTIONS

In this chapter, software bug prediction is performed using three classifiers with wolf search feature selection. The results showed that all prediction models are performing almost similarly. Some important implications have been observed. While various measures such as MCC, Accuracy, F-score with built-in model time provide a basic understanding of bug prediction models, it is imperative to know that not a single classifier can detect all kinds of bugs that are present in a software product. Furthermore, it's observed that the selection of features can improve performance as different classifiers use a different subset of features. We suggest new ways to develop either ensemble classifiers or a new meta-heuristic search to improve the model's predictive performance. These opportunities may attract future researchers to use some tools to address the performance issues currently experienced in software bug prediction.

REFERENCES

Aboamer, M. A., Azar, A. T., Wahba, K., Schumann, A., & Bär, K. J. (2019). Nonlinear single-input single-output model-based estimation of cardiac output for normal and depressed cases. *Neural Computing & Applications, 31*(7), 2955–2978. doi:10.100700521-017-3245-8

Aboamer, M. M., Azar, A. T., Mohamed, A. S. A., Bär, K. J., Berger, B. S., & Wahba, K. (2014b). Nonlinear Features of Heart Rate Variability in Paranoid Schizophrenic. *Neural Computing & Applications, 25*(7-8), 1535–1555. doi:10.100700521-014-1621-1

Aboamer, M. M., Azar, A. T., Wahba, K., & Mohamed, A. S. A. (2014a). Linear model-based estimation of blood pressure and cardiac output for Normal and Paranoid cases. *Neural Computing & Applications, 25*(6), 1223–1240. doi:10.100700521-014-1566-4

Adak, M. F. (2018), Software Defect Detection by using Data Mining based Fuzzy Logic. *2018 Sixth International Conference on Digital Information, Networking, and Wireless Communications (DINWC)*, 65-69. 10.1109/DINWC.2018.8356997

Ajeil, F. H., Ibraheem, I. K., Azar, A. T., & Humaidi, A. J. (2020). Grid-Based Mobile Robot Path Planning Using Aging-Based Ant Colony Optimization Algorithm in Static and Dynamic Environments. *Sensors (Basel), 20*(7), 1880. doi:10.339020071880 PMID:32231091

Alsaeedi, A., & Khan, M. Z. (2019). Software Defect Prediction Using Supervised Machine Learning and Ensemble Techniques: A Comparative Study. *Journal of Software Engineering and Applications, 12*(5), 85–100. doi:10.4236/jsea.2019.125007

Amara, K., Malek, A., Bakir, T., Fekik, A., Azar, A. T., Almustafa, K. M., Bourennane, E., & Hocine, D. (2019). Adaptive neuro-fuzzy inference system based maximum power point tracking for stand-alone photovoltaic system. Int. J. Modelling. *Identification and Control, 33*(4), 311–321. doi:10.1504/IJMIC.2019.107480

Ammar, H.H., Azar, A.T., Shalaby, R., & Mahmoud, M. I. (2019). Metaheuristic Optimization of Fractional Order Incremental Conductance (FO-INC) Maximum Power Point Tracking (MPPT). *Complexity*. doi:10.1155/2019/7687891

Anter, A. M., Abu ElSoud, M., Azar, A. T., & Hassanien, A. E. (2014). A Hybrid Approach to Diagnosis of Hepatic Tumors in Computed Tomography Images. *International Journal of Rough Sets and Data Analysis, 1*(2), 31–48. doi:10.4018/ijrsda.2014070103

Anter, A. M., Azar, A. T., & Fouad, K. M. (2020). Intelligent Hybrid Approach for Feature Selection. In *The International Conference on Advanced Machine Learning Technologies and Applications (AMLTA2019). AMLTA 2019. Advances in Intelligent Systems and Computing*, (vol. 921). Springer. 10.1007/978-3-030-14118-9_8

Anter, A. M., Hassanien, A. E., Abu ElSoud, M., & Azar, A. T. (2015). Automatic Liver Parenchyma Segmentation System from Abdominal CT Scans using Hybrid Techniques. *International Journal of Biomedical Engineering and Technology*, *17*(2), 148–168. doi:10.1504/IJBET.2015.068052

Asad, A. H., Azar, A. T., & Hassaanien, A. E. (2012). Integrated Features based on Gray-level and Hu moment-invariants with Ant Colony System for Retinal Blood Vessels Segmentation. *International Journal of Systems Biology and Biomedical Technologies*, *1*(4), 60–73. doi:10.4018/ijsbbt.2012100105

Asad, A. H., Azar, A. T., & Hassanien, A. E. (2013). Ant Colony-based System for Retinal Blood Vessels Segmentation. *Proceedings of Seventh International Conference on Bio-Inspired Computing: Theories and Applications (BIC-TA 2012) Advances in Intelligent Systems and Computing*, *201*, 441-452. 10.1007/978-81-322-1038-2_37

Asad, A. H., Azar, A. T., & Hassanien, A. E. (2014a). A New Heuristic Function of Ant Colony System for Retinal Vessel Segmentation. *International Journal of Rough Sets and Data Analysis*, *1*(2), 15–30. doi:10.4018/ijrsda.2014070102

Asad, A. H., Azar, A. T., & Hassanien, A. E. (2014b). A Comparative Study on Feature Selection for Retinal Vessel Segmentation Using Ant Colony System. *Recent Advances in Intelligent Informatics Advances in Intelligent Systems and Computing*, *235*, 1–11. doi:10.1007/978-3-319-01778-5_1

Azar, A. T., El-Said, S. A., & Hassanien, A. E. (2013). Fuzzy and Hard Clustering Analysis For Thyroid Disease. *Computer Methods and Programs in Biomedicine*, *111*(1), 1–16. doi:10.1016/j.cmpb.2013.01.002 PMID:23357404

Aziz, A. S. A., Azar, A. T., Hassanien, A. E., & Hanafy, S. E. (2012). Continuous Features Discretizaion for Anomaly Intrusion Detectors Generation. *The 17th Online World Conference on Soft Computing in Industrial Applications (WSC17).*

Aziz, A. S. A., Hassanien, A. E., Azar, A. T., & Hanafy, S. E. (2013). Genetic Algorithm with Different Feature Selection Techniques for Anomaly Detectors Generation. *2013 Federated Conference on Computer Science and Information Systems (FedCSIS).*

Aziz, A. S. A., Hassanien, A. E., Azar, A. T., & Hanafy, S. E. (2013). Machine learning techniques for anomalies detection and classification. *Advances in Security of Information and Communication Networks Communications in Computer and Information Science*, *381*, 219–229.

Banu, P. K. N., Azar, A. T., & Inbarani, H. H. (2017). Fuzzy firefly clustering for tumor and cancer analysis. *Int. J. Modelling Identification and Control*, *27*(2), 92–103. doi:10.1504/IJMIC.2017.082941

Banu, P. K. N., Inbarani, H. H., Azar, A. T., Hala, S., Own, H. S., & Hassanien, A. E. (2014). Rough Set Based Feature Selection for Egyptian Neonatal Jaundice. In *Advanced Machine Learning Technologies and Applications: Second International Conference, AMLTA 2014, Cairo, Egypt, November 28-30, 2014. Proceedings, Communications in Computer and Information Science*, (*Vol. 488*). Springer-Verlag GmbH Berlin/Heidelberg. 10.1007/978-3-319-13461-1_35

Barakat, M. H., Azar, A. T., & Ammar, H. H. (2020). Agricultural Service Mobile Robot Modeling and Control Using Artificial Fuzzy Logic and Machine Vision. In *The International Conference on Advanced Machine Learning Technologies and Applications (AMLTA2019). AMLTA 2019. Advances in Intelligent Systems and Computing*, (vol. 921, pp. 453-465). Springer. 10.1007/978-3-030-14118-9_46

Bashir, K., Li, T., & Yohannese, C. W. (2019). An Empirical Study for Enhanced Software Defect Prediction Using a Learning-Based Framework. *International Journal of Computational Intelligence Systems, 12*(1), 282–298. doi:10.2991/ijcis.2018.125905638

Ben Smida, M., Sakly, A., Vaidyanathan, S., & Azar, A. T. (2018). *Control-Based Maximum Power Point Tracking for a Grid-Connected Hybrid Renewable Energy System Optimized by Particle Swarm Optimization. Advances in System Dynamics and Control.* IGI-Global. doi:10.4018/978-1-5225-4077-9.ch003

Bishnu, P. S., & Bhattacherjee, V. (2012). Software Fault Prediction Using Quad Tree-Based K-Means Clustering Algorithm. *IEEE Transactions on Knowledge and Data Engineering, 24*(6), 1146–1150. doi:10.1109/TKDE.2011.163

Breiman, L. (2001). Random Forests. *Machine Learning, 45*(1), 5–32. doi:10.1023/A:1010933404324

Challagulla, V. U. B., Bastani, F. B., Yen, I-L., & Paul, R. A. (2005). Empirical Assessment of Machine Learning based Software Defect Prediction Techniques. *Proceedings of the 10th IEEE International Workshop on Object-Oriented Real-Time Dependable Systems*, 263-270. 10.1109/WORDS.2005.32

Chantar, H., Mafarja, M., Alsawalqah, H., Heidari, A. A., Aljarah, I., & Faris, H. (2019). Feature selection using binary grey wolf optimizer with elite-based crossover for Arabic text classification. *Neural Computing & Applications*, 1–20. doi:10.100700521-019-04368-6

Chengying, M., Xinxin, Y., & Jifu, C. (2012). Swarm Intelligence-Based Test Data Generation for Structural Testing. *2012 IEEE/ACIS 11th International Conference on Computer and Information Science*, 623-628. 10.1109/ICIS.2012.103

Choudhary, G. R., Kumar, S., Kumar, K., Mishra, A., & Catal, C. (2018). Empirical analysis of change metrics for software fault prediction. *Computers & Electrical Engineering, 67*, 15–24. doi:10.1016/j.compeleceng.2018.02.043

COLREGS. (2013). Convention on the International Regulations for Preventing Collisions at Sea. International Maritime Organization, 8(7), 14-20.

Czibula, I.-G., Czibula, G., Marian, Z., & Ionescu, V.-S. (2016). A Novel Approach Using Fuzzy Self-Organizing Maps for Detecting Software Faults. *Studies in Informatics and Control, 25*(2), 207–216. doi:10.24846/v25i2y201608

D'Ambros, M., Lanza, M., & Robbes, R. (2012). Evaluating defect prediction approaches a benchmark and an extensive comparison. *Empirical Software Engineering, 17*(4), 531–577. doi:10.100710664-011-9173-9

ElBedwehy, M. N., Ghoneim, M. E., Hassanien, A. E., & Azar, A. T. (2014). A Computational Knowledge Representation Model for Cognitive Computers. *Neural Computing & Applications, 25*(7-8), 1517–1534. doi:10.100700521-014-1614-0

Elkholy, H. A., Azar, A. T., Magd, A., Marzouk, H., & Ammar, H. H. (2020a). Classifying Upper Limb Activities Using Deep Neural Networks. In *Proceedings of the International Conference on Artificial Intelligence and Computer Vision (AICV2020). AICV 2020. Advances in Intelligent Systems and Computing,* (vol. 1153, pp. 268-282). Springer. 10.1007/978-3-030-44289-7_26

Elkholy, H. A., Azar, A. T., Shahin, A. S., Elsharkawy, O. I., & Ammar, H. H. (2020b) Path Planning of a Self Driving Vehicle Using Artificial Intelligence Techniques and Machine Vision. In *Proceedings of the International Conference on Artificial Intelligence and Computer Vision (AICV2020). AICV 2020. Advances in Intelligent Systems and Computing,* (vol. 1153, pp. 532-542). Springer. 10.1007/978-3-030-44289-7_50

Elshazly, H. I., Azar, A. T., Elkorany, A. M., & Hassanien, A. E. (2013a). Hybrid System based on Rough Sets and Genetic Algorithms for Medical Data Classifications. *International Journal of Fuzzy System Applications*, *3*(4), 31–46. doi:10.4018/ijfsa.2013100103

Elshazly, H. I., Elkorany, A. M., Hassanien, A. E., & Azar, A. T. (2013b). Ensemble Classifiers for Biomedical Data: Performance Evaluation. *IEEE 8th International Conference on Computer Engineering & Systems (ICCES),* 184-189. DOI: 10.1109/ICCES.2013.6707198

Emary, E., Zawbaa, H., Hassanien, A. E., Schaefer, G., & Azar, A. T. (2014a). Retinal Vessel Segmentation based on Possibilistic Fuzzy c-means Clustering Optimised with Cuckoo Search. *IEEE 2014 International Joint Conference on Neural Networks (IJCNN 2014).* 10.1109/IJCNN.2014.6889932

Emary, E., Zawbaa, H., Hassanien, A. E., Schaefer, G., & Azar, A. T. (2014b). Retinal Blood Vessel Segmentation using Bee Colony Optimization and Pattern Search. *IEEE 2014 International Joint Conference on Neural Networks (IJCNN 2014).*

Emary, E., Zawbaa, H. M., & Hassanien, A. E. (2016a). Binary ant lion approaches for feature selection. *Neurocomputing*, *213*, 54–65. doi:10.1016/j.neucom.2016.03.101

Emary, E., Zawbaa, H. M., & Hassanien, A. E. (2016b). Binary grey wolf optimization approaches for feature selection. *Neurocomputing*, *172*, 371–381. doi:10.1016/j.neucom.2015.06.083

Faris, H., Mafarja, M. M., Heidari, A. A., Aljarah, I., Ala'M, A. Z., Mirjalili, S., & Fujita, H. (2018). An efficient binary salp swarm algorithm with crossover scheme for feature selection problems. *Knowledge-Based Systems*, *154*, 43–67. doi:10.1016/j.knosys.2018.05.009

Fekik, A., Denoun, H., Azar, A. T., Zaouia, M., Benyahia, N., Hamida, M. L., Benamrouche, N., & Vaidyanathan, S. (2018a). *Artificial Neural Network for PWM Rectifier Direct Power Control and DC Voltage Control. Advances in System Dynamics and Control.* IGI-Global. doi:10.4018/978-1-5225-4077-9.ch010

Fekik, A., Denoun, H., Hamida, M. L., Azar, A. T., Atig, M., & Zhu, Q. M. (2018b) Neural Network Based Switching State Selection for Direct Power Control of Three Phase PWM-Rectifier. In *2018 10th International Conference on Modelling, Identification and Control (ICMIC).* 10.1109/ICMIC.2018.8529997

Giove, S., Azar, A. T., & Nordio, M. (2013). Fuzzy Logic Control for Dialysis Application. In A. T. Azar (Ed.), *Biofeedback Systems and Soft Computing Techniques of Dialysis. Springer-Verlag GmbH Berlin/ Heidelberg*. doi:10.1007/978-3-642-27558-6_9

Gondra, I. (2008). Applying machine learning to software fault-proneness prediction. *Journal of Systems and Software*, *81*(2), 186–195. doi:10.1016/j.jss.2007.05.035

Gong, L., Jiang, S., & Jiang, L. (2019). Tackling Class Imbalance Problem in Software Defect Prediction Through Cluster-Based Over-Sampling with Filtering. *IEEE Access: Practical Innovations, Open Solutions*, *7*, 145725–145737. doi:10.1109/ACCESS.2019.2945858

Gorripotu, T. S., Samalla, H., Jagan Mohana Rao, C., Azar, A. T., & Pelusi, D. (2019). TLBO Algorithm Optimized Fractional-Order PID Controller for AGC of Interconnected Power System. In J. Nayak, A. Abraham, B. Krishna, G. Chandra Sekhar, & A. Das (Eds.), *Soft Computing in Data Analytics. Advances in Intelligent Systems and Computing* (Vol. 758). Springer. doi:10.1007/978-981-13-0514-6_80

Grishma, B. R., & Anjali, C. (2015). Software root cause prediction using clustering techniques: A review. *2015 Global Conference on Communication Technologies (GCCT)*. 10.1109/GCCT.2015.7342714

Harrington, P. (2012). *Machine Learning in Action*. Manning.

Hassanien, A. E., & Azar, A. T. (2015). Brain Computer Interfaces: Current Trends and Applications. In Intelligent Systems Reference Library, (Vol. 74). Springer-Verlag GmbH Berlin/Heidelberg.

Hassanien, A. E., Azar, A. T., Gaber, T., Bhatnagar, R., & Tolba, M. F. (2019a). The International Conference on Advanced Machine Learning Technologies and Applications (AMLTA2019). In *Advances in Intelligent Systems and Computing*, (Vol. 921). Springer.

Hassanien, A. E., Azar, A. T., Gaber, T., Oliva, D., & Tolba, M. F. (2020). Proceedings of the International Conference on Artificial Intelligence and Computer Vision (AICV2020). In *Advances in Intelligent Systems and Computing*, (Vol. 1153). Springer.

Hassanien, A. E., Azar, A. T., Snasel, V., Kacprzyk, J., & Abawajy, J. H. (2015). Big Data in Complex Systems: Challenges and Opportunities. Springer-Verlag GmbH Berlin/Heidelberg.

Hassanien, A.E., Moftah, H.M., Azar, A.T., & Shoman, M. (2014b). MRI Breast cancer diagnosis hybrid approach using adaptive Ant-based segmentation and Multilayer Perceptron neural networks classifier. *Applied Soft Computing, 14*(Part A), 62-71.

Hassanien, A. E., Tolba, M., & Azar, A. T. (2014a). Advanced Machine Learning Technologies and Applications. In *Second International Conference, AMLTA 2014, Cairo, Egypt, November 28-30, 2014. Proceedings, Communications in Computer and Information Science*, (*Vol. 488*). Springer-Verlag GmbH Berlin/Heidelberg. 10.1007/978-3-319-13461-1

Hassanien, A. E., Tolba, M. F., Shaalan, K., & Azar, A. T. (2019b). Proceedings of the International Conference on Advanced Intelligent Systems and Informatics 2018. In *Advances in Intelligent Systems and Computing*, (Vol. 845). Springer.

Hussien, A. G., Hassanien, A. E., Houssein, E. H., Amin, M., & Azar, A. T. (2020). New binary whale optimization algorithm for discrete optimization problems. *Engineering Optimization*, *52*(6), 945–959. doi:10.1080/0305215X.2019.1624740

Ibraheem, G. A. R., Azar, A. T., Ibraheem, I. K., & Humaidi, A. J. (2020). A Novel Design of a Neural Network based Fractional PID Controller for Mobile Robots Using Hybridized Fruit Fly and Particle Swarm Optimization, Complexity, vol. 2020. *Article ID*, *3067024*, 1–18. doi:10.1155/2020/3067024

Ibrahim, H. A., Azar, A. T., Ibrahim, Z. F., & Ammar, H. H. (2020). A Hybrid Deep Learning Based Autonomous Vehicle Navigation and Obstacles Avoidance. In *Proceedings of the International Conference on Artificial Intelligence and Computer Vision (AICV2020)*. AICV 2020. Springer. 10.1007/978-3-030-44289-7_28

Inbarani, H. H., Azar, A. T., & Jothi, G. (2014). Supervised hybrid feature selection based on PSO and rough sets for medical diagnosis. *Computer Methods and Programs in Biomedicine*, *113*(1), 175–185. doi:10.1016/j.cmpb.2013.10.007 PMID:24210167

Inbarani, H. H., Azar, A. T., & Jothi, G. (2020). Leukemia Image Segmentation using a Hybrid Histogram based Soft Covering Rough K-Means Clustering Algorithm. *Electronics (Basel)*, *9*(1), 188. doi:10.3390/electronics9010188

Inbarani, H. H., Bagyamathi, M., & Azar, A. T. (2015b). A novel hybrid feature selection method based on rough set and improved harmony search. *Neural Computing & Applications*, *26*(8), 1859–1880. doi:10.100700521-015-1840-0

Inbarani, H. H., Banu, P. K. N., & Azar, A. T. (2014). Feature selection using swarm-based relative reduct technique for fetal heart rate. *Neural Computing & Applications*, *25*(3-4), 793–806. doi:10.100700521-014-1552-x

Inbarani, H. H., Kumar, S. S., Azar, A. T., & Hassanien, A. E. (2015a). Hybrid TRS-PSO Clustering Approach for Web2.0 Social Tagging System. *International Journal of Rough Sets and Data Analysis*, *2*(1), 22–37. doi:10.4018/ijrsda.2015010102

Inbarani, H. H., Kumar, S. S., Azar, A. T., & Hassanien, A. E. (2015c). Hybrid TRS-PSO Clustering Approach for Web2.0 Social Tagging System. *International Journal of Rough Sets and Data Analysis*, *2*(1), 22–37. doi:10.4018/ijrsda.2015010102

Inbarani, H. H., Kumar, S. U., Azar, A. T., & Hassanien, A. E. (2018). Hybrid Rough-Bijective Soft Set Classification system. *Neural Computing & Applications*, *29*(8), 67–78. doi:10.100700521-016-2711-z

Jia, C., & Chan, W. K. (2013). Which compiler optimization options should I use for detecting data races in multithreaded programs. *2013 8th International Workshop on Automation of Software Test (AST)*, 53-56. 10.1109/IWAST.2013.6595791

Jia, L. (2018). A Hybrid Feature Selection Method for Software Defect Prediction. *IOP Conf. Ser.: Mater. Sci. Eng., ACMME, 394*(3). doi:10.1088/1757-899X/394/3/032035

Jothi, G., Inbarani, H. H., & Azar, A. T. (2013). Hybrid Tolerance-PSO Based Supervised Feature Selection For Digital Mammogram Images. *International Journal of Fuzzy System Applications*, *3*(4), 15–30. doi:10.4018/ijfsa.2013100102

Jothi, G., Inbarani, H. H., Azar, A. T., & Devi, K. R. (2019). Rough set theory with Jaya optimization for acute lymphoblastic leukemia classification. *Neural Computing & Applications*, *31*(9), 5175–5194. doi:10.100700521-018-3359-7

Jothi, G., Inbarani, H. H., Azar, A. T., Koubaa, A., Kamal, N. A., & Fouad, K. M. (2020). Improved Dominance Soft Set based Decision Rules with Pruning for Leukemia Image Classification. *Electronics (Basel)*, *9*(5), 794. doi:10.3390/electronics9050794

Jothi, G., Inbarani, H. H., Azar, A. T., Koubaa, A., Kamal, N. A., & Fouad, K. M. (2020). Improved Dominance Soft Set based Decision Rules with Pruning for Leukemia Image Classification. *Electronics (Basel)*, *9*(5), 794. doi:10.3390/electronics9050794

Jothi, G., Inbarani, H. H., Azar, A. T., & Polat, K. (2017). Tolerance rough set firefly-based quick reduct. *Neural Computing & Applications*, *28*(10), 2995–3008. doi:10.100700521-016-2514-2

Jureczko, M., & Madeyski, L. (2010). Towards identifying software project clusters with regard to defect prediction. In *Proceedings of the 6th International Conference on Predictive Models in Software Engineering, ser. PROMISE '10*. New York, NY: ACM. Available: https: //doi.acm.org/10.1145/1868328.1868342

Kaen, E., & Algarni, A. (2019). Feature Selection Approach for Improving the Accuracy of Software Bug Prediction. *JKAU: Comp. IT. Sci.*, *8*(1), 35–44.

Kamal, N. A., Azar, A. T., Elbasuony, G. S., Almustafa, K. A., & Almakhles, D. (2020). PSO-based Adaptive Perturb and Observe MPPT Technique for Photovoltaic Systems. In *The International Conference on Advanced Intelligent Systems and Informatics AISI 2019*. Springer.

Koubaa, A., Ammar, A., Alahdab, M., Kanhouch, A., & Azar, A.T. (2020). DeepBrain: Experimental Evaluation of Cloud-Based Computation Offloading and Edge Computing in the Internet-of-Drones for Deep Learning Applications. *Sensors 2020, 20*(18), 1-25. doi:10.339020185240

Kumar, L., Hota, C., Tummalapalli, S., & Murthy Neti, L. B. (2020). Selection of Significant Metrics for Improving the Performance of Change-Proneness Modules. In *Automated Software Engineering: A Deep Learning-Based Approach. Learning and Analytics in Intelligent Systems* (Vol. 8). Springer. doi:10.1007/978-3-030-38006-9_1

Kumar, S., Inbarani, H.H., Azar, A.T., Own, H.S., & Balas, V.E. (2014). Optimistic Multi-Granulation Rough set based Classification for Neonatal Jaundice Diagnosis. *Soft Computing Applications, 356*, 307-317. DOI doi:10.1007/978-3-319-18296-4_26

Kumar, S. S., Inbarani, H. H., Azar, A. T., & Hassanien, A. E. (2015a). Rough Set Based Meta-Heuristic Clustering Approach for Social E-Learning Systems. *International Journal of Intelligent Engineering Informatics*, *3*(1), 23–41. doi:10.1504/IJIEI.2015.069098

Kumar, S. S., Inbarani, H. H., Azar, A. T., & Polat, K. (2017). Covering-based rough set classification system. *Neural Computing & Applications*, *28*(10), 2879–2888. doi:10.100700521-016-2412-7

Kumar, S. U., Azar, A. T., Inbarani, H. H., Liyaskar, O. J., & Almustafa, K. M. (2019). Weighted Rough set Theory for Fetal Heart Rate Classification. *Int. J Sociotechnology and Knowledge Development*, *11*(4), 1–19. doi:10.4018/IJSKD.2019100101

Kumar, S. U., Inbarani, H. H., & Azar, A. T. (2015b). Hybrid Bijective soft set - Neural network for ECG arrhythmia classification. *International Journal of Hybrid Intelligent Systems*, *12*(2), 103–118. doi:10.3233/HIS-150209

Lamamra, K., Azar, A. T., & Ben Salah, C. (2017). *Chaotic system modelling using a neural network with optimized structure. Studies in Computational Intelligence* (Vol. 688). Springer-Verlag.

Lamba, T., Kavita, & Mishra, A. K. (2019). Optimal Machine learning Model for Software Defect Prediction. *I. J. Intelligent Systems and Applications*, *2*(2), 36–48. doi:10.5815/ijisa.2019.02.05

Lessmann, S., Baesens, B., Mues, C., & Pietsch, S. (2008). Benchmarking classification models for software defect prediction: A proposed framework and novel findings. *IEEE Transactions on Software Engineering*, *34*(4), 485–496. doi:10.1109/TSE.2008.35

Lingden, P., Alsadoon, A., Prasad, P.W.C., Alsadoon, O. H., Ali, R. S. & Nguyen. (2019). A novel modified undersampling (MUS) technique for software defect prediction. *Computational Intelligence*, 1–18. DOI: doi:10.1111/coin.12229

Liu, L., Ma, D., Azar, A. T., & Zhu, Q. (2020). Neural computing Enhanced Parameter Estimation for Multi-Input and Multi-Output Total Nonlinear Dynamic Models. *Entropy (Basel, Switzerland)*, *22*(5), 510. doi:10.3390/e22050510

Malhotra, R. (2016). An empirical framework for defect prediction using machine learning techniques with Android software. *Appl. Soft Comput.*, *49*(C), 1034–1050.

Manjula, C., & Florence, L. (2018). Deep neural network-based hybrid approach for software defect prediction using software metrics. *Cluster Computing*, *22*(S4), 9847–9863. doi:10.100710586-018-1696-z

Mikyeong, P., & Hong, E. (2014). Software Fault Prediction Model using Clustering Algorithms Determining the Number of Clusters Automatically. *International Journal of Software Engineering and Its Applications*, *8*, 199–204.

Mirjalili, S., Mirjalili, S. M., & Lewis, A. (2014). Grey wolf optimizer. *Advances in Engineering Software*, *69*, 46–61. doi:10.1016/j.advengsoft.2013.12.007

Mishra, D., & Mishra, A. (2009). Software process improvement in SMEs: A comparative view. *Computer Science and Information Systems*, *6*(1), 111–140. doi:10.2298/CSIS0901111M

Mjahed, S., Bouzaachane, K., Azar, A. T., Hadaj, S. E., & Raghay, S. (2020). Hybridization of Fuzzy and Hard Semi-Supervised Clustering Algorithms Tuned with Ant Lion Optimizer Applied to Higgs Boson Search. *CMES-Computer Modeling in Engineering & Sciences*, *125*(2), 459–494. doi:10.32604/cmes.2020.010791

Mohamed, N. A., Azar, A. T., Abbas, N. E., Ezzeldin, M. A., & Ammar, H. H. (2020) Experimental Kinematic Modeling of 6-DOF Serial Manipulator Using Hybrid Deep Learning. In *Proceedings of the International Conference on Artificial Intelligence and Computer Vision (AICV2020). AICV 2020*. Springer.

Moustafa, S., ElNainay, M. Y., El Makky, N., & Abougabal, M. S. (2018). Software bug prediction using weighted majority voting techniques. *Alexandria Engineering Journal*, *57*(4), 2763–2774. doi:10.1016/j.aej.2018.01.003

Najm, A. A., Ibraheem, I. K., Azar, A. T., & Humaidi, A. J. (2020). Genetic Optimization-Based Consensus Control of Multi-Agent 6-DoF UAV System. *Sensors (Basel)*, *20*(12), 3576. doi:10.339020123576 PMID:32599862

Palomba, F., Zanoni, M., Fontana, F. A., Lucia, A. D., & Oliveto, R. (2019). Toward a smell-aware bug prediction model. *IEEE Transactions on Software Engineering*, *45*(2), 194–218. doi:10.1109/TSE.2017.2770122

Panda, M. (2019). Software defect prediction using hybrid distribution base balance instance selection and radial basis function classifier. *International Journal of System Dynamics Applications*, *8*(3), 53–75. doi:10.4018/IJSDA.2019070103

Panda, M., & Patra, M. R. (2009). Evaluating machine learning algorithms for detecting network intrusions. *International Journal of Recent Trends in Engineering*, *1*(1), 472–477.

Pandey, B., Rhmann, W., Ansari, G. & Pandey, D. K. (2019). Software fault prediction based on change metrics using hybrid algorithms: An empirical study. *Journal of King Saud University – Computer and Information Sciences*. doi:10.1016/j.jksuci.2019.03.006

Pilla, R., Azar, A. T., & Gorripotu, T. S. (2019). Impact of Flexible AC Transmission System Devices on Automatic Generation Control with a Metaheuristic Based Fuzzy PID Controller. *Energies*, *12*(21), 4193. doi:10.3390/en12214193

Pilla, R., Botcha, N., Gorripotu, T. S., & Azar, A. T. (2020). Fuzzy PID Controller for Automatic Generation Control of Interconnected Power System Tuned by Glow-Worm Swarm Optimization. In J. Nayak, V. Balas, M. Favorskaya, B. Choudhury, S. Rao, & B. Naik (Eds.), *Applications of Robotics in Industry Using Advanced Mechanisms. ARIAM 2019. Learning and Analytics in Intelligent Systems* (Vol. 5, pp. 140–149). Springer. doi:10.1007/978-3-030-30271-9_14

Premalatha, H. M., & Srikrishna, C. V. (2018). Software Fault Prediction and Classification using Cost based Random Forest in Spiral Life Cycle Model. *International Journal of Intelligent Engineering and Systems*, *11*(2), 1–8. doi:10.22266/ijies2018.0430.02

Rajkumar, G., & Alagarsamy, K. (2013). The Most Common Factors for The Failure of Software development Project. International Journal of Computer. *Science and Applications*, *1*(11), 74–77.

Rana, Z. A., Mian, M. A., & Shamail, S. (2015). Improving Recall of software defect prediction models using association mining. *Knowledge-Based Systems*, *90*, 1–13. doi:10.1016/j.knosys.2015.10.009

Ren, J -H & Liu, F. (2019). Predicting Software Defects Using Self-Organizing Data Mining. *IEEE Access, 7*(2019), 122796 – 122810. DOI-. doi:10.1109/ACCESS.2019.2927489

Rhmann, W. (2018). Cross-project defect prediction using hybrid search-based algorithms. *Int. J. Inf. Tecnol.* doi:10.100741870-018-0244-7

Roy, P., Goswami, S., Chakraborty, S., Azar, A. T., & Dey, N. (2014). Image Segmentation Using Rough Set Theory: A Review. *International Journal of Rough Sets and Data Analysis*, *1*(2), 62–74. doi:10.4018/ijrsda.2014070105

Sallam, O. K., Azar, A. T., Guaily, A., & Ammar, H. H. (2020). Tuning of PID controller using Particle Swarm Optimization for cross flow heat Exchanger based on CFD System Identification. In *The International Conference on Advanced Intelligent Systems and Informatics AISI 2019*. Springer. 10.1007/978-3-030-31129-2_28

Santoro, A., Mancini, E., & Azar, A. T. (2013). Biofeedback Systems and Their Application in the Hemodialysis Therapy. In A. T. Azar (Ed.), *Biofeedback Systems and Soft Computing Techniques of Dialysis,* (Vol. 405, pp. 1081–1107). doi:10.1007/978-3-642-27558-6_6

Sayed, A. S., Azar, A. T., Ibrahim, Z. F., Ibrahim, H. A., Mohamed, N. A., & Ammar, H. H. (2020). Deep Learning Based Kinematic Modeling of 3-RRR Parallel Manipulator. In *Proceedings of the International Conference on Artificial Intelligence and Computer Vision (AICV2020). AICV 2020*. Springer.

Sayed, G. I., Hassanien, A. E., & Azar, A. T. (2019). Feature selection via a novel chaotic crow search algorithm. *Neural Computing & Applications, 31*(1), 171–188. doi:10.100700521-017-2988-6

Shepperd, M., Bowes, D., & Hall, T. (2014). Researcher bias: The use of machine learning in software defect prediction. *IEEE Transactions on Software Engineering, 40*(6), 603–616. doi:10.1109/TSE.2014.2322358

Soliman, M., Azar, A. T., Saleh, M. A., & Ammar, H. H. (2020) Path Planning Control for 3-Omni Fighting Robot Using PID and Fuzzy Logic Controller. In *The International Conference on Advanced Machine Learning Technologies and Applications (AMLTA2019). AMLTA 2019*. Springer.

Tang, R., Fong, S., Yang, X.-S., & Deb, S. (2012). Wolf search algorithm with ephemeral memory. In *IEEE Seventh International Conference on Digital Information Management (ICDIM 2012)* (pp.165-172). IEEE.

Taradeh, M., Mafarja, M., Heidari, A. A., Faris, H., Aljarah, I., Mirjalili, S., & Fujita, H. (2019). An evolutionary gravitational search-based feature selection. *Inf Sci, 497*, 219–239. doi:10.1016/j.ins.2019.05.038

Xu, Z., Khoshgoftaar, T. M., & Allen, E. B. (2000). Prediction of software faults using fuzzy nonlinear regression modelling. In *Fifth IEEE international symposium on high assurance systems engineering*, (pp. 281–290). IEEE.

Yamany, W., Emary, E., & Hassanien, A. E. (2014). Wolf search algorithm for attribute reduction in classification. *2014 IEEE Symposium on Computational Intelligence and Data Mining (CIDM)*, 351–358. 10.1109/CIDM.2014.7008689

Zhu, Q., & Azar, A. T. (2015). *Complex system modelling and control through intelligent soft computations. Studies in Fuzziness and Soft Computing* (Vol. 319). Springer-Verlag.

Chapter 14
Brushless Motor and Wireless Recharge System for Electric Vehicle Design Modeling and Control

Mohamed Naoui
National School of Engineering of Gabes, Tunisia

Flah Aymen
https://orcid.org/0000-0002-3463-6096
National School of Engineering of Gabes, Tunisia

Ben Hamed Mouna
ISSIG, Tunisia

Lassaad Sbita
https://orcid.org/0000-0002-6589-3976
National School of Engineering of Gabes, Tunisia

ABSTRACT

This chapter deals with the problem of energy storage inside an electric vehicle. The main source of energy is based on a wireless system. This recharge tool regroups inside several components as the storage system, which consists of an ensemble of batteries and serving as the main power source, a special electronic converter that is based on the buck-boost principle and a coil receiver placed undo the vehicle. From the other side, one or more than coil transmitters are placed on the road, where the vehicle is. Modeling all of these components and expressing their mathematical models seems interesting for defining the possible control method that can guarantee a high autonomy when the vehicle is moving. So, taking into account if the care is driving or stopped, this recharge system is studied for verifying the effectiveness of this recharge system and showing the relationship between the vehicle situation and the quantity of received power. Using the platform Matlab/Simulink the results were shown and discussed.

DOI: 10.4018/978-1-7998-5788-4.ch014

INTRODUCTION

After years of debate over whether or not there is global warming, most experts agree that the earth's temperature is indeed rising. This increase is directly linked to greenhouse gases. CO_2 has seen its concentration increase by 30% since 1750, while methane has increased by 150%.

However, it is CO_2 which constitutes the main contributor among greenhouse gases, this being linked to human activities: 90% of its emissions come from the combustion of fossil fuels (petroleum products, coal, natural gas) (Hamidi, Weber, & Nasiri, 2013), and in particular to 34% of the transport sector, which remains the first sector in terms of greenhouse gas emissions (Sarkar & Bhattacharyya, 2012) . The particles released by heat engines are also very harmful to humans: they enter the lungs and can cause many respiratory and cardiovascular diseases. Mainly rejected by diesel engines, most of them are captured by particulate filters (Sudimac, Ugrinović, & Jurčević, 2020). However, according to some studies, the latter would manage to effectively remove a large number of particles, but in return release smaller particles, which would go deeper into the lungs (Mousavi Agah & Abbasi, 2012).

In this context, the electrified transportation infrastructure calls for the development of a wide range of charging networks to promote the worldwide adoption of electric vehicles (Spichkova & Margaret Hamilton, 2016). As there are many charging systems and modes, it seems important to find a useful solution to help solve the issue of autonomy for an electric vehicle (Chau, Chan, & Liu, 2008). The plugin recharging mode is currently the most useful technic, based on the concept of connecting the vehicle to the grid through a recharging cable (R. G. Mohamed, Ebrahim, & Bendary, 2017). This type has two versions, the high-speed recharge mode, which is generally based on DC voltage, and the low-speed version, which is based on the home version of 220 voltages. The high-speed charging system can take 30 minutes for the vehicle to be fully charged, but it can take 8 to 10 hours for the low-voltage mode to do this. Nevertheless, whether it is in motion or halted, the only difficulty that characterizes this recharge mode is attached to the vehicle situation. Thus, if the vehicle battery has decreased to the low region on a high-speed road or a simple long road, using these techniques may not be effective. This scenario poses a frightening situation for electric vehicle users. It is safe to assume that this is why the world does not use electric cars as much (C. S. Lee, Jeong, Lee, & Hur, 2011).

Concentrating on haw help resolving the vehicle autonomy problem, this chapter deals with the field of energy storage and how it is possible to facilitate the recharge phase for an electric vehicle on the road. The idea is to use the wireless recharge system for charging the vehicle when it is in movement. In the future, the wireless charging systems for electric vehicles (Wr) will be a potential alternative technology to charge electric vehicles. Basing on the existing literature, various researchers were focused on this field to improve the global system efficiency. These works have studied the relationship between various parameters as the quantity of transmitted power (Rosu, S. G 2016), the needed recharge times, the related battery SOC, the vehicle speed, and the user control technic. Concerning this field, different results are compared to highlight the benefits and the disadvantages of each method (Flah. A, 2018).

According to those researches, the major problem is how to estimate the recharge time of the vehicle when it is in movement and haws the vehicle speed can affect the efficiency of this recharging method.

Generally, the efficiency of such a recharge method is maximum when the vehicle is stopped. Then the time recharge duration can be founded or estimated easily.

The reliability of the recharge system is guaranteed only in home-use applications, such as garages fitted with wireless recharge systems (Abdelwahab & Shaaban, 2019). However, for the bus applications, this is not the case, where they stopped time is limited relative to the car in a garage. The main problem is that when the EV is in motion, two cases are available for the recharge delay. The first is when the EV stops for a short period, such as at a traffic light. The second case is connected, where the vehicle is on, to the rechargeable whet lane. To accurately determine the quantity of the obtained power from this recharging process, this specification must be expressed correctly and then the battery SOC value and the new vehicle autonomy value can be calculated.

According to the objective of this chapter, the interior design of this manuscript composes, an introduction section, which describes the problem and the exposed solution concerning the wireless recharge vehicle method. The next section gives a general overview of the electric vehicle, which needs an electric recharge system and where the wireless recharge solution can be used. Four types of electric vehicles are discussed. The proportional electronic converter that can be used for the traction part of the recharge system is also expressed in the next section. The corresponding mathematical models and equations are also given. The next section is concentrated on the wireless recharge solution. All the necessary information is exposed and the corresponding mathematical model is given to. The interior composition is also exposed. This is by taken into consideration all the necessary equipment of such a recharge system as the battery and sup capacitor models. Section four takes into consideration the traction system which is based on the PMSM machine and expresses its corresponding mathematical model and gives the interior design of the control loop. These blocks are then combined in the simulation steps at the end to evaluate the recharge system when the vehicle is in motion and this by supervising the battery SOC. The corresponding results are then discussed. The conclusion at the end of this chapter concludes this work and gives the future endeavors of this work.

TYPES OF ELECTRIC VEHICLES

Electric vehicles can be classified into two groups in general, the first being the pure electric vehicle and the second is the hybrid electric vehicle (HEV). This model blends electricity with some other source. There is an intrinsic advantage of hybrid electric vehicles (HEVs) that can further stretch the fuel economy by incorporating the best of both an engine and a battery. Thus, the vehicle could be driven on a battery when in an urban/populated area and then could turn to the engine while outside the city. Thus, hybrids can be subdivided into plug-in hybrid electric vehicles (PHEVs) and fuel cell electric vehicles (FCEVs). Thus, EVs may be classified into hybrid electric vehicles (HEVs), battery-electric vehicles (BEVs), plug-in hybrid electric vehicles (PHEVs), and electric fuel cell vehicles (FCEVs).

1. Hybrid Electric Vehicles (HEVs)

The HEV was described by the International Technical Committee for Electric Road Vehicles as vehicles using two or more sources of energy or storage in such a way that at least one supplies electrical energy (Sreedhar. V, 2006). An electric motor with a battery and an engine with a fuel tank is widely available for the HEV. Energy is extracted solely from petrol and regenerative braking. HEVs may also be categorized according to the drivetrain configuration into four types: series hybrids, parallel hybrids, series-parallel hybrids, and complex hybrids.

2. Battery Electric Vehicles (BEVs)

The battery-electric vehicles that rely only on batteries for power are called BEV; they are without traditional ICE and must be connected to the battery from an external energy source. The battery's capacity directly affects the vehicles' power. The modern Battery Electric Vehicles (BEV) can typically cover a single charge of 100-400 km (Hwang, 2013). High-range battery models for electric vehicles are also available, but the price is the trade-off, as these are considered luxury models and have a large price tag typically. The Battery Electric Vehicles (BEV) can recharge its battery with regenerative brakings, like all other electric vehicles, which slows the vehicle down using the engine and then recovers some energy that is converted to heat (Mohrehkesh & Nadeem, 2011). The avantages of BEV include easy installation, ease of use, and total noise-free use. It is also made environmentally sustainable by the lack of any greenhouse gas pollution. The only downside is a shorter per-charge range that makes it an attractive choice for the urban household, coupled with its advantages.

3. Plug-in Hybrid Electric Vehicles (PHEVs)

The plug-in hybrids were developed to increase the performance of HEVs (Yilmaz & Krein, 2013). Using a battery, Plug-in Hybrid Electric Vehicles (PHEVs) can be charged through the recharger system and the electric motor. The battery is also provided by a recharging device that can repair or replace the battery while the vehicle operates at a low speed. As plug-in hybrid electric cars use energy directly from the recharger grid, fuel savings are higher compared to hybrid electric vehicles.

4. Fuel Cell Electric Vehicles (FCEVs)

In this mode an ICE generates electricity from an electrochemical reaction and it can run indefinitely if it is supplied with hydrogen motor similar to an ICE (Un-Noor, Padmanaban, Mihet-Popa, Mollah, & Hossain, 2017). The industry uses the PEMFC (proton exchange membrane fuel cell) as its fuel type.

Water is produced by Fuel Cell Electric Vehicles (FCEVs) as a by-product which is ejected from the tailpipes. Poorly designed refueling infrastructure and the absence of an inexpensive and sustainable method of processing hydrogen make the fuel cell costly. Hydrogen also ignites faster than gasoline and it is very hard to detect the flames. However, researchers are discussing these issues, and hopefully, they will be eradicated shortly.

PROPORTIONAL CONVERTERS STRUCTURE

Figure (1) displays the most used composition for electric (Thounthong, Raël, & Davat, 2007)(Khaligh, Member, Li, & Member, 2010)(Zandi, 2011). It consists of a 0-500V DC link supplied by a 10 kW of the wireless charging system used as the main source, complete a current nonreversible DC/DC integrated converters, a battery used as a source which is connected to the DC link through a current reversible DC/DC boost-buck converter and the load consisted of an inverter driving the electric motor PMSM of power around 50Kw. The battery is used as a power source that supplies power demand and peak current required during acceleration and deceleration stages.

Figure 1. Composition for electric vehicles: recharge and traction systems.

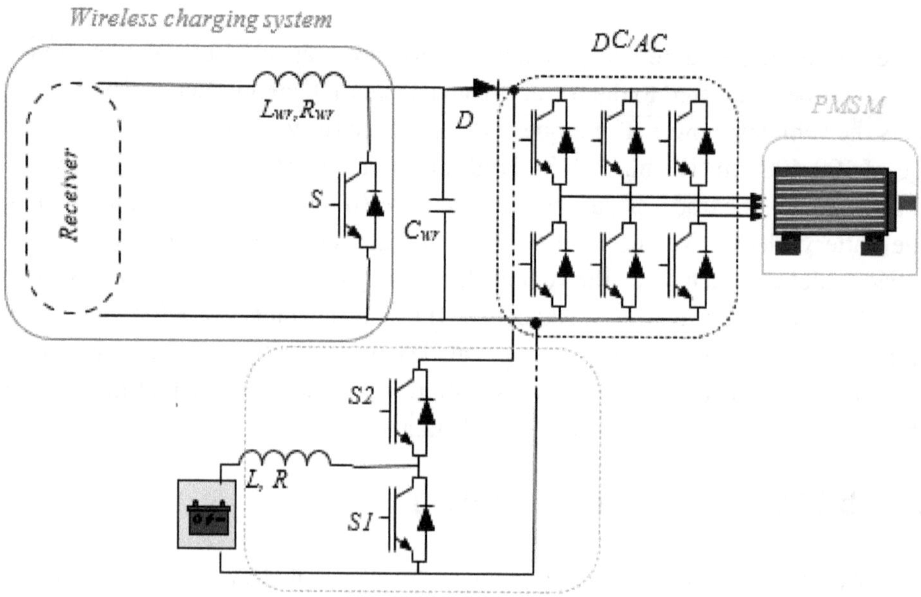

Due to the ease of use and the good robustness of the wireless recharge system, this part will be concentrating on (N. Mohamed, Flah, & Ben Hamed, 2018). Figure (2), expresses the global architecture of this recharge principle this system involves in transferring energy from the source to the vehicle without physical contact, this is by three different converters and two transmitter coils. A general block diagram for the contactless charger for EV batteries is shown in Figure (2). This system is based on the resonant circuit, four topologies, SS (series-series), SP (series-parallel), PS (parallel series), and PP (parallel - parallel) (Bi, 2016). The overall system is equipped with two parts, fixed transmitters, and a moving receiver: the contactless coupler and the power electronics system connected to it. Before the wireless stage, there are two conversion steps: The low-frequency AC from the grid is converted to DC, and then the DC to AC high-frequency. These conversion stages allow adjusting the power level by controlling the input voltage and the frequency. After the wireless stage, a final conversion from high frequency 50 kHz AC to DC is done to provide energy to the battery (Yilmaz & Krein, 2013).

Figure 2. Composition of the wireless charging system

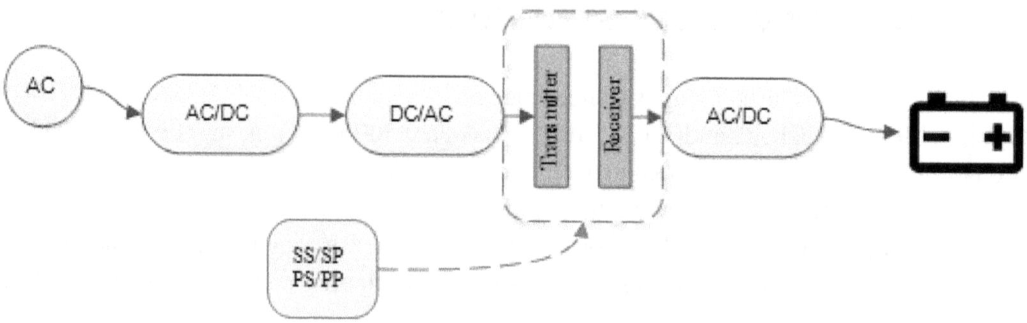

MODELING OF THE WIRELESS RECHARGE SYSTEM

1. Recharge System Composition

Basing on the literature, the recharge systems that can be found inside an electric vehicle are numerous, as the plugin method, the wireless method, the hybrid method, and other solutions. Each system has a unique form regarding its interior architecture. Concentrating on the wireless method, the work steps start by showing the related interior composition and then be given the proportional mathematical model.

This system has two different modes, all of them is related to the vehicle situation if it is stopped or in movement. The mathematical model change from a situation to another. We have called the first model, "the static model", which represents that the vehicle is stopped, and we have mentioned by the dynamic model for the second case, where the vehicle is in motion.

As it is known, the wireless recharge system for an electric vehicle or any other electric system composes of two essential parts. The first one is related to the first coil, it is nominated as the primary coil or the transmitter, and the second part which is related to the secondary coil or the receiver part.

a. The Static Model

This part exposes the simplified architecture of the first and the second coils inside the transmitter and the receiver respectively. Figure (3). Exposes the interior components of this model, and we note by V_P and V_S the input and output voltages (N. Mohamed, Aymen, Ben Hamed, & Lassaad, 2019)(Auvigne, Germano, Civet, & Perriard, 2014).

Figure 3. The mutual inductance coupling model

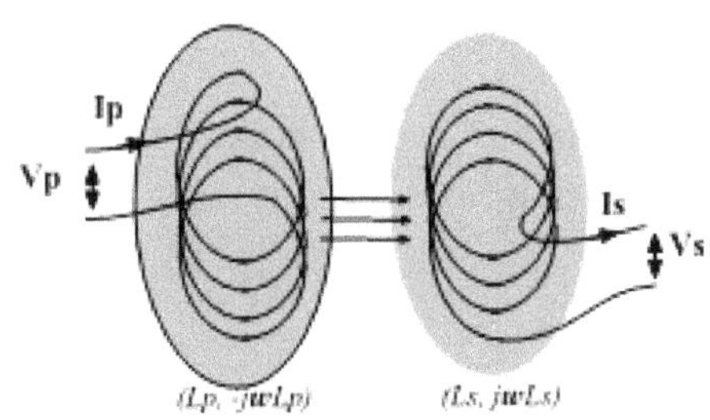

If searching the relationship between the first and the second voltages coming from the two coils, a mutual inductance "M", an operational frequency "ω", appears in the voltage equations as it is in equation (2). We note that the mutual inductance is related to the magnetic coupling coefficient and expressed in (1).

$$\begin{cases} V_p = j\omega L_p I_p - j\omega M I_s \\ V_s = j\omega M I_p - j\omega L_s I_s \end{cases} \tag{1}$$

$$k = \frac{M}{\sqrt{L_p L_s}} \tag{2}$$

On the other side, it is important to express the reflected impedance from the secondary to the primary coils, for facilitating haw power losses can be estimated. Equation (3), expresses this relation. We note by Zs the impedance of the secondary network and this depends on the selected compensation topology.

$$Z_r = \frac{\omega^2 M^2}{Z_s} \tag{3}$$

The current flowing through the secondary winding can be expressed in equation (4).

$$I_s = \frac{j\omega M I_p}{Z_s} \tag{4}$$

Also, knowing the primary and secondary resonant frequencies will be important next, so the corresponding mathematical relation can be visualized in equation (5).

$$\begin{cases} \omega_0 = \dfrac{1}{\sqrt{C_p L_p}} \\ \omega_0 = \dfrac{1}{\sqrt{C_s L_s}} \end{cases} \tag{5}$$

Finally, and for knowing how much power can be obtained in the second part of this wireless transmitter, equation (6), express the relationship.

$$\begin{cases} P_p = V_p I_p \\ P_s = V_s I_s \end{cases} \tag{6}$$

b. The Dynamic Model

For explaining the process of the dynamic model, Figure (4), show that the flux distribution between two parallel coils depends on the primary face of the secondary coil position. Even the distance between the center of the coil is minimum (D1=0), the quantity of the transmitted flux is high. This is the case of the static model.

Figure 4. The shape of the transmitted magnetic flux in both cases of primary and secondary coils position

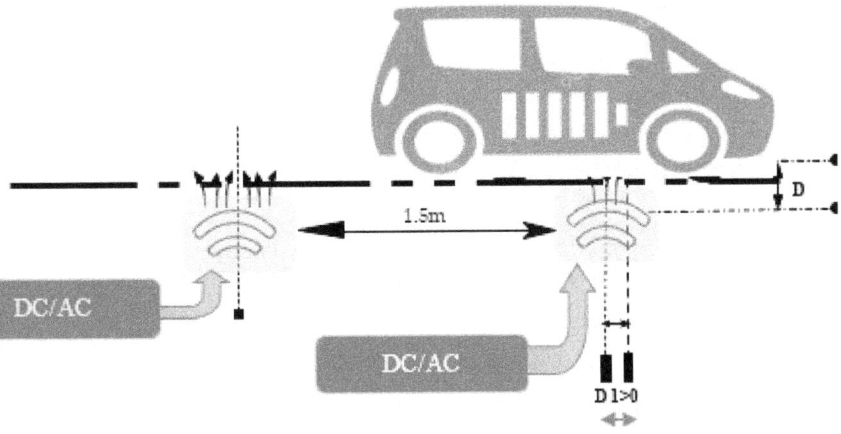

The coil form and the distance between the transmitter and the receiver will be responsible for the quantity of the injected power (Naoui, Mohamed, Flah, & Ben hamed, 2020). Also, the coils overlap percentage present an important factor that should be taken into consideration to improve the quantity of the receiver power. Equations (7) to (10) expose the corresponding mathematical models.

Starting with the corresponding physical-mathematical equation of the flux in the primary coil, which is notated in equation (7). With is the length of the solenoid, N number of turns, S the surface, current, the magnetic field, and the magnetic constant.

$$\phi_1(t) = N.S.B(t) = \left(\frac{\mu_0.S.N^2}{\ell}\right) i_1(t) \tag{7}$$

However, in a mobile vehicle, the two coils are not confused, then we notate that they obtained flux in the secondary coil "attached to the vehicle", is expressed as it is in equation (8). Where k is the percentage of the obtained flux in the secondary coil.

$$\phi_2 = k.\phi_1 \tag{8}$$

This is related to several factors as, vehicle speed and coils parameters.

$$\begin{cases} 0 < k < 1 & if\ 0 < D1 < \text{coil radius} \\ k = 1 & if\ D1 = 0 \end{cases} \tag{9}$$

The portioned outputted voltage in the secondary coil will be as it is in equation (10)

With "R_2" presents the resistance of the secondary coil. « L » is the coil inductance," w" is the width of winding, "r" is the average winding radius, and" N "is the number of turns.

$$\begin{cases} U_2(t) = R_2.i_2(t) + k.L.\dfrac{di_1(t)}{dt} \\ L(\mu\,\mathrm{H}) = \dfrac{\mathrm{r}^2.N^2}{8r + \pi w} \end{cases} \tag{10}$$

2. Battery and Supercapacitor Model

In the literature, various battery models were used. Each type can be presented by several methods as in (Tremblay & Dessaint, 2009)(C. S. Lee, Jeong, Lee, & Hur., 2011). The lithium-ion battery model can be expressed as it is in Figure (5).

Figure 5. Equivalent electrical diagram of a battery element

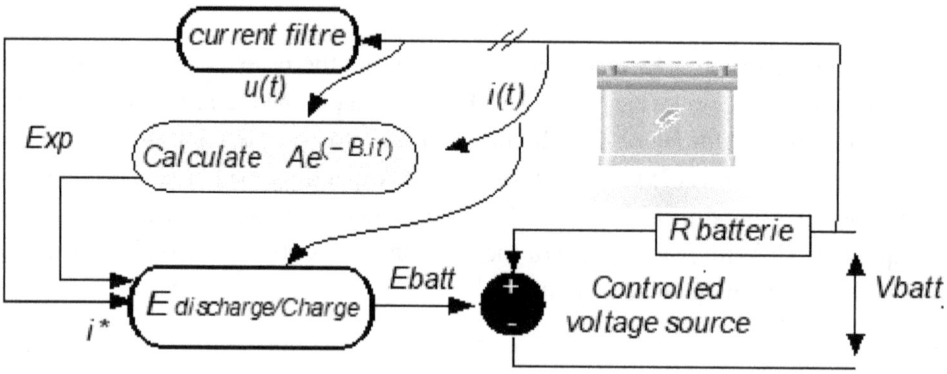

The charge/discharge model for the lithium-ion battery (Mahmoudi, Flah, & Sbita, 2017) can be described as follows the equation (11). Where i* is the current of low-frequency dynamics (A), K is the polarization constant or polarization resistance (Ω), Q is the battery capacity (Ah), "A" is the exponential zone amplitude (V), "B" is the exponential zone time constant inverse $(\mathrm{Ah})^{-1}$,"E_0 " is the standard voltage (V).

$$\begin{cases} E_{charge} = E_0 - K \cdot \dfrac{Q}{Q-it} \cdot i^* - K\dfrac{Q}{Q-it} \cdot it + A \cdot e^{(-B \cdot it)} \\ E_{discharge} = E_0 - K \cdot \dfrac{Q}{0.1 \cdot Q + it} \cdot i^* - K\dfrac{Q}{Q-it} \cdot it + A \cdot e^{(-B \cdot it)} \end{cases} \tag{11}$$

The State of Charge (SOC) is the ratio of the amount of electricity remaining at time t. (Q(t)) it's the nominal capacity (Q_{Max}). SOC is often given in%, η_b battery performance, and battery current.

$$SOC = 100\left(1 - \frac{1}{Q}\int_0^t i(t)dt\right) \tag{12}$$

The battery pack voltage depends on the number of parallel/ serial used cells. The battery resistance can be found also in equation (13). We note by "Ro", the battery cell charging or discharging resistance.

$$R_b = (R_0 + R_{st}.\frac{I_{st}}{I_I} + R_{It}.\frac{I_{It}}{I})\frac{N_s}{N_p} \tag{13}$$

Basing on these equations, it is possible to express the battery heat generation noted "Q_{bp}" in equation (14). "ξ" is the charge efficiency.

$$Q_{bp} = I_b R_b^2 + (1-\xi) I_b V_b \tag{14}$$

$$V_b = \frac{N_s}{N_p} V_{b/cell} \tag{15}$$

Then it is simple to express the battery pack temperature noted "Tb" as it is in equation (16). "Qbc" is the heat losses and "Cpb" is the battery heat capacity (Kroeze & Krein, 2008), (S. Lee, Cherry, Safoutin, McDonald, & Olechiw, 2018).

$$T_b = \int \frac{Q_{bp} - Q_{bc}}{m_{ba} C_{pb}} dt + T_0 \tag{16}$$

The supercapacitor model, for one cell, can be visualized in equation (17). The capacity value depends," the number of layers in the electrode "Ne. "ϵ_0" and "ϵ" is the permittivity of the free space and the electrolyte material. "c" presents the molar concentration and "A_i" expresses the joint area between electrodes and electrolytes. The Helmholtz layer length is "d" and "Qc" presents the cell electric charge (C). The final capacitance value depends, the number of parallel and serial cells, noted respectively by "Np" and "Ns", and it can be calculated as it is in equation (18). Then the outputted voltage value for the supercapacitor module can be expressed as it is in equation (19) (Zhou, Gao, Ravey, Al-Durra, & Simões, 2017), (Zhang, Tao, & Zhou, 2019).

$$\frac{1}{C_{\text{sup}/cell}} = \left[\frac{d}{N_e \varepsilon_0 A_1} + \frac{2N_e RT}{FQ_C \sinh(\frac{Q_C}{N_e^2 A_1 \sqrt{8RT\varepsilon_0 c}})} \right] \tag{17}$$

$$C_{\text{sup}} = \frac{N_p}{N_s} C_{\text{sup}/cell} \tag{18}$$

$$V_{sup} = \frac{N_p Q_c}{C_{sup}} - R_{sup} i_{sup} \tag{19}$$

3. The Recharge System Proportional to the Boost–Buck Converter

The proportional converter for such a recharge system is the buck converter or the boost converter. Indeed, in charging mode ($i_{bat} < 0$), it operates as a buck converter and in discharging mode ($i_{bat} > 0$), the converter operates as a boost converter. As the goal is to enforce the battery current i_{bat} to track its reference i_{batref} (provided by the energy management system), one can define a binary variable X as follows:

$$X = \begin{cases} 1 & if\ i_{batref} > 0 \\ 0 & if\ i_{batref} < 0 \end{cases} \tag{20}$$

Hence, the following boost–buck converter model can be easily obtained in:

$$\begin{cases} \frac{di_{bat}}{dt} = -\left[X(1-u_1)+(1-X)u_2\right]\frac{v_{dc}}{L_1} - \frac{R_1}{L_1} i_{bat} + \frac{v_{bat}}{L_1} \\ i_2 = \left[X(1-u_1)+(1-X)u_2\right] i_{bat} \end{cases} \tag{21}$$

MODELING OF THE TRACTION SYSTEM

1. Permanent Magnet Machine for the Traction Part

In the traction bloc, the permanent magnet synchronous machine is used. It is the main charge inside the vehicle. For assuring the global loop function, modeling this system is necessary and the corresponding mathematical model can be supervised in equation (22).

But for staring showing the corresponding model, it is important to indicate that a set of simplifying assumptions as mentioned in the following points (Hussein, Ammar, & Hassan, 2017).

- Neglecting iron losses.
- The permeability of magnets is considered to be close to that of air.
- The excitation flux of the permanent magnet is considered constant.

The voltage Equations of voltages are in equation (22).

$$[v_s] = [R_s][i_s] + \frac{d[\varphi_s]}{dt} \tag{22}$$

With:

$$\left[\varphi_f\right]=\varphi_{sf}\begin{bmatrix}\cos(\theta)\\ \cos(\theta-\frac{2\pi}{3})\\ \cos(\theta-\frac{4\pi}{3})\end{bmatrix} \tag{23}$$

Equation (23) illustrates the flux expression generated by the magnets. In which φs_f the peak (constant) value created by the magnet through the stator windings.

Note that equation (22) is non-linear and coupled. To eliminate this problem we adopt variable changes and transformations that reduce the complexity of the system. In this case, we proceed to the transformation of Park, which consists of transforming the immobile windings (a, b, c) by winding (d, q) which turns with the rotor. Using Park's transformation, we pass real stator quantities (voltage, current flows) to their dummy components called the d-q components (Sbita & Flah, 2013).

Figure 6. Diagram of a PMSM machine

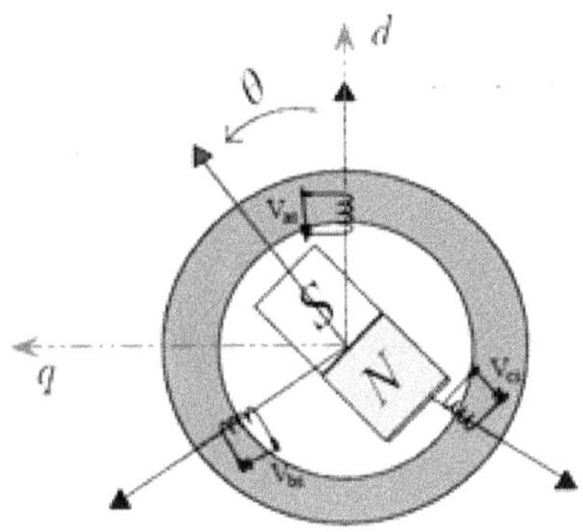

With: $[P(\theta)]$ the matrix of the Park transforms which allows the passage of the stator quantities [vs], [φs] and [Is] to their relative components [Vdq], [Idq] and [φdq].

$$\left[P(\theta)\right]=\sqrt{\frac{2}{3}}\begin{bmatrix}\cos(\theta) & \cos(\theta-\frac{2\pi}{3}) & \cos(\theta+\frac{2\pi}{3})\\ -\sin(\theta) & -\sin(\theta-\frac{2\pi}{3}) & -\sin(\theta+\frac{2\pi}{3})\\ \frac{1}{\sqrt{2}} & \frac{1}{\sqrt{2}} & \frac{1}{\sqrt{2}}\end{bmatrix} \tag{24}$$

The dynamic properties of the PMSM can be described by a set of nonlinear differential equations linking the stator and rotor currents and voltages with the mechanical quantities; torque, speed, and angular position (Xu, Wang, Xie, Xu, & Feng, 2004),(An & Sun, 2008).The application of Park's transformation to equation (23) gives after development the necessary equation:

The electric equations:

$$\dot{i}_d = \frac{1}{L_d}\left(v_d - Ri_d + L_q\omega_e i_q\right) \quad (25)$$

$$\dot{i}_q = \frac{1}{L_q}\left(v_q - Ri_q - L_d\omega_e i_d - \varphi_f p\Omega\right) \quad (26)$$

After the Park transformation the voltage expressions in the (d,q) axis are presented in equation (27). where v_d, v_q, i_d, i_q, L_d, L_q are respectively the direct and the quadrature voltages, currents, and the stator inductances, R_s is the stator resistance, ω is the mechanical speed and λm_t he permanent magnet flux linkage.

$$\begin{aligned} v_d &= R_s i_d + L_d\frac{di_d}{dt} - \omega L_q i_q \\ v_q &= R_s i_q + L_q\frac{di_q}{dt} + \omega L_d i_d + \omega\lambda_m \end{aligned} \quad (27)$$

The electromechanical torque can be expressed as follows:Where $\lambda d_= Ld_i d_+ \lambda m_a$nd $\lambda q = L_q iq_.$

$$T_e = \left(\frac{3}{2}\right)\left(\frac{P}{2}\right)\left(\lambda_d i_q - \lambda_q i_d\right) \quad (28)$$

For a no salient poles PMSM model, $L_d = L_q = L_s$ then the modified torque expression becomes:

$$T_e = \left(\frac{3}{2}\right)\left(\frac{P}{2}\right)\left(\lambda_m i_q\right) \quad (29)$$

The mechanical motor in equation (30). where T_l, P and J are respectively the load torque, the poles number, and the rotor inertia coefficient.

$$\left(T_e\_T_l\right) = \left(\frac{P}{2}\right)\left(J\frac{d\omega}{dt} + f\omega\right) \quad (30)$$

2. Proportional Inverter for the Traction Part

The voltage inverter which supplies the PMSM can be ideally represented according to the figure 7, where K_i and K_i^* (i = a, b, c) are IGBT transistors mounted in antiparallel with diodes, S_i and S_i^* are the logic commands associated with them such as:

- if S_i=1: the switch K_i is on and K_i^* is open,
- if S_i=0: the switch K_i is open and K_i^* is on.

Figure 7. Equivalent diagram of the inverter

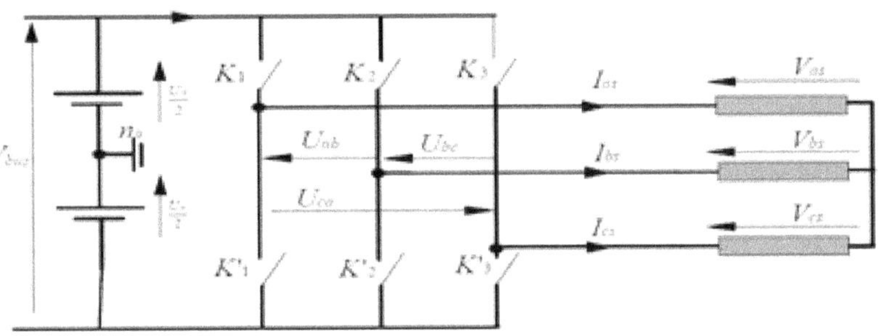

The inverter consists of three arms, each arm consists of two transistors whose control is complementary. Each arm of the inverter can be presented by a two-position switch as shown in figure 7.

The line voltages delivered by the inverter are:

$$\begin{cases} U_{ab} = V_{as} - V_{bs} = V_{bus}(S_a - S_b) \\ U_{bc} = V_{bs} - V_{cs} = V_{bus}(S_b - S_c) \\ U_{ca} = V_{cs} - V_{as} = V_{bus}(S_c - S_a) \end{cases} \tag{31}$$

Or

$$V_{ac} + V_{bc} + V_{cs} = 0 \tag{32}$$

$$\begin{cases} V_{as} = \frac{V_{bus}}{3}(2S_a - S_b - S_c) \\ V_{bs} = \frac{V_{bus}}{3}(2S_b - S_a - S_c) \\ V_{cs} = \frac{V_{bus}}{3}(2S_c - S_b - S_a) \end{cases} \tag{33}$$

CONTROL TECHNIQUE FOR THE TRACTION SYSTEM

The vector control technique is used to establish a linear model and transform the synchronous magnet machine PMSM to a DC machine with separate excitation.

Consequently, the rotor flux is located at an angle θ with respect to the real axis of phase α. The application of vector control requires that the axis of the current i_q be in quadrature with respect to the rotor flux, which leads to the component i_d of the stator current being collinear with the rotor flux. If the current i_d, is in the same direction as the rotor flux, the stator flux along the axis d is added to the fluxes of the magnets, which gives an increase to the airgap flux. On the other hand, if the current i_d is negative the stator flux will in opposition to that of the rotor which gives a reduction in the air gap flow. Normally the current i_d, must be zero when the system works at constant torque. Therefore, if the current i_d is forced to zero id equal to i_q, the component φ_d of the stator flux becomes:

$$\varphi_d = \varphi_f \tag{34}$$

The form of electromagnetic torque will be:

$$C_e = p\,\varphi_f\, i_q = K\, i_q \tag{35}$$

The reference of the direct component of the stator currents is equal to zero ($i_d^*= 0$). This component is zero to obtain a unit power factor (the reactive power created, and the associated losses are therefore zero). Also, the inductances on the axes q and d being equal ($L_d = L_q$), the component i_d does not affect the calculation of the electromagnetic torque.

The speed regulator output i_q^* constitutes the torque reference C_e^*. The references of the currents "id*" and "iq*" are compared separately with the measured currents of the machine i_d and i_q. Current errors are applied to the input of conventional PI type regulators. A decoupling block generates the reference voltages v_d^* and v_q^*. The outputs of the decoupling block v_d^* and v_q^* transfer to through a two-phase to the three-phase transformation which gives us the three reference voltages.

- **Decoupling**

As you can see, the i_d and i_q of the current depend simultaneously on the input quantities v_d and v_q. Here we have a multi-variable system with two inputs and two outputs. In order to be able to set up mono variable controls, we will seek from the equations which govern the dynamic speed of the engine feedback which decouples the system.

The electrical equations (27) of the PMSM and putting:

$$\begin{cases} v_d' = v_d + e_q \\ v_q' = v_q + e_d \end{cases} \tag{36}$$

With:

$$e_q = \omega_e L_q i_q \tag{37}$$

$$e_d = \omega_e L_d i_d + \omega_e \varphi_f \tag{38}$$

- **Calculation of regulator**

As already mentioned, the regulation is carried out using PI type regulators (proportional, integral). Even the best algorithms are always a combination of these actions. We have adopted a PI regulator whose integral coefficient 4 'is used to reduce the difference between the setpoint and the regulated quantity therefore reducing overshoot. As the proportional term 4 allows the adjustment of the speed of the system and therefore the response time (Alain, Romanic, & Bertrand, 2017).

In an onboard system such as a vehicle, one does not expect that it will have oscillations, otherwise, one will seek a regulation adapted to a critical regime. The situation became similar to that of the DC machine. The speed is measured directly and it is possible to train its regulation using a PI regulator. At the level of the calculation of the regulator parameters (Mousa, Ebrahim, & Hassan, 2015). The process to be ordered is broken down into two subsystems.

- Current adjustment system, therefore, i_q torque also.
- Mechanical system subsystem.

Figure 8. Speed control loop

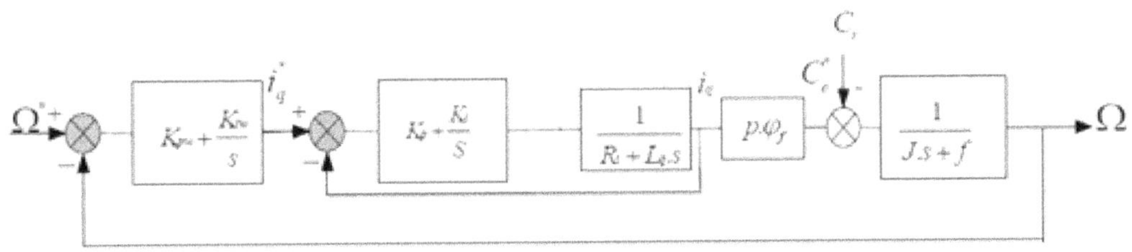

The speed control loop is shown in figure (8):

$$G(p) = \frac{1 + \tau.s}{1 + \tau.s + \frac{J}{K_{iw}} s^2} \tag{39}$$

With:

$$\tau = \frac{K_{pw}}{K_{iw}} \tag{40}$$

Knowing that the transfer function of a second-order system is:

$$F(p)=\frac{1+\tau.s}{1+\frac{2\xi}{\omega_0}s+\frac{J}{\omega_0}s^2} \quad (41)$$

By analogy, we will try to find the parameters of the regulator (PI)

$$\frac{K_{iw}}{J}=\omega_0, \tau=\frac{2\xi}{\omega_0} \quad (42)$$

For critical damping

$$K_{iw}=\frac{4J}{\tau^2}, K_{pw}=K_{iw}.\tau \quad (43)$$

We will take τ=0.8s, knowing that the electric time constant must be less than the mechanical time constant.

The closed-loop transfer function gives us a 1st order type system with an electric time constant:

$$\tau_e=\frac{R_s}{K_i} \quad (44)$$

$$G(s)=\frac{1}{\tau_e.s+1} \quad (45)$$

The response time t_{rip} is defined by the instant when the output quantity is 95% of the final value given when the time becomes infinite, that is:

$$t_{rep}=3.\tau_e \quad (46)$$

$$K_p=3.\frac{L_d}{t_{rep}} \quad (47)$$

$$K_i=3.\frac{R_s}{t_{rep}} \quad (48)$$

Results Discussions

1. Simulation Discretion

The simulations are performed considering a vehicle with the following specifications: acceleration of 0.2,0.7,01,0.3 in 15s. The build EV model is implemented in Matlab/Simulink. The present model will be verified under various conditions related to the vehicle status and according to algorithm control. Therefore, in this section, we expose and discuss the obtained results and we concentrate on the battery SOC evolution in the function of algorithm control. Table 1 describes the wireless recharge system parameters. Here, the objective is to check the behavior of the electric vehicle and the performance of the technic recharge. So, in the two next subsections, the PMSM results, according to the used-control technic, will show. Also, the power performance of the recharge system according to the vehicle speeds, will be exposed and discussed.

Table 1. Recharge system dimensions

Parameter	Value
i_1	32 A
U_{ref}	500V
Coil diameter	50 cm
Distance between coil	150 cm
Width of winding "w"	21 cm
Average winding radius "r"	14.5 cm
Number of turns "N"	15 Turns

2. Traction Control Loop Results

To global control scheme of the PMSM, based on the FOC control principle is illustrated in figure (9). This control loop is tested using Matlab/Simulink software package. The obtained results were depicted using the motor and inverter parameters as it is in follow. For the inverter parameter, the DC bus voltage U_{dc}= 700 V, and the switching frequency is 10Khz. Table (2), illustrates all the necessary information about the machine parameters.

Table 2. Motor specifications

Motor parameters	*value*
Motor Power	400 W
Nominal current	2A
Maximum current	4A
Nominal speed	1750 rpm
Nominal Torque	1.5 N.M

Figure 9. Block diagram of the proposed control system of the PMSM

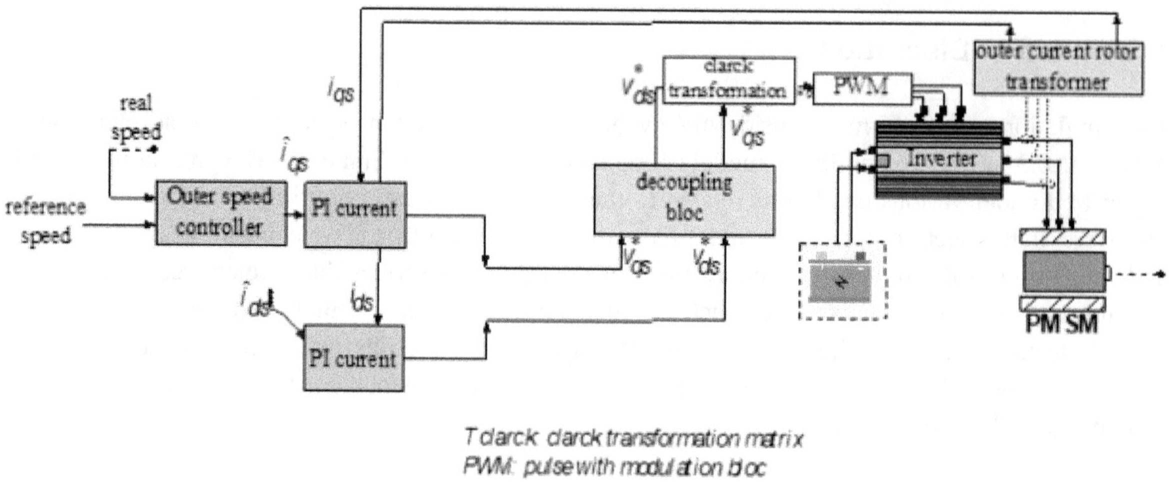

In this simulation, we have supposed that the motor torque is constant and then we have depicted the related results as the motor speed, currents, and battery performance as the battery power.

Figure (10), shows the motor speed and proves the robustness of the control loop. The decoupling principle is also verified by the current results. The direct components are always nothing.

Figure 10. PMSM speed evolution under the FOC control technology

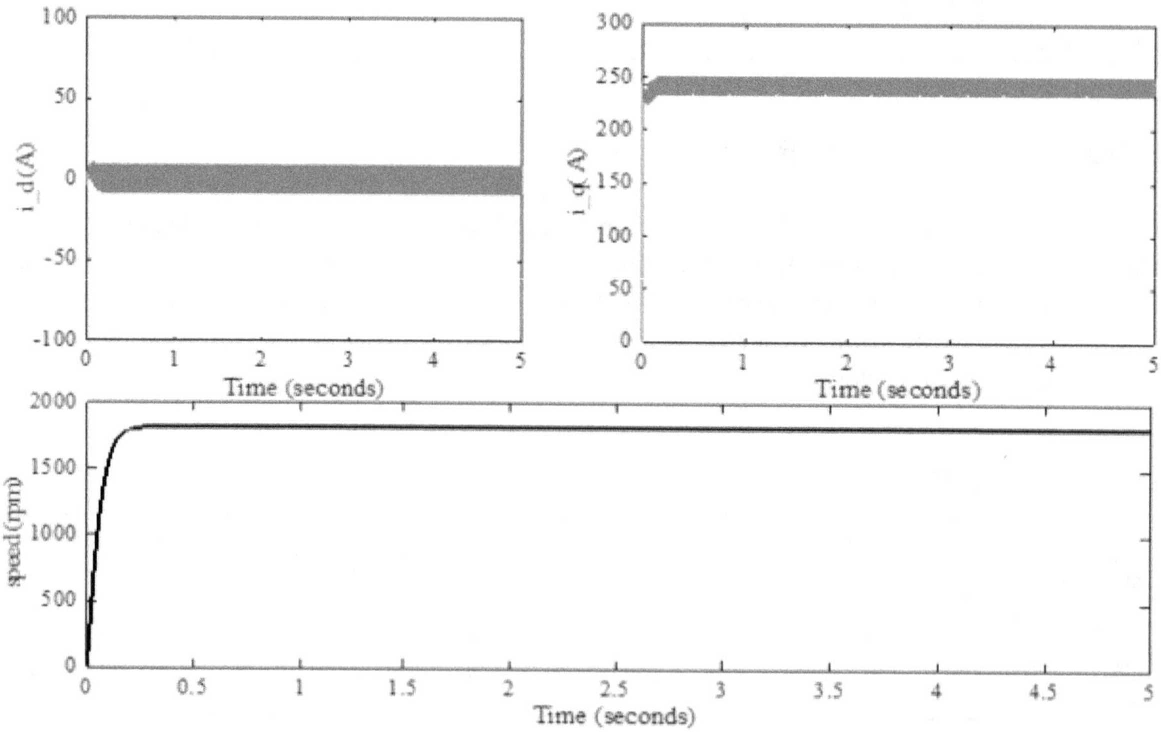

3. WIRELESS RECHARGE METHOD: RESULTS

In order to test the efficacity of the used wireless recharge system, we have made a special test, which takes a various acceleration form for obtaining a various car speed. So, we have supposed that the vehicle is moved on a road contains an unlimited wireless transmitter coils and the vehicle is equipped with only one receiver. Then, the simulation will concentrate on the obtained power from the wireless system in order to charge the battery inside the vehicle.

Figure 11. Power situation using the wireless recharge method under different vehicle speeds.

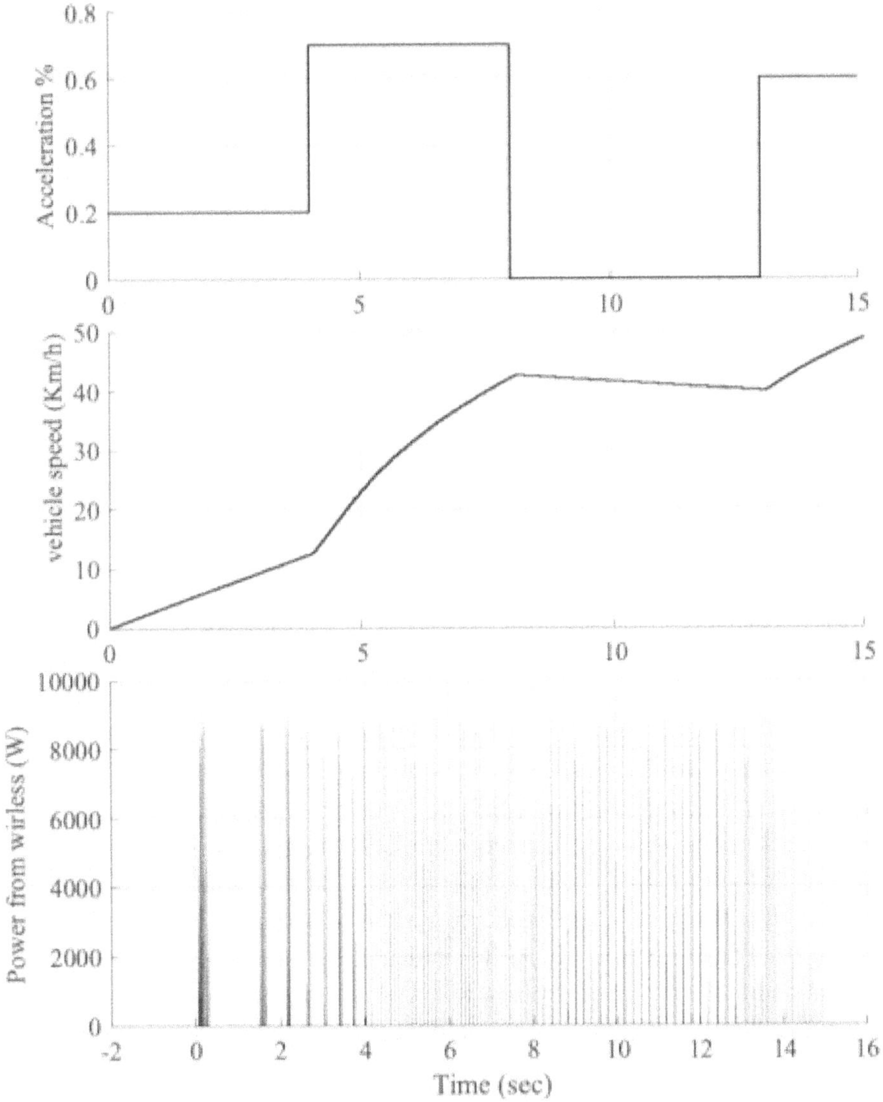

The battery SOC will be then changed and this is will be according to the vehicle speed. This is due to the quantity of the demand motor power according to the vehicle speed. Figure (11), shows the acceleration form and the vehicle speed, which touch 40km/h for an acceleration of 70%. In the same figure (11), the obtained power from the wireless system is shown, and it is clear that if the vehicle speed increases the number to the touched coils will be also increased. In the time zone from 4 seconds to 8 seconds, the maximum of touched coils is done. However, for the lowest speed it is the inverse.

So, the battery SOC evolution will be as it is in figure (12), where the value increase for the lowest speed and the inverse in the high speeds as the demands motor power is higher from the obtained from the recharge system. Therefore, it is recommended conducting the vehicle with a low speed if the road is equipped with wireless recharge technic.

Figure 12. Battery SOC evolution according the vehicle speed under the wireless recharge method

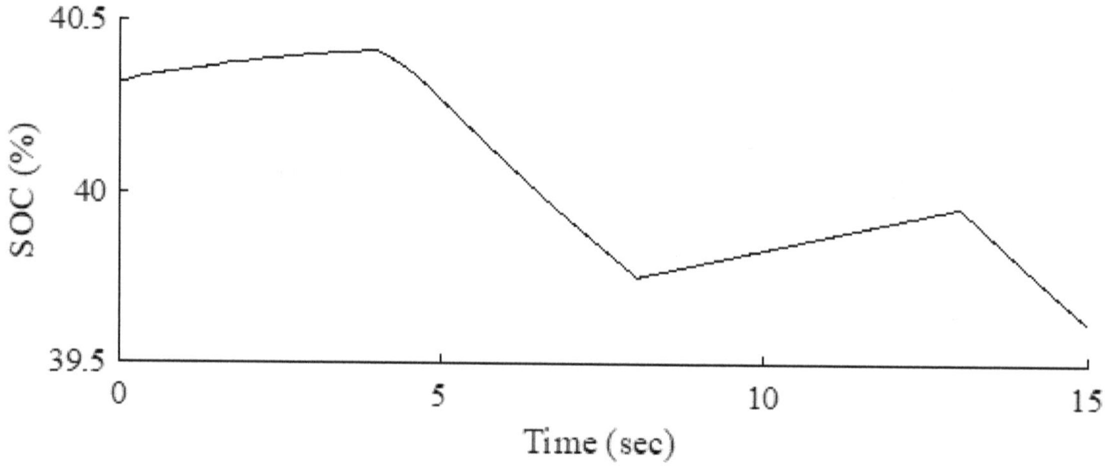

CONCLUSION

This chapter was formulated to express the mathematical models of the interior components inside an electric vehicle, taken into consideration the wireless recharge method as a solution for improving vehicle autonomy on road. The traction system and its components were defined and modeled, and then the overall mathematical model was tested under a simulation test. Basing on a brushless machine as the permanent magnet synchronous motor, the corresponding vector control loop was defined and implemented in the simulation phase. Also, the recharge system was defined and detailed on mathematical equations and then tested on the simulation. As it is indicated before, the wireless recharge method is the main recharge solution for this case, the exposed results show the influence of the vehicle speed on the performance of this recharging method. Basing on the obtained results, several future endeavors were appeared as searching a better and easier control solutions as the Direct torque control loop, which can operate the vehicle in high speed region. This is will give a better global performance for some sports cars. But also, concentrating on the problem of recharge phase and as they obtained power from this recharge solution is not enough, having a double wireless receiver can be a useful solution. This is for a high way and with a higher speeds, contrary to this case which is efficient only in low speed regions.

REFERENCES

Abdelwahab, O. M., & Shaaban, M. F. (2019). PV and EV charger allocation with V2G capabilities. *Proceedings - 2019 IEEE 13th International Conference on Compatibility, Power Electronics and Power Engineering, CPE-POWERENG.IEEE, 2019*, 1–5. 10.1109/CPE.2019.8862370

Alain, K. S. T., Romanic, K., & Bertrand, F. H. (2017). Dynamics and Improved Robust Adaptive Control Strategy for the Finite Time Synchronization of Uncertain Nonlinear Systems. *International Journal of System Dynamics Applications*, *6*(4), 34–62. doi:10.4018/IJSDA.2017100103

An, Q., & Sun, L. (2008). On-line parameter identification for vector controlled PMSM drives using adaptive algorithm. *2008 IEEE Vehicle Power and Propulsion Conference, VPPC 2011*, 59(11), 4169-4178. 10.1109/VPPC.2008.4677634

Auvigne, C., Germano, P., Civet, Y., & Perriard, Y. (2014). Design considerations for a contactless battery charger. *2014 16th European Conference on Power Electronics and Applications, EPE-ECCE Europe 2014*. 10.1109/EPE.2014.6910704

Bi, Z., Kan, T., Mi, C. C., Zhang, Y., Zhao, Z., & Keoleian, G. A. (2016). A review of wireless power transfer for electric vehicles: Prospects to enhance sustainable mobility. *Applied Energy*, *179*, 413–425. doi:10.1016/j.apenergy.2016.07.003

Chau, K. T., Chan, C. C., & Liu, C. (2008). Overview of permanent-magnet brushless drives for electric and hybrid electric vehicles. *IEEE Transactions on Industrial Electronics*, *55*(6), 2246–2257. doi:10.1109/TIE.2008.918403

Flah, A. (2018). Internal Fuzzy Hybrid Charger System for a Hybrid Electrical Vehicle. *ASME. Journal of Energy Resources Technology*, *140*(1), 12003–12008. doi:10.1115/1.4037352

Hamidi, A., Weber, L., & Nasiri, A. (2013). EV charging station integrating renewable energy and second-life battery. *Proceedings of 2013 International Conference on Renewable Energy Research and Applications, ICRERA 2013*, 1217–1221. 10.1109/ICRERA.2013.6749937

Hussein, H. A. T., Ammar, M. E., & Hassan, M. A. M. (2017). Three Phase Induction Motor's Stator Turns Fault Analysis Based on Artificial Intelligence. *International Journal of System Dynamics Applications*, *6*(3), 1–19. doi:10.4018/IJSDA.2017070101

Hwang, J. J., Kuo, J. K., Wu, W., Chang, W. R., Lin, C. H., & Wang, S. E. (2013). Lifecycle performance assessment of fuel cell/battery electric vehicles. *International Journal of Hydrogen Energy*, *38*(8), 3433–3446. doi:10.1016/j.ijhydene.2012.12.148

Khaligh, A., Member, S., Li, Z., & Member, S. (2010). Battery, Ultracapacitor, Fuel Cell, and Hybrid Energy Storage Systems for Electric, Hybrid Electric, Fuel Cell, and Plug-In Hybrid Electric Vehicles : State of the Art. *IEEE Transactions on Vehicular Technology*, *59*(6), 2806–2814. doi:10.1109/TVT.2010.2047877

Kroeze, R. C., & Krein, P. T. (2008). Electrical battery model for use in dynamic electric vehicle simulations. *IEEE Power Electronics Specialists Conference, 2008. PESC 2008*, 1336–1342. 10.1109/PESC.2008.4592119

Lee, C. S., Jeong, J. B., Lee, B. H., & Hur, J. (2011). Study on 1.5 kW battery chargers for neighborhood electric vehicles. *2011 IEEE Vehicle Power and Propulsion Conference. VPPC, 1–4*. Advance online publication. doi:10.1109/VPPC.2011.6043129

Lee, S., Cherry, J., Safoutin, M., McDonald, J., & Olechiw, M. (2018). Modeling and Validation of 48V Mild Hybrid Lithium-Ion Battery Pack. *SAE International Journal of Alternative Powertrains, 7*(3), 273–288. doi:10.4271/2018-01-0433

Mahmoudi, C., Flah, A., & Sbita, L. (2017). Prototype design of a compact plug in solar Electric Vehicle. *International Conference on Green Energy Conversion Systems (GECS)*, 1-4. http://batteryuniversity.com/learn/article/electric_vehicle

Mohamed, N., Aymen, F., Ben Hamed, M., & Lassaad, S. (2019). Analysis of battery-EV state of charge for a dynamic wireless charging system. *Energy Storage, 2020, 2*(2). doi:10.1002/est2.117

Mohamed, N., Aymen, F., & Mouna, B. H. (2018). Wireless Charging System for a Mobile Hybrid Electric Vehicle. *2018 International Symposium on Advanced Electrical and Communication Technologies (ISAECT)*, 1–5. 10.1109/ISAECT.2018.8618829

Mohamed, R. G., Ebrahim, M. A., & Bendary, F. M. (2017). *Transient Stability Enhancement for 20 MW PV Power Plant via Incremental Conductance Controller*. doi:10.4018/IJSDA.2017100106

Mohrehkesh, S., & Nadeem, T. (2011). Toward a wireless charging for battery electric vehicles at traffic intersections. *IEEE Conference on Intelligent Transportation Systems, Proceedings, ITSC*, 113-118. 10.1109/ITSC.2011.6083137

Mousa, M. E., Ebrahim, M. A., & Hassan, M. A. M. (2015). Stabilizing and Swinging-Up the Inverted Pendulum Using PI and PID Controllers Based on Reduced Linear Quadratic Regulator Tuned by PSO. *International Journal of System Dynamics Applications, 4*(4), 52–69. doi:10.4018/IJSDA.2015100104

Mousavi Agah, S. M., & Abbasi, A. (2012). The impact of charging plug-in hybrid electric vehicles on residential distribution transformers. *2012 2nd Iranian Conference on Smart Grids, ICSG 2012, 25*(1), 371–380.

Naoui, M. (2019). Inductive charger efficiency under internal and external parameters variation for an electric vehicle in motion. *International Journal of Powertrains, 8*(4), 343–358. doi:10.1504/IJPT.2019.104674

Rosu, S. G. (2016). A Dynamic Wireless Charging System for Electric Vehicles Based on DC / AC Converters with SiC MOSFET-IGBT Switches and Resonant Gate-Drive. IEEE.

Sarkar, J., & Bhattacharyya, S. (2012). Application of graphene and graphene-based materials in clean energy-related devices Minghui. *Archives of Thermodynamics, 33*(4), 23–40. doi:10.2478/v10173-012-0026-8

Sbita, L., & Flah, A. (2013). High speed PMSM control without speed encoder based on Q-MRAS observer. *CEIT-13EIT-13*, 174–178.

Spichkova, M., & Hamilton, M. (2016). Dynamic Decision Making System for Public Transport Routes. *International Journal of System Dynamics Applications, 5*(3), 47-70. doi:10.4018/IJSDA.2016070103

Sreedhar, V. (2006). Plug-in hybrid electric vehicles with full performance. *2006 IEEE Conference on Electric and Hybrid Vehicles, ICEHV*, 1-2. 10.1109/ICEHV.2006.352291

Sudimac, B., Ugrinović, A., & Jurčević, M. (2020). The application of photovoltaic systems in sacred buildings for the purpose of electric power production: The case study of the Cathedral of St. Michael the Archangel in Belgrade. *Sustainability (Switzerland), 12*(4), 1408. doi:10.3390u12041408

Thounthong, P., Raël, S., & Davat, B. (2007). Control strategy of fuel cell and supercapacitors association for a distributed generation system. *IEEE Transactions on Industrial Electronics, 54*(6), 3225–3233. doi:10.1109/TIE.2007.896477

Tremblay, O., & Dessaint, L. (2009). *Experimental Validation of a Battery Dynamic Model for EV Applications*. Academic Press.

Un-Noor, F., Padmanaban, S., Mihet-Popa, L., Mollah, M. N., & Hossain, E. (2017). A comprehensive study of key electric vehicle (EV) components, technologies, challenges, impacts, and future direction of development. *Energies, 10*(8), 1217. doi:10.3390/en10081217

Xu, J., Wang, F., Xie, S., Xu, J., & Feng, J. (2004). A new control method for permanent magnet synchronous machines with observer. *PESC Record - IEEE Annual Power Electronics Specialists Conference, IEEE*, 1404-1408. 10.1109/PESC.2004.1355628

Yilmaz, M., & Krein, P. T. (2013). Review of battery charger topologies, charging power levels, and infrastructure for plug-in electric and hybrid vehicles. *IEEE Transactions on Power Electronics, 28*(5), 2151-2169. doi:10.1109/TPEL.2012.2212917

Zandi, M., Payman, A., Martin, J., Pierfederici, S., Davat, B., & Meibody-Tabar, F. (2011). Energy Management of a Fuel Cell / Supercapacitor / Battery Power Source for Electric Vehicular Applications. *IEEE Transactions on Vehicular Technology, 2010, 60*(2), 433-443. doi:10.1109/TVT.2010.2091433

Zhang, R., Tao, J., & Zhou, H. (2019). Fuzzy Optimal Energy Management for Fuel Cell and Supercapacitor Systems Using Neural Network Based Driving Pattern Recognition. *IEEE Transactions on Fuzzy Systems, 27*(1), 45-57. doi:10.1109/TFUZZ.2018.2856086

Zhou, D., Gao, F., Ravey, A., Al-Durra, A., & Simões, M. G. (2017). Online energy management strategy of fuel cell hybrid electric vehicles based on time series prediction. *2017 IEEE Transportation Electrification Conference and Expo (ITEC)*, 113-118. 10.1109/ITEC.2017.7993256

APPENDIX: NOMENCLATURE

vd, vq: Direct and quadrature stator voltages
λd, λq: Direct and quadrature stator field
id, iq: Direct and quadrature stator currents
Ld, Lq: Direct and quadrature stator inductances
Ls: Stator inductance
λm: Magnet flux
Rs: Stator resistance
Tl: Load torque
np: Poles numbers
J: Rotor inertia coefficient
e: Error between the estimated and the real current

Section 2
Control Design and Nonlinear Dynamics

Chapter 15
Robust Control Methods for Finite Time Synchronization of Uncertain Nonlinear Systems

Kammogne Soup Soup Tewa Alain
University of Dschang, Cameroon

Fotsin Hilaire Bertrand
University of Dschang, Cameroon

ABSTRACT

This chapter addresses the dynamic analysis and two different control strategies for the synchronization of new topology of Colpitts oscillator submitted to uncertainties and external disturbances. The diagrams obtained reveal precisely spirals bifurcation and chaos when for a specific values of the system parameters. Based on the relevant control, the authors have controlled this striking phenomenon in the system. The first (control) deals with the sliding mode control (SMC) method. Some important aspects of the design and implementation are considered to reach a suitable controller for the applications. The second presents an adaptive robust tracking control strategy based on a modified polynomial observer which tends to follow exponentially the chaotic Colpitts circuits brought back to a topology of the Chua oscillator with perturbations. To highlight the contribution, they also present some simulation results with the purpose to compare the proposed method to the classical polynomial observer.

1. INTRODUCTION

The study of nonlinear systems is very relevant because the nature and all its phenomena are intrinsically nonlinear. The understanding of nonlinear systems is a continuous challenge and requires an interdisciplinary approach that brings together methodologies and tools from different research areas, and also inspires the development of new approaches based on the theory of dynamical systems. Many nonlinear phenomena are described by differential equations, especially by partial differential equations (PDEs) and ordinary differential equations (ODEs). Note that, symmetry group analysis of a differential equa-

DOI: 10.4018/978-1-7998-5788-4.ch015

tion appears as a powerful method to analyze PDEs (Ibragimov, 1985; Olver, 1993). Among its many applications, we highlight the fact that they allow us to obtain exact solutions of a PDE, directly or by using similarity solutions (Avdonina et al., 2013); classify invariant equations; reduce the number of independent variables and construct conservation laws. Many approaches to analyze the ODEs are developed in literature (Kudryashov et al., 2005, Kudryashov, 2013). Due to implementation of sophisticated algorithms, the approach is basically qualitative through computer-based methods. Many electronic circuits such as Colpitts oscillators, van der Pol oscillators, Chen systems, Chua circuit, Harley oscillator, Jerk oscillator and so on depicts nonlinear phenomena. Amongst theses topologies, the dynamic of Chua's circuit is better understood due to its piecewise nonlinear continuous function. The impact of nonlinearity on circuit analysis for the design of new electronic devices have generated much interest in nonlinear circuit theory (Vaidyanathan et al., 2017a, 2017b; Volos et al., 2016; Azar et al., 2017a; Pham et al., 2017; Vaidyanathan and Azar, 2016a, 2016b)

The various models associated with the Chua's circuit have been recently found helpful in demonstrating and explaining the various facets of chaos (Chua and Lin, 1900). Attempts are made in the literature to discover new members of the large family of Chua's circuits and to find their close relatives (Chua, 1993). The present communication deals with the well-known Colpitts oscillator, which was shown to be topologically similar to Chua's circuit (Sarafian and Kaplan, 1995). The nonlinearity of the active device in the Colpitts oscillator is modified to be purely odd, then the circuit exhibits chaotic phenomena closely related to those exhibited by the classical Chua's circuit. There exists a robust relationship between the Colpitts oscillator and the Chua's oscillator (Kennedy, 1995), one can show that both systems are not simply close relatives, but they are even strongly related and can be regarded as being conjugate one to another. The dynamics behavior of the Colpitts oscillator that results from this kind of modeling is reported in this paper and is related to the well-known chaotic behavior of Chua's circuit. In the past, these works was essentially theoretical and the technological aspect was limited (Kennedy, 1995). It ought to be mentioned that the relationship between Colpitts and Chua oscillators with a symmetric nonlinearity leads to a singular phenomenon with only very few cases reported. The research on this model is motivated by the discovery of a simple third-order ordinary differential equations of the form $\dddot{x} = J(x, \dot{x}, \ddot{x})$ whose solution are chaotic. The nonlinear function J is called a *Jerk,* because it describes the third-time derivative of acceleration in a mechanical system (Schot, 1978). Sprott (Sprott, 2000) proposed many new jerk systems with several nonlinearities that show chaotic behavior with easy electronic implementation. In one of the chapter of "Elegant Chaos: Algebraically Simple flow", book published in 2010, Sprott proposed a list of 16 autonomous jerk oscillators with different nonlinearities called memory oscillators (MO_0 to MO_{15}) (Sprott, 2010). A simple chaotic jerk circuit was used in a sound encryption scheme (Volos, 2017). Especially, a three-dimensional novel jerk chaotic oscillator with two hyperbolic sinusoidal nonlinearities was reported in (Vaidyanathan, 2014). We should note that, the representation of the Colpitts oscillator as jerky dynamics has another useful and important advantage: based on their jerky dynamics, a classification of different dynamical systems is possible, because the transformation of certain functionally different three-dimensional dynamical systems can lead to the same jerky dynamics (Kammogne et al., 2017; Eichhorn Ralf, et al., 2002).

Synchronization of chaotic systems is a current challenge in control theory (Azar and Vaidyanathan, 2017; Azar and Vaidyanathan, 2015a, 2015b, 2015c; Wang et al., 2017). Researches on the coupled nonlinear oscillators constitute an excellent framework for understanding the various complex collective dynamics that spontaneously emerge in real-life systems (Hossein, et al., 2018; Pikovsky and Rosenblum,

2001) and (Balochian et al., 2018; Lakshmanan and Senthilkumar, 2010). Master-slave synchronization has many applications in technology such as in secured telecommunication (Chen, 2009) and (Kammogne et al., 2015; Ouannas et al., 2017a, 2017b, 2017c; Ouannas et al., 2016a, 2016b; Boulkroune et al., 2016a; Vaidyanathan and Azar, 2016f, 2016g). Specifically, motivated by the vulnerability to attacks of certain architectures, the use of chaotic systems for data scrambling has thoroughly been investigated. It is well known that some researches today are most oriented to neural network synchronization (Ma et al., 2015) and (Tosi et al., 2016) but the master-slave synchronization remains the benchmark of all methods encountered in the literature. The observer-based method has been naturally proposed for chaos synchronization, where the outputs of the master system are to control the slave system so that they can oscillate in a synchronized manner. A variety of observer-based approaches have been proposed for the synchronization of chaotic systems, which include the exponential polynomial observer (Mata et al., 2010), sliding observer (Ghabi et al., 2018; Feki, 1990), higher-order sliding observer (Mounir et al., 2018; Rodriguez, 2005), fuzzy disturbance observer (Yoo, 2010), etc. Lately J.L Mata proposed a chaotic systems synchronization via high order observer design (Mata et al., 2011).

A practical system is often subjected to external unknown disturbances, i.e, in the modeling process, there exist a vast amount of uncertainties caused by modeling error. The disturbance may lead to oscillations and even increase the instability of the system. In the field of traditional control, it is well known that feedforward control provides an effective disturbance compensation method which can achieve prompt disturbance attenuation which may not only degrade the performance of the control system but also even lead to instability of the dynamics system. Therefore, the uncertainty should be considered in the control design to improve the closed-loop system performance of fractional-order systems. However, the disturbance has to be measured by sensors for the implementation of traditional feedforward control. Unfortunately, disturbances are usually difficult or even impossible to be measured physically using sensors. Since disturbance observers can estimate external disturbances from known information of the controlled plants, the output of disturbance observers can be used to design the control law. As a result, disturbance rejection is guaranteed to improve the performance and robustness of the closed-loop system. Due to the simplicity in implementation, low sensitivity to external perturbations, robustness against parameter variation and high stability attributes, the sliding mode control strategy has achieved more popularity among the proposed controllers for nonlinear industrial systems.

It is well known that sliding-mode control is an effective robust control scheme and has the features of fast global convergence and high robustness to external disturbances (Sabanovic, 2011). In recent years, sliding-mode control has been investigated for integer-order linear and nonlinear systems (Shi et al., 2006; Byungkook, 1998) and many important results have been reported for the control of chaotic systems using the sliding-mode technique (Vaidyanathan and Azar, 2015e, 2015f; Bouzaida and Sakly, 2018). However, most of the traditional sliding mode controllers use at least the upper bounds of the system uncertainties and external noises for handling their undesirable effects. However, in practical situations, there is usually no information about the externally perturbed terms and parameter variations in advance. Since the bounds of the uncertain parameters and external disturbances have been used as coefficients for the sign function in the switching control rule, and the upper bounds are unknown, the switching gains have been commonly selected to be large enough to respond a wide variety of uncertain terms. Such large switching gains may be foundation of chattering and the robot manipulator will oscillate around the sliding surface due to the physical limitations of actuators and relay delays (Lee et al., 2007). To respond to this drawback, the adaptive laws are considered which is a valid control technique for linear and nonlinear systems with parameter uncertainty. So far, integer-order linear have been extensively investigated based on the adaptive control method (Kammogne et al., 2019; Kammogne et al., 2020). In addition, another major problems in the design of sliding mode controller is chattering phenomenon. So far, many methods have been presented to attenuate chattering. One of the common ways to reduce chattering is to use boundary layer method (Slotine, 1991). In Lee et al., (2001), an improvement was reported in control performance emanating from the utilization of fuzzy boundary layer approach in SMC design. Another approach is to design high-frequency second-order SMC (Huangfu et al., 2012; Shtessel et al. 2003). Furthermore, state observer (Shi et al. 2018; Eng and Sci., 2011) and adaptation algorithm (Yang and Yan 2016) are two effective strategies to alleviate chattering problem.

New robust theories for chaotic synchronization are needed in order to model and control highly complex systems such as those involving extremely large number of control loops, to control nonlinear, hybrid and stochastic systems and to handle very large model uncertainties. It is a challenge to develop powerful new identification tools that are needed to model large scale systems. In this case, a polynomial strategy is considered here as an alternative solution. Besides, a new exponential observer for a class of nonlinear oscillators is proposed by J.L Mata and coworkers [Mata-Machuta et al. 2011]. These authors demonstrated that the construction of the observer can be made by means of simple algebraic manipulations. Lately, the same approach was investigated by Rafael Martinez-Guerra et al., [Martinez-Guerra et al. 2013] in which the properties of non-decreasing functions, non-increasing functions, linear matrix inequalities and the Lyapunov method was used to prove that the estimation error converges to zero exponentially. The works in the previous references are limited to the specific system configuration without environmental perturbations consideration. One limitation of the proposed scheme was the evaluation of the control performance of the scheme because the authors did not consider the uncertainties. It's worth noting that uncertainties occur commonly in many real-world control systems. It has been noted that when chaotic synchronization is applied to engineering such as communications, these chaos-based synchronization schemes are rather sensitive to noise and distortions in channels. This is an undesirable feature, as it is one of the main factors leading to a failure of the intended synchronization. This effect should be taken into account when we want to evaluate the performance of a practical chaos synchronization scheme. It's important to note that the results obtained in the works previously mentioned let us conjecture that there are some unexplored performances aspects of the polynomial observer such as the robustness and the time duration of the control process. Such type of analysis has not yet been studied in the universal Chua circuit in spite of the huge amount of related research works in control theory when a complex environment is concerned and thus represents a meaningful contribution to the understanding of the behaviour of robust exponential polynomial observer in general.

This work considers the master–slave synchronization problem via generalized sliding mode control method and a robust modified exponential polynomial observer (RMEPO) based on differential algebraic techniques. An additional term is added to the well-known controller which enable to establish the observability condition, to overcome the perturbations on the system's parameters and to control the time response of the system. An observable system in this sense can be regarded as a system whose state variables can be expressed in terms of the input and output variables and a finite number of their time derivatives. Thus, chaos synchronization problem can be posed as an observer design procedure, where the coupling signal is viewed as output and the slave system is regarded as observer. The strategy consists of proposing a RMEPO which exponentially reconstructs the unknown states of an uncertain nonlinear system.

The aim of this chapter is as follows:

a- Analyse the dynamical of the Colpitts brought back to the Chua circuit with common chaotic indicator.
b- design a robust sliding mode control with an modified approach proposed in by
c- design modified exponential polynomial observer (RMEPO) for chaotic synchronization.
d- make a simple comparison of the proposed controller with that encountered in the literature.

As it is known, the observer method is widely used in many applications (Kammogne et al., 2011) and (Kammogne et al., 2013).

Moreover, to deal with uncertainties and disturbances, the adaptive method is used and the RMEPO is derived in order to achieve robust output synchronization for an uncertain Colpitts oscillator brought back to the Chua topology. Based on Lyapunov theory and some algebraic criteria, we get a new sufficient condition that can guaranteed the observer is robust against the effect of disturbances. Numerical results show its satisfactory performance.

The rest of the paper is organized as follows. In Section 2, the nonlinear system is presented and the dynamical behaviors of the circuit are identified with the help of bifurcation diagram and one parameter Lyapunov exponent diagram. The problem is formulated in section 3. The finite-time synchronization issue with the help of the RMEPO is presented in Section 4. The simulation results are shown in Section 5. In the last section, the conclusions are summarized.

2. THE PROPOSED CHAOTIC SYSTEM AND ITS ANALYSES

2.1 System description

Let us consider the two models represented by Figure. 1(a) and Figure. 1(b). The slight difference between them resides at the level of the nonlinear element. Whereas the passive network is the same in both cases. The states equations of the Colpitts oscillator are:

$$\begin{aligned} & c_1 \frac{dv}{dt} - i_L = 0 \\ & Ri_L + L\frac{di_L}{dt} + v + v_2 = 0 \\ & -i_L + c_2 \frac{dv_2}{dt} - f(v) = 0 \end{aligned} \tag{1}$$

With the aim of accentuating the chaotic behavior, we model the nonlinearity by omitting the quadratic term (Chua and Lin, 1900). That is to say:

$$f(v) = \beta_1 v + \beta_3 v^3 \tag{2}$$

Figure 1. **(a)** *Topology of Colpitts oscillator* **(b)** *Topology of the Chua oscillator*

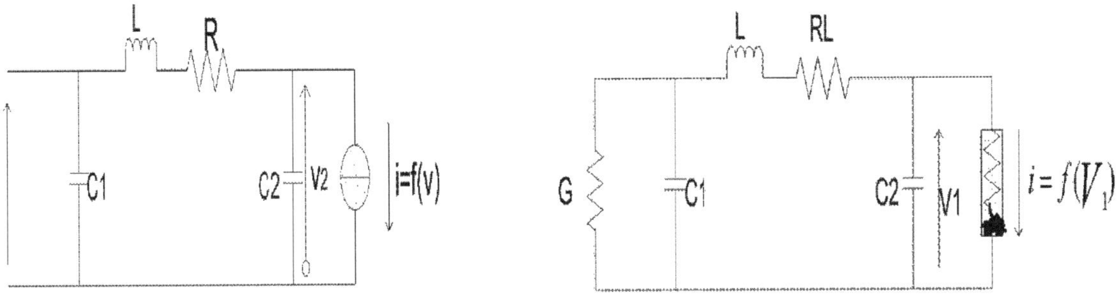

By combining the equations of the system (1) and by taking into account of (2) we obtain the following third order differential equation:

$$\frac{d^3v}{dt^3}+\alpha_1\frac{d^2v}{dt^2}+\alpha_2\frac{dv}{dt}+\alpha_3 v+k_3v^3=0 \tag{3}$$

with:

$$\alpha_1=\frac{R}{L},\alpha_2=\frac{c_1+c_2}{Lc_1c_2},\alpha_3=\frac{\beta_1}{Lc_1c_2}, \text{ and } k_3=\frac{\beta_3}{Lc_1c_2}$$

These various parameters are those of the topology of Chua, deduced from the Standard Colpitts oscillator. The reference (Kennedy, 1995) gives the value of its coefficients and proposes the following standardization made on the case of a transistor of reference 2N2222:

$\frac{1}{c_2}=2.2, \frac{1}{c_1}=2.4$, L=1, R_L=0.249, β1_1, β3=-0.2

From equation (3) we deduce the following system:

$$\begin{aligned}\frac{dx_1}{dt}&=x_3\\ \frac{dx_2}{dt}&=x_1\\ \frac{dx_3}{dt}&=-\alpha_2x_1-\alpha_1x_3-\alpha_3x_2-k_3x_2^3\end{aligned} \tag{4}$$

2.2 Existence and Uniqueness of the Solution

System (4) can be written in the following form:

$$Dx(t) = H_1x(t) + x_2^2H_2x(t), \quad \forall t \in (0,T] \tag{5}$$

with initial conditions of the system given by

$$x(0) = x_0 \tag{6}$$

where

$$x = \begin{bmatrix} x_1 \\ x_2 \\ x_3 \end{bmatrix}, x_0 = \begin{bmatrix} x_{10} \\ x_{20} \\ x_{30} \end{bmatrix}, H_1 = \begin{bmatrix} 0 & 0 & 1 \\ 1 & 0 & 0 \\ -\alpha_2 & -\alpha_3 & -\alpha_1 \end{bmatrix}, \text{and } H_2 = \begin{bmatrix} 0 & 0 & 0 \\ 0 & 0 & 0 \\ 0 & 0 & -k_3 \end{bmatrix} \tag{7}$$

The supremum norm utilized in the following analysis is defined for the class of continuous function $C[0,T]$ by

$$\|\grave{E}\| = \sup_{t\in(0,T]} |\grave{E}(t)|, \quad \grave{E}(t) \in C[0,T], \tag{8}$$

and for a matrix $M=[m_{ij}(t)]$, it is defined by

$$\|M\| = \sum_{i,j} \sup_{t\in(0,T]} |m_{ij}(t)| \tag{9}$$

Theorem 1. *The sufficient condition for existence and uniqueness of the solution of system (4) with initial conditions* $x(0)=x_0$ *in the region* $\Omega\times J$ *is:*

$$k_s = \|B_1\| + \|B_2\|(2x_0 + 4\varepsilon_0^4) \tag{10}$$

Proof. The existence and uniqueness of the solution is studied in the region $\Omega\times J$ where $J=(0,T]$ and $\Omega=\{(x_1,x_2,x_3)\}$. We consider the function

$$H(x) = H_1x(t) + x_2^2H_2x(t), \quad \forall t \in (0,T] \tag{11}$$

continuous and bounded on any interval $[x_0-\varepsilon, x0\_\varepsilon]$ for any $\varepsilon \succ 0$ from the Lipschitz equivalence of the equation (11) and the system (5)-(6), denoting the right hand side of (11) by $H(x)$, then

$$\begin{aligned} H(x)-H(\omega) &= \left\| H_1x(t)+x_2^2H_2x(t)-H_1\omega(t)-\omega_2^2H_2\omega(t) \right\| \\ &= \left\| H_1x(t)-H_1\omega(t)+x_2^2H_2x(t)-\omega_2^2H_2\omega(t) \right\| \end{aligned} \tag{12}$$

The following inequality holds for in the region $\Omega \times J$

$$\begin{aligned} \left| x_2^2-\omega_2^2 \right| = \left| x_2-\omega_2 \right| \left| x_2+\omega_2 \right| &\le 2\varepsilon_0 \left| x_2-\omega_2 \right| \\ &\le 2\varepsilon_0 \left| x-\omega \right| \end{aligned} \tag{13}$$

therefore

$$\begin{aligned} \left| x_2^2H_2x(t)-\omega_2^2H_2\omega(t) \right| &\le \left| H_2(x_2^2(t)-\omega_2^2(t))x(t)+H_2\omega_2^2(x(t)-\omega(t)) \right| \\ &\le \left| 2\varepsilon_0H_2 \left| x(t)-\omega(t) \right| x(t)+H_2\omega_2^2(x(t)-\omega(t)) \right| \\ &\le \left\| H_2 \right\| \left| 2\varepsilon_0 \left| x(t) \right| + \left| \omega_2^2 \right| \right| \left| x(t)-\omega(t) \right| \\ &\le \left\| H_2 \right\| \left| 2\varepsilon_0 \left| 2x_0 \right| + \left| 2\varepsilon_0 \right|^2 \right| \left| x(t)-\omega(t) \right| \end{aligned} \tag{14}$$

which implies that

$$\begin{aligned} \left| H(x)-H(\omega) \right| &\le \left\| H_1 \right\| \left| x(t)-\omega(t) \right| + \left\| H_2 \right\| \left| 2\varepsilon_0 \left| 2x_0 \right| + \left| 2\varepsilon_0 \right|^2 \right| \left| x(t)-\omega(t) \right| \\ &\le \left(\left\| H_1 \right\| + 2\varepsilon_0 \left\| H_2 \right\| \left| \left| 2x_0 \right| + 2\varepsilon_0 \right| \right) \left| x(t)-\omega(t) \right| \\ &\le k_s \left| x(t)-\omega(t) \right| \end{aligned} \tag{15}$$

where

$$k_s = \left(\left\| H_1 \right\| + 2\varepsilon_0 \left\| H_2 \right\| \left| \left| 2x_0 \right| + 2\varepsilon_0 \right| \right) \tag{16}$$

The following consideration can be made if $k_s<1$, then then mapping $\dot{x} = H(x)$ is a contraction mapping. The proof of *theorem 1* is complete.

2.3 Dissipativity and Stability Analysis

2.2.1 Dissipativity

An important step towards the stability analysis could be to evaluate the volume contraction of the oscillator modeled by system (4) in order to get preliminary insights of the types of attractors (stable or unstable) which might coexist in the system.

We recall that the volume contraction rate of the system (4) is given by

$$\Pi = \nabla . F(X) = \frac{\partial \dot{x}_1}{\partial x_1} + \frac{\partial \dot{x}_2}{\partial x_2} + \frac{\partial \dot{x}_3}{\partial x_3} \quad (17)$$

Referring to our model, it is straightforward to show that:

$$\Pi = -\alpha 1 < 0 \quad (18)$$

Therefore, the dissipativity is always negative, implying that all orbits are ultimately confined to a specific subset of zero volume, and the asymptotic motion settles onto an attractor.

2.2.2 Stability of the Equilibrium Points.

It easily shown that the system (4) has three equilibriums points E_0= (0,0,0); $E_+ = \left(\sqrt{5}, 0, 0\right)$ and $E_- = \left(-\sqrt{5}, 0, 0\right)$.

The jacobian matrix of system (4) about the equilibrium point $\bar{E} = (\bar{x}, \bar{y}, \bar{z})$ provide the characteristic equation

$$\lambda^3 + \beta_1 \lambda^2 + \beta_2 \lambda + \beta_3 = 0 \quad (19)$$

where $\beta_1 = -\alpha_1$, $\beta_2 = -\alpha_2$, $\beta_3 = -\alpha_3 - 3k_3\bar{y}^2$

If $\bar{y} = 0$ (or $y = \pm\sqrt{5}$), then $\beta 1 < 0$ and $\beta 3 < 0$. According to Routh-Hurwitz criteria the equilibrium point E0= (0,0,0) is unstable. The singularity is saddle point. Since the equilibrium points of system (4) are unstable. The control problem takes place.

If the parameters are taken as $\alpha 1$=0.249, $\alpha 2$=4.6, $\alpha 3$=5.28, k3= -1.056, the systems specified by (4) exhibit chaotic dynamics. Figures. 2(a) and 2(b) give the time history and phase portrait of the systems. Figures. 3(a) and 3(b) show a one-parameter bifurcation diagram obtained with the variation of $\alpha 1$ and the maximal Lyapunov exponent.

Remark 1:

The circuit on figure (1a) is more practical and less boring in practice more than those propose in the literature. The nonlinearity is a simply odd function of one variable. This prototype topology derived from the original Colpitts oscillator will find a general application in control theory in the purpose to build robust and efficient controllers.

Figure 2. **(a)** *Chaotic attractor z-y.* **(b)** *Chaotic attractor z-x.*

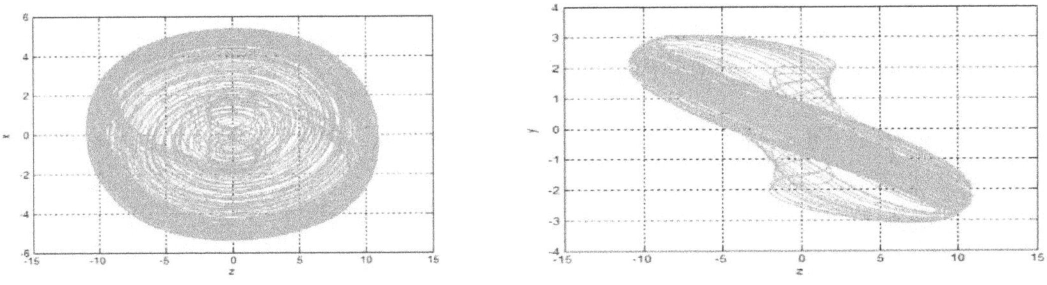

Figure 3. **(a)** *one-parameter bifurcation diagram of x when α1 varies.* **(b)** *Maximum Lyapunov exponent.*

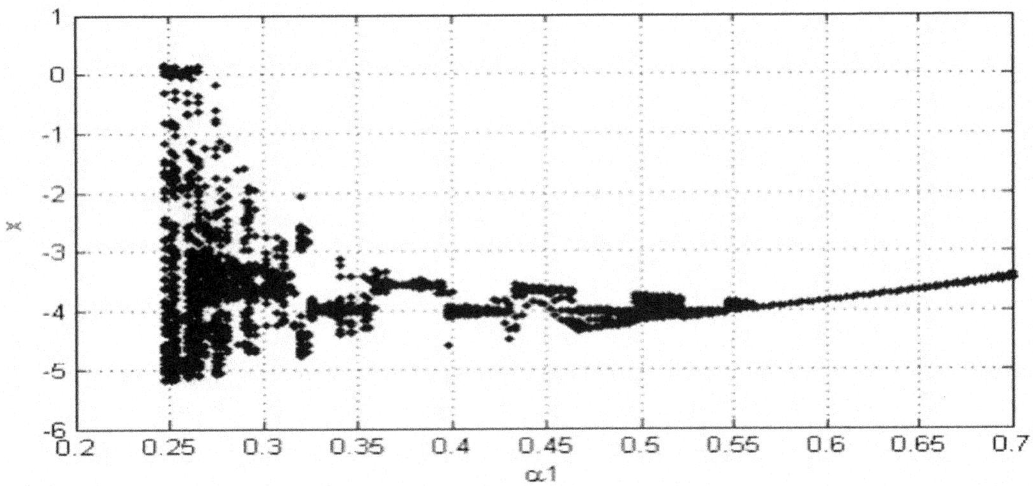

(a)

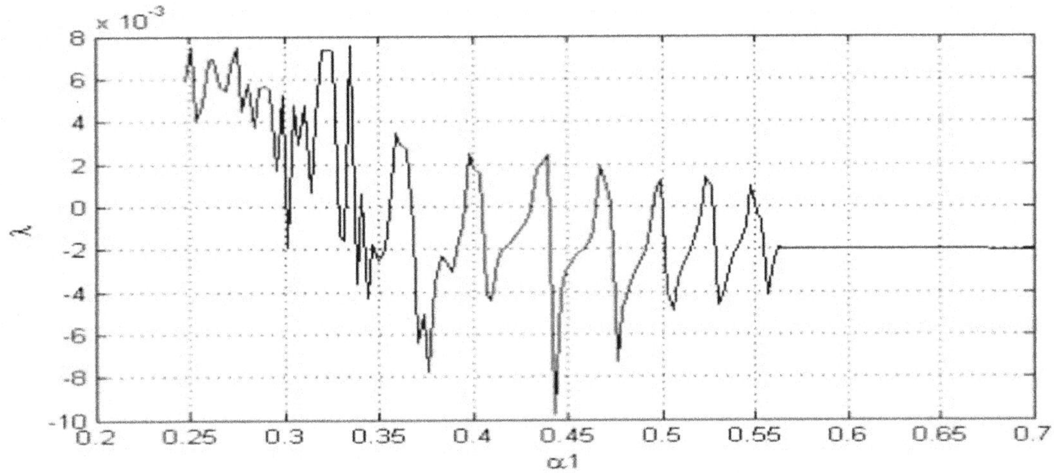

(b)

3. SLIDING CONTROL APPROACH FOR THE ROBUST SYNCHRONIZATION OF TWO CHAOTIC SYSTEMS.

3.1 System Presentation and Transformation

At this step, we introduce a relevant linear transformation of the coordinates of the system (21). From Eqs. (4), we let

$$[x_1, x_2, x_3] = [kx_1', kx_2', kx_3'] \tag{20}$$

where k is the positive constant. The Eqs. (4) becomes

$$\begin{aligned}\frac{dx_1'}{dt} &= x_3' \\ \frac{dx_2'}{dt} &= x_1' \\ \frac{dx_3'}{dt} &= -\alpha_2 x_1' - \alpha_1 x_3' - \alpha_3 x_2' - k^2 k_3 x_2'^3\end{aligned} \tag{21}$$

Remark 1: It's worth noting that the great sensibility of the system parameters can destroy the performance of the proposed sliding mode controller. An additional option is needed to handle with good accuracy the convergence of the problem and increase the degree of freedom of the system.

By letting $k^2k_3=k_0$, the following three-dimensional uncertain equations of the master system in it canonical form is presented as follows:

$$\begin{aligned}\frac{dx_{1m}}{dt} &= x_{3m} \\ \frac{dx_{2m}}{dt} &= x_{1m} \\ \frac{dx_{3m}}{dt} &= -\alpha_2 x_{1m} - \alpha_1 x_{3m} - \alpha_3 x_{2m} + \Delta f(x_{1m}, x_{2m}, x_{3m}, t) - k_0 x_{2m}^3\end{aligned} \tag{22}$$

where $x_{1m}(t)$, $x_{2m}(t)$, $x_{3m}(t)$, represent the states of the master system. The model of the slave chaotic system is specified as

$$\begin{aligned}\frac{dx_{1s}}{dt} &= x_{3s} + u_1(t)\\ \frac{dx_{2s}}{dt} &= x_{1s} + u_2(t)\\ \frac{dx_{3s}}{dt} &= -\alpha_2 x_{1s} - \alpha_1 x_{3s} - \alpha_3 x_{2s} - k_0 x_{2s}^3 + \Delta g(x_{1s}, x_{2s}, x_{3s}, t) + u_3(t)\end{aligned} \tag{23}$$

where $x_{1s}(t)$, $x_{2s}(t)$, $x_{3s}(t)$, represent the states of the master system. $u_1(t)$, $u_2(t)$, and $u_3(t)$ are the control inputs and $\Delta f(x_{1s}, x_{2s}, x_{3s}, t)$ is the external disturbance.

We define the synchronization error as follows

$$\begin{aligned}e_1(t) &= x_{1s}(t) - x_{1m}(t)\\ e_2(t) &= x_{2s}(t) - x_{2m}(t)\\ e_3(t) &= x_{3s}(t) - x_{3m}(t)\end{aligned} \tag{24}$$

The dynamical errors leads to the following expressions

$$\dot{e}_1(t) = \dot{x}_{1s}(t) - \dot{x}_{1m}(t) = e_3(t) + u_1(t) \tag{25}$$

$$\dot{e}_2(t) = \dot{x}_{2s}(t) - \dot{x}_{2m}(t) = e_1(t) + u_2(t) \tag{26}$$

$$\dot{e}_3(t) = -\alpha_2 e_1 - \alpha_1 e_3 - \alpha_3 e_2 - k_0\left(x_{2s}^3 - x_{2m}^3\right) + \Delta g(x_{1s}, x_{2s}, x_{3s}, t) + u_3(t) \tag{27}$$

Assumption 1: The uncertainties terms $\Delta f(x_{1m}, x_{2m}, x_{3m}, t)$ and $\Delta g(x_{1s}, x_{2s}, x_{3s}, t)$ are bounded such that:

$$\left|\Delta f(x_{1m}, x_{2m}, x_{3m}, t)\right| \prec \Omega_1, \quad \Delta g(x_{1s}, x_{2s}, x_{3s}, t) \prec \Omega_2 \tag{28}$$

As a result, one has:

$$\left|\Delta f(x_{1m}, x_{2m}, x_{3m}, t) - \Delta g(x_{1s}, x_{2s}, x_{3s}, t)\right| \prec \Omega \tag{29}$$

where Ω_1, Ω_2, and Ω are the known positive constants. The main purpose of this paper is design a suitable robust nonsingular controller for synchronization of two chaotic systems (22) and (23) in finite-time.

Lemma (Moulay and Perruquetti, 2006). Assume that the continuous positive-define function $V(t)$ fulfills the following differential inequality:

$$\dot{V}(t) \le -\delta_1 V(t)^{\eta} - \delta_2 V(t) \qquad \forall t \ge t_0, V(t_0) \ge 0 \tag{30}$$

where δ1 and δ2 are two positive constants, and is a ratio of two odd positive integers ($1 \succ \eta \succ 0$). Then, for the initial time *t0* the positive-definite function V*(t)* approaches to the origin at least in a finite time t*s*:

$$t_s = t_0 + \frac{1}{\delta_2(1-\eta)} \ln \frac{\delta_1 + \delta_2 V(t_0)^{1-\eta}}{\delta_1} \tag{31}$$

3.2 Sliding Controller Design

In the following, the sliding control approach is used to design a robust control law which ensures synchronization objective. Design the control law is divided into two phases: first, choosing an adequate terminal sliding surface to achieve the control objective. Second, design of a discontinuous control law which forces the system trajectories.

The sliding surface is defined as

$$s(t) = \sum_{i=1}^{3} h_i \left(e_i(t) - e_i(0) \tanh(-p_i t) \right) \tag{32}$$

where h_1,h_2,h_3 are the gain constants and p_i are the appropriate positive coefficients.

Theorem 2: Considering the master-slave equations (20)-(21) and adopting the control inputs $u_1(t)$, $u_2(t)$, and $u_3(t)$ as

$$\begin{aligned} u_1(t) &= -e_3(t) - h_1 e_1(0) \tanh(-p_1 t) - \rho_1 s(t) \\ u_2(t) &= -e_1(t) - h_2 e_2(0) \tanh(-p_2 t) - \rho_2 \operatorname{sgn}(s(t)) |s(t)|^q \\ u_3(t) &= \alpha_2 e_1 + \alpha_1 e_3 + \alpha_3 e_2 + k_3 \left(x_{2s}^3 - x_{2m}^3 \right) - h_3 e_3(0) \tanh(-p_3 t) - \rho_3 \operatorname{sgn}(s(t)) \end{aligned} \tag{33}$$

then the error dynamics given by equations (23)–(25) are forced to move from initial conditions to the sliding surface of equation (30) in a finite time T, determined by (32)

$$T = t_0 + \frac{1}{\delta_{20}(1-\eta')} \ln \frac{\delta_{10} + \delta_{20} V(t_0)^{1-\eta'}}{\delta_{10}} \tag{34}$$

where $\delta_{10} = 2^{\frac{\eta+1}{2}} \rho_1 h_2$, $\delta_{21} = 2 h_1 \rho_2$ and $\eta' = \frac{\eta+1}{2}$

Proof. The derivative of the surface *s(t)* is obtained from the equation (30)

$$\dot{s}(t) = \sum_{i=1}^{3} h_i \left(\dot{e}_i(t) - e_i(0) \tanh(-p_i t) \right) \tag{35}$$

Define a Lyapunov function as follows

$$V(t)=\frac{1}{2}s(t)^2 \tag{36}$$

Taking the time derivative of *V(t)*, one has:

$$\begin{aligned}\dot{V}(t)&=\dot{s}(t)s(t)\\ &=\dot{s}(t)\left\{\sum_{i=1}^{3}h_i\left(\dot{e}_i(t)-e_i(0)\tanh(-p_it)\right)\right\}\end{aligned} \tag{37}$$

Now, applying the control laws of equations (33) in equation (37), it follows that:

$$\begin{aligned}\dot{V}(t)&=s(t)\left\{-h_1\rho_1 s(t)-h_2\rho_2\operatorname{sgn}(s(t))\left|s(t)\right|^q-h_3\rho_3\operatorname{sgn}(s(t))+h_3(\Delta f(x_{1m},x_{2m},x_{3m},t)-\Delta g(x_{1s},x_{2s},x_{3s},t))\right\}\\ &\leq -h_1\rho_1 s(t)^2-h_2\rho_2\left|s(t)\right|^{q+1}-h_3\rho_3 s(t)+h_3 s(t)\Omega\end{aligned}$$

we suppose that $\rho_3 \gg \|\Omega\|$, one can obtain

$$\begin{aligned}\dot{V}(t)&\leq -2h_1\rho_1 V(t)-2^{\frac{q+1}{2}}h_2\rho_2 V(t)^{q+1}-h_3\left(\rho_3-h_3\Omega\right)s(t)\\ &\leq -\delta_{10}V(t)^{\eta'}-\delta_{20}V(t)\end{aligned} \tag{38}$$

The result of this proof is that the Lyapunov *V*(*t*) function decreases gradually and the sliding surface (30) converges to the origin in a finite time. This means that the tracking errors of equation (22)-(23) converges to zero in a finite time. Now, According to Lemma 1, the error trajectories reach the sliding surface in finite time T calculated as follows:

$$T=t_0+\frac{1}{\delta_2(1-\eta)}\ln\frac{\delta_1+\delta_2 V(t_0)^{1-\eta}}{\delta_1} \tag{39}$$

where $V(t_0)=\frac{1}{2}s(t_0)$. This completes the proof.

Some simulations are executed to evaluate the reliability and feasibility of the proposed adaptive sliding-mode controller. The initial conditions of the master and slave systems are assumed to be

$$\left[x_{1m},x_{2m},x_{3m}\right]=\left[10^{-7},10^{-7},10^{-7}\right] \text{ and } \left[x_{1s},x_{2s},x_{3s}\right]=\left[2.10^{-9},3.10^{-9},3.10^{-9}\right]$$

respectively. The system parameters are as follows, h_1=1.5, h_2=2, h_3=1, p_1=2, p_2=1.2, p_3=1.2, η=0.6, k=0.1, Ω=0.8. Fig. 4a, 4b, 4c depicts the synchronization error of the scheme. One can observe that after the transient period, the master and the slave systems asymptotically converge. Sine $|e_i|\leq\Omega$ for i=1,2,3 the proposed sliding mode control appear robust against the uncertainties. The designer can acts only on the parameter k to accelerate the convergence of the trajectories on the sliding surface which prove the better performance of the controller compared to other similar results encountered in the literature (See refs (Mobayen et al., 2016) and references therein) . The control synthesis is well understood and easily implementable in practice which is in good agreement with our theoretical analysis.

4. ROBUST SYNCHRONIZATION APPROACH USING POLYNOMIAL OBSERVER

4.1 Preliminaries

In this section, to synchronize two Colpitts, we will use the adaptive method based on the states observed. Let put the perturbed chaotic system (4) in the following form

$$\begin{aligned} \dot{x}(t) &= (A+\Delta A_1)x(t)+Bf(x(t))+d(t) \\ y(t) &= Cx(t)+Ed(t) \end{aligned} \tag{40}$$

where $x\in R^n$ and $\tilde{x}\in R^n$ are the systems states, $y\in R^{n\times 1}$ and $\tilde{y}\in R^{n\times 1}$ are the output vectors of the systems (4). $A,B\in R^{n\times n}$ and $C,E\in R^{n\times 1}$ are known matrices such that (A,C) is observable. f: $R^n\rightarrow R^n$ is nonlinear continuously differentiable vector function. $d(t)\in\Im_\infty[0,\infty)$ is an unknown exogenous disturbance, and $\Delta A_1(t)$ is the unknown matrices representing time-varying parameter uncertainties. The transmitted signal is affected by noise and disturbances denoted by $Ed(t)$.

Assumption 2: One state only is available
Assumption 3: (A,B) is stabilizable

The structured uncertainties ($\Delta A_i (i=1,2)$) satisfy the following assumption:

Assumption 4:

$$\Delta A_1 = F((x),t)H_1\,, \Delta A_2 = F((\tilde{x}),t)H_2\,,$$

where H_1 and H_2 are uncertain matrices function with appropriate dimensions which are Lebesque measurable in t ant satisfy:

$$(H_1-H_2)^T(H_1-H_2)\leq I \tag{41}$$

Figure 4. Synchronization errors; **(a)**: $e_1(t)$, **(b)**: $e_2(t)$, **(c)** : $e_3(t)$,

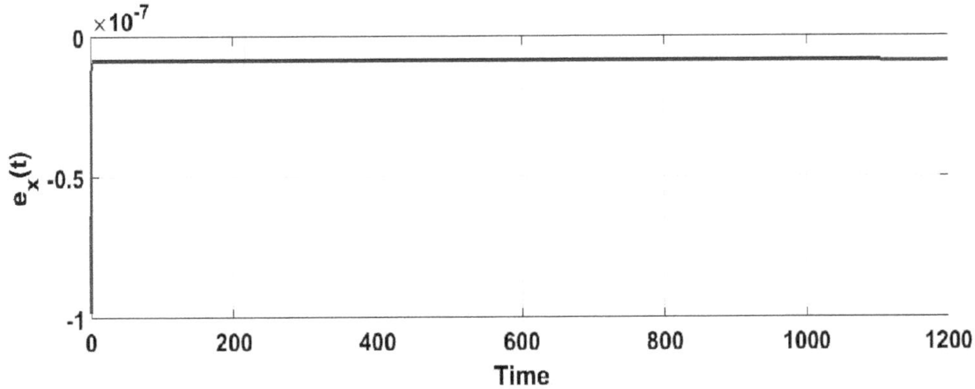

(a)

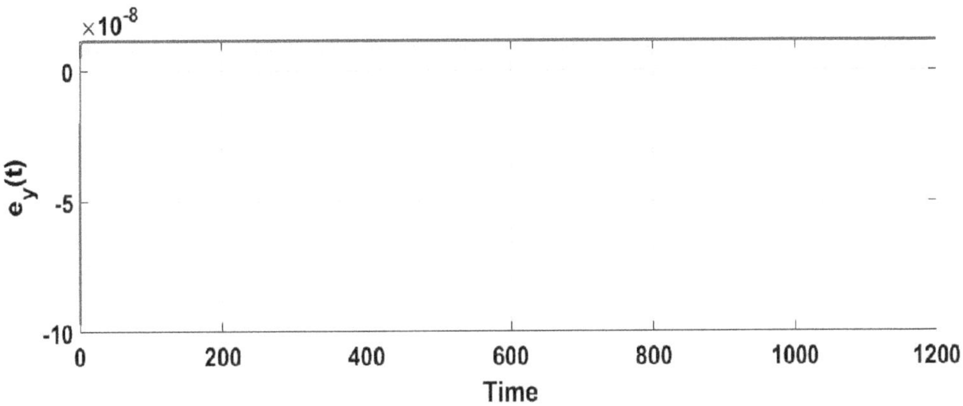

(b)

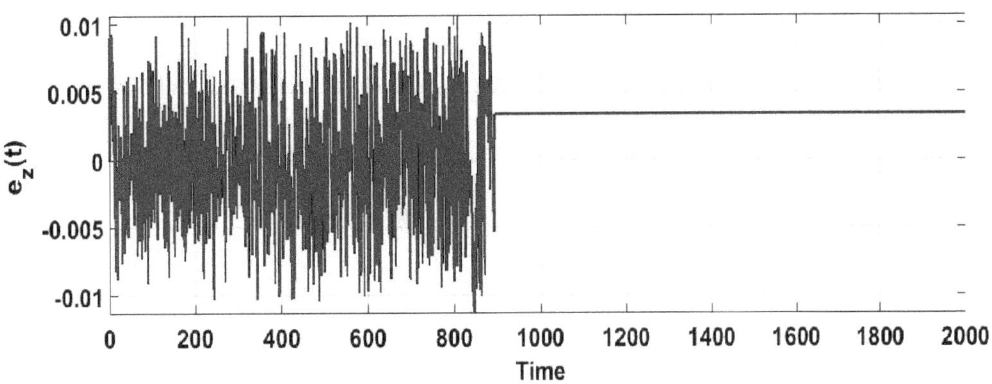

(c)

where I denotes the identity matrix with appropriate dimensions. Furthermore, the following inequality holds:

$$\|F((\bullet),t)\| \le \omega \tag{42}$$

For any known positive constant ω.

Assumption 5: Let $\omega i_{>}0$ be arbitrarily given for each i∈{1,…,n}. Define the compact $\Omega \subset R^n$

$$\Omega = \{x \in R^n : |x_i| \le \omega_i\} \tag{43}$$

For each Ω there exist a nondecreasing, that is, for all $a,b \in R^n$ and $a>b$, it satisfies

$$0 \le \frac{f(a)-f(b)}{a-b} \tag{44}$$

4.2 Problem Statement

The synchronization problem consists in designing a slave dynamical system, based on the observer method:

$$\begin{aligned} \dot{\tilde{x}}(t) &= (A+\Delta A_2)\tilde{x}(t) + Bf(\tilde{x}(t)) + u(t) \\ \tilde{y}(t) &= C\tilde{x}(t) \end{aligned} \tag{45}$$

$u(t)$ is any admissible nonlinear control signal.

Due to the uncertainties in the master system (40), the slave system (45) is usually difficult to achieve complete synchronization. Therefore a concept of practical synchronization, also called synchronization with uniform ultimate boundedness in order literatures, is introduced, in this paper.

Definition 1. *The master-slave systems (40) and (45) achieve practical synchronization if for any initial state* $x(0)$ and $\tilde{x}(0)$, there exist constants $h>0$ and $T_0>0$ such that the trajectory $x(t,x(0))$ of the system (6) and trajectory $\tilde{x}(t,\tilde{x}(0))$ of the system (7) satisfy

$$\|x(t,x(0)) - \tilde{x}(t,\tilde{x}(0))\| \le h, \text{ for any } t > T_0 \tag{46}$$

h indicates the error bound.

In order word, our goal is to design an appropriate controller $u(t)$ such that the trajectory of the slave system (40) asymptotically approaches the master system (45) and finally achieves synchronization.

Define the synchronization error vector as

$$e(t) = x(t) - \tilde{x}(t), y_e(t) = y(t) - \tilde{y}(t) \tag{47}$$

So the error dynamical system is

$$\begin{aligned} \dot{e}(t) &= Ae(t) + B\big(f(x(t)) - f(\hat{x}(t))\big) + d(t) + \Delta A_1 x + \Delta A_2 \hat{x} + u(t) \\ y_e(t) &= Ce(t) + Ed(t) \end{aligned} \tag{48}$$

Suppose that $y_e(t)=He(t)$ stands for the controller output for error state where H is a known matrix. Our purpose is to design the controller $u(t)$ such that the observer error dynamics is asymptotically stable with the following specified H_∞ norm upper bound is guaranteed.

$$\|y_e(t)\| \le \mu \|d(t)\| \tag{49}$$

μ is a positive constant.

Definition 2: (H_∞ synchronization): The error system (27) is H_∞ synchronized if the synchronization error $e(t)$ satisfies.

$$\int_0^\infty e^T(t)Qe(t)dt < \gamma^2 \int_0^\infty \left(\omega^T(t)\omega(t) - n(t)\frac{dK}{dt} \right) dt \tag{50}$$

where $n(t)$ and $K(t)$ are continuous functions. For a given level $\gamma>0$ under zero initial condition, where Q is a positive symmetric matrix. The parameter γ is called the H∞ $_n$orm bound or the disturbance attenuation level which is defined for $n(t)=0$ by

$$\gamma = \frac{\sqrt{\int_0^\infty y^T{}_e(t) y_e(t)dt}}{\sqrt{\int_0^\infty \omega^T(t)\omega(t)dt}}. \tag{51}$$

5 MAIN RESULTS

Assumption 5: We can choose a constant matrix $L \in R^{3\times 3}$ and two positive definite matrices $P=P^T>0$ and $Q=Q^T$ such that

$$\begin{aligned} &(A - LC + K_1C)^T P + P\big(A - LC + K_1C\big) + \big(\delta_1 + \gamma^{-2}(1+r^2) - (2m-1)\varepsilon_1\big)PP \\ &+2PLC + \delta_1^{-1}\rho^2 I + -(2m-1)\varepsilon_1^{-1}h_{\max}^2 = -Q \end{aligned} \tag{52}$$

and

$$B^T P = C \tag{53}$$

where

$$\Omega^2 = \delta_1 + \gamma^{-2}(1 + r^2) \tag{54}$$

Assumption 6. We can choose the controller in this form

$$u(t) = \sum_{i=1}^{m} KB\left(y - C\hat{x}(t)\right)^{2i-1} + \mu(t) \text{ for } 1 \leq i \leq m \tag{55}$$

where $m \in Z^{+}$, *m odd*, $m > 1$ and K is the estimated feedback gain which is updated according to the following adaptation algorithm:

$$\dot{K} = \rho \left\| y - C\hat{x}(t) \right\|^2 \tag{56}$$

with ρ and ε are positive constants.

For the sake of robustness μ(t) is introduced in the controller and have the following form

$$\mu(t) = \begin{cases} \dfrac{B^T Pe}{\left\| B^T Pe \right\| + \varepsilon} \left\| Ce \right\| m'(t) \text{ if } \left\| B^T Pe \right\| + \varepsilon \neq 0 \\ 0, \quad \textit{otherwise} \end{cases} \tag{57}$$

The control objective is stated in the mathematical form in (55). One advantage of this type of controller is that it can be easily constructed through time varying resistors, capacitors, or operational amplifiers and their combinations or using a digital signal processor together with the appropriate converters.

Proposition 1: For an initial value $K(0)$ of $K(t)$, there exist a smooth function $\eta(t) \left(\left\| \eta \right\| = m'_{\max} \right)$ such that the following relation holds:

$$K(t) - K(0) \leq -\eta(t)\gamma^3 \tag{58}$$

Theorem 3: *System (40) asymptotically synchronizes system (45) under condition (55) and (56)*

Proof: The closed loop error system with the control (48) can be written in the following form

$$\dot{e}(t) = \left(A - LC\right)e(t) + LCe(t) + \phi(e) - \sum_{i=1}^{m} KB\left(y - C\hat{x}(t)\right)^{2i-1} - \frac{B^T Pe}{\left\| B^T Pe \right\| + \varepsilon} \left\| Ce \right\| m'(t) + \Delta A_1 x - \Delta A_2 \hat{x} + d(t) \tag{59}$$

where $\phi(e) = f(x) - f(\hat{x})$

It follows from (23) that each component ϕ(e) satisfies

$$0 \le \frac{\phi_i(e)}{e_i}, \forall e_i \ne 0, \tag{60}$$

which leads to the following relationship

$$e^T\phi(e) = \sum_{i=1}^{n} e_i\phi_i(e_i) = \sum_{i=1}^{n} e_i^2 \frac{\phi_i(e_i)}{e_i} \tag{61}$$

By using (39) we have the following condition

$$0 \le e^T\phi(e) \tag{62}$$

Consider the Lyapunov function

$$V(t) = e^T(t)Pe(t) + \frac{1}{2\rho}\left(K - K_0\right)^2. \tag{63}$$

Differentiating V with respect to time t one obtains

$$\begin{aligned}
\dot{V}(t) &= \dot{e}^T(t)Pe(t) + e^T(t)P\dot{e}(t) + \frac{1}{\rho}\left(K - K_0\right)\dot{K} = (A - LC)^T Pe(t) + e^T(t)P\left(A - LC\right)e(t) \\
&+ 2e^TPLCe - 2e^TP\sum_{i=1}^{m} KB\left(Ce + Ed(t)\right)^{2i-1} - 2e^TP\frac{B^TPe}{\left\|B^TPe\right\| + \varepsilon}\left\|Ce\right\|m'(t) + 2e^TP\phi(e) \\
&+ \frac{1}{\rho}\left(K_i - K_0\right)\dot{K}_i + 2e^TP(t)(\Delta A_1x + \Delta A_2 \hat{x})e(t) + e^T(t)Pd(t) + d^T(t)Pe(t)
\end{aligned} \tag{64}$$

Proposition 2: For any matrices $Q_1 \in R^{n\times m}$, $Q_2 \in R^{n\times m}$, $\Theta = \Theta^T > 0$, and $\Theta \in R^{n\times m}$ we have the following inequalities:

$$Q_1^TQ_2 + Q_2^TQ_1 \le Q_1^T\Theta Q_1 + Q_2^T\Theta^{-1}Q_2 \tag{65}$$

These propositions hold:

$$\begin{aligned}
& e^T(t)Pd(t) + d^T(t)Pe(t) \le \gamma^2 d^T(t)d(t) + \frac{1}{\gamma^2}e^T(t)PPe(t)2e^TP(\Delta A_1x - \Delta A_2 \hat{x})e(t) \\
& \le \delta_1 e^T(t)P^2e(t) + \delta_1^{-1}\rho^2 e(t)e^T(t) + \gamma^{-2}r^2e^T(t)P^2e(t) + \gamma^2\rho^2(H_1 - H_2)^T(H_1 - H_2) \\
& 2e^TP\sum_{i=1}^{m} KB\left(Ed(t)\right)^{2i-1} \le (2m-1)\left(\varepsilon_1 e^T(t)P^2e(t) + \varepsilon_1^{-1}h_{\max}^2 e^T(t)e(t)\right)
\end{aligned} \tag{66}$$

$\delta 1$ and $\varepsilon 1$ are the positive constants. Based on the bounded property of chaotic attractor, we consider that $\|x(t)\| \le r$ for t<0 and $\|Ed(t)\|\left[\left\|I+(Ed)^{2m-2}\right\|\right] \le h_{\max}$

Using Ineq. (40), we obtain

$$\begin{aligned}
\dot{V} &\le e^T(t)\left((A-LC+K_1C)^T P + P(A-LC+K_1C) + \frac{1}{\gamma^2}PP + 2PLC\right)e(t) - 2\|Ce(t)\|^2 m'(t) \\
&+2e^T P\phi(e) - 2e^T P\sum_{i=1}^{m} KB(Ce)^{2i-1} - 2e^T P\sum_{i=1}^{m} KB(Ed(t))^{2i-1} - \delta_1 e^T(t)P^2 e(t) + \delta_1^{-1}\rho^2 e(t)e^T(t) \\
&+\gamma^{-2}r^2 e^T(t)P^2 e(t) + \gamma^2\rho^2 I + \gamma^2 d^T(t)d(t) + \frac{1}{\rho}(K-K_0)\dot{K} \le e^T(t)\left((A-LC+K_1C)^T P\right. \\
&+P(A-LC+K_1C) + \left(\delta_1 + \gamma^{-2}(1+r^2)\right)PP + 2PLC + \delta_1^{-1}\rho^2 I\big)e(t) - 2\|Ce(t)\|^2 m'(t) \\
&+2e^T P\phi(e) + \gamma^2 d^T(t)d(t) + \frac{1}{\gamma^2}e^T(t)PPe(t) + \frac{1}{\rho}(K-K_0)\dot{K} - \sum_{i=2}^{m}(Ce)^{2i-2}e^T\left(M_i + M_i^T\right)e \\
&-(2m-1)\left(\varepsilon_1 e^T(t)P^2 e(t) + \varepsilon_1^{-1}h_{\max}^2 e^T(t)e(t)\right) + \gamma^2\rho^2(H_1-H_2)^T(H_1-H_2)
\end{aligned} \tag{67}$$

$$\begin{aligned}
&= e^T(t)\left((A-LC+K_1C)^T P + P(A-LC+K_1C) + \left(\delta_1 + \gamma^{-2}(1+r^2) - (2m-1)\varepsilon_1\right)PP\right. \\
&+2PLC + \delta_1^{-1}\rho^2 I + -(2m-1)\varepsilon_1^{-1}h_{\max}^2\big)e(t) - \sum_{i=2}^{m}(Ce)^{2i-2}e^T\left(M_i + M_i^T\right)e \ + 2e^T P\phi(e) \\
&+\gamma^2\left(\rho^2 I + d^T(t)d(t) - \frac{\dot{K}}{\rho}\right) - 2\|Ce\|^2 m'(t)
\end{aligned}$$

with

$$M = PKC \tag{68}$$

such that $\lambda m_{in(}M+MT^{)} \ge 0$, i=2,…,m,

Finally we obtain

$$\dot{V} \le -e^T(t)Qe(t) + 2e^T P\phi(e) - \sum_{i=2}^{m}(Ce)^{2i-2}e^T\left(M_i + M_i^T\right)e + \gamma^2\left(\omega^T(t)\omega(t) - \frac{\dot{K}}{\rho}\left(\eta + \frac{2}{\gamma^2}m'(t)\right)\right) \tag{69}$$

In the view of (61) and (62), we obtains

$$\dot{V} \le -e^T(t)Qe(t) + \gamma^2\left(\omega^T(t)\omega(t) - \frac{\dot{K}}{\rho}\left(\eta + \frac{2}{\gamma^2}m'(t)\right)\right) \tag{70}$$

with $\omega^T(t)\omega(t) = \rho^2 I + d^T(t)d(t)$

It is easy to find that when the disturbances are not considered, we have

$$\dot{V}(t) \le -e^T Qe \le -\lambda_{\min}(Q)\|e(t)\|^2 < 0 \qquad (e(t) \ne 0) \tag{71}$$

where $\lambda m_{in(}Q)$ is the eigenvalue of Q, then

$$\lim_{t\to\infty} \|e(t)\| = 0 \tag{72}$$

It is obvious that the asymptotic stability of the overall system is guaranteed.

By using (71), we get

$$\lambda_{\min}(Q)\int_0^t e^T(s)e(s)ds \le V(e(t)) - V(e(0)) \tag{73}$$

This proves that $\int_0^\infty e^T(s)e(s)ds$ is bounded.

If $d(t)\neq 0$ and $\mu(t)\neq 0$, integrating both sides of Ineq.(70) gives

$$V(\infty) - V(0) < -\int_0^\infty e(t)^T Qe(t)dt + \gamma^2\int_0^\infty\left(\omega^T(t)\omega(t) - \frac{\dot{K}}{\rho}\left(\eta + \frac{2}{\gamma^2}m'(t)\right)\right)dt \tag{74}$$

We now start to prove that $m'(t)$ is bounded by contradiction. If $m'(t)$ is unbounded and we know by its definition that it is increasing and positive, we conclude that $m'(t) \to \infty$. Or Q is positive definite, we know that V is bounded and decreasing. Then $\int_0^\infty e^T(s)e(s)ds$ is bounded. This implies that $m'(t) = \int_0^t e^T(s)e(s)ds + m'(0)$ is bounded, which leads to a contradiction to the assumption that $m(t)$ is not bounded. Therefore, we have proved that $m(t)'$ has to be bounded and there exists a positive constant $\tilde{m}$ such that $\lim_{t\to\infty} m'(t) = \tilde{m}$.

We let $n(t) = \frac{1}{\rho}\left(\eta + \frac{2}{\gamma^2}m(t)\right)$. Since $V(\infty)\geq 0$ and $V(0)=0$, we have the relation in Ineq. (53). This completes the proof.

Remark 2: The "control" function $n(t)$ describes the performance of the synchronization. The incremental term $\frac{1}{\gamma^2}\frac{2}{\rho}m'(t)$ increase the rate of convergence of the synchronization error in the perturbed environment. The faster $m'(t)$ grows, the shorter the time for the estimation of the feedback gain to track the uncertainty.

It is left to analyze the behavior of the error when the system is perturbed in presence of the variable structure controller. For this purpose, let's consider the following theorem:

Theorem 4 If *Ineq. (48) holds, then the adaptive robust control law (35) guarantees that; all signals are bounded and the dynamic error is bounded above by*

$$\|e(t)\|^2 \leq \exp(-k't)\|e(0)\|^2 + \frac{\varepsilon'}{k'} \tag{75}$$

Where k' and ε' are positives constants

Proof. From (70), we obtain

$$\dot{V}(t) \leq -\lambda_{\min}(Q)\|e(t)\|^2 + \gamma^2\left(\omega^T(t)\omega(t) - n(t)\dot{K}(t)\right) \tag{76}$$

Since P is positive matrix, the error norm $\|e(t)\|$ can be estimated using Rayleigh inequality:

$$\lambda_{\min}(P)\|e(t)\|^2 \leq V(t) \leq \lambda_{\max}(P)\|e(t)\|^2 \tag{78}$$

$$\dot{V}(t) + \frac{\lambda_{\min}(Q)}{\lambda_{\max}(P)} V(t) \leq \gamma^2\left(\omega^T(t)\omega(t) - n(t)\dot{K}(t)\right) \tag{79}$$

where $\lambda m_{ax(}P)$ is the maximum eigenvalue of matrix P

Solving the differential inequation (79) for $V(t)$ when $t \geq 0$ results in

$$V(t) \leq \exp(-\lambda t)V(0) + \gamma^2 \int_0^t \exp(-\lambda(t-\tau))\left(\omega^T(\tau)\omega(\tau) - n(\tau)\dot{K}(\tau)\right)d\tau \tag{80}$$

where $\lambda = \dfrac{\lambda_{\min}(Q)}{\lambda_{\max}(P)} > 0$. We obtain using relation (78)

$$\begin{aligned}
\|V(t)\| &\leq \left\| V(0)\exp(-\lambda t) + \gamma^2 \int_0^t \left(\omega^T(\tau)\omega(\tau) - n(\tau)\dot{K}(\tau)\right)\exp\left(-\lambda(t-\tau)\right)d\tau \right\| \\
&\leq \exp(-\lambda t)V(0) + \gamma^2 \exp\left(-\lambda(t)\right)\int_0^t \left(\omega^T(\tau)\omega(\tau) - n(\tau)\dot{K}(\tau)\right)\exp\left(\lambda\tau\right)d\tau \\
&\leq \exp(-\lambda t)V(0) + \gamma^2 \exp\left(-\lambda(t)\right)\int_0^t \left(\int_0^\infty e^T(\tau)Qe(\tau)dt\right)\exp\left(\lambda\tau\right)d\tau \\
&\leq \exp(-\lambda t)V(0) + \gamma^2 \lambda_{\min}(Q)\exp\left(-\lambda t\right)\int_0^t \left(\|e(\tau)\|^2\right)\exp\left(\lambda\tau\right)d\tau \\
&\leq \exp(-\lambda t)V(0) + \gamma^3 \lambda_{\min}(Q)\exp\left(-\lambda t\right)\int_0^t \exp\left(\lambda\tau\right)d\tau \\
&\leq \exp(-\lambda t)V(0) + \frac{\gamma^3 \lambda_{\min}(Q)}{\lambda}\exp\left(-\lambda t\right)\left(\exp\left(\lambda t\right)-1\right) \\
&\leq \exp(-\lambda t)V(0) + \frac{\gamma^3 \lambda_{\min}(Q)}{\lambda}\left(1-\exp\left(-\lambda t\right)\right)
\end{aligned} \tag{81}$$

The upper bound is defined as follows

$$\begin{aligned}
\lambda_{\min}(P)\|e(t)\|^2 &\leq \exp(-\lambda t)V(0) + \frac{\gamma^3 \lambda_{\min}(Q)}{\lambda}\left(1-\exp\left(-\lambda t\right)\right) \\
\|e(t)\|^2 &\leq \frac{V(0)}{\lambda_{\min}(P)}\exp(-\lambda t) + \frac{\gamma^3}{\lambda}\frac{\lambda_{\min}(Q)}{\lambda_{\min}(P)}\left(1-\exp\left(-\lambda t\right)\right) \\
&= \left(\frac{V(0)}{\lambda_{\min}(P)} - \frac{\gamma^3}{\lambda}\frac{\lambda_{\min}(Q)}{\lambda_{\min}(P)}\right)\exp\left(-\lambda t\right) + \frac{\gamma^3}{\lambda}\frac{\lambda_{\min}(Q)}{\lambda_{\min}(P)}
\end{aligned} \tag{82}$$

Letting

$$V'(0) = \frac{V(0)}{\lambda_{\min}(P)} - \frac{\gamma^3}{\lambda}\frac{\lambda_{\min}(Q)}{\lambda_{\min}(P)} \text{ and } \frac{\varepsilon'}{k'} = \frac{\gamma^3}{\lambda}\frac{\lambda_{\min}(Q)}{\lambda_{\min}(P)},$$

i.e., the tracking error exponentially decays to a ball. The exponential converging rate k' and the size of the final tracking error $\left(\|e(\infty)\| \leq \sqrt{\frac{\varepsilon'}{k'}} \leq \gamma\right)$ can be freely adjusted by the controller parameters ε' and k' in a known form. The proof is completed.

After a while of synchronization process, the error asymptotically approaches a constant . We can easily found the maximum amplitude of error given by

$$\lim_{t\to\infty}\|e(t)\| = \frac{\gamma^3}{\lambda}\frac{\lambda_{\min}(Q)}{\lambda_{\min}(P)} \tag{83}$$

From relation (83), the master-slave systems (40) and (45) achieve practical synchronization under high constraint perturbations.

Theorem 4. *For any nonzero scaling constant* η adapted atomically from the previous section, the master system (40) can synchronize the slave system (45) with controller (55) in a finite time given by:

$$t_r \leq \frac{\rho c^2 \|e(0)\|^2}{\lambda\left(\gamma_0 - \rho c^2 \|e(0)\|^2\right)}\left[1-\gamma^3\left(\eta + \frac{\rho c^2}{\lambda^2}\frac{\lambda_{\min}(Q)}{\lambda_{\min}(P)}\right)\right] \tag{62}$$

The time duration of the synchronization process is given by:

a. When only the exponential polynomial observer is activated: $\eta(t)=0$

$$t_{EPO} \leq \frac{\rho c^2 \|e(0)\|^2}{\lambda\left(\gamma_0 - \rho c^2 \|e(0)\|^2\right)}\left[1-\gamma^3 \frac{\rho c^2}{\lambda^2}\frac{\lambda_{\min}(Q)}{\lambda_{\min}(P)}\right] \tag{63}$$

b. When only the exponential polynomial observer is activated: $\eta(t)=m(t)$

$$t_{RMEPO} \leq \frac{\rho c^2 \|e(0)\|^2}{\lambda\left(\gamma_0 - \rho c^2 \|e(0)\|^2\right)}\left[1-\gamma^3\left(\eta + \frac{\rho c^2}{\lambda^2}\frac{\lambda_{\min}(Q)}{\lambda_{\min}(P)}\right)\right] \tag{64}$$

The accurate of each observer is obtained by evaluating the discrepancy between (63) and (64), whence

$$T_{d\max} = -\gamma^3\eta \frac{\rho c^2 \|e(0)\|^2}{\lambda\left(\gamma_0 - \rho c^2 \|e(0)\|^2\right)} < 0 \tag{65}$$

One can deduce that

$$t_{EPO} > t_{RMEPO} \tag{66}$$

6. SIMULATION INVESTIGATION

In this section, we will give simulation results of the chaotic Colpitts circuit brought back to a topology of Chua system to demonstrate that the above criterion is effective. Let's rewrite the master system (40)

$$\begin{aligned} \dot{x}(t) &= Ax(t) + Bf(x(t)) + d(t) \\ y(t) &= Cx(t) \end{aligned} \tag{67}$$

where

$$x=\begin{bmatrix}x_1\\x_2\\x_3\end{bmatrix},A=\begin{bmatrix}0&0&1\\1&0&0\\-\alpha_2&-\alpha_3&-\alpha_1\end{bmatrix},B=\begin{bmatrix}0&0&0\\0&0&0\\0&0&-k_3\end{bmatrix},f(x(t))=\begin{bmatrix}0\\0\\x_2^3\end{bmatrix},d(x,t)=\begin{bmatrix}d_1(t)\\d_2(t)\\d_3(t)\end{bmatrix} \tag{68}$$

For convenience, we select the following matrices:

$$C=\Lambda\begin{bmatrix}0\\0\\-1\end{bmatrix},L=\begin{bmatrix}1&0&0\\0&1&0\\0&0&1\end{bmatrix},\text{ and }E=\begin{bmatrix}0&0&0\\0&0&0\\-0.24&0&0.65\end{bmatrix} \tag{65}$$

where $\Lambda\neq 0$

The controller can be expressed as

$$\begin{aligned}u(t)&=\sum_{i=1}^{m}KB\left(y-C\hat{x}(t)\right)^{2i-1}+\mu(t)\\&=\begin{bmatrix}K_{11}(x_3-\hat{x}_3)+K_{21}(x_3-\hat{x}_3)^3+\dots K_{(2m-1)1}(x_3-\hat{x}_3)^{2m-1}\\K_{12}(x_3-\hat{x}_3)+K_{22}(x_3-\hat{x}_3)^3+\dots K_{(2m-1)2}(x_3-\hat{x}_3)^{2m-1}\\K_{13}(x_3-\hat{x}_3)+K_{23}(x_3-\hat{x}_3)^3+\dots K_{(2m-1)3}(x_3-\hat{x}_3)^{2m-1}\end{bmatrix}-\frac{\Lambda}{k_3|e_3|+\varepsilon}\begin{bmatrix}0\\0\\|e_3|m'(t)\end{bmatrix}\end{aligned} \tag{66}$$

If we let Λ= -1; then the estimated feedback gain is updated according to the following adaptation algorithm:

$$\dot{K}_{1j}=\rho_1(x_3-\hat{x}_3)^2,\ \ j=1,2,3 \tag{67}$$

The structure variable controller gain is given by

$$\dot{m}(t)=c_0e^T(t)e(t) \tag{68}$$

We consider the following parameters

$$m=2,\ \rho_1=10^{-4},\ c_0=10^{-4},\ \ K_{2j}=\begin{bmatrix}K_{21}\\K_{22}\\K_{23}\end{bmatrix}=10^{-5}\begin{bmatrix}2\\20\\2\end{bmatrix}.$$

Let us select parameter perturbation applied to the master system (6) in the following form

$$d(x,t) = \begin{bmatrix} \lambda \sin(2t).(10x_1) \\ \lambda \cos t.x_2 \\ \lambda \sin(3t).\frac{8}{3}x_3 \end{bmatrix} \tag{69}$$

Where λ represents the perturbed strength compared with the magnitude of state variables $x_i (i=1,2,3)$. The parametric uncertainties are given as follows:

$$F((\bullet),t) \leq \omega = 1.4, H_1 = \begin{bmatrix} \sin(t) & 1 & 0 \\ 0 & 0 & 0 \\ 0 & -1 & 0 \end{bmatrix} \text{ and } H_2 = \begin{bmatrix} 0 & -1 & 0 \\ 0 & \cos(t) & 0 \\ 0 & 1 & \cos(t) \end{bmatrix} \tag{70}$$

The controller without compensation is applied, that is $\mu(t)=0$, ∀t. Here the initial conditions for both master and slave system are respectively adopted as $(x1_{0,}\ x2_{0,}\ x3_{0,}) = (1\times10^{-10}, 1.5\times10^{-10}, 3\times10^{-10})$ and $(\hat{x}_{10}, \hat{x}_{20}, \hat{x}_{30}) = (2\times10^{10}, 12\times10^{10}, 4\times10^{10})$. In addition, we select $K(0)=0.016$ and $\lambda = 1\%$. In order to verify the convergence of the trajectory, let us introduce the quadratic gap of the error system: $e_q(t) = \|e(t)\|$. For the sake of clarity, let's analyze the influence of the additional term $\mu(t)$ in the controller. Figures 5(a), 5(b) and 5(c) respectively present the synchronization errors and the error norm $eq_(t)$ using the exponential polynomial observer, it's clearly appears that after the transient oscillatory period, the synchronization error converges to zero in the both case. The graphs in blue represents the time evolution of the error when the control $\mu(t)$ is activated. Those ones in red described the evolution of the system when $\mu(t)=0$. Two major observations emerge from these analyzes: the *response time of the controller and* the rob*ustness of* the scheme against the perturbations. The net effect is a large time denoted Td which is cancel by the presence of $\mu(t)$. *A* very good agreement is obtained and complete the proof of (64). Figure 6 shows the time evolution of the estimated parameter K in the both cases. One can observe that K increases slowly with time and then saturates at its optimal value, which is consistent with our theory analysis.

The performance index (quadratic gap of synchronization error) of the corresponding synchronization process is calculated as

$$J(t) = \frac{1}{t+0.001}\int_0^t \left|e_q(t)\right|_{Q'}, \quad Q' = I \tag{71}$$

Figure. 7 shows the performance index of the synchronization, which is bounded and has tendency to decrease, globally Fig.7 depicts an exponential behaviour.

Figure 5. Time evolution of the synchronization errors for the adaptive feedback synchronization. (a) . (b) . (c) (Color online)

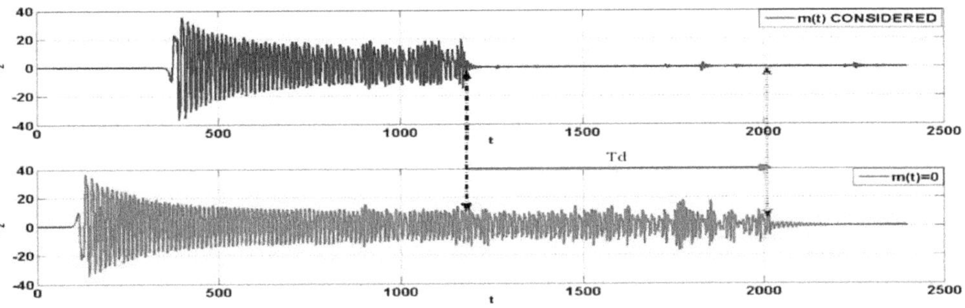

(a)

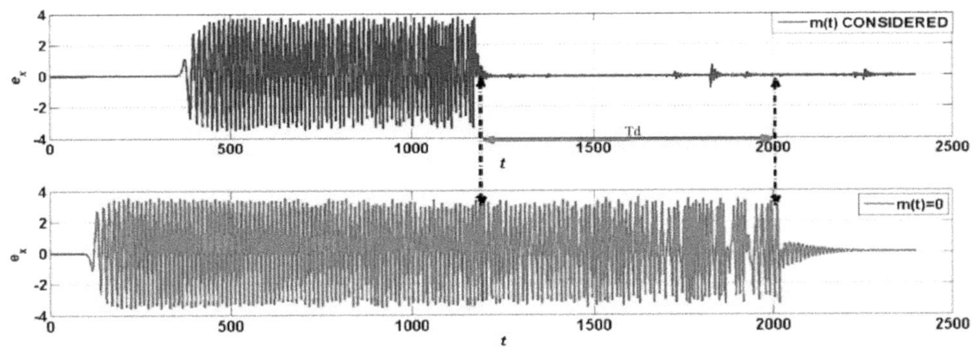

(b)

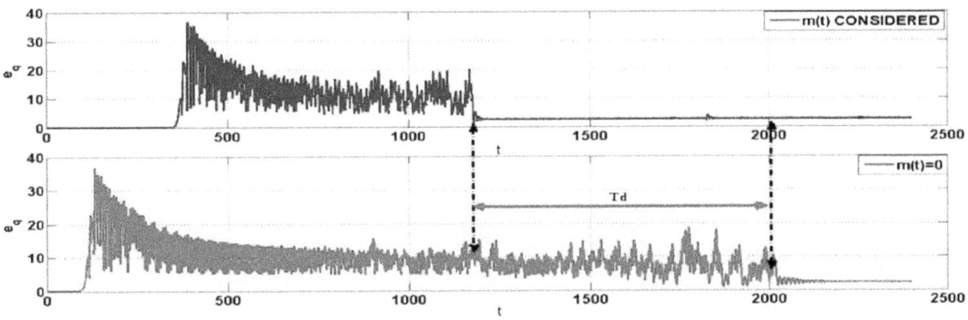

(c)

Figure 6. The parameter K(t) or the adaptive feedback synchronization. (Color online)

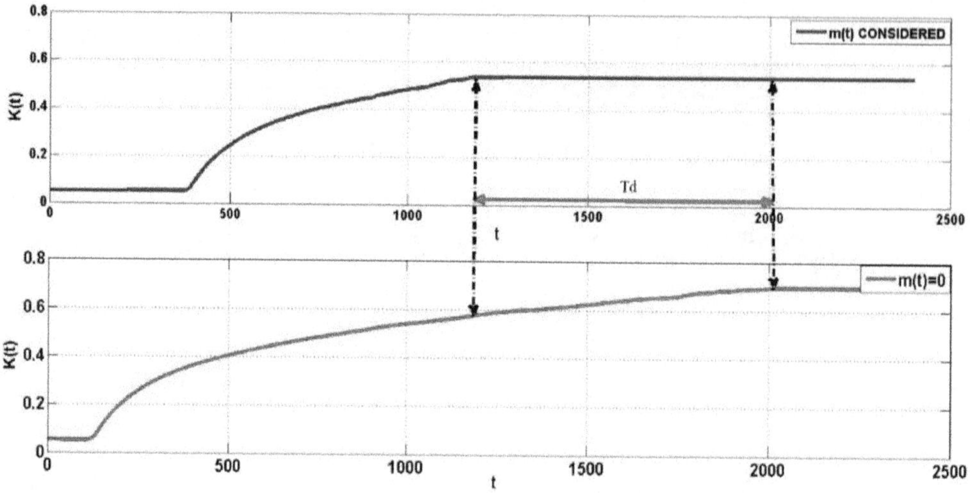

Figure 7. Performance index (Color online)

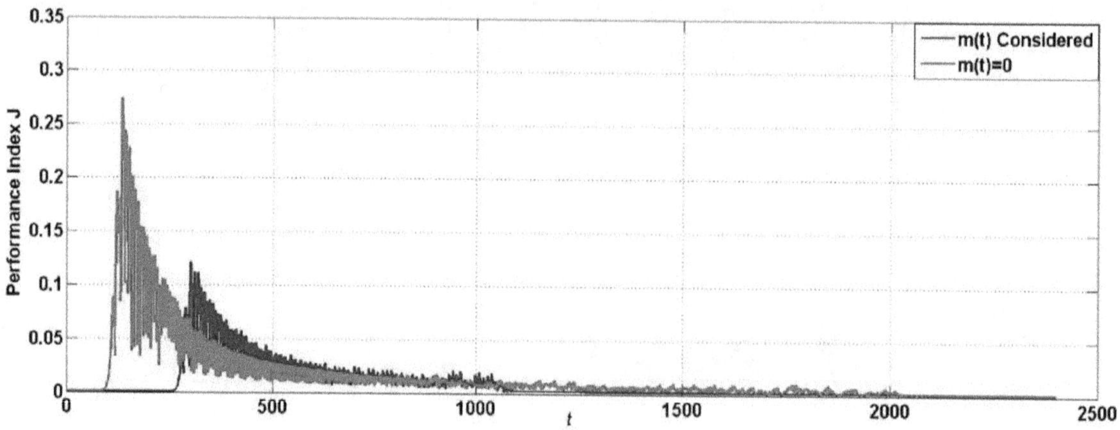

6. CONCLUDING AND REMARKS

The dynamics of a three-dimensional Colpitts system with smooth nonlinearity was studied by numerical solution methods. We have shown that, this topology of Colpitts circuit, though simple it is, can exhibit a chaotic behaviour. Such type of jerky dynamics has not yet been reported in the universal Colpitts circuit in spite of the huge amount of related research works and thus represents a meaningful contribution part to the understanding of the behaviour of nonlinear dynamical system in general. The second part of this work demonstrates that chaos in Colpitts circuit brought back to a topology of Chua system can be easily controlled using adaptive control techniques. We have proposed a novel sliding mode-based robust finite-time control technique for the synchronization of a class of master–slave chaotic systems with uncertainties. The chattering phenomenon in the control input is eliminated using the adaptive finite-time controller, and the unknown upper bound of the external disturbance is estimated. The approach used shows better performance for more complex chaotic systems. Secondly, due to the existence of exoteric perturbation, an extended state robust observer, based on Lyapunov theory, is designed for the output synchronization of a class of uncertain systems when the full state vector is not available. We proposed a structure variable polynomial observer, and by means of properties of nondecreasing and with the help of the Lyapunov method we proved that the estimation error exponentially converges to zero at finite time. An adaptive parameter is designed for the control gain to improve the suitability for different control situations and reduce the chattering in the control signal, and the extended state observer is adopted to estimate the system uncertainties. Finally, we have presented some simulations to illustrate the effectiveness of the suggested approach, which shows some robustness properties against noisy measurements. The simulations demonstrate its good performance in terms of simplicity, feasibility, and robustness, which indicate that it has better practical significance for real world applications.

REFERENCES

Avdonina, E. D., & Ibragimov, N. H. (2013). Conservation laws and exact solutions for nonlinear diffusion in anisotropic media. *Communications in Nonlinear Science and Numerical Simulation*, *18*(10), 2595–2603. doi:10.1016/j.cnsns.2013.02.009

Azar, A. T., & Vaidyanathan, S. (2015a). *Handbook of Research on Advanced Intelligent Control Engineering and Automation*. IGI Global. doi:10.4018/978-1-4666-7248-2

Azar, A. T., & Vaidyanathan, S. (2015b). *Computational Intelligence applications in Modeling and Control, Studies in Computational Intelligence* (Vol. 575). Springer-Verlag.

Azar, A. T., & Vaidyanathan, S. (2015c). *Chaos Modeling and Control Systems Design, Studies in Computational Intelligence* (Vol. 581). Springer-Verlag.

Azar, A. T., Vaidyanathan, S., & Ouannas, A. (2017). *Fractional Order Control and Synchronization of Chaotic Systems, Studies in Computational Intelligence* (Vol. 688). Springer-Verlag., doi:10.1007/978-3-319-50249-6

Azar, A.T., Volos, C., Gerodimos, N.A., Tombras, G.S., Pham, V.T., Radwan, A.G., Vaidyanathan, S., Ouannas, A., & Munoz-Pacheco, J.M. (2017a). *A novel chaotic system without equilibrium: Dynamics, synchronization and circuit realization.* . doi:10.1155/2017/7871467

Balochian, S., & Rajaee, N. (2018). Fractional-order optimal control of fractional-order linear vibration systems with time delay. *International Journal of System Dynamics Applications*, *7*(3), 72–93. doi:10.4018/IJSDA.2018070104

Boulkroune, A., Bouzeriba, A. B., Azar, A.T. (2016a). Fuzzy Adaptive Synchronization of Uncertain Fractional-order Chaotic Systems. In Advances in Chaos Theory and Intelligent Control, Studies in Fuzziness and Soft Computing (Vol. 337). Springer-Verlag.

Bouzaida, S., & Sakly, A. (2018). Adaptive neuro-fuzzy sliding mode controller. *International Journal of System Dynamics Applications*, *7*(2), 34–54. doi:10.4018/IJSDA.2018040103

Byungkook, Y., & Ham, W. (1998). Adaptive fuzzy sliding mode control of nonlinear systems. *IEEE Transactions on Fuzzy Systems*, *6*(2), 315–321. doi:10.1109/91.669032

Chen, H. H. (2009). Chaos control and global synchronization of Liu chaotic systems using linear Balanced feedback control. *Chaos, Solitons, and Fractals*, *40*(1), 466–473. doi:10.1016/j.chaos.2007.07.098

Chou, H. G., Chuang, C. F., Wang, W. J., & Lin, J. C. (2013). A Fuzzy-Model-Based Chaotic Synchronization and Its Implementation on a Secure Communication System. *IEEE Trans. Signal Processing Society*, *8*, 2177–2185.

Chua, L. O. (1993). Article. *IEICE Transactions on Fundamentals of Electronics, Communications and Computer Science*, *E76-A*(5), 704–734.

Chua, L. O., & Lin, G.-N. (1990). Canonical realization of Chua's circuit family. *IEEE Transactions on Circuits and Systems*, *37*(7), 885–902. doi:10.1109/31.55064

Eichhorn, R., Stefan, J. L., & Hanggi, P. (2002). Simple polynomial classes of chaotic jerky dynamics. *Chaos, Solitons, and Fractals*, *13*(1), 1–15. doi:10.1016/S0960-0779(00)00237-X

Eng, T. J. E., & Sci, C. (2011). Sliding mode controller gain adaptation and chattering reduction techniques for DSP-based PM DC motor. *Turkish Journal of Electrical Engineering and Computer Sciences*, *19*(4), 531–549.

Ghabi, J., Rhif, A., & Vaidyanathan, S. (2018). Discrete Time Sliding Mode Control Scheme for Nonlinear Systems With Bounded Uncertainties. *International Journal of System Dynamics Applications*, *7*(2), 15–33. doi:10.4018/IJSDA.2018040102

Hossein, S. (2018). The Stability of a Class of Fractional Order Switching System with Time-Delay Actuator. *International Journal of System Dynamics Applications*, *7*(1), 85–96. doi:10.4018/IJSDA.2018010105

Huangfu, Y., Ma, R., & Miraoui, A. (2012). Avoidance high-frequency chattering second-order sliding mode controller design: Buck converter in wind power system. *International Journal of Antennas and Propagation*, *2012*, 176830. doi:10.1155/2012/176830

Ibragimov, N. K. (1985). Transformation groups applied to mathematical physics. Dordrecht: Kammogne ST, Ahmad TA, Kengne R, Fotsin HB. (2020). Stability analysis and robust synchronization of fractional or- der modified colpitts oscillators. *International Journal of Automation and Control*, *14*(1), 52–79.

Kammogne, S. T., Azar, A. T., Fotsin, H. B., & Kengne, R. (2019). Robust observer-based synchronization of chaotic oscillators with structural perturbations and input nonlinearity. *International Journal of Automation and Control*, *13*(4), 387–412. doi:10.1504/IJAAC.2019.100467

Kammogne, S. T., & Fotsin, H. B. (2011). Synchronization of modified Colpitts oscillators with structural perturbations. *Physica Scripta*, *83*(6), 65011–65018. doi:10.1088/0031-8949/83/06/065011

Kammogne, S. T., & Fotsin, H. B. (2015). Robust adaptive exponential synchronization of two different stochastic perturbed chaotic systems with structural perturbations. *International Journal of Engineering Mathematics., 2015*, 1–10. doi:10.1155/2015/535317

Kammogne, S.T., Fotsin, H.B., Kounchou, M. & Louodop, P., (2013) A robust observer design for passivity-based synchronization of uncertain modified Colpitts oscillators and circuit simulation. *Asian Journal of Science and Technology, 5*, 29-41.

Kammogne, S. T., Kengne, R., & Fotsin, H. (2017). Dynamics and improve robust adaptive control strategy for the finite time synchronization of uncertain nonlinear system. *International Journal of System Dynamics Applications*, *6*(4), 32–62.

Kudryashov, N. A. (2005). Simplest equation method to look for exact solutions of nonlinear differential equations. *Chaos, Solitons, and Fractals*, *24*, 1217–1231.

Kudryashov, N. A., & Loguinova, N. B. (2008). Extended simplest equation method for nonlinear differential equations. *Applied Mathematics and Computation*, *205*(1), 396–40. doi:10.1016/j.amc.2008.08.019

Lakshmanan, M., & Senthilkumar, D. V. (2010). *Dynamics of Nonlinear Time-Delay Systems*. Springer.

Lee, H., Kim, E., Kang, H.-J., & Park, M. (2001). A new sliding-mode control with fuzzy boundary layer. *Fuzzy Sets and Systems*, *120*(1), 135–143. doi:10.1016/S0165-0114(99)00072-X

Lee, H., & Utkin, V. I. (2007). Chattering suppression methods in sliding mode control systems. *Annual Reviews in Control*, *31*(2), 179–188. doi:10.1016/j.arcontrol.2007.08.001

Ma, J., Song, X., Jin, W., & Wang, C. (2015). Autapse-induced synchronization in a coupled neuronal network. *Chaos, Solitons, and Fractals*, *80*, 31–38. doi:10.1016/j.chaos.2015.02.005

Martínez-Guerra, R., Corona-Fortunio, D. M. G., & Mata-Machuca, J. L. (2013). Synchronization of chaotic Liouvillian systems: An application to Chua's oscillator. *Applied Mathematics and Computation*, *219*(23), 10934–10944. doi:10.1016/j.amc.2013.04.056

Mata-Machuca, J., Martinez-Guerra, R., & Aguilar-Lopez, R. (2010). An exponential polynomial observer for synchronization of chaotic systems. *Communication and Nonlinear Sciciences Numerical Simulation*, *15*(12), 4114–4130. doi:10.1016/j.cnsns.2010.01.040

Mata-Machuca, J. L., Martínez-Guerra, R., & Aguilar-López, R. (2011). Chaotic systems synchronization via high order observer design. *Journal of Applied Research and Technology*, *9*(1), 57–68. doi:10.22201/icat.16656423.2011.9.01.448

Mobayen, S., Baleanu D., & Tchier, F. (2016). Second-order fast terminal sliding mode control design based on LMI for a class of non-linear uncertain systems and its application to chaotic systems. *Journal of Vibration and Control*. . doi:10.1177/1077546315623887

Moulay, E., & Perruquetti, W. (2006). Finite time stability and stabilization of a class of continuous systems. *Journal of Mathematical Analysis and Applications*, *323*(2), 1430–1443. doi:10.1016/j.jmaa.2005.11.046

Mounir, D., Azar, A. T., Drid, S., & Mehdi, D. (2018). Higher Order Sliding Mode Control for Blood Glucose Regulation of Type 1 Diabetic Patients. *International Journal of System Dynamics Applications*, *7*(1), 65–84. doi:10.4018/IJSDA.2018010104

Novic, A. (2011). Variable structure systems with sliding modes in motion control—A survey. *IEEE Transactions on Industrial Informatics*, *7*(2), 212–223. doi:10.1109/TII.2011.2123907

Ouannas, A., Azar, A. T., & Abu-Saris, R. (2016a). A new type of hybrid synchronization between arbitrary hyperchaotic maps. *International Journal of Machine Learning and Cybernetics*. Advance online publication. doi:10.100713042-016-0566-3

Ouannas, A., Azar, A. T., Vaidyanathan, S., & Radwan, A. G. (2016b). New Control Schemes of Synchronization for Incommensurate and Commensurate Fractional Order Chaotic Systems. *Proceedings of the 4th International Conference on Control Engineering & Information Technology CEIT '16.*

Ouannas, A., Azar, A. T., Ziar, T., & Radwan, A. G. (2017c). Study on coexistence of different types of synchronization between different dimensional fractional chaotic systems. In Fractional Order Control and Synchronization of Chaotic Systems, Studies in Computational Intelligence (Vol. 688, pp. 637-6690). Springer Verlag.

Ouannas, A., Azar, A. T., Ziar, T., & Vaidyanathan, S. (2017a). Fractional inverse generalized chaos synchronization between different dimensional systems. *Fractional Order Control and Synchronization of Chaotic Systems Studies in Computational Intelligence*, *688*, 525–551.

Ouannas, A., Azar, A. T., Ziar, T., & Vaidyanathan, S. (2017b). A new method to synchronize fractional chaotic systems with different dimensions. *Fractional Order Control and Synchronization of Chaotic Systems Studies in Computational Intelligence*, *688*, 581–611.

Pikovsky, A., Rosenblum, M., & Kurths, J. (2001). *Synchronization: A universal concept in nonlinear sciences*. Cambridge University Press.

Reidel. Olver, P. (1993). *Applications of Lie groups to differential equations*. Springer. doi:10.1007/978-1-4612-4350-2

Rodriguez, A., De Leon, J., & Fridman, L. (2005). Synchronization in reduced-order of chaotic systems via control approaches based on high-order sliding-mode observer. *Chaos Soliton Fractals*, *42*, 3219–3233. doi:10.1007/978-3-319-50249-6_22

Ruth, B., Patrizio, C., Dugard, L., Allgöwer, F., Kleimenov, A., & Scherer, C. (2008). Trends in Theory of Control System Design. *Proceedings of the 17th World Congress, the International Federation of Automatic Control.*

Sarafian, G., & Kaplan, B. Z. (1995). Article. *IEEE Trans. Circ. Systems I*, *42*(6), 376–376. doi:10.1109/81.390276

Schot, S. H. (1978). The time rate of change of acceleration. *American Journal of Physics*, *46*(11), 1090–1094. doi:10.1119/1.11504

Shi, P., Xia, Y. Q., Liu, G. P., & Rees, D. (2006). On designing of sliding-mode control for stochastic jump systems. *IEEE Transactions on Automatic Control*, *51*(1), 97–103. doi:10.1109/TAC.2005.861716

Shi, S. L., Li, J. X., & Fang, Y. M. (2018). Extended-state-observer-based chattering free sliding mode control for nonlinear systems with mismatched disturbance. *IEEE Access: Practical Innovations, Open Solutions*, *6*, 22952–22957. doi:10.1109/ACCESS.2018.2828868

Shtessel, Y. B., Shkolnikov, I. A., & Brown, M. D. J. (2003). An asymptotic second-order smooth sliding mode control. *Asian Journal of Control*, *5*(4), 498–504. doi:10.1111/j.1934-6093.2003.tb00167.x

Slotine, J. E. (1991). *Applied nonlinear control*. Prentice-Hall.

Sprott, J. C. (2000). Simple chaotic systems and circuits. *American Journal of Physics*, *68*(8), 758–763. doi:10.1119/1.19538

Sprott, J. C. (2010). *Elegant chaos: algebraically simple flow*. Singapore: World Scientific Publishing.

Tosi, Z., & Yoshimi, J. (2016). Simbrain 3.0: A flexible, visually-oriented neural network simulator. *Neural Networks*, *83*, 1–10. doi:10.1016/j.neunet.2016.07.005 PMID:27541049

Vaidyanathan, S., & Azar, A. T. (2016a). *Dynamic Analysis, Adaptive Feedback Control and Synchronization of an Eight-Term 3-D Novel Chaotic System with Three Quadratic Nonlinearities*. Springer-Verlag. doi:10.1007/978-3-319-30340-6_7

Vaidyanathan, S., & Azar, A. T. (2016b). *Qualitative Study and Adaptive Control of a Novel 4-D Hyperchaotic System with Three Quadratic Nonlinearities*. Springer-Verlag. doi:10.1007/978-3-319-30340-6_8

Vaidyanathan, S., & Azar, A. T. (2016f). *Adaptive Control and Synchronization of Halvorsen Circulant Chaotic Systems. Advances in Chaos Theory and Intelligent Control*. Springer-Verlag.

Vaidyanathan, S., & Azar, A. T. (2016g). *Generalized Projective Synchronization of a Novel Hyperchaotic Four-Wing System via Adaptive Control Method*. Springer-Verlag. doi:10.1007/978-3-319-30340-6_12

Vaidyanathan, S., Azar, A. T., & Ouannas, A. (2017a). An Eight-Term 3-D Novel Chaotic System with Three Quadratic Nonlinearities, its Adaptive Feedback Control and Synchronization. *Fractional Order Control and Synchronization of Chaotic Systems Studies in Computational Intelligence*, *688*, 719–746.

Vaidyanathan, S., Azar, A. T., & Ouannas, A. (2017b). Hyperchaos and Adaptive Control of a Novel Hyperchaotic System with Two Quadratic Nonlinearities. *Fractional Order Control and Synchronization of Chaotic Systems Studies in Computational Intelligence*, *688*, 773–803.

Vaidyanathan, S., Volos, C., Pham, V.-T., Madhavan, K., & Idowu, B. A. (2014). Adaptive backstepping control, synchronization and circuit simulation of a 3-D novel jerk chaotic system with two hyperbolic sinusoidal nonlinearities. *Archives of Control Sciences*, *24*(3), 375–403. doi:10.2478/acsc-2014-0022

Volos, C., Akgul, A., Pham, V.-T., Stouboulos, I., & Kyprianidis, I. (2017). A simple chaotic circuit with a hyperbolic sine function and its use in a sound encryption scheme. *Nonlinear Dynamics*, *89*(2), 1047–1106. doi:10.100711071-017-3499-9

Wang, Z., Volos, C., Kingni, S. T., Azar, A. T., & Pham, V. T. (2017). Four-wing attractors in a novel chaotic system with hyperbolic sine nonlinearity. *Optik (Stuttgart)*, *131*, 1071–1078. doi:10.1016/j.ijleo.2016.12.016

Xiaojian, X., Saleh, M., Haipeng, R., & Sajad, J. (2017). Robust finite-time synchronization of a class of chaotic systems via adaptive global sliding mode control. *Journal of Vibration and Control*, *0*(0), 1–13.

Yang, Y., & Yan, Y. (2016). Attitude regulation for unmanned quadrotors using adaptive fuzzy gain-scheduling sliding mode control. *Aerospace Science and Technology*, *54*, 208–217. doi:10.1016/j.ast.2016.04.005

Yoo, W., Ji, D., & Won, S. (2010). Synchronization of two different non-autonomous chaotic systems using fuzzy disturbances observer. *Physics Letters. [Part A]*, *374*(11-12), 1354–1361. doi:10.1016/j.physleta.2010.01.023

Chapter 16
Synchronization and Anti-Synchronization of Unidirectional and Bidirectional Coupled Chaotic Systems by Terminal Sliding Mode Control

Ahmad Taher Azar
https://orcid.org/0000-0002-7869-6373
Faculty of Computers and Artificial Intelligence, Benha University, Benha, Egypt & College of Computer and Information Sciences, Prince Sultan University, Riyadh, Saudi Arabia

Fernando E. Serrano
Universidad Tecnologica Centroamericana (UNITEC), Honduras

Nashwa Ahmad Kamal
Faculty of Engineering, Cairo University, Giza, Egypt

ABSTRACT

In this chapter, the synchronization and anti-synchronization of coupled unidirectional and bi-directional chaotic systems by terminal sliding mode control strategy are shown. The unidirectional synchronization consists in establishing a drive chaotic system and a response chaotic system in order to synchronize the variables of the response system in finite time. The unidirectional and bi-directional anti-synchronization consist in anti-synchronizing mutually a coupled chaotic system in both directions. For these purposes, terminal sliding mode control techniques are implemented. Three systems considered for experimental purposes in this study, a Lorenz, Rossler, and Ikeda systems are used for analysis and experimentation of synchronization and anti-synchronization. Three numerical experiments are shown to test the performance of the obtained proposed strategy.

DOI: 10.4018/978-1-7998-5788-4.ch016

1 INTRODUCTION

Synchronization and anti-synchronization in chaotic systems have been thoroughly studied in recent years (Moaddy et al, 2011; Li et al., 2015; Mishra et. al., 2016; Khan et al., 2020a; Wang et al., 2017). These topics have many applications in the physical system, such as electrical, mechanical, optical and semiconductor applications and superconductivity. The synchronization of coupled chaotic systems, in its identical and non-identical forms, consists of a chaotic drive system and a chaotic response system in which the output of the second tracks the response of the first system. The case explained above is related to unidirectional synchronization, in bi-directional synchronization both systems are considered to be drive and response systems at the same time. Something similar happens in unidirectional and bi-directional anti-synchronization in which, in the first case, anti-synchronization is achieved in order to maintain the variables of the drive and response systems in the opposite direction. In the second case, anti-synchronization is achieved at the same time in both systems.

There are many examples in the literature concerning the synchronization of chaotic, unidirectional and bi-directional chaotic systems in their identical and non-identical form. For example, in Shahverdiev & Shore (2009), the synchronization of chaotic laser diodes in uni-directional and bi-directional forms is shown defining a dynamic model that describes the experimental setup and, later, the synchronization is performed using the appropriate technique. Shahverdiev et al. (2013) proposed the inverse chaos synchronization of coupled time delay chaotic systems. In Shahverdiev (2019), coupled junctions of Josephson and powered by a central junction are synchronized providing excellent application in the field of superconductivity. Another interesting example of chaos synchronization in bi-directional systems is found in Ji et al. (2008) where generalized chaos synchronization for bi-directional coupled differential equations is shown. In Khennaoui et al. (2019), an interesting example is given for the stabilization and synchronization of fractional order discrete time chaotic systems. An interesting application can be found in Sang et al. (2019) where a bi-directional synchronization control for an electro-hydraulic servo system is shown to prove another application of the theoretical context required for the design of synchronizers for chaotic systems. In Tarai et al. (2009), a synchronization of generalized linearly coupled two-way chaotic systems is shown in which a unified chaotic framework is implemented to solve this problem.

The synchronization and stabilization of various forms of chaotic systems such as fractional order or complex can be found also in some studies (Khennaoui et al., 2020; Ouannas et al., 2020a; Mittal et al., 2021). Vishal & Agrawal (2017) proposed a dynamic analysis control and synchronization of a new complex chaotic system. In Wembe & Yamapi (2009) synchronization of nonlinear bi-directional chaotic and hyperchaotic systems is achieved, given that hyperchaotic systems are more difficult to stabilize, an effective synchronization strategy is implemented in this analysis. Other examples can be found in Yuan et. al. (2012) where discrete time chaotic system synchronization is done bidirectionally. Another interesting application for chaos synchronization of chaotic systems can be found in Chen et al. (2020) where bi-directional synchronization of two Kuramoto systems is proposed. Also, in Deng et al. (2009), a bi-directional synchronization of semiconductor lasers by opto-electronic feedback, is proposed. Finally, in Zhou et al. (2007), hyper-chaotic systems in their linear and non-linear forms are synchronized.

There are many examples found in anti-synchronization literature. For example, the anti-synchronization of two non-identical chaotic systems is proposed by Al-Sawalha (2009). In Beregov & Melkikh (2015), two inductively coupled Van der Pol oscillators are de-synchronized. In Bao et al. (2020), a bi-directional chaotic communication configuration scheme based on wavelet-division-multiplexing WDM is implemented for semiconductor laser systems with time delay concealment. In Volos et al. (2011), some synchronization phenomena of chaotic systems are investigated where two bi-directional coupled double scrolling circuits are synchronized and inverted.

Sliding mode control has become one of the most commonly used techniques for many control applications, synchronization, anti-synchronization, stabilization of chaotic and hyper-chaotic systems of different forms (Vaidyanathan et al., 2019; Kumar et al., 2018; Singh et al., 2017; Mekki et al., 2015). Numerous sliding mode techniques, such as higher order, integral and terminal sliding mode solutions have been applied for the design of chaotic system synchronization and anti-synchronization controllers noting that modern sliding mode techniques have not been so widely introduced in recent years (Vaidyanathan et al., 2015a; Djouima et al., 2017; Meghni et al., 2017; Azar et al., 2020). Due to the importance of sliding mode control, some important results and contributions are worth mentioning because they provide a useful theoretical framework for this study, some results can be found in the literature. For example, Gurumurthy & Das (2020) proposed an adaptive supertwisting controller with a terminal sliding mode observer designed for a class of non-linear systems. Another example can be found in Wang et. al. (2020) where a new error configuration function is designed for the position and attitude of a fully controlled six-degree freedom rigid body, so that the terminal sliding mode controller is configured for stabilization purposes on the basis of this configuration error. In Rabiee et al. (2019), there is a non-singular terminal sliding mode controller built along with a sliding mode observer for unknown non-linear systems. This technique consists of obtaining a continuous and non-singular terminal sliding mode controller based on an adaptive sliding mode disturbance observer, given that the first derivative of the disturbance is bound. An adaptive, quick, non-singular terminal sliding mode controller for a flexible aircraft is then shown in Miao et al. (2019). This control system consists of the use of the Lagrange process for the transfer, rotation of the aircraft. The quick terminal sliding mode controller is therefore obtained by designing an appropriate sliding manifold. Finally, a guideline law based on terminal sliding mode control is introduced in Hou et al. (2019). The first part is obtained for the missile to intercept the target and the second part is to obtain the desired impact time.

This chapter demonstrates the synchronization and anti-synchronization of a coupled unidirectional and bidirectional chaotic system by terminal sliding mode control. The synchronization of the unidirectional chaotic system is accomplished by first considering a drive system and a response system that tracks the trajectory of the drive system at a finite time. The anti-synchronization of unidirectional and bidirectional chaotic processes is performed to drive the variables in the opposite direction. Just one terminal sliding variable is designed for unidirectional synchronization and anti-synchronization, whereas two terminal sliding mode variables are chosen for bi-directional anti-synchronization. The synchronization and anti-synchronization control laws are effectively established by selecting the appropriate Lyapunov functionality. Three numerical examples are shown with the Lorenz chaotic system, the Rossler system and the Ikeda system to validate theoretical results using synchronizer and anti-synchronizer strategies. The anti-synchronization control strategy presented in this study can be considered for identical and non-identical chaotic systems, an advantage in comparison with other strategies found in the literature in where only the identical case is considered.

The rest of the chapter is organized as follows. Section 2 presents theoretical framework and preliminaries. Section 3 discusses unidirectional synchronization of non-identical chaotic systems. The unidirectional and bidirectional anti-synchronization of chaotic systems is presented in Section 4. Numerical experiments are presented in Section 5 followed by a discussion in Section 6. Finally, the chapter is concluded in Section 7 with future scope of work.

2 THEORETICAL FRAMEWORK AND PRELIMINARIES

Synchronization plays a key role in achieving the optimal output of various types of physical chaotic systems, such as mechanical, electrical, optical and semiconductor systems (Alain et al., 2020, 2019, 2018; Kammogne et al, 2020; Ouannas et al., 2016, 2017a,b,c,d,e,f,g,h, 2020b,c,d,e,f; Vaidyanathan et al., 2017a,b,c, 2015b,c, 2018a,b,c,d; Khan et al., 2020b; Singh et al., 2018a,b,c). As described in the previous section, synchronization happens when a chaotic system, called a response system, uses the drive system reference to ensure that the response system follows the desired trajectory (Pham et al., 2017a,b, 2018; Shukla et al., 2018; Khettab et al., 2018; Volos et al., 2018). The theoretical principles of synchronization are based on finding an equilibrium point where the drive and response variables are synchronized while minimizing the error variable before the origin is reached (Singh & Azar, 2020a,b; Grassi et al., 2017; Vaidyanathan & Azar, 2016a,b,c,d,e,f,g; Ouannas et al., 2019a,b,c). As mentioned in the literature, there are several techniques for the synchronization of chaotic systems in their identical and non-identical form (Munoz-Pacheco et al., 2017). Many of these techniques are based on robust, efficient and sliding mode control. The difficulty in synchronizing two chaotic systems is due to certain properties and characteristics of chaotic systems, such as entropy and ergodicity, among others, and there are other characteristics such as input constraints, time delays, uncertainties and disturbances that make it difficult to achieve this goal (Vaidyanathan & Azar, 2015a,b,c,d). Robust synchronization includes a wide variety of techniques when instability is detected in the drive or in the response system or both at the same time. Some techniques, such as optimum control, provide certain benefits for synchronization purposes as a fixed time for synchronization can be chosen to minimize the error variable before the origin has been reached. It is important to note that sliding mode control techniques for synchronization systems have now been increased for synchronization purposes. The sliding mode synchronization techniques in this study provide a fast, oscillation-free capability to deal with uncertainties, attributes that are desirable given the above-mentioned characteristics of chaotic or hyper-chaotic systems. The design of the sliding manifold ensures a quick reaching time while maintaining the error variable between the drive and the response variables within the reaching phase. For this analysis, it is important to remember that only one-way synchronization is considered, which means that synchronization is only performed in one direction, so that a special synchronization law is designed.

It is well known that uni-directional anti-synchronization is achieved when two chaotic systems, one acting as a drive system, and the other acting as a response system, track the first system in the opposite direction of the second system, allowing the difference or error variable to reach zero in the finite time. Bi-directional anti-synchronization is a more complicated task compared to unidirectional anti-synchronization, as in this case both systems operate as a response and drive mechanism at the same time. There are several techniques for the bi-directional anti-synchronization of chaotic systems, starting with the basics and finalizing with the more complex and advanced. Simple state feedback has been shown to be useful for bi-directional synchronization purposes, and these methods have the advantage of being easy to incorporate in many types of physical systems, such as optical systems. Given this, the drawback of a bi-directional state feedback anti-synchronizer is that, when the complexity of the chaotic system is increased, this technique is no longer effective, so advanced complex techniques such as sliding mode control are required. The application of the sliding mode technique for synchronization purposes is a real problem that is not difficult to overcome, but is effectively solved by applying the right mathematical structure. First, it is important to develop two anti-synchronization laws for bi-directional anti-synchronization, so that both variables are shifted in the opposite direction by choosing an appropriate error variable. Something that is important to remember is that even if the unilateral synchronization case is very similar, it is not so obvious that appropriate control laws are being devised that act simultaneously in both chaotic systems. There are two ways to achieve bi-directional synchronization by sliding mode technique, one by selecting a unique sliding manifold and the other by selecting two sliding manifolds. Two sliding variables were selected for the purpose of this study. In this way, two different anti-synchronization laws are implemented for both systems.

Today, various types or configurations of sliding mode control introduced for synchronization and anti-synchronization of uni-directional and bi-directional chaotic systems have been increased. Although though these technical solutions are used widely, they have not been applied as they must be. The discovery of new types of physical systems and the increase in applications in recent years have led to the development of new synchronization and anti-synchronization techniques based on sliding mode control. The design of the correct sliding manifolds with their sliding synchronizer and anti-synchronizer laws provides a fast reaching phase, which means that the convergence of the error variable, in the case of synchronization and anti-synchronization, is zero in the finite time. Another thing to remember is that sliding mode control offers on chattering without oscillating action in synchronizer and anti-synchronizer laws, which is relevant when such laws are applied in real physical systems. Terminal sliding mode control has recently been introduced and applications for stabilization of various types of chaotic systems have been introduced due to rapid integration and no chatting among other characteristics or attributes of this control strategy. The sliding manifold designed for the terminal sliding mode synchronization and anti-synchronization techniques used in this study provides the desired closed loop stability thus reducing the error variable to zero in the finite period, which is accomplished by selecting the appropriate laws. This approach offers an ability to deal with complexities that have certain advantages over other approaches as an optimum control, something that is necessary to consider because of the complexity of the chaotic system and its attributes. It is important to note that terminal sliding mode control techniques for the synchronization and anti-synchronization of chaotic systems can be effectively applied to various types of chaotic systems such as fractional order or complex form.

3 UNIDIRECTIONAL SYNCHRONIZATION OF NON-IDENTICAL CHAOTIC SYSTEMS

Unidirectional synchronization of chaotic processes is done by choosing a drive and a response system. A unique terminal sliding manifold is designed to ensure that the error variable, which is the difference between the state variables of the drive and the response system, reaches zero at a finite time. It is accomplished by driving the sliding manifold to zero with a fast-reaching phase while preserving the sliding phase as time reaches infinity. It is achieved by selecting the appropriate synchronizer law and the correct Lyapunov function to ensure the closed loop stability of the response system when tracking its variables to the drive system variables. The kind of chaotic systems considered for this study are non-identical but of the same dimension, and can obviously be adapted for the same systems. Terminal sliding mode control techniques prove to be effective for synchronization of non-identical systems, and it is analyzed later that this control / synchronization strategy is appropriate for synchronization purposes.

For unidirectional synchronization of non-identical chaotic systems consider the drive system:

$$\begin{aligned} \dot{x}_1 &= f_1(x) \\ \dot{x}_2 &= f_2(x) \\ &\vdots \\ \dot{x}_n &= f_n(x) \end{aligned} \tag{1}$$

where $X \in \mathbb{R}^n$ is the state vector and $f_n(x)$ is a nonlinear vector field. The response system is given by:

$$\begin{aligned} \dot{y}_1 &= g_1(y) + u_1 \\ \dot{y}_2 &= g_2(y) + u_2 \\ &\vdots \\ \dot{y}_n &= g_n(y) + u_n \end{aligned} \tag{2}$$

where $y \in \mathbb{R}^n$ is the state vector, $g_n(y)$ is a nonlinear vector field and u_i is the synchronization input. Now consider the error variables $e_i = x_i - y_i$ and its derivative $\dot{e}_i = \dot{x}_i - \dot{y}_i$ and the following sliding variable:

$$s_i(t) = k_1 e_i + k_2 \int_0^t \left[e_i + sign(e_i)|e_i|^\rho \right] d\tau$$

$$\dot{s}_i(t) = k_1 \dot{e}_i + k_2 e_i + k_2 sign(e_i)|e_i|^\rho \tag{3}$$

Where $k_1, k_2 \in \mathbb{R}$ with $k_1, k_2 > 0$ are the gain constants meanwhile ρ is a constant with appropriate values. In order to achieve the synchronization between the drive and response system the following theorem is necessary:

Theorem 1: *The response system (2) is synchronized driving the error variable e_i to zero as time goes to infinity implementing the following control law:*

$$u_i = \dot{x}_i - g_i(y) + k_1 sign(e_i) sign(s_i) \dot{e}_i + k_2 sign(e_i) sign(s_i) e_i + k_2 sign(s_i)|e_i|^\rho \tag{4}$$

obtained by selecting an appropriate Lyapunov functional.

Proof: Consider the following Lyapunov function

$$V = \sum_{i=1}^{n} |s_i| + \sum_{i=1}^{n} |e_i| \tag{5}$$

so, the derivative of (5) is:

$$\dot{V} = \sum_{i=1}^{n} sign\left(s_i\right)\dot{s}_i + \sum_{i=1}^{n} sign\left(e_i\right)\dot{e}_i \tag{6}$$

Now substituting the derivative of the sliding variable (3) into (6) yields:

$$\dot{V} = \sum_{i=1}^{n} sign\left(s_i\right)\left[k_1\dot{e}_i + k_2 e_i + k_2 sign\left(e_i\right)\left|e_i\right|^{\rho}\right] + \sum_{i=1}^{n} sign\left(e_i\right)\left[\dot{x}_i - \dot{y}_i\right] \tag{7}$$

now substituting systems (1) and (2) into (7) we obtain:

$$\dot{V} = k_1\sum_{i=1}^{n} sign\left(s_i\right)\dot{e}_i + k_2\sum_{i=1}^{n} sign\left(s_i\right)e_i + k_2\sum_{i=1}^{n} sign\left(s_i\right)sign\left(e_i\right)\left|e_i\right|^{\rho}$$

$$+\sum_{i=1}^{n} sign\left(e_i\right)\dot{x}_i - \sum_{i=1}^{n} sign\left(e_i\right)g_i\left(y\right) - \sum_{i=1}^{n} sign\left(e_i\right)u_i \tag{8}$$

obtaining the following synchronization law:

$$u_i = \dot{x}_i - g_i\left(y\right) + k_1 sign\left(e_i\right)sign\left(s_i\right)\dot{e}_i + k_2 sign\left(e_i\right)sign\left(s_i\right)e_i + k_2 sign\left(s_i\right)\left|e_i\right|^{\rho} \tag{9}$$

with this the derivative of the Lyapunov function shown in (8) becomes in:

$$\dot{V} \leq 0 \tag{10}$$

so, the synchronization is achieved due to the Lyapunov stability is assured.

4 UNIDIRECTIONAL AND BIDIRECTIONAL ANTI-SYNCHRONIZATION OF CHAOTIC SYSTEMS

Unidirectional and bi-directional anti-synchronization of chaotic systems is achieved to ensure that both coupled systems drive system variables in opposite directions as time goes to infinity. In the unidirectional anti-synchronization case, one chaotic system is known to be a drive system, while the other is chosen as a response system, and by designing the appropriate error variables, it ensures that the response system variables track the drive system variable in the opposite direction in the finite time. In the case of bi-directional synchronization, both systems control their respective variables in the opposite direction by choosing two anti-synchronization input laws for both systems. Only one sliding variable is selected for unidirectional anti-synchronization, so that the unidirectional anti-synchronization input law is selected by designing the appropriate Lyapunov function. Then, for the bi-directional anti-synchronization case, two sliding variables are selected to identify the two anti-synchronization laws for coupled chaotic systems, and this objective is achieved by designing the appropriate Lyapunov functionality. The results obtained for unidirectional anti-synchronization are quite close to the synchronization case, so the deductions are obtained by specifying the appropriate mathematical framework. In the meantime, for the bi-directional anti-synchronization case, a theorem is developed to define the two anti-synchronization laws that are needed.

The unidirectional and bidirectional synchronization of chaotic systems is presented in this section. Unidirectional synchronization is achieved by selecting the terminal sliding mode variables shown in the previous section but by changing the direction of the error variable. For bidirectional synchronization, two terminal sliding mode variables are designed for two error variables and synchronization laws are obtained by selecting the appropriate Lyapunov function.

4.1 Unidirectional Anti-Synchronization

Consider the de-synchronizer error variable $e_i=x_i+y_i$ and its derivative $\dot{e}_i = \dot{x}_i + \dot{y}_i$ taking into consideration the sliding variable shown in (3) and the Lyapunov functional and its derivative shown in (5) and (6) obtaining the following result:

$$\dot{V} = k_1\sum_{i=1}^{n} sign(s_i)\dot{e}_i + k_2\sum_{i=1}^{n} sign(s_i)e_i + k_2\sum_{i=1}^{n} sign(s_i)\,sign(e_i)|e_i|^{p}$$

$$+\sum_{i=1}^{n} sign(e_i)\dot{x}_i + \sum_{i=1}^{n} sign(e_i)g_i(y) + \sum_{i=1}^{n} sign(e_i)u_i \tag{11}$$

now by selecting the following anti-synchronization law:

$$u_i = -g_i(y) - \dot{x}_i - k_1 sign(e_i)\,sign(s_i)\dot{e}_i - k_2 sign(e_i)\,sign(s_i)e_i - k_2 sign(e_i)|e_i|^{p} \tag{12}$$

with this the derivative of the Lyapunov function shown in (11) becomes in:

$$\dot{V} \le 0 \tag{13}$$

4.2 Bidirectional Anti-Synchronization

For the bidirectional anti-synchronization consider the following systems:

$$\begin{aligned}\dot{x}_1 &= f_1(x) + u_1\\ \dot{x}_2 &= f_2(x) + u_2\\ &\vdots\\ \dot{x}_n &= f_n(x) + u_n\end{aligned} \tag{14}$$

where $X \in \mathbb{R}^n$ is the state vector, $f_n(x)$ is a nonlinear vector field and u_i is the control input. The response system is given by:

$$\begin{aligned}\dot{y}_1 &= g_1(y) + v_1 \\ \dot{y}_2 &= g_2(y) + v_2 \\ &\vdots \\ \dot{y}_n &= g_n(y) + v_n\end{aligned} \tag{15}$$

where $y \in \mathbb{R}^n$ is the state vector, $g(y)$ is a nonlinear vector field and v_i is the synchronization input. Now consider the following two sliding variables:

$$s_{ai}(t) = k_{a1}e_{ai} + k_{a2}\int_0^t \left[e_{ai} + sign(e_{ai})\left|e_{ai}\right|^{\rho}\right] d\tau$$

$$s_{bi}(t) = k_{b1}e_{bi} + k_{b2}\int_0^t \left[e_{bi} + sign(e_{bi})\left|e_{bi}\right|^{\rho}\right] d\tau \tag{16}$$

$k_{ai}, k_{bi} \in \mathbb{R}$ where $k_{ai}, k_{bi} > 0$ are the sliding gains and defining to $e_{ai} = y_i - x_i$ and $e_{bi} = x_i - y_i$ so the derivatives of (16):

$$\dot{s}_{ai}(t) = k_{a1}\dot{e}_{ai} + k_{a2}e_{ai} + k_{a2}sign(e_{ai})\left|e_{ai}\right|^{\rho}$$

$$\dot{s}_{bi}(t) = k_{b1}\dot{e}_{bi} + k_{b2}e_{bi} + k_{b2}sign(e_{bi})\left|e_{bi}\right|^{\rho} \tag{17}$$

Consider the following theorem to find the anti-synchronization bidirectional laws:

Theorem 2: *The systems (14) and (15) are bidirectionally anti-synchronized by the following laws:*

$$u_i = \dot{y}_i - sign(e_{ai})\,sign(e_{bi})\,g_i(y) + sign(e_{ai})\,sign(s_{bi})\,\dot{s}_{bi}$$

$$v_i = \dot{x}_i - sign(e_{bi})\,sign(e_{ai})\,f_i(x) + sign(e_{bi})\,sign(s_{ai})\,\dot{s}_{ai} \tag{18}$$

by selecting an appropriate Lyapunov functional in order to anti-synchronize the systems in both directions.

Proof: Consider the following Lyapunov functional

$$V = \sum_{i=1}^{n}\left|s_{ai}\right| + \sum_{i=1}^{n}\left|s_{bi}\right| + \sum_{i=1}^{n}\left|e_{ai}\right| + \sum_{i=1}^{n}\left|e_{bi}\right| \tag{19}$$

so, the derivative of (19) is:

$$\dot{V} = \sum_{i=1}^{n} sign(s_{ai})\,\dot{s}_{ai} + \sum_{i=1}^{n} sign(s_{bi})\,\dot{s}_{bi} + \sum_{i=1}^{n} sign(e_{ai})\,\dot{e}_{ai} + \sum_{i=1}^{n} sign(e_{bi})\,\dot{e}_{bi} \tag{20}$$

Now substituting the error variable and its derivative in (20) yields:

$$\dot{V} = \sum_{i=1}^{n} sign\left(s_{ai}\right)\dot{s}_{ai} + \sum_{i=1}^{n} sign\left(s_{bi}\right)\dot{s}_{bi} + \sum_{i=1}^{n} sign\left(e_{ai}\right)\dot{y}_i - \sum_{i=1}^{n} sign\left(e_{ai}\right)f_i\left(x\right)$$

$$-\sum_{i=1}^{n} sign\left(e_{ai}\right)u_i + \sum_{i=1}^{n} sign\left(e_{bi}\right)\dot{x}_i - \sum_{i=1}^{n} sign\left(e_{bi}\right)g_i\left(y\right) - \sum_{i=1}^{n} sign\left(e_{bi}\right)v_i \quad (21)$$

now by substituting the following bi-directional anti-synchronization laws:

$$u_i = \dot{y}_i - sign\left(e_{ai}\right)sign\left(e_{bi}\right)g_i\left(y\right) + sign\left(e_{ai}\right)sign\left(s_{bi}\right)\dot{s}_{bi}$$

$$v_i = \dot{x}_i - sign\left(e_{bi}\right)sign\left(e_{ai}\right)f_i\left(x\right) + sign\left(e_{bi}\right)sign\left(s_{ai}\right)\dot{s}_{ai} \quad (22)$$

yields $\dot{V} \le 0$ so the anti-synchronization is achieved and the proof is completed.

5 NUMERICAL EXPERIMENTS

This numerical experiment section consists of three numerical experiments, the first of which is the unidirectional synchronization of the Lorenz chaotic system and the Rossler chaotic system. The second numerical experiment consists of the unidirectional anti-synchronization of the system described above, and the last experiment consists of the bi-directional synchronization of the Ikeda system. As mentioned in this section, the theoretical results obtained in this chapter are effectively validated by the three empirical examples given in this section. Three chaotic systems were introduced for these numerical cases, first the Lorenz chaotic system, then the Rossler system, and finally the Ikeda system. The scheme of Lorenz shall be defined by:

$$\dot{x}\left(t\right) = a\left(y - x\right)$$

$$\dot{y}\left(t\right) = cx - xz - y$$

$$\dot{z}\left(t\right) = xy - bz \quad (23)$$

where the constants and initial conditions can be found in the literature. *The variables are considered in this study as* $x=x_1$, $y=x_2$ *and* $z=x_3$. Meanwhile the Rossler system is given by:

$$\dot{x}_1\left(t\right) = x_2 + ax_1$$

$$\dot{x}_2 = -x_1 - x_3$$

$$\dot{x}_3 = x_2 x_3 - b x_3 + c \tag{24}$$

With the appropriate constants and initial conditions. The Ikeda system used in this study is (Shahverdiev et. al., 2013):

$$\frac{dx}{dt} = -\pm x - m_1 sin\left(x\right) - m_2 sin\left(x\right)$$

$$\frac{dy}{dt} = -\pm y - m_1 sin\left(y\right) - m_2 sin\left(y\right) \tag{25}$$

With its appropriate constants as found in Shahverdiev et. al. (2013). It is important to mention that for the system (23), (24) and (25) are represented as uncontrolled system and the necessary input variables are considered.

In Fig. 1, Fig. 2 and Fig. 3, the phase portraits for the Lorenz attractor are shown. Meanwhile, in Fig. 4, Fig. 5 and Fig. 6, the phase portraits of the Rossler system are illustrated.

Figure 1. Phase portrait of the variables $\mathbf{x}_1$ and $\mathbf{x}_2$ for the Lorenz attractor

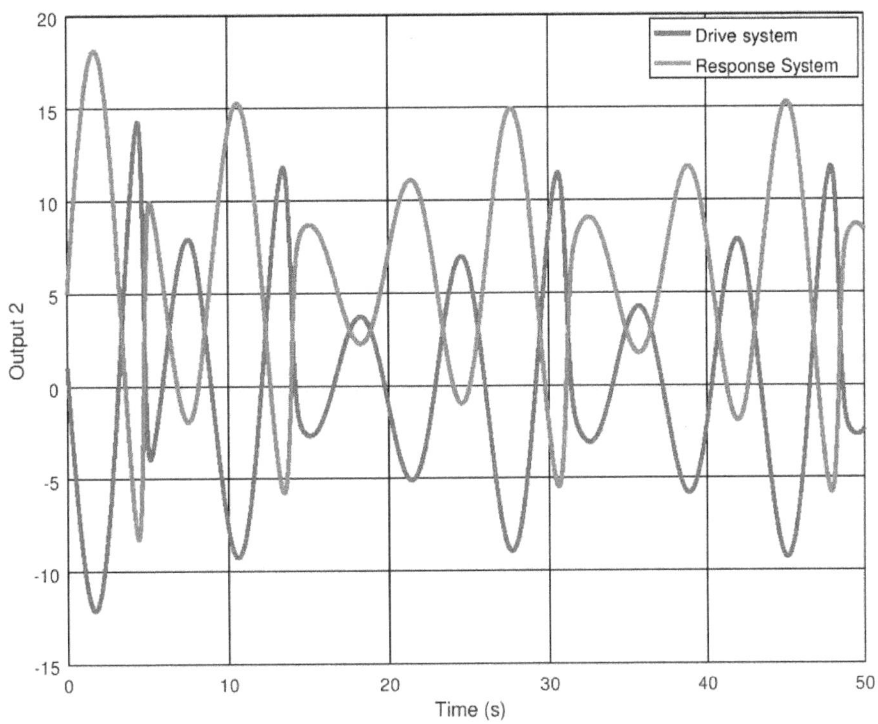

Figure 2. Phase portrait of the variables $\mathbf{x}_2$ *and* $\mathbf{x}_3$ *for the Lorenz attractor*

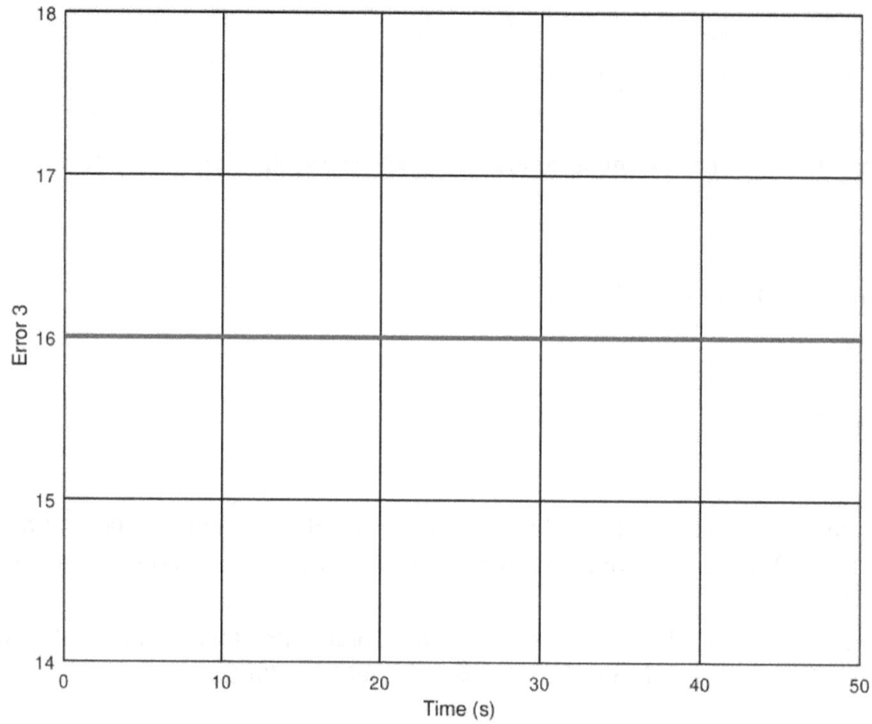

Figure 3. Phase portrait of the variables $\mathbf{x}_1$, $\mathbf{x}_2$ *and* $\mathbf{x}_3$ *for the Lorenz attractor*

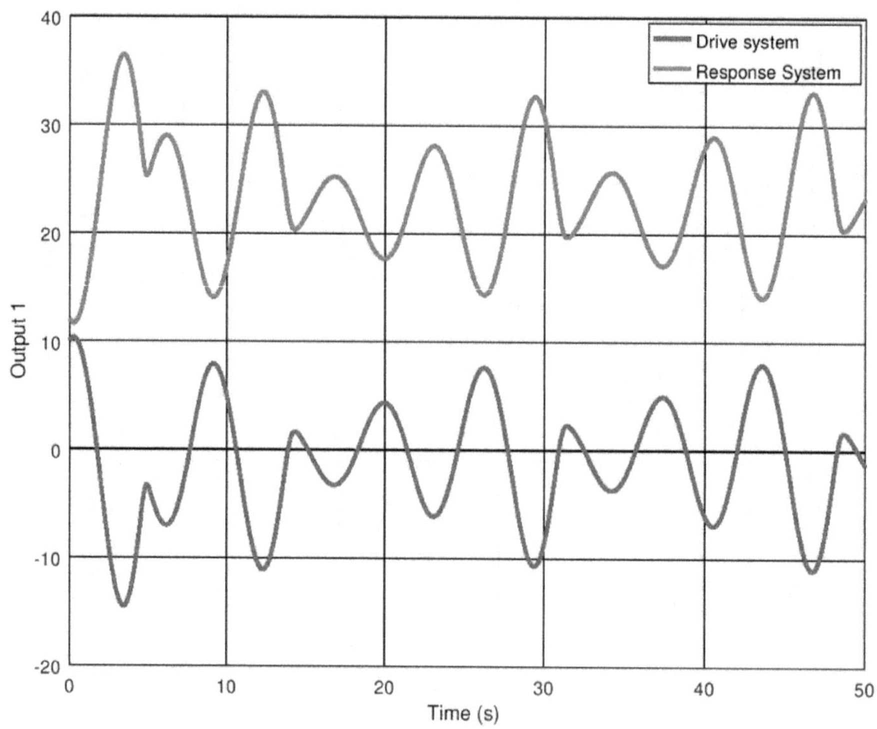

Figure 4. Phase portrait of the variables $\mathbf{x}_1$ and $\mathbf{x}_2$ for the Rossler system

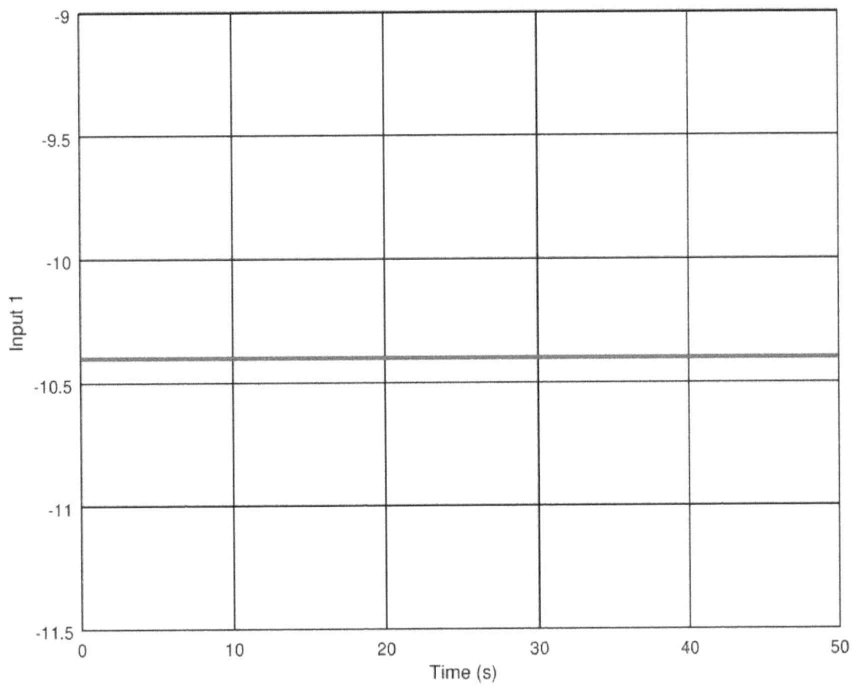

Figure 5. Phase portrait of the variables $\mathbf{x}_2$ and $\mathbf{x}_3$ for the Rossler system

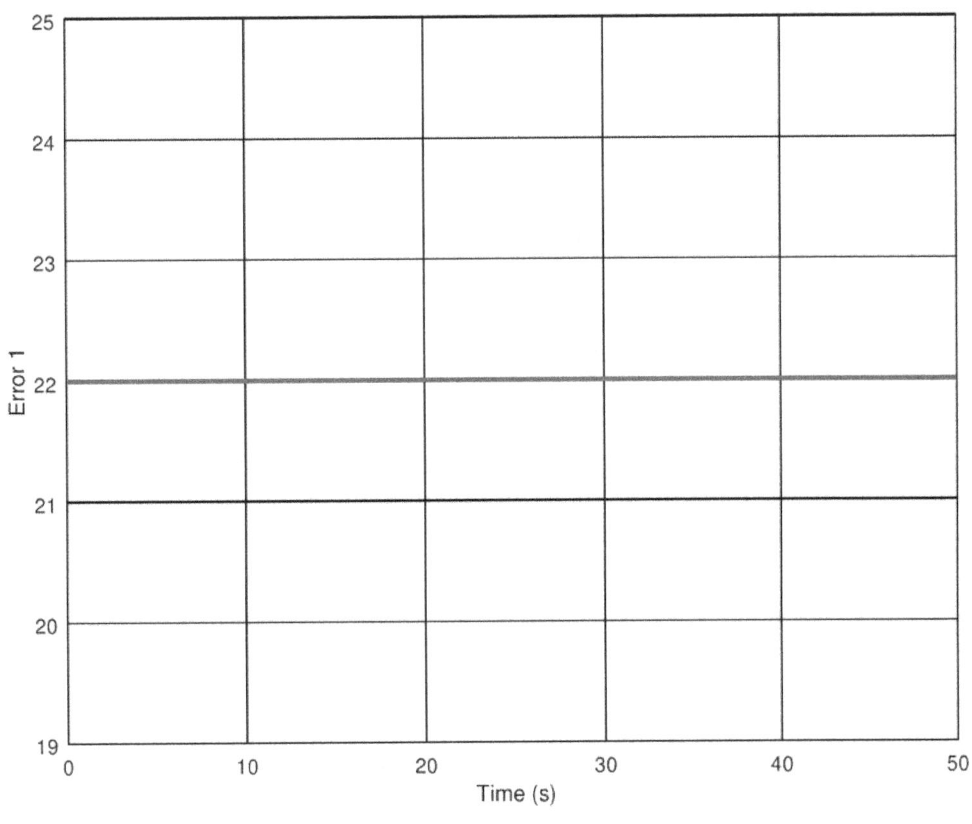

Figure 6. Phase portrait of the variables $\boldsymbol{x}_1$, $\boldsymbol{x}_2$ *and* $\boldsymbol{x}_3$ *for the Rossler system*

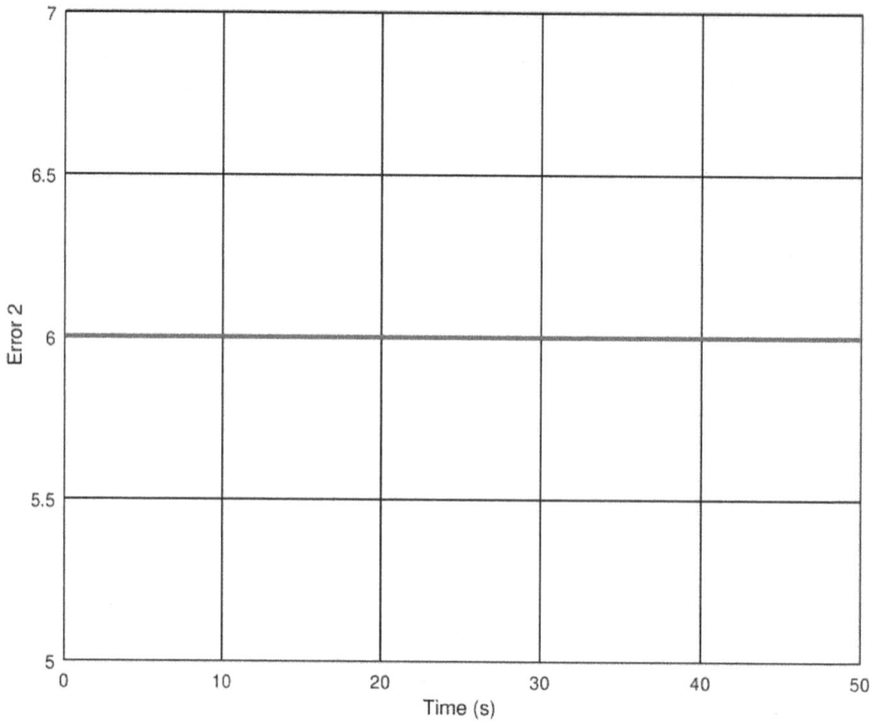

5.1 Numerical Experiment 1

For this experiment, chaos synchronization is accomplished by the use of the Rossler system (24) as a drive system and the Lorenz system (24) as a response system. The synchronization laws obtained in Section 2 shall be applied with the following results. It is verified that the terminal sliding mode chaotic system provides quicker, no chatter and small control effort for the unidirectional synchronization of chaotic systems. The theoretical framework used in this study is validated in this experiment, which shows how the response system variables are effectively driven by terminal sliding mode control techniques, contributes to the implementation of efficient synchronization techniques compared to other strategies found in the literature, and takes into account that sliding mode has not been extensively used.

In Fig. 7, Fig. 8 and Fig. 9, the for response of the variables x_1, x_2 and x_3 for experiment 1 indicates that it is a perfect synchronization between the drive and the response system. As later confirmed, the errors are driven to zero by the action of the terminal sliding mode synchronizer, which demonstrates the effectiveness of the proposed strategy. It can be seen how the action of the synchronization law drives the response system according to the desired path generated by the drive system for the three variables in finite time, with zero error as time goes to infinity. Almost perfect synchronization is achieved, taking into account that the action of the terminal sliding mode synchronizer drives the response system variables with a fast convergence time and no chatter during the reaching phase and sliding phase, which is advantageous compared to other synchronization techniques found in the literature. However, if the two systems used in this experiment are not identical, it is shown that accurate synchronization is accomplished by reducing the synchronization of the error variable to zero, taking into account the characteristics of the drive and the response system due to its complexity.

Figure 7. Response of the variable $\mathbf{x}_1$ *for experiment 1*

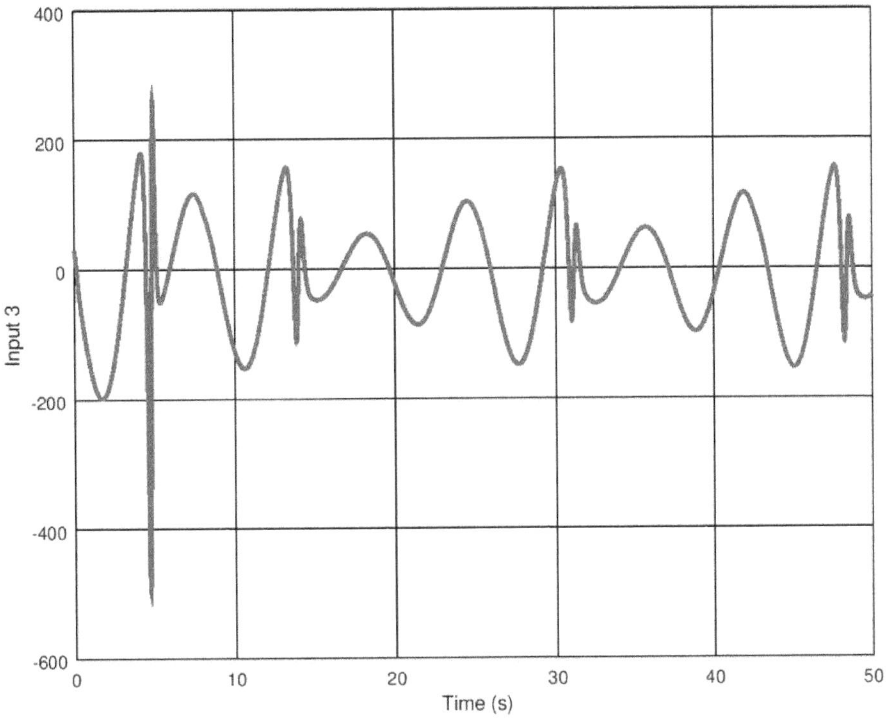

Figure 8. Response of the variable $\mathbf{x}_2$ *for experiment 1*

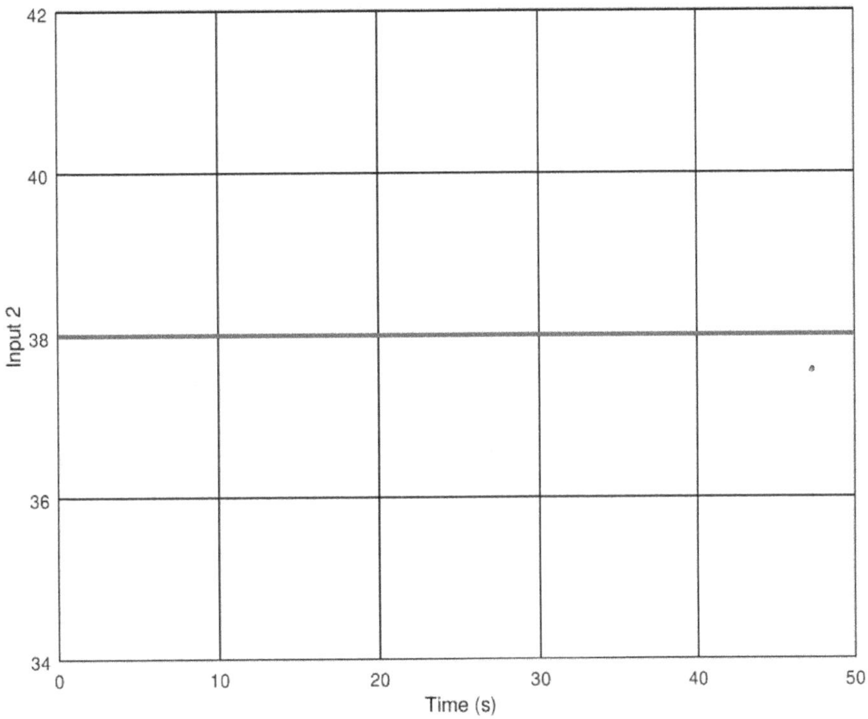

Figure 9. Response of the variable $\mathbf{x}_3$ *for experiment 1*

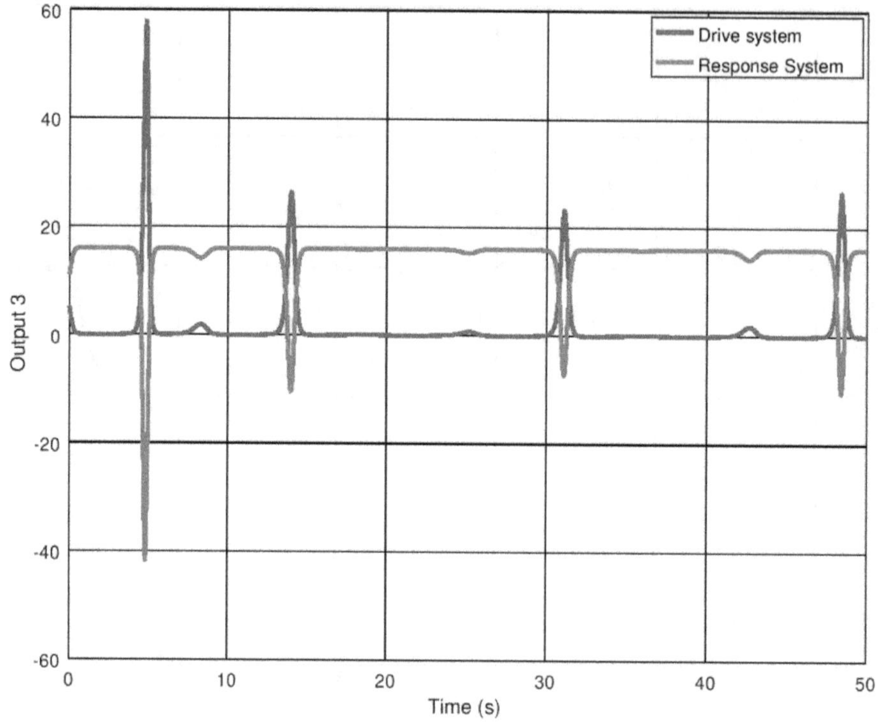

In Fig. 10, Fig. 11 and Fig. 12 can be observed the error variables e_1, e_2 and e_3 taking into consideration that these variables reach the origin by the terminal sliding mode synchronizer action, proving the effectiveness of the proposed strategy. The terminal sliding mode synchronizer action, as proved in the response tracking plots, drive the error system variables to zero in finite time, this is done by realizing an appropriate synchronization law control design in order that the global asymptotic stability is obtained by designing an appropriate Lyapunov function and meeting the required stability conditions in order to reduce the error between the drive and response system variables to zero in finite time. The theoretical results are corroborated in such a way that this results can be implemented of course in identical chaotic systems of any dimension considering the generalized theoretical results obtained in this study.

Finally, in Fig. 13, Fig. 14 and Fig. 15 the input variables u_1, u_2 and u_3 obtained by terminal sliding mode synchronizer action indicate that the proposed strategy results in a small controller action avoiding input saturation and no chatting. This limited control action is provided by the operation of the terminal sliding mode synchronizer, which, due to an appropriate theoretical design, provides a control input to the response system without the saturation of the synchronizer and without chattering, which is helpful when this synchronizer is used in actual physical systems such as mechanical, electrical and optical systems. Due to the rapid convergence of the sliding manifold, a fast synchronization is achieved with the appropriate synchronization input needed to track the response system variables to the drive system variables at a finite time, reducing the error variable to zero at a finite time. As a conclusion of this first experiment, apart from corroborating the theoretical results obtained in this analysis, a novel, quick and no chattering synchronization law is provided which has more advantages compared to other types of strategies found in the literature given that sliding mode technique has not been implemented as needed for chaotic system synchronization.

Figure 10. Error variable e_1 for experiment 1

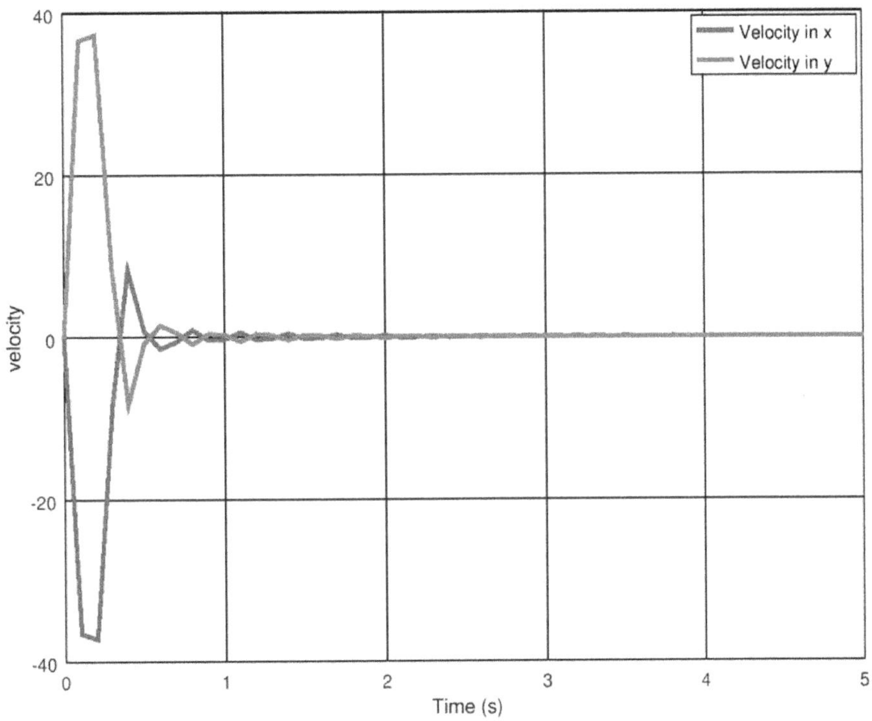

Figure 11. Error variable e_2 for experiment 1

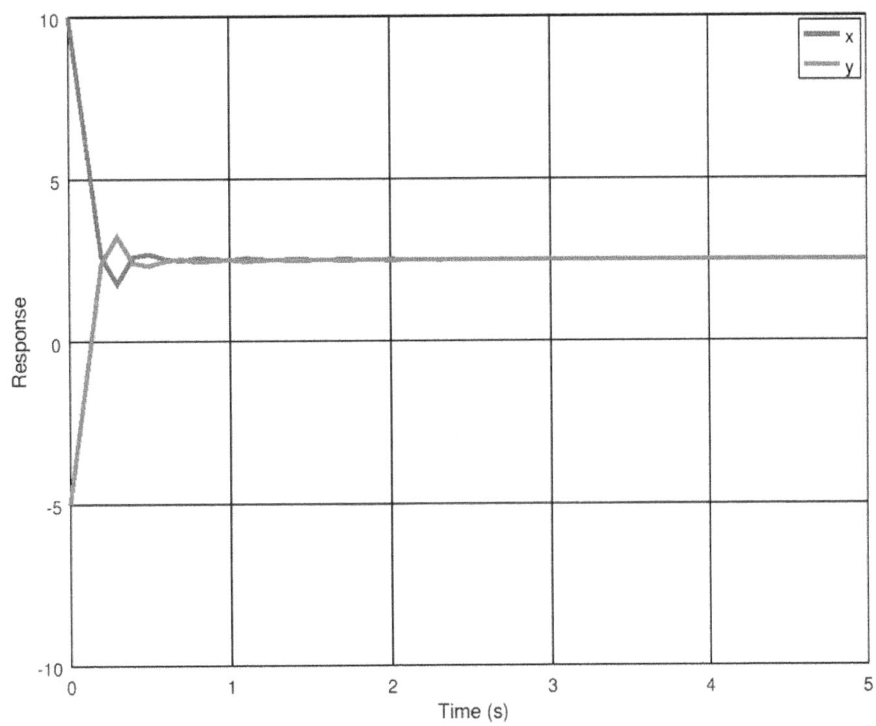

Figure 12. Error variable e_3 for experiment 1

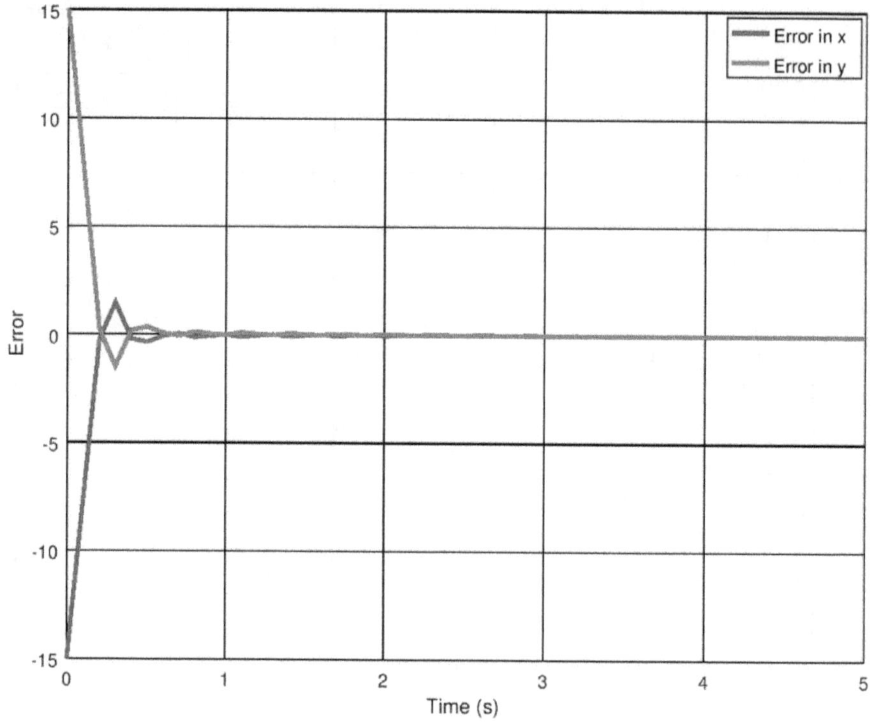

Figure 13. Input variable u_1 for experiment 1

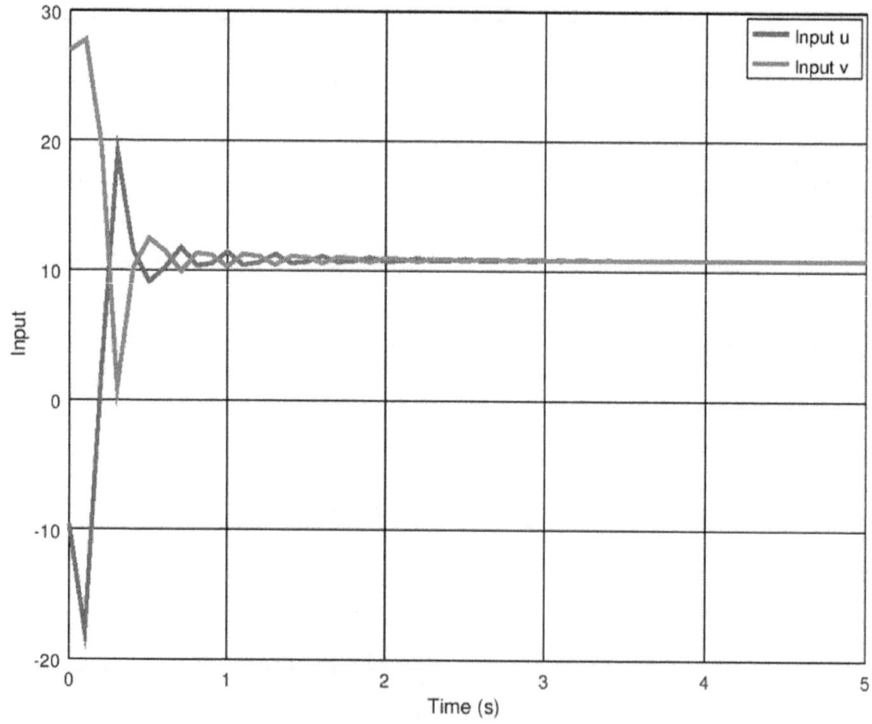

Figure 14. Input variable u_2 for experiment 1

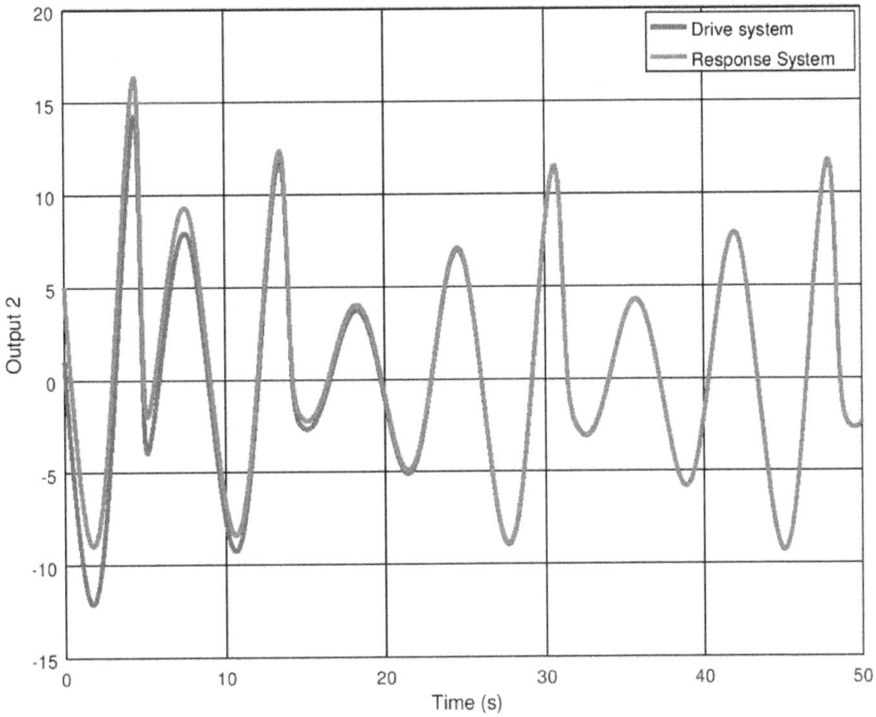

Figure 15. Input variable u_3 for experiment 1

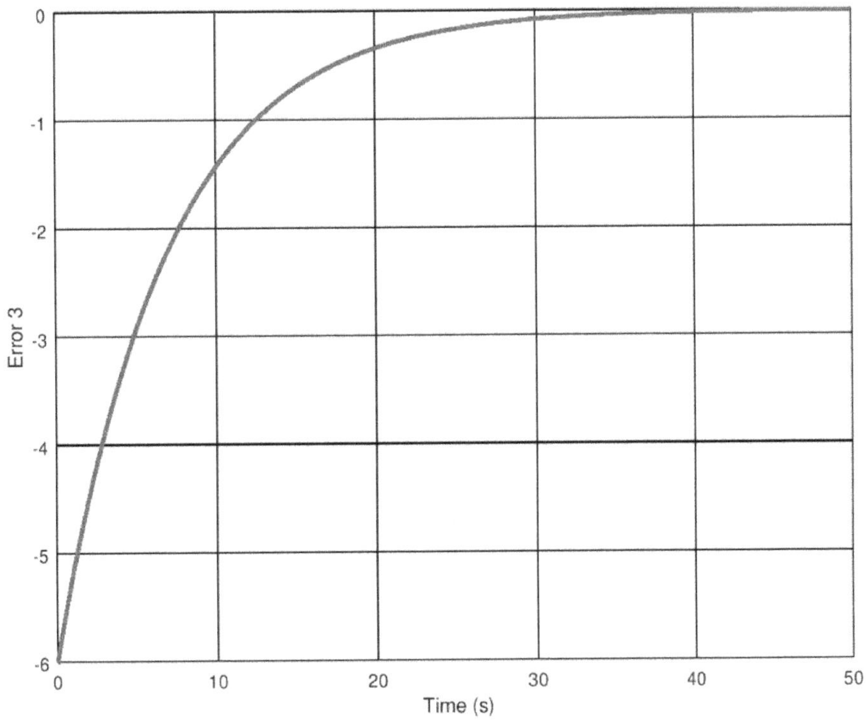

5.2 Numerical Experiment 2

An anti-synchronization between the Rossler system, which acts as a drive system, and the Lorenz attractor is shown in this experiment. The terminal sliding mode controller de-synchronizer law is used for this purpose showing the effectiveness of the proposed strategy. Similar to the unidirectional synchronization case, the theoretical results obtained in this study and validated numerically in this experiment prove that the anti-synchronization of two non-identical chaotic systems is effectively carried out by means of a single terminal sliding mode control law. It is confirmed how the chaotic system variables are effectively driven in the opposite direction while obtaining less control effort, no chatter and fast controller response, which is very useful for many types of coupled chaotic physical systems. Experiments are carried out on the basis of the correct initial conditions that make the system enter a chaotic regime in order to carry out the respective specified numerical experiments. The performance of the terminal sliding mode strategy for synchronization purposes is validated by selecting a single sliding manifold along with the respective Lyapunov functionality, designing an appropriate error variable to drive both system variables in the opposite direction in the finite time.

In Figs. 16, 17 and 18, the output variables x_1, x_2 and x_3 are displayed, indicating the anti-synchronization in the opposite direction between the drive system and the response system, so the de-synchronizer action controller drives the variables in the desired direction. The terminal sliding mode uni-directional synchronization law drives the system response variables in the opposite direction, as previously theoretically designed. The terminal sliding mode uni-directional anti-synchronizer provides fast, accurate and non-synchronous results that are validated in this experiment by selecting the appropriate error variable while minimizing the error due to the complexity of the uni-directional anti-synchronization function, taking into account the characteristics of the chaotic systems used as drive and response systems. The anti-synchronization control inputs are shown in the following figures.

Finally, in Fig. 19, Fig 20 and Fig. 21, the de-synchronizer action of the input variable u_1, u_2 and u_3 shows that a small controller action is required and that the terminal sliding mode de-synchronizer does not generate unwanted chattering. The terminal sliding mode synchronizer action provides a fast and accurate response while minimizing error by selecting the correct Lyapunov feature to obtain anti-synchronization laws. The proposed terminal sliding mode control provides a small control action for anti-synchronization purposes, so that input saturation is avoided provided that these results can be implemented in real physical systems as indicated in the previous numerical experiment. Similar to the unidirectional synchronization case, an effective control action is eliminated in the anti-synchronization chattering phase, which is very necessary given the characteristics of the chaotic drive system's anti-synchronized response.

5.3 Numerical Experiment 3

For this numerical experiment, the Ikeda system is used for bi-directional anti-synchronization of bi-directional coupled chaotic systems implementing the terminal sliding mode strategy. Similar to the previous experiment, the system variables are driven in the opposite direction, proving the effectiveness of the proposed strategy. The bi-directional anti-synchronization experiment seen in this section describes two one-dimensional chaotic systems that can obviously be used in identical and non-identical anti-synchronization as long as the dimensions of both systems coincide. It is validated how the two sliding manifolds, together with their corresponding Lyapunov functions, provide the necessary control laws for each coupled chaotic system. The proposed control strategy provides a small control effort, no chatter and rapid response, similar to the previous two numerical experiments. It is also observed in this numerical experiment that the velocity variable matches the response of both systems and that it is not too high to prove the effectiveness of the proposed control strategy for anti-synchronization purposes.

In Fig. 22 and Fig. 23 the variables x and y along with the input variables u and v are shown, proving in the first case, that the state variables are driven in the opposite direction, and in the second case, the input variables are bound to show that there is no saturation and the chattering effect is reduced to zero. In the case of the response of the variables x and y, it is noted that the input variables are anti-synchronized in opposite directions by the operation of the two anti-synchronization laws. The appropriate design of the two sliding manifolds to obtain the two control laws by sliding mode control, along with the appropriate selection of the Lyapunov function, provides rapid convergence in comparison with other strategies found in the literature. The control inputs obtained in this experiment provide a limited control effort that is very necessary and desirable while suppressing the chattering effect. It is important to note that the sliding mode technique for bi-directional anti-synchronization, in comparison to other techniques such as state feedback, provides a quicker and more efficient strategy given the various types of bi-directional coupled chaotic systems found in actual physical applications.

Figure 16. Variable x_1 for experiment 2

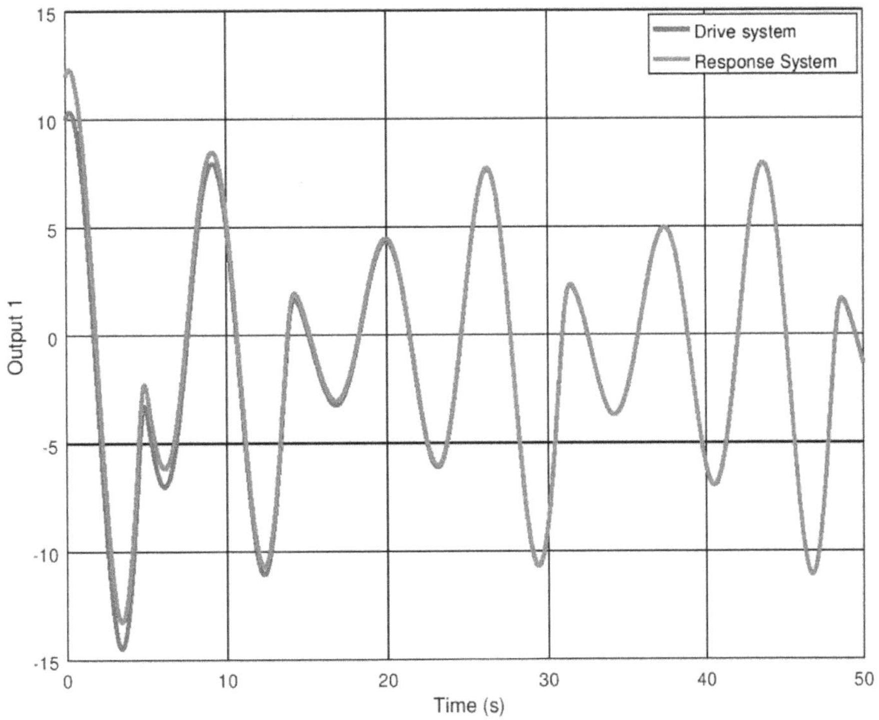

Figure 17. Variable x_2 for experiment 2

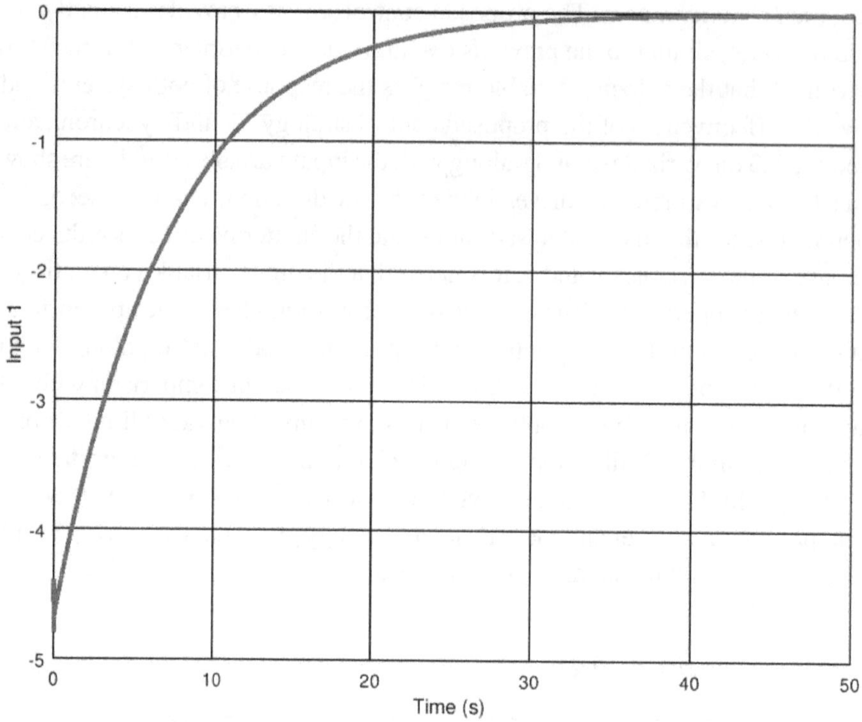

Figure 18. Variable x_3 for experiment 2

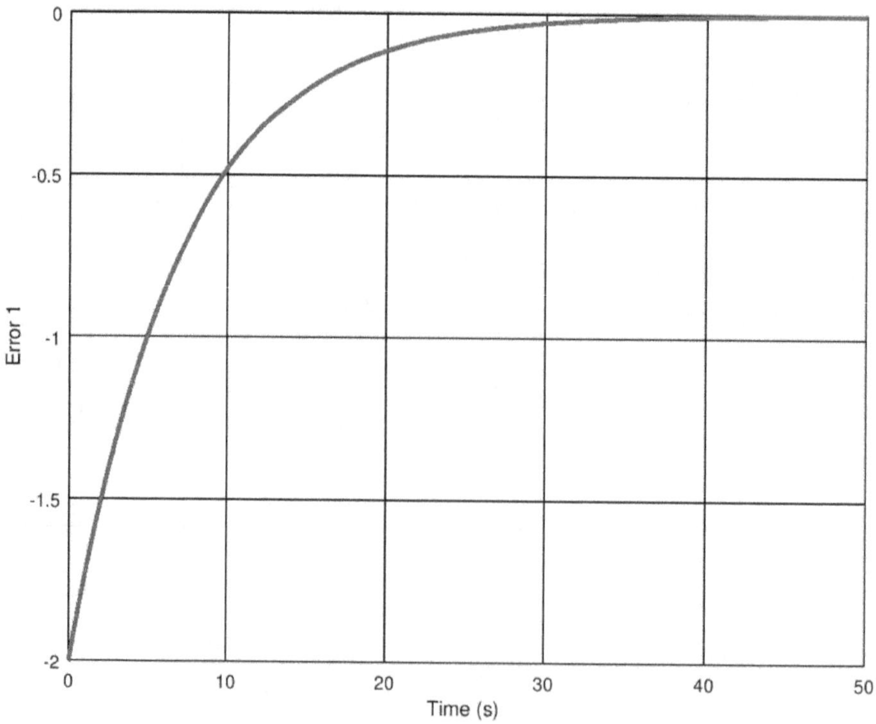

Figure 19. Input variable u_1 for experiment 2

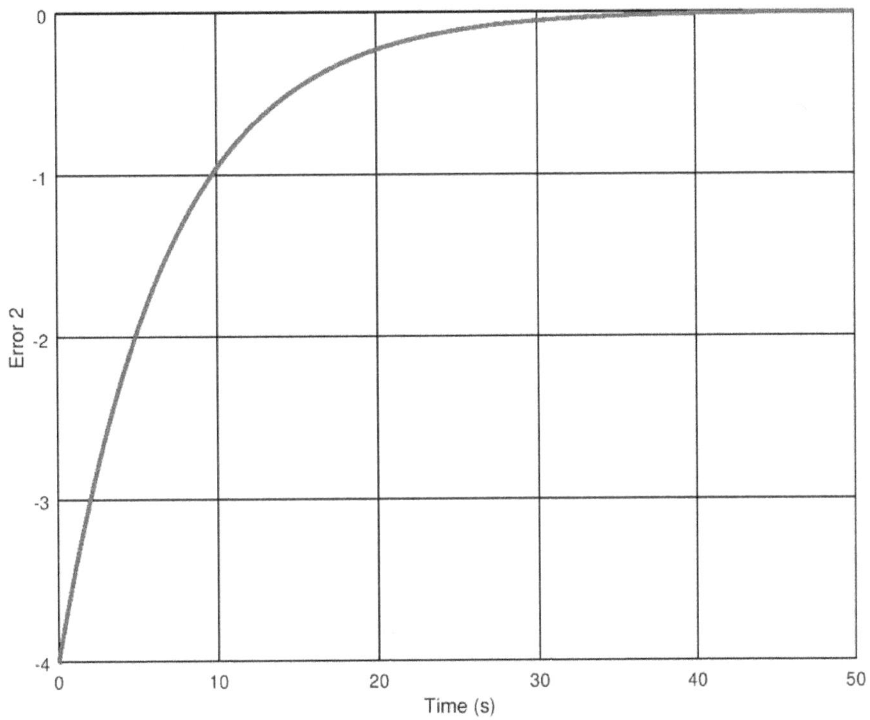

Figure 20. Input variable u_2 for experiment 2

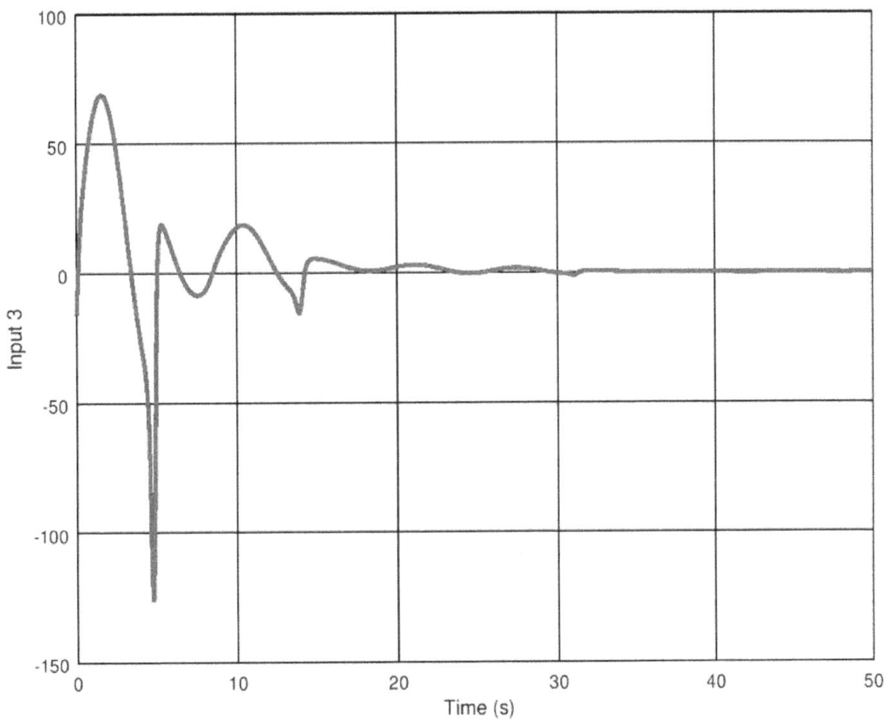

Figure 21. Input variable u_3 for experiment 2

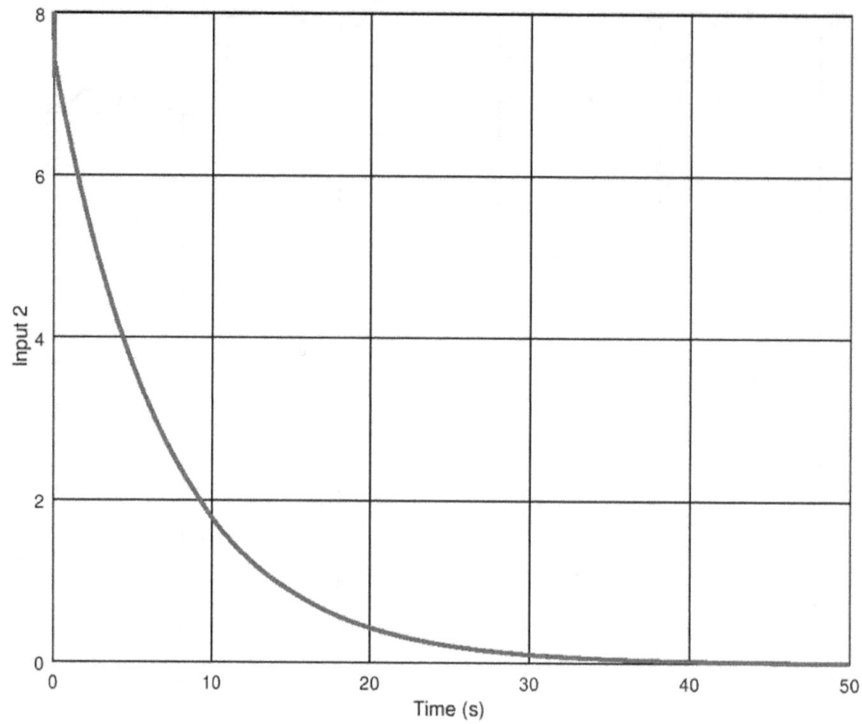

Figure 22. Response of the variables x and y for experiment 3

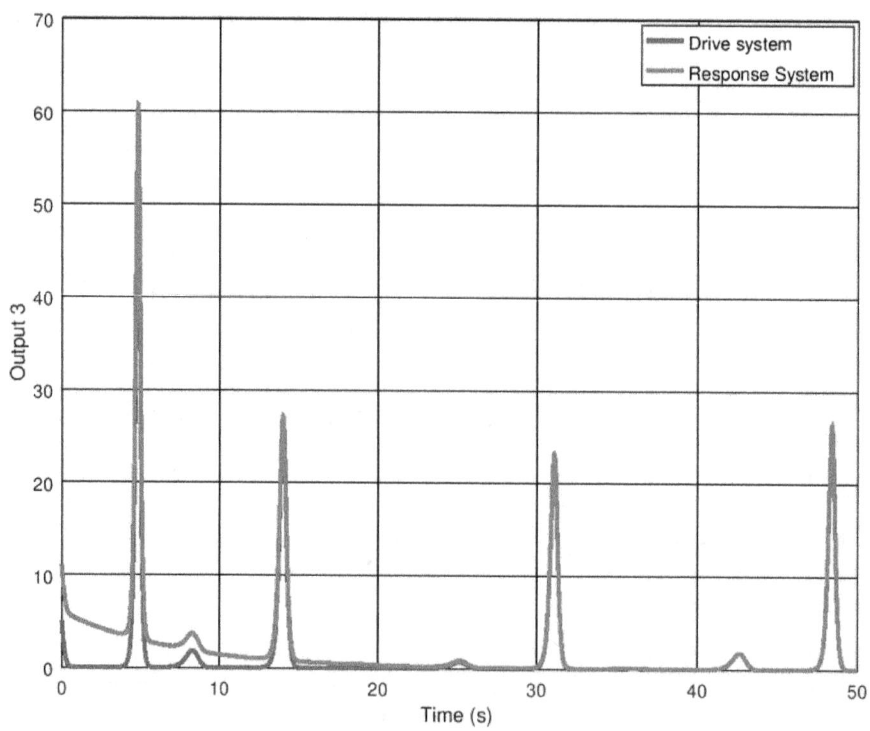

Figure 23. Input variables u and v for experiment 3

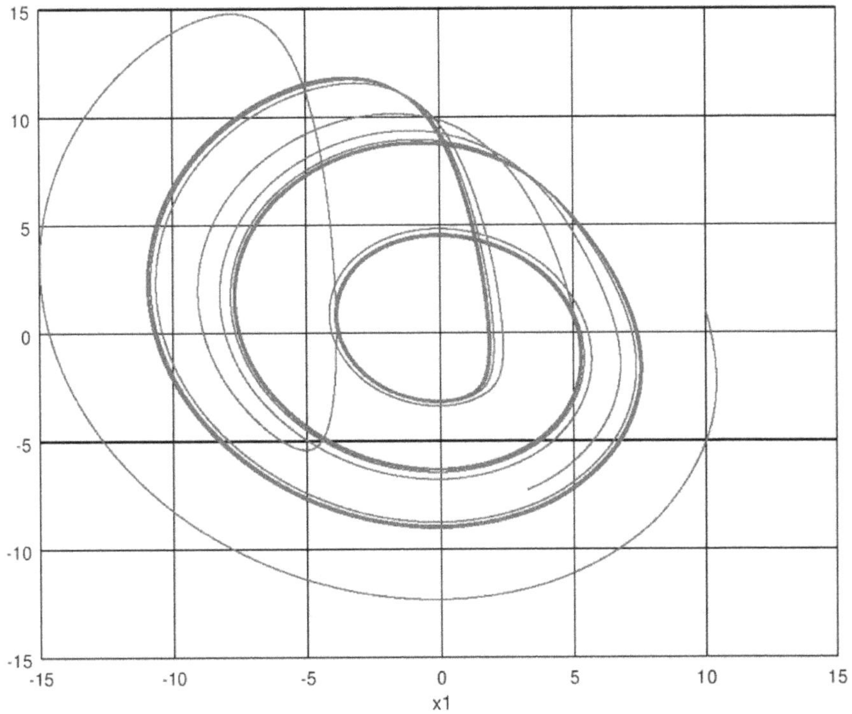

Figure 24. Velocities in x and y for experiment 3

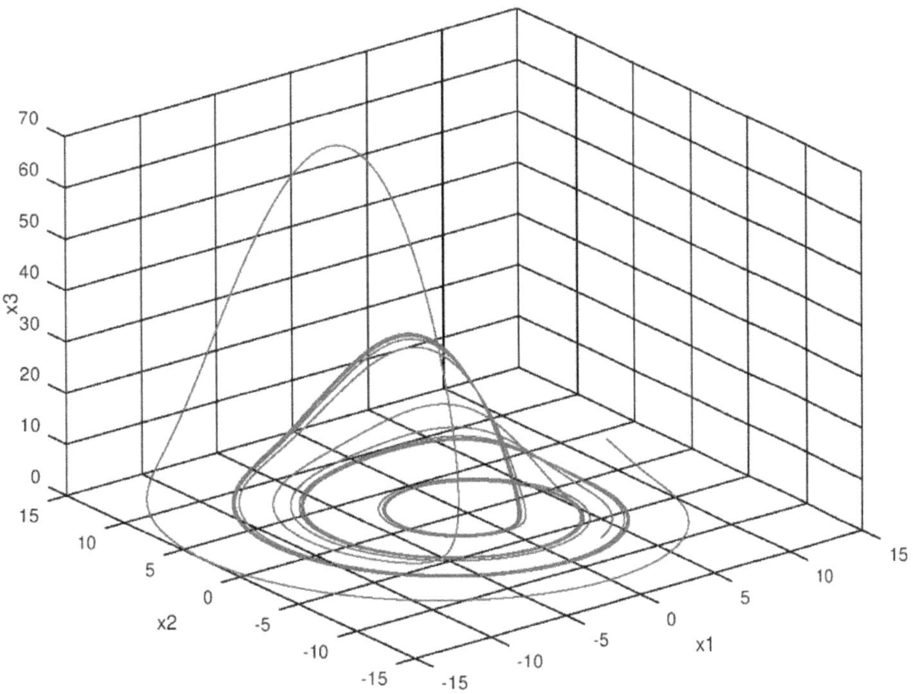

Finally, in Fig. 24, the velocity shown in x and y are also indicates the de-synchronization, similar to the previous figures. Velocity is also driven in the opposite direction. Despite the faster characteristics of the terminal sliding mode bi-directional anti-synchronization technique, the anti-synchronization velocity is not excessively high, provided that the overshoot is much smaller than that which can affect the response of both chaotic systems that have been anti-synchronized. As explained above, the response of the chaotic systems coincides with their velocity in direction and the chattering effect is removed. Excessively high oscillations are removed, taking into account the complex features of both chaotic systems to be anti-synchronized, surpassing the performance of other strategies.

6 DISCUSSIONS

As theoretically and numerically demonstrated, the results obtained in this study contribute to the synchronization and anti-synchronization of uni-directional and bi-directional chaotic systems. Theoretically and experimentally, it is shown that the synchronization and anti-synchronization strategies designed provide an effective alternative for uni-directional and bi-directional synchronization and anti-synchronization of coupled chaotic systems in their non-identical and identical form, provided, in the first case, that the systems must be of the same dimension. This contribution provides a novel strategy suitable for different types of coupled chaotic physical systems due to its effectiveness, given that the results found in the literature on sliding mode synchronization and anti-synchronization of chaotic systems are very limited. Terminal sliding mode ensures a fast convergence or, in other words, a fast reaching time of the sliding variable in finite time, along with other characteristics related to the control effort in synchronization and anti-synchronization cases. For this reason, this study has shown that both unidirectional and bi-directional synchronizer and anti-synchronizer laws are valid. The Lyapunov approach design of an effective control rule for the implementation of a terminal sliding manifold guarantees stability and zero convergence of error variables for the synchronization and anti-synchronization of uni-directional systems. In the case of bi-directional anti-synchronization, both error signals originate in finite time due to the definition of two terminal sliding variables. Anti-synchronization of the bi-directional system is an important contribution of this research, given that similar findings have not been recorded in the literature, anti-synchronization is accomplished by selecting very novel sliding variables to move system variables in opposite directions. For this purpose, two sliding manifolds, along with their respective Lyapunov, are designed to ensure stability during the anti-synchronization phase and, as confirmed in the numerical experiment section, a rapid convergence is achieved with the proposed control strategy due to the appropriate sliding mode manifold design. Another benefit is that chattering is not noticed and a limited control effort is achieved which makes it suitable for implementing bi-directional anti-synchronization in many types of physical systems. The numerical experiment example section validates how the two de-synchronizer laws drive system variables in the opposite direction, which is important for the real implementation of anti-synchronization in physical chaotic systems.

7 CONCLUSION

This chapter of this book addresses the synchronization and anti-synchronization of unidirectional and bi-directional coupled chaotic systems by terminal sliding mode control. In the case of unidirectional synchronization and anti-synchronization, appropriate control laws are obtained by selecting suitable terminal sliding variables and selecting Lyapunov functions to ensure synchronization and anti-synchronization of the origin variable, in the first case by selecting an appropriate unique sliding manifold and a suitable Lyapunov function. In the case of anti-synchronization, the state variables are driven in the opposite direction for all chaotic systems. For the uni-directional anti-synchronization, a unique sliding manifold, similar to the synchronization case, is designed to make the respective Lyapunov functional and to find the anti-synchronization law. In the case of the anti-synchronization of coupled bi-directional systems, two terminal sliding variables are designed to drive the state variables of the two systems in the opposite direction. Anti-synchronization of bi-directional chaotic systems is a novel strategy that has not been reported in the literature and has made an excellent contribution to the system. It is important to note that the control effort generated by the terminal sliding mode control strategies provides a small control effort, no chatter and a quick response in the three designed approaches. This research offers innovative theoretical results that can be used in the synchronization and anti-synchronization of various forms of real physical chaotic systems, such as electrical, mechanical and biological systems, among others.

REFERENCES

Al-Sawalha, A. (2009). Chaos anti-synchronization of two non-identical chaotic systems with known or fully unknown parameters. *Chaos, Solitons, and Fractals*, *42*(3), 1926–1932. doi:10.1016/j.chaos.2009.03.100

Alain, K. S. T., Azar, A. T., Fotsin, H. B., & Romanic, K. (2019). Robust Observer-based Synchronization of Chaotic Oscillators with Structural Perturbations and Input Nonlinearity. *Int. J. Automation and Control*, *13*(4), 387–412. doi:10.1504/IJAAC.2019.100467

Alain, K. S. T., Azar, A. T., Kengne, R., & Bertrand, F. H. (2020). Stability analysis and robust synchronisation of fractional-order modified Colpitts oscillators. *Int. J. Automation and Control*, *14*(1), 52–79. doi:10.1504/IJAAC.2020.103806

Alain, K. S. T., Romanic, K., Azar, A. T., Vaidyanathan, S., Bertrand, F. H., & Adele, N. M. (2018). *Dynamics Analysis and Synchronization in Relay Coupled Fractional Order Colpitts Oscillators. Advances in System Dynamics and Control*. IGI-Global. doi:10.4018/978-1-5225-4077-9.ch011

Azar, A. T., Serrano, F. E., Rossell, J. M., Vaidyanathan, S., & Zhu, Q. (2020). Adaptive self-recurrent wavelet neural network and sliding mode controller/observer for a slider crank mechanism. *International Journal of Computer Applications in Technology*, *63*(4), 273–285. doi:10.1504/IJCAT.2020.110404

Bao, X., Li, Q., Wu, T., Deng, Y., Hu, M., & Zeng, R. (2020). WDM-based bidirectional chaotic communication for semiconductor lasers system with time delay concealment. *Optics Communications*, *472*, 472. doi:10.1016/j.optcom.2020.125868

Beregov, R. Y., & Melkikh, A. V. (2015). De-synchronization and chaos in two inductively coupled Van der Pol auto-generators. *Chaos, Solitons, and Fractals*, *73*, 17–28. doi:10.1016/j.chaos.2014.12.017

Chen, S. H., Sia, C. H., & Shiue, M. C. (2020). On mathematical analysis of synchronization to bidirectionally coupled Kuramoto oscillators. *Nonlinear Analysis Real World Applications*, *56*, 103169. doi:10.1016/j.nonrwa.2020.103169

Deng, T., Xia, G. Q., Cao, L. P., Chen, J. G., Lin, X. D., & Wu, Z. M. (2009). Bidirectional chaos synchronization and communication in semiconductor lasers with optoelectronic feedback. *Optics Communications*, *282*(11), 2243–2249. doi:10.1016/j.optcom.2009.02.040

Djouima, M., Azar, A. T., Drid, S., & Mehdi, D. (2018). Higher Order Sliding Mode Control for Blood Glucose Regulation of Type 1 Diabetic Patients. *International Journal of System Dynamics Applications*, *7*(1), 65–84. doi:10.4018/IJSDA.2018010104

Grassi, G., Ouannas, A., Azar, A. T., Radwan, A. G., Volos, C., Pham, V. T., Ziar, T., Madian, A., Kyprianidis, I. M., & Stouboulos, I. N. (2017). *Chaos Synchronisation Of Continuous Systems Via Scalar Signal.* The 6th International Conference on Modern Circuits and Systems Technologies (MOCAST), Thessaloniki Greece. 10.1109/MOCAST.2017.7937629

Gurumurthy, G., & Das, D. K. (2020). Terminal sliding mode disturbance observer based adaptive super twisting sliding mode controller design for a class of nonlinear systems. *European Journal of Control.* Advance online publication. doi:10.1016/j.ejcon.2020.05.004

Hou, Z., Yang, Y., Liu, L., & Wang, Y. (2019). Terminal sliding mode control based impact time and angle constrained guidance. *Aerospace Science and Technology*, *93*, 1270–9638. doi:10.1016/j.ast.2019.04.050

Ji, Y., Liu, T., & Min, L. (2008). Generalized chaos synchronization theorems for bidirectional differential equations and discrete systems with applications. *Physics Letters. [Part A]*, *372*(20), 3645–3652. doi:10.1016/j.physleta.2008.02.022

Kammogne, A. S. T., Kountchou, M. N., Kengne, R., Azar, A. T., Fotsin, H. B., & Ouagni, S. T. M. (2020). Polynomial Robust Observer Implementation based-passive Synchronization of Nonlinear Fractional-Order Systems with Structural Disturbances. *Frontiers of Information Technology & Electronic Engineering.*, *21*(9), 1369–1386. doi:10.1631/FITEE.1900430

Khan, A., Singh, S., & Azar, A. T. (2020a). Combination-Combination Anti-Synchronization of Four Fractional Order Identical Hyperchaotic Systems. In *The International Conference on Advanced Machine Learning Technologies and Applications (AMLTA2019). AMLTA 2019. Advances in Intelligent Systems and Computing*, (vol 921, pp. 406-414). Springer. 10.1007/978-3-030-14118-9_41

Khan, A., Singh, S., & Azar, A. T. (2020b). Synchronization between a Novel Integer-Order Hyperchaotic System and a Fractional-Order Hyperchaotic System Using Tracking Control. In *The International Conference on Advanced Machine Learning Technologies and Applications (AMLTA2019). AMLTA 2019. Advances in Intelligent Systems and Computing*, (vol 921, pp. 382-391). Springer. 10.1007/978-3-030-14118-9_38

Khennaoui, A. A., Ouannas, A., Bendoukha, S., Grassi, G., Lozi, R. P., & Pham, V. T. (2019). On fractional–order discrete–time systems: Chaos, stabilization and synchronization. *Chaos, Solitons, and Fractals*, *119*, 150–162. doi:10.1016/j.chaos.2018.12.019

Khennaoui, A. A., Ouannas, A., Grassi, G., & Azar, A. T. (2020). Dynamic Analysis of a Fractional Map with Hidden Attractor. In *Proceedings of the International Conference on Artificial Intelligence and Computer Vision (AICV2020). AICV 2020. Advances in Intelligent Systems and Computing*, (vol 1153, pp. 731-739). Springer. 10.1007/978-3-030-44289-7_68

Khettab, K., Bensafia, Y., Bourouba, B., & Azar, A. T. (2018). Enhanced Fractional Order Indirect Fuzzy Adaptive Synchronization of Uncertain Fractional Chaotic Systems Based on the Variable Structure Control: Robust H? Design Approach. In Mathematical Techniques of Fractional Order Systems, Advances in Nonlinear Dynamics and Chaos (ANDC) series, (pp. 559-595). Elsevier.

Kumar, J., Azar, A. T., Kumar, V., & Rana, K. P. S. (2018) Design of Fractional Order Fuzzy Sliding Mode Controller for Nonlinear Complex Systems. In Mathematical Techniques of Fractional Order Systems, Advances in Nonlinear Dynamics and Chaos (ANDC) series, (pp. 249-282). Elsevier. doi:10.1016/B978-0-12-813592-1.00009-X

Li, Z., Chen, D., Zhu, J., & Liu, Y. (2015). Nonlinear dynamics of fractional order Duffing system. *Chaos, Solitons, and Fractals*, *81*, 111–116. doi:10.1016/j.chaos.2015.09.012

Meghni, B., Dib, D., & Azar, A. T. (2017). A Second-order sliding mode and fuzzy logic control to Optimal Energy Management in PMSG Wind Turbine with Battery Storage. *Neural Computing & Applications*, *28*(6), 1417–1434. doi:10.100700521-015-2161-z

Mekki, H, Boukhetala, D., & Azar, A.T. (2015). Sliding Modes for Fault Tolerant Control. *Studies in Computational Intelligence, 576*, 407-433. DOI doi:10.1007/978-3-319-11173-5_15

Miao, Y., Hwang, I., Liu, M., & Wang, F. (2019). Adaptive fast nonsingular terminal sliding mode control for attitude tracking of flexible spacecraft with rotating appendage. *Aerospace Science and Technology*, *93*, 1270–9638. doi:10.1016/j.ast.2019.105312

Mishra, V., Das, S., Jafari, H., & Ong, S. H. (2016). Study of fractional order Van der Pol equation. *Journal of King Saud University - Science, 28*(1), 55-60.

Mittal, S., Azar, A. T., & Kamal, N. A. (2021). Nonlinear Fractional Order System Synchronization via Combination-Combination Multi-switching. In *Proceedings of the International Conference on Advanced Intelligent Systems and Informatics 2020. AISI 2020. Advances in Intelligent Systems and Computing*, (vol 1261). Springer. 10.1007/978-3-030-58669-0_75

Moaddy, K., Hashim, I., & Momani, S. (2011). Non-standard finite difference schemes for solving fractional-order Rössler chaotic and hyperchaotic systems. *Computers & Mathematics with Applications (Oxford, England)*, *62*(3), 1068–1074. doi:10.1016/j.camwa.2011.03.059

Munoz-Pacheco, J. M., Volos, C., Campos-Canton, E., Azar, A. T., Pham, V. T., & Radwan, A. G. (2017). New Trends on Modeling, Design, and Control of Chaotic Systems. *Mathematical Problems in Engineering*, *2017*, 3737681. Advance online publication. doi:10.1155/2017/3737681

Ouannas, A., Azar, A. T., & Abu-Saris, R. (2017h). A new type of hybrid synchronization between arbitrary hyperchaotic maps. *International Journal of Machine Learning and Cybernetics*, *8*(6), 1887–1894. doi:10.100713042-016-0566-3

Ouannas, A., Azar, A. T., & Radwan, A. G. (2016). On Inverse Problem of Generalized Synchronization Between Different Dimensional Integer-Order and Fractional-Order Chaotic Systems. In *The 28th International Conference on Microelectronics*. IEEE. 10.1109/ICM.2016.7847942

Ouannas, A., Azar, A. T., & Vaidyanathan, S. (2017a). New hybrid synchronisation schemes based on coexistence of various types of synchronisation between master-slave hyperchaotic systems. *International Journal of Computer Applications in Technology*, *55*(2), 112–120. doi:10.1504/IJCAT.2017.082868

Ouannas, A., Azar, A. T., & Vaidyanathan, S. (2017g). On A Simple Approach for Q-S Synchronization of Chaotic Dynamical Systems in Continuous-Time. Int. J. *Computing Science and Mathematics*, *8*(1), 20–27.

Ouannas, A., Azar, A. T., & Ziar, T. (2019c). Control of Continuous-Time Chaotic (Hyperchaotic) Systems: F-M Synchronization. *Int. J. Automation and Control*, *13*(2), 226–242. doi:10.1504/IJAAC.2019.098215

Ouannas, A., Azar, A. T., & Ziar, T. (2020f). On Inverse Full State Hybrid Function Projective Synchronization for Continuous-time Chaotic Dynamical Systems with Arbitrary Dimensions. *Differential Equations and Dynamical Systems*, *28*(4), 1045–1058. doi:10.100712591-017-0362-x

Ouannas, A., Azar, A. T., Ziar, T., & Radwan, A. G. (2017e). *A Study On Coexistence of Different Types of Synchronization Between Different dimensional Fractional Chaotic Systems. Studies in Computational Intelligence* (Vol. 688). Springer-Verlag.

Ouannas, A., Azar, A. T., Ziar, T., & Radwan, A. G. (2017f). *Generalized Synchronization of Different Dimensional Integer-order and Fractional Order Chaotic Systems. Studies in Computational Intelligence* (Vol. 688). Springer-Verlag.

Ouannas, A., Azar, A. T., Ziar, T., & Vaidyanathan, S. (2017b). *On New Fractional Inverse Matrix Projective Synchronization Schemes. Studies in Computational Intelligence* (Vol. 688). Springer-Verlag.

Ouannas, A., Azar, A. T., Ziar, T., & Vaidyanathan, S. (2017c). *Fractional Inverse Generalized Chaos Synchronization Between Different Dimensional Systems. Studies in Computational Intelligence* (Vol. 688). Springer-Verlag.

Ouannas, A., Azar, A. T., Ziar, T., & Vaidyanathan, S. (2017d). *A New Method To Synchronize Fractional Chaotic Systems With Different Dimensions. Studies in Computational Intelligence* (Vol. 688). Springer-Verlag.

Ouannas, A., Grassi, G., & Azar, A. T. (2020b). A New Generalized Synchronization Scheme to Control Fractional Chaotic Systems with Non-identical Dimensions and Different Orders. In *The International Conference on Advanced Machine Learning Technologies and Applications (AMLTA2019). AMLTA 2019. Advances in Intelligent Systems and Computing*, (vol 921, pp. 415-424). Springer. 10.1007/978-3-030-14118-9_42

Ouannas, A., Grassi, G., & Azar, A. T. (2020c). Fractional-Order Control Scheme for Q-S Chaos Synchronization. In *The International Conference on Advanced Machine Learning Technologies and Applications (AMLTA2019). AMLTA 2019. Advances in Intelligent Systems and Computing*, (vol 921, pp. 434-441). Springer. 10.1007/978-3-030-14118-9_44

Ouannas, A., Grassi, G., Azar, A. T., & Gasri, A. (2019b). A New Control Scheme for Hybrid Chaos Synchronization. In *Proceedings of the International Conference on Advanced Intelligent Systems and Informatics 2018. AISI 2018. Advances in Intelligent Systems and Computing*, (vol 845). Springer. 10.1007/978-3-319-99010-1_10

Ouannas, A., Grassi, G., Azar, A. T., & Khennaoui, A. A. (2020e). Synchronization Control in Fractional Discrete-Time Systems with Chaotic Hidden Attractors. In A. Hassanien, R. Bhatnagar, & A. Darwish (Eds.), *Advanced Machine Learning Technologies and Applications. AMLTA 2020. Advances in Intelligent Systems and Computing* (Vol. 1141, pp. 661–669). Springer.

Ouannas, A., Grassi, G., Azar, A. T., Khennaouia, A. A., & Pham, V. T. (2020a). Chaotic Control in Fractional-Order Discrete-Time Systems. In *AISI 2019. The International Conference on Advanced Intelligent Systems and Informatics AISI 2019. Advances in Intelligent Systems and Computing*, (vol 1058, pp. 207-217). Springer. 10.1007/978-3-030-31129-2_20

Ouannas, A., Grassi, G., Azar, A. T., Khennaouia, A. A., & Pham, V. T. (2020d). Synchronization of Fractional-Order Discrete-Time Chaotic Systems. In *The International Conference on Advanced Intelligent Systems and Informatics AISI 2019. Advances in Intelligent Systems and Computing*, (vol 1058, pp. 218-228). Springer. 10.1007/978-3-030-31129-2_21

Ouannas, A., Grassi, G., Azar, A. T., Radwan, A. G., Volos, C., Pham, V. T., Ziar, T., Kyprianidis, I. M., & Stouboulos, I. N. (2017g). Dead-Beat Synchronization Control in Discrete-Time Chaotic Systems. *The 6th International Conference on Modern Circuits and Systems Technologies (MOCAST)*. 10.1109/MOCAST.2017.7937628

Ouannas, A., Grassi, G., Azar, A. T., & Singh, S. (2019a). New Control Schemes for Fractional Chaos Synchronization. In *Proceedings of the International Conference on Advanced Intelligent Systems and Informatics 2018. AISI 2018. Advances in Intelligent Systems and Computing*, (vol 845). Springer. 10.1007/978-3-319-99010-1_5

Pham, V. T., Gokul, P. M., Kapitaniak, T., Volos, C., & Azar, A. T. (2018). Dynamics, Synchronization and Fractional Order Form of a Chaotic System With Infinite Equilibria. In Mathematical Techniques of Fractional Order Systems, Advances in Nonlinear Dynamics and Chaos (ANDC) series, (pp. 475-502). Elsevier.

Pham, V. T., Vaidyanathan, S., Volos, C. K., Azar, A. T., Hoang, T. M., & Yem, V. V. (2017a). *A Three-Dimensional No-Equilibrium Chaotic System: Analysis, Synchronization and Its Fractional Order Form. Studies in Computational Intelligence* (Vol. 688). Springer-Verlag.

Pham, V. T., Volos, C. K., Vaidyanathan, S., & Azar, A. T. (2017b). Dynamics, Synchronization and Fractional Order Form of a Chaotic System without Equilibrium. In C. K. Volos (Ed.), *Nonlinear Systems: Design, Applications and Analysis*. Nova Science Publishers.

Rabiee, H., Ataei, M., & Ekramian, M. (2019). Continuous nonsingular terminal sliding mode control based on adaptive sliding mode disturbance observer for uncertain nonlinear systems. *Automatica, 109*.

Sang, Y., Sun, W., Duan, F., & Zhao, J. (2019). Bidirectional synchronization control for an electrohydraulic servo loading system. *Mechatronics*, *62*, 62. doi:10.1016/j.mechatronics.2019.102254

Shahverdiev, E. M. (2019). Effect of parameter mismatches on chaos synchronization between Josephson junctions coupled in series and driven by a central junction. *Physica C: Superconductivity and its Applications, 31*(1), 230-235.

Shahverdiev, E. M., Bayramov, P. A., & Qocayeva, M. V. (2013). Inverse chaos synchronization between bidirectionally coupled variable multiple time delay systems. *Optik (Stuttgart)*, *124*(18), 3427–3429. doi:10.1016/j.ijleo.2012.10.055

Shahverdiev, E. M., & Shore, K. A. (2009). Synchronization of chaos in unidirectionally and bidirectionally coupled multiple time delay laser diodes with electro-optical feedback. *Optics Communications*, *282*(2), 310–316. doi:10.1016/j.optcom.2008.10.009

Shukla, M. K., Sharma, B. B., & Azar, A. T. (2018). Control and Synchronization of a Fractional Order Hyperchaotic System via Backstepping and Active Backstepping Approach. In Mathematical Techniques of Fractional Order Systems, Advances in Nonlinear Dynamics and Chaos (ANDC) series, (pp. 597-624). Elsevier. doi:10.1016/B978-0-12-813592-1.00019-2

Singh, S., & Azar, A. T. (2020a). Multi-switching Combination Synchronization of Fractional Order Chaotic Systems. In *Proceedings of the International Conference on Artificial Intelligence and Computer Vision (AICV2020). AICV 2020. Advances in Intelligent Systems and Computing*, (vol 1153, pp. 655-664). Springer. 10.1007/978-3-030-44289-7_61

Singh, S., & Azar, A. T. (2020b). Controlling Chaotic System via Optimal Control. In *The International Conference on Advanced Intelligent Systems and Informatics AISI 2019. Advances in Intelligent Systems and Computing*, (vol 1058, pp. 277-287). Springer. 10.1007/978-3-030-31129-2_26

Singh, S., Azar, A. T., Bhat, M. A., Vaidyanathan, S., & Ouannas, A. (2018b). *Active Control for Multi-Switching Combination Synchronization of Non-Identical Chaotic Systems. Advances in System Dynamics and Control*. IGI-Global. doi:10.4018/978-1-5225-4077-9.ch005

Singh, S., Azar, A. T., Ouannas, A., Zhu, Q., Zhang, W., & Na, J. (2017). Sliding Mode Control Technique for Multi-switching Synchronization of Chaotic Systems. *Proceedings of 9th International Conference on Modelling, Identification and Control (ICMIC 2017)*, 880-885. 10.1109/ICMIC.2017.8321579

Singh, S., Azar, A. T., Vaidyanathan, S., Ouannas, A., & Bhat, M. A. (2018a). Multiswitching Synchronization of Commensurate Fractional Order Hyperchaotic Systems Via Active Control. In Mathematical Techniques of Fractional Order Systems, Advances in Nonlinear Dynamics and Chaos (ANDC) series, (pp. 319-345). Elsevier. doi:10.1016/B978-0-12-813592-1.00011-8

Singh, S., Azar, A. T., & Zhu, Q. (2018c). Multi-switching Master-Slave Synchronization of Nonidentical Chaotic Systems. In Q. Zhu, J. Na, & X. Wu (Eds.), *Innovative Techniques and Applications of Modelling, Identification and Control. Lecture Notes in Electrical Engineering* (Vol. 467, pp. 321–330). Springer. doi:10.1007/978-981-10-7212-3_20

Tarai, A., Poria, S., & Chatterjee, P. (2009). Synchronization of generalised linearly bidirectionally coupled unified chaotic system. *Chaos, Solitons, and Fractals*, *40*(2), 885–892. doi:10.1016/j.chaos.2007.08.039

Vaidyanathan, S., & Azar, A.T. (2015a). Anti-Synchronization of Identical Chaotic Systems using Sliding Mode Control and an Application to Vaidyanathan-Madhavan Chaotic Systems. *Studies in Computational Intelligence, 576*, 527-547. DOI . doi:10.1007/978-3-319-11173-5_19

Vaidyanathan, S., & Azar, A.T. (2015b). Hybrid Synchronization of Identical Chaotic Systems using Sliding Mode Control and an Application to Vaidyanathan Chaotic Systems. *Studies in Computational Intelligence, 576*, 549-569. DOI . doi:10.1007/978-3-319-11173-5_20

Vaidyanathan, S., & Azar, A.T. (2015c). Analysis, Control and Synchronization of a Nine-Term 3-D Novel Chaotic System. *Studies in Computational Intelligence, 581*, 3-17. DOI . 63. doi:10.1007/978-3-319-13132-0_1

Vaidyanathan, S., & Azar, A.T. (2015d). Analysis and Control of a 4-D Novel Hyperchaotic System. *Studies in Computational Intelligence, 581*, 19-38. DOI . doi:10.1007/978-3-319-13132-0_2

Vaidyanathan, S., & Azar, A. T. (2016a). *Dynamic Analysis, Adaptive Feedback Control and Synchronization of an Eight-Term 3-D Novel Chaotic System with Three Quadratic Nonlinearities. Studies in Fuzziness and Soft Computing* (Vol. 337). Springer-Verlag.

Vaidyanathan, S., & Azar, A. T. (2016b). *Qualitative Study and Adaptive Control of a Novel 4-D Hyperchaotic System with Three Quadratic Nonlinearities. Studies in Fuzziness and Soft Computing* (Vol. 337). Springer-Verlag.

Vaidyanathan, S., & Azar, A. T. (2016c). *A Novel 4-D Four-Wing Chaotic System with Four Quadratic Nonlinearities and its Synchronization via Adaptive Control Method. Advances in Chaos Theory and Intelligent Control. Studies in Fuzziness and Soft Computing* (Vol. 337). Springer-Verlag.

Vaidyanathan, S., & Azar, A. T. (2016d). *Adaptive Control and Synchronization of Halvorsen Circulant Chaotic Systems. Advances in Chaos Theory and Intelligent Control. Studies in Fuzziness and Soft Computing* (Vol. 337). Springer-Verlag.

Vaidyanathan, S., & Azar, A. T. (2016e). *Adaptive Backstepping Control and Synchronization of a Novel 3-D Jerk System with an Exponential Nonlinearity. Advances in Chaos Theory and Intelligent Control. Studies in Fuzziness and Soft Computing* (Vol. 337). Springer-Verlag.

Vaidyanathan, S., & Azar, A. T. (2016f). *Generalized Projective Synchronization of a Novel Hyperchaotic Four-Wing System via Adaptive Control Method. Advances in Chaos Theory and Intelligent Control. Studies in Fuzziness and Soft Computing* (Vol. 337). Springer-Verlag.

Vaidyanathan, S., & Azar, A. T. (2016g). Takagi-Sugeno Fuzzy Logic Controller for Liu-Chen Four-Scroll Chaotic System. *Int. J. of Intelligent Engineering Informatics, 4*(2), 135–150. doi:10.1504/IJIEI.2016.076699

Vaidyanathan, S., Azar, A. T., Akgul, A., Lien, C. H., Kacar, S., & Cavusoglu, U. (2019). A memristor-based system with hidden hyperchaotic attractors, its circuit design, synchronisation via integral sliding mode control and voice encryption. *Int. J. Automation and Control, 13*(6), 644–667. doi:10.1504/IJAAC.2019.102665

Vaidyanathan, S., Azar, A. T., & Boulkroune, A. (2018d). A novel 4-D hyperchaotic system with two quadratic nonlinearities and its adaptive synchronization. *Int. J. Automation and Control, 12*(1), 5–26. doi:10.1504/IJAAC.2018.088612

Vaidyanathan, S., Azar, A. T., & Ouannas, A. (2017a). *An Eight-Term 3-D Novel Chaotic System with Three Quadratic Nonlinearities, its Adaptive Feedback Control and Synchronization. Studies in Computational Intelligence* (Vol. 688). Springer-Verlag.

Vaidyanathan, S., Azar, A. T., & Ouannas, A. (2017c). *Hyperchaos and Adaptive Control of a Novel Hyperchaotic System with Two Quadratic Nonlinearities. Studies in Computational Intelligence* (Vol. 688). Springer-Verlag.

Vaidyanathan, S., Azar, A. T., Rajagopal, K., & Alexander, P. (2015b). Design and SPICE implementation of a 12-term novel hyperchaotic system and its synchronization via active control. *Int. J. Modelling Identification and Control, 23*(3), 267–277. doi:10.1504/IJMIC.2015.069936

Vaidyanathan, S., Azar, A. T., Rajagopal, K., Sambas, A., Kacar, S., & Cavusoglu, U. (2018c). A new hyperchaotic temperature fluctuations model, its circuit simulation, FPGA implementation and an application to image encryption. *Int. J. Simulation and Process Modelling, 13*(3), 281–296. doi:10.1504/IJSPM.2018.093113

Vaidyanathan, S., Azar, A. T., Sambas, A., Singh, S., Alain, K. S. T., & Serrano, F. E. (2018a). *A Novel Hyperchaotic System With Adaptive Control, Synchronization, and Circuit Simulation. Advances in System Dynamics and Control.* IGI-Global. doi:10.4018/978-1-5225-4077-9.ch013

Vaidyanathan, S., Idowu, B.A., & Azar, A.T. (2015c). Backstepping Controller Design for the Global Chaos Synchronization of Sprott's Jerk Systems. *Studies in Computational Intelligence, 581*, 39-58. DOI . doi:10.1007/978-3-319-13132-0_3

Vaidyanathan, S., Jafari, S., Pham, V. T., Azar, A. T. E., & Alsaadi, F. (2018b). A 4-D chaotic hyperjerk system with a hidden attractor, adaptive backstepping control and circuit design. *Archives of Control Sciences., 28*(2), 239–254.

Vaidyanathan, S., Sampath, S., & Azar, A. T. (2015a). Global chaos synchronisation of identical chaotic systems via novel sliding mode control method and its application to Zhu system. *Int. J. Modelling Identification and Control, 23*(1), 92–100. doi:10.1504/IJMIC.2015.067495

Vaidyanathan, S., Zhu, Q., & Azar, A. T. (2017b). *Adaptive Control of a Novel Nonlinear Double Convection Chaotic System. Studies in Computational Intelligence* (Vol. 688). Springer-Verlag.

Vishal, K., & Agrawal, S. K. (2017). On the dynamics, existence of chaos, control and synchronization of a novel complex chaotic system. *Zhongguo Wuli Xuekan, 55*(2), 519–532. doi:10.1016/j.cjph.2016.11.012

Volos, C. K., Kyprianidis, I. M., & Stouboulos, I. N. (2011). Various synchronization phenomena in bidirectionally coupled double scroll circuits. *Communications in Nonlinear Science and Numerical Simulation, 16*(8), 3356–3366. doi:10.1016/j.cnsns.2010.11.015

Volos, C. K., Pham, V. T., Azar, A. T., Stouboulos, I. N., & Kyprianidis, I. M. (2018). Synchronization Phenomena in Coupled Dynamical Systems with Hidden Attractors. In V. T. Pham, S. Vaidyanathan, C. Volos, & T. Kapitaniak (Eds.), *Nonlinear Dynamical Systems with Self-Excited and Hidden Attractors. Studies in Systems, Decision and Control* (Vol. 133, pp. 375–401). Springer. doi:10.1007/978-3-319-71243-7_17

Wang, Y., Hong, H., Guo, J., Wang, X., & Shang, W. (2020). Configuration error function design and application to fixed-time geometric terminal sliding-mode control on SE(3). *Acta Astronautica, 174*.

Wang, Z., Volos, C., Kingni, S. T., Azar, A. T., & Pham, V. T. (2017). Four-wing attractors in a novel chaotic system with hyperbolic sine nonlinearity. *Optik (Stuttgart)*, *131*, 1071–1078. doi:10.1016/j.ijleo.2016.12.016

Wembe, E. T., & Yamapi, R. (2009). Linearly and nonlinearly bidirectionally coupled synchronization of hyperchaotic systems. *Chaos, Solitons, and Fractals*, *14*(4), 1439–1453.

Yuan, Z., Xu, Z., & Guo, L. (2012). Generalized synchronization of two bidirectionally coupled discrete dynamical systems. *Communications in Nonlinear Science and Numerical Simulation*, *17*(2), 992–1002. doi:10.1016/j.cnsns.2011.07.014

Zhou, J., Lu, J. A., & Wu, X. (2007). Linearly and nonlinearly bidirectionally coupled synchronization of hyperchaotic systems. *Chaos, Solitons, and Fractals*, *31*(1), 230–235. doi:10.1016/j.chaos.2005.09.056

Chapter 17
Soft-Computing-Based Real-Time Control of Two Wheel Mobile Robot (TWMR)

Ashwani Kharola
Tulas Institute, India

Suyashi Raiwani
EXL Services, India

Shristi Kharola
Graphic Era University, India

ABSTRACT

This chapter considers various soft-computing techniques for control of self-balancing two wheel mobile robot (TWMR). Initially, a mathematical model of the system was developed using Newton's second law. Thereafter, a simulink of the proposed system was developed in Matlab Simulink environment. Two different controllers, namely fuzzy logic controller and ANFIS controller, were used for control of proposed system. Finally, a real-time model of TWMR was designed which was controlled using Arduino Uno microcontroller, and its results were used for training of ANFIS controller.

1.0 INTRODUCTION

Robotic systems comprises of elements of complex configurations which are keen source of interest for researchers in the past few decades (Yasuda, 2016). Two-wheel mobile robot (TWMR) is a complex dynamical configuration of robot in which a single rotating wheel is replaced by two identical rotating wheels (Pham *et al.,* 2017). The configuration of the system increases its nonlinearity thereby making it difficult to control (Melkou & Hamerlain, 2014). The objective remains same to stabilise chassis and wheels at upright angle and at particular position in presence of external uncertainities and disturbances (Ghabi *et al*., 2018). Watanabe (1993) recommended a fuzzy gaussian neural control for position and

DOI: 10.4018/978-1-7998-5788-4.ch017

azimuth angle regulation of a two-wheel movable machine. The author designed the proposed controller using gaussian shape neural networks having different configuration and weighted connections. A simplified model of two-input and two-output fuzzy neural controller having 49 fuzzy rules was considered in the study. The study can be further extended to a balance control problem. In a study by Ruan *et al.* (2008) reinforcement learning based fuzzy neural controller was successfully applied for motion control of a double-wheel portable robotic vehicle. The outcomes illustrated that above technique was capable of controlling robot in different positions quickly compared to conventional PID control. There is a scope for further improvement of settling time below 5.0 sec by increasing the number of hidden neurons. A real-time stabilisation of double-wheel mutable robotic assembly was proposed by Abeygunawardhana and Murakami (2008). The authors designed a self-tuned PD controller which was capable of adjusting their gains automatically for control of proposed system.

An adaptive fuzzy based approach for regulating motion of two-wheeled standing robot was proposed by Ruan *et al.* (2009). The controller was capable of stabilising the system in large initial angles and provides better stability. As an extension to future work real-time control of proposed system can be achieved for better validation. Goher *et al.* (2010) presented a strategy for stabilisation of a double-wheel robotic manipulator carrying different loads. The study considered a PD and PD-fuzzy controller for control of the proposed system. Results highlighted improved performance of fuzzy-PD approach in contrast to PD controller. The fuzzy controller was designed using five triangular shape memberships and the settling time obtained to control robot position was around 100 sec. The performance can be surely improved by either changing the fuzzy control rules or changing the shape and number of memberships. In a study by Wu and Zhang (2011), pole placement feedback technique was compared to fuzzy control in order to stabilse a double-wheel self-balancing robotic manipulator. A structural model of proposed system was built using kinetic equations derived using Newton dynamics. Simulations confirmed that fuzzy control resulted in lesser stabilisation time whereas pole placement control resulted in lower overshoot compared to fuzzy control.

Ahmad *et al.* (2011) constructed a modular fuzzy controller for regulating an extremely unstable two-wheel wheelchair configuration. A two-link model of typical wheelchair system was developed in Visual Nastran software and its control was confirmed in Matlab. Vermeiren *et al.* (2011) presented a study which considered modeling and control of a two-wheeled Segway called B2. The authors derived a parallel distribution compensation algorithm based on takagi-sugeno fuzzy model to stabilise above system in real-time. The steady state error of the system can to be improved to yield better stabilisation. A two-wheel robotic vehicle carrying a payload was controlled using a PID and fuzzy controller by Almeshal *et al.* (2013). The fuzzy controller was designed using five bell shape membership functions. The results showed that fuzzy controller minimised the overshoot and settle time with lesser disturbances compared to other controller. A real-time model of proposed system can be further constructed for better validation and control. A comparison study for balancing of a two-wheel transportable robot was proposed by Wen *et al.* (2013). This study also compared fuzzy and PID controllers for stabilisation of proposed system in terms of balanced efficacy and disturbance rejection ability. Results showed that both the control scheme were able to stabilise the proposed system but fuzzy control again provides better results. Bature *et al.* (2014) performed comparison study of three dissimilar balancing approaches namely fuzzy reasoning, LQR and PID for regulation of a double-wheel inverted pendulum system. The outcomes of the controllers were validated and compared through experiments conducted in real-time. The fallouts illustrated enhanced performance of fuzzy controller when compared to other two control approaches. The PID controller gives the maximum overshoot and settling time response compared to other two controllers.

Efforts could be made to further improve the percentage overshoot responses of all the three controllers. Unluturk & Aydogdu (2017) discussed about an artificial neural network control for two-wheel robotic system. The controller was designed to stabilise the system in different orientations and terrains. Kalman filter was used for estimating the tilt angle of the robot. The study compared the proposed controller to the classical one and results showed better performance of the former. Esmaeili *et al.* (2017) proposed a sliding mode technique for position and path control of two wheel robot. Lagrangian technique was adopted to develop a mathematical model of above system. The authors considered a two layer sliding mode control for controlling angle of tilt and linear velocity of robot. Simulation results validated the effectiveness of proposed control scheme.

TWMR considered in this study encompasses of a chassis of mass (M_c=1kg) and height (2H=0.2m) mounted on two identical wheels of mass (M_w=0.5kg) and radius (r_w=0.05m). The system is capable of rotating and travelling on difficult terrains very easily. Robot chassis is inclined at an angle θ from vertical axis. The objective is same as that of single wheel mobile robot i.e. to stabilise chassis at vertical upright position when the wheels are stationary at desired location. A schematic view of two wheel mobile robot is shown in Figure 1. It is apparent from the diagram that a control force (F) is required to move the system straight under influence of gravity (g=9.81m/s^2).

Figure 1. Schematic view of TWMR

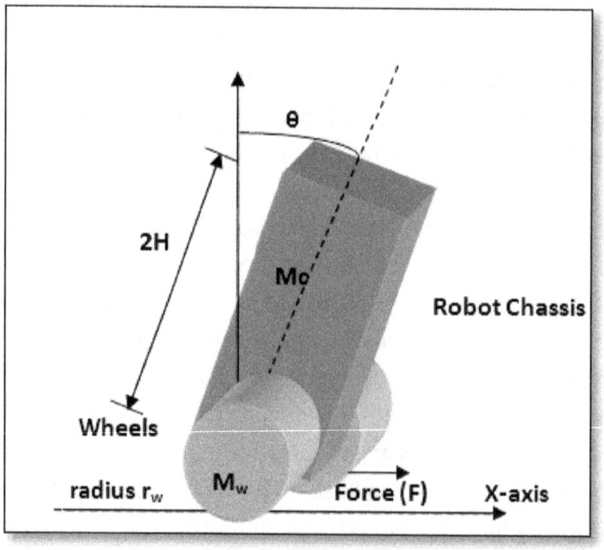

2.0 MATHEMATICAL MODEL OF TWMR

The expressions derived for angular acceleration of chassis ($\ddot{\theta}$) and linear acceleration of wheels ($\ddot{x}$) for the proposed system are as follows,

$$\ddot{\theta} \cdot \frac{g\sin\theta + \cos\theta\left[\dfrac{-F - M_c H \ \sin\theta\dot{\theta}^2}{M_c + M_w}\right]}{H\left[\dfrac{4}{3} - \dfrac{M_c \ \cos^2\theta}{M_c + M_w}\right]} \tag{1}$$

$$\ddot{x} \cdot \frac{F + M_c H\left(\dot{\theta}^2 \sin\theta - \ddot{\theta}\cos\theta\right)}{M_c + M_w} \tag{2}$$

Where $\dot{\theta}$.and $\dot{x}$.are expressions for linear and angular velocity of robot chassis and wheel respectively. The above two equations were further employed to construct a simulink of proposed system as depicted in Figure 2.

Figure 2. Simulink model of TWMR

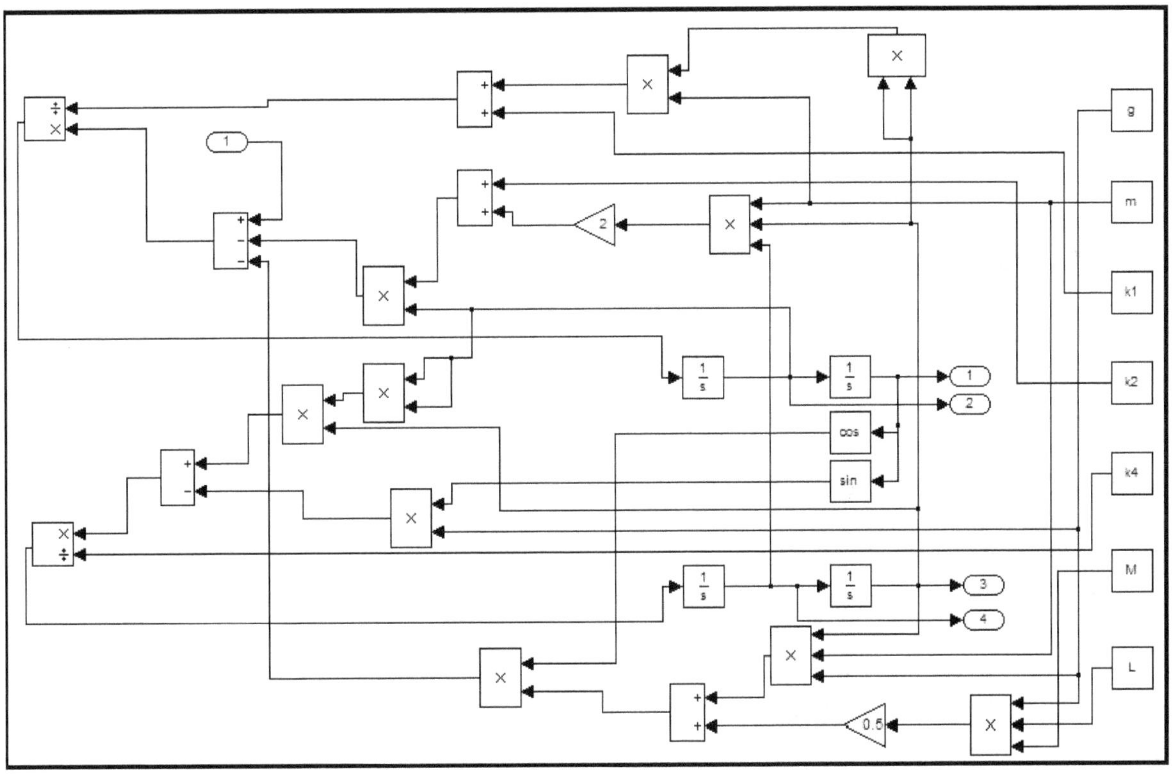

Figure 3. Gbell shape fuzzy memberships for wheel position and chassis angle

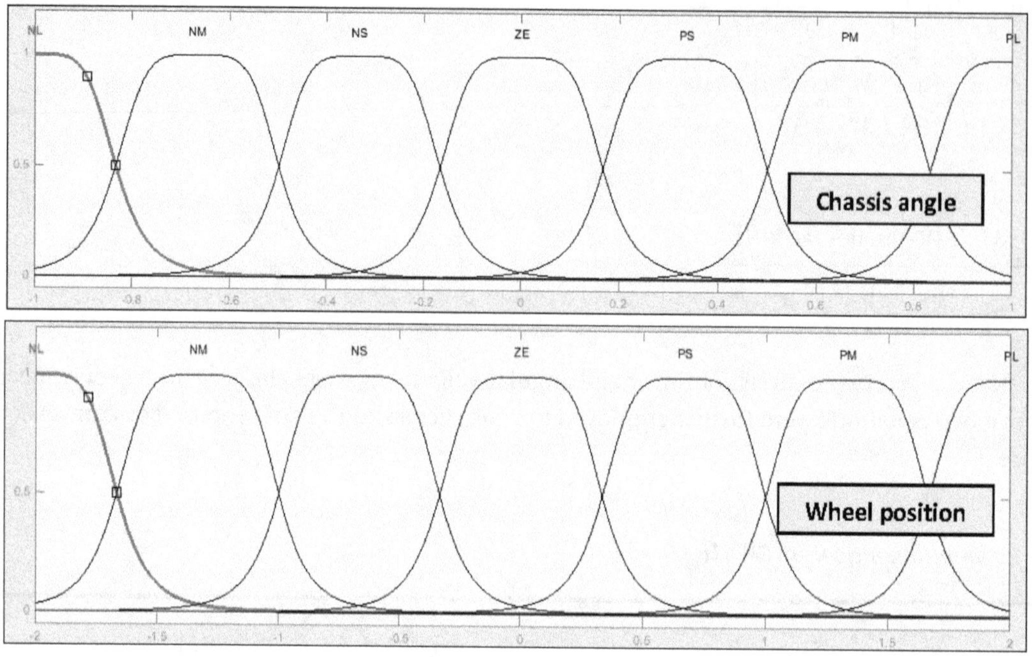

3.0 FUZZY CONTROL OF TWMR

The fuzzy controllers for chassis and wheel sub-system were designed using seven gbell shape membership functions as shown in Figure 3. Gbell Membership functions were used for defining linguistic variables of proposed controller. The initial fuzzy inference structure was developed using mamdani based inference system. If-then fuzzy rules were derived for control of single wheel mobile robot were further used for control of TWMR. A three-dimensional view of surface viewer obtained for robot wheel controller is represented in Figure 4.

3.1 Fuzzy based ANFIS control of TWMR

The ANFIS controller for TWMR has been trained using 211 data samples received after simulation of fuzzy controller described in the previous section (Bouzaida *et al.*, 2018). Two different controllers were designed for controlling chassis and wheel sub-systems respectively. The training epochs considered for training of chassis and wheel controllers were 20 and 50 respectively and learning error was put to zero. A pictorial representation of loading and training process of controllers is demonstrated in Figure 5 and Figure 6 respectively.

The error tolerances obtained after training through 16 gbell shape memberships were 0.0538 and 0.02861 respectively. It is clear from the above figure that generalization for above controller's halts after 18 and 39 epochs respectively. The memberships and surface viewer received after learning of above controllers are given in Figure 7 and Figure 8 respectively.

Figure 4. Surface viewer obtained for wheel controller

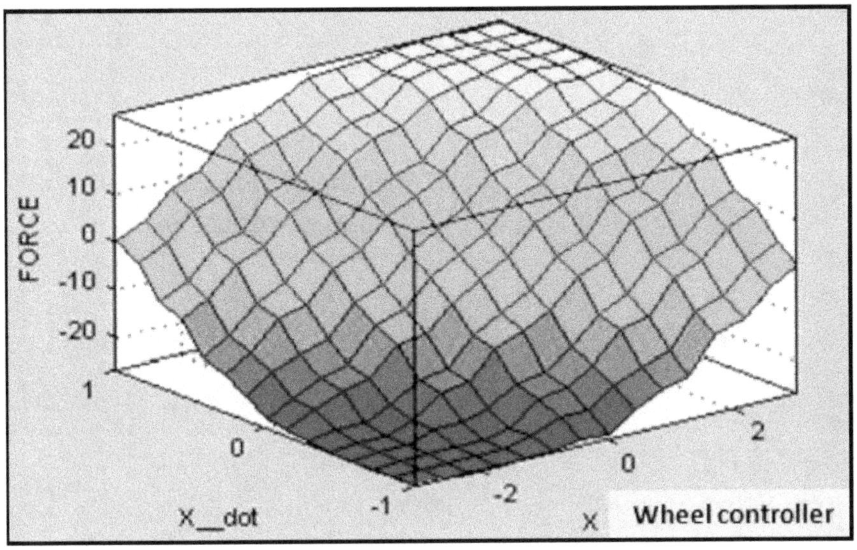

Figure 5. Loading of samples for controllers

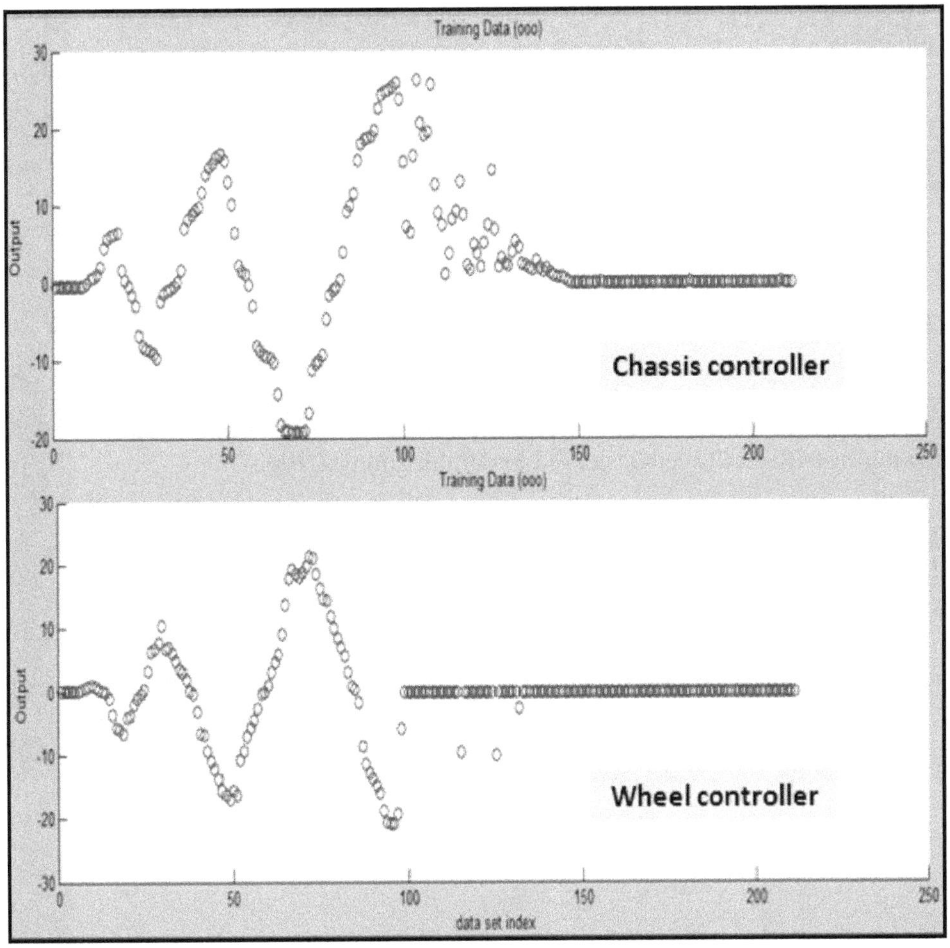

Figure 6. Training of samples for controllers

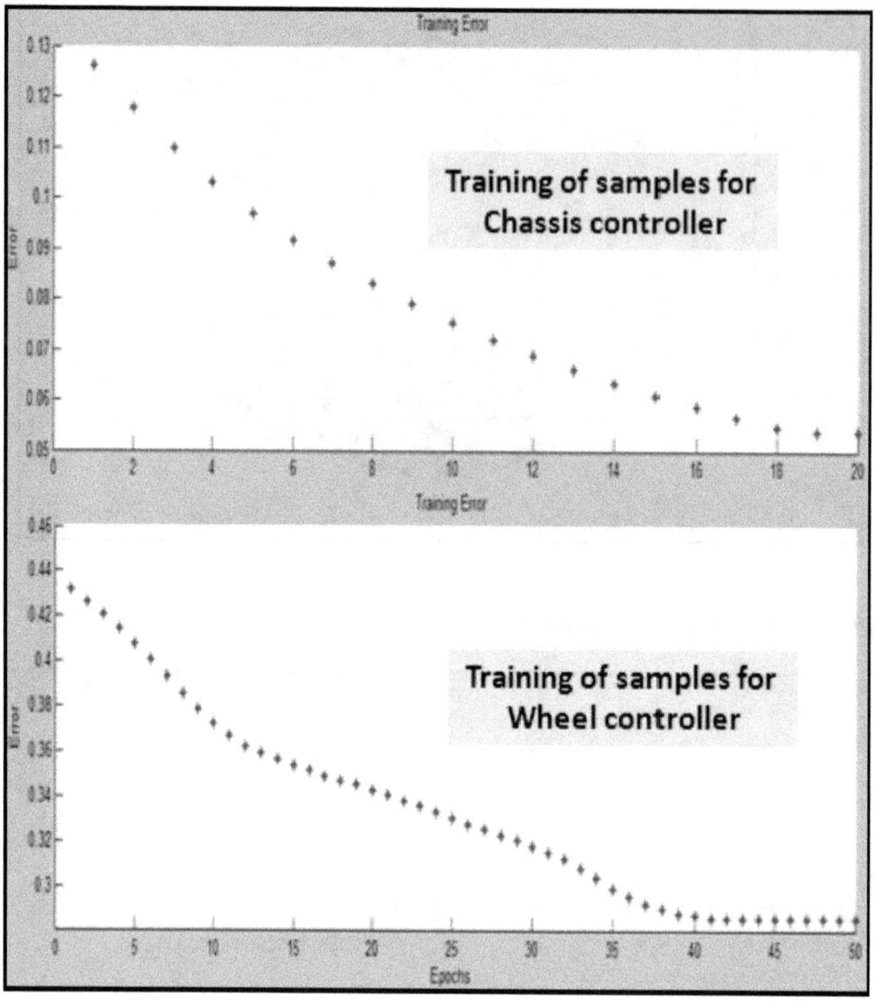

The simulation responses for TWMR using both fuzzy and ANFIS controller are shown with the help of following figures. The time considered for simulation was 100 sec.

(a) Simulation response: Chassis angle (Figure 9)
(b) Simulation response: chassis angular velocity (Figure 10)
(c) Simulation response: wheel position (Figure 11)

A comparison of simulation results obtained using different controllers for TWMR is given in Table 1.

Figure 7. Tuned memberships for chassis angle and wheel position

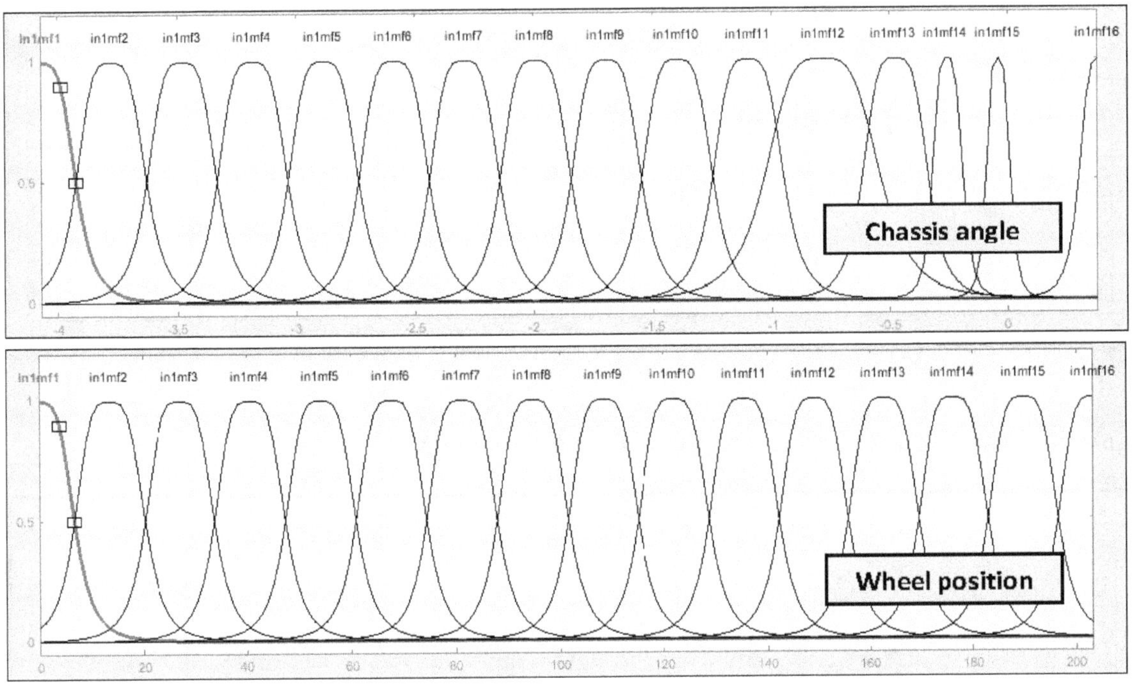

Figure 8. Surface viewer obtained for wheel controller

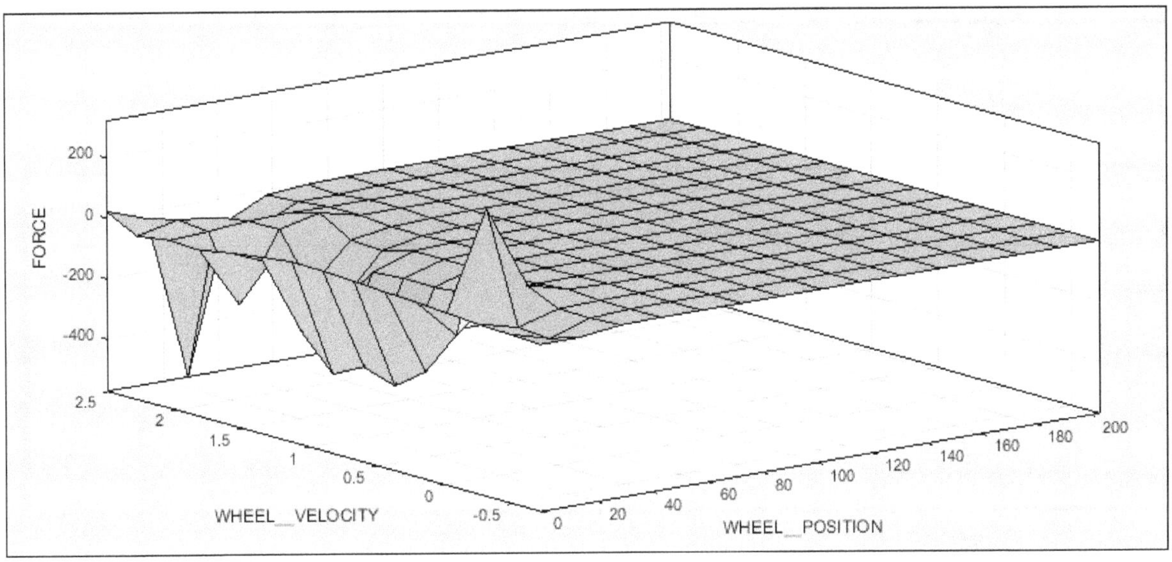

Figure 9. Chassis angle response using fuzzy and ANFIS controllers

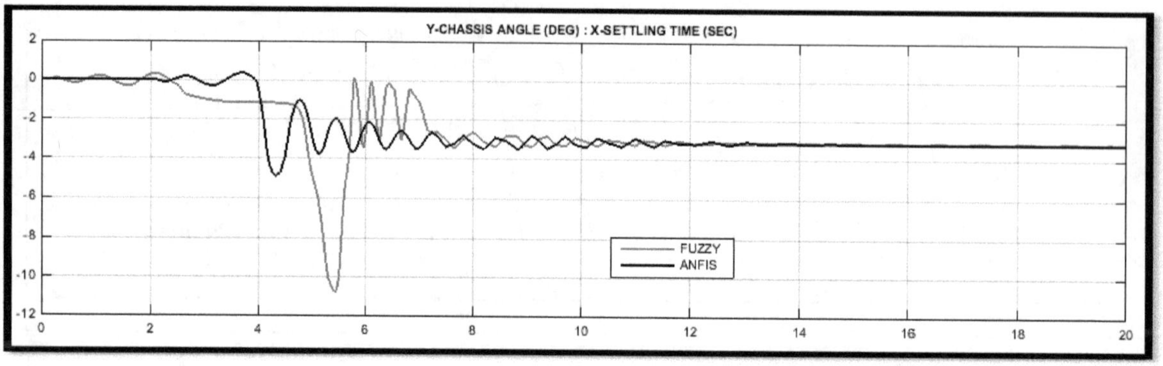

Figure 10. Chassis angular velocity response using fuzzy and ANFIS controllers

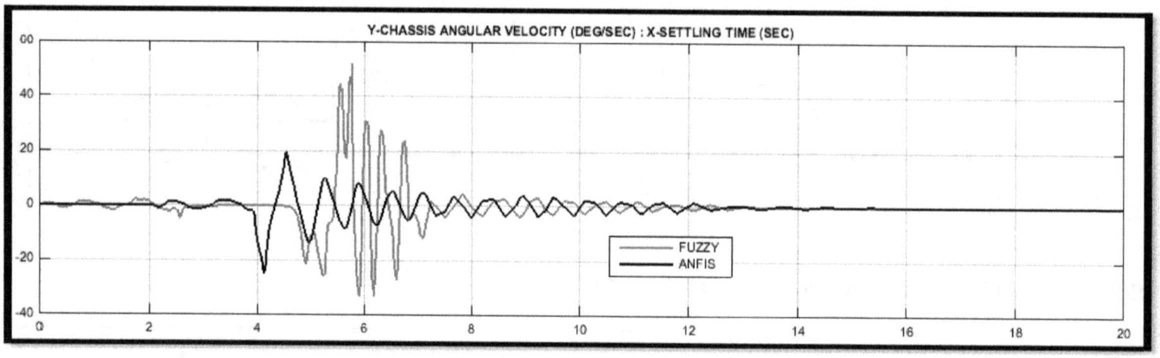

Figure 11. Wheel position response using fuzzy and ANFIS different controllers

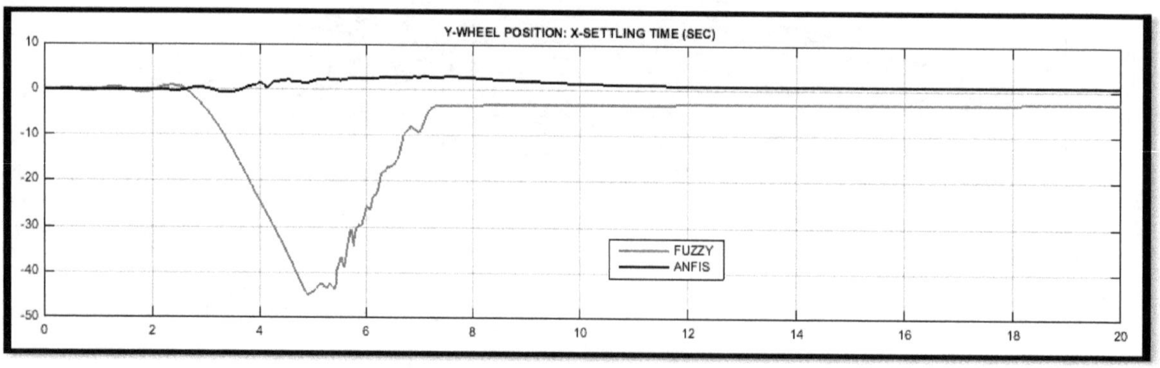

Table 1. Simulation results of TWMR using different controllers

Controller	Settling time (seconds)	Max. Overshoot (degree)	Steady state error
Chassis angle			
ANFIS	6.0 sec	0.2° to -5.0°	0
Fuzzy	8.0 sec	0.1° to -11.0°	0
Chassis angular velocity			
ANFIS	7.0 sec	20.0° to -25.0°	0
Fuzzy	8.0 sec	50.0° to -30.0°	0
Wheel position			
ANFIS	7.0 sec	2.5° to -1.0°	0
Fuzzy	7.0 sec	1.0° to -45.0°	0

4. REAL-TIME CONTROL OF TWMR

The dynamics and behaviour of TWMR discussed earlier were further analysed for its real-time implementation. A physical model of two wheel robot has been developed using Arduino microcontroller and real-time data collected after simulation was used for training of ANFIS controller. The system comprises of a MPU-6050 sensor having a single axis accelerometer and gyroscope for determining rotational velocity and tilt angle of robot (Al-Dahan *et al.*, 2016; Kim & Kwon, 2015). The sensor comprises of 16-bits analog to digital conversion hardware and uses 12C-bus for interfacing with Arduino microcontroller. The response of the accelerometer is proportional to acceleration of the system. Digital motion processing system was used for processing the real-time data obtained from MPU-6050 sensor.

This study considers an Arduino Uno microcontroller board based on ATmega 328P controller having 14 digital input/output pins and 6 analog pins (D'Ausilio, 2012). We can connect the board to a computer via USB or powered it with an external battery. The model also considers a dual H-bridge L298N motor driver for driving a DC brushed motors of 12V (200 rpm). The front and top view of real-time model of two wheel robot constructed for control is shown in Figure 12 and Figure 13 respectively.

Initially a total of 496 data sets were collected from result of real-time simulation. These data sets were further reduced to a number of 197 by identifying and removing the noises and disturbances that occurred during the control process.

4.1 ANFIS Based Supervised Learning and Control of TWMR

The ANFIS controller was designed with the help real-time data sets collected from simulation of two-wheel robot designed in the previous section. The network was trained using hybrid learning algorithm and seven gbell shape memberships (Taher & Abdeljawad, 2013). The number of iteration and training error during learning was set to 300 and 0 respectively. The process of loading and training of samples is shown with the help of Figure 14 and Figure 15 respectively.

One can clearly observe from above figure that training error stops improving after 35 epochs and least error obtained after learning was 0.013591 which is very much within acceptable limits. The learning process results in tuning of if-then fuzzy and the memberships obtained after training is shown in Figure 16.

Figure 12. Front view of real-time model of TWMR

Figure 13. Top view of real-time model of TWMR

Figure 14. Loading of data sets for training in ANFIS

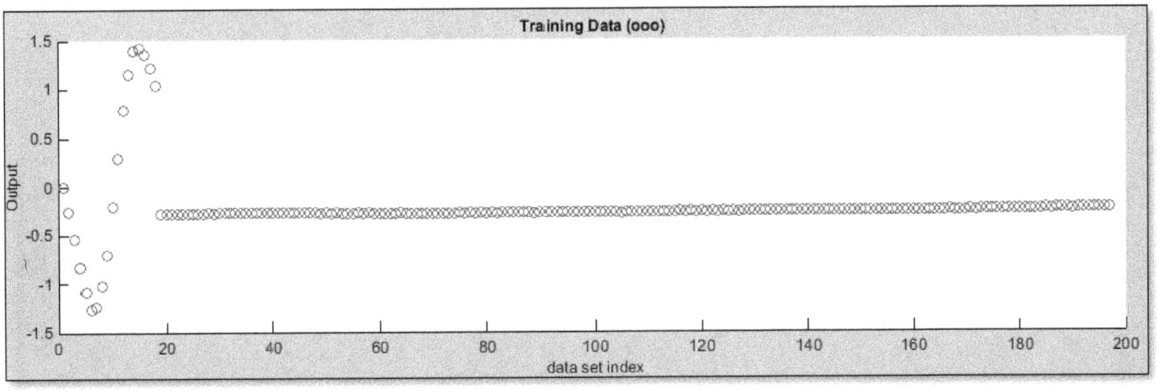

Figure 15. Training of data sets for in ANFIS

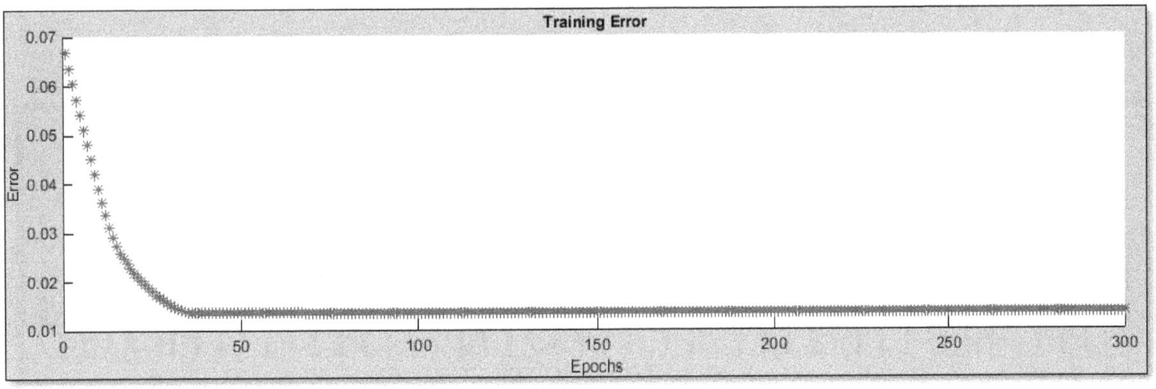

Figure 16. Tuned gbell memberships obtained after training

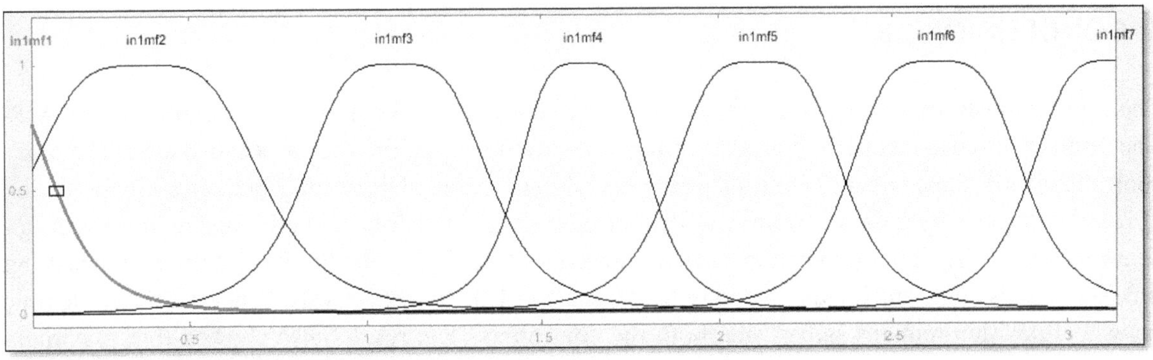

The simulation results obtained for chassis angle and angular velocity after learning of real-time data by ANFIS controller are shown in Figure 17 and Figure 18 respectively. The time considered for simulation is 50 sec for both the attributes of two wheel robot.

(a) Simulation responses: chassis angle (Figure 17)
(b) Simulation responses: chassis angular velocity (Figure 18)

Figure 17. Chassis angle response using ANFIS controller

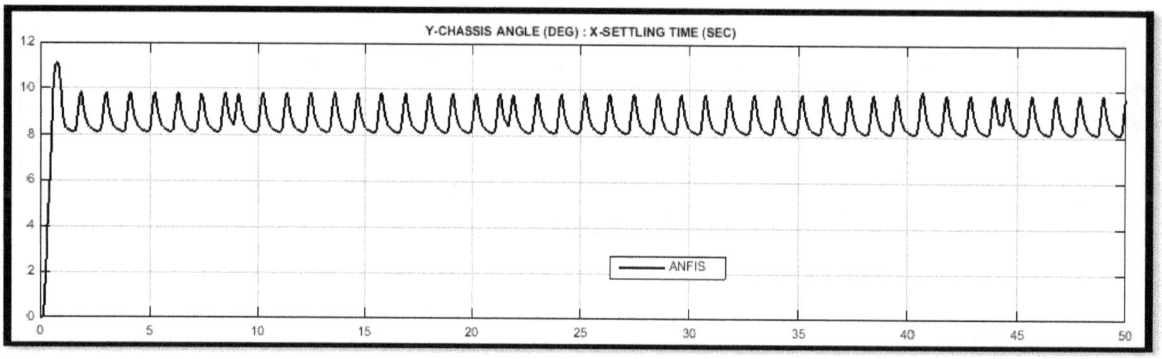

Figure 18. Chassis angular velocity response using ANFIS controller

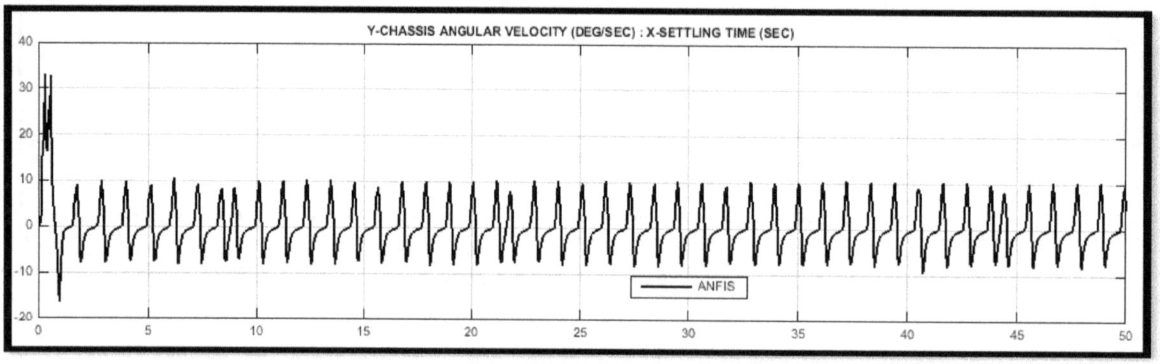

5. CONCLUSION

The chapter illustrated comparison of fuzzy and fuzzy based ANFIS controllers for control of TWMR. The results showed that both the control techniques were able to stabilise the proposed system. The fuzzy controller was designed using 7 gbell shape membership functions. The results obtained after simulation of fuzzy controller were used to train ANFIS controller designed using 16 gbell shape memberships for better training. The simulation results clearly showed that ANFIS controller has improved the settling time for chassis angle and angular velocity by 2.0 sec and 1.0 sec respectively. Whereas, the settle time for wheel position remains same using both the controllers. The results also showed that maximum overshoot ranges are improved using ANFIS controllers. Finally, it is experiential that both the controllers have acceptable steady state response. The performance of proposed fuzzy based ANFIS controller can be further improved by either optimising fuzzy control rules or by changing the shape or number of memberships for ANFIS controller.

This chapter also presents a successful real-time position and angle control of TWMR with the help of an Arduino Uno microcontroller. The chapter considers a MPU-6050 sensor for sensing and collecting the real-time of proposed system. The results showed that controller was able to stabilise the complete system within 2.0 sec within acceptable range of overshoots even when subjected to external disturbances and noises. The real-time data collected after simulation was used for training of ANFIS controller which makes the control more reliable and robust as demonstrated by simulations.

REFERENCES

Abeygunawardhana, P. K. W., & Murakami, T. (2008). Stability improvement of two wheel driven mobile manipulator using nonlinear PD controller. *IEEE Transactions on Industry Applications*, *128*(10), 1149–1156. doi:10.1541/ieejias.128.1149

Ahmad, S., Siddique, N. H., & Tokhi, M. O. (2011). A Modular fuzzy control approach for two-wheeled wheelchair. *Journal of Intelligent & Robotic Systems*, *64*(3), 401–426. doi:10.100710846-011-9541-0

Al-Dahan, Z. T., Bachache, N. K., & Bachache, L. N. (2016). Design and implementation of fall detection system using MPU 6050 Arduino. In *Proceedings of 14th International Conference on Inclusive Smart Cities and Digital Health*, (pp. 180-187). 10.1007/978-3-319-39601-9_16

Almeshal, M., Tokhi, M. O., & Goher, K. O. (2013). Stabilization of a new configurable Two-wheeled machine using a PD-PID and a hybrid FL control strategies: A Comparative study. *International Scholarly and Scientific Research & Innovation*, *6*(10), 61–66.

Bature, A. A., Buyamin, S., Ahmad, M. N., & Muhammad, M. (2014). A comparison of controllers for balancing two wheeled inverted pendulum robot. *International Journal of Mechanical & Mechatronics Engineering*, *14*(3), 62–68.

Bouzaida, S., & Sakly, A. (2018). Adaptive neuro-fuzzy sliding mode controller. *International Journal of System Dynamics Applications*, *7*(2), 1–21. doi:10.4018/IJSDA.2018040103

D'Ausilio, A. (2012). Arduino: A low cost multipurpose lab equipment. *Behavior Research Methods*, *44*(2), 305–313. doi:10.375813428-011-0163-z PMID:22037977

Esmaeili, N., Alfi, A., & Khosravi, H. (2017). Balancing and trajectory tracking of two-wheeled mobile robot using backstepping sliding mode control: Design and experiments. *Journal of Intelligent & Robotic Systems*, *87*(3-4), 601–613. doi:10.100710846-017-0486-9

Ghabi, J., Rhif, A., & Vaidyanathan, S. (2018). Discrete time sliding mode control scheme for nonlinear systems with bounded uncertainities. *International Journal of System Dynamics Applications*, *7*(2), 1–19. doi:10.4018/IJSDA.2018040102

Goher, K. M., Tokhi, M. O., & Siddique, N. H. (2010). Dynamic modeling and control of a two wheeled robotic vehicle with a virtual payload. *Journal of Engineering and Applied Sciences (Asian Research Publishing Network)*, *6*(3), 7–41.

Kharola, A., & Patil, P. (2017). Position and tilt control of two-wheeled robot (TWR): A neuro-fuzzy approach. *International Journal of System Dynamics Applications*, *6*(4), 1–17. doi:10.4018/IJSDA.2017100102

Kim, S., & Kwon, S. J. (2015). Dynamic modeling of a two-wheeled inverted pendulum balancing mobilr robot. *International Journal of Control, Automation, and Systems*, *13*(4), 926–933. doi:10.100712555-014-0564-8

Melkou, L., & Hamerlain, L. (2014). Classical sliding and generalized variable structure controls for a manipulator robot arm with pneumatic artificial muscles. *International Journal of System Dynamics Applications*, *3*(1), 1–24. doi:10.4018/ijsda.2014010103

Mnasser, A., Bounani, F., Manar, T. E., & Ksouri, M. (2014). Neural networls predictive controller using an adaptive control rate. *International Journal of System Dynamics Applications*, *3*(3), 1–21. doi:10.4018/ijsda.2014070106

Pham, V. T., Volos, C., & Vaidyanathan, S. (2017). A chaotic time-delay system with saturation nonlinearity. *International Journal of System Dynamics Applications*, *6*(3), 1–19. doi:10.4018/IJSDA.2017070106

Ruan, X., Cai, J., & Chan, J. (2008). Learning to control two-wheeled self-balancing robot using reinforcement learning rules and fuzzy neural networks. In *Proceedings of 4th IEEE International Conference on Natural Computation*, (pp. 395-398). 10.1109/ICNC.2008.361

Ruan, X., Chan, J., Cai, J., & Dai, L. (2009). Balancing control of the two-wheeled upstanding robot using adaptive fuzzy control method. In *Proceedings of IEEE International Conference on Intelligent Computing and Intelligent Systems*, (pp 95-98). 10.1109/ICICISYS.2009.5358152

Tahre, M. A., & Abdeljawad, M. (2013). A new modular strategy for action sequence automation using neural networks and hidden markov models. *International Journal of System Dynamics Applications*, *2*(3), 1–18. doi:10.4018/ijsda.2013070102

Tsai, C. C., Lin, S. C., & Lin, B. C. (2010). Intelligent adaptive motion control using fuzzy basis function networks for self-balancing two-wheeled transporters. In *Proceedings of IEEE International Conference on Fuzzy Systems*, (pp. 1-6). 10.1109/FUZZY.2010.5584070

Unluturk, A., & Aydogdu, O. (2017). Adaptive control of two-wheeled mobile balance robot capable to adapt different surfaces using a novel artificial neural network based real-time switching dynamic controller. *International Journal of Advanced Robotic Systems*, *14*(2), 1–9. doi:10.1177/1729881417700893

Vermeiren, L., Dequidt, A., Guerra, T. M., Rago-Tirmant, H., & Parent, M. (2011). Modeling, control and experimental verification on a two-wheeled vehicle with free inclination: An urban transportation system. *Control Engineering Practice*, *19*(7), 744–756. doi:10.1016/j.conengprac.2011.04.002

Watanabe, K., Tang, J., Nakamura, M., Koga, S., & Fukuda, T. (1993). Mobile robot control using fuzzy gaussian neural networks. In *Proceedings of IEEE/RSJ International Conference on Intelligent Robots and Systems (IROS)*, (pp. 919-925). 10.1109/IROS.1993.583253

Wen, Y. H., Lin, Y. S., & Leu, Y. G. (2013). Design and implementation of the balance of two-wheeled robots. In *Proceedings of IEEE International Conference on Advanced Robotics and Intelligent Systems (ARIS)*, (pp. 130-134). 10.1109/ARIS.2013.6573547

Wu, J., & Zhang, W. (2011). Design of fuzzy logic controller for two-wheeled self-balancing robot. In *Proceedings of 6th International Forum on Strategic Technology (IFOST)*, (pp. 1266-1270). 10.1109/IFOST.2011.6021250

Yasuda, G. (2016). Design and Implementation of Distributed Autonomous Cordinators for cooperative multi-robot systems. *International Journal of System Dynamics Applications*, *5*(4), 1–15. doi:10.4018/IJSDA.2016100101

Chapter 18
Non-Linear Adaptive Control of Induction Motor Drive for Standalone Photovoltaic Water Pumping System

Salima Meziane
Souk Ahras University, Algeria

Riad Toufouti
Souk Ahras University, Algeria

Loubna Atarsia
Souk Ahras University, Algeria

ABSTRACT

The integration in the isolated areas and rural sectors is a better solution for producing the electrical energy needed for applications such as pumping systems. The rural water demand for crop irrigation and domestic water supplies is increasing. For this, one of the most conceived solutions is the photovoltaic water pumping technology which has the advantage of being sustainable and respectful of the environment to supply water to rural areas. This chapter describes a robust control of a standalone photovoltaic water pumping system using induction motor drive coupled with a centrifugal hydraulic pump. The induction motor is controlled by algorithm called an adaptive nonlinear control uses a combination of the adaptive observer for rotor flux and nonlinear control technique. The variables to be controlled are the rotor speed and the rotor flux norm required to implement the nonlinear control algorithm is estimated by adaptive flux observer. Simulations are carried out in order to show the effectiveness of the drive and the robustness to parameters variations.

DOI: 10.4018/978-1-7998-5788-4.ch018

INTRODUCTION

For over fifty years, Direct Current (DC) motors have been widely used in variable speed drives applications principally due to their fast torque response, high precision of regulation and the possibility to use these motors in whichever mode of operation (Cruz & Rivas, 2000). However, DC motors with drawbacks of spark, corrosion and necessity of maintenance, have been replaced by AC induction motors (IMs) (Wai & Lin, 2005). Since its discovery, induction motor is considered as actuator privileged in the in all industries with a reported 90% utilization in electrical motor applications, since it has many advantages such as low cost, high efficiency, good self-starting, simplicity of design, absence of the collector brooms system and a small inertia (Hussein et al., 2017; Cruz & Rivas 2000; Archana et al., 2012). The inventions of power electronics and advances in computing have made a radical revolution in developing control strategies for induction motors. Many control strategies have been developed and used for IM drives namely Scalar Control Technique, Direct Torque Control (DTC) and Field-Oriented Control (FOC), (Muchande et al, 2013). FOC schemes, proposed by F. Blaschke in 1972 (Blaschke, 1972), can provide at least the same performance, for an inverter-driven induction motor, as it is available for a DC motor (Wade et al., 1997), by decoupling the stator flux and electromagnetic torque. However, this technique is very sensitive to the deviation of motor parameters, particularly the rotor time-constant (Wai & Lin, 2005) and it requires a mechanic sensor to correctly determine the orientation of the rotor flux vector (Cruz & Rivas, 2000). This mechanic sensor increases the cost of the drive equipment and makes the driving system much larger and unstable (Casadei et al., 2002).

Trying to reduce the complexity of the algorithms involved in a FOC, researchers presented new techniques used for estimation of the motor speed and fluxes without a need of speed and flux sensors that can reduce the cost and having the features of precision, rejection of disturbance and fast torque response (Zhu et al., 2013). One of the most popular control methods is the feedback linearization which is based on the transformation of the nonlinear model into a linear one which makes the linear feedback control easy to be applied in respect to trajectory reference (Zhu et al.; Enev, 2007). The feedback linearization based on differential geometry control theory is used to uncouple and linearize the model, put it under canonical form and then make control using linear control techniques (Enev, 2007; Raumer et al., 1993). The advantage of the Input-Output Linearization Control (IOLC) over the FOC is the fact that by applying the linearization transformation, system stability and a complete decoupling of the torque and flux in a stator fixed reference frame is achieved, which enables the optimization of motor performances without degrading the mechanical output regularization (Raumer et al., 1993).

To improve the Field Oriented Control, many control strategies have been proposed (Wade et al., 1997). Several feedback linearization based solutions were proposed. It is, by now, well known that the induction motor model with a linear magnetic circuit is not feedback linearizable by static feedback (Ismail, 2012). In (Enev, 2007) the authors have proposed a controller designed to track torque and rotor flux references, in (Marino et al., 1993) have developed an input-output decoupling controller which decouples the regulation of the rotor flux and the rotor speed. To solve the influence of large dependencies on the parameters variation and need accurate cancellation the dynamic when using exact linearization method to nonlinear system. The authors in (Meng et al., 2011; Rashed et al., 2006) introduced Nonlinear decoupling control of induction motor based on parameter adaptive identification to observe the load torque changes. In (Enev, 2007) the author proposed an input-output linearizing control for induction motor, along with a simple scheme enabling the calculation of the rotor resistance and the load torque using steady-state speed and stator current information. In (Moutchou,2014) a structure on input-output linearization based Model reference adaptive system (MRAS), allowing simultaneous observation of stator resistance and rotor speed, in order to estimate rotor flux using a sliding mode observer has been proposed. Water supply, livestock and domestic needs are principal problems in the development of isolated regions and rural remote. Generally, the grid network is not present and the extension of the conventional electric network to the remote areas remains very hard to do (Chikh & Chandra, 2009; Meziane, 2009). Therefore, it is necessary to look for other energy resources to fulfill their energy requirements. A recently proposed solution to this problem is the usage of autonomous systems using inexhaustible resources (Camocardi et al, 2008; Hirpara, 2020). These sources include wind energy, geothermal energy, bio energy and solar energy. PV is renewable and clean energy source because it has no contaminated waste, noise, disturbing effects, radiation, does not even need to the fuel and is converted directly to direct current (Mohamed et al.,2017; Biji, 2012; Ramya, 2012).

Recently, the application of a directly coupled photovoltaic panel to the electromechanical system for water pumping has received increasing devotion because of the expected cost reduction in PV array (Kolhe et al, 2004). Generally, a stand-alone PV pumping system consists of a PV array, a storage element, a control/power processing components and a motor coupled to a hydraulic pump (Chikh & Chandra, 2009). In this paper we present a robust control of a standalone photovoltaic water pumping system using induction motor drive, the algorithm control called a nonlinear control of induction motor coupled with a centrifugal hydraulic pump, powered by a photovoltaic array speeding system. The proposed system is designed for usage in rural or remote areas where the grid network is not present in order to reduce both cost, while concurrently guaranteeing optimal utilization of the PV array.The PV array is directly connected to the dc bus of the voltage source inverter (VSI), with intermediate DC-DC Boost Converter.

The VSI is controlled by space vector modulation (SVM). A perturbation based nonlinear controller adjusts the reference speed of the (IM) to attain the maximum power point (MPPT) operation of the photovoltaic array. In this study we will prove a stability of the proposed adaptive flux observer by the Lyapunov's theorem, the parameters identified adaptively are stator and rotor resistance which vary with motor temperature. Therefore, this study proposes an easy but effective way to compensate the voltage drop on the stator resistance so that the stator flux can be constructed without identifying the stator resistance as done by most authors. Simulation results shows validate the effectiveness of the proposed method. The remainder of this paper is organized as follows. The mathematical model of the induction motor is developed in Section 2, while Section 3 describes a brief overview of the field oriented control theory. In Section 4, Input-Output linearization control of the induction motor will be presented.

Sections 5 describe a proposed adaptive flux observer. In Section 6 we extend the previous scheme control applied to a photovoltaic water-pumping system. Section 6.1 section presents the photovoltaic generator in operation at the point of maximum power (MPPT), hence the architecture of the proposed MPPT, as well as the basic modeling necessary to the understanding of its principle of operation. The performance of the proposed control of the water pumping system is verified by simulation against rotor resistance variations, load torque solar radiation variations. Finally, the conclusions and future works are reported in section 8.

NONLINEAR MODEL OF INDUCTION MOTOR

Using nonlinear feedback allows to contort the model in the stator fixed (α,β) reference frame avoiding the transformation in a rotating reference frame. The complete model in stator fixed (α,β) reference frame is given by (1) for a control in speed and flux (Casadei et al., 2002; Moutchou & Mahmoudi, 2014 Pham, 2017; Meziane, 2018):

$$\begin{cases} \dot{x} = f(x) + g(x).u \\ y_1 = h_1(x); \quad y_2 = h_2(x) \end{cases} \tag{1}$$

Where $x = [I_{s\pm} \;\; I_{s^2} \;\; \varphi_{r\pm} \;\; \varphi_{r^2} \;\; ©]^T$ is an n dimension state vector, and $u = [u_{s\pm} \;\; u_{s^2}]^T$ is an m dimension control signal, $y = \left[© \quad (\varphi_{r\pm}^2 + \varphi_{r^2}^2)\right]$ is an p dimension measurement vector.

The state x belongs to the set $\left\{x \in \Re^5 : \varphi_{r\pm}^2 + \varphi_{r^2}^2 \neq 0\right\}$ and the vector function *f(x)* is defined as follows:

$$f(x) = \begin{bmatrix} -\gamma I_{s\alpha} + \frac{K}{Tr}\varphi_{r\alpha} + p\Omega K\varphi_{r\beta} \\ -\gamma I_{s\beta} - p\Omega K\varphi_{r\alpha} + \frac{K}{Tr}\varphi_{r\beta} \\ \frac{M}{Tr} I_{s\alpha} - \frac{1}{Tr}\varphi_{r\alpha} - p\Omega\varphi_{r\beta} \\ \frac{M}{Tr} I_{s\beta} + p\Omega\varphi_{r\alpha} - \frac{1}{Tr}\varphi_{r\beta} \\ p\frac{M}{JL_r}(\varphi_{r\alpha} I_{s\beta} - \varphi_{r\beta} I_{s\alpha}) - \frac{1}{J}(T_L + f) \end{bmatrix}$$

$$[g] = \begin{bmatrix} g_1 \\ g_2 \end{bmatrix}^T = \begin{bmatrix} \frac{1}{\sigma L_S} & 0 & 0 & 0 & 0 \\ 0 & \frac{1}{\sigma L_S} & 0 & 0 & 0 \end{bmatrix}^t$$

Let us introduce the following definitions:

$$Tr = \frac{L_r}{R_r}; \sigma = 1 - \frac{M^2}{L_s L_r}; K = \frac{M}{\sigma L_s L_r}; \gamma = \frac{Rs}{\sigma L_s} + \frac{R_r M^2}{\sigma L_s L_r^2}$$

$i_{s\alpha}$, $i_{s\beta}$ denote the stator currents, $\varphi_{r\alpha}$, $\varphi_{r\beta}$ the rotor fluxes, $u_{s\alpha}$, $u_{s\beta}$ the stator voltages, L_s, L_r the stator and rotor inductance, R_s, R_r the stator and rotor resistance, Ω the speed, *J* is the inertia of the machine, *M* the mutual inductance, *f* the friction coefficient, *p* the poles pair number, T_L the load torque, and finally T_r is the rotor time constant. From the above representation, we can see that the dynamic model of an induction motor is a strongly coupled nonlinear multivariable system (Fekih & Chowdhury, 2004).

FIELD ORIENTED CONTROL

One particular approach for the control of induction motors is the Field Oriented Control (FOC). This control strategy is very popular in industrial applications. There are two basic categories: The direct oriented control proposed by Blaschke and the indirect field oriented control proposed by (Fekih & Chowdhury, 2004). In both methods, it is necessary to determine correctly the orientation of the rotor flux vector, lack of which leads to degradation in the speed control of the motor. In next study we present the indirect field-orientation method (Cruz & Rivas, 2000; Casadei et al., 2002). Under ideal IFOC conditions, the rotor flux linkage is oriented along the d-axis of the motor, we have:

$$\begin{cases} \varphi_r = \varphi_{rd} \\ \varphi_{rq} = 0 \end{cases} \tag{2}$$

Substituting (2) into complete model in the rotator reference frame yields the following equations:

$$\frac{d\varphi_{rd}}{dt} + \frac{1}{Tr}\varphi_{rd} = \frac{M}{T_r} I_{sd} \tag{3}$$

$$\acute{E}_s = \acute{E} + \frac{M_{sr} I_{sq}}{Tr\varphi_r} \tag{4}$$

$$Te = \frac{3}{2} p \frac{M}{L_r} \varphi_{rd} I_{sq} \tag{5}$$

These equations represent the basic principle of the field oriented control in the rotor flux reference frame. A decoupled control of rotor flux magnitude and electromagnetic torque can be achieved acting on the q and axis stator current components, respectively. Which is equivalent to a separately excited DC motors (Casadei et al, 2002). Because of the variation of T_r, the desired field orientation condition cannot always be maintained and the drive performance can be significantly affected.

INPUT-OUTPUT LINEARIZATION TECHNIQUE

The input-output linearization (IOL) is a common approach used in controlling nonlinear systems. This approach is based on exact cancellation of the system's nonlinearities using a nonlinear state feedback in order to obtain a linear relationship between the control inputs and the system outputs (Marino et al., 1993; Merabet, 2007; Meziane, 2018). Consider the following multiple-input multiple-output (MIMO) nonlinear system:

$$\begin{cases} x = f(x) + \sum_{i \in m} g_i(x).u_i \\ y = h(x) \end{cases} \tag{6}$$

Where *x*ÎÂ^{n} is the state vector, uÎUÌÂ^{m} represents the control vector. *y* Î Â^{m} is the output. The functions *f(x):* Â^{n} àÂ^{n}, g*(x):* Â^{n} àÂ^{nxm} and h*(x):* Â^{n} àÂ^{mxn}

We consider the system model as described in (6). We try to find a special nonlinear state-space transformation:

$$z = h(x) \tag{7}$$

Where z has a linear dynamic, (Lee et al., 2008):

$$\dot{z} = Az + Bv \qquad z \in \Re^{n}, v \in \Re^{m} \tag{8}$$

With A and B two matrices to be determined, v is the new control input. Linear techniques can then be applied in the design of the control *v* (Fekih & Chowdhury, 2004). The outputs to be controlled are $[h_1(x)\ldots\ldots h_m(x)]^T$. The manipulated quantities which will be controlled are differentiated with respect to time until the input u appears and the derivatives of the state variables are eliminated using the state-space representation of (6) (Chouiref et al., 2013) . The derivatives of the output may be expressed as follows:

$$y_i^{(r_i)} = L_f^{ri} h_i(x) + \sum_{j=1}^{m} L_{gj} L_f^{r_i - 1} h_i(x) u_j \tag{9}$$

With r_i is the relative degree associated to the ith output where i ∈{1, ... m}.The relative degree of system:

$$r = \sum_{i=1}^{m} r_i \tag{10}$$

The following notation is used for the Lie derivative of state function $h_j(x)$ along a vector field $f(x) = (f_1(x)\ldots f_n(x))$.

$$L_f h(x) = \sum_{i=1}^{n} \frac{\partial h(x)}{\partial x_i} f_i(x) \tag{11}$$

$$L_g L_f h_i = \frac{\partial L_f h_i}{\partial x_i} g(x) \tag{12}$$

Iteratively, we have $L_f^i h(x) = L_f\left(L_f^{(i-1)} h(x)\right)$ That is, once the system is linearized, one can use a linear controller in the design of the control signal input u for the system in the new reference. The decoupling via input-output linearization can be applied to a system (9), of *n* order if r≤n, the following system of differential equations is obtained (Lee et al., 2008; Pham, 2017; Meziane, 2018):

$$\begin{bmatrix} y_1^{(r1)} \\ \vdots \\ y_m^{(rm)} \end{bmatrix} = \begin{bmatrix} L_f^{r1} h_1(x) \\ \vdots \\ L_f^{r_m} h_m(x) \end{bmatrix} + D(x).\begin{bmatrix} u_1 \\ \vdots \\ u_m \end{bmatrix} \tag{13}$$

Or equivalently

$$y = A(x) + D(x)u \tag{14}$$

Where the *mxm* decoupling matrix *E(x)* was defined as:

$$E(x) = \begin{bmatrix} L_{g_1} L_f^{r_1-1} h_1(x) & \cdots & L_{g_m} L_f^{r_1-1} h_1(x) \\ \vdots & \ddots & \vdots \\ L_{g_1} L_f^{r_m-1} h_m(x) & \cdots & L_{g_m} L_f^{r_m-1} h_m(x) \end{bmatrix} \tag{15}$$

And then, if *E(x)* is non-singular, the new transformed system with input-output linearization can now be written as (Fekih & Chowdhury; Rashed et al., 2006; Chouiref et al, 2013):

$$u = D^{-1}(x)(v - A(x)) \tag{16}$$

Substituting equation (16) into (14) results in a linear differential relation between the output y and the new input v that is given by:

$$v_i = y_i^{(r_i)} \tag{17}$$

Furthermore, we can see each input v_i only affects its correspondent output $y_i^{(r_i)}$, so it is decoupled linear relation (Lee et al, 2008). In order to exactly track desired reference signals, the new input signals v is designed as (Chouiref et al., 2013):

$$v_i = k_1\left(z_{iref} - z_i\right) + k_2\left(z_{iref}^{(1)} - z_i^{(1)}\right) + \ldots + k_r\left(z_{iref}^{(r_i-1)} - z_{iref}^{(r_i-1)}\right) + z_{iref}^{(r_i)} \tag{18}$$

The k_i are obtained using pole placement technique to configure the system closed loop poles in the desired position in order to acquire satisfactory dynamic quality and stability (Chouiref et al., 2013).

Linearization to Induction Motor

The control problem is to choose V_{sd}, V_{sq} in such a way as to control independently the rotor speed and the square of the rotor flux to follow the desired control signals in a second-order system response form. For the application of linearization technique of induction motor, we considering the speed and the rotor flux magnitude as outputs (so $m = 2$), the following equations can be derived using equation (9) (Even, 2007; Merabet, 2007; Meziane, 2009; Pham, 2017; Meziane, 2018):

Thus the derivatives of the outputs are given by

- **For the first output** $y_1=\Omega$

$$\begin{cases} y_1 = L_f^0 h_1(x) = h_1(x) = \copyright \\ \dot{y}_1 = L_f h_1(x) = \dfrac{pM}{JL_r}\left(\text{Æ}_{r\pm} I_{s^2} - \text{Æ}_{r^2} I_{s\pm}\right) - \dfrac{f_r}{J}\copyright - \dfrac{C_r}{J} \\ \ddot{y}_1 = L_f^2 h_1(x) + L_{g1}L_f h_1(x).u_{s\pm} + L_{g2}L_f h_1(x).u_{s^2} \end{cases} \tag{20}$$

- **For the second output** $y_1 = \varphi_r^2$

$$\begin{cases} y_2 = L_f^0 h_2\left(x\right) = h_2\left(x\right) = \varphi_r^2 = \varphi_{r\pm}^2 + \varphi_{r^2}^2 \\ \dot{y}_2 = \dot{h}_2(x) = L_f h_2(x) \\ \ddot{y}_2 = \ddot{h}_2(x) = L_f^2 h_2(x) + L_{g1}L_f h_2(x).u_{s\pm} + L_{g2}L_f h_2(x).u_{s^2} \end{cases} \tag{21}$$

The global relative degree is lower than the order n of the system $r=r_1+r_2=4$ is lower than the n=5 degree system. The system is siding partly linear zed; we obtain a not observable dynamics of order 2. (Even, 2007; Meng et al., 2013; Merabet, 2007; Meziane, 2009; Pham, 2017; Meziane, 2018) define the change of coordinates:

$$\begin{cases} z_1 = y_1 = h_1\left(x\right) \\ z_2 = L_f h_1\left(x\right) \\ z_3 = y_2 = h_2\left(x\right) \\ z_4 = L_f h_2\left(x\right) \\ z_5 = y_3 = \tan^{-1}(\varphi_{r^2}/\varphi_{r\pm}) \end{cases} \tag{22}$$

Where z_1, z_2, z_3, z_4 and z_5 are nonlinear state functions.

The choice of the variable (Z_5) is arbitrary, this variable represent the angle of the rotor flux corresponding to the same approach (Meziane, 2009; Merabet, 2007; Pham, 2017; Meziane, 2018). The dynamic model of induction motor presented in equation (9) can be expressed by the following new coordinates:

$$\begin{cases} \dot{z}_1 = \dot{y}_1 = L_f h_1(x) \\ \dot{z}_2 = \ddot{y}_1 = L_f^2 h_1(x) + L_g L_f h_1(x) u_{s\pm} + L_g L_f h_1(x) u_{s^2} \\ \dot{z}_3 = \dot{y}_2 = L_f h_2(x) \\ \dot{z}_4 = \ddot{y}_2 = L_f^2 h_2(x) + L_g L_f h_2(x) u_{s\pm} + L_g L_f h_2(x) u_{s^2} \\ \dot{z}_5 = \dot{y}_3 = L_f h_3(x) \end{cases} \tag{23}$$

The return status is then applied according to linearize accepting the input-output behavior of the equation system (9). Derived outputs are given by:

$$\begin{bmatrix} \ddot{y}_1(x) \\ \ddot{y}_2(x) \end{bmatrix} = \begin{bmatrix} L_f^2 h_1(x) \\ L_f^2 h_2(x) \end{bmatrix} + D(x).\begin{bmatrix} u_{s\pm} \\ u_{s^2} \end{bmatrix} \tag{24}$$

The decoupling matrix D(x) is determined by:

$$D(x) = \begin{bmatrix} -\frac{PM}{J\tilde{A}L_s L_r}\varphi_{r^2} & \frac{PM}{J\tilde{A}L_s L_r}\varphi_{r\pm} \\ \frac{2M}{\tilde{A}L_s T_r}\varphi_{r\pm} & \frac{2M}{\tilde{A}L_s T_r}\varphi_{r^2} \end{bmatrix} \tag{25}$$

The matrix *D(x)* is not singular as the determinant is given by:

$$\det D(x) = 2.\frac{pM^2}{(L_s.\tilde{A})^2.T_r.L_r.J}.(\varphi_{r\pm}^2 + \varphi_{r^2}^2) \neq 0 \tag{26}$$

The dynamics of the rotor flux angle z_5 is made unobservable by linearizing state feedback is derived by:

$$\dot{z}_5 = \ddot{y}_3 = P© + \frac{M}{T_r}.\frac{(\varphi_{r\pm} I_{s^2} - \varphi_{r^2} I_{s\pm})}{\varphi_{r^2}^2 + \varphi_{r\pm}^2} = Pz_1 + \frac{R_r}{Pz_3}(J_{z2} + C_r) \tag{27}$$

The difference between the two speeds $(\dot{z}_5 - p©)$, respectively, the flow velocity and the angular rotor speed, corresponds to the slip speed expressed by:

$$\dot{z}_5 - P© = \frac{M}{T_r}.\frac{(\varphi_{r\pm} I_{s^2} - \varphi_{r^2} I_{s\pm})}{\varphi_{r^2}^2 + \varphi_{r\pm}^2} = \frac{R_r}{P}\frac{C_e}{\varphi_r^2} \tag{28}$$

When rotor flux does not equal to zero, *D(x)* is nonsingular everywhere. In order to avoid the singularity problem, the control algorithm is activated when the rotor-flux magnitude is greater than zero.

The nonlinear Sate feedback control is given by (Marino et al., 1993; Raumer et al, 1993).

$$\begin{bmatrix} u_{s\pm} \\ u_{s^2} \end{bmatrix} = \left[D(x)\right]^{-1} \cdot \begin{bmatrix} -L_f^2 h_1(x) + v_1 \\ -L_f^2 h_2(x) + v_2 \end{bmatrix} \tag{29}$$

Substituting (28) into (23) yields the following two decoupled and linear systems [16]:

$$\begin{cases} \ddot{h}_1(x) = v_1 = \ddot{y}_1 \\ \ddot{h}_2(x) = v_2 = \ddot{y}_2 \end{cases} \tag{30}$$

$$\begin{cases} \dot{z}_1 = z_2; \quad \dot{z}_2 = v_1; \quad \dot{z}_3 = z_4 \\ \dot{z}_4 = v_2; \quad \dot{z}_5 = Pz_1 + \dfrac{R_r}{Pz_3}\left(Jz_2 + C_r\right) \end{cases} \tag{31}$$

The closed loop system (32) is input-output decoupled and linear. To ensure perfect tracking of speed and flux references, v_1 and v_2 are chosen as follows:

$$\begin{cases} v_1 = -k_{11}\left(\copyright - \copyright^*\right) - k_{12}\left(\dfrac{pM}{JL_r}\left(\varphi_{r\pm} I_{s^2} - \varphi_{r^2} I_{s\pm}\right) - \dfrac{C_r}{J} - \dot{\copyright}^*\right) + \ddot{\copyright} * \\ v_2 = -k_{21}\left(\varphi_r^2 - \varphi_r^{*2}\right) + k_{22}\left(\dfrac{2}{T_r}\left(M\left(\varphi_{r\pm} I_{s^2} + \varphi_{r^2} I_{s\pm}\right) - \left(\varphi_{r\pm}^2 + \varphi_{r^2}^2\right)\right) - \varphi_r^{*2}\right) + \varphi_r^{*2} \end{cases} \tag{32}$$

Where k_{11}, k_{12} and k_{21}, k_{22} are positive non-zero constant which specify the transient response characteristics of the system nonlinear transformed states, they are determined in order to make the closed loop system (32) stable and to have fast response in variable tracking according that the following order decoupled 2 (Merabet, 2007; Meziane,2009; Fekih & Chowdhury,2004):

$$\begin{cases} \ddot{\copyright} - \ddot{\copyright}^* = -k_{11}\left(\dot{\copyright} - \dot{\copyright}^*\right) - k_{12}\left(\copyright - \copyright^*\right) \\ \ddot{\varphi}_{r\pm}^2 + \ddot{\varphi}_r^{*2} = -k_{21}\left(\dot{\varphi}_r^2 - \dot{\varphi}_r^{*2}\right) + k_{22}\left(\varphi_r^2 - \varphi_r^{*2}\right) \end{cases} \tag{33}$$

In closed loop tracking error is given by:

$$\begin{cases} \ddot{e}_1 + k_{11}.\dot{e}_1 + k_{12}.e_1 = 0 \\ \ddot{e}_2 + k_{21}.\dot{e}_2 + k_{22}.e_2 = 0 \end{cases} \tag{34}$$

With:

$$\begin{cases} e_1 = ©^* - © \\ e_2 = \varphi_r^* - \varphi_r \end{cases} \tag{35}$$

The proposed control strategy is illustrated by feedback control block diagram reported in (Figure 1).

Figure 1. Control feedback control block

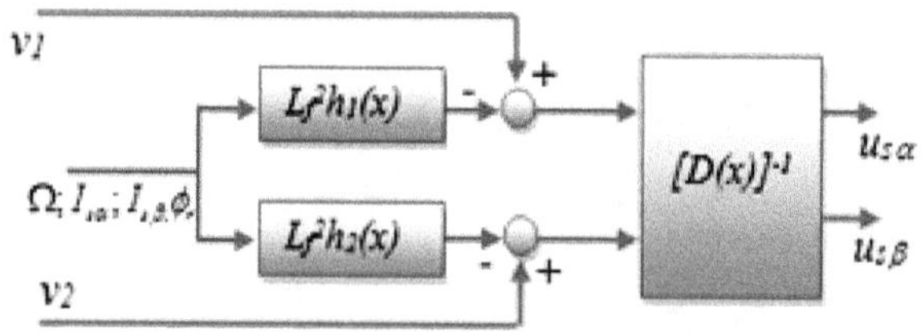

Space Vector Modulation

Space Vector Modulation (SVM) is one of the most widely utilized techniques to generate sinusoidal line-to-line voltages and currents with a VSI. The SVM based on the concept of approximating a rotating reference voltage vector V_{ref} using a combination of two out of the eight possible vectors that can be generated from a VSI, (Cruz & Rivas, 2000; Toufouti, 2009). This process is illustrated in (Figure 2).

The magnitude of this vector is related to the magnitude of the output voltage and the time this vector takes to complete one period is the same as the fundamental time period of the output voltage (Cruz & Rivas, 2000; Toufouti & al.,2009).

Figure 2. Space vector diagram (Toufouti, 2009).

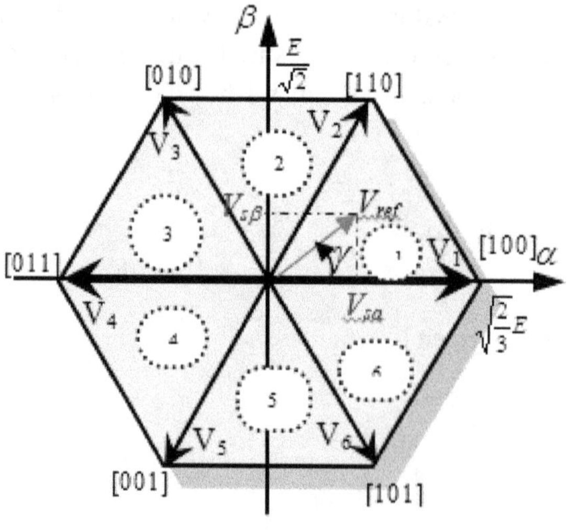

Figure 3. Commutation sequence for the inverter (Toufouti, 2009).

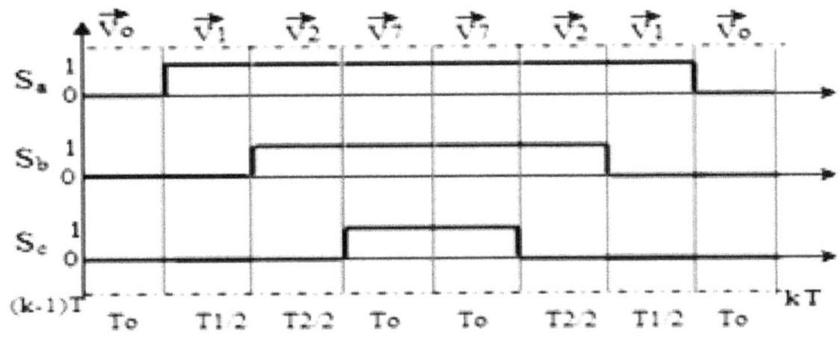

However the general idea is based on sequential switching of active and zero (V_0 or V_7) vectors (Figure 3). Equating the volt-sec integral of reference vector V_{ref} obtained from the controller with the inverter output voltage vectors over a single space vector modulation cycle gives (Cruz & Rivas, 2000):

$$\int_0^{T_s} \bar{V}ref\,dt = \int_0^{T_1} \bar{V}_k\,dt + \int_{T_1}^{T_1+T_2} \bar{V}_{k+1}\,dt + \int_{T_1+T_2}^{T_s} \bar{V}_0\,dt \tag{36}$$

V_0 does not affect the magnitude of the voltage since it is a null vector. Assuming that the vectors V_k or V_{k+1} are constants and given that V_0 is a null vector:

$$\bar{V}ref\,T_s = \bar{V}_k T_1 + \bar{V}_{k+1} T_2 \tag{37}$$

Expressing (37) in the d-q reference plane and considering the region $0<\gamma<\pi/3$, it follows that.

$$\frac{2}{3}T_1 E\begin{bmatrix}1\\0\end{bmatrix} + \frac{2}{3}T_2 E\begin{bmatrix}cos(\pi/3)\\sin(\pi/3)\end{bmatrix} = T_s V_{ref}\begin{bmatrix}cos(\gamma)\\sin(\gamma)\end{bmatrix} \tag{38}$$

The amplitude of the vector is depended on the duration times, which for discrete system with sampling time T_s, can be calculated by the equations.

$$T_1 = \frac{\sqrt{3}T_s V_{ref}}{E}.\sin(\frac{\pi}{3}-\gamma) \tag{39}$$

$$T_2 = \frac{\sqrt{3}T_s V_{ref}}{E}.\sin(\gamma) \tag{40}$$

$$T_3 = T_0 - (T_1 + T_2) \tag{41}$$

In order to determine the duration of the pulses in the other regions, the reference vector is rotated -nπ/3 radians until its angle is equal to $0<\gamma<\pi/3$.

ADAPTIVE FLUX OBSERVER

The parameters are updated only in a powering operation. Figure.4 shows a block diagram of the proposed flux observer with the parameter adaptive scheme (S. Belkacem, 2005; Hisao Kubota, 1990).

Figure 4. Block Diagram of Proposed Flux Observer (Meziane., 2009).

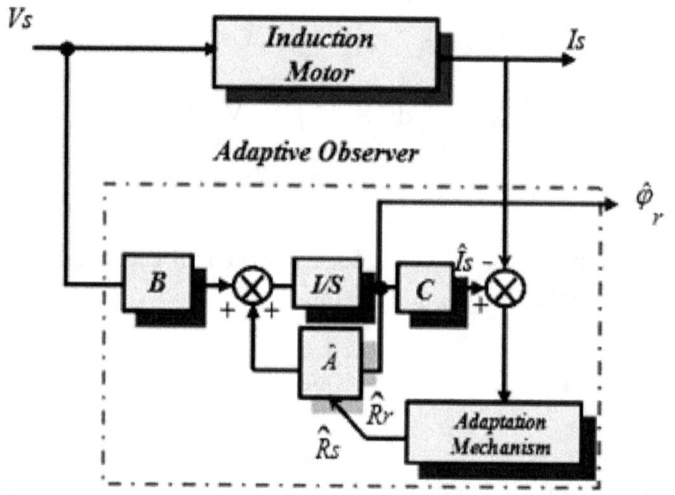

The full order state observer which estimates the stator current and the rotor flux is written by following equations (H. Kubota, 1993). The state observer, which estimates the stator current and the rotor flux together, is written as the following equation.

$$\begin{bmatrix} \dot{\hat{I}}_s \\ \dot{\hat{\phi}}_r \end{bmatrix} = \begin{bmatrix} A_{11} & A_{12} \\ A_{21} & A_{22} \end{bmatrix} \begin{bmatrix} \hat{I}_s \\ \hat{\phi}_r \end{bmatrix} + \begin{bmatrix} B_1 \\ 0 \end{bmatrix} V_s + G(\hat{I}_s - I_s) \tag{42}$$

Where:

$$V_s = \begin{bmatrix} V_{s\alpha} & V_{s\beta} \end{bmatrix}^T, \hat{I}_s = \begin{bmatrix} \hat{I}_{s\alpha} & \hat{I}_{s\beta} \end{bmatrix}^T, \hat{\phi}_r = \begin{bmatrix} \hat{\phi}_{r\alpha} & \hat{\phi}_{r\beta} \end{bmatrix}^T$$

$$G = \begin{bmatrix} g_1 & g_2 & g_3 & g_4 \\ -g_2 & g_1 & -g_4 & g_3 \end{bmatrix}^T$$

$$g_1 = (k-1)(-a_{r11} - a_{r22}) \,, g_2 = (k-1)(-a_{i22})$$
$$g_3 = (k-1)(ca_{r11} - a_{r21}) + (k-1)(-a_{r11} - a_{r22})$$
$$g_4 = c(k-1)(-a_{r22}) \quad , c = -\frac{\sigma L_s L_r}{M}$$

k is the proportional constant

Rotor Flux Observer with Parameter Adaptation

The Influence of the parameter variations on the flux estimation will be investigated in this section. We propose the addition of a parameter adaptive scheme to the flux observer described by equation (43) in order to solve the problem of the parameter variations. We propose the following update law (Hisao Kubota, 1990; Alain, 2017):

$$\begin{cases} \dot{\hat{R}}_s = -\lambda_1 (e_{is\alpha} \hat{I}_{s\alpha} + e_{is\beta} \hat{I}_{s\beta}) \\ \dot{\hat{R}}_r = \lambda_2 \left\{ e_{is\alpha} (\hat{\phi}_{r\alpha} - M\hat{I}_{s\alpha}) + e_{is\beta} (\hat{\phi}_{r\beta} - M\hat{I}_{s\beta}) \right\} \\ e_{is\alpha} = (\hat{I}_{s\alpha} - I_{s\alpha}), \quad e_{is\beta} = (\hat{I}_{s\beta} - I_{s\beta}) \end{cases} \tag{43}$$

λ_1, λ_2: arbitrary positive gain.

Stability of Flux Observer

A stability of the proposed flux observer with the parameter adaptation scheme is proved by the Lyapunov's theorem. For the simplification of the proof, the induction motor and the observer are expressed by (E.M. Chady, 2002, Hisao Kubota, 1990; H. Kubota, 1993).

$$\begin{bmatrix} \dot{I}_s \\ \dot{\phi}_s \end{bmatrix} = \begin{bmatrix} a_{r11} & a_{r12} + z.a_{112} \\ a_{r21} & a_{r22} + z.a_{122} \end{bmatrix} \begin{bmatrix} I_s \\ \phi_s \end{bmatrix} + \begin{bmatrix} B_1 \\ 0 \end{bmatrix} V_s \tag{44}$$
$$\dot{\hat{x}} = (A + \Delta A)\hat{x} + bV_s$$

Where ΔA is an error matrix caused by the parameter variation. An error equation lowing can be expressed by:

$$\dot{e} = (A(x - \hat{x}) - \Delta A\hat{x}) = (Ae - \Delta A\hat{x}) \tag{45}$$

We define a following Lyapunov's function candidate:

$$V = e^T e + \frac{(\Delta R_s)^2}{\lambda_1 \sigma L_s} + \frac{(\Delta R_r)^2 (1/L_2)^2 M}{\lambda_1 \sigma} \tag{46}$$

Computing the time derivative of V and using equation (8) gives following equation:

$$\dot{V} = e^T (A^T + A) e \tag{47}$$

Equation (47) is negative-semi definite, because the matrix (AT+A) negative-semi definite. Therefore the flux observer with the parameter adaptive scheme. The stability in the case of k≠1 can be proved similarly (the observer should have stable poles) (Hisao Kubota, 1990).

DESCRIPTION OF THE PHOTOVOLTAIC PUMPING SYSTEM

The general structure of photovoltaic pumping system studied is shown in (Figure 5). This system is composed by a photovoltaic panel coupled to the induction motor via VSI control thrue space vector modulation. The considered load of PV is a pumping unit composed of an induction motor driving a centrifugal pump. The speed and the flux of the induction motor are controlled by nonlinear control (Jaziri & Jemli, 2013).

Figure 5. Diagram of the photovoltaic pumping system. (Meziane, 2018).

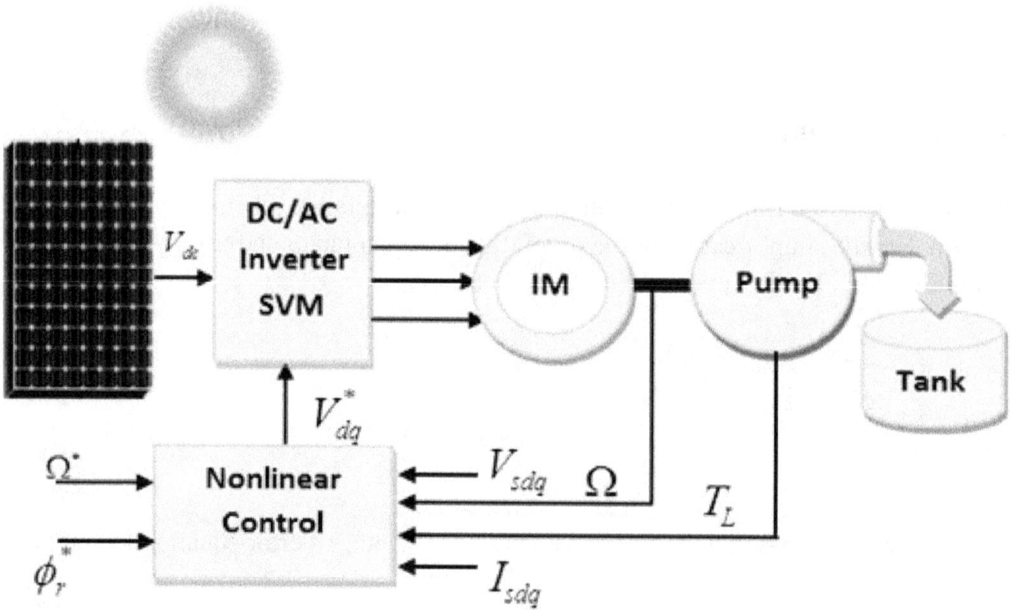

Photovoltaic Source

The equivalent circuit of the PV device can be represented as a current source in parallel with diode and a resistor *(Rp)*, expressing a leakage current, and a series resistor *(Rs)* describing an internal resistance. The model is shown in (Figure 6).

Figure 6. Equivalent circuit of the PV

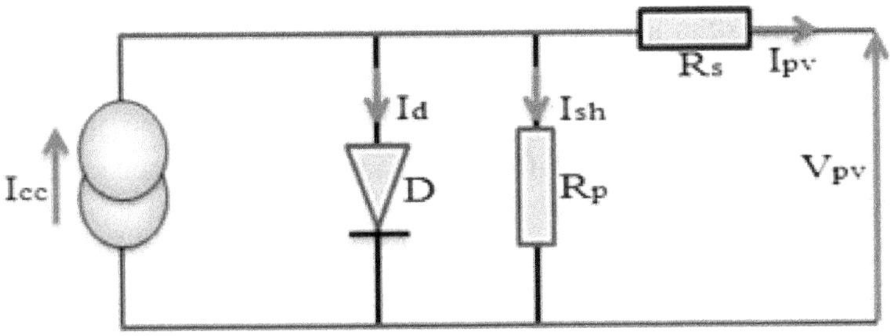

The characteristic (V-I) of a photovoltaic module is nonlinear and depends on solar radiation and cell temperature. The output voltage-current relationship is given by the following expression (Chikh & Chandra, 2009; Mohamed, 2017):

$$I = I_{ph} - \underbrace{I_o \left\{ \exp\left[\frac{q(V + R_s.I)}{aVt} \right] - 1 \right\}}_{I_d} - \underbrace{\frac{V + R_s I}{R_P}}_{I_{sh}} \tag{48}$$

Where

- I_{ph} and I_d are respectively the photocurrent and the junction diode current,
- I_0 is the reverse saturation current, *a* is the diode ideality constant, q[C]: Electron charge.
- $V_t = N_s k/q$ is the thermal voltage of the array with N_s cells connected in series.
- *k* is the Boltzmann's constant, *T[K]*: Temperature of the p-n junction. *k [J/K]*: Boltzmann constant,

DC-DC Boost Converter

The mathematical modeling of the Li-Ion battery bank used in the simulation program is been introduced in this partition. The open voltage source is calculated with a non- linear equation based on the actual SOC of the battery as follows (O. Tremblay, 2007):

$$V_{batt} = E_{batt} - RI \tag{49}$$

The controlled voltage source is described by equation (1):

$$E_{batt} = E_0 - K\frac{Q}{Q-it} + A.\exp(-b.it) \quad (50)$$

Where:

E = no-load voltage (V), Eo = battery constant voltage (V)
K = polarisation voltage (V), Q = battery capacity (Ah), R = internal resistance (Ω)
i. = actual battery charge (Ah), A = exponential zone amplitude (V),
B = exponential zone time constant inverse (Ah)−1, Vbatt = battery voltage (V),

In order to extend the life of battery, it is also important that the battery does not stay in the high-voltage stage for a long time and is allowed to drop after full charge has been reached. This could be achieved by keeping the state of charge (SOC) within the safe range. In the recent year, DC-DC converters are widely used in photovoltaic panel to increased there reliability, efficiency, and controllability and it can reduce such problems like manufacturing cost, efficiency and production (Pandey, S, 2016). The circuit diagrams of the boost converters concerning this study are shown in Figure.8. It contains one commutation cell made up by transistor type and a diode in series with load (Lahoti, G, 2016).

Figure 7. Circuit of electrical elements of the boost converters

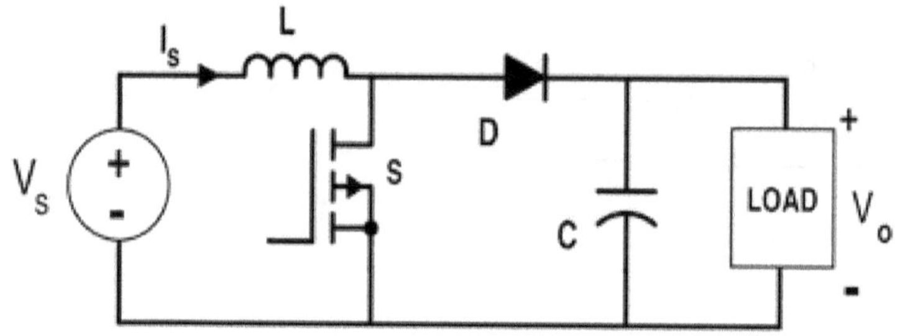

The output voltage of DC/DC Converter is given by:

$$V_o = \frac{1}{1-\alpha} V_s \quad (51)$$

Where α is the duty cycle of the DC/DC converter

Centrifugal Pump

The hydraulic pump is an important part in the photovoltaic water pumping systems. The centrifugal pump type is a commonly used technology because of its operation simplicity and modular flexibility. (Chikh & Chandra, 2009). The pump model relates between the flow rate and the manometric head, for a given speed. It can be approximated as (Karampuri, 2014)

$$H_{pump} = k_0 + k_1 Q + k_2 Q^2 \tag{52}$$

Where, k_0, k_1, k_2 are the constant of the pump, *H[m]* and *Q[L/s]* are respectively, the head and the flow rate of the pump. (Karampuri et al., 2014; Belarbi,2006). The IM is mechanically coupled to a centrifugal pump. In the steady state, the electromagnetic torque of (IM) equilibrates the mechanical torque of the load torque. The combined torque dynamics of IM and pump can be described by the equation (53) as (Jain & Umashankar, 2016).

$$J\frac{d©}{dt} = T_{em} - T_{pump} \tag{53}$$

The centrifugal pump is designed using torque- speed relation as,

$$T_{pump} = k_{T0} + k_{T1}\Omega + k_{T2}\Omega^2 \tag{54}$$

Where, k_{T0}, k_{T1}, k_{T2} are the empirical constants of the pump which can be obtained from the measurements.

Figure 8. Stator Resistance Variation

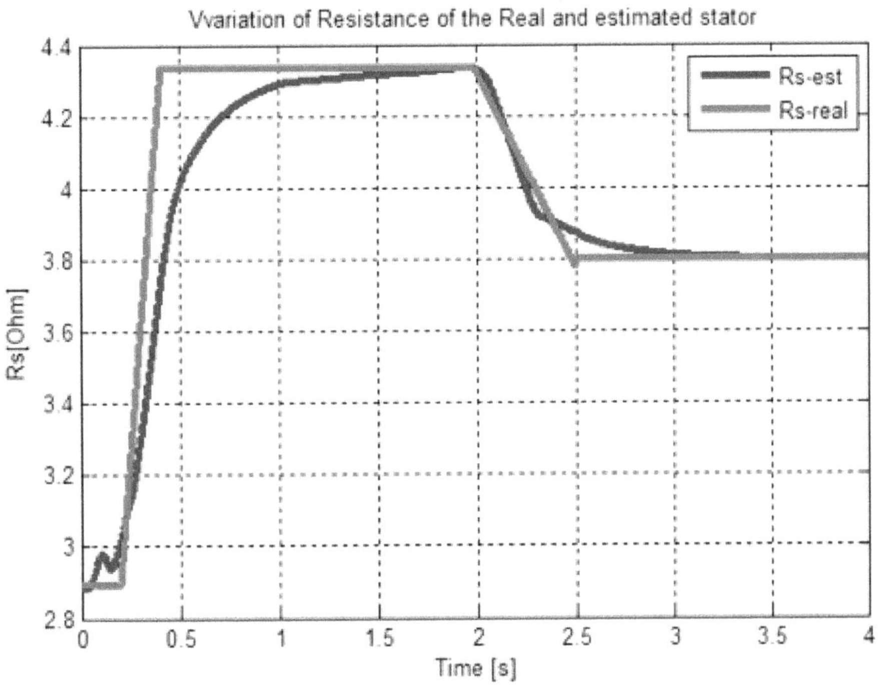

SIMULATION RESULTS

The proposed Photovoltaic Water Pumping (PWP) based Nonlinear adaptive Control (NAC) is tested in Matlab/Simulink in we provide a series of simulations to validate the performances of the proposed control strategy, including rotor and stator resistances and solar insolation variations.

A Nonlinear controller and a DC-DC Boost converter adjusts the speed and Dc output voltage of PV panel to attain the maximum power point (MPPT) in the case of insolation decreasing. The simulations are conducted by a 3kW induction motor driving a centrifugal pump whose data are listed in the Appendix. In order to show the robustness of the proposed nonlinear adaptive control of PV water pumping scheme to provide a decouple control of flux and speed; in the interval of time 0–4s the simulation results including rotor and stator resistances and solar insolation variations are respectively shown in Figures (9-18).

From the simulation result we remark a good tracking performance for proposed technique and show that adaptive flux observer is capable of tracking the stator the stator and rotor resistances variation very well, this is clearly demonstrated in figures.(8-9) shows the tracking performance of the estimator to the actual change in Rs and Rs.

Figure. 11, shows the output DC voltage of the DC-DC boost converter. As it may be observed from the waveform; at t =1.2s as soon as the insolation decreases (Figure.10), the DC voltage decreases and the feed water flow decreases. At some point, the pump will no longer be able to pump water. To solve this problem the nonlinear controller adjusts the reference speed of the IM to compensate the decreasing insolation and the water pumping is restarted in times t =2.7s. (Figure 12 and Figure 13).

Figure 9. Rotor Resistance Variation

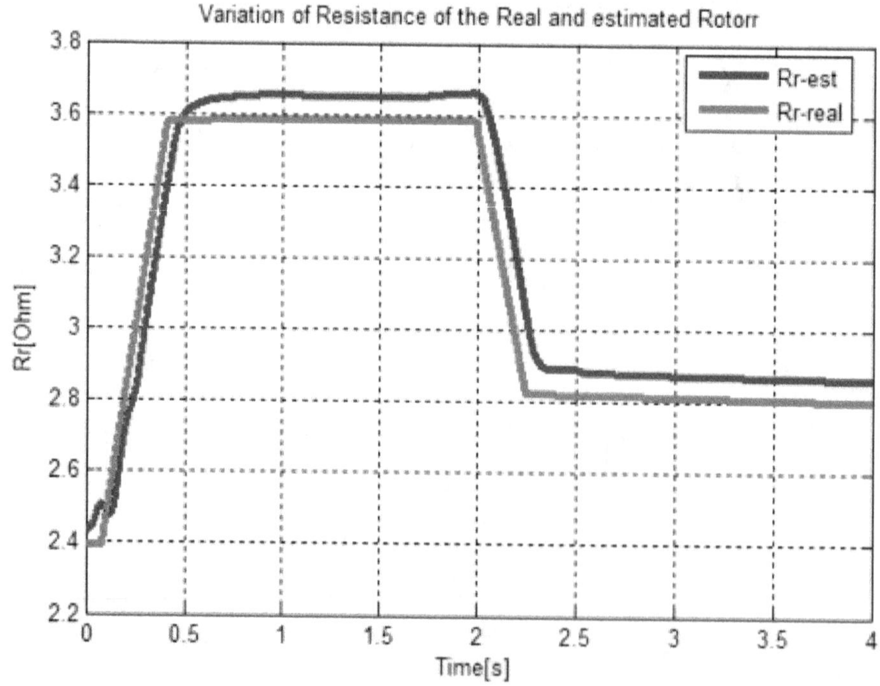

Figure 10. Variation of the Insolation

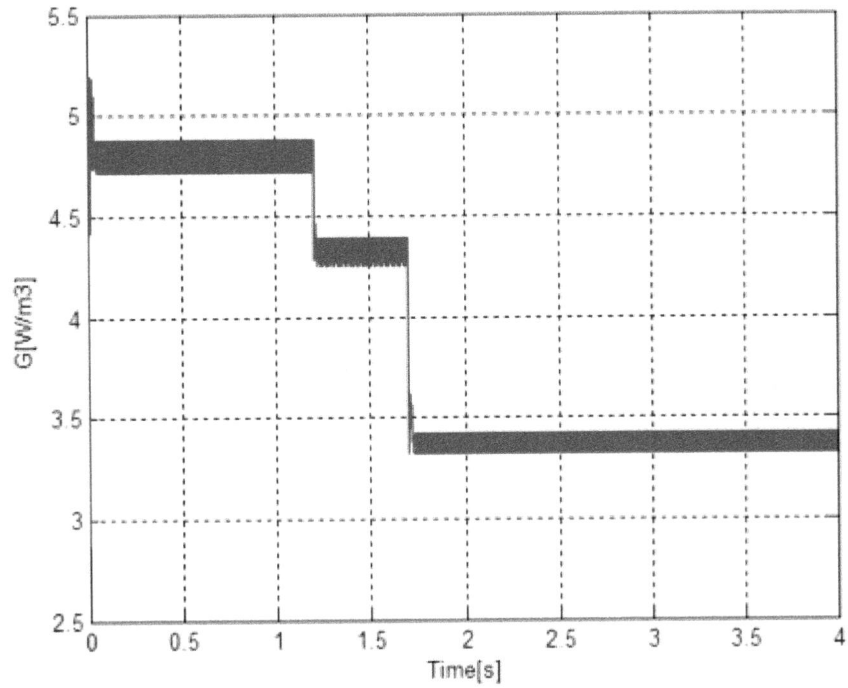

Figure 11. Output DC Voltage of Pc panel

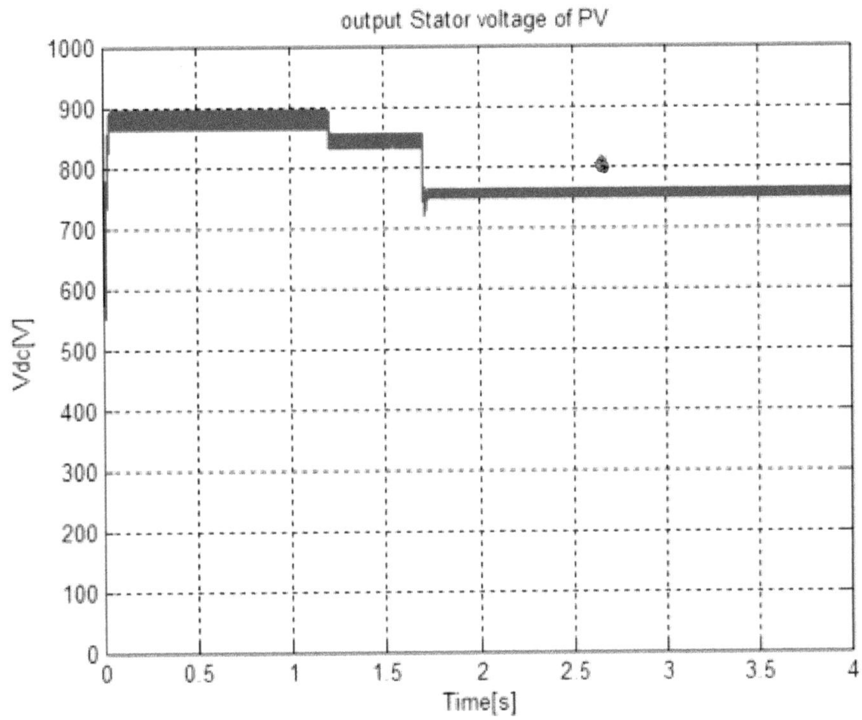

Figure 12. Flow Rate of the Water Pumping

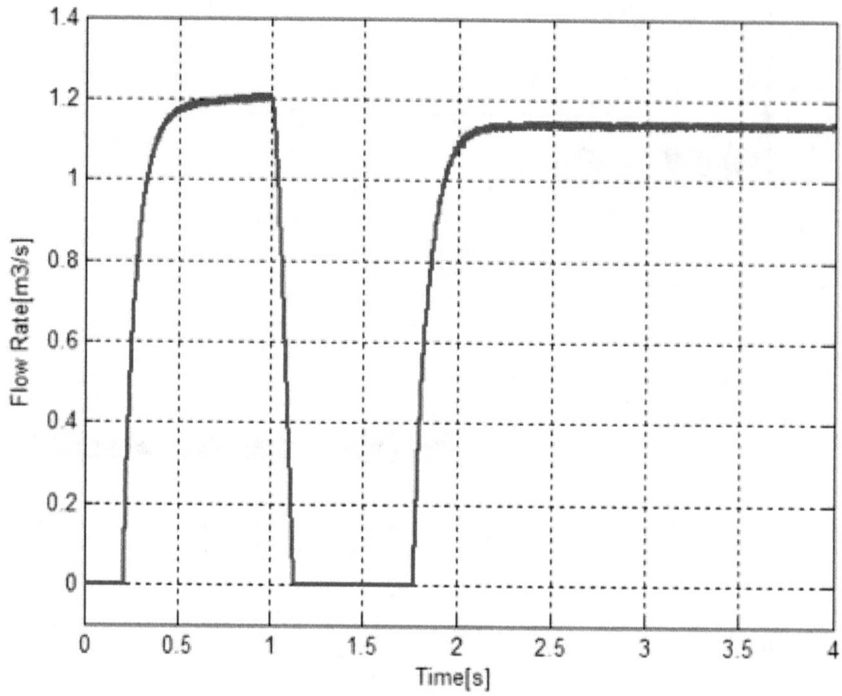

Figure 13. Total Head monomeric

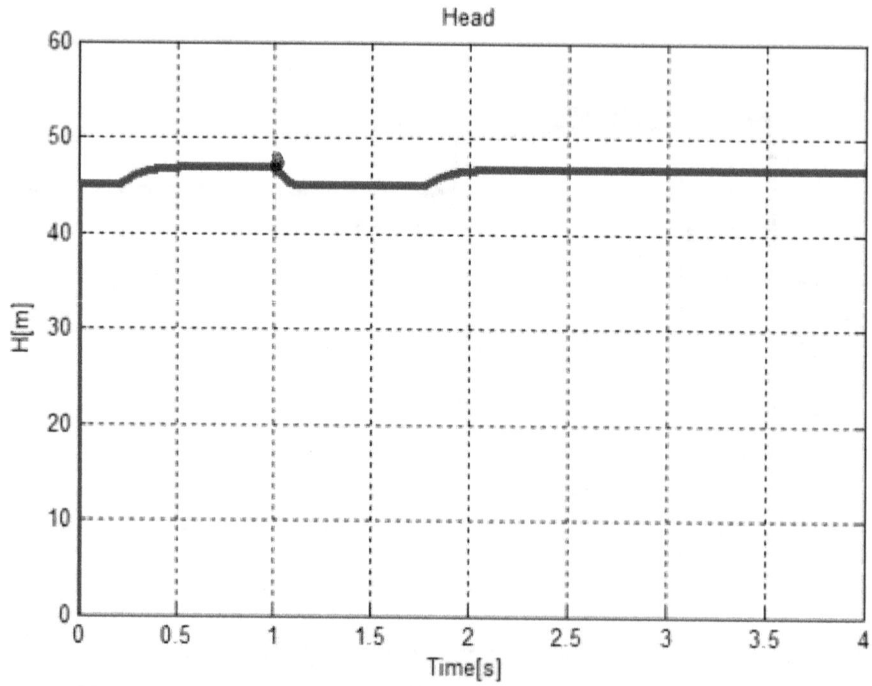

Figure. 14, shows the desired rotor speed and the real rotor speed. As it may be observed from the waveform, the motor speed tracks the change in reference speed quickly and steady state and become unaffected by the variation of the stator, rotor and insolation. Furthermore, comparison of Figure.15 shows good torques responses in steady state. It can be observed that the torque track the torque pump (load torque tracking) at 0.2s and is not affected by the rotor speed change; but the motor torque present the oscillation around load torque (effect the SVM).

Figure 14. Rotor and Reference Speed

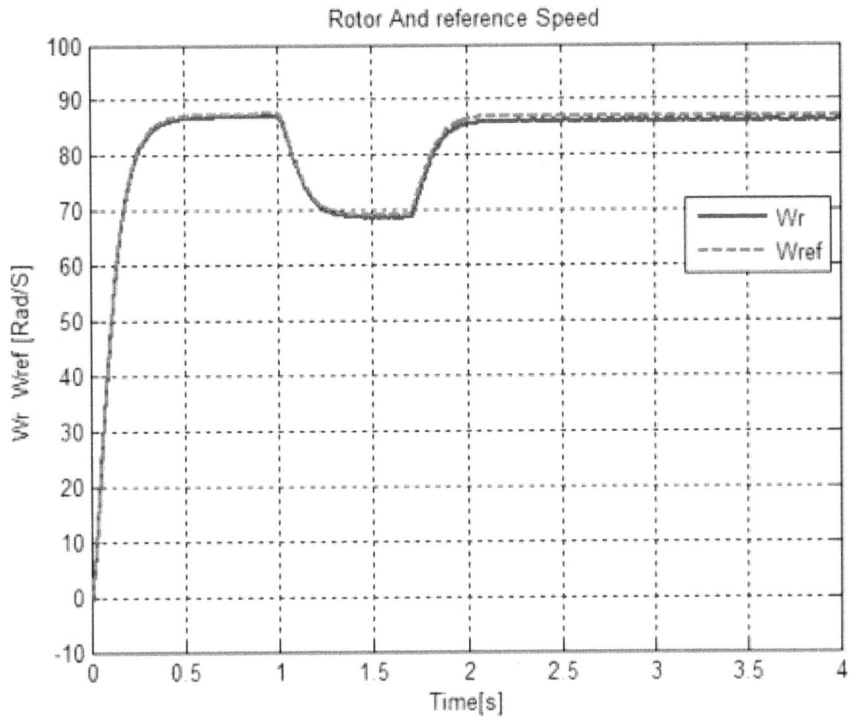

Those figures show that the waveform real and estimated phase stator current is sinusoidal within acceptable limits of the nominal current of the motor; (Figure.16).

Figure.17 shows that the fast response of the estimated and real rotor flux in the transient state is very tracking to the reference flux and is not affected by the disturbances due to resistances variation flux. They show clearly that the decoupling control is more maintained for the nonlinear control design ($\phi_r=\phi_{ref}$ and $\phi_{rq}=0$). The decoupling of torque and flux is obtained.

From these simulation results we remark that a good tracking performance is achieved, those results demonstrate that the one step ahead nonlinear adaptive control has strong robustness properties in face of uncertainties including motor parameters and decreasing insolation. Thus, the proposed Water Pumping System can reduce the cost, while concurrently guaranteeing optimal utilization of the PV array.

Figure 15. Electromagnetic and pump torques

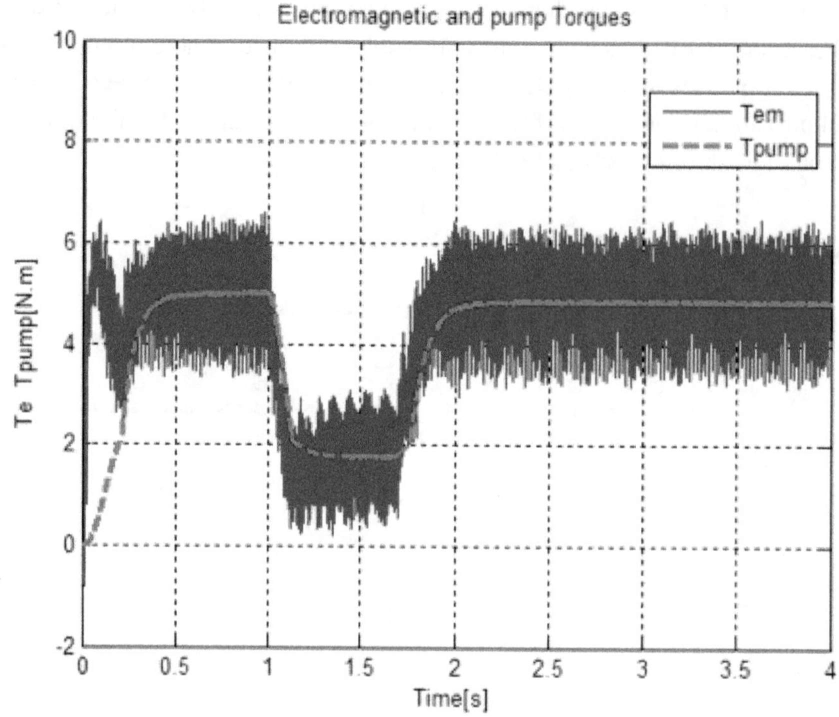

Figure 16. Phase Stator current

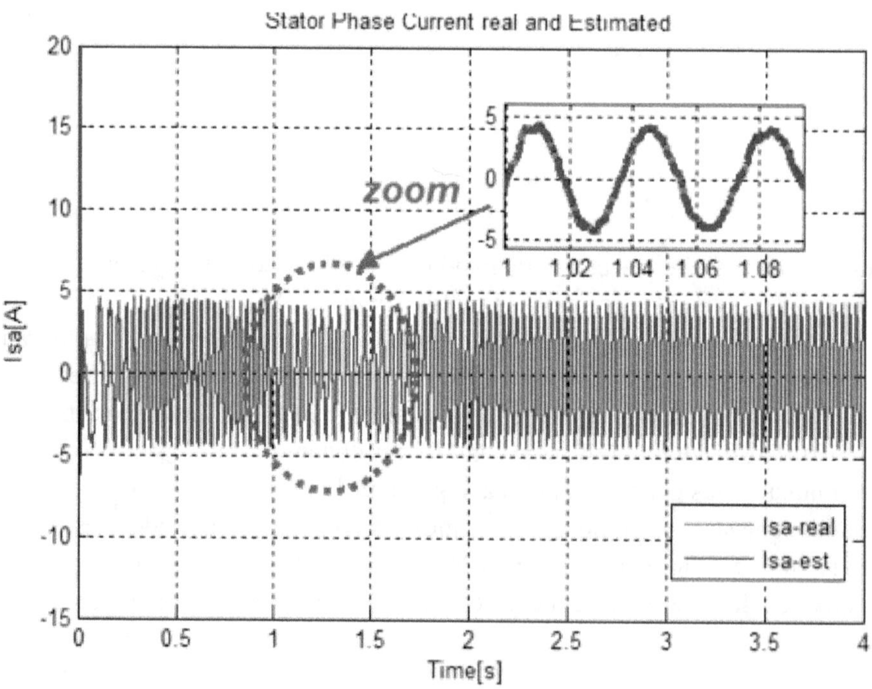

Figure17. Estimated rotor flux for IOLC

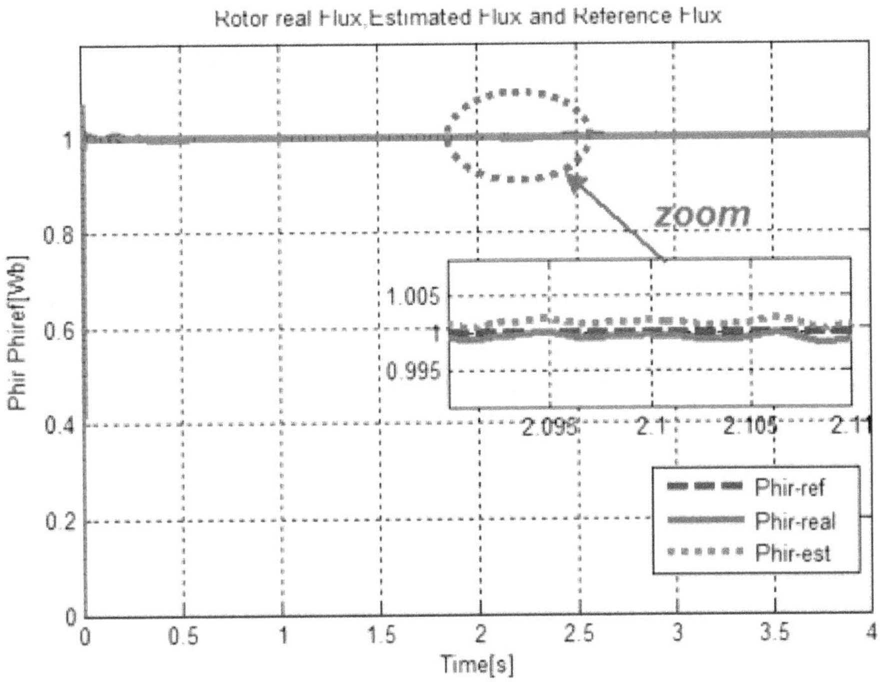

CONCLUSION

In this paper we have presented the application of Nonlinear Adaptive Control (NAC) for a Photovoltaic Water Pumping (PWP) system. The application of such NAC controller allows the regulation of the flux and the speed of the induction motor by eliminating system's nonlinearity and coupling effect control. Thus, nonlinear controller adjusts the reference speed of the IM to attain the maximum power point (MPPT) operation of the photovoltaic array.

From the results obtained it can be noted that the proposed control provides an excellent decouple control of flux and speed in face of uncertainties including motor parameters and decreasing insolation. Thus, the proposed Water Pumping System can reduce the cost, while concurrently guaranteeing optimal utilization of the PV array. As future work, we plan to validate the proposed photovoltaic pumping system by the experimental tests;

REFERENCES

Alain, K. S. T., Romanic, K., & Bertrand, F. H. (2017). Dynamics and Improved Robust Adaptive Control Strategy for the Finite Time Synchronization of Uncertain Nonlinear Systems. *International Journal of System Dynamics Applications*, *6*(4), 34–62.

Archana, D., Kalyani, K., & Prasad, B. S. (2012). Efficiency optimization control of induction motor using Fuzzy logic. *International Journal of Soft Computing and Engineering*, *2*(3), 327–332.

Belarbi, M. (2006). *Modélisation et Simulation d'un Système de Pompage Photovoltaïque (Master's)*. University of Oran.

Belkacem & Naceri. (2005). Speed Sensorless DTC of Induction Motor Based on an improved Adaptive Flux Observer. *IEEE International Conference on Industrial Technology,* 1192-1197.

Blaschke, F. (1972). The principle of field orientation as applied to the new transvector closed-loop system for rotating-field machines. *Siemens Review, 34*(3), 217-220.

Camocardi, P., Battaiotto, P., & Mantz, R. (2008). Wind generator with double stator induction machine. Control strategy for a water pumping application. In *University Power Engineering Conference, UPEC.* September. *43rd International* (pp. 1-5). IEEE.

Casadei, D., Profumo, F., Serra, G., & Tani, A. (2002). FOC and DTC: Two viable schemes for induction motors torque control. *IEEE Transactions on Power Electronics, 17*(5), 779–787. doi:10.1109/TPEL.2002.802183

Chady, Mendes, & Razek. (2002). Decoupled Direct Control for PWM Inverter-Fed Induction Motor Drives. *IEEE Transactions on Industry Applications, 38*(5).

Chikh, A., & Chandra, A. (2009, October). Optimization and control of a photovoltaic powered water pumping system. In Electrical Power & Energy Conference (EPEC), 2009 IEEE (pp. 1-6). IEEE.

Cruz, P. P., & Rivas, J. R. (2000). Induction motor space vector control using adaptive reference model direct and indirect methods. In *Industrial Electronics, Proceedings of the IEEE International Symposium on* (Vol. 1, pp. 300-305). IEEE.

Enev, S. (2007, June). Input-output decoupling control of induction motors with rotor resistance and load torque identification. In *IEEE Mediterranean Conference on Control & Automation 2007. MED'07* (pp. 1-5). IEEE.

Hirpara & Sharma. (2020). An analysis of a Wind Turbine-Generator System In The Presence of Stochasticity and Fokker-Planck Equations. *International Journal of System Dynamics Applications, 9*(1), 18–43.

Hussein, H. T., Ammar, M. E., & Hassan, M. M. (2017). Three Phase Induction Motor's Stator Turns Fault Analysis Based on Artificial Intelligence. *International Journal of System Dynamics Applications, 6*(3), 1–19. doi:10.4018/IJSDA.2017070101

Ismail, M. M. (2012). Applications of anfis and fuzzy algorithms for improvement of the DTC performance for the three phase saturated model of induction motor. *International Journal of System Dynamics Applications, 1*(3), 54–83. doi:10.4018/ijsda.2012070102

Ismail, M. M., Fattah, H. A. A., & Bahgat, A. (2012). Adaptive output feedback voltage-based control of magnetically-saturated induction motors. *International Journal of System Dynamics Applications, 1*(3), 1–53. doi:10.4018/ijsda.2012070101

Jain, G., & Umashankar, S. (2016, April). Modelling and simulation of solar photovoltaic fed induction motor for water pumping application using perturb and observer MPPT algorithm. In *Energy Efficient Technologies for Sustainability (ICEETS), 2016 International Conference on* (pp. 250-254). IEEE.

Jaziri, S., & Jemli, K. (2013, May). Optimization of a photovoltaic powered water pumping system. In *Control, Decision and Information Technologies (CoDIT), International Conference on* (pp. 422-428). IEEE.

Karampuri, R., Jain, S., & Somasekhar, V. T. (2014, December). A single-stage solar PV power fed Open-End Winding Induction Motor pump drive with MPPT. In *Power Electronics, Drives and Energy Systems (PEDES), 2014 IEEE International Conference on* (pp. 1-6). IEEE.

Kubota, Matsuse, & Nakano. (1990). New adaptive flux observer of induction motor for wide speed range motor drives. *Industrial Electronics Society, 1990. IECON APOS, 90. 16th Annual Conference of IEEE, 2,* 921 – 926.

Kubota & Matsuse. (1993). DSP based speed adaptive flux observer of induction motor. *IEEE Transactions on Industry Applications*, *29*(2), 344–348.

Lahoti, G., Shah, U., & Kadam, P. (2016). Robust control of boost converter for flexible operation in PV based systems. *Energy Efficient Technologies for Sustainability (ICEETS), International Conference on,* 345-349.

Marino, R., Peresada, S., & Valigi, P. (1993). Adaptive input-output linearizing control of induction motors. *IEEE Transactions on Automatic Control*, *38*(2), 208–221. doi:10.1109/9.250510

Meng, Z., Sun, C., An, Y., & Chen, R. (2011, August). Nonlinear decoupling control of induction motor based on parameter adaptive identification. In *Intelligent Human-Machine Systems and Cybernetics (IHMSC), 2011 International Conference on* (Vol. 2, pp. 98-102). IEEE. 10.1109/IHMSC.2011.94

Meziane, S. (2009). *Commande Adaptative et Prédictive de la Machine Asynchrone* (Doctoral dissertation). Constantine University.

Mohamed, R. G., Ebrahim, M. A., Bendary, F. M., & Osman, S. A. A. (2017). Transient Stability Enhancement for 20 MW PV Power Plant via Incremental Conductance Controller. *International Journal of System Dynamics Applications*, *6*(4), 102–123.

Mohamed, R. G., Ebrahim, M. A., Bendary, F. M., & Osman, S. A. A. (2017). Transient Stability Enhancement for 20 MW PV power plant via incremental conductance controller. *International Journal of System Dynamics Applications*, *6*(4), 102–123.

Muchande, S., Kadam, A., Unni, K., & Thale, S. (2013). Design and Implementation of a Direct Torque Control Space Vector Modulated Three Phase Induction Motor Drive. In *Advances in Computing, Communication, and Control* (pp. 659–672). Springer. doi:10.1007/978-3-642-36321-4_61

Pandey, S., Dwivedi, B., & Tripathi, A. (2016). Performance analysis of SMC controlled PV fed boost converter. *7th India International conference on In Power Electronics (IICPE),* 1-4.

Pham, V. T., Volos, C., & Vaidyanathan, S. (2017). A Chaotic Time-Delay System With Saturation Nonlinearity. *International Journal of System Dynamics Applications*, *6*(3), 111–129.

Rashed, M., MacConnell, P. F., & Stronach, A. F. (2006). Nonlinear adaptive state-feedback speed control of a voltage-fed induction motor with varying parameters. *IEEE Transactions on Industry Applications*, *42*(3), 723–732. doi:10.1109/TIA.2006.872951

Raumer, T. V., Dion, J. M., & Dugard, L. (1993). Adaptive non-linear speed control of induction motors. *International Journal of Adaptive Control and Signal Processing*, *7*(5), 435–455. doi:10.1002/acs.4480070511

Salima, M., Loubna, A., & Riad, T. (2018). A Global Stability of Linearizing Control of Induction Motor for PV Water Pumping Application. *International Journal of System Dynamics Applications*, *7*(3), 31–56.

Toufouti, R., Meziane, S., & Benalla, H. (2009). New direct torque neuro-fuzzy control based SVM for dual two level inverter-fed induction motor. *Journal of Control Engineering and Applied Informatics*, *11*(2), 3–13.

Tremblay, D., & Dekkiche. (2007). A Generic Battery Model for the Dynamic Simulation of Hybrid Electric Vehicles. *IEEE Vehicle Power and Propulsion Conference,* 284-289.

Wade, S., Dunnigan, M. W., & Williams, B. W. (1997). Modeling and simulation of induction machine vector control with rotor resistance identification. *IEEE Transactions on Power Electronics*, *12*(3), 495–506. doi:10.1109/63.575677

Wai, R. J., & Lin, K. M. (2005). Robust decoupled control of direct field-oriented induction motor drive. *IEEE Transactions on Industrial Electronics*, *52*(3), 837–854. doi:10.1109/TIE.2005.847585

Zhu, J., Xu, B., Wang, X., Feng, H., & Xu, X. (2013). The research of sensorless vector control for permanent magnet linear synchronous motor. *Journal of Computers*, *8*(5), 1184–1191. doi:10.4304/jcp.8.5.1184-1191

Chapter 19
Modeling, Analysis, and Control of Wide Distributed Large-Scale PV Power Plant Using Recent Optimization Techniques

Rania Gamal Mohamed
Marg High Institute for Engineering and Modern Technology, Cairo, Egypt

Mohamed Ahmed Ebrahim
https://orcid.org/0000-0001-8100-5206
Faculty of Engineering at Shoubra, Benha University, Egypt

ABSTRACT

Atom search optimization algorithm (ASOA) has recently been explored to develop a novel algorithm for distributed optimization and control. This chapter proposes the ASOA-based design of maximum power point tracking controllers (MPPTCs) for controlling the boost converter voltage to harvest the maximum power and enhance the damping of oscillations in the output power of the photovoltaic power plants. The proposed ASOA-based MPPTCs are PI and fractional-order PI controllers. ASOA is utilized to search for optimal controller parameters by minimizing a candidate time-domain based objective function. The performance of the proposed ASOA-based MPPTCs has been compared to the MPPTCs optimized by grey wolf optimizer (GWO) to demonstrate the superior efficiency of the ASOA-based MPPTCs. Simulation results emphasis on the better performance of the proposed MPPTCs compared to MPPTCs and GWO-based PI- FOPI controllers over a wide range of operating conditions.

INTRODUCTION

Like many other countries, Egypt faces increasing challenges in the provision of energy services. Therefore, presenting a comprehensive strategy for growth in the energy sector to cope with local market requirements becomes crucial. The primary strategy objectives are how to achieve energy efficiency and

DOI: 10.4018/978-1-7998-5788-4.ch019

sustainability at competitive costs by reducing carbon emissions under the availability of energy resources constraints. Also, how to improve the quality of the existing networks, as well as the extent to which this strategy, economic efficiency, and environmental sustainability. Egypt is one of the countries which lie inside the Sunbelt region and enjoy the high direct solar radiation. So, the Egyptian market has become a promising market in the field of solar energy especially, after the availability of modern technologies enabling customers to rely on solar energy within the electricity network or outside it (Hanane Yatimi, et al., 2018), (Nayara Teixeira Santos, et al., 2020). Egypt has a great asset to renewable energy, which paves the way for more current researches to exploit them economically (NREA, 2019). Egypt, one of the countries which lie inside the Sunbelt region and enjoy the high direct solar radiation. As well as solar energy falling on the deserts of Egypt is enough to cover the world's consumption of electricity. Figure (1) presents Egypt's average solar radiation (NREA, 2005). The only drawback to solar energy technology is the high cost of production units produced from solar energy compared to conventional sources. However, nanotechnology has recently achieved a significant reduction in the cost of generating electricity from the sun, and it deduces importance directing scientific research in Egypt, especially to improve the economics of solar energy (Egyptian State Information service,2019). Photovoltaic (PV) is one of the latest technologies used in electrical power generation to convert sunlight directly into electricity; it uses a semiconductor material such as silicon, which is extracted from pure sand (World Energy Council), (Hanane Yatimi, et al., 2018). PV is a renewable and clean energy source because it has no contaminated waste, noise, radiation and does not even need the fuel. However, the initial cost is high compared with other energy sources (Renewable Energy policy network for the twenty-first century, 2017). The main benefit of PV is that there are no moving parts exposed to malfunction. For this, PV works over satellite with high efficiency and no need for maintenance or repairs or fuel (Regional center for Renewable Energy and Energy Efficiency, 2019), (Seyed Hossein, et al., 2018).

The biggest problem with PV system utilization is the weather uncertainties such as solar radiation variation, continuous temperature change, and partial shading (Brian J. Galli, et al., 2020). Therefore, it is crucial to run PV at Maximum Power Point (MPP) to extract the highest power (Mohamed. A. Ebrahim et al., 2017). Nowadays Maximum Power Point Tracking (MPPT) system is considered the central part of any PV system to achieve the maximum power at all times under the change in weather conditions (Jalel Ghabim et al., 2018). There are many MPPT algorithms discussed in scientific research such as Incremental Conductance (IC), Fuzzy Logic Control (FLC), Artificial Neural Networks (ANN), Fractional Open-Circuit Voltage (FOCV), Sliding Mode Control (SMC), Particle Swarm Optimization (PSO), Perturb and Observe (P&O), Fractional Short-Circuit Current (FSCC) (Hadjer Bounechba, et al., 2016). However, these algorithms are differing from each other according to sensors number, easy or complexity implementation and cost. The best algorithm is selected according to its accuracy, fast-tracking performance and minimum error due to changing the countries weather conditions. IC is introduced in this research work due to its ease of implementations and low-cost. Most of the researchers presented the IC-based MPPT algorithms based on I-controller as a classical MPPT algorithm (Hadjer Bounechba, et al., 2016). However, this work offers two modified MPPT algorithms based on Proportional-Integral control (PI), Fractional Order Proportional-Integral control (FOPI), The main problem of FOPI and PI controllers is how to attain the optimum value of the gains to achieve the best result (Mousa, M.E, et al., 2017).

For this purpose, two recent optimization techniques, namely; Atom search optimization (ASO) and Grey Wolf Optimizer (GWO) are used to achieve the optimal gains of PI and FOPI controllers

Figure 1. Egypt's average solar radiation.

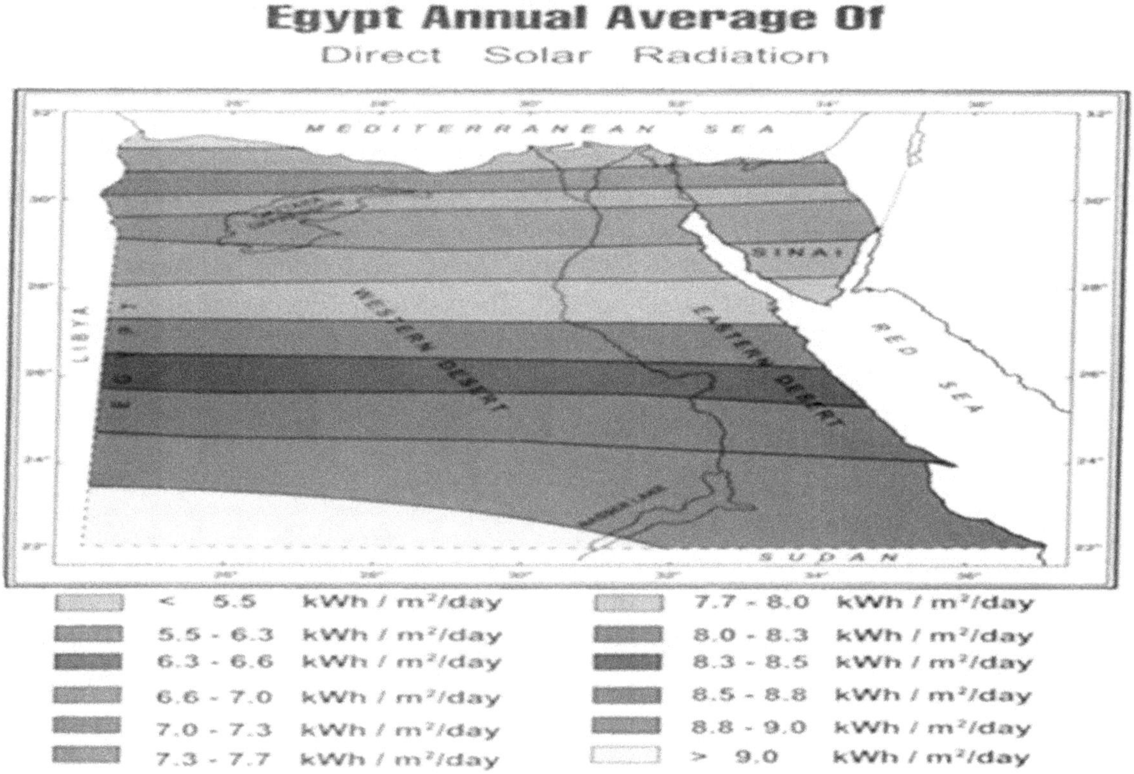

The main contributions of this research work are:

- Complete design for Komombo PV power plant (including sizing for each component like PV array, a boost converter, MPPT, inverter, transformer, and filter) is presented.
- Simulink model for 20 MW PV solar power plant located at Komombo, Aswan, Egypt (Practical Case Study) is constructed.
- This research work proposed two modified MPPT techniques based on FOPI and PI controllers tuned using new optimization techniques called GWO and ASO, aiming to enhance the transient stability of MPPT controllers.
- A comprehensive comparative study between the modified MPPT controllers is conducted to verify the effectiveness of the proposed GWO and ASO based FOPI and PI controllers.

PRACTICAL CASE STUDY

This study presents a 20 MW PV solar power plant connected to the national grid and installed in Komombo, Aswan, Egypt. Komombo locates at E 32.8° longitude North and N 24.6° latitude Eastern at time zone GMT+2. The project site is a 500,000 m^2 area approximately 80 km north of Aswan, 9 km west of Farris village, and 13 km west of Nile. The site is a trapezium of sides approximately 240 m by

2,100 m, with the long side running approximately north-south. This project is the largest PV project planned in Egypt. In May 2012, Grant Agreement was signed between the Ministry of International Cooperation and the French Development Agency (AFD) to finance the feasibility study. It is planned to start the operation of the project in 2017. There are four substations in the region of the site Kalabsha, Siwa, ALNorka, and Komombo. The project site is away from Kalabsha substation by 13 Km, 16.3 km from Komombo substation, 26 km from Siwa substation and 31.8 km from ALNorka substation. Connection to the nearest substation, Kalabsha, is impractical. The loads at the substation are not sufficient to consume the plant output. It is not reasonable to connect to the substation only as an entry point to the grid. Siwa appears to be the most attractive substation for connection to the grid. The substation has the capacities to support connection at 220 kV or 66 kV. The path from the project to Siwa substation is clear desert, unimpeded by agricultural land or buildings. Connecting to the Siwa substation also avoids the need to cross the Nile.

KOMOMBO PV SOLAR PLANT SIZING

The sizing of Komombo PV solar power plant is done using the author-friendly MATLAB software package based on the equations in references (Rania. G Mohamed et al., 2014; Rania. G. Mohamed et al., 2012; Adel. A. Elbasetet al., 2015). Figure (2) shows a flowchart for the sizing steps of Komombo PV solar plant. According to the authors' design, the total size of Komombo PV solar plant results from the developed MATLAB program is 20 MW.

- The numbers of series modules and parallel strings are 18 and 3643, respectively (according to Yingli Energy (China) YL305P 35b _PV module).
- The total number of modules is 65574. There are 33 PV similar PV substations; each PV substation produces 615.3 kW.
- The total size of the inverter is 16.5 MW with 33 inverters each inverter has a capacity of 500kW (according to ABB PVS800_500 kw_inverter).

The size of the boost converter for 615.3 kW substations is shown in Table 1.

Table 1. Sizing results of the boost converter.

Boost Converter Design	
Input voltage	666V
output voltage	825V
duty cycle ratio	0. 192
Switching frequency	100 kHz
Inductor value	0.0043H
Capacitor value	38.545μ F

- The size of the transformer is 1250 kVA for each PV substation.

Table 2 shows Komombo PV solar plant sizing result for one substation (615.3 kW).

Table 2. Sizing results of Komombo PV Solar plant

Komombo PV Solar Plant615.3 kW	
Number of Inverters	1
Inverter Capacity (kW)	500
PV Module Power (W)	305.25
Total number of Module per string	18
Total number of string per Inverter	112
Total rated power (D.C) (kW)	615.3
Capacitor bank filter (kVAR)	61.53
Transformer rating (kVA)	1250

PV MODULE AND CHARACTERISTICS

Figure 3 represents the equivalent circuit for Yingli Energy (China) YL305P-35b _PV module to express the voltage-current (V-I) characteristic of this PV module (Mohamed. A. Ebrahim et al., 2016; Mohamed.J, Ahmedet al., 2016). The values of PV module elements will be calculated using the Matlab program developed by authors. The developed program is based on equations (1:11).

$$I = I_{Ph} - \exp\left(\frac{V + IR_{se}}{aV_T}\right) - 1 - \frac{V + IR_{se}}{R_{sh}} \tag{1}$$

$$V_T = n\frac{KT}{q} \tag{2}$$

$$I_s = I_{os}\left[\frac{T_k}{T_{re}}\right]^3 \exp\left\{\frac{qE_{GO}}{aK}\left\{\frac{1}{T_{re}} - \frac{1}{T_k}\right\}\right\} \tag{3}$$

$$I_L = \left[I_{sct} + K_i\left(T_{ce} - 25\right)\right] \times \frac{H}{1000} \tag{4}$$

$$T_{ce} = T_{air} + 0.2 \times H\% \tag{5}$$

$$R_{se} = A[\omega_{-1}(Be^c) - (D + C)] \tag{6}$$

$$A = \frac{aV_T}{I_{mpp}} \tag{7}$$

$$B = -\frac{V_{mpp}\left(2I_{mpp} - I_{sct}\right)}{\left(\left(V_{mpp}I_{sct}\right) + V_{oc}\left(I_{mpp} - I_{sct}\right)\right)} \tag{8}$$

$$C = -\frac{2V_{mpp} - V_{oc}}{aV_T} + \frac{\left(V_{mpp}I_{sct} - V_{oc}I_{mpp}\right)}{\left(V_{mpp}I_{sct} + V_{oc}\left(I_{mpp} - I_{sct}\right)\right)} \tag{9}$$

$$D = \frac{V_{mpp} - V_{oc}}{aV_T} \tag{10}$$

$$R_{sh} = \frac{\left(V_{mpp} - I_{mpp}R_{se}\right)\left(V_{mpp} - R_{se}\left(I_{sct} - I_{mpp}\right) - aV_T\right)}{\left(V_{mpp} - I_{mpp}R_{se}\right)\left(I_{sct} - I_{mpp}\right) - aV_T I_{mpp}} \tag{11}$$

Where,

I: YL305P-35b _PV module output current (A).
V: YL305P-35b _PV module output voltage (V).
I_{ph}: The photocurrent due to an incident of solar irradiation (A).
R_{se}: The YL305P-35b _PV module series resistance (0.44 ohm).
R_{sh}: The YL305P-35b _PV module shunt resistance (294.1069 ohm).
V_T: The diode thermal voltage.
K: The constant of Boltzmann (1.38×10^{-23} joule/°K).
q: The charge of the electron (1.602×10^{-19} coulomb).
n: The number of cells in series.
T: The normal temperature (°C).
I_s: The YL305P-35b_PV module reverse saturation current (A).
I_{os}: The reverse saturation current of the cell ($19.97 *10^{-6}$A).
T_k: The temperature of the cell in Kelvin (°K).
T_{re}: The reference temperature (301°K).
a: The ideality factor (1.92).
E_{GO}: The energy band gap of silicon (1.11 e.v).
I_L: The YL305P-35b_PV module light generated current (8.8857 A).
Isct: The YL305P-35b module short-circuit current at STC (8.87 A).
K_i: The coefficient of the short-circuit current (0.047001 A / °C).
T_{ce}: The temperature of the cell in Celsius (°C).

Figure 2. Komombo PV solar plant sizing flowchart

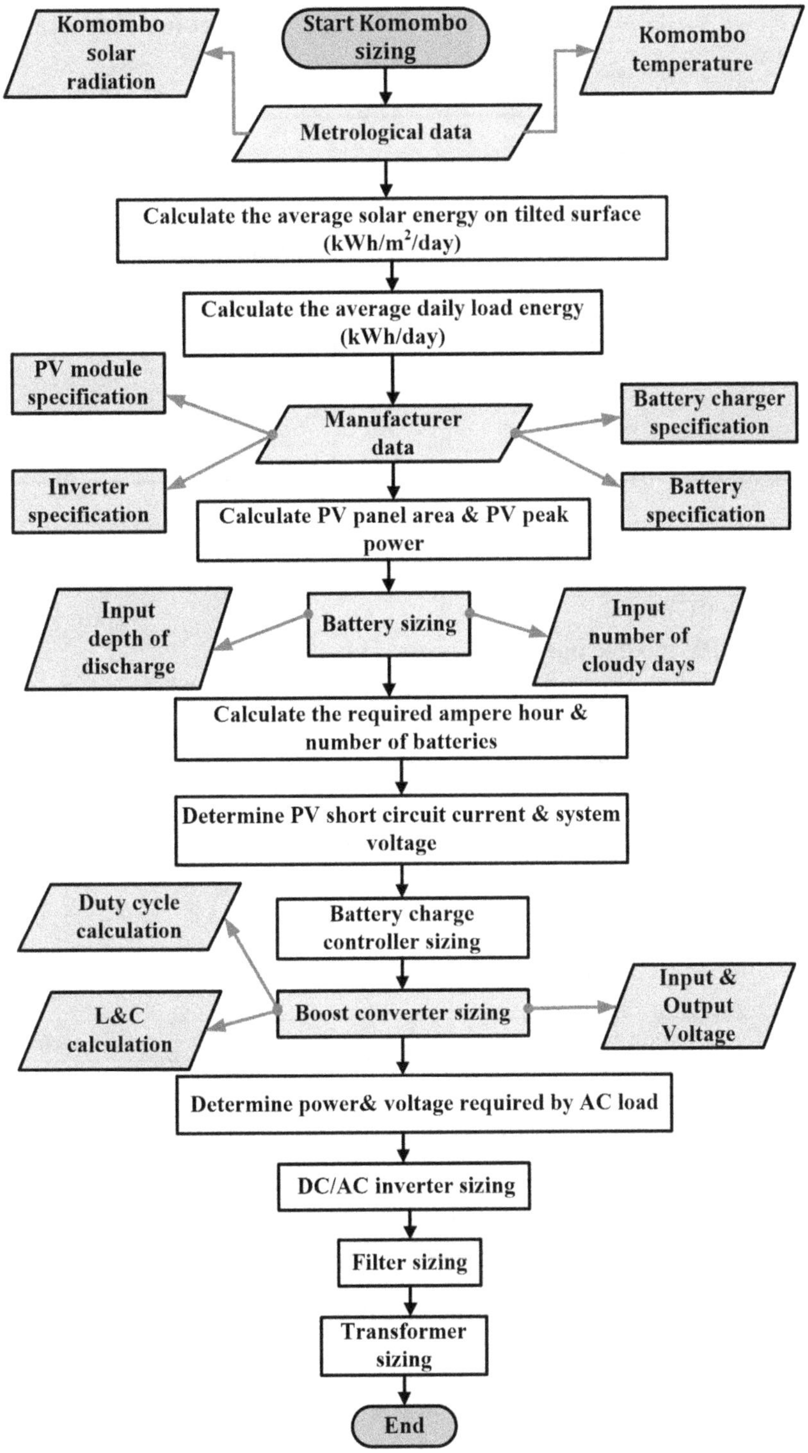

Figure 3. YL305P-35b _PV module equivalent circuit

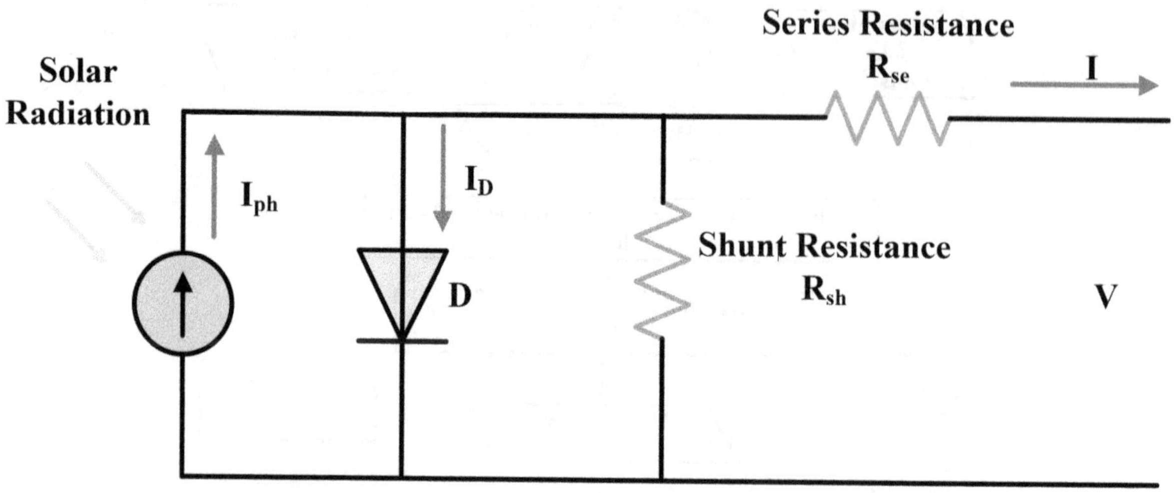

H: The radiation incident on the cell (W/m^2).
T_{air}: The air temperature (ºC).
I_{mpp}: The YL305P-35b _PV module operating current at its MPP.
ω_{-1}: The expression of Lambert W-function [W (z)], where z = (z) e(z)·
Voc: The YL305P-35b _PV module open-circuit voltage at STC.
Vm_{pp}: The YL305P-35b _PV module output voltage at its MPP.

The electrical parameters of the YL305P-35b_PV module using the author developed Matlab program are giving in Table (3).

Table 3. Electrical parameters of the YL305P-35b_PV module.

Electrical Parameter	Rating
Light generated current: I_L	8.8857 A
Reverse saturation current:I_s	3.8623e^{-10} A
Series resistance: R_{se}	0.44 ohm
Shunt resistance: R_{sh}	294.1069 ohm

Figure (4) illustrates the response of the current and power against the voltage variation for the simulated YL305P-35b_PV module at STC.

Figure 4. Simulated YL305P-35b_PV module Current and Power at S.T.C.

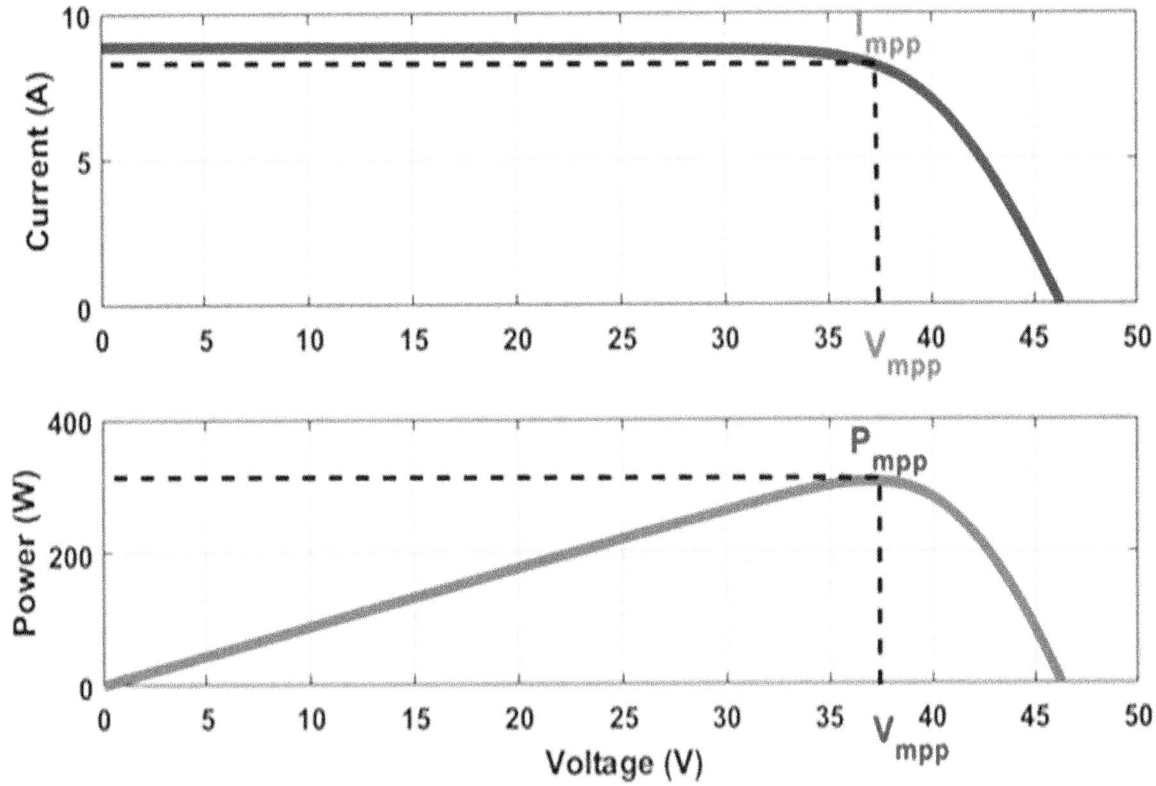

BATTERY MATHEMATICAL MODEL

In general, the battery model describes the relationship between its voltage, current and State of Charge (SoC). When the battery voltage is higher than the Low Battery Voltage (LBV), the switch is on, and the power can be delivered by the battery when it must supply power. In this research work, the lead-acid battery securing energy source for operating the monitoring and control devices for the station during the absence of the sun. Figure (5) shows the equivalent circuit for battery (Mohamed. E. Sawada et al., 2015). The terminal voltage (V_{TB}) of the battery is given by the following equation:

$$V_{TB} = V_{OB} + I_{BC} \times R_{IB} \tag{12}$$

Where,

V_{OB}: is the battery open-circuit voltage (V).
I_{BC}: is the battery current (A).
R_{IB}: is the internal battery resistance (Ω).
V_{OB} and R_{IB} are estimated by a set of equations according to the battery operation mode.

Figure 5. Equivalent circuit for Battery

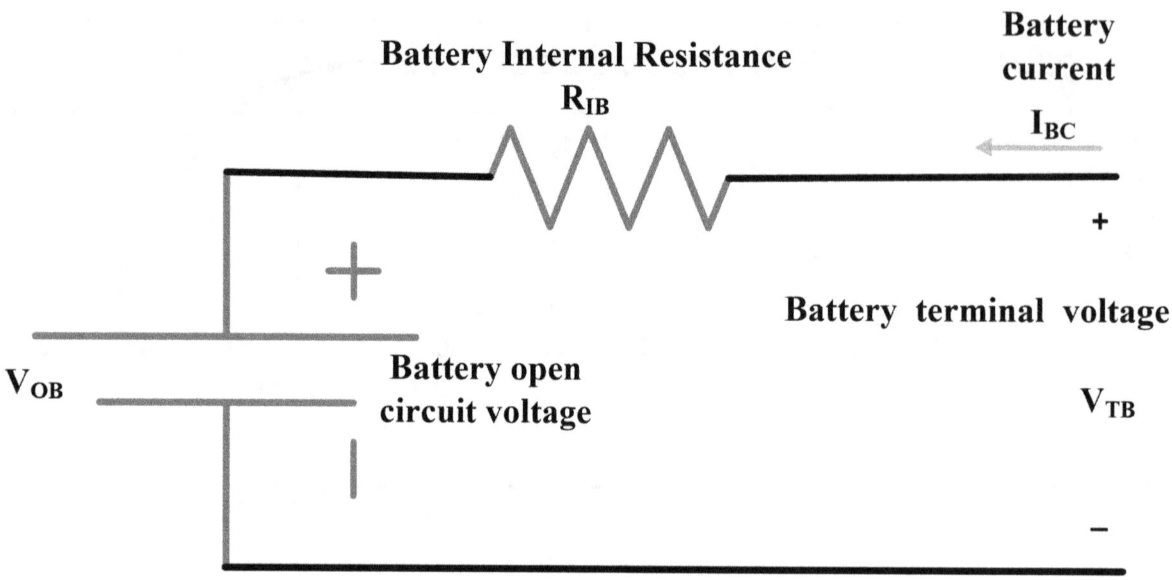

Charge Mode

According to the following equations, the battery voltage and SoC can be estimated in charging mode (Mohamed. E. Sawada et al., 2015)1:

$$V_{OB} = V_{ch} = [2 + 0.148 \times SOC(t)] \times n_s \tag{13}$$

$$R_{IB} = R_{ch} = \frac{0.758 + 0.1309 / \left[1.06 - SOC(t)\right] \times n_s}{Q_m} \tag{14}$$

Discharge Mode

In discharging mode, the battery voltage and SoC relationship is given by (Mohamed. E. Sawada et al., 2015):

$$V_{OB} = V_{dch} = [1.926 + 0.124 \times SOC(t)] \times n_s \tag{15}$$

$$R_{IB} = R_{dch} = \frac{0.19 + 0.1037 / \left[SOC(t) - 0.14\right] \times n_s}{Q_m} \tag{16}$$

Where,

SOC(t): is the state of charge of the battery.

n_s: is the number of 2V battery series cells

Q_m: is the maximum battery capacity (Wh).

The battery SOC(t) can be estimated using the following equation.

$$SOC(t) = SOC(t-1) + \left(\int_{t-1}^{t} \frac{K_b \times V_1 \times I_{BC}}{Q_m} - SOC(t-1) \times D \right) dt \tag{17}$$

Where,

kb: is the battery charge/discharge efficiency.
D: is the battery self-discharge rate (h-1).

Technical data for the battery used in this study given in Table (4).

Table 4. Technical data for battery

Battery type	Lead-acid battery
Capacity	12 kWh
Nominal Capacity Ah	500 Ah
Voltage	24 V
Min. SOC	30%
Efficiency	85%
Lifetime	5 years

KOMOMBO CONTROL STRATEGIES

Komombo PV Solar plant will face two major challenges,

- The first challenge is how to control the boost converter to get the highest power from PV array all the time under any temperature and radiation conditions.
- The last one is how to operate the monitoring and control devices for the station during the absence of the sun.

For this purpose, two modified MPPT controllers are employed to overcome these challenges.

INCREMENTAL CONDUCTANCE MPPT CONTROL

Conventional IC technology overcomes the disadvantages of the classical P&O technique. The classical IC technique, which based on the integral controller and genetic algorithm, depends on the position of the YL305P-35b_PV module slope, as shown in Figure (6). At MPP, the slope is zero, on the left of MPP, the slope is positive, and at the right of MPP, the slope is negative (P. Srinivas et al., 2016).

Figure (7) showed a flowchart of the modified IC-MPPT algorithm. Most research of IC-MPPT disregarded zero division problems, which represents a major problem and led to stopping the technique (Mohamed. Salah El-Din et al., 2017; Atef. Gharbiet al., 2017). This paper presents a solution to avoid the zero division problems by inserting a tiny number instead of zero.

$$\begin{cases} \frac{dP}{dV} = 0, \text{ at MPP} \\ \frac{dP}{dV} > 0, \text{ left of MPP} \\ \frac{dP}{dV} < 0, \text{ right of MPP} \end{cases} \tag{18}$$

Since,

$$\frac{dP}{dV} = \frac{d(IV)}{dV} = I + V\frac{dI}{dV} = I + V\frac{\Delta I}{\Delta V} \tag{19}$$

$$\begin{cases} \frac{\Delta I}{\Delta V} = -\frac{I}{V}, \text{ at MPP} \\ \frac{\Delta I}{\Delta V} > -\frac{I}{V}, \text{ left of MPP} \\ \frac{\Delta I}{\Delta V} < -\frac{I}{V}, \text{ right of MPP} \end{cases} \tag{20}$$

The biggest problem of the classical IC technique is that it has a relatively slow transient stability and dynamic response time (Hanane Yatimi, et al., 2017). Therefore, this work presents two modified IC-MPPT algorithms based on FOPI, PI controller with ASO and GWO, aimed at providing an enhanced performance as compared to the classical IC technique. The main function of FOPI and PI controllers is to eliminate or minimize error to help IC MPPT algorithm in transient stability enhancement (Mohamed. Salah El-Din et al., 2017; Atef. Gharbiet al., 2017).

$$\text{Error} = \frac{\Delta I}{\Delta V} + \frac{I}{V} \tag{21}$$

Figure 6. The power-voltage curve of YL305P-35b _PV module at STC

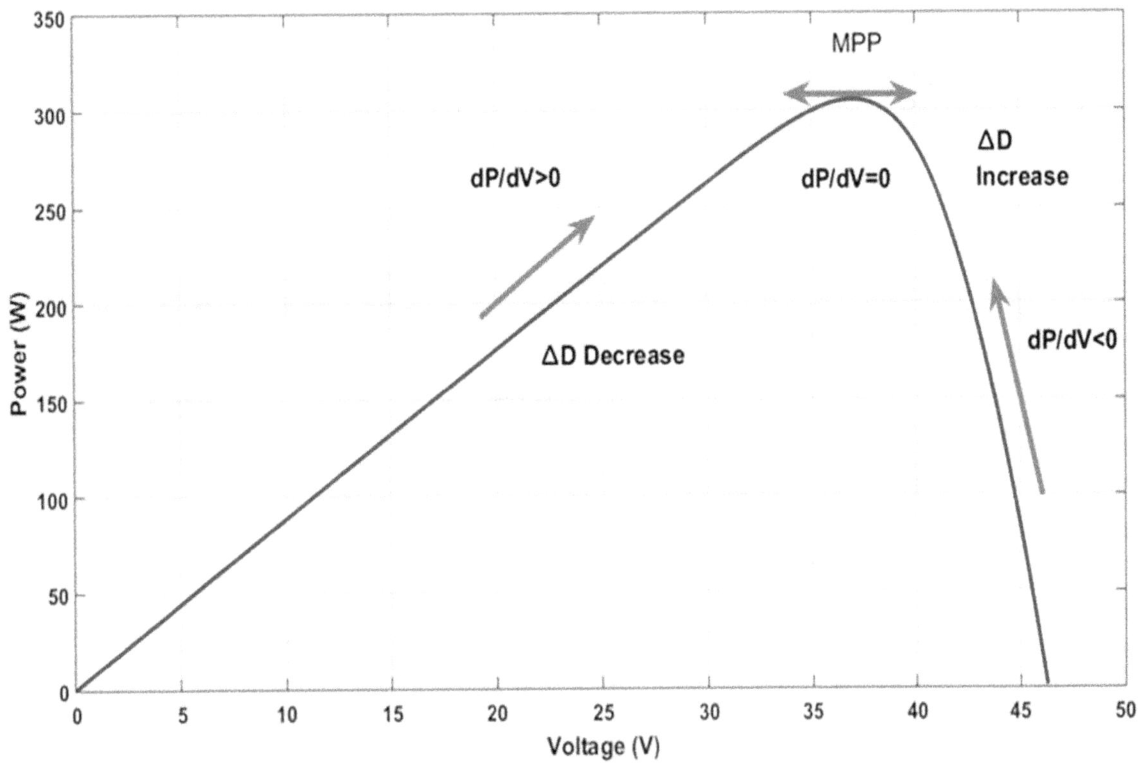

BATTERY CONTROL STRATEGY

Although the main advantages of PV technology, that does not even need to fuel, it is not able to supply the load demand during the absence of the sun. So, the battery is a crucial element in any PV solar power plant, especially in existing critical loads. The main functions of the battery storage system are:

- Securing energy source for operating the monitoring and control devices of the station during the absence of the sun.
- Improving the steady-state and dynamic behavior of the generation system.

OPTIMIZATION TECHNIQUES (GWO&ASO)

The main contribution of using GWO and ASO in this work to attain the optimum gains of FOPI and PI controllers to minimize or eliminate the error.

Figure (8) shows the flow chart that describes the steps of the GWO algorithm. Figure (9) shows the flow chart that describes the steps of ASO algorithm

Figure 7. Modified IC flowchart

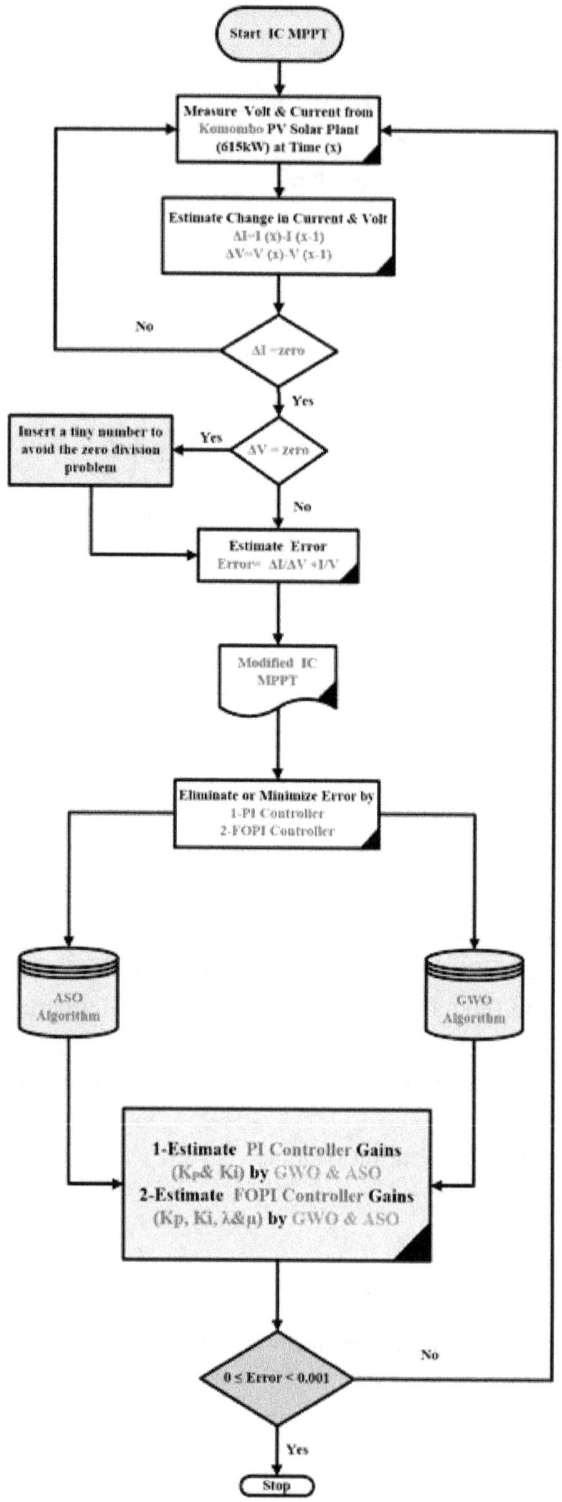

Figure 8. Flow-chart of GWO Algorithm

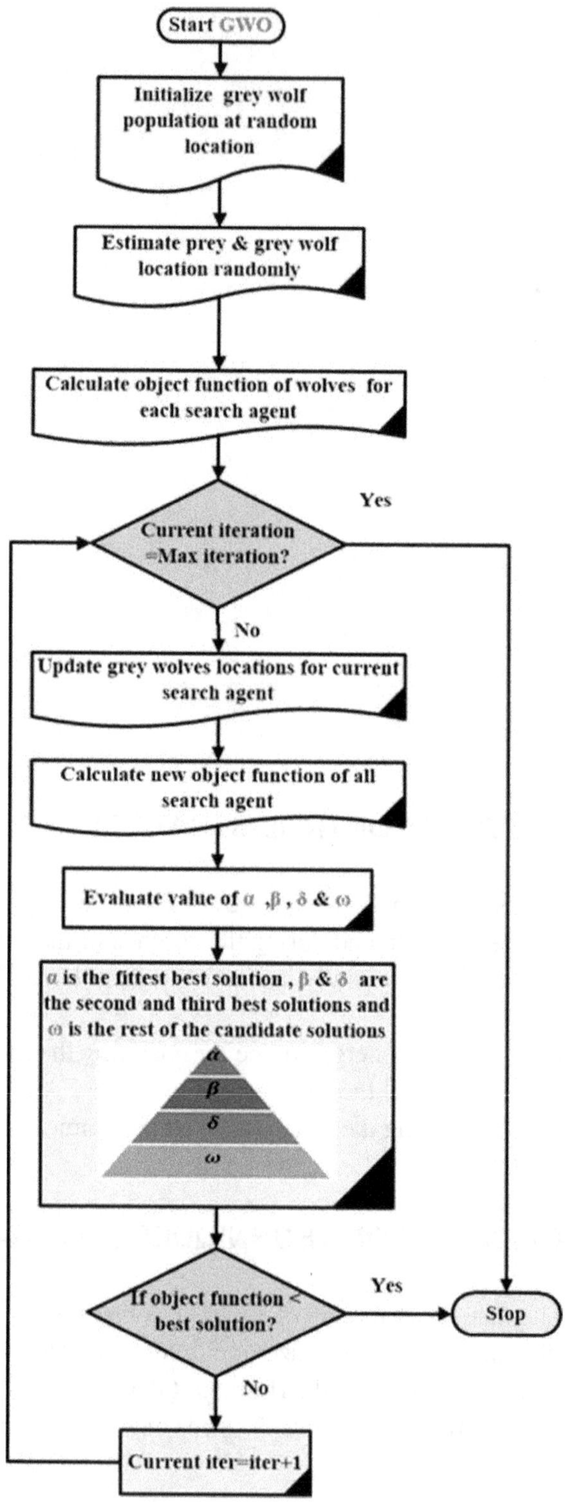

Figure 9. Flow-chart of ASO Algorithm

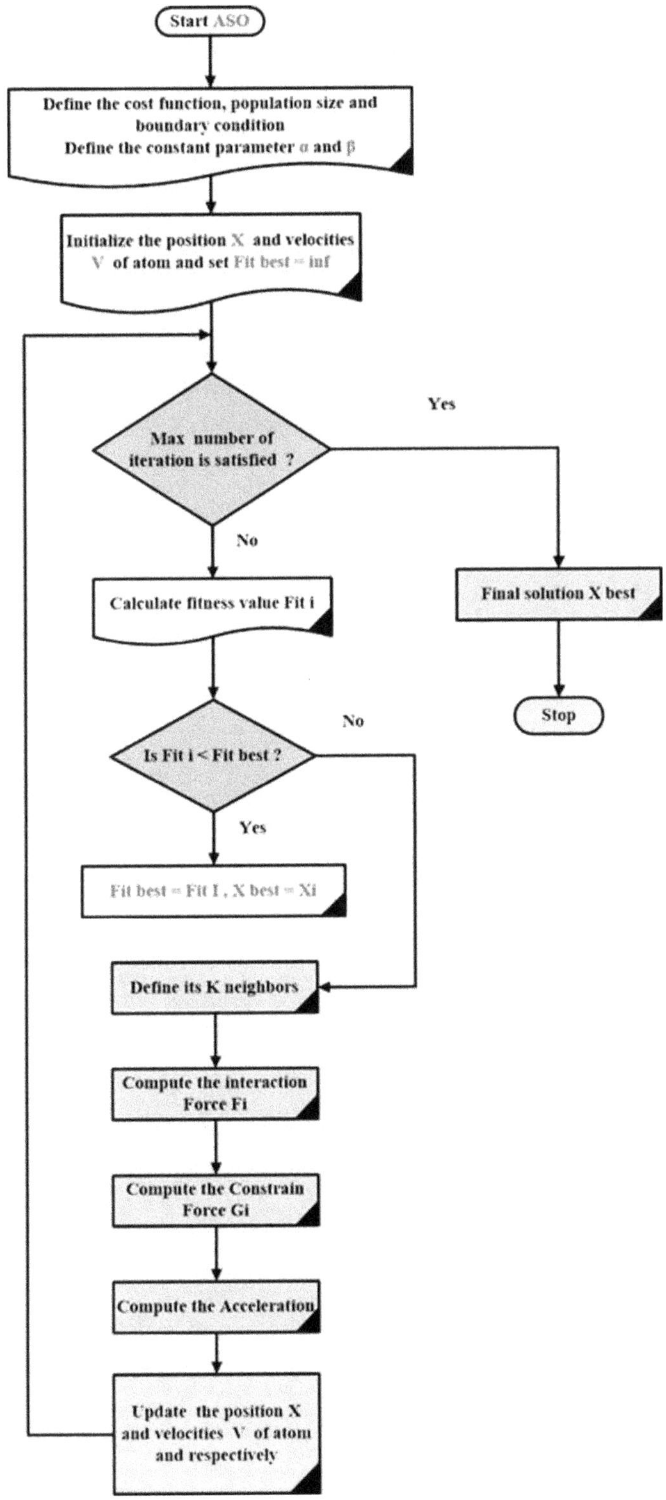

SIMULATION RESULTS AND DISCUSSION

Komombo PV solar plant Simulink model contains various simulation blocks such as PV array, three-phase Voltage Source Inverter (VSI), MPPT controllers, a filter circuit, load, and the utility grid. Figure (10) shows Komombo PV solar plant simulation system. In this system, PV array connected to an 11-kV network via a DC-DC boost converter and a three-phase three-level VSI. In this paper, PV array generates a voltage of 666 V DC for a solar irradiance of 1000W/m^2. 100-kHz DC-DC boost converter increases the voltage from PV nominal voltage (666 V DC at maximum power) to 825 V DC. The optimization of the duty cycle is done using standard and two modified IC MPPT controllers This MPPT method automatically varies the duty cycle to generate the required voltage to extract the maximum power. 1980-Hz 3-level 3-phase voltage source inverter is used to convert the 825 V DC link voltage to 300 V AC. There is 61.5 kVAR capacitor bank filtering harmonics produced by VSI, 1250-kVA, 300V/11kV three-phase coupling transformer to step up the voltage from 300V to 11 kV, and the utility grid 11-kV.

Figure 10. Komombo PV solar plant simulation system (615.3kW)

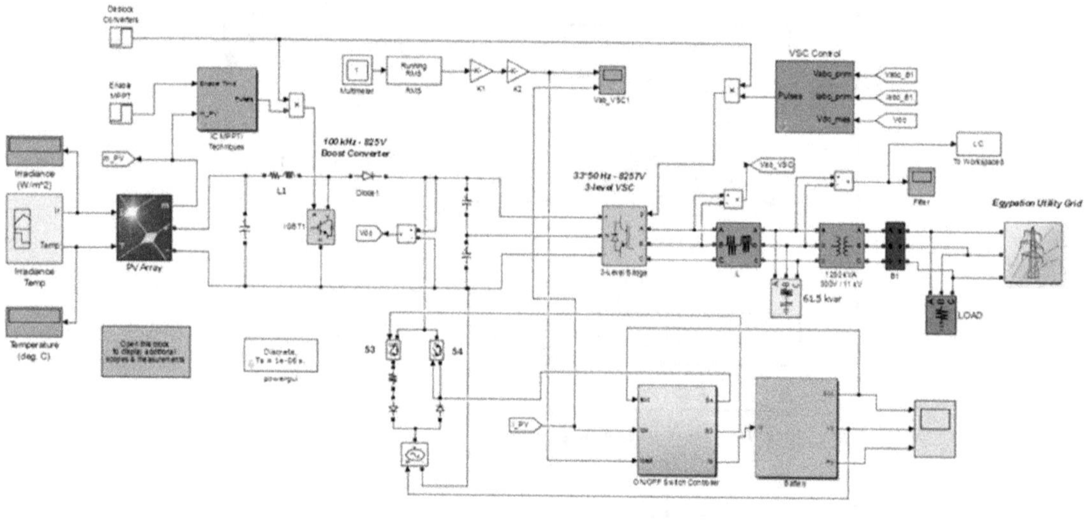

Figure 11. The radiation and temperature waveform for Komombo PV solar plant. Plot (a) shows case.1 at S.T.C. Plot (b) shows case.2 on a winter day in the year at Komombo, Aswan, Egypt (Realistic Case).

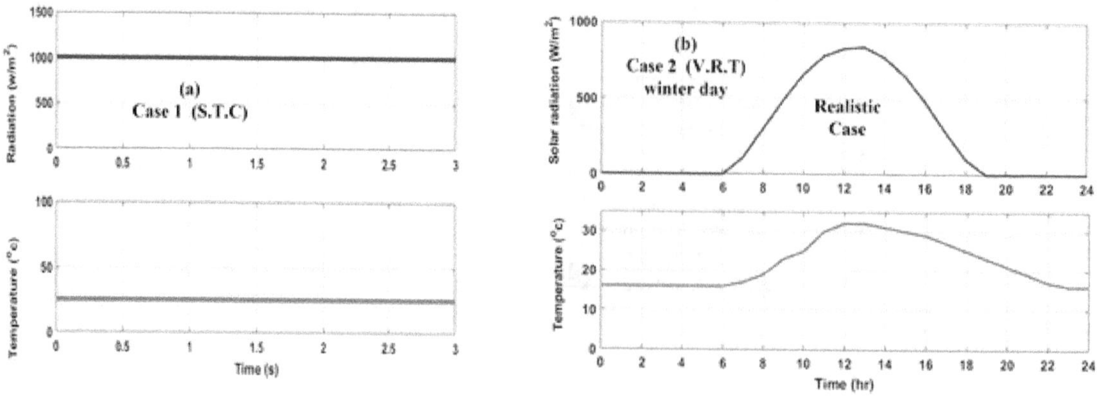

The comparison between different MPPT algorithms requires performance criteria to adjust each algorithm appropriately according to the application, so the average power, transient behavior, Energy Availability (E.A), and Array Fill Factor (A.F.F) are used as the performance criteria for each MPPT algorithm. Komombo simulation model is tested under the following:

Case.1: In this case, the model is tested at Standard Test Condition (S.T.C) 1000 W/m^2&25°C, as shown in Figure.11 (a), where the maximum output from our PV array is 615.3 kW, which satisfied load demand.

Case 2 (Realistic Case): In this case, the model is tested for realistic solar radiation and temperature for a winter day in the year at Komombo, Aswan, Egypt is shown in Figure.12 (b).

After implementing the two modified IC MPPT algorithms on two simulation cases. It can note that IC based on FOPI Controller and ASO performs best results than IC, which based on PI controller & ASO and IC based on FOPI &PI Controllers and GWO according to the transient behavior response analysis as shown in Figure (12). Also, IC-based on FOPI Controller and ASO perform best results according to Average Power (A.P) Array Fill Factor (A.F.F) and Energy Availability (E.A) criteria as shown in Table (5). At STC, the battery storage system does not work because of load demand power equal PV array generation power, as illustrated in the figure.13 (a). However, at variable radiation and temperature level, the output power of PV array does not satisfy the load and the battery storage system will maintain the system power balance by discharging its power as shown in Figure.13 (b).

The optimal values of FOPI and PI controller's gains using GWO and ASO are given in table (6). For the IC_PI technique, the optimum gains of the PI controller achieved by Integral Time Square Error (ITSE), but for the IC_FOPI technique, the optimum gains of the FOPI controller obtained by Integral Time Absolute Error (ITAE). Results from Table (6) perform that IC_FOPI_ASO has the lowest over-shoot compared to IC_FOP_GWO and IC_PI_GWO. Also, the IC_FOPI_ASO arrives at Maximum Power Point (MPP) faster than both IC_FOPI_GWO and IC_PI_GWO.

The comparative study showed that ASO is more efficient than GWO in achieving MPPT because it has,

1- Low number of iteration.
2-low computation time.
3-high speed to reach the best solution.
4- And more stable.

Table 5. A.P, A.F.F and E.A for, two modified IC MPPT algorithms

		IC_PI_GWO	IC_FOPI_GWO	IC_PI_ASO	IC_FOPI_ASO
simulation case study.1 S.T.C	**A.P (kW)**	587.8198	588.1765	588.3644	589.1253
	F.F	0.9552	0.9558	0.9561	0.9856
	E.A (kWh)	1763.5	1764.5	1765.1	1767.3
simulation case study.2 Realistic Case	**A.P (kW)**	158.7952	159.5906	159.9335	160.2365
	F.F	0.2580	0.2593	0.2599	0.2798
	E.A (kWh)	381.1086	383.0177	383.8406	383.9987

Figure 12. The output power with dynamic and transient responses. Plot (a) shows a comparison between two modified IC MPPT at standard test condition. Plot(b) illustrates the comparison between two modified IC MPPT at Realistic Case.

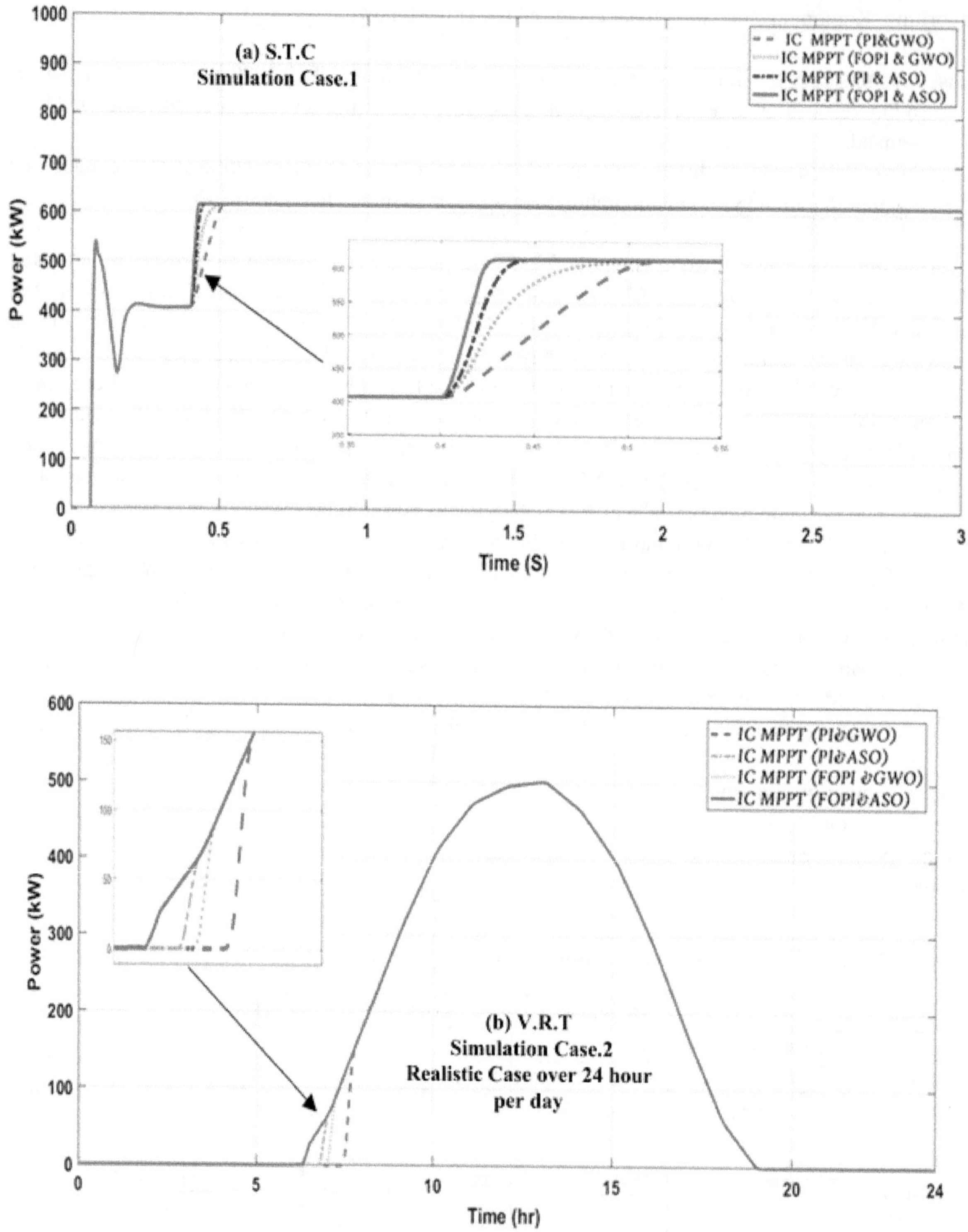

Figure 13. Battery behavior at the two-simulation case study. Plot (a) shows battery output voltage, power and state of charge at standard test condition. Plot (b) shows Battery output voltage, power, and state of charge Realistic Case.

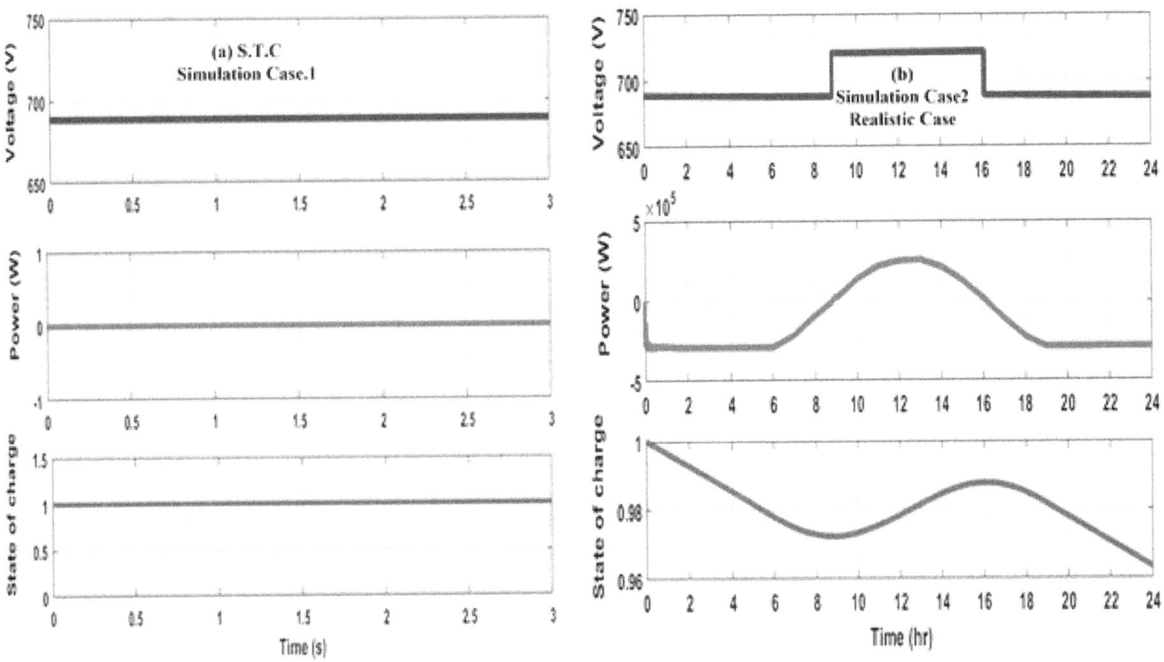

Table 6. IP I&FOPI controllers gains for IC_PI and IC_FOPI using GWO and ASO

	IC_PI_GWO	IC_FOPI_GWO	IC_PI_ASO	IC_FOPI_ASO
KP	0.002958	0.008217	0.00698699	0.0046235
KI	4.23145	5.14423	8.89984	9.2156
λ	1	0.013485	1	0.0373718
Rise time μ sec	0.789	0.789	0.799	0.811
Settling time μ sec	1.0556	0.927	0.892	0.821
Maximum Overshoot	1.055	0.956	0.832	0.825

CONCLUSION

This study presents a real PV solar power plant with a battery backup system connected to the medium voltage distribution network (practical case study). This paper presents two control techniques, namely PI and FOPI, for controlling the IC MPPT algorithm. The results confirm that the modified IC technique based on FOPI control is better than the modified IC technique based on PI control according to transient behavior, Energy Availability (E.A), and array Fill factor (A.F.F). Also, the FOPI controller eliminated forced oscillations and steady-state error, resulting in enhancing the dynamic behavior of the system as compared with the PI controller. This work proposed a new optimization algorithm inspired by grey wolves for determining the optimal controller's parameters. The results showed that ASO can provide highly competitive results compared to a well-known GWO Algorithm. As well as, battery helps in improving the steady-state and dynamic behavior of our system to support PV array in stable and reliable output for our load.

REFERENCES

Abu-Seada, H. F., Mansor, W. M., Bendary, F. M., Emery, A. A., & Moustafa Hassan, M. A. (2013). Application of Particle Swarm Optimization in Design of PID Controller for AVR System. *International Journal of System Dynamics Applications*, *3*(2), 1–17. doi:10.4018/ijsda.2013070101

Azar, A. T., & Serrano, F. E. (2016). Stabilization of Mechanical Systems with Backlash by PI Loop Shaping. *International Journal of System Dynamics Applications*, *5*(3), 20–47. doi:10.4018/IJSDA.2016070102

Bounechba, H., Bouzid, A., Snani, H., & Lashab, A. (2016). Real-time simulation of MPPT algorithms for PV energy system. *Electrical Power and Energy Systems, Elsevier.*, *3*(8), 201–212. doi:10.1016/j.ijepes.2016.03.041

Ebrahim, M. A., Mousa, M. E., & Moustafa Hassan, M. A. (2016). Stabilizing and Swinging-Up the Inverted Pendulum Using PI and PID Controllers Based on Reduced Linear Quadratic Regulator Tuned by PSO. *International Journal of System Dynamics Applications*, *4*(4), 52–69.

Ebrahim, AbdelHadi, & Mahmoud. (2016). Optimal Design of MPPT Controllers for Grid Connected Photovoltaic Array System. *International Journal of Emerging Electric Power Systems*, *4*(17), 511–517.

Ebrahim, Elyan, Wadie, & Abd-Allah. (2017). Optimal design of RC snubber circuit for mitigating transient overvoltage on VCB via hybrid FFT/Wavelet Genetic approach. *Electric Power Systems Research*, *3*(4), 451-461.

El-Din, Aziz. Elsamahy, Hassan, & Bendary. (2017). Enhancement of Turbo-Generators Phase Backup Protection Using Adaptive Neuro-Fuzzy Inference System. *International Journal of System Dynamics Applications*, *4*(6), 58–76.

Elbaset & Hassan. (2015). Design and Implantation of 1oo kW Roof Top Grid Connected PV System. *3rd International Conference on Energy Systems and Technologies*, 16–19.

Galli. (2020). The Application of Systems Engineering to Project Management A Review of Their Relationship. *International Journal of System Dynamics Applications*, *9*(1), 81-95.

Ghabim, Rhif, & Vaidyanathan. (2018). Discrete Time Sliding Mode Control Scheme for Nonlinear Systems With Bounded Uncertainties. *International Journal of System Dynamics Applications*, *7*(2), 15-23.

Gharbi, Gharsellaoui, & Khalgui. (2017). Real-Time Reconfigurations of Embedded Control Systems. *International Journal of System Dynamics Applications*, *4*(6), 71–93.

Jangholi, M., & Sedaghati, A. (2016). Design and Optimization of MPPT Routing in Solar Car Cells with the Maximum Power Point Tracking. *International Journal of Advanced Research in Electrical. Electronics and Instrumentation Engineering*, (5), 4.

Ministry of Electricity & Energy New & Renewable Energy Authority (NREA). (n.d.). *Annual Report 2019/ 020*. Author.

Mohamed, J., & Ahmed, S. (2016). Design and Optimization of MPPT Routing in Solar Car Cells with the Maximum Power Point Tracking. *International Journal of Advanced Research in Electrical. Electronics and Instrumentation Engineering*, *1*(5), 354–363.

Mohamed & Ibrahim. (2012). Optimal Sizing and Economic Analysis of Stand-Alone Photovoltaic System. *The 2012 World Congress on Power and Energy Engineering (WCPEE'12), 4*(5), 188-195.

Mohamed, Ibrahim, Youssef, & Rakha. (2014). Optimal Sizing and Economic Analysis of Different Configurations of Photovoltaic Systems. *International Review of Electrical Engineering, 4*(9), 111-121.

Montoya, D. G., Carlos, A. R. P., & Giral, R. (2016). Maximum power point tracking of photovoltaic systems based on the sliding mode control of the module admittance. *Electric Power Systems Research, Elsevier*, *136*, 125–134. doi:10.1016/j.epsr.2016.02.001

Mousa, M. E., Ebrahim, M. A., & Hassan, M. M. (2017). Optimal Fractional Order Proportional Integral Differential Controller for Inverted Pendulum with Reduced Order Linear Quadratic Regulator. In *Fractional Order Control and Synchronization of Chaotic Systems* (pp. 225–252). Springer International Publishing. doi:10.1007/978-3-319-50249-6_8

Nabavi, Branch, Balochian, & Branch. (2018). The Stability of a Class of Fractional Order Switching System with Time-Delay Actuator. *International Journal of System Dynamics Applications, 7*(1), 85-99.

Naglaa, K. (2014). Load Frequency Control in Power System via Improving PID Controller Based on Particle Swarm Optimization and ANFIS Techniques. *International Journal of System Dynamics Applications*, *3*(3), 1–24. doi:10.4018/ijsda.2014070101

Santos, Santos, Silva, & Ferreira. (2020). A System Dynamics Model for Sales and Operations Planning: An Integrated Analysis for the Lime Industry. *International Journal of System Dynamics Applications, 9*(1), 1-15.

Sawada. (2015). Energy Storage Management in Grid Connected Solar Photovoltaic System. *Journal of Engineering Research and Applications, 5*, 1-5.

Srinivas, Lakshmi, & Ramesh. (2016). Simulation of Incremental Conductance MPPT Algorithm for PV Systems using Lab View. *International Journal of Innovative Research in Electrical, Electronics. Instrumentation and Control Engineering*, *1*(4), 34–38.

Yatimi & Aroudam. (2018). Standalone Photovoltaic System with Maximum Power Point Tracking: Modeling and Simulation. *International Journal of System Dynamics Applications, 7*(3), 94-117.

Chapter 20
Modeling, Analysis, of Induction Motor's Stator Turns Fault Using Neuro-Fuzzy

Hussein. A. Taha
https://orcid.org/0000-0001-9456-4795
Department of Electrical Engineering, Sohag University, Egypt

M. E. Ammar
Electric Power Department, Cairo University, Egypt

M. A. Moustafa Hassan
https://orcid.org/0000-0001-6116-6707
Electric Power Department, Cairo University, Egypt

ABSTRACT

This chapter discusses modeling and analysis methods for fault detection and diagnosis of stator inter-turn short circuit in three-phase induction machines. dq frame was used to model the induction motor for both health and fault cases to facilitate recognition of motor current and simulate motor environment. Fault diagnosis system was designed with adaptive neuro-fuzzy inference system (ANFIS) to provide an efficient online diagnostic tool. ANFIS diagnostic tool was trained with simulated data that generated by induction motor healthy and faulty models. Approached tool is verified online with a motor under different loading conditions. It determines the fault severity values using the motor current signature analysis (MCSA). Developed tool performance is investigated with a case study of two HP three-phase induction motor using Matlab/Simulink® software.

DOI: 10.4018/978-1-7998-5788-4.ch020

1. INTRODUCTION

Induction machines are most common use machines in all industries. It is utilized with ninety percent over electrical motor industrial applications(Abdel, Hassan, & Bendary, 2018; Hussein, Ammar, & Hassan, 2017; Joshi & Talange, 2016). Induction machines are low price and reliability. Analysis of induction machines' faults is crucial to minimize downtime and cost of damages in additions to early detections (Abdel, El Samahy, Hassan, & El Bendary, 2016; Abdel et al., 2018; Milić, 2020). Machine faults can be divided into two categories, internal and external. The external faults can be determined as electrical (transient voltage, unbalanced voltage, voltage fluctuations…), mechanical (pulsating load, overloading, and poor mounting…) and environmental (Humidity, temperature, cleanliness…) as discussed in (Atig, Bouheraoua, & Fekik, 2018; Geethanjali & Ramadoss, 2019; He, Xie, & Luo, 2020). The induction machine internal faults contain two types of faults mechanical such as rotor strikes, bearing faults, and Eccentricity and electrical as Stator winding faults, Broken Rotor Bar, Broken End ring, and Rotor winding faults for the wound rotor machine.

Induction machine faults include winding faults, unbalanced stator and rotor, broken rotor bars, eccentricity and bearing faults. The failure due to the stator and bearing fault breakdown are considering the dominant fault incidents(Abdel et al., 2016; Joshi & Talange, 2016; Lázaro & Pinilla, 2020). The main goal of fault diagnosis is to determine the fault as early as possible and classify it correctly to facilitate localization. In order to ensure reliable diagnosis, advanced strategies must be used. Fault detection and diagnosis methods fall under three categories model-based fault detection and history-based method to build up an online signal based model (Enshaei, Hammad, & Naderkhani, 2020; Lázaro & Pinilla, 2020; Lei et al., 2020). The model-based technique requires a system modeling that is derived from the physics of the system to describe the motor in both quantitative and qualitative (Lei et al., 2020; Rahimi, Kumar, & Alighanbari, 2020).

Model-based fault diagnosis methods built on motor observer design, based on induction motor model (Han, Li, Dahhou, Cabassud, & He, 2020). Figure 1 shows schematic of model-based fault detection method. The observer estimates the system states in healthy operating conditions though the system state vector is measured. The error signal has a value when there is a difference between the measured and estimated system states. When the motor is working under healthy mode of operation the difference between actual and estimated states is almost zero given that the observer is well designed. However with inter-turn fault happen the real motor environment will be changed, but the observer keeps as it is. Hence the error signal will have a significant value that indicates an inter-turn fault. Model-based fault diagnosis method can detect the severity of the fault independently of loading condition and supply voltage quality. But it needs an exact mathematical model of the faulty motor with variable fault severity. Genetic algorithms were used in (Raie & Rashtchi, 2002) with classical model of the induction motor to estimate the severity of the stator inter-turn fault, where they used fitness function depends on the motor current and speed signals.

However model-based methods needless for a priori data. But it has a complexity of observer design. Besides the availability of a priori healthy and faulty operation data can support in the fine-tuning of the model-based observers.

The process history-based method requires enough history of measurements or system data. Parameter estimation is based on the input-output data where a mathematical equation could be realized using the measurements and decisions. Signal based models depend on online process system measurements. Online diagnosis use highly efficient sensors and Data Acquisition Card (DAC) to identify system fault

(Hashemi & Rahideh, 2019; Jain & Raj, 2018; Karner, Glawar, Sihn, & Matyas, 2019). The main scope of fault diagnosis is to determine the fault as early as possible and diagnosis it in a correct way to facilitate the localization of the fault. To verify the safe and reliable process, advance fault diagnosis strategies must be used. Different methods of fault detection model base and knowledge base. Model-based technique is considered to have a mathematical model that describes the process system. A mathematical model is derived from physics of the system. This model can describe the system in both quantitative and qualitative models (Abdel et al., 2018; Boudinar, Bendiabdellah, & Benouzza, 2019; Joshi & Talange, 2016; Kumar, Ramya, & Irfan, 2019).

Figure 1. Model-Based Fault Detection Schematic

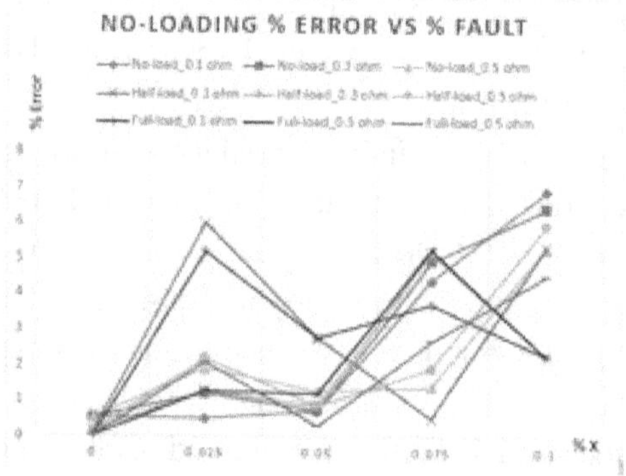

Process history-based methodology does not depend on any analytical relations to refer to the physical system. Process history also called in literature as data driven approach. It is depending on feature extraction from the measured data in order to build a model detect motor faults. Data driven models could be categorized to statistically based and artificial intelligence based (Ellefsen, Æsøy, Ushakov, & Zhang, 2019; Enshaei et al., 2020). Figure 2 shows data-driven process.

Historical data Statistical recognizes any abnormalities comparing of measured data mean and variance relative to healthy data conditions mean and variance. Faulty data has a different means and/or variance than healthy data. The most common statistical way to detect faults is outlier detections. The other historical data model is built on AI. Researchers dominate the two methods of process history and called as expert systems.

Figure 2. Data Driven Model Process

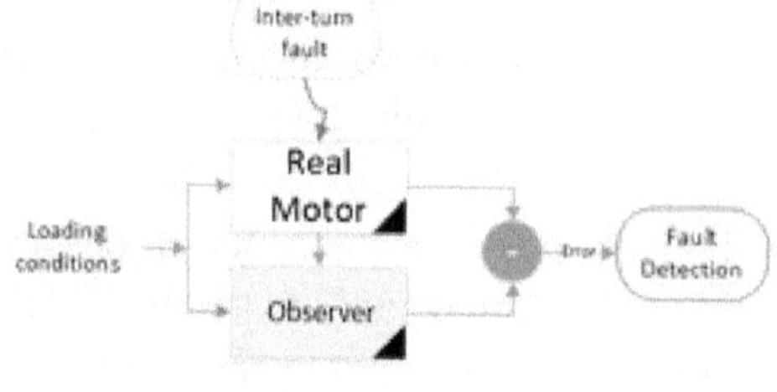

Expert system uses the knowledge-based I/O data to provide the decision of fault diagnosing. It uses different algorithms to detect the faults such as statistical learning, neural network, fuzzy systems, Neuro Fuzzy and deep learning (Abdullah & Iqbal, 2020; Alain, Azar, Kengne, & Bertrand, 2020; Saufi, Ahmad, Leong, & Lim, 2020; Srinivasan & LD, 2020; Zhou, Yang, Fujita, Chen, & Wen, 2020). Knowledge-based data should contain variable states of process systems for both faulty and healthy cases that will lead to accurate diagnosis. It uses the logical rules (IF condition THEN decision) rules to determine the fault as reviewed in (Atig et al., 2018; Boudinar et al., 2019; Joshi & Talange, 2016; Kumar et al., 2019).

The experimental or simulation measurements can give input output data for the process system. System characteristics are obtained based on the measurements (Boudinar et al., 2019; Kumar et al., 2019). Spectral Analysis and Signal Based of Process System Measurements parameters should have a specific frequency spectral at the normal operation. At the fault state, the process measurement for the same parameter deviates from the healthy case. This eases the fault determination. The spectral analysis is used for vibration and current signal analysis. It uses signal processing mathematical operations of Discrete Fourier Transform (DFT), Fast Fourier Transform (FFT) and Short Fourier Transform (SFT) as reviewed by (Atig et al., 2018). The machine's signals of vibration analysis, motor current signature, instantaneous partial and total power, wavelet analysis and electrical torque harmonic analysis are used for detecting the faults. Signal based approach requires high efficiency sensors. This technique can be divided into pattern recognition and decision-making to identify the fault as mentioned in (Atig et al., 2018). Statistical analysis uses the stochastic methods to distinguish the machine faults. It uses Principle Component Analysis (PCA) Partial Least Square method (PLS) and Fisher Discriminate Analysis (FDA) to distinguish process system fault. The PCA has been used in the fault diagnosis research area, but the partial least square method has better configuration for small scale classification. From a theoretical point of view, the FDA has an advantage over the other methods. Those techniques are low-dimensional techniques. The PCA is most accurate low-dimensional technique leading to abnormal fault analysis as explained in (Shi et al., 2016). FDA: It is an optimal lower-dimensional technique. It represents fault diagnosis through data class classification. Each class of data is collected from process system output in faulty and healthy cases. It is one of the pattern classifications methods.

PLS: It is a type of data decomposing. The covariance between predictor of the system and the predicted value is used as the fault detection criteria. After computing the predicted block based PLS1 &PLS2.The fault analysis is done based on detecting the column to predict block which has maximum value among each observation index. Moreover, support vector machine, linear regression, KNN and classification statistical learning methods are used to diagnosis and detections of faults for industrial systems (Nguyen & Park, 2020; Taha, Sakr, & Yacout, 2019). Figure 3 shows a schematic of detection and diagnosis methods. Block of (*1): represent the quantitative model base methods, it depends on the mathematical modeling of the process system like parameter estimation and state space observer. Block of (*2): represent the qualitative model base methods, it depends on physical property of the process system like the causal relation between input and output of the system and graphical tools. Block of (*3): represent the qualitative database methods, it did not depend on the mathematical modeling. It uses the model historical data and hypothesis trends. Block of (*2): represent the quantitative database methods, it depends on process system data classification and recognition use of statistical methods or neural network. The second method depends on enough system history of measurement. It doesn't need any parameter estimation based on the process system physics, mainly the input-output data. Depending on the measured and decision (actual and systems theoretical), a mathematical equation could be realized.

This equation is used for pattern recognition (Gangsar & Tiwari, 2019; Nguyen & Park, 2020; Saufi et al., 2020; Zhou et al., 2020)

The simplified physical method is a method to estimate the system parameters and state variables. This method depends on the Lumped parameters assumption. The state variables estimation requires analytical knowledge of the system process and linearization of the process around the operating point(Chen, Fink, & Sansavini, 2017; Liang, Liu, Sheng, & Jiang, 2019)

Figure 3. Fault Detection and Diagnosis Methods

The most common condition monitoring techniques as reviewed by (Atig et al., 2018; Geethanjali & Ramadoss, 2019) are thermal monitoring, vibration monitoring and electrical monitoring. Thermal measure is the local or bulk temperature of the machine, it requires the installation of thermocouples. Small machines do not use this technique. It is considered as an indirect machine fault diagnosis method for detecting inter-turn faults and bearing. It is a low efficiency method for inter-turn fault detection as temperature raise rate is too small. The turn fault will lead to progressive faults and total phase turns will be damaged. Also for bearing fault detection the machine temperature increases due to friction as presented in (Atig et al., 2018; Geethanjali & Ramadoss, 2019).Vibration monitoring is mechanical vibration monitoring at a specific frequency which can be detectable in abnormalities. It is used for detecting and monitoring gear faults, bearing faults and unbalanced rotors. This technique uses three types of measurement sensors: displacement, velocity or acceleration. It is an optimal methodology for mechanical fault detection, but the measurement sensors are costly as discussed by (Choudhury, 2019; Kumar et al., 2019). It is a commonly used monitoring technique for detection of several of machine and inverter faults. Electrical monitoring depends on current, voltage and flux monitoring. The analysis of motor current signature analysis (MCSA) detects the inter-turn faults, unbalanced voltage, air gaps and mechanical vibration as well. Through current spectral analysis by FFT, the current magnitude and/or harmonic magnitude have different values according to each fault. Mechanical vibration leads to air gap uniformity change, which causes inductance change and finally the stator current change as given in (Atig et al., 2018; Geethanjali & Ramadoss, 2019; Hussein et al., 2017).

The failure due to stator winding breakdown is occurred once out of three induction motor faults as shown in (Boudinar et al., 2019; Joshi & Talange, 2016; Kumar et al., 2019) The prediction of stator faults will save the high maintenance cost. There are a lot of approaches to diagnose the stator turns fault. Some methods are based on the stator currents and Fast Fourier Transform (FFT) while other methods use the torque profile analysis and vibration study. Recent research work investigated the use of intelligent control, Fuzzy Logic (FL), Neural Network (NN), combination of FL and NN and adaptive control in fault analysis.

Expert systems use knowledge-based I/O data to provide the decision of fault diagnosing. Different algorithms are used to detect the faults (Neural Network, fuzzy system, Neuro Fuzzy, etc.). Knowledge-based data should contain the various states of the process for both faulty and healthy cases to lead to accurate diagnosis. The fuzzy expert system uses logical rules (IF condition THEN decision) to detect and diagnosis the fault as discussed in (Alain et al., 2020; Azar, Serrano, Flores, Vaidyanathan, & Zhu, 2020; Fatma, Ghabi, & Dhouibi, 2020; Srinivasan & LD, 2020). Some fault diagnosis methods are based on stator currents and fast Fourier transforms (FFT) while other methods use torque profile analysis and a vibration study (Hashemi & Rahideh, 2019; Jain & Raj, 2018; Joshi & Talange, 2016). Recent research work investigated the use of intelligent control, Fuzzy logic (FL), Artificial Neural Network (ANN), combination of FL and NN and adaptive control in fault analysis (Das & Ghosh, 2020; Lucke, Stief, Chioua, Ottewill, & Thornhill, 2020; Rahimi et al., 2020)

ANN has data classification and pattern recognition capabilities. ANN's accuracy depends on the network structure as well as training data (Pourpanah et al., 2019; Siddique & Adeli, 2013; Tambouratzis, Giannatsis, Kyriazis, & Siotropos, 2020). For accurate fault detection, the network should be trained with input output data for healthy and faulty cases. Fuzzy logic uses the IF antecedent the consequent rules. The fuzzy rules are the core of fuzzy logic and must be developed by an expert (Siddique & Adeli, 2013). The rules verify the main knowledge base to have an efficient fuzzy logic system for the applied process (Guedes, Castoldi, Goedtel, Agulhari, & Sanches, 2019; Hussein et al., 2017) . In addition, a combination between ANN and fuzzy system was proposed to introduce learning ability to fuzzy systems. The linguistic rules and/or membership functions are permitted to learn from training data. The fuzzy system is an expert system based on rules, on the other side the NN knowledge is based on training with historical data (Abdullah & Iqbal, 2020; Siddique & Adeli, 2013). In a Neuro-Fuzzy system, an implementation of FL, fuzzy rule parameters and structure are identified based on ANN learning and training data. Adaptive Neuro-Fuzzy Inference System (ANFIS) use ANN learning to identify FL system parameters providing advantageous features such as easier implementation, fast learning, and accuracy (Alain et al., 2020; Fatma et al., 2020; Srinivasan & LD, 2020). The next industrial generation is going deeply for historical data analysis, where it is lower computational time for online possessing (Karner et al., 2019; Milić, 2020; Rahimi et al., 2020). Industrial Internet of Things (IIOT) and digitalization facilitate connectivity of all industrial systems besides data availability. New Industrial Revolution concepts dominated with approached diagnostic tool and it could support maintenance and decision-making strategies.

2. MAIN FOCUS OF THE CHAPTER

This chapter discusses fault detection and diagnosis for induction motors stator turn faults within acceptable accuracy and detection time. It shows the two methods of fault detection's model-based and historical data and then verification of proposed neuro-fuzzy model. Chapter is organized as solutions and recommendations including mathematical modelling of a three-phase induction motor, ANFIS proposed model, and system performance analysis are developed in section 3. Conclusion and future research are conducted by section 4.

3. SOLUTIONS AND RECOMMENDATIONS

3.1 Modelling of A Three Phase Induction Motor

A dq frame is used to reduce the complexity of differential equations where the original stator and rotor frames are transformed to a common frame that rotates with arbitrary angular velocity (Ortega, Pyrkin, Bobtsov, Efimov, & Aranovskiy, 2020; Pourpanah et al., 2019).

3.1.1 Healthy Case

The three phases of a healthy motor are symmetrical as all phases have the same number of turns (Das & Ghosh, 2020; P. Krause, Wasynczuk, Sudhoff, & Pekarek, 2013; P. C. Krause, Wasynczuk, Sudhoff, & Pekarek, 2002; Ortega et al., 2020) The rotor is a balanced star connection cage rotor.

Stator and cage rotor voltage equations in ABC frame are

$$v^{s}_{abc} = [v^{s}_{a}\ v^{s}_{b}\ v^{s}_{c}\ ]^{T} \quad (1)$$

$$v^{s}_{abc} = r^{s}_{abc}\ i^{s}_{abc} + p\lambda^{s}_{abc} \quad (2)$$

Where v^{s}_{abc}, r^{s}_{abc} and λ^{s}_{abc} stator voltage, resistance and flux linkage in three phase ABC frame respectively

The cage rotor is symmetrical with shorted ring

$$v^{r}_{abc} = [\ v^{r}_{a}\ v^{r}_{b}\ v^{r}_{c}\ ]^{T} \quad (4)$$

$$0 = r^{r}_{abc} i^{r}_{abc} + p\lambda^{r}_{abc} \quad (5)$$

Where v^{r}_{abc}, r^{r}_{abc} and λ^{r}_{abc} rotor voltage, resistance and flux linkage in three phase ABC frame respectively

The parameters with subscripts a, b, c refer to the abc phases and the superscription of s, r refers to the stator and rotor respectively. Figure 4 shows symmetrical star connected three-phase induction motor and rotor and stator abc phases relations.

The stator and rotor inductances and flux linkage relation in ABC frame are

$$\begin{bmatrix} \lambda^{s}_{abc} \\ \lambda^{r}_{abc} \end{bmatrix} = \begin{bmatrix} L(\theta)^{s}_{abc} & L(\theta)^{sr}_{abc} \\ L(\theta)^{sr\ T}_{abc} & L(\theta)^{rr}_{abc} \end{bmatrix} \cdot \begin{bmatrix} i^{s}_{abc} \\ i^{r}_{abc} \end{bmatrix} \quad (6)$$

Where the $L(\theta)^{s}_{abc}, L(\theta)^{sr}_{abc}, L(\theta)^{rr}_{abc}$, and Θ are stator self-inductance, stator to rotor mutual inductance, rotor self-inductance and rotor position respectively (P. Krause et al., 2013).

The Stator Voltage Equations in dq frame voltage equations of stator and rotor are derived in equation (7) as given in (P. Krause et al., 2013). Based on balanced Y 3ph induction motor, the neutral current has zero value $I^{s}_{0}=I^{r}_{0}=0$. According to balanced stator condition, the turns of each phase are equal (Na= Nb= Nc= Ns).

Figure 4. Star Connected Symmetrical Three Phase Induction Motor, Adapted from (P. Krause et al., 2013)

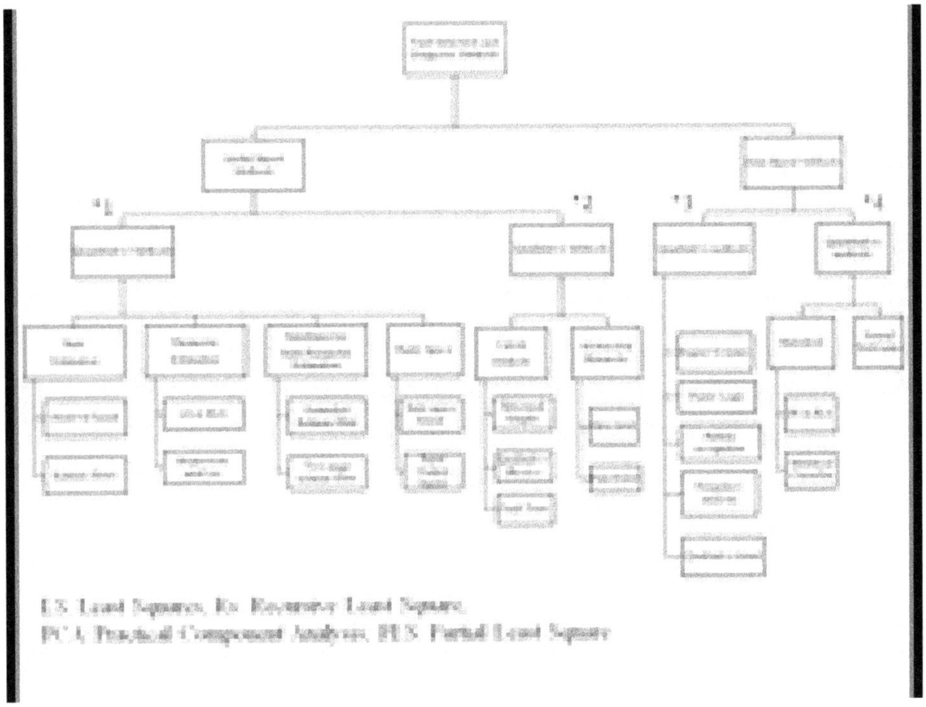

The supply voltage in the dq frame is:

$V_q^s=2/3[V_a^s -0.5(V_b^s+V_c^s)]$, $V_d^s=1/\sqrt{3}\,(-V_b^s+V_c^s)$,

$$V_0^s=1/3\,(V_a^s + V_b^s+ V_c^s) \tag{7}$$

The induction motor inductance in dq frame is given by:

$$L^{ss}_{qd0} = \begin{bmatrix} L^{ss}_{11} & L^{ss}_{12} & L^{ss}_{13} \\ L^{ss}_{21} & L^{ss}_{22} & L^{ss}_{23} \\ L^{ss}_{31} & L^{ss}_{32} & L^{ss}_{33} \end{bmatrix}, L^{sr}_{qd0} = \begin{bmatrix} L^{sr}_{11} & L^{sr}_{12} & 0 \\ L^{sr}_{21} & L^{sr}_{22} & 0 \\ L^{sr}_{31} & L^{sr}_{32} & 0 \end{bmatrix}, L^{rr}_{qd0} = \begin{bmatrix} L^{rr}_{11} & 0 & 0 \\ 0 & L^{rr}_{22} & 0 \\ 0 & 0 & L^{rr}_{33} \end{bmatrix} \tag{8}$$

The Rotor resistance is

$$R^r_{qd0} = r^r\, I_{3x3} \tag{9}$$

The Flux Equations are:

$$\begin{bmatrix} \lambda_q^s \\ \lambda_d^s \\ \lambda_q^r \\ \lambda_d^r \end{bmatrix} = \begin{bmatrix} L_{11}^{rs} & L_{12}^{ss} & L_{11}^{sr} & L_{12}^{sr} \\ L_{21}^{ss} & L_{22}^{ss} & L_{21}^{sr} & L_{22}^{sr} \\ L_{11}^{rs} & L_{12}^{rs} & L_{11}^{sr} & 0 \\ L_{21}^{rs} & L_{22}^{rs} & 0 & L_{22}^{rr} \end{bmatrix} \begin{bmatrix} i_q^s \\ i_d^s \\ i_q^r \\ i_d^r \end{bmatrix} \tag{10}$$

Where $\lambda^s_{dq,}$ λ^r_{dq} are flux linkage in dq frame for stator and rotor respectively

Electromechanical Torque and speed are given by equation (11) in dq frame (P. Krause et al., 2013; Pourpanah et al., 2019) (Krause et al., 2013; Arkan et al.., 2005):

$$T_d = (3/2)\,(P)\,(\lambda^s_d i^s_q - \lambda^s_q i^s_d) \tag{11}$$

$$P\, w_r = P/\,(J)\,(T_d - T_L - T_{damp}) \tag{12}$$

Where P is machine pair poles

3.1.2 Inter-Turn Fault Case

Under inter-turn fault in phase a, motor parameters (stator resistance, inductance and mutual inductance between all phases and the faulty phase) change as shown in Figure 5.

$$X\ (\text{fault }\%) = N_a\ (\text{healthy turns})/N_s\ (\text{phase turns}) \tag{13}$$

$$r_{sh} = X\, r_{a)\,healthy} = r_{a)\,f},$$

$$L_{a1a1)\,f} = (1-X)^2\, L_{asas)\,healthy} = L'_{asas}, \tag{14}$$

$$L_{a2a2)\,f} = X^2\, L_{m)\,healthy} = L_{shsh},$$

$$L_{a1a2)\,f} = (1-X)\, X\, L_{m)\,healthy} = L_{assh},$$

$$L_{a1sr)\,f} = (1-X)\, Lm_{)\,healthy},\ L_{m)\,healthy} = L'_{asr} + L_{shar} \tag{15}$$

Where:

r_{sh} and $L_{a1a1)f}$ are faulty phase resistance and self-inductance for faulty area of phase A
$L_{a1a2)f}$ is mutual inductance between faulty area and healthy area of phase A
$L_{a1sr)f}$ is mutual inductance between phase A and any rotor phases

Figure 5. Three Phase Induction Motor Stator Winding Fault Schematic.

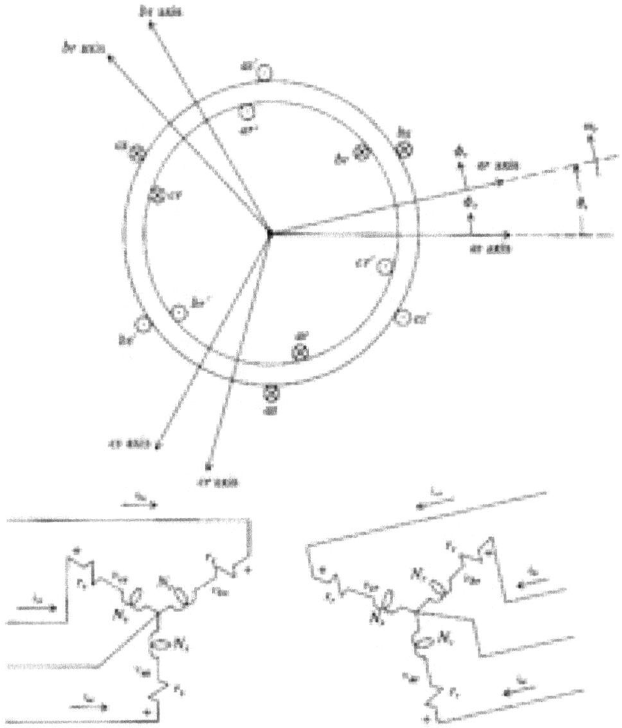

Figure 6. Three-phase Induction Motor In dq Frame with Turn Fault In q Phase Represent Phase a Fault

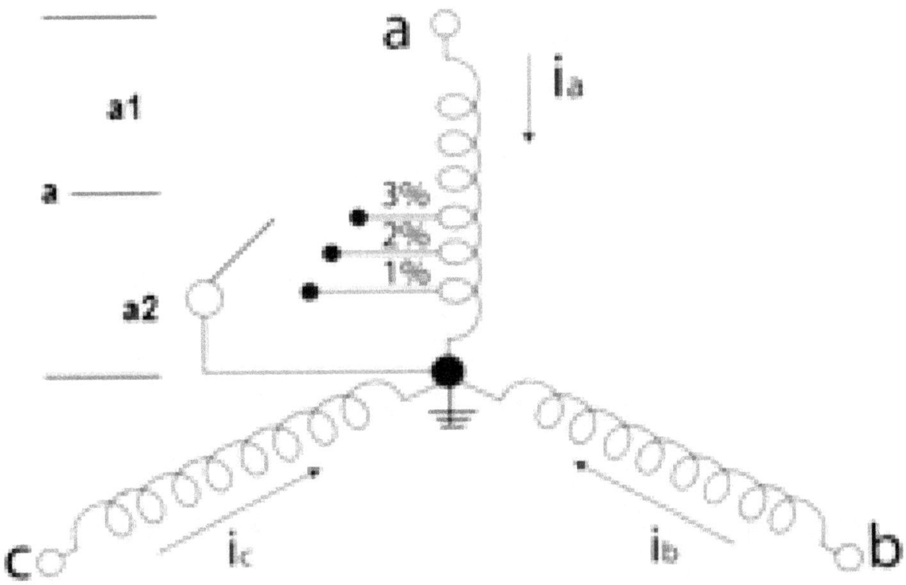

Faulty phase is represented by q axis as depicted by Figure 6.

The flux equation in dq frame after taking the shorted turns in consideration is:

$$\begin{bmatrix} \text{»}_q^{sh} \\ \text{»}_q^{s} \\ \text{»}_d^{s} \\ \text{»}_q^{r} \\ \text{»}_q^{r} \end{bmatrix} = \begin{bmatrix} L_q^{sh} & L_q^{ssh} & 0 & L_q^{shr} & 0 \\ L_q^{ssh} & L_q^{s} & 0 & L_q^{sr} & 0 \\ 0 & 0 & L_d^{s} & 0 & L_q^{sr} \\ L_q^{shr} & L_q^{sr} & 0 & L_q^{r} & 0 \\ 0 & 0 & L_d^{sr} & 0 & L_d^{r} \end{bmatrix} \begin{bmatrix} I_q^{sh} \\ I_q^{s} \\ I_d^{s} \\ I_q^{r} \\ I_q^{r} \end{bmatrix}$$

$$\begin{bmatrix} \text{»}_q^{sh} \\ \text{»}_q^{s} \\ \text{»}_d^{s} \\ \text{»}_q^{r} \\ \text{»}_q^{r} \end{bmatrix} = \begin{bmatrix} L_q^{sh} & L_q^{ssh} & 0 & L_q^{shr} & 0 \\ L_q^{ssh} & L_q^{s} & 0 & L_q^{sr} & 0 \\ 0 & 0 & L_d^{s} & 0 & L_q^{sr} \\ L_q^{shr} & L_q^{sr} & 0 & L_q^{r} & 0 \\ 0 & 0 & L_d^{sr} & 0 & L_d^{r} \end{bmatrix} \begin{bmatrix} I_q^{sh} \\ I_q^{s} \\ I_d^{s} \\ I_q^{r} \\ I_q^{r} \end{bmatrix}$$

$$\begin{bmatrix} \text{»}_q^{sh} \\ \text{»}_q^{s} \\ \text{»}_d^{s} \\ \text{»}_q^{r} \\ \text{»}_q^{r} \end{bmatrix} = \begin{bmatrix} L_q^{sh} & L_q^{s} & 0 & L_q^{shr} & 0 \\ L_q^{ssh} & L_q^{s} & 0 & L_q^{sr} & 0 \\ 0 & 0 & L_d^{s} & 0 & L_q^{sr} \\ L_q^{shr} & L_q^{sr} & 0 & L_q^{r} & 0 \\ 0 & 0 & L_d^{sr} & 0 & L_d^{r} \end{bmatrix} \begin{bmatrix} I_q^{sh} \\ I_q^{s} \\ I_d^{s} \\ I_q^{r} \\ I_q^{r} \end{bmatrix} \quad (16)$$

Stator resistance is given by:

$$\begin{bmatrix} r_q^{sh} \\ r_q^{s} \\ r_d^{s} \end{bmatrix} = \begin{bmatrix} \frac{2}{3} r_q^{sh} & 0 & 0 \\ 0 & r_{11}^{s} & r_{12}^{s} \\ 0 & r_{21}^{s} & r_{22}^{s} \end{bmatrix}$$

$$\begin{bmatrix} r_q^{sh} \\ r_q^{s} \\ r_d^{s} \end{bmatrix} = \begin{bmatrix} \frac{2}{3} r_q^{sh} & 0 & 0 \\ 0 & r_{11}^{s} & r_{12}^{s} \\ 0 & r_{21}^{s} & r_{22}^{s} \end{bmatrix} \quad (17)$$

Where

L_q^{ssh} Mutual inductance between faulty area and healthy area of phase A in dq frame

L^{shr}_q Mutual inductance between phase A faulty area and any rotor phases in dq frame
L^{sr}_q Mutual inductance between phase A healthy area and any rotor phases in dq frame
L^{sh}_q Self-inductance of phase A faulty area in dq frame
L^{s}_q Self-Inductance for healthy area of phase A in dq frame

Flux linkage derived from equation (10) is:

$P\lambda^{sh}_q = V_q^{sh} - r^{sh} i^{sh}_q$,

$P\lambda^{s}_q = V_q^{s} - V_q^{sh} - r^{s}_{11} i^{s}_q - r^{s}_{12}\, i^{s}_d$,

$P\lambda^{s}_d = V_d^{s} - r^{s}_{21} i^{s}_q - r^{s}_{22}\, i^{s}_d$,

$$P\lambda^{r}_q = -r^{r}_r i^{r}_q + w_r \lambda^{r}_d,\ p\lambda^{r}_d = -r^{r}_r i^{r}_d - w_r \lambda^{r}_q \quad (18)$$

$$V_q^{sh} = I_f * R_f \quad (19)$$

Where Rf is the fault resistance, in case of solid fault equal to zero.

Equations (13-18) show induction motor dq modelling with fault conditions and the effect of fault severity on motor parameters.

3.2 Historical Data Generation

Data was generated according to motor mathematical model discussed in section 3.1. A Simulink model was built to simulate induction motor environment in case of health and fault as given by figure 7. Motor current, loading condition and injected fault percentages were recorded to have a simulated historical data.

Figure 7. Three-phase Induction Motor In dq Overall Simulink Model

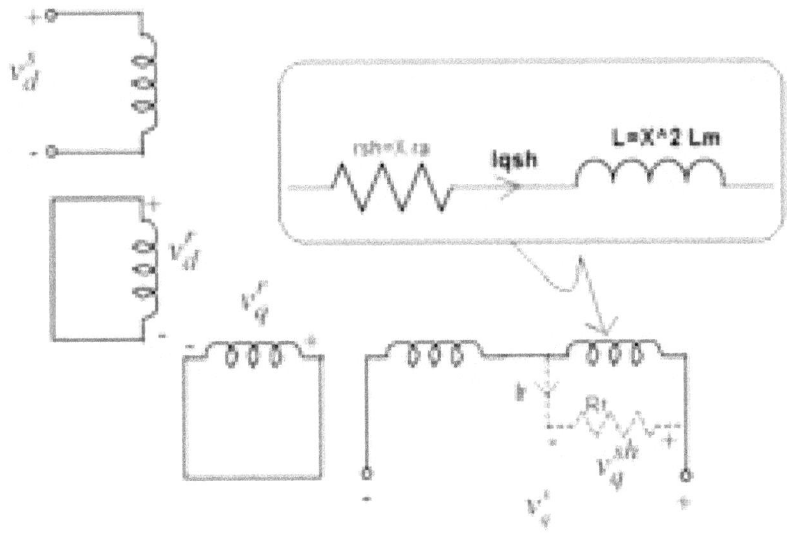

3.3 Adaptive Neuro-Fuzzy Inference Diagnostic Tool System

The Adaptive Neuro-Fuzzy inference system (ANFIS) depends on two main systems fuzzy logic (FL) and artificial neural networks (ANN). The ANN has the ability of data classification and pattern recognition. For accurate fault detection, the network should be trained with various input-output data of health and fault data. The ANN accuracy depends on the network structure as well as training data (Azar et al., 2020; Siddique & Adeli, 2013). The supervised ANN is trained offline at first with the input-output data of the faulty and healthy cases. The training changes the network weights. The inputs data for training the NN are the system measured data (e.g. stator currents) and the desired output of NN as fault severity (Sharma, Verma, Mathew, & Chatterji, 2018). While unsupervised ANN use unsupervised method of learning is not supervised. This means that target is unknown. It has internal representation to respond for different input patterns with the different parts of NN. This learning technique is based on input signal feature, regularities and trends for network configuration and weight adjustment. The unsupervised learning NN is suitable for data clustering and signal categorization as presented in (Siddique & Adeli, 2013).

The Fuzzy Logic (FL) emulates the human brain logic. It is a type of expert system. The fuzzy linguistic variable is used to present the human compares logic instead of rigorous mathematical formulas. The fuzzy logic is capable of transforming the linguistic variable to numerical values based on the fuzzy rules and membership functions. The fuzzy logic structure main items are the input/output membership functions, linguistic variable and fuzzy rules as proposed in(Abdel et al., 2018; Azar et al., 2020; Das & Ghosh, 2020; Fatma et al., 2020; Jain & Raj, 2018; Milić, 2020; Pourpanah et al., 2019; Siddique & Adeli, 2013; Srinivasan & LD, 2020). Both FL and ANN increase the system efficiency and decrease the mathematical equations compared to other detection methods (Abdullah & Iqbal, 2020; Zhu & Azar, 2015). The system is widely used for many applications of systems modelling, control systems and forecasting predictions (Hussein et al., 2017; Jain & Raj, 2018). ANFIS consists of IF-then rules, training and learning algorithms (Siddique & Adeli, 2013) .

For fuzzy inference system, consider a system with two inputs (X, Y) and one output (Z). Fuzzy rules based on 1st order Sugeno type (Siddique & Adeli, 2013) are:

Rule1: IF X is A1 and Y is B1 Then $f1=p_1X+q_1Y+r_1$,
Rule2: IF X is A2 and Y is B2 Then $f2=p_2X+q_2Y+r_2$,

A_i,B_i are fuzzy set, f_i is the system outputs within the specified fuzzy rules and pi, qi and ri are the design parameters based on the ANFIS training (Fatma et al., 2020; Jain & Raj, 2018; Siddique & Adeli, 2013).

Adaptive Neuro-Fuzzy inference (ANFIS) network consists of five layers with a normalization layer added to the Neuro-Fuzzy network (Abdullah & Iqbal, 2020; Hussein et al., 2017; Zhu & Azar, 2015) as presented by Figure 8.

Layer 1 is fuzzification layer adaptive nodes with bell membership function:

$$MF1,\ i= \mu A_i(X)\ \&\ MF1,\ i= \mu B_i(Y),\ \text{for } i=1, 2, 3 \tag{19}$$

The A_i and B_i are the linguistic variable of X and Y

Figure 8. ANFIS Structure for Two I/P with Three MMF And One O/P

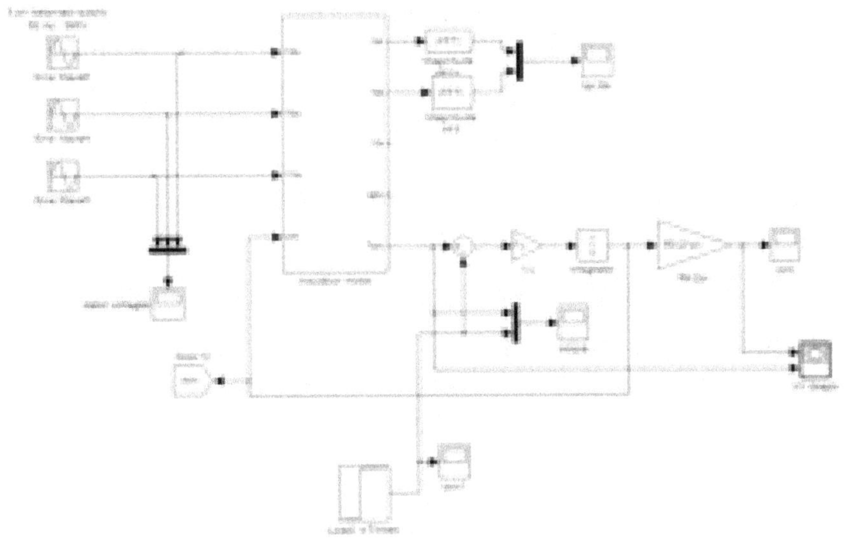

Layer 2 is the rules layer where its output is considered as fire strength of each node

$$W_i = \mu A_i(X)*\mu B_i(Y),\ i=1, 2, 3 \tag{20}$$

Layer 3 is the normalization layer and its output is normalized fire strength

$$\bar{W}_i = \frac{Wi}{(W1+W2+\ldots+W9)} \tag{21}$$

Layer 4 is the consequent layer where each node is an adaptive node and its output is the product of the consequent polynomial of fuzzy rules and normalized firing strength

$$\bar{W} f_i = \bar{W}\,(p_iX+q_iY+r_i),\ i=1, 2, 3\ldots\ldots 9 \tag{22}$$

Layer 5 is defuzzification layer which has only one node (output node) and its output is the overall ANFIS output, summation of the layer 4 output. The proposed ANFIS defuzzification is average weight.

$$f=\sum_{1}^{i}\bar{W}1\ f1+\bar{W}2\ f2+\ldots+\bar{W}i\ fi \tag{23}$$

ANFIS modelling is a trade of complexity and accuracy. More rules and input/output membership functions provide a more accurate system but results in high emulation cost of the system (Fatma et al., 2020; Siddique & Adeli, 2013; Srinivasan & LD, 2020). The structure of Neuro-Fuzzy systems is adapted based on learning algorithm where the rules change to track input output training data to have the optimal ANFIS structure. The initial structure of ANFIS requires selecting the number of rules which depends on the system knowledge. As the system is unknown, rules number range (maximum and minimum) can be validated by measurements of the system. Another approach is to start with a high rules number and iteratively reduce it by merging compatible rules to achieve the required accuracy (Siddique & Adeli, 2013) .

The conventional fault diagnosis computing techniques were recently replaced by artificial intelligence (NN, Neuro Fuzzy, and ANFIS). Because simple technique, it is used for online diagnosis and doesn't require large memory. The Neuro Fuzzy System is capable to adapt with the system and update the knowledge base online

3.4 ANFIS Detection Model Algorithm

The developed fault analysis technique is investigated through MATLAB simulations. An induction motor with inter-turn stator fault is modelled in SIMULINK based on equations of section 3.1. Fault detection technique uses an ANFIS network to estimate the inter turn fault percentage. Learning rules of ANN change the FL parameters. Figure 9.a chart shows algorithms of the off-line stage of proposed diagnostic model. At the beginning, construction the neuro-fuzzy system network architecture, defining inputs and outputs, fuzzy membership functions, and fuzzy rules. Secondly, dividing healthy and faulty motor data to train, validation, and testing data. The final stage is to ingest data to constricted model and checking model training performance before going to the online stage.

While motor is running, motor current signals are measured and transformed with Fast Fourier Transformation to be sent to ANFIS diagnostic model proceed detection to clarify fault severity. This online process is shown by flowchart given by Figure 9.b.

Figure 9. ANFIS diagnostic model design and implementation flowcharts
(a) Off-line process (b) On-liner Process

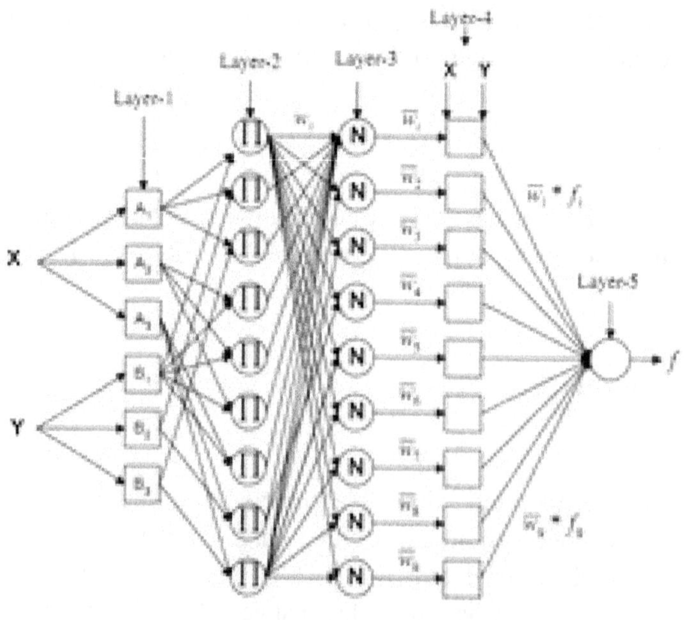

Training and testing data are generated from the SIMULINK induction motor model. Motor loading condition was varied to simulate no-load, 25%, 50%, 75%, full load and 110% loading. The inter turn fault percentage was varied to span a range of 0~16% with steps of 0.005. The total points are 199. Fault severity was varied from 0% till 16% for this study. However, an actual fault will be limited to 10% fault only at 110% loading based on induction motor overload protection setting Io. L (overload current setting) =1.5 x I rated (motor rated current). The proposed fault detection technique is evaluated under the stator inter turn faults with and without contact resistance. Figures 10 (a, b) show the ANFIS initial parameter rules and membership functions. Figures 11 (a, b) classify trained parameters. It is clear view how ANFIS changes initial rules and membership functions to be adapted with motor turns fault detection.

The ANFIS network was trained with 66% of the total data and checked/tested with the remaining 34%. The design is based on three inputs fuzzy membership functions. It was observed that the learning phase was completed in the first 120 Epochs out of 300 iterations. Figure 12 illustrates the fault analysis system procedure and how it can be implemented to online system. Proposed ANFIS fault detector is tested under solid fault i.e., with zero fault resistance and resisted faults as given by figure 10 and induction motor parameters are given by table 1.

3.5 Solid Fault

The qd current provides better resolution for fault detection. It increases as the fault percentage increases as shown in Figure 13. Figure 14 shows proposed fault analyzer percentage error for motor different loading with respected to targeted values of fault percentage Simulation results illustrate the ANFIS fault detection accuracy even for cases that were not included in the training data. The fault percentage at 25% loading case gives the highest error as ANFIS network was not trained with this fault data. The maximum % error is 6.83% at 25% loading and 16% fault.

Figure 15 shows the ANFIS error for the different loading and different fault percentages when trained the 25% loading data.

The network's initial configuration influences the accuracy of the fault diagnosis system. Table (2) shows the errors for a two input and three input membership functions. The three-input membership function has a lower error for testing and checking data.

3.6 Inter-Turn Fault with Resistance

The fault detection technique was designed for inter turn faults without contact resistance which is the worst case. In order to fully investigate the technique's effectiveness, it was tested for inter turn faults with a contact resistance. The ANFIS detector was tested for varying motor loading and various fault resistance. Figure 16 shows the ANFIS detector percentage error under R_f of 0.1, 0.3 and 0.5 Ω at no-loading, half-loading and full-loading conditions. The maximum percentage error in fault detection was found to be 7%.

The simulations results show that the ANFIS analyzer performed better in investigating turn to turn stator faults with contact resistance as the accuracy of detection is around 93%.

Figure 10. Initial parameters of FL of ANFIS system
(b) Initial membership function (b) Initial Rules

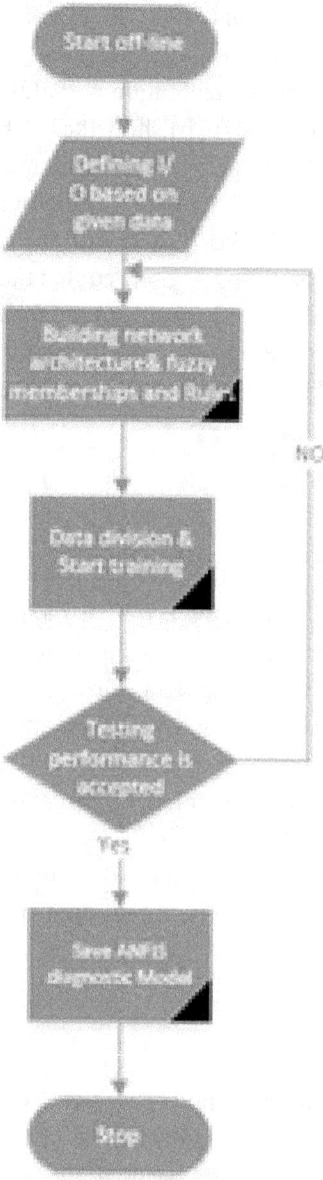

Figure 11. Adapted parameters of FL of ANFIS system
(a) adapted membership function (b) adapted Rules

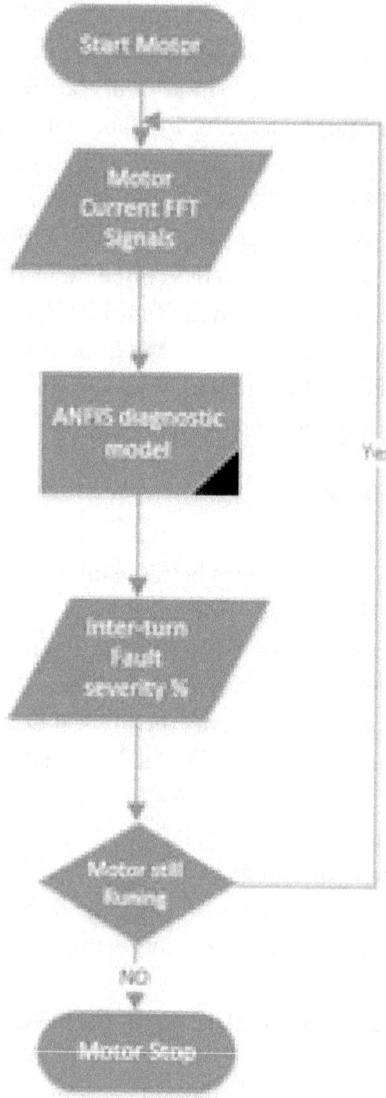

Table 1. Case study motor parameters

parameter	power	no. poles	R_s	L_{ls}	R_r	L_{lr}	L_m	I_{rated}	Vs	f
Value	2hp~1.5kw	4	4.05 ohm	0.014H	2.6 ohm	0.014H	0.5387H	2.81A	220V	50Hz

Figure 12. Online Fault Analysis System for Induction Motor with dq Modelling

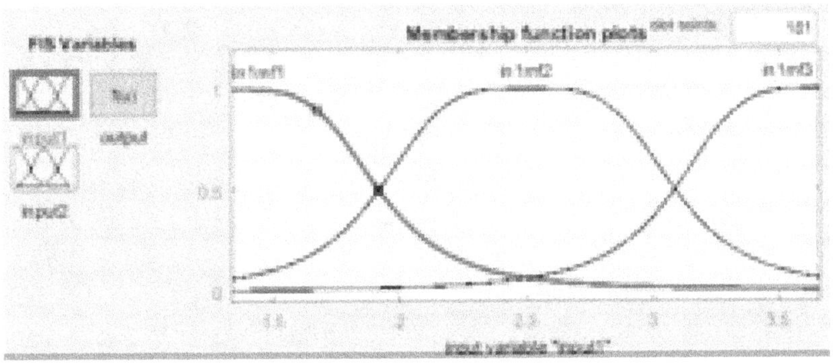

Figure 13. d Current VS. Fault Percentages (X %) At Different Loading

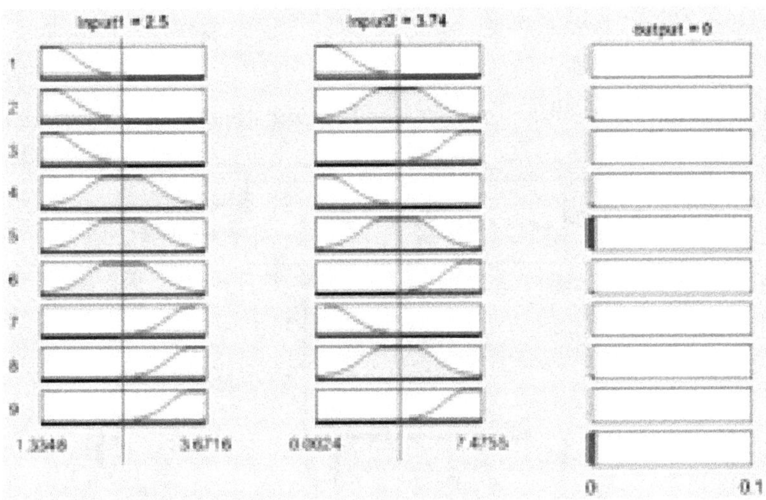

Figure14: % Error VS. Fault Percentages at Different Loading

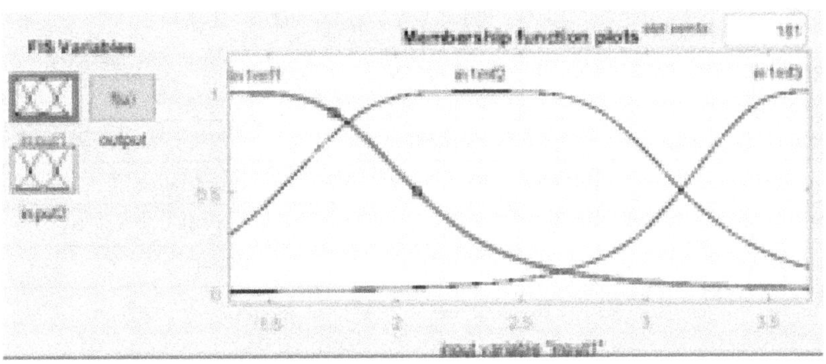

Figure15. % Error VS. Fault Percentages at Different Loading with adding 25% loading to training data

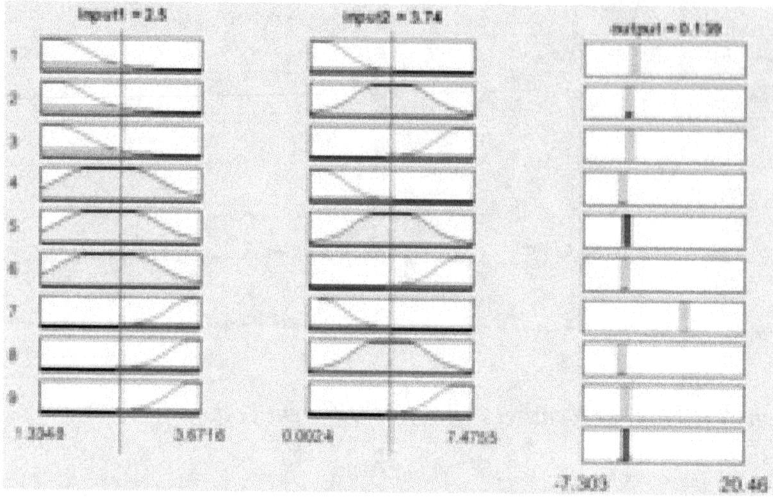

Figure16: % Error VS. Fault Percentages at different-Loading conditions for different fault resistance

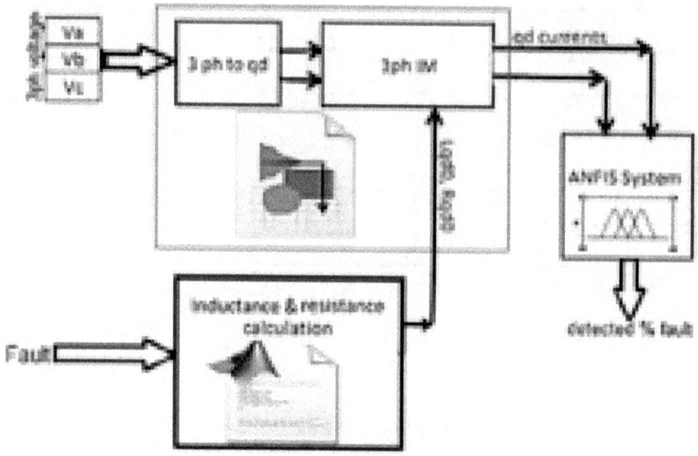

Table 2. The Error Comparison for the Two and Three Membership

Error	Two MMF	Three MMF
Training data	9*10-4	6*10-4
Testing	6*10-3	4*10-3
Checking	9*10-3	3*10-2

4. CONCLUSION AND FUTURE RESEARCH

This chapter study the fault diagnosis of induction motor stator winding faults, for safe and high-performance machines. This was achieved through two main points:

(a). A model of the induction motor was created based on dq components for both cases healthy and faulty cases. (b) Fault detection system was designed based on ANFIS system. It is capable of detecting the fault and analysis its severity online. (c) Fault severity detection is based on stator currents Isq and Isd. This algorithm can be used for online inter-turn fault detection and diagnosis. Simulations results show the ability to distinguish the fault severity, and the accuracy of the estimated values compared to the actual case. This chapter provides an inter turn fault diagnosis technique for induction motors based on an artificial network system and evaluations for solid and resistance faults. The ANFIS network detects the inter turn stator faults with high accuracy even for low fault percentages. The average ANFIS error is 1% among all data training, testing and checking for solid inter-turn faults and maximum of 7% with fault resistance. ANFIS initial structure influences fault detection system accuracy. This algorithm can be used for online inter-turn fault detection and diagnosis as it provided promising results. The use of ANFIS for fault diagnosis achieved great results. It has a good accuracy for low levels turns short circuit and for small changes has better sensitivity. The obtained results are promising for the new fault detection technique.

For the same research area, there are several researching points could enhance the presented results and increase the added value. Practical tests and experiments are needed for further evaluation of the winding fault detection with different motor ratings. Experiments are needed to evaluate the rotor and bearing fault detection sensitivity. The use of the presented algorithm for fault-tolerant control research, as input for controller. For monitoring, supervision and online reporting of induction machines group faults and status require further research through control network. Experimental data should be collected to enhance the fault diagnosis technique. Practical tests are needed for further evaluation of the winding and bearing fault detection with different motor ratings.

Funding Information

This research received no specific grant from any funding agency in the public, commercial, or not-for-profit sectors

REFERENCES

Abdel, M. S. E.-D. A., El Samahy, M., Hassan, M. A. M., & El Bendary, F. (2016). Applications of ANFIS in Loss of excitation faults detection in hydro-generators. *International Journal of System Dynamics Applications*, *5*(2), 63–79.

Abdel, M. S. E.-D. A., Hassan, M. A. M., & Bendary, F. M. (2018). Loss of Excitation Protection of Medium Voltage Hydro Generators Using Adaptive Neuro Fuzzy Inference System. In *Advances in System Dynamics and Control* (pp. 253–285). IGI Global.

Abdullah, M., & Iqbal, S. (2020). The Dawn of Paternal HR: An Exploratory Study Using System Dynamics Approach. In Occupational Stress: Breakthroughs in Research and Practice (pp. 217-237). IGI Global.

Alain, K. S. T., Azar, A. T., Kengne, R., & Bertrand, F. H. (2020). Stability analysis and robust synchronisation of fractional-order modified Colpitts oscillators. *International Journal of Automation and Control*, *14*(1), 52–79.

Atig, M., Bouheraoua, M., & Fekik, A. (2018). An Experimental Investigation of Heating in Induction Motor under Open Phase Fault. *Iranian Journal of Electrical and Computer Engineering*, *8*(3), 1288.

Azar, A. T., Serrano, F. E., Flores, M. A., Vaidyanathan, S., & Zhu, Q. (2020). Adaptive neural-fuzzy and backstepping controller for port-Hamiltonian systems. *International Journal of Computer Applications in Technology*, *62*(1), 1–12.

Boudinar, A. H., Bendiabdellah, A., & Benouzza, N. (2019). Diagnosis of the Broken Rotor Bars Faults by Root-MUSIC Method. In *Advanced Condition Monitoring and Fault Diagnosis of Electric Machines* (pp. 59–88). IGI Global.

Chen, Y. Q., Fink, O., & Sansavini, G. (2017). Combined fault location and classification for power transmission lines fault diagnosis with integrated feature extraction. *IEEE Transactions on Industrial Electronics*, *65*(1), 561–569.

Choudhury, A. (2019). Vibration Monitoring of Rotating Electrical Machines: Vibration Monitoring. In Advanced Condition Monitoring and Fault Diagnosis of Electric Machines (pp. 163-188). IGI Global.

Das, M., & Ghosh, S. K. (2020). Advanced Bayesian Network Models with Fuzzy Extension. In Enhanced Bayesian Network Models for Spatial Time Series Prediction (pp. 101-113). Springer.

Ellefsen, A. L., Æsøy, V., Ushakov, S., & Zhang, H. J. I. T. o. R. (2019). *A comprehensive survey of prognostics and health management based on deep learning for autonomous ships*. Academic Press.

Enshaei, N., Hammad, A., & Naderkhani, F. (2020). A Comprehensive Review on Advanced Maintenance Strategies for Smart Railways. In *Handbook of Research on Interdisciplinary Approaches to Decision Making for Sustainable Supply Chains* (pp. 433–457). IGI Global.

Fatma, L., Ghabi, J., & Dhouibi, H. (2020). Applying Interval Fuzzy Petri Net to Failure Analysis. *International Journal of Service Science, Management, Engineering, and Technology*, *11*(1), 14–30.

Gangsar, P., & Tiwari, R. (2019). Online Diagnostics of Mechanical and Electrical Faults in Induction Motor Using Multiclass Support Vector Machine Algorithms Based on Frequency Domain Vibration and Current Signals. *ASCE-ASME Journal of Risk and Uncertainty in Engineering Systems, Part B. Mechanical Engineering*, *5*(3).

Geethanjali, M., & Ramadoss, H. (2019). Fault Diagnosis of Induction Motors Using Motor Current Signature Analysis: A Review. In Advanced Condition Monitoring and Fault Diagnosis of Electric Machines (pp. 1-37). IGI Global.

Guedes, J. J., Castoldi, M. F., Goedtel, A., Agulhari, C. M., & Sanches, D. S. (2019). Differential evolution applied to line-connected induction motors stator fault identification. *Soft Computing*, *23*(21), 11217–11226.

Han, X., Li, Z., Dahhou, B., Cabassud, M., & He, M. (2020). Nonlinear Observer Based Fault Diagnosis for an Innovative Intensified Heat-Exchanger/Reactor. *Proceedings of the 11th International Conference on Modelling, Identification and Control (ICMIC2019)*.

Hashemi, Z., & Rahideh, A. (2019). Rotor Electrical Fault Detection of Wind Turbine Induction Generators Using an Unscented Kalman Filter. *Iranian Journal of Science and Technology, Transactions of Electrical Engineering*, 1-10.

He, F., Xie, G., & Luo, J. (2020). Electrical bearing failures in electric vehicles. *Friction*, *8*(1), 4–28.

Hussein, H. T., Ammar, M., & Hassan, M. M. (2017). Three Phase Induction Motor's Stator Turns Fault Analysis Based on Artificial Intelligence. *International Journal of System Dynamics Applications*, *6*(3), 1–19.

Jain, V., & Raj, T. (2018). An adaptive neuro-fuzzy inference system for makespan estimation of flexible manufacturing system assembly shop: A case study. *International Journal of System Assurance Engineering and Management*, *9*(6), 1302–1314.

Joshi, S., & Talange, D. (2016). Fault tolerant control of an auv using periodic output feedback with multi model approach. *International Journal of System Dynamics Applications*, *5*(2), 41–62.

Karner, M., Glawar, R., Sihn, W., & Matyas, K. (2019). An industry-oriented approach for machine condition-based production scheduling. *Procedia CIRP*, *81*, 938–943.

Krause, P., Wasynczuk, O., Sudhoff, S. D., & Pekarek, S. D. (2013). *Analysis of Electric Machinery and Drive Systems* (Vol. 75). John Wiley & Sons.

Krause, P. C., Wasynczuk, O., Sudhoff, S. D., & Pekarek, S. (2002). *Analysis of electric machinery and drive systems* (Vol. 2). Wiley Online Library.

Kumar, K. V., Ramya, K., & Irfan, M. (2019). Advanced Fault Diagnosis Monitoring Scheme in Asynchronous Motor Using Soft Computing Method. In Advanced Condition Monitoring and Fault Diagnosis of Electric Machines (pp. 89-97). IGI Global.

Lázaro, J. G. M., & Pinilla, C. B. (2020). Detection and Classification of Wear Fault in Axial Piston Pumps: Using ANNs and Pressure Signals. In Pattern Recognition Applications in Engineering (pp. 286-316). IGI Global.

Lei, Y., Yang, B., Jiang, X., Jia, F., Li, N., & Nandi, A. K. (2020). Applications of machine learning to machine fault diagnosis: A review and roadmap. *Mechanical Systems and Signal Processing*, *138*, 106587.

Liang, H., Liu, Y., Sheng, G., & Jiang, X. (2019). Fault-cause identification method based on adaptive deep belief network and time–frequency characteristics of travelling wave. *IET Generation, Transmission & Distribution*, *13*(5), 724–732.

Lucke, M., Stief, A., Chioua, M., Ottewill, J. R., & Thornhill, N. F. (2020). Fault detection and identification combining process measurements and statistical alarms. *Control Engineering Practice*, *94*, 104195.

Milić, S. D. (2020). Fuzzy-Decision Algorithms for Cyber Security Analysis of Advanced SCADA and Remote Monitoring Systems. In Cyber Security of Industrial Control Systems in the Future Internet Environment (pp. 131-155). IGI Global.

Nguyen, S., & Park, A. (2020). A Comparison of Machine Learning Algorithms of Big Data for Time Series Forecasting Using Python. In Open Source Software for Statistical Analysis of Big Data: Emerging Research and Opportunities (pp. 197-218). IGI Global.

Ortega, R., Pyrkin, A., Bobtsov, A., Efimov, D., & Aranovskiy, S. (2020). A Globally Convergent Adaptive Indirect Field-Oriented Torque Controller for Induction Motors. *Asian Journal of Control*, *22*(1), 11–24.

Pourpanah, F., Wang, R., Lim, C. P., Wang, X., Seera, M., & Tan, C. J. (2019). An improved fuzzy ARTMAP and Q-learning agent model for pattern classification. *Neurocomputing*, *359*, 139–152.

Rahimi, A., Kumar, K. D., & Alighanbari, H. (2020). Fault detection and isolation of control moment gyros for satellite attitude control subsystem. *Mechanical Systems and Signal Processing*, *135*, 106419.

Raie, A., & Rashtchi, V. J. E. E. (2002). *Using a genetic algorithm for detection and magnitude determination of turn faults in an induction motor*. Academic Press.

Saufi, M. S. R. B. M., Ahmad, Z. A. B., Leong, M. S., & Lim, M. H. (2020). Gearbox fault diagnosis using a deep learning model with limited data sample. *IEEE Transactions on Industrial Informatics*.

Sharma, A., Verma, P., Mathew, L., & Chatterji, S. (2018). *Using Motor Current Analysis for Broken Rotor Bar Fault Detection in Rotary Machines*. Paper presented at the 2018 3rd International Conference on Communication and Electronics Systems (ICCES).

Shi, H., Liu, J., Wu, Y., Zhang, K., Zhang, L., & Xue, P. (2016). Fault diagnosis of nonlinear and large-scale processes using novel modified kernel Fisher discriminant analysis approach. *International Journal of Systems Science*, *47*(5), 1095–1109.

Siddique, N., & Adeli, H. (2013). *Computational intelligence: synergies of fuzzy logic, neural networks and evolutionary computing*. John Wiley & Sons.

Srinivasan, S., & LD, D. B. (2020). A Neuro-Fuzzy Approach to Detect Rumors in Online Social Networks. *International Journal of Web Services Research*, *17*(1), 64–82.

Taha, H. A., Sakr, A. H., & Yacout, S. (2019). *Aircraft Engine Remaining Useful Life Prediction Framework for Industry 4.0*. Paper presented at the 4th North America conference on Industrial Engineering and Operations Management, Toronto, Canada.

Tambouratzis, T., Giannatsis, J., Kyriazis, A., & Siotropos, P. (2020). Applying the Computational Intelligence Paradigm to Nuclear Power Plant Operation: A Review (1990-2015). *International Journal of Energy Optimization and Engineering*, *9*(1), 27–109.

Zhou, F., Yang, S., Fujita, H., Chen, D., & Wen, C. (2020). Deep learning fault diagnosis method based on global optimization GAN for unbalanced data. *Knowledge-Based Systems*, *187*, 104837.

Zhu, Q., & Azar, A. T. (2015). *Complex system modelling and control through intelligent soft computations* (Vol. 319). Springer.

Chapter 21
Modeling and Control of Hybrid Electric Vehicle

Flah Aymen
https://orcid.org/0000-0002-3463-6096
National School of Engineering of Gabes, Tunisia

Habib Kraiem
College of Engineering, Northern Border University, Saudi Arabia

Lassaad Sbita
https://orcid.org/0000-0002-6589-3976
National School of Engineering of Gabes, Tunisia

ABSTRACT

The transportation systems have become more electrified, and the major countries of the world program using electric scooters, electric bicycles, electric trains, electric buses, and electric vehicles for their transport. The traditional energy resource stocks are still decreasing rapidly, which makes the world afraid about the future of the transport sector. Therefore, several international restrictions and laws have limited using this kind of energy in relation to the transport sector by encouraging public transport and making a high taxes for the highly energy-consuming cars. The robustness and the efficiency of transportation systems designs are related especially to the internal electric motor and to the battery capacity used. From the other side, the energy management problem presents a serious factor that must be optimized in order to guarantee the overall efficiency and rentability. This chapter explores the modeling and control of hybrid electric vehicles.

1. INTRODUCTION

The world is evolving, and the physical systems are becoming increasingly smarter and somewhat successful towards the present. Every day of order to promote human life, new technology or a new approach or procedure were learnt to make it more comfortable. The area of transportation has also benefited from

DOI: 10.4018/978-1-7998-5788-4.ch021

the introduction of the vehicles due to the evolution steps (Bajracharya, 2016). Its development started in 1769 with the steam-driven automobile, in 1808 with the car powered by an internal combustion engine using hydrogen, and in 1870 with the combustion engine powered by petrol. Most of it was in the 19th century. ANYOS Jedlik was invented during the same period, and precisely in 1828, the tiny model car focused on the electric motor (Rajashekara, 1994), (Salima, Atarsia Loubna, & Toufouti Riad, 2018). Ten years later, the inventor Vermont Thomas, built on a DC machine, unveiled the electric car, and then the new ones appeared. After such dates, the electrified models have performed, and the standard versions have restricted areas of speeds ranging from 4 to 6 km/h. Because of the issue of limited rates and the battery recharge problems, such versions were not successful. Therefore the gasoline-based internal combustion engine was found to be more convenient even with the pollution problem. It is actually due to many factors these as the lowest fuel price, the high touched speeds, and the high torque is given. Recently, the growth of the ICE figure based on cars has raised the pollution issue, and the major environmental governments and organizations have been asked to limit the use of these model cars by using other ecological solutions. Hence, the electric vehicle was returned, and special laws were introduced by the major governments, such as tax minimization and other incentives, promoting the use of its form of the transport system (Kharola & Patil, 2017). This is officially started after 2008, and the then mass-produced electric car was in development (Bajracharya, 2016). First, with this policy, the U.S. government began, then China and the European countries followed. Recent facts have shown that more than 25 percent of the world's current vehicle is wholly or partially based on electric propulsion and this number is still growing (Duff & Ricketts, 2015). This propulsion device has been developed and updated throughout different times in history. These propulsion systems, in relation to the primary power source used, were planned. Many of them are linked to the energy from combustion and others were based on renewable energy sources. Others were combined to solve the problems in the other models, more than energy sources. The combustion-based propulsion systems have various issues related to the high gas emission level and the low-speed zones when high torque is required. The engine used also needs regular maintenance and this enables the use of this type of machine. The propulsion systems based on clean energy as electrical energy or hydrogen energy were defined from the other side by numerous advantages related in particular to the absence of any type of gas emission. Such propulsion systems have been covered by the mentioned difficulties, which characterized the previous system based on combustion energy. However, these two types are distinguished by two particular problems; the first is related to the high noises in the exhaust pipes, the reliability of the engine and the health of the user with the hydrogen version. For the electric variant, these problems do not occur and the underlying problem is related to the autonomy of the system and the lack of a useful transportable energy source adaptable for a similar application. In Table (1), by mentioning the problems and advantages, reveals and classifies the propulsion systems concerning the type of energy sources.

Basing on the table (1), it is clear that the propulsion system which uses the electric energy, have various advantageous and this is why the electrical vehicle number is still increasing. Many new names appear in the international vehicle market and produce vehicles basing on electric energy. The electric vehicle term was used for any kind of car using electric energy as the main power source. However, some versions were combined with two types of energy sources as the fuel energy and electrical energy. This type of vehicle was entitled to the hybrid electric vehicle (HEV) and it is exposed in more than architecture. Those architectures regrouped, an or more than an electric motor, an internal combustion engine, two kinds of energy blocs, electronic equipment and an electronic control unit. In what way

Table 1. Advantageous and weakness for propulsion systems concerning the used energy source

	Combustion energy	Hydrogen energy	Electric energy
Advantageous	°High speeds for high torque, °low cost of the used energy, °low noise	°zero gas emission °high given torque °very low cost of the used energy	°zero gas emission °constant torque for low and high speeds °simplicity of the propulsion system architecture °simplicity of problems identification °low noise °absence of a gearbox
Weaknesses	°Low speeds for high torque, °presence of gearbox, °gas emission, °periodic maintenance for the engine °complex technology for manufacturing	°high level of noise in the combustion phase °complex internal architecture	°low autonomy marge °battery technology not evolved yet °complex control algorithms °technology not standardized yet

Figure 1. hybrid electric vehicle: two versions (a): serial version, (b): a parallel version

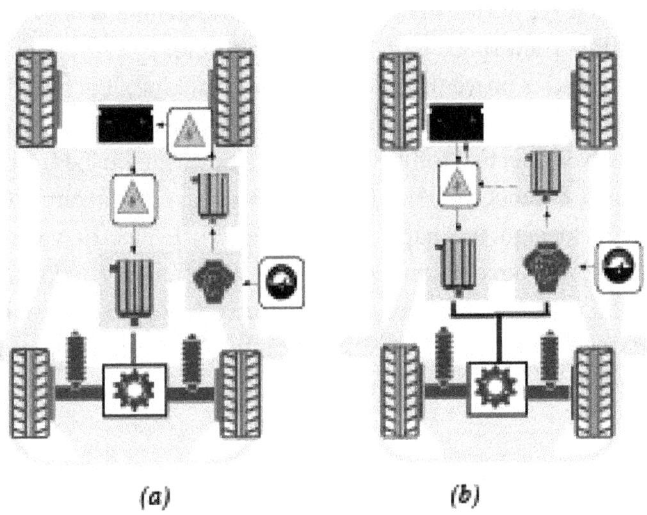

those systems are regrouped, makes the correct nomination, if it is serial or parallel. Figure (1), shows the two architectures and explains the relationship between the main blocs.

By these two versions, it is clear that two electrical motors were used inside the HEV architecture. One of those electrical machines is used for the powertrain phase and it is connected to the wheels (Salima, Loubna, & Riad, 2018). The other machine is connected to the ICE motor (Un-Noor, Padmanaban, Mihet-Popa, Mollah, & Hossain, 2017). These two machines are connected to an electronic bloc too, which is connected to the electric source of power. Basing on these descriptions, the two electric machines architecture where found hardness in the productions and make an excessive weight at the final vehicle weight, which increases the vehicle cost and increases the totality of consumed electric power (Wei, 2013). This combination of motors having a unique objective, related to the autonomy problem. As the battery capacity is limited and then the vehicle autonomy, using permanent charger was a solution for this difficulty. This recharge system composes an electrical motor connected to an internal combustion engine, which has power from a fuel tank. The function loop of the system is related to the battery status and the fuel tank. Generally, if the battery condition is limited, the recharge mode will be in action by turning On the ICE engine, which makes the electrical motor in generator mode (Flah, 2017). with this function mode, it is clear that the two existing motors can be used at the same time. One of them is as motor mode and the other as a generator mode.

Researchers were searching about the possible electrical motor that can be used for this kind of application, and the recent results were shown that the permanent magnet motor based on, is the most suitable for the specification of this application (Santiago et al., 2012)(Singh, Bansal, & Singh, 2019).

With these running mode specifications, researchers were worked on electrical machines and search how it is possible to combine all of the two machines in only one prototype. The goal was to minimize the space inside the car, boost the propulsion system and reduce the inside components. Therefore, a special electrical machine based on the double rotor was exposed. This machine has merged the two roles of two machines in the parallel HEV version. The first motor basing two rotors appeared in 2008. This original version is based switched reluctance model and it is for HEV version too (Cui, Yuan, & Wang, 2008). As the magnet materiel showed its efficiency, another version based on is exposed as it is in (Pišek, Štumberger, Marčič, & Virtič, 2013). This kind of machine was evolved rapidly and the new generation has shown an important space gain inside the vehicle.

In the table (2), some statistics prove that with a special machine, it is possible to have a weight and Power gain inside the vehicle. This table gives the related statistics if using two conventional machines or using a special machine, based on two rotors. Those statistics were raised from those references (Jahns et al., 2006), (Engineering & Sciences, 2018), (Tutelea, Boldea, & Deaconu, 2012) .

With this special machine design, this paper exposes the hybrid electric vehicle architecture with this machine by given the global mathematical models, testing its efficiency and showing the related results. These results were obtained using Matlab Simulink platform. Also, the control loop of the overall blocs is formulated, described and implemented for simulating the function mode of the HEV taken into account the recharge and the action mode (Bouzaida & Sakly, 2018).

Table 2. Gain if using special machine face the conventional machine for HEV model

	Conventional machines		Special machine
	First machine	Second machine	Only one machine
Dimension (mm)/(mm)	L.120/ Ø.245	L.320/ Ø.142	L.50 /Ø.394
The total weight (kg)	**126**	**120**	**133**
Torque (N.m)	85	562	138/438
Speed 10^2 rot/min	45	80	55/120
Power gain (kW)	--	--	15
Power (kW)	100	50	55/110
Space gain (liter)	**--**	**--**	**4.6**
Needed Space (Liter)	**5.43**	**5.07**	**5.6**

Therefore, this paper regroups five sections. After the introduction part, the used blocs for building the HEV model were exposed, described and their mathematical models were given. Next, the control loop is described and discussed in order to define all the steps of control. The fourth section is formulated in order to test the built blocs with their control loop using a Matlab tool. In this simulation, a special drive cycle is used to demonstrate the energetic performance of this system and its vehicle efficiency. In the end, a conclusion resumes this work and gives some perspectives for this work.

2. HEV BLOCS

As it is exposed in the introduction section, various components, form the hybrid electric version. The principle blocs are the main electric motor, the internal combustion engine, the system of batteries and the used electronics inverter. The wheels and the gear system models are also important for simulating the real comportment of a vehicle. As the vehicle system is exposed to external forces like wind and road slopes, taken into consideration, all of those parameters is important to have a real close model.

The mathematical model for each bloc is given as it is in the next.

1. Propulsion system components
    a. Electric motor model: special design for double rotor machine

For this special machine, formed by two rotors, the system of equations of this electrical machine composes two parts. This machine regroups an inner rotor and an outer rotor, the two of them is inside the stator bloc. Figure (2), shows it is an internal form and indicates all of these parts inside. As it is shown in this figure, one rotor is covered by magnets and the other is formed by windings.

Figure 2. special machine design

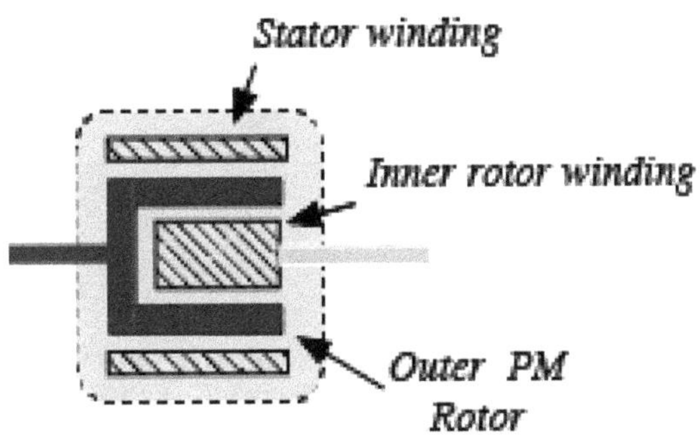

Generally, the stator form has the same present in all AC machines. The used variables are nominated ωm for the outer rotor speed and by ωrm the inner rotor speed. υds $_{\upsilon q}$s a$_{re}$ respectively the direct and transversal voltage components of the stator. The direct and transversal voltage components of the outer rotor are υdr $\upsilon q_{r,}$ re$_{sp}$ectively.

The inner and the outer rotor voltage equations are respectively in equation (1) and (2).

$$v_{ds} = R_s i_{ds} + \frac{d\lambda_{ds}}{dt} - \omega_m \lambda_{qs}$$
$$v_{qs} = R_s i_{qs} + \frac{d\lambda_{qs}}{dt} + \omega_m \lambda_{ds} \tag{1}$$

$$v_{dr} = R_r i_{dr} + \frac{d\lambda_{dr}}{dt} - (\omega_m - \omega_{rm})\lambda_{qr}$$
$$v_{qr} = R_r i_{qr} + \frac{d\lambda_{qr}}{dt} + (\omega_m - \omega_{rm})\lambda_{dr} \tag{2}$$

The flux expressions are in equations (3) and (4). It is important to indicate that the used magnet on the outer rotor in the two sides cannot be equivalent. Therefore, two different flux parameters noted λm1 and λm2 exist.

$$\lambda_{ds} = L_{ds} i_{ds} + L_{dm} i_{dr} + \lambda_{m1}$$
$$\lambda_{qs} = L_{qs} i_{qs} + L_{qm} i_{qr} \tag{3}$$

$$\lambda_{dr} = L_{dr} i_{dr} + L_{dm} i_{ds} + \lambda_{m2}$$
$$\lambda_{qr} = L_{qr} i_{qr} + L_{qm} i_{qs} \tag{4}$$

For each rotor, the corresponding electromagnet torque as it is in equation (5) and (6). Tout is the outer rotor torque, and Tin is the inner rotor torque.

$$T_{out} = \frac{3}{2} P\left[\lambda_{m1} i_{qs} + (L_{ds} - L_{qs}) i_{ds} i_{qs} + L_{md} i_{qs} i_{dr} - L_{mq} i_{ds} i_{qr}\right] \tag{5}$$

$$T_{in} = -\frac{3}{2} P(\lambda_{m2} i_{qr} + L_{md} i_{ds} i_{qr} - L_{mq} i_{qs} i_{dr} - (L_{qr} - L_{dr}) i_{dr} i_{qr}) \tag{6}$$

Then, speeds equations can be expressed. The inner rotor speed is noted as "ωrm". The outer rotor speed noted "ωm". "Tic$_e$" is the home torque of this rotor. "TL" is the load torque. Equation (7) show them.

$$\frac{d\omega_{rm}}{dt} = \frac{1}{J_r}(T_{in} - T_{ice} - B_r \omega_{rm}) \tag{7}$$

b. electronic converter for double rotor machine

The use of static converters in manufacturing has become an extremely broad field, as variable speed drive is increasingly being used in industrial equipment. Voltage inverters are commonly used for asynchronous motor control.

The output of the analog or digital control algorithms represents the voltages or currents desired at the terminals of the machine. The technique of Pulse Width Modulation (Pulse Width Modulation) allows these outputs to be reconstructed from a DC voltage source by means of a direct converter. Controlling AC machines with a voltage inverter generally uses pulse width modulation techniques to regulate the power switches. By controlling the switchers of the power transistors, it is possible to minimize the switching losses. This is remarkably alters on the applied voltages on the electric motor. The pulse width modulation techniques are responsible on the quality on the outputted voltages to the machine. Multiple choices exist and the select of one of them depends on the type of command that can be applied to the machine, the modulation frequency of the inverter and the harmonic constraints set by the user. The principle of modulation can be done by various approaches, conventionally by comparing references to a triangular function or using a real-time calculation satisfying a criterion. The traction control system of the vehicle allows it to endure alternating voltages of variable frequency and amplitude. Three-phase voltage converters allow a system of three-phase voltages to supply the electric motor, obtained from a DC input signal. For variable-speed applications, inverters run at switching frequencies of a few kHz on small and medium power machines. The inverter used in this work is a three-phase inverter with two voltage levels, having eight switching cells (IGBT's) and eight covering diodes. Each inverter leg is made up of two switching cells, each consisting of the switch with its diode in reverse. The output corresponding to the midpoint of the leg. The control signals of the switches on each arm must be complementary so as not to short-circuit the continuous supply to the inverter. For guarding against a short circuit, it is necessary to introduce a waiting time for the closing of the switch, usually called dead time. The consequence of this dead time causes discontinuities on the voltage resulting in distortions of the currents and an increase in the amplitude of the corresponding harmonics when changing the direction of the line currents.

In this paper, the four-leg voltage inverter used to supply this special AC machine based on double rotor and a single stator that operates with a variable frequency. To optimize motor control while avoiding magnetic coupling and reducing stator losses, used the principle of two separate stator windings with only a four-leg inverter is the fixed choice. Figure (3), shows the converter architecture.

During operation, the voltage equations of this type of inverter will be different from those produced by the conventional three-phase inverter. In fact, all the voltages in each arm are defined with reference to the midpoint (between the two capacities).

The classic method for determining the switching times (and the application of positive or negative voltages) is entitled Pulse width modulation (PWM). It is based on the comparison of a modulation wave "V_p"(generally a triangular signal of high frequency "f_p"), with a reference wave "V_{ref}"(representing the desired output voltage with frequency "f_{ref}") .

The switching signal is determined from the intersection of these two signals. In the case of the three-phase inverter with four arms, there will be four control signals (one for each arm of the inverter [12]. The interactive modulation is characterized by its modulation index "r" and its modulation depth "m_a", such as:

Figure 3. electronic converter design

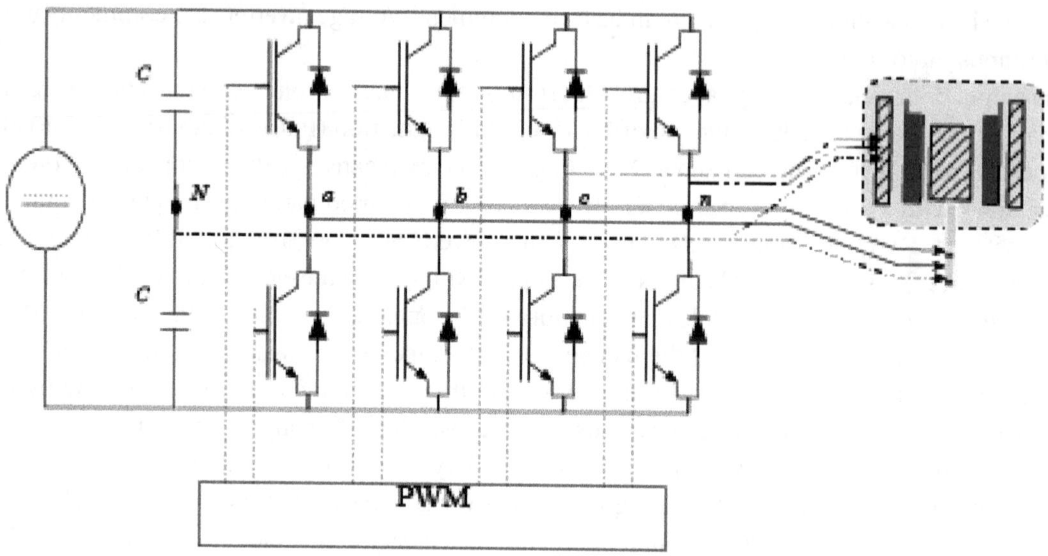

$$m_a = \frac{V_{ref}}{V_p} \quad r = \frac{f_p}{f_{ref}}$$

$$\left.\begin{array}{l} V_{i\_ref} < V_p \Rightarrow S_i = 0,\ V_{iN} = -\frac{V_{dc}}{2} \\ V_{i\_ref} > V_p \Rightarrow S_i = 1,\ V_{iN} = \frac{V_{dc}}{2} \end{array}\right\} V_{iN} = \frac{V_{dc}}{2}\left(2S_i - 1\right), i = a,b,c,n \tag{8}$$

The relationship between the output voltages and the state of the switchers can be defined as it is in equation (9).

$$\begin{bmatrix} V_{aN} \\ V_{bN} \\ V_{cN} \\ V_{nN} \end{bmatrix} = \frac{V_{dc}}{2} \times \begin{bmatrix} 2S_a - 1 \\ 2S_b - 1 \\ 2S_c - 1 \\ 2S_1 - 1 \end{bmatrix} \tag{9}$$

The phase voltage can be given by equation (10).

$$\begin{bmatrix} V_{an} \\ V_{bn} \\ V_{cn} \\ V_{n} \end{bmatrix} = \begin{bmatrix} 1 & 0 & 0 & -1 \\ 0 & 1 & 0 & -1 \\ 0 & 0 & 1 & -1 \\ 0 & 0 & 0 & 0 \end{bmatrix} \begin{bmatrix} V_{aN} \\ V_{bN} \\ V_{cN} \\ V_{nN} \end{bmatrix} \tag{10}$$

2. Batteries and supercapacitors models

The electrical transportation systems are based on energy storage devices such as the battery or the supercapacitors, connected to the propulsion system inside the electric drive. These energy sources can feed the vehicle at or out of departure by the required power. These tools are largely used for the pure electric vehicle or for the hybrid-electric model. This offers a primordial instrument for the main transportation systems. Generally, those mobile sources have specific characteristics regarding the charge or the discharge mode. For the batteries case, the problem appears in the recharge time face the supercapacitor. Actually, the recharge period is very higher for the battery case and very low for the case of the supercapacitor. Similar to the discharge mode, there is a discharge cycle number for the batteries, and it is unlimited for the capacitors, which make the essential weaknesses of the battery. Parallel, the energy density can be more interesting for the case of batteries faces the capacitors. This is the opposite of the power density factor. For all of those reasons the price of the supercapacitors is higher than the batteries. The performances of these sources depend on various parameters and variables. Equation (11) exposes the battery outputted voltage, by one cell, noted "V_b." Actually, the voltage expression depends on the « Voc » which is the open-circuit voltage. « R_b » is the ohmic resistance of a cell. « R_{st} » and « C_{st} » are the resistances and the capacitances of the electro-magnetic short-time-interval double layer effects, respectively and « R_{lt} » and « C_{lt} » are the resistances and capacitances of the electro-chemical long-time-interval mass transport effects. « I_b » is the cell load current. As this element can be discharged or charged « I_{st} » will be positive or negative respectively

$$V_{b/cell} = V_{oc} + R_b I_b + \int \frac{I_b - V_{st} / R_{st}}{C_{st}} dt + \int \frac{I_b - V_{lt} / R_{lt}}{C_{lt}} dt \tag{11}$$

The battery pack voltage depends on the number of serial and parallel of used cells. Equation (12), explains that relation. The battery resistance can also be found in equation (13). "Ro", indicates the battery cell charging or discharging resistance. With these equations, it is possible to express the battery heat generation noted "Q_{bp}" in equation (14). "ξ" is the charge efficiency. So based on the battery heat generation and basing from the initial battery temperature "T0„, it is possible to express the battery pack temperature noted "Tb" and expressed in (15). "Qb_c„ is the heat losses and "Cp_b„ is the battery heat capacity (Kroeze & Krein, 2008), (Lee, Cherry, Safoutin, McDonald, & Olechiw, 2018).

$$R_b = (R_o + R_{st}{*}I_{st}/I_l + R_{lt}{*}I_{lt}/I)\frac{N_s}{N_p} \tag{12}$$

$$V_b = \frac{N_s}{N_p} V_{b/cell} \tag{13}$$

$$Q_{bp} = I_b R_b^2 + (1-\xi) I_b V_b \tag{14}$$

$$T_b = \int \frac{Q_{bp} - Q_{bc}}{m_{ba} C_{pb}} dt + T_0 \tag{15}$$

The supercapacitor model, for one cell, can be visualized in equation (16). The capacity value depends, "Ne" which is the number of layers in the electrode. "є0" and "є" is the permittivity's of the free space and the electrolyte material. "c" presents the molar concentration and "Ai" expresses the joint area between electrodes and electrolyte. The Helmholtz layer length is "d" and "Qc" presents the cell electric charge (C). the final capacitance value depends, the number of parallel and serial cells, noted respectively by "Np" and "Ns", and it can be calculated as it is in equation (17). Then the outputted voltage value for the supercapacitor module can be expressed as it is in equation (18).

$$\frac{1}{C_{sup/cell}} = \left[\frac{d}{N_e \varepsilon \varepsilon_0 A_i} + \frac{2 N_e R T}{F Q_c \sinh\left(\frac{Q_c}{N_e^2 A_i \sqrt{8 R T \varepsilon \varepsilon_0 c}} \right)} \right] \tag{16}$$

$$C_{\sup} = \frac{N_p}{N_s} C_{\sup/cell} \tag{17}$$

$$V_{\sup} = \frac{N_p Q_c}{C_{\sup}} - R_{\sup} i_{\sup} \tag{18}$$

Equation (19) can be used for supervising the batteries and the capacitors state of charge, noted « $SOC_B(t)$ » and « $SOC_U(t)$ ». « η_B » is the charge and discharge efficiency of the battery pack (Zhou, Gao, Ravey, Al-Durra, & Simões, 2017), (Zhang, Tao, & Zhou, 2019).

$$\begin{cases} SOC_B(t) = SOC_I - \eta_B \int \frac{I_b(t)}{3600 Q_b} \\ SOC_U(t) = \frac{\left(V_{\sup} - 2 R_{\sup} i_{\sup}(t)\right)^2}{\left(V_{\sup,\max}\right)^2} \end{cases} \tag{19}$$

3. Mechanical Vehicle model

The car vehicle model regroups various equations, related to the speed form. This equation depends on the wheel radius, noted "r_w" and the mechanical gear factor, noted "G" and expressed in equation (20). Firstly it is necessary to express the vehicle speed formula, which can be presented as it is in equation (20). This equation is limited by the vehicle and motor maximum seeds noted respectively, $V_{v,max}$ and $\omega m_{,max}$.

$$V_v = \frac{w_a}{\underbrace{\frac{\omega_{m,\max}}{V_{v,\max}} \frac{2\pi}{60} \frac{r_w}{1.1}}_{G}} r_w \tag{20}$$

Also, expressed the vehicle torque equation "C_v" is necessary in order to calculate the real applied force on the vehicle. This expression regroups various parameters and variables, as the total vehicle weight "M_v" and the road grade "θ", the road gravity noted "g" and the permanent friction force noted "fr,,.This is explained in equation (21). Regarding the vehicle weight, the vehicle motor weight and the batteries pack weight are added to the chassis vehicle weight as it is expressed in equation (22).

$$M_v = m_{v0} + m_{ba} + m_{em} \tag{21}$$

$$C_v = r_w \left[f_r M_v g \cos(\theta) + M_v g \sin(\theta) + \frac{1}{2} C_d \rho A_v V_v^2 \right] \tag{22}$$

The vehicle torque exposed previously can help to form the traction force, which can be expressed as it is in equation (23). This equation depends on the frontal vehicle area, the air density and the given acceleration factors, noted Av, ρ and Acc.

$$F_t = M_v \left[g\left(\sin(\theta) + f_r \cos(\theta)\right) + \rho A_v \left(\frac{V_v - V_w}{2M_v} \right) + f_m Acc \right] \tag{23}$$

This bloc will simulate the external space comportment influence in relation to vehicle speed and acceleration.

4. Internal combustion engine model

Inside the hybrid electric vehicle and especially in the parallel version, the internal combustion engine can be used for powering the vehicle by the traction power. Similarly, it can be used for powering the vehicle by electric energy if it is used inside the serial version. However, in some specific applications, this system can be just used in the recharge phase. This machine is mechanically coupled to an electric generator and makes this machine ON or OFF is related to the energy management algorithm decision. Actually, this machine will give the necessary torque for the electrical machine, which converts this mechanical power to an electrical power transferring to the batteries pack. The obtained torque "T_{ICE}" can be expressed in function of the maximum possible given torque noted "T_{max}" and power "P_{max}". The engine speed appears too in the mathematical model as it is expressed in equation (24) (Flah, 2018). "wp,, and "wt" are the Speed at maximum power and torque, respectively.

$$T_{ICE} - T_{\max} = \frac{0.5(\omega - \omega_t)^2 P_{\max}}{\omega_p^2(\omega_t - \omega_p)} \tag{24}$$

The obtained torque from the ICE engine will be transferred to the electrical generator and then the given electromagnet torque for the machine will be in function of the ICE and machine inertia noted "J_{ICE}" and "J_M" respectively. The gear ratio parameter appears too in this equation, also the friction force, too, as a linear relationship between the vehicle vector force and the wheel radius. This is valid for the case of a parallel HEV version. For the serial version, this parameter can be eliminated. Equation (25), express this final obtained electromagnet torque.

$$T_e = \left(T_{ICE}/G\right) - F_t r_w \frac{d\omega_m}{dt}\left(J_{ICE}/G + GJ_M\right) \tag{25}$$

3. CONTROL ARCHITECTURE

The HEV model regroups many blocs, as it is exposed previously. Arrange all of the related parameters, and assure the correct flow of information, present the important factor for controlling the vehicle correctly. Two basic systems englobe the overall function of the HEV. The first one is related to the traction phase and the second one is related to the recharge phase. A mutual element related to these two parts and it is the battery pack.

For the traction phase, the main active element is the electrical motor bloc. As this one is attached through a gearbox to the wheels, this one seems responsible for controlling the vehicle acceleration and deceleration. With the literature, various motor models and designs were exposed and used for doing this touch. The conventional DC machine, the conventional AC machines are the basic versions presented for this application. Actually, each machine model is characterized by a special performance which is improved in the new motors version as the Brushless DC machine "BLDC" and the permanent magnet synchronous machine "PMSM". Basing on the PMSM version, the new versions have appeared as the double stator or the double rotor machine. In this study, the new design based on the double rotor version is implemented.

This machine has a special design that makes its control loop special and unique face the conventional machine. There are a double rotor and a single stator permanent magnet inside. Actually, its double rotor will be used in parallel if it is necessary. The primary rotor is attached to the wheels and the secondary rotor is attached to the ICE engine, as it is exposed previously. Figure (4) gives in what way the wheels and the ICE engine are coupled with this machine. As it is exposed in this figure, two electronic converters are used for moving the power from and to the battery pack. When the ICE makes the inner rotor running, the obtained electric power is used for charging the battery. This is possible even the first

Figure 4. special machine attached to the traction and recharge systems

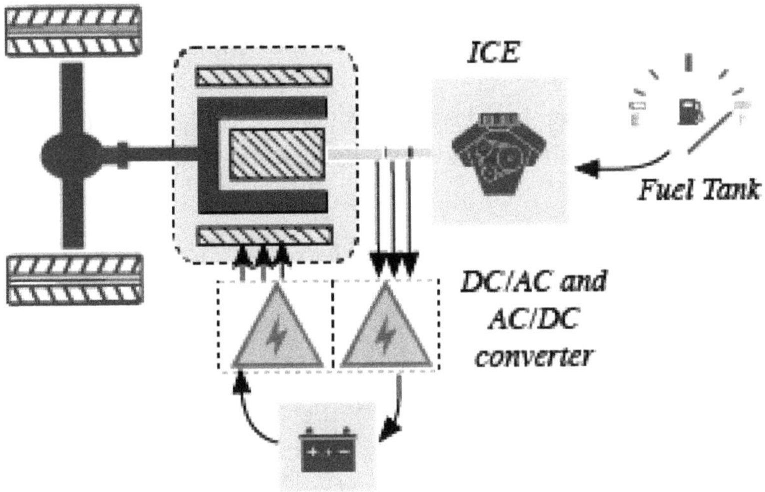

rotor "or the external rotor" is stopped (Njajra, Flah, Elhak Chariag, & Sbita, 2017), (Pišek et al., 2013), (Xiao, Yang, Zhu, & Cao, 2009).

As this machine is used for both traction and recharge phases, the corresponding control loops will be merged into only one version as the outputted currents and voltages, and the inner currents and voltages will be supervised, respectively to the outer and inner rotors.

The control of this machine is very special, as it is possible to have two different currents from two different sources in the motor stator and in the inner rotor windings. This special running mode demands a decoupling control method. In literature, man strategies have resolved this problem for the conventional machines as the Direct torque control loop DTC. This technic was explained and ameliorated in several works using intelligent technics as it is in (Mohamed, 2012), (Abu-Seada, Mansor, Bendary, Emery, & Hassan, 2013). This technique was considered appropriate for this type of machine and had numerous stability and robustness-related advantages. However, it was found not suitable in low-speed level, where chattering problems appear. Another strategy was also applied for this kind of machine and based on the field-oriented vector control loop. This method was found easier than the DTC and had acceptable performances in all region speeds. Actually, this control method was applied for all kinds of applications based on AC machines. This reference (Salima, Atarsia Loubna, et al., 2018) explores this method. Basing on the statistics, which classify the two conventional control loops and cite their weaknesses and advantages, the choice was concentrated on the FOC version. Therefore a modification loop was built for this machine and based on this control strategy.

Figure (5), explores the control loop architecture in relation to the electric motor. Actually, the user control loop is based on the field-oriented vector control strategy. The first stator control loop is used for controlling the rotor speed of the outer rotor. This one is attached to the wheels' systems. However, the inner rotor speed and torque are controlled through the related windings into this rotor. In this case, it is possible to simulate that the related stator for this rotor is based on permanent magnet material. This is not the case for the outer rotor.

It is important to indicate that for succeeding the running function and the independence of the two sections as recharge and traction phases, the used control strategy must assure the independence between the inner and the outer rotor. The used strategy eliminates the coupling problem between the two parts (Cai & Xu, 2015).

As it is presented in this control loop, various PI controllers are used and for tuning their parameters, the trial error method is used. Other recommendations will be shared next for improving the PI performances.

Figure 5. control loop of the used special electric machine

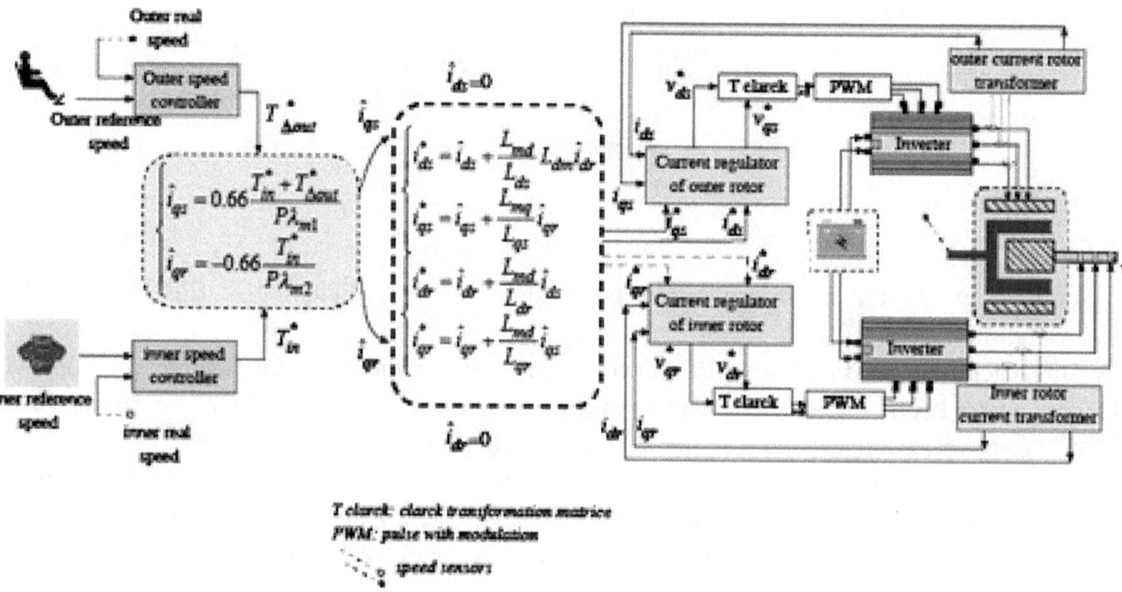

The recharge action starts by making the ICE engine ON. This is can be by various solutions and methods. Actually, the simple architecture is the relay control or ON/OFF control method. Its principle is to supervise the battery SOC and take the decision of the recharge or not. After fixing the battery SOC minimum and maximum limit, the engine can be turned ON or OFF. However, only one ICE reference speed is applied and it is the maximum possible engine speed.

For this case, the used-control method gives the recharge system a variety of reference speeds. Three regions were used, the lowest, the middle and the highest reference speeds. Each zone is limited too by two limits. In the exposed flowchart in figure (6), the recharge method is explained, were a hyperbolic function is used as a reference parameter. It is important to indicate that a limit on the ICE speeds appear if the fuel tank level is critical.

4. CASE STUDY AND RESULTS DISCUSSION

All the mathematical models outlined in the previous sections have been implemented for all the necessary blocks. The global vehicle model has been implemented in the Simulink framework of Matlab and is built into two critical parts, the first one is related to the traction stage and the second one is related to

Figure 6. control loop of the recharge phase

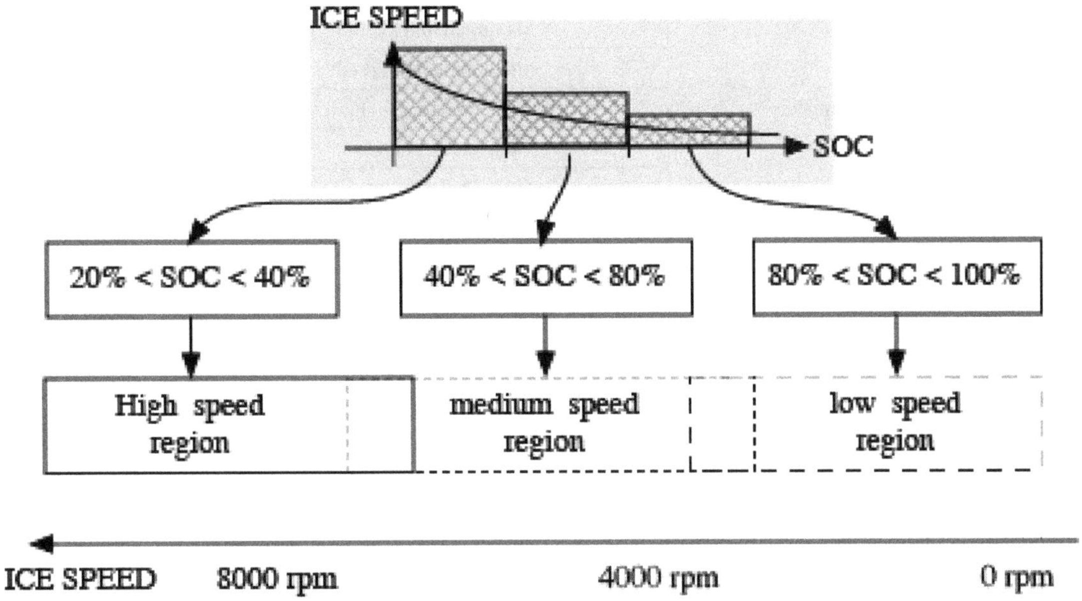

the recharge step. The blocks are linked, as shown in figure (7). The given type of acceleration is taken as the overall model's input signal. The vehicle's actual speed is regulated by overseeing the electric motor speed and this is for the traction process. The battery charging status is an essential element that must be monitored for the recharging cycle. This is achieved by an operation on a computer speed generator. This one is connected to the main motor of the ICE.

Figure 7. blocs association for HEV simulation

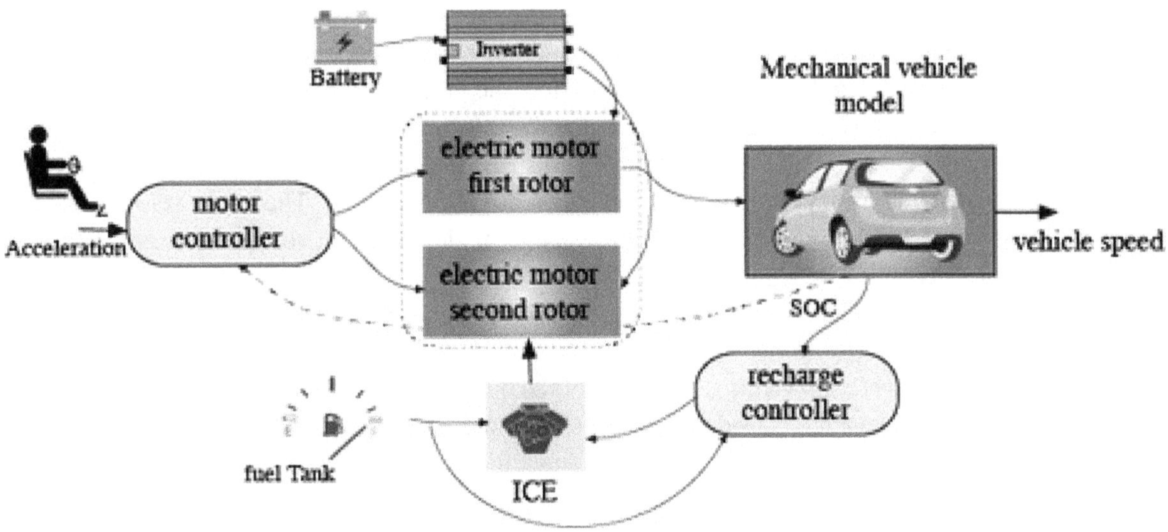

Table 3. Vehicle specifications

Electric machine parameters	**Maximum inner speed**	**10000 rpm**
	Maximum outer speed	**1500 rpm**
	Maximum Outputted power	**20000 W**
	Maximum current	**25/100 A**
	Voltage	**500 V**
	voltage	210V
Battery pack characteristics		
	Initial state of charge	79%
ICE engine parameters	Maximum speed	8000 rpm
	Maximum torque	300
	Vehicle weight	700 kg
Vehicle and road specification	Road slope	0
	Maximum speed	65 km/h

Table (3) resumes the vehicle specifications and precisely the electric motor, ICE and battery pack parameters. Also, the road condition is given inside.

If using the parameters in the table (3), the simulation condition was based on the reference form of acceleration, as it is in figures (8) applied in 250 seconds. The given acceleration was started from zero and touch 30% as a maximum acceleration value. Then it decreases again to 15% at the end of this simulation. The proportional vehicle speed follows the outer rotor speed form of the electrical machine. The gear ratio transforms the outer rotor speed to the vehicle and for 710 rpm, the vehicle touch 61 km/h. It is clear that the motor speed evolution is smooth, which validate the motor control method.

With this drive cycle and based on the initial battery state of charge, which is 79%, the battery evolution will be as it is in figure (9). Actually, the battery SOC is still decreasing for the lowest speed value, as it is in the periods 0 to 40 seconds. However, since the speed becomes more than 100 rpm, the battery SOC becomes increasing. This is due to the internal combustion engine, which is starting and makes the inner rotor in generator mode. In this situation, the inner rotor speed touches 2000 rpm, and this is due to the given torque by the ICE, which is equal to 50 Nm. Figure (10), shows the evolution of the recharging equipment as the ICE engine torque and speed and shows the obtained power from this generator. As the generator power exists, the battery SOC is still increasing. The cases were the battery voltage is higher than 210 V, indicate that it is in recharge mode. Even the voltage increase more, the battery SOC increases rapidly.

From the other side and if concentrating on the motor speeds in figures (8) and (10), it is remarkable the independence of each rotor, which validates the used motor control strategy.

The obtained results were shown the satisfaction of the built model for simulating the hybrid electric vehicle running mode. Taken into consideration the traction mode and the recharge mode and by using a special electric motor for both steps, the built model was validated by the obtained results, which shone all the related parameters.

Figure 8. vehicle speed according to the given acceleration and the obtained outer rotor speed

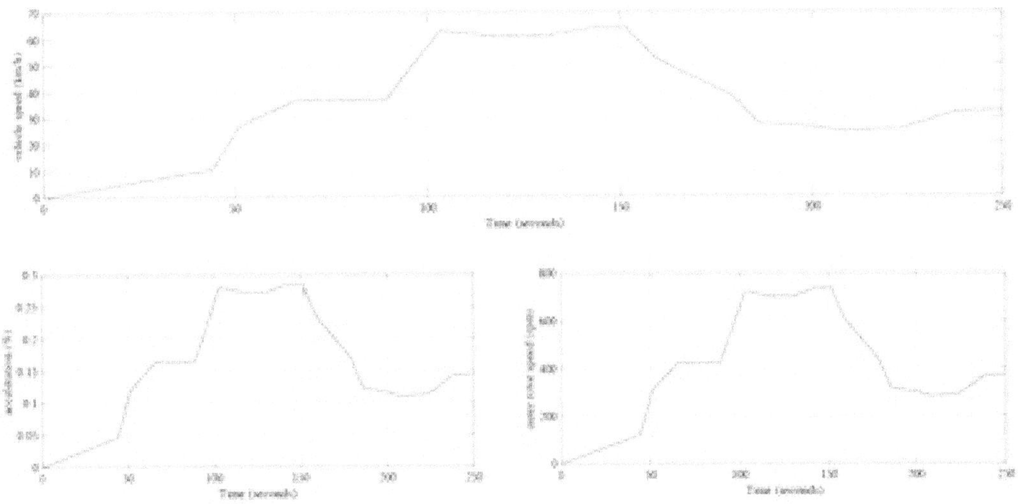

Figure 9. Battery statistics: power, voltage and state of charge

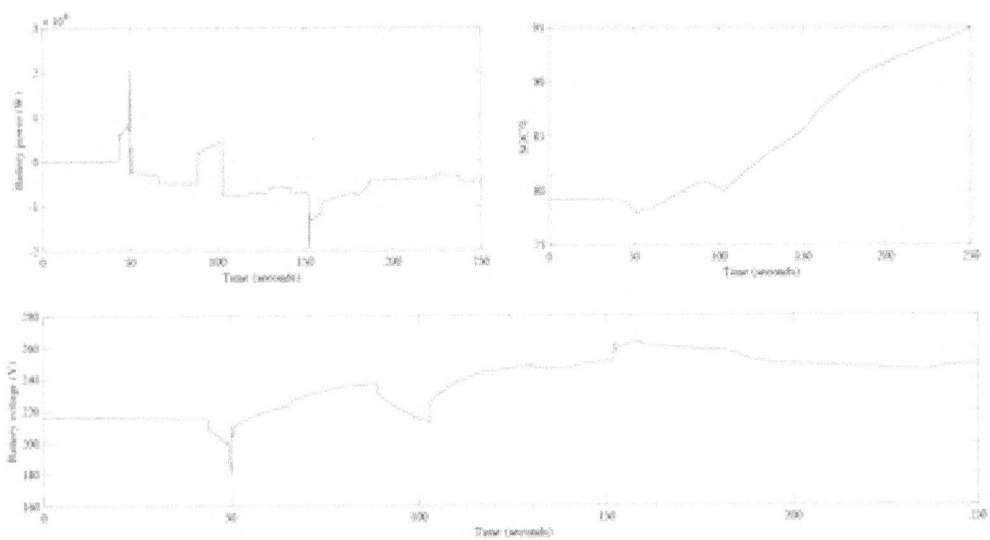

FUTURE RESEARCH DIRECTIONS

The research discussed in this chapter was expanded to use mathematical analysis and then translated in Simulink / Matlab into blocs. Certain parameters were tracked and obtained after running the overall system as vehicle speed, motor speed, and others. A lot of PI controllers within using the exposed control loops inside the overall system are also exposed. The problem of PI parameters modification was addressed in that related section. The used PI parameters were actually changed using trial and error method, and the results obtained were acceptable. Nevertheless, these results can be improved by giving in this part a list of recommendations for the future use of methods and resources to improve the

Figure 10. recharge mode system: related parameters

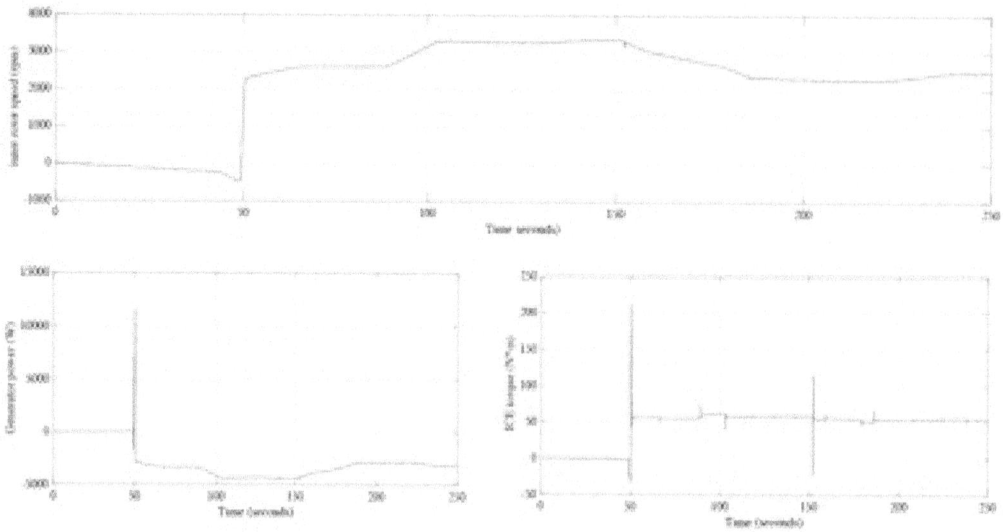

vehicle's global efficiency. These forms of guidance provide a list of possible contributions and future research. Using the optimization algorithm to adjust the PI parameters and we're referencing the PSO program and BFO tool. There is some useful reference in (Salima et al., 2018), (Flah & Sbita, 2013) A direct torque control methodology can also be applied to control the inner rotor inside the system and to keep the FOC for the outer. This will increase the charger system performance. On the other hand, and by using the fuzzy approach, it is possible to improve the charger controller. This will also increase the charger system performance and give the ICE engine more flexibility by precision fuel injection. Reference is made to some applications based on the fuzzy technique (Azar & Vaidyanathan, 2018). Any work related to the field of power management may support this report and may boost the vehicle's energetic performance.

CONCLUSION

At the end of this chapter, the exposed application is exposed so that the contribution can be explained shortly. In addition, this chapter takes into consideration the study of the transport system and concentrates in particular on the version of the hybrid electric car. This addressed and demonstrated the internal components. The mathematical models were also provided and detailed in order to have information about the inner parameters which operate the overall The battery pack and the internal combustion engine were used precisely because inside the vehicle, and the essential parts are in the recharge mode. Since this experiment is based on a special electrical machine, an important part of this report was reserved to illustrate the concept of electrical machines and their specifications. The corresponding mathematical model has been given and described in detail in it's working mode. The second section of this revision describes the control behavior and processes used inside the hybrid vehicle for the traction and recharge phases to be efficient. For controlling the machine, a special control method is designed and implemented based on field-oriented vector control. For making the ICE function versatile, a special control architecture for the ICE engine has been specified from the other side. A global scheme was developed using Matlab Simulink at the end of this study and tested using a real HEV prototype parameter. The principal supervised parameters are vehicle speed, motor speed, torque, and battery circumstances. A list of improvement measures was listed at the end to open up a future research project.

ACKNOWLEDGMENT

There is no funding for this project.

REFERENCES

Abu-Seada, H. F., Mansor, W. M., Bendary, F. M., Emery, A. A., & Hassan, M. A. M. (2013). Application of Particle Swarm Optimization in Design of PID Controller for AVR System. *International Journal of System Dynamics Applications*, *2*(3), 17. doi:10.4018/ijsda.2013070101

Bajracharya, A. (2016). Public Transportation and Private Car: A System Dynamics Approach in Understanding the Mode Choice. *International Journal of System Dynamics Applications*, *5*(2), 18. doi:10.4018/IJSDA.2016040101

Bouzaida, S., & Sakly, A. (2018). Adaptive Neuro-Fuzzy Sliding Mode Controller. *International Journal of System Dynamics Applications*, *7*(2), 34–54. doi:10.4018/IJSDA.2018040103

Cai, H., & Xu, L. (2015). Modeling and Control for Cage Rotor Dual Mechanical Port Electric Machine - Part II: Independent Control of Two Rotors. *IEEE Transactions on Energy Conversion*, *30*(3), 966–973. doi:10.1109/TEC.2015.2400971

Cui, S., Yuan, Y., & Wang, T. (2008). Research on switched reluctance double-rotor motor used for hybrid electric vehicle. *2008 International Conference on Electrical Machines and Systems*.

de Santiago, J., Bernhoff, H., Ekergård, B., Eriksson, S., Ferhatovic, S., Waters, R., & Leijon, M. (2012). Electrical Motor Drivelines in Commercial All-Electric Vehicles: A Review. *IEEE Transactions on Vehicular Technology*, *61*(2), 475–484. Advance online publication. doi:10.1109/TVT.2011.2177873

Engineering, C., & Sciences, N. (2018). Optimal Design of High-Performance Interior Pm. *The International Journal of Energy & Engineering Sciences*, *3*(2), 26–35.

Flah, A. (2017). HEV recharge battery algorithm using a fuzzy controller. *International Conference on Green Energy Conversion Systems (GECS)*.

Flah, A. (2018). Internal Fuzzy Hybrid Charger System for a Hybrid Electrical Vehicle. *ASME. Journal of Energy Resources Technology*, *140*(1), 12003–12008. doi:10.1115/1.4037352

Jahns, T. M., Han, S., El-Refaie, A. M., Baek, J., Aydin, M., Guven, M. K., & Soong, W. L. (2006). Design and Experimental Verification of a 50 kW Interior Permanent Magnet Synchronous Machine. In *Conference Record of the 2006 IEEE Industry Applications Conference Forty-First IAS Annual Meeting* (Vol. 4, pp. 1941–1948). 10.1109/IAS.2006.256801

Kharola, A., & Patil, P. P. (2017). Position and Tilt Control of Two-Wheeled Robot (TWR). *International Journal of System Dynamics Applications*, *6*(4), 17–33. doi:10.4018/IJSDA.2017100102

Kroeze, R. C., & Krein, P. T. (2008). Electrical battery model for use in dynamic electric vehicle simulations. In *IEEE Power Electronics Specialists Conference, 2008. PESC 2008.* (pp. 1336–1342). 10.1109/PESC.2008.4592119

Lee, S., Cherry, J., Safoutin, M., McDonald, J., & Olechiw, M. (2018). Modeling and Validation of 48V Mild Hybrid Lithium-Ion Battery Pack. *SAE International Journal of Alternative Powertrains*, *7*(3), 1–11. doi:10.4271/2018-01-0433

Mohamed, M. I. (2012). Applications of ANFIS and Fuzzy Algorithms for Improvement of the DTC Performance for the Three Phase Saturated Model of Induction Motor. *International Journal of System Dynamics Applications*, *1*(3), 30. doi:10.4018/ijsda.2012070102

Njajra, Z., Flah, A., Elhak Chariag, D., & Sbita, L. (2017). A PMDRM decoupling vector control strategy for EVT drive in hybrid vehicles. *International Conference on Green Energy Conversion Systems (GECS)*. 10.1109/GECS.2017.8066227

Pišek, P., Štumberger, B., Marčič, T., & Virtič, P. (2013). Design analysis and experimental validation of a double rotor synchronous PM machine used for HEV. *IEEE Transactions on Magnetics*, *49*(1), 152–155. doi:10.1109/TMAG.2012.2220338

Salima, M., Loubna, A., & Riad, T. (2018). A Global Stability of Linearizing Control of Induction Motor for PV Water Pumping Application. *International Journal of System Dynamics Applications*, *7*(3), 26. doi:10.4018/IJSDA.2018070102

Singh, K. V., Bansal, H. O., & Singh, D. (2019). A comprehensive review on hybrid electric vehicles: Architectures and components. *Journal of Modern Transportation*, *27*(2), 77–107. doi:10.100740534-019-0184-3

Tutelea, L. N., Boldea, I., & Deaconu, S. I. (2012). Optimal design of dual rotor single stator PMSM drive for automobiles. *2012 IEEE International Electric Vehicle Conference, IEVC 2012*. 10.1109/IEVC.2012.6183224

Un-Noor, F., Padmanaban, S., Mihet-Popa, L., Mollah, M. N., & Hossain, E. (2017). A comprehensive study of key electric vehicle (EV) components, technologies, challenges, impacts, and future direction of development. *Energies*, *10*(8), 1–82. doi:10.3390/en10081217

Wei, L. (2013). Hybrid Vehicle System Modeling. In *Introduction to Hybrid Vehicle System Modeling and Control*. Academic Press.

Xiao, R., Yang, X., Zhu, W., & Cao, J. (2009). Modeling and simulation of dual-rotor permanent-magnet synchronous motor based on MATLAB. *Proceedings - The 12th International Conference on Electrical Machines and Systems, ICEMS 2009*, 1–4. 10.1109/ICEMS.2009.5382660

Zhang, R., Tao, J., & Zhou, H. (2019). Fuzzy Optimal Energy Management for Fuel Cell and Supercapacitor Systems Using Neural Network Based Driving Pattern Recognition. *IEEE Transactions on Fuzzy Systems*, *27*(1), 45–57. doi:10.1109/TFUZZ.2018.2856086

Zhou, D., Gao, F., Ravey, A., Al-Durra, A., & Simões, M. G. (2017). Online energy management strategy of fuel cell hybrid electric vehicles based on time series prediction. In 2017 IEEE Transportation Electrification Conference and Expo. ITEC. doi:10.1109/ITEC.2017.7993256

Chapter 22
Discrete Time Chaotic Maps With Application to Random Bit Generation

Lazaros Moysis
https://orcid.org/0000-0002-5652-2532
Laboratory of Nonlinear Systems – Circuits & Complexity, Physics Department, Aristotle University of Thessaloniki, Greece

Ahmad Taher Azar
https://orcid.org/0000-0002-7869-6373
Robotics and Internet-of-Things Lab (RIOTU), Prince Sultan University, Riyadh, Saudi Arabia

Aleksandra Tutueva
https://orcid.org/0000-0002-9747-1962
Youth Research Institute, Saint Petersburg Electrotechnical University "LETI", Russia

Denis N. Butusov
Youth Research Institute, Saint Petersburg Electrotechnical University "LETI", Russia

Christos Volos
Laboratory of Nonlinear Systems – Circuits & Complexity, Physics Department, Aristotle University of Thessaloniki, Greece

ABSTRACT

Chaotic behavior is a term that is attributed to dynamical systems whose solutions are highly sensitive to initial conditions. This means that small perturbations in the initial conditions can lead to completely different trajectories in the solution space. These types of chaotic dynamical systems arise in various natural or artificial systems in biology, circuits, engineering, computer science, and more. This chapter reports on some new chaotic discrete time two-dimensional maps that are derived from simple modifications to the well-known Hénon, Lozi, Sine-Sine, and Tinkerbell maps. Numerical simulations are carried out for different parameter values and initial conditions, and it is shown that the mappings either diverge to infinity or converge to attractors of many different shapes. The application to random bit generation is then considered using a collection of the proposed maps by applying a simple rule. The resulting bit generator successfully passes all statistical tests performed.

DOI: 10.4018/978-1-7998-5788-4.ch022

1. INTRODUCTION

By chaos and chaotic systems, we generally refer to dynamical systems whose behavior is highly sensitive to initial conditions. That is, small changes in the initial conditions of such a system will lead to completely different trajectories in the solution space, a phenomenon that is widely referred to as the butterfly effect, see for example Alligood et al. (1996). Thus, despite such systems being deterministic in nature with no stochastic parameters, their complex nature constitutes their behavior hard to predict and control. This happens due to the presence of nonlinear terms in the differential equations that describe the system.

Chaotic systems were introduced in the seminal work of Lorenz (1963), that introduced a three-dimensional weather model that described forced dissipative hydrodynamic flow. Since then, chaos theory as a discipline has vastly expanded and chaotic systems have been extensively studied over the last 40 years. The advancements in this area have been greatly influenced by the advancement of technology and personal computers, which makes it easier to follow the trajectories of chaotic systems for a longer time period and with a higher precision. Thus, the theoretical analysis of chaos has been assisted by the numerical simulations of such chaotic systems, made possible by the use of computers.

Out of the many chaotic systems, the ones that seem to be the most interesting are those that exhibit an attracting behavior. By that, we refer to systems whose trajectories converge with time to a set of values, called an *attractor* (Wang et al., 2017; Volos et al., 2018). This can be a single point, a curve, a surface or a manifold. For example, in a system with a spherical bowl and a rolling marble, the bottom of the bowl is called a *fixed-point* attractor. The attractor is called *strange* if it is a set with fractal structure, see Grassberger and Procaccia (2004); Ott (1981). Examples include the aforementioned Lorenz system, as well as the *Hénon, Lozi* and *Tinkerbell* maps that will be introduced in the next sections. It should also be noted that a system may have different attractors depending on the choice of initial conditions.

Chaotic systems have found their place in applications spanning a wide variety of natural and artificial systems. These include oscillators (Heagy, 1992; Kengne et al. 2012; Sharma et al. 2012), secure communications (Murali & Lakshmanan, 1998; Zaher, & Abu-Rezq, 2011), controller design (dos Santos Coelho & Mariani, 2012), lasers (Li et al. 2014; Yuan et al. 2014), chemical reactions (Gaspard, 1999), pseudo-random number generation (Stoyanov & Kordov, 2015), optical ring phase resonators (Aboites et al. 2009), biology (Das et al. 2014; Kyriazis, 1991) medicine (Qu, 2011, Witte & Witte, 1991), finance (Guegan, 2009; Nosoohi & Parvizian, 2013) and more.

A major task in the control of chaotic systems, is the one of finding a feedback control of the general form $u(t)=k(x(t))$, where u the input, x the state of the system and k the feedback law, in order to stabilize the system around its unstable equilibrium. Techniques include active control, adaptive control, sliding mode control, fuzzy logic control (Alain et al., 2017; Vaidyanathan & Azar, 2015a, 2016a,b,c).

Chaos synchronization is another significant problem, which rises when two or more chaotic systems are coupled, or when a chaotic system, called the *master system*, drives another chaotic system, called the *slave system* (El-Ghazaly; 2012; Azar et al., 2017; Ouannas et al., 2017a,b,c,d,e; Singh et al., 2017, 2018a,b,c; Khan et al., 2020a,b; Longge & Xiangjie, 2013;). Due to the aforementioned butterfly effect in chaotic systems which causes the exponential divergence of the trajectories of two identical systems with nearly the same initial conditions, their synchronization is a difficult task and is in itself a subject of extensive research. As with the feedback control, various techniques have been established for chaos synchronization. The choice of the technique used depends on the type of the system and the available resources. For example, the *active control method* (Rafikov & Balthazar, 2008; Longe et al., 2013) is used when all the system parameters are available for measurement and other more robust methods like *adaptive control* (Vaidyanathan & Azar, 2016a,b,c), *sliding mode control* (Vaidyanathan et al., 2019; Vaidyanathan & Azar, 2015b,c,d,e) and fractional order control (Ouannas et al., 2020a,b,c,d, 2019; Alain et al., 2018; Khettab et al., 2018; Ouannas et al., 2017f,g; Pham et al., 2018, 2017) are preferred when some or all the system parameters are not available for measurement and estimates of them are used.

A special case of chaotic systems is hyperchaotic systems. They are defined as chaotic systems having more than one positive Lyapunov exponents and where first introduced by Rössler (1979). The minimum dimension for a hyperchaotic continuous time system is four, while for discrete time chaotic maps this is not necessary. As with chaotic systems, hyperchaotic systems have applications in oscillators (Zou et al. 2005), secure communications (Filali et al. 2014; Li et al. 2005; Aguilar-Bustos & Cruz-Hernández, 2009; encryption (Zhang et al. 2013) and more. Similar to chaotic systems, various methods have been developed for their synchronization .

The main aim of this work is to introduce new chaotic two-dimensional discrete time maps with chaotic behavior. These are derived from previously introduced and well-studied maps like the Hénon map, the Lozi map, the Tinkerbell map and the recently introduced Sine-sine map. The new maps are derived from adding nonlinear terms on these systems, or by multiplying any nonlinear term in the equations by a scaling factor. Thus, new chaotic maps are introduced, taking different shapes for different parameter values and different initial conditions (Balibrea-Iniesta et al., 2015; Meranza-Castillón et al., 2019). The next step of this work would be to study various qualitative properties of these maps, like their basin of attraction, dissipativity, their Lyapunov exponents and dimension, as well as the problem of adaptive control and synchronization. In addition, the construction of novel discrete hyperchaotic maps from the ones introduced here is a subject of interest.

This book chapter extends the work of Moysis & Azar (2017) by considering the application of the proposed maps to the problem of random bit generation (RBG), see Boyacı & Tantuğ, (2017); Stoyanov & Kordov (2015); Huang et al. (2019), Wang et al. (2016), Wang & Cheng, (2019); Murillo-Escobar et al. (2017), Patidar et al., (2009), Hamza (2017), Irfan et al., (2020), Nepomuceno et al., (2019), Tutueva et al., (2020), Alawida et al., (2020), Rezk et al. (2020), Liu et al. (2016), Volos et al. (2016), Meranza et al., (2019). The problem posed is to design a simple rule for generating a bit sequence, based on the values of a chaotic map. Using chaotic maps to generate a bit sequence classifies this technique as pseudo random bit generation, since a deterministic system is used, rather than a random natural process, as in true random bit generators. The resulting bit sequence should have the characteristics of a random sequence, which is tested using the National Institute of Standards and Technology statistical test suite (NIST 800-22), see Rukhin et al. (2001). This package consists of 15 tests that test the randomness of a bitstream, and it is seen that the proposed generators pass all the tests successfully. Moreover, autocorrelation and cross-correlation tests are performed, as well as sensitivity simulations. Finally, the key space for the proposed RBG is computed and it is seen that is above the desired threshold value.

This chapter is organized as follows. In each section, a known map is presented and novel maps that are derivations of it are introduced. In Section 2, the Hénon and Lozi maps are presented first and the Hénon and Lozi like maps are introduced. In section 3 the same is done with the Sine-sine map and the maps that are derived from it. Section 4 deals with the Tinkerbell map and its similar novel maps. In Section 5, the application to random bit generation is considered. The last Section concludes the chapter with future directions. At the end of the chapter, an Appendix with the Matlab simulation codes is provided. Each map presented in this work is accompanied by multiple simulations for varying parameter values.

2. HÉNON-LOZI LIKE MAPS

In this section we begin by providing the known *Hénon* and *Lozi* maps, and then we provide some new maps that are derived by simple modifications of the above.

2.1 The Hénon Map

The *Hénon Map*, was first introduced in Hénon (1976) and is a two dimensional and invertible map that is one of the most known and studied examples of a dynamical system with a strange attractor. It is a discrete time system that maps a point (x_n, y_n) using the following formula

$$\begin{aligned} x_{n+1} &= 1 - ax_n^2 + y_n \\ y_{n+1} &= bx_n \end{aligned} \tag{1}$$

where a and b are the system parameters. The map was initially developed as a simplified discrete system that could exhibit similar properties to the Lorenz attractor (1963), which was developed by Lorenz in 1963 as a system of differential equations that models forced dissipative hydrodynamic flow.

Numerical investigations of the Hénon map, along with studies on its various properties, like its fractal structure, its Lyapunov exponents, its basin of attraction, its Kaplan-Yorke dimension, its correlation dimension, have been carried out in numerous works, see for example Akhmet & Fen (2016); BalibreaIniesta et al. (2015); Benedicks & Carleson (1991); Cvitanović et al. (1988); Farmer et al. (1983); Galias (2016); Grassberger & Procaccia (1983, 2004); Guckenheimer & Holmes (2013); Lichtenberg & Lieberman (2013); Lorenz (2008); Michelitsch & Rössler (1989); Ott (1981); Ouannas & Al-sawalha (2016); Parlitz (2016); Richter (2002); Sprott & Xiong (2015); Strogatz (2014); Tanase (2016); Wolf et al. (1985), Meranza et al., (2019); Valendar et al. (2020), and the references therein. These properties are used extensively in studying the behavior of a dynamical system and identifying its chaotic structure. For example, the Lyapunov exponent measures the sensitivity of a system to small changes in the initial conditions. The basin of attraction gives the set of initial conditions for which the system converges to the attractor as time approaches infinity. Kaplan-Yorke and correlation dimensions are other dimensional measures that are used in studying fractal structures.

Physical interpretations of the Hénon map have also been explored. For example, in Heagy (1992) the map is linked to the period one return map of a periodically kicked harmonic oscillator and in Biham & Wenzel (1989) it is shown that there is a one-to-one correspondence between the trajectories of this map and the static extrema of a Hamiltonian describing an infinite chain of atoms interacting with a potential and among themselves.

For different values of the parameters the attractor can take various shapes. The classic map is the one with $a = 1.4$, $b = 0.3$, given in Figure 1. Zooming into a particular part of the map reveals that each curved line consists of multiple components. This is done in Figure 2 where the first 5000000 points are computed, which reveals the fractal nature of this map. In contrast though to the Lorenz system whose trajectories are all bounded, for different initial conditions the Hénon map may either converge to the attractor or diverge to infinity. In Figure 3, we plot the first 100000 points of this mapping for different initial conditions and different parameter values.

Figure 1. The classic Hénon map

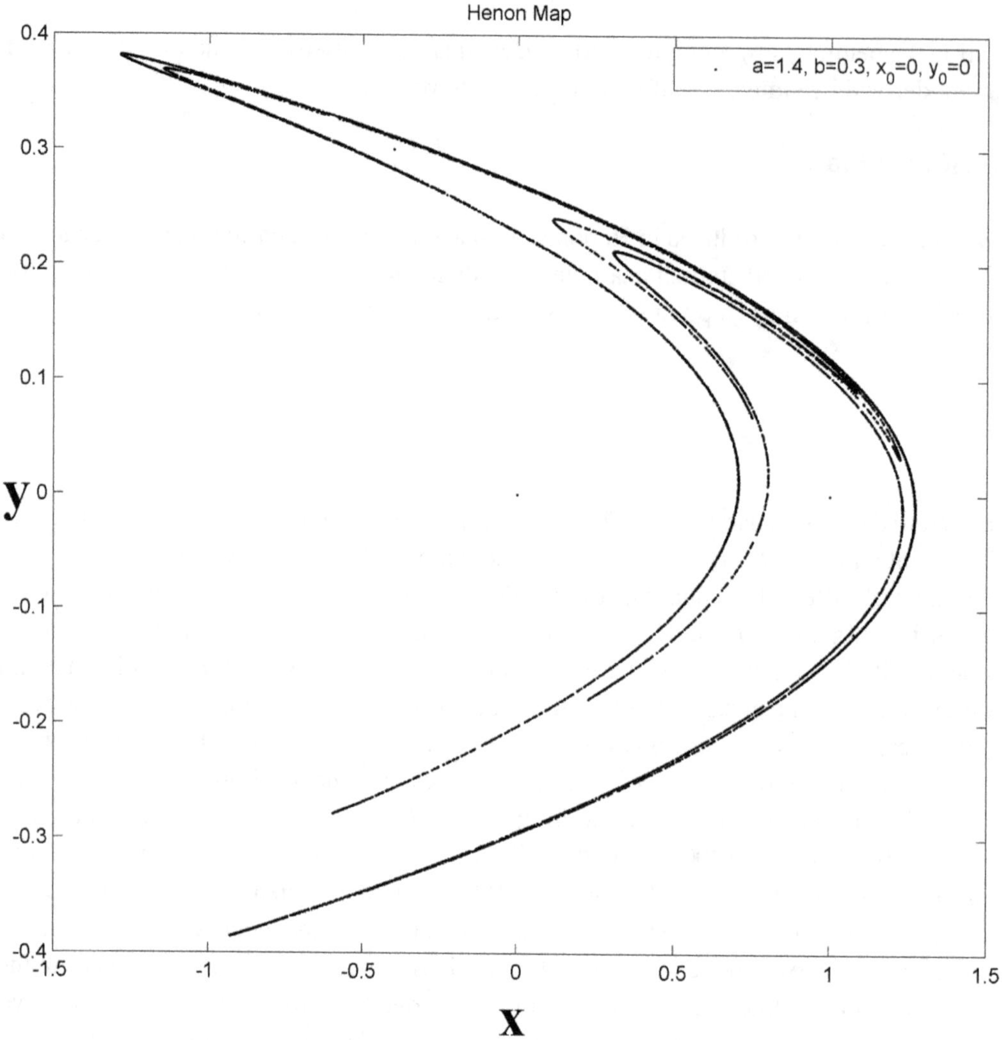

2.2 The Lozi Map

Similar to the Hénon map is the *Lozi map*, first introduced in Lozi (1978) as an ever simpler map compared to the Hénon map. It is the discrete time dynamical system

$$\begin{aligned} x_{n+1} &= 1 - a|x_n| + by_n \\ y_{n+1} &= bx_n \end{aligned} \tag{2}$$

where *a, b* are the system parameters. Its basic difference to the Hénon map is that its derivative has a discontinuity. The Lozi map has been analyzed in various works, see for example Aziz-Alaoui et al. (2001); Cvitanović et al. (1988); Guckenheimer & Holmes (2013); Lopesino et al. (2015); Sprott & Xiong (2015); Strogatz (2014); Merah et al. (2017) and the references therein. A detailed collection of

Figure 2. Close up look of the Hénon map

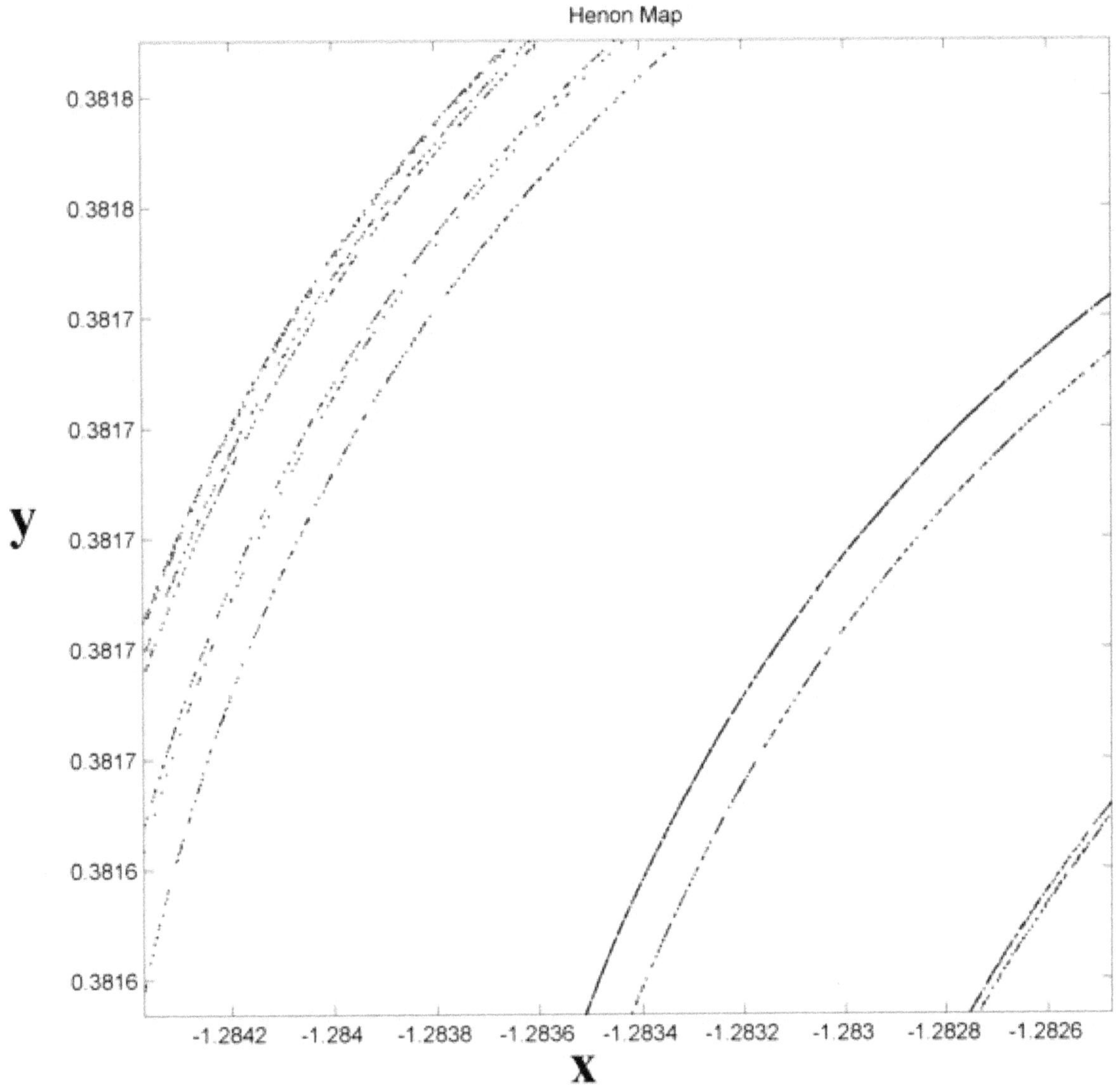

Figure 3. Hénon map for different parameter values

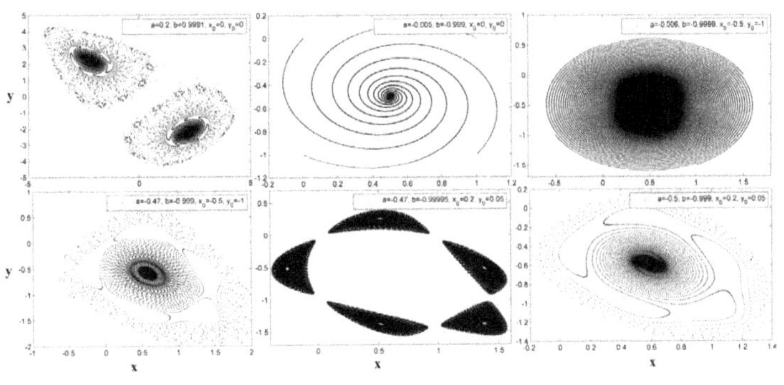

the works around Lozi mappings is the book of Elhadj (2013), which provides an analysis on all aspects and properties of Lozi mappings, as well as provide an extensive (over a thousand) reference list on the ongoing work around the Lozi map that has taken place over the last 35 plus years.

Applications of the Lozi map span multiple fields, like that of control systems, synchronization theory and secure communications. For a detailed presentation of various applications see (Elhadj, 2013, Chapter 5).

What is historically notable about this map is that although introduced later than the Hénon map on which it was based, the Lozi map was proven first to possess a strange attractor in Misiurewicz (1980), while for the Hénon map it was proven in Benedicks and Carleson (1991), around ten years later.

The classic map is the one with $a = 1.7, b = 0.5$, as shown in Figure 4. Again as with the Hénon map, zooming into a particular part of the map reveals that each straight line consists of multiple components. This is done in Figure 5 where the first 5000000 points are computed, revealing the fractal nature of this map. In Figure 6, we compute and plot the first 100000 points of this map for different initial conditions and parameter values. As with the Hénon map, for different parameters and initial conditions the map may diverge to infinity.

Figure 4. The classic Lozi map

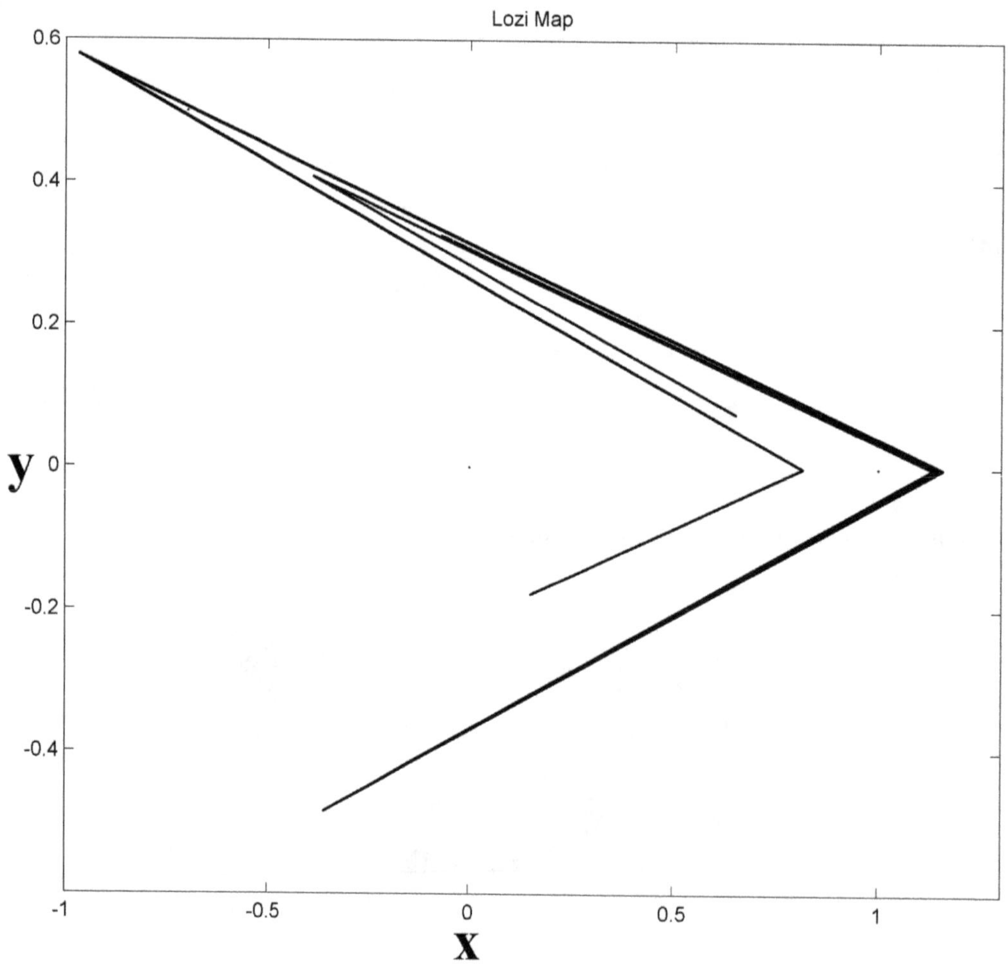

Figure 5. Close up look of the Lozi map

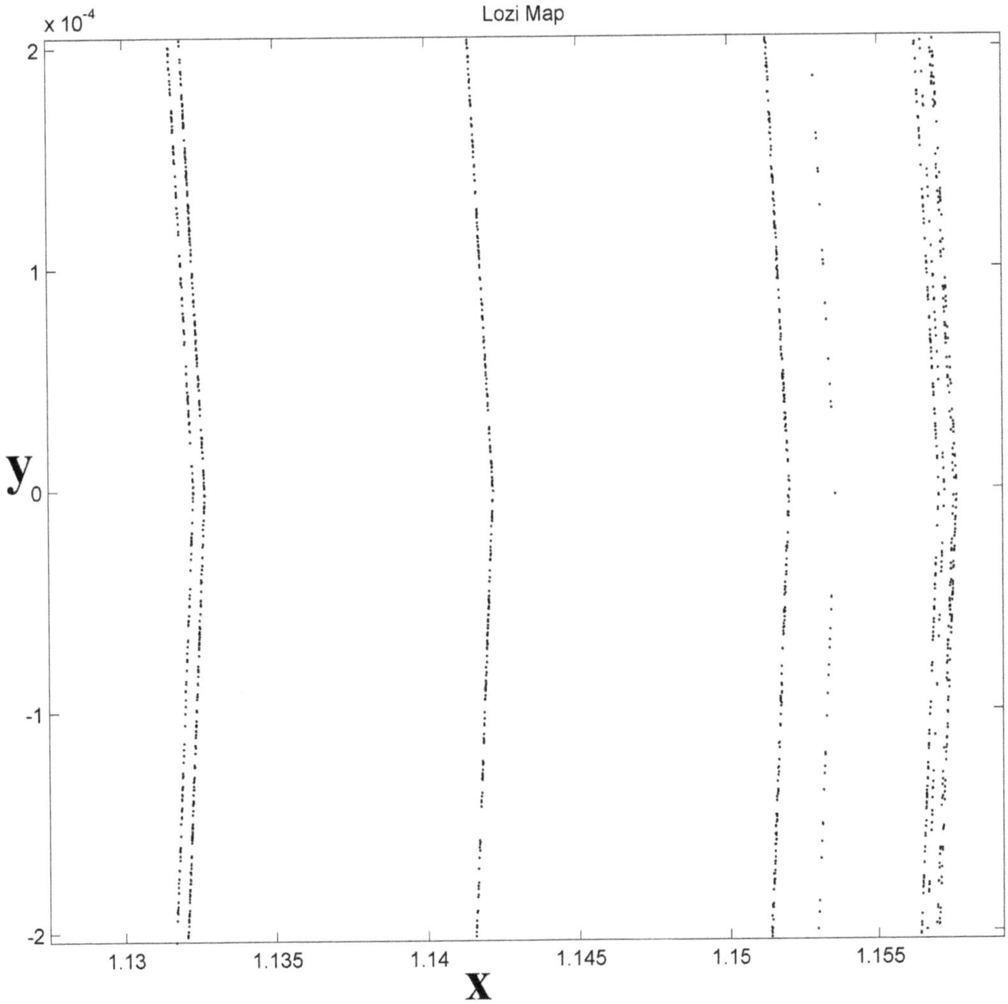

Figure 6. Lozi map for different parameter values

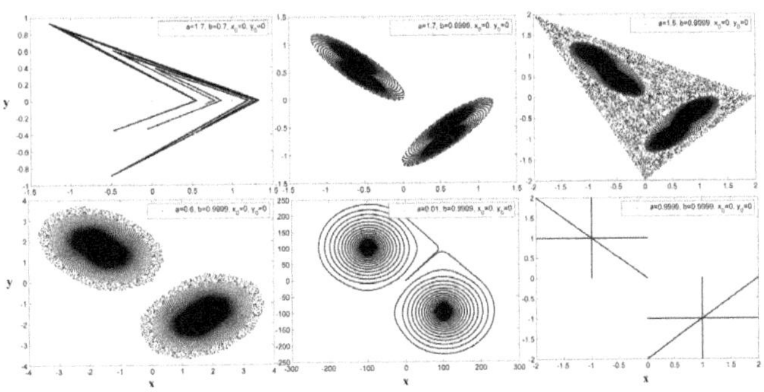

2.3 Hénon and Lozi Type Maps

Similar to the Hénon map and the Lozi map, consider the following four Hénon and Lozi like maps, that include the absolute value of x_n and/or the power of x_n, y_n.

$$\begin{aligned} x_{n+1} &= 1 - a x_n^2 + y_n^2 \\ y_{n+1} &= b x_n \end{aligned} \tag{3}$$

$$\begin{aligned} x_{n+1} &= 1 - a | x_n | + y_n^2 \\ y_{n+1} &= b x_n \end{aligned} \tag{4}$$

$$\begin{aligned} x_{n+1} &= 1 - a (x_n + y_n)^3 \\ y_{n+1} &= b | x_n |; \end{aligned} \tag{5}$$

$$\begin{aligned} x_{n+1} &= 1 - a x_n^6 + y_n^5 \\ y_{n+1} &= b x_n; \end{aligned} \tag{6}$$

For the above maps, we plot the first 100000 points for zero initial conditions and different parameter values.

Figure 7. The Hénon-Lozi like map (3)

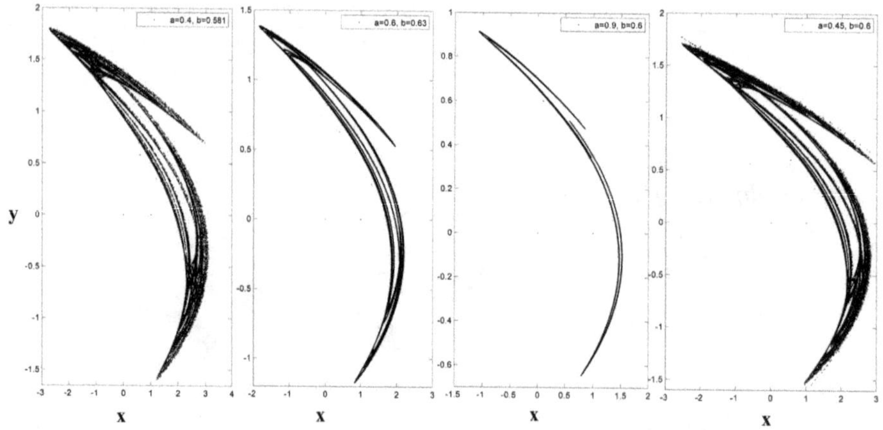

3. THE SINE-SINE LIKE MAPS

In this section we begin by providing the recent *Sine-Sine* map and then provide some new maps that are derived by simple modifications of it.

3.1 The Sine-Sine Map

The sine-sine map is a discrete time dynamical system introduced recently in Sprott and Xiong (2015) and is given by

$$\begin{aligned} x_{n+1} &= \sin x_n - \sin 2y_n \\ y_{n+1} &= x_n \end{aligned} \tag{7}$$

This system was introduced as an example of a map that has a basin which includes the complete state space, except from a set of points with zero measure. This is why it is called a global attractor. In Sprott and Xiong (2015) its Luapunov exponents and Kaplan-Yorke dimensions where also studied. Another recently introduced chaotic nonlinear map that includes sinusoidal functions is the one in Bi et al. (2013). For the sine-sine map, we plot the first 100000 points of this map with initial conditions x(0) = y(0) = 1.

Figure 8. The Hénon-Lozi like map (4)

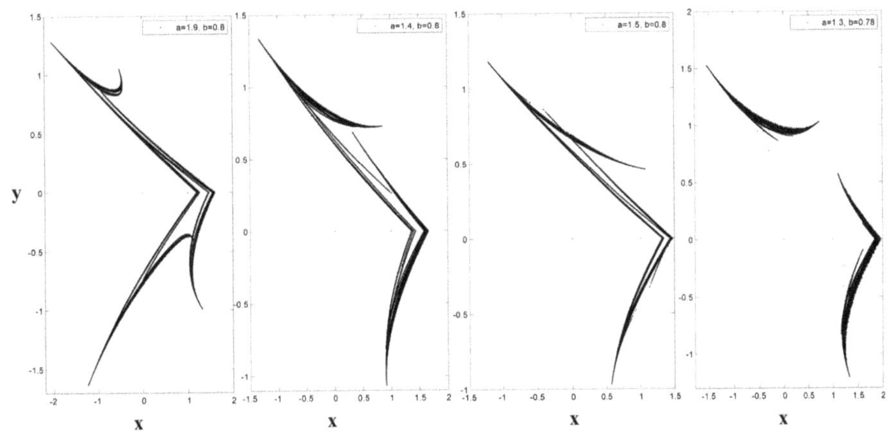

Figure 9. The Hénon-Lozi like map (5)

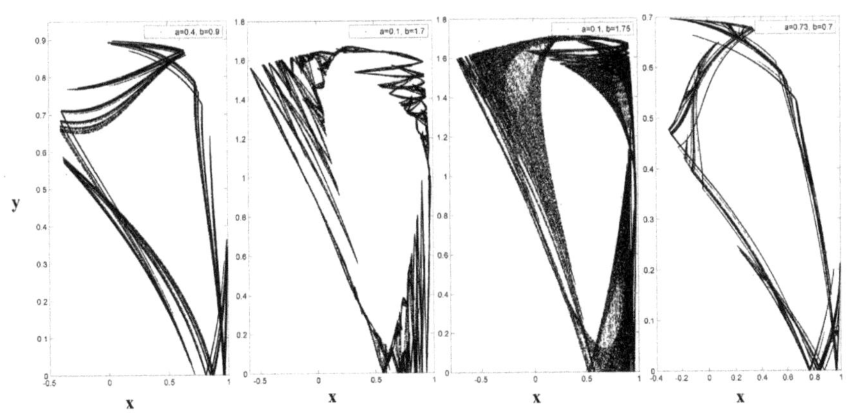

Figure 10. The Hénon-Lozi like map (6)

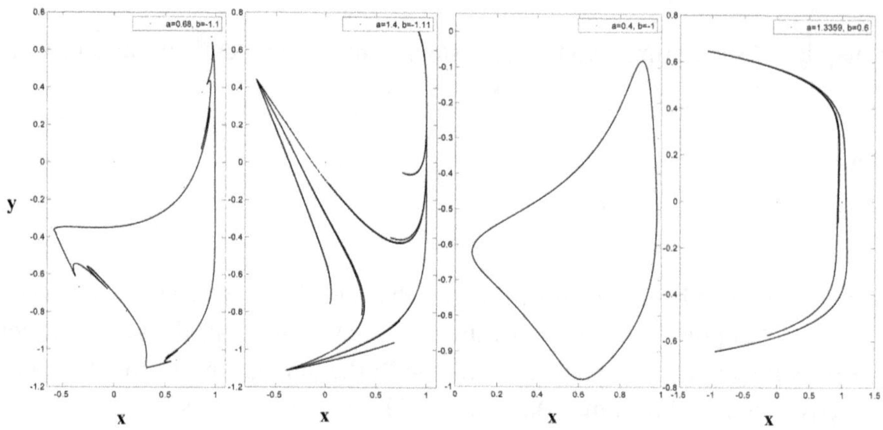

3.1 Sine-Sine Like Maps

Similar to the sine-sine map, we can construct different maps similar to it with peculiar chaotic behavior.
First, consider the following two Sine-sine maps with scaling factors

$$\begin{aligned} x_{n+1} &= a\sin x_n - \sin by_n \\ y_{n+1} &= cx_n \end{aligned} \tag{8}$$

$$\begin{aligned} x_{n+1} &= a\sin x_n - c\sin by_n \\ y_{n+1} &= x_n \end{aligned} \tag{9}$$

For these maps, we compute the first 400000 points of these maps with initial conditions $x(0) = y(0) = 1$.

Now, define the following three *Abs-Sine-Sine* maps that involve the absolute values of the sine functions

$$\begin{aligned} x_{n+1} &= |\sin x_n - \sin 2y_n| \\ y_{n+1} &= x_n \end{aligned} \tag{10}$$

$$\begin{aligned} x_{n+1} &= |\sin x_n| - |\sin 2y_n| \\ y_{n+1} &= x_n \end{aligned} \tag{11}$$

$$\begin{aligned} x_{n+1} &= |\sin x_n| - |\sin 2y_n| \\ y_{n+1} &= |x_n| \end{aligned} \tag{12}$$

Again, we compute the first 400000 points of these maps with initial conditions $x(0) = y(0) = 1$.

Figure 11. The sine-sine map

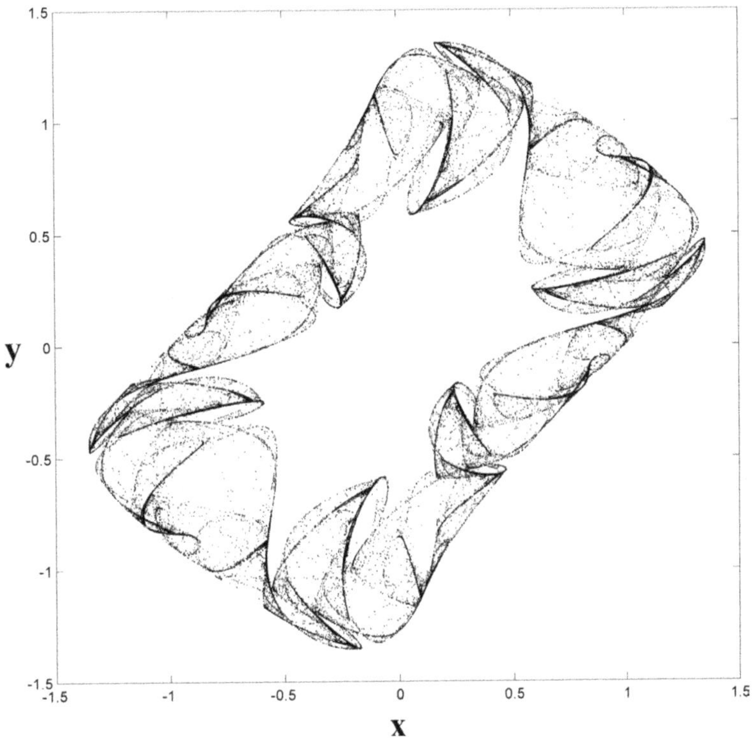

Figure 12. Scaled Sine-sine map (8)

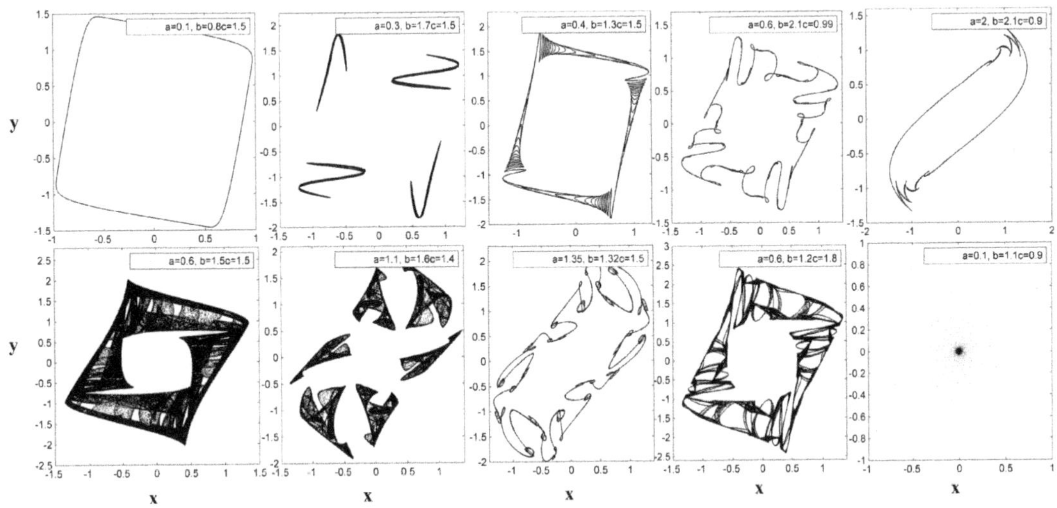

Figure 13. Scaled Sine-sine map (9)

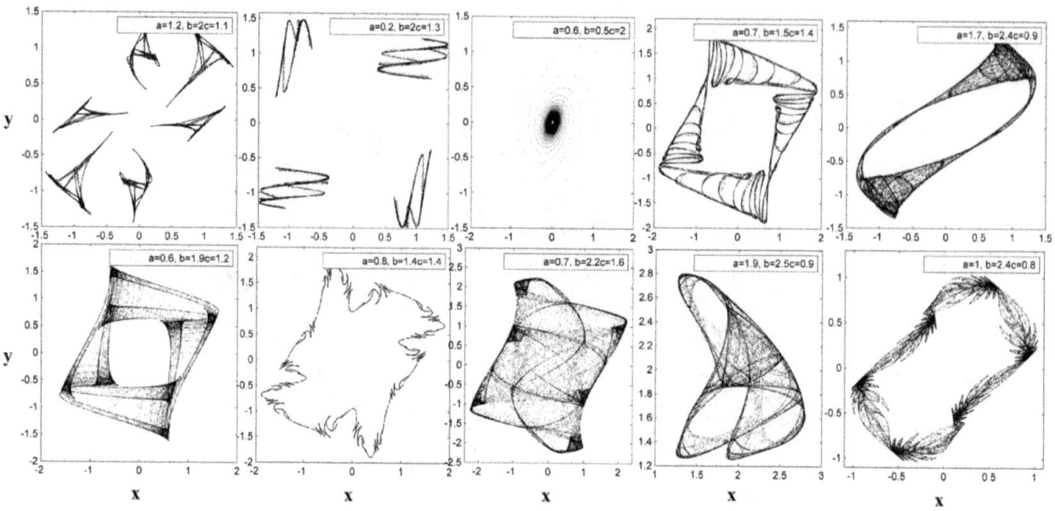

Figure 14. The Abs-sine-sine maps (10)-(12)

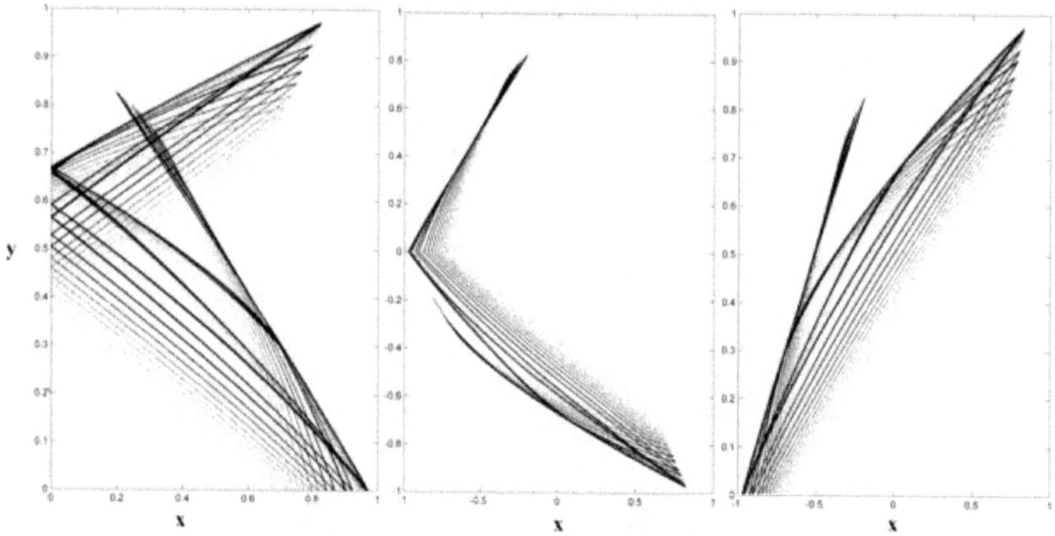

Now, consider the *Square-Sine-Sine* map that includes the square of a sine function

$$x_{n+1} = j(\sin x_n)^2 - \sin 2y_n$$
$$y_{n+1} = x_n \tag{13}$$

where the initial conditions are x(0)=1, y(0)=1. The value of *j* can take values, yielding different and interesting looking maps each time. For each value of *j* we plot the first 400000 points of the map. As the parameter *j* grows larger, the map takes the form of a vertical sine-like shape. For large values of *j*, we can see that the points of the map lie in a square shaped region whose bottom left point starts at the origin, with length sides close to, but not equal, to *j*. This can be seen in Figures 15, 16 and 17.

Figure 15. The Square-sine-sine map (13)

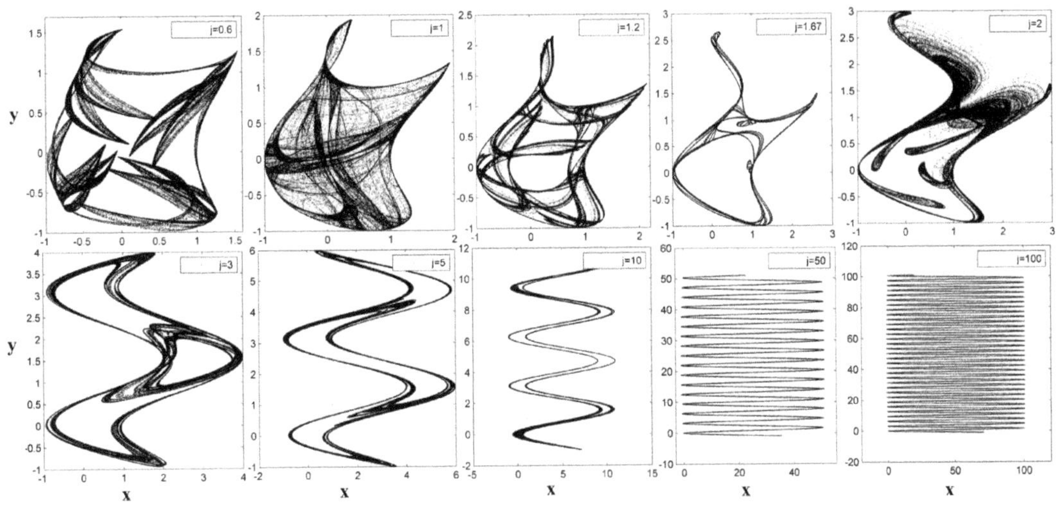

Figure 16. Square-sine-sine map for j=1000

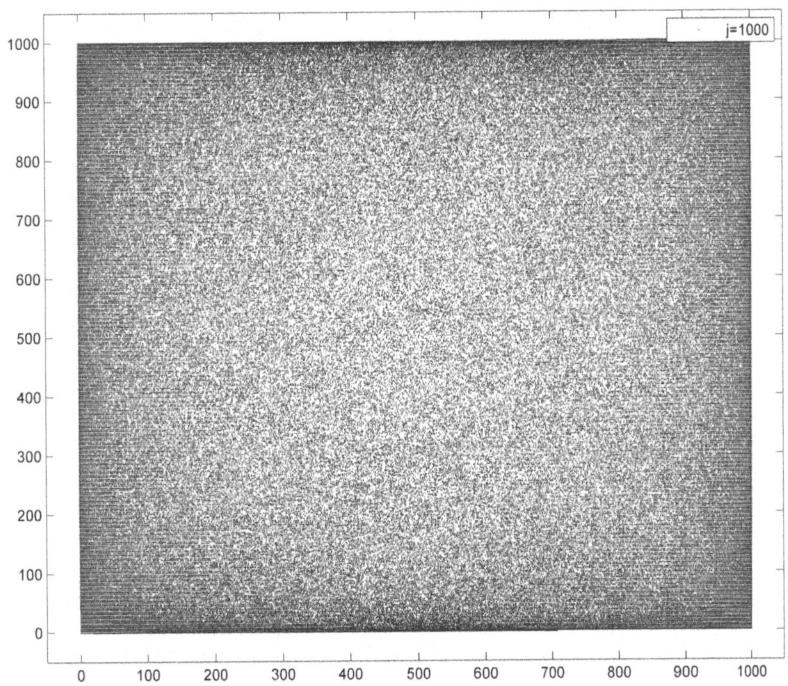

Similar to the Square-Sine-Sine map, define the *Cubic-Sine-Sine map* that includes the cubic of a sine function

$$x_{n+1} = j(\sin x_n)^3 - \sin 2y_n \qquad (14)$$
$$y_{n+1} = x_n$$

where the initial conditions are $x(0)=1$, $y(0)=1$. For different values of the parameter j the mapping yields completely different results, as can be seen in the following figure where we plot the first 400000 points of each map. It is clearly seen that this map is very peculiar in its behavior. As the parameter j grows larger, the map takes the form of a vertical sine-like shape, as with the previous map. For large values of j, we can see that the points of the map lie in a square shaped region centered at the origin, with length sides close to, but not equal to $2j$.

Figure 17. Close up of the Square-sine-sine map

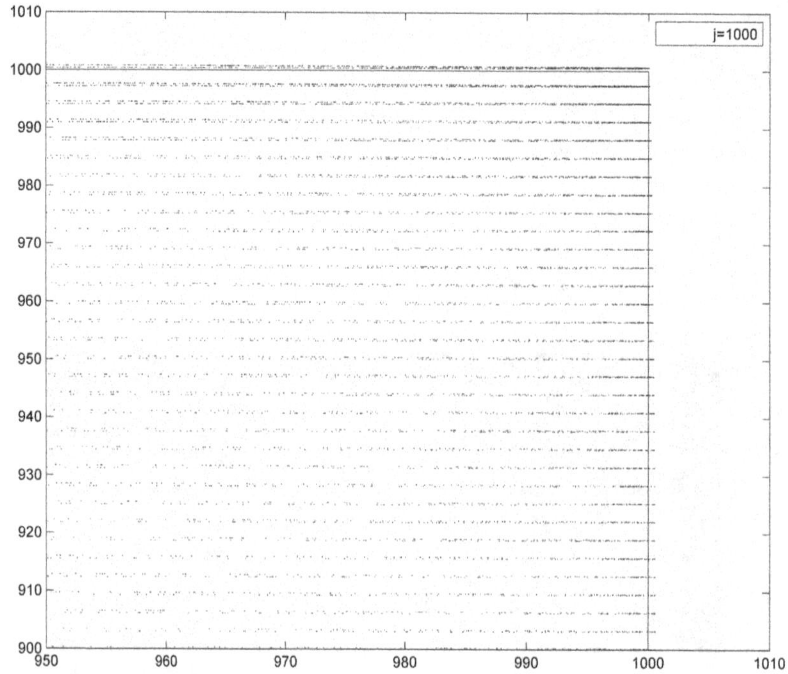

Figure 18. The Cubic-sine-sine map (14)

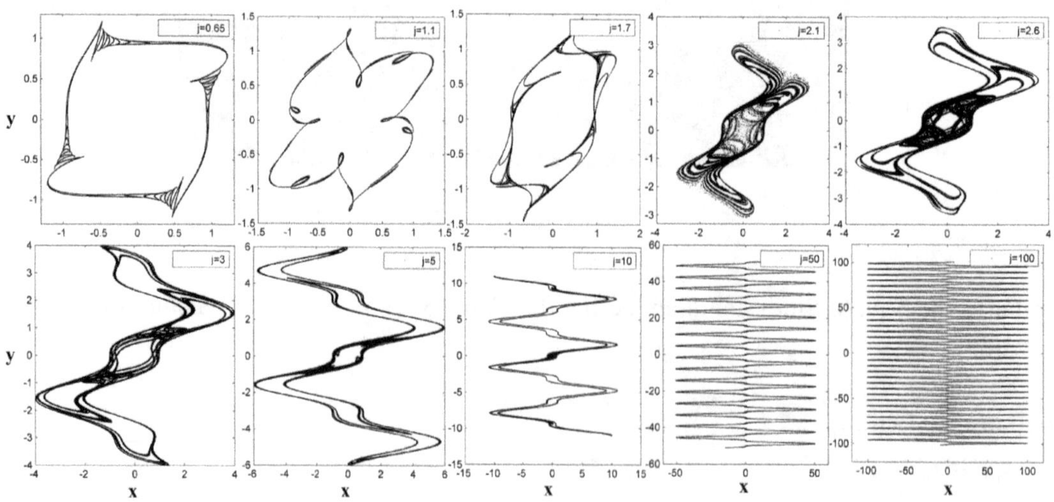

Figure 19. The Higher order sine-cos map (15)

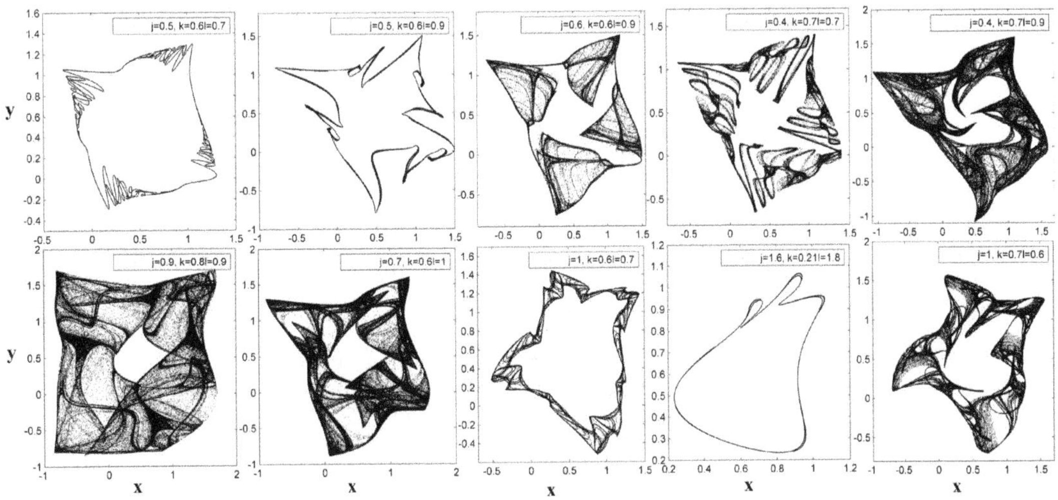

Lastly, consider the following *Higher order Sine-Cos map* that has tree cubic terms

$$
\begin{aligned}
x_{n+1} &= j(\sin x_n)^3 + k(\cos 2y_n)^3 + l(cos(x_n + y_n))^3 \\
y_{n+1} &= x_n
\end{aligned} \tag{15}
$$

where the initial conditions are $x(0)=1$, $y(0) = 1$. Since this matrix has three parameters, it can take various different shapes. Again though, as with the previous maps, for higher values of the parameters *j, k, l* the map will take a square form similar to Figure 16. The map for different values of the parameters for the first 400000 points is given in Figure 19. We can see that for specific parameters the map closely resembles the three winged chaotic attractor in Alpar (2015), which also happens for many of the maps presented here, see for example Figures 12, 13, 18, 19 and 23.

4. THE TINKERBELL LIKE MAPS

In this section we begin by providing the known *Tinkerbell* maps and then provide some new maps that are derived by simple modifications of the above

4.1 The Tinkerbell Map

The *Tinkerbell map* Alligood et al. (1996); Davidchack et al. (2001); Goldsztejn et al. (2011); Nusse and Yorke (2012); Sprott and Xiong (2015); Yuan et al. (2011) is another example of a chaotic map. It has been extensively analyzed numerically, as well as found its place in various applications, for example in an optical ring phase conjugated resonator Aboites et al. (2009), in PID controller design in dos Santos Coelho and Mariani (2012) and in pseudo-random number generators in Stoyanov and Kordov (2015); Todorova et al. (2018). The map is given by

$$x_{n+1} = x_n^2 - y_n^2 + ax_n + by_n$$
$$y_{n+1} = 2x_n y_n + cx_n + dy_n \quad (16)$$

The origin of the name has never been confirmed, though by looking at the plot of the map, it should be safe to assume that it is a reference to the Disney character of Tinker Bell as she flies over the castle in the company's opening logo. For this map we plot the first 100000 points for different values of the parameters, with initial conditions $x(0) = y(0) = -0.1$.

Figure 20. The Tinkerbell map

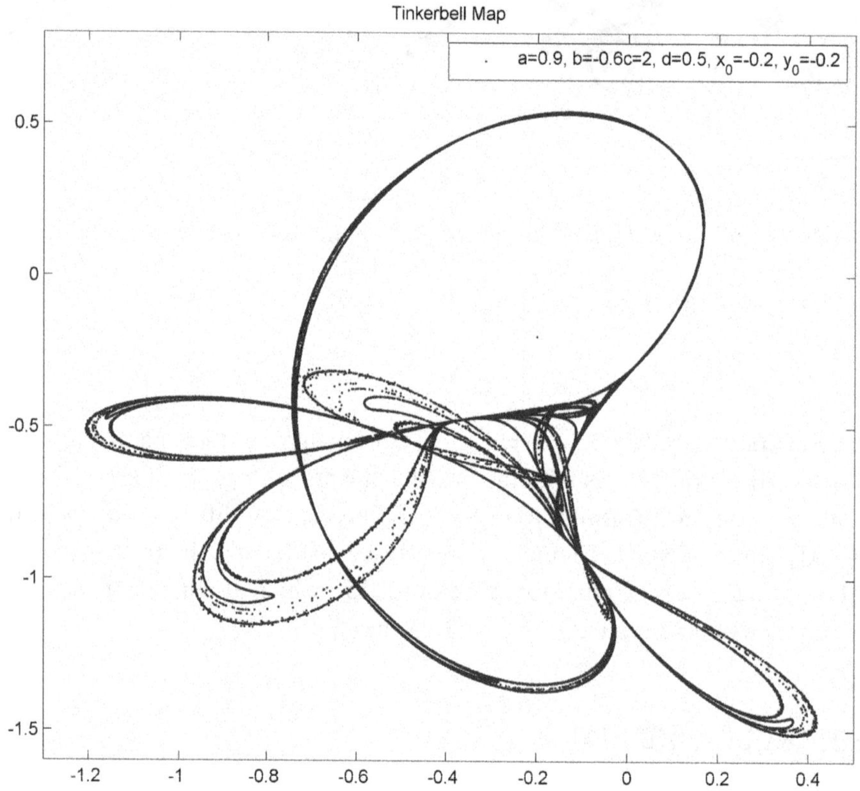

4.2 Tinkerbell Like Maps

Now, similar to the Tinkerbell map, consider the following maps that include cubic terms of x and y

$$x_{n+1} = x_n^2 - y_n^2 + ax_n + by_n^3$$
$$y_{n+1} = 2x_n y_n + cx_n + dy_n \quad (17)$$

$$x_{n+1} = x_n y_n + ax_n^3 + by_n$$
$$y_{n+1} = jx_n y_n + cx_n + dy_n + gx_n^3 \quad (18)$$

$$x_{n+1} = x_n y_n + a x_n^3 + b y_n$$
$$y_{n+1} = j x_n^2 y_n + c x_n + d y_n + g x_n^3 \tag{19}$$

where for each map the initial conditions are taken close to the origin. If the initial conditions are taken far from the origin the maps diverge to infinity. Since each map has three parameters, again as with the previous examples, the shapes can vary significantly. For each of these maps we plot the first 100000 points for different values of the parameters and initial conditions.

Figure 21. The Tinkerbell map for different parameter values

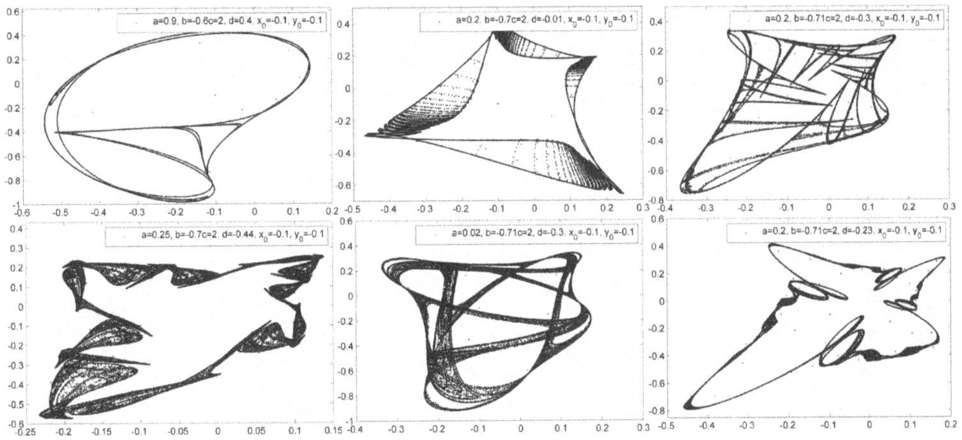

Figure 22. Tinkerbell like map (17)

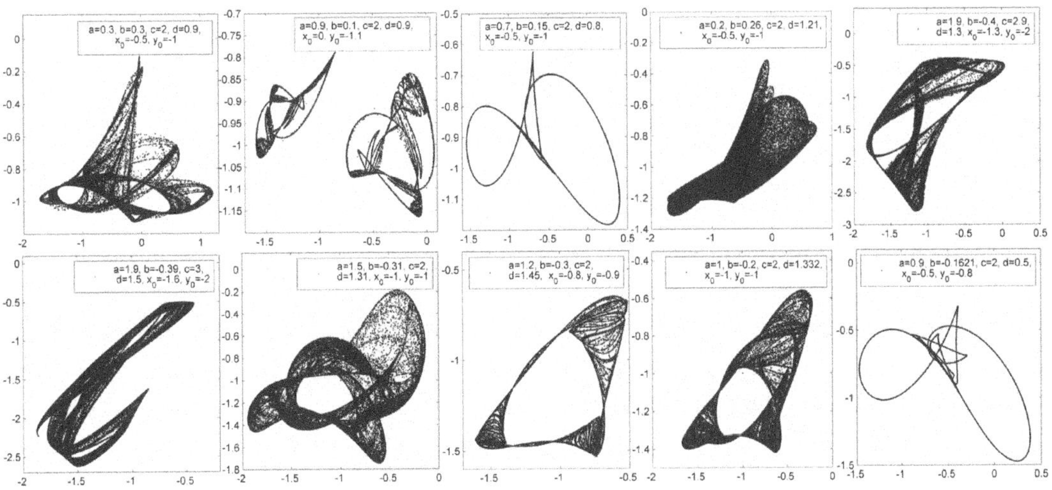

Figure 23. Tinkerbell like map (18)

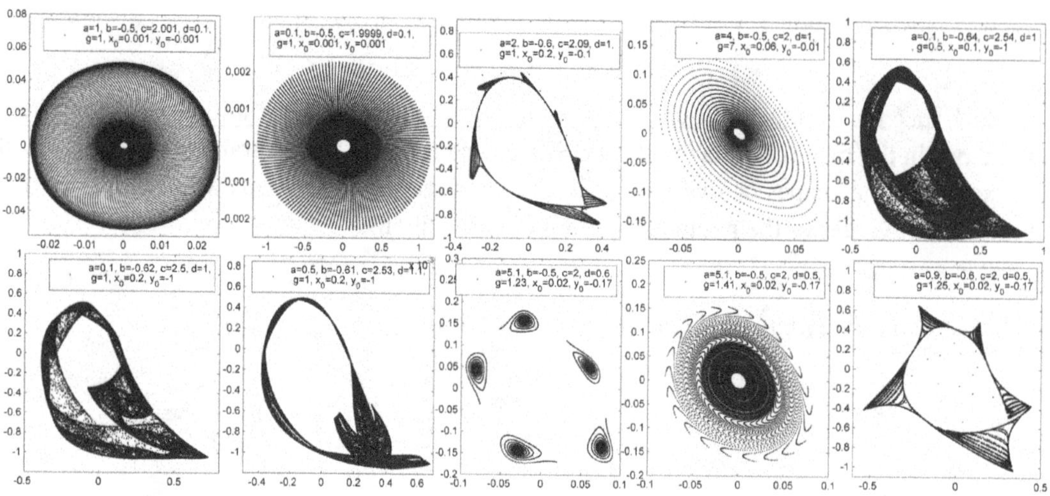

Figure 24. Tinkerbell like map (19)

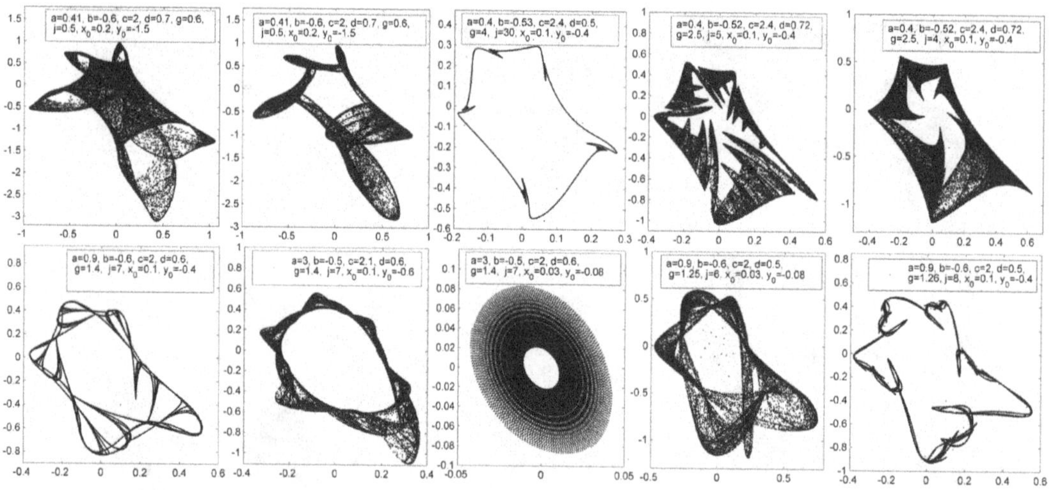

5. APPLICATION TO RANDOM BIT GENERATION

In this section, a collection of the above chaotic maps will be applied to the problem of random bit generation. Specifically, the maps considered are the scaled sine-sine map (9), the squared sine-sine map (13) and the cubic map (14). The proposed RBG consists of the following simple rule of comparing two chaotic maps. First, two chaotic maps (x_i, y_i) and (z_i, w_i) are initialized, using different initial conditions (or parameter values). Then, in each iteration, the first 11 decimal digits of the values of x_i, z_i are discarded. The resulting numbers b_i, q_i are compared and a "0" or "1" is generated depending on the result. This procedure is outlined by the following equations

Initialize x_0, y_0, z_0, w_0 and parameter values.

$$\begin{aligned} &\text{In each iteration } i: \\ &d_i = 10^{11} x_i \bmod 1 \\ &q_i = 10^{11} z_i \bmod 1 \\ &b_i = \begin{cases} 1, & d_i \geq q_i \\ 0, & d_i < q_i \end{cases} \end{aligned} \tag{16}$$

The bifurcations diagrams for the maps (9), (13) and (14) are shown in Figs. 25, 27 and 29, and the corresponding diagrams of the maximum Lyapunov exponent in Figs. 26, 28 and 30 respectively. For the map (9), the bifurcation diagram with respect to parameter a shows a rich dynamical behavior, with phenomena like quasiperiodic behavior, a period doubling route to chaos, and also crisis, where the system exits abruptly from chaos, These chaotic phenomena are verified from the diagram of maximum Lyapunov exponent shown in Fig. 26.

Similar phenomena are encountered in the bifurcation diagram and diagram of maximum Lyapunov exponent of systems (13) and (14) with respect to parameter j, *as seen in Figs.* 27,29 and 28,30 respectively. The phenomenon of antimonotonicity also appears, where the system enters into chaos through a period doubling route, and also exits from it following the reverse period halving route.

Figure 25. Bifurcation diagram of the scaled sine-sine map with respect to parameter a, and x_0=1, y_0=1, b=2.3, c=0.8.

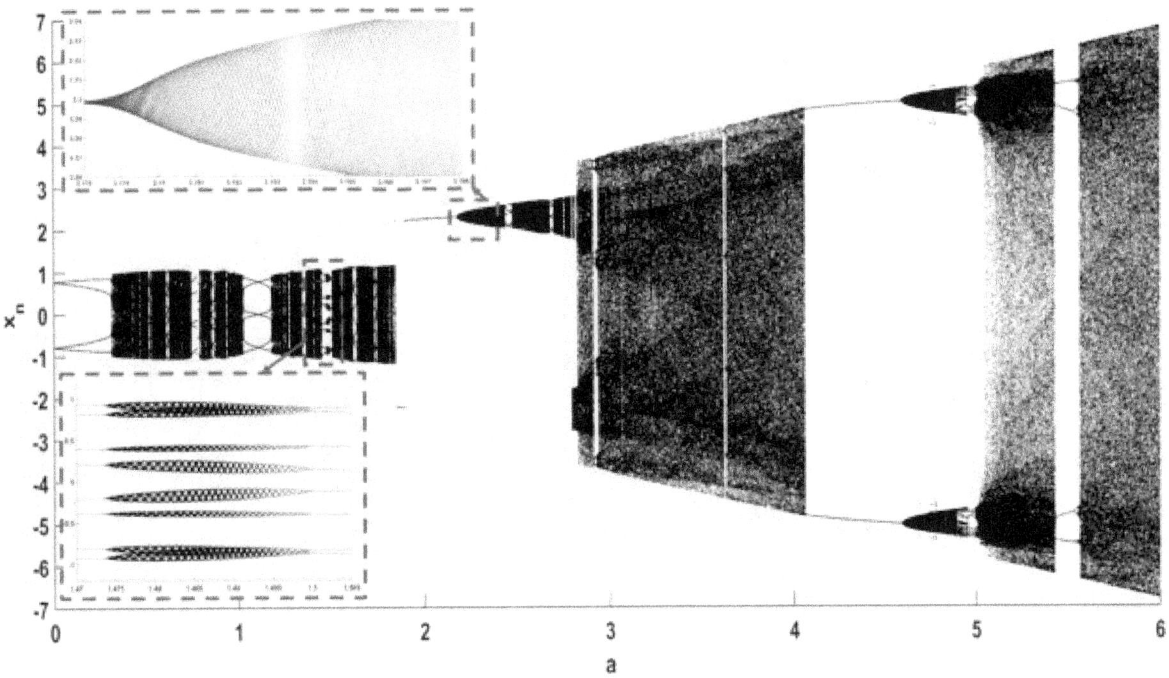

Figure 26. Diagram of the maximum Lyapunov exponent of the scaled sine-sine map with respect to parameter a, and $x_0=1$, $y_0=1$, $b=2.3$, $c=0.8$.

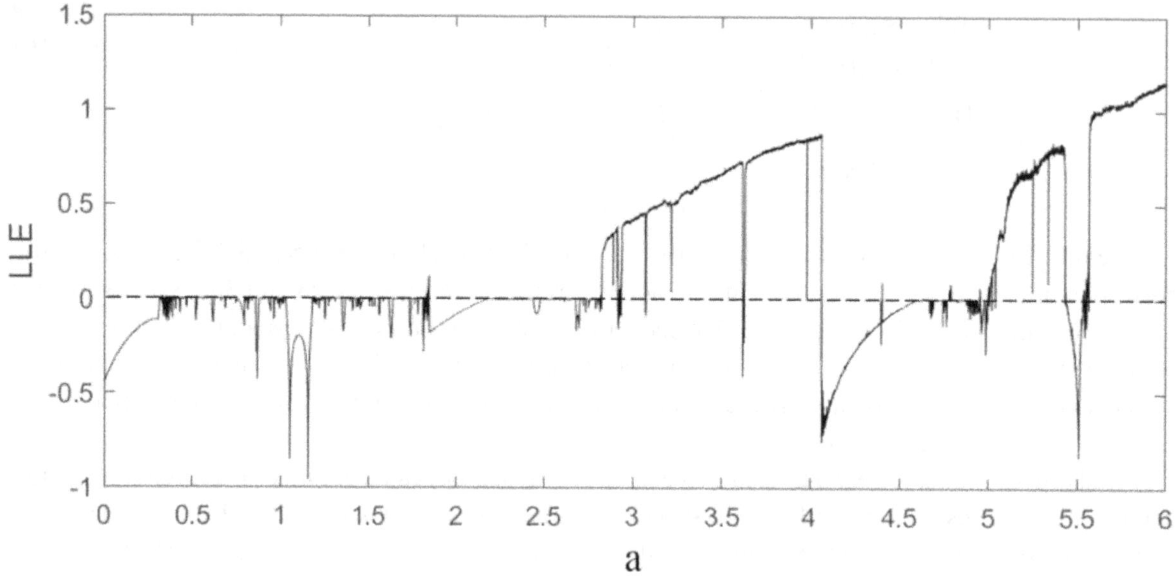

Figure 27. Bifurcation diagram of the squared sine-sine map (13) with respect to parameter j, and $x_0=1$, $y_0=1$.

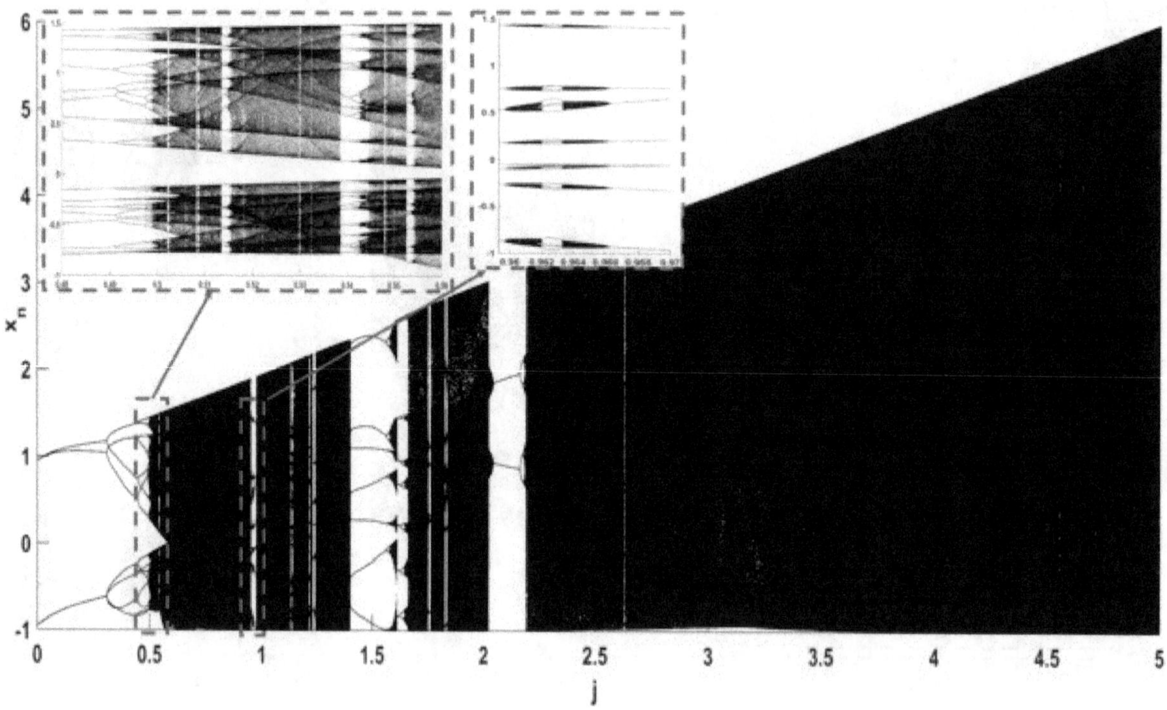

Figure 28. Diagram of the maximum Lyapunov exponent of the squared sine-sine map (13) with respect to parameter j, and $x_0=1$, $y_0=1$.

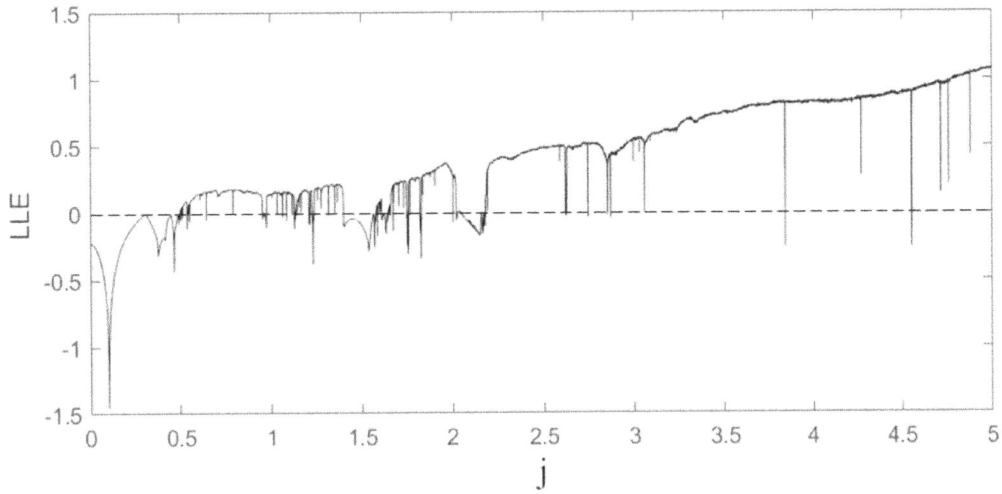

Based on the bifurcation diagrams, the bit generators are tested for parameter values $a=6$, $b=2.3$, $c=0.8$ for map (9) and $j=5$, 6 for maps (13) and (14). For the NIST tests, the significance level for each test is taken as the default value $s=0.01$. A set of 100 bitstreams of length 10^6 is tested, and the results are shown in Tables 1, 2 and 3.It is seen that for all three cases the tests are passed successfully.

To further study the randomness of the sequence, the distribution of the correlation coefficient for the obtained sequences for map (9) is shown in Fig. 31. The absolute value of the calculated measure does not exceed 0.005. This corresponds to a weak correlation, which the desired result. For the two other studied maps, the obtained results are similar.

Figure 29. Bifurcation diagram of the cubic sine-sine map (14) with respect to parameter j, and $x_0=1$, $y_0=1$

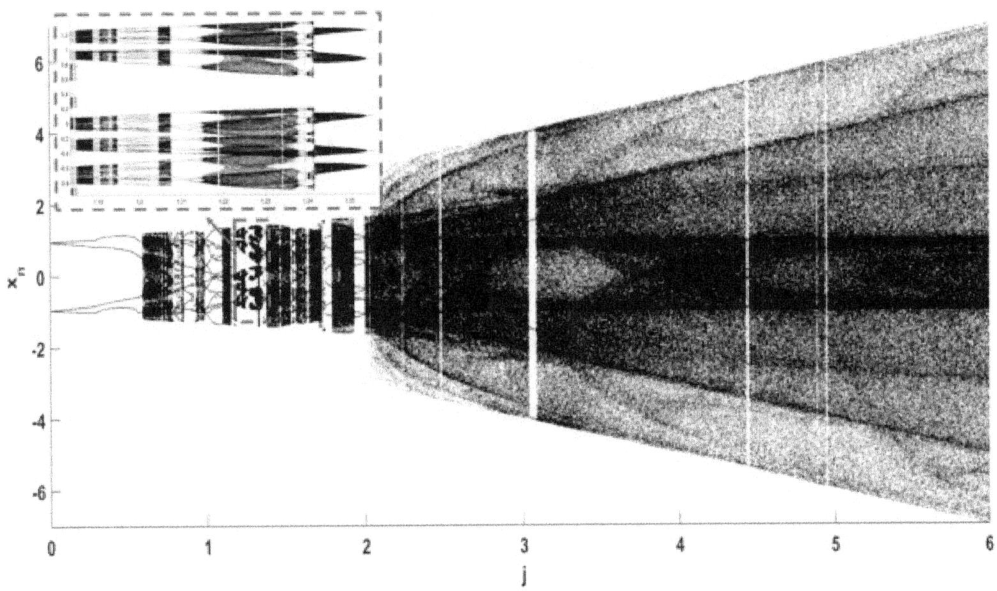

Figure 30. Diagram of the maximum Lyapunov exponent of the cubic sine-sine map (14) with respect to parameter j, and $x_0=1$, $y_0=1$.

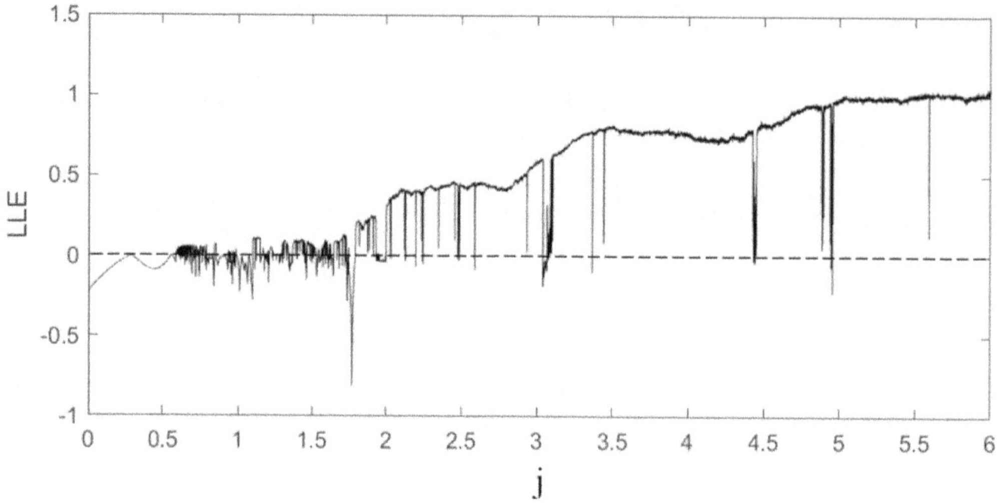

Figure 31. The distribution of the correlation coefficient for the sequences generated by the PRBG based on map (9).

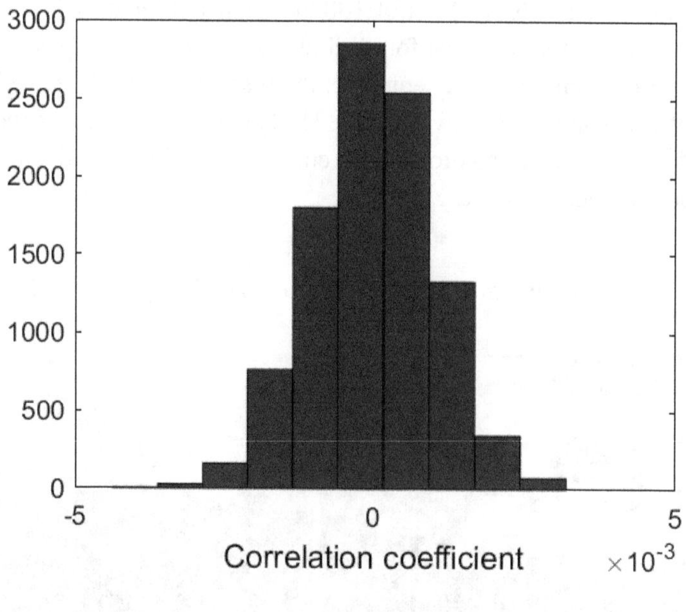

Also, to showcase the sensitivity of the design to initial conditions and parameter changes, a simulation is performed for small variations of them, shown in Figures 32 and 33. In each simulation, the initial conditions or parameter values of the first map (9) in (16) are changed, while the second map in (16) remains the same. It is clearly seen from the Figures that small changes lead to a different state trajectory and bit sequence after a very short number of iterations. Similar results are obtained for the other maps as well.

Figure 32. Sensitivity to initial conditions for the bit generation design (16) using system (9), with system parameters $y_0=1$, $z_0=0.1$, $w_0=0.1$, $a=6$, $b=2.3$, $c=0.8$.

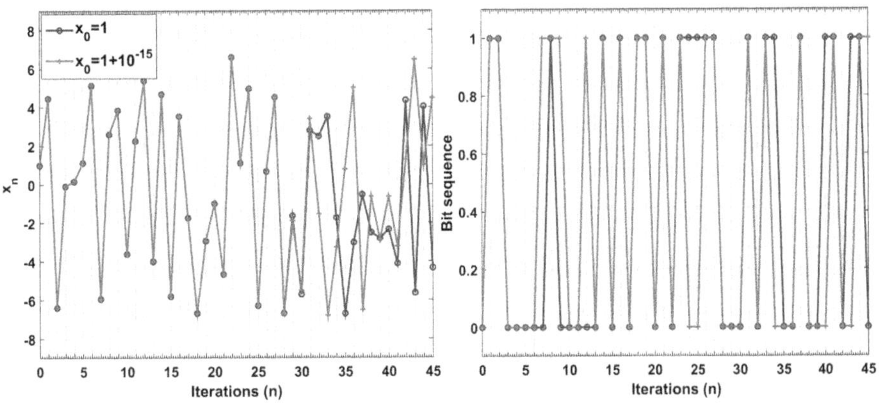

Figure 33. Sensitivity to parameter values for the bit generation design (16) using system (9), with system parameters $x_0=1$, $y_0=1$, $z_0=0.1$, $w_0=0.1$, $b=2.3$, $c=0.8$.

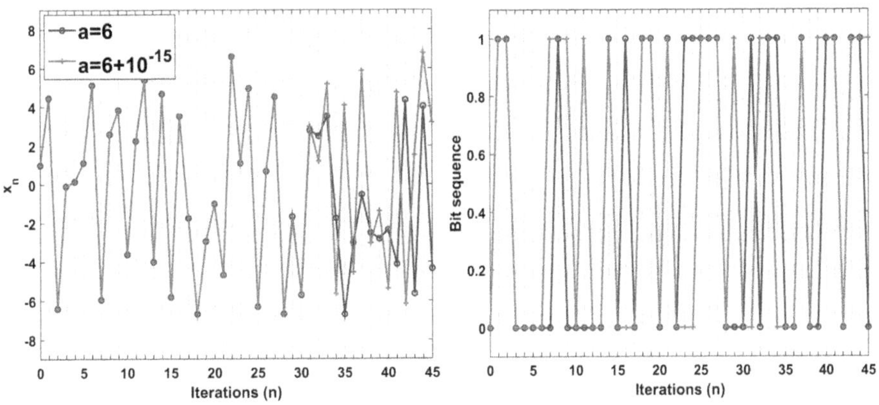

The key space of the considered generator depends on the used chaotic map. In the simplest case, a two-dimensional map possesses two pairs of initial conditions and one nonlinearity parameter. To compute a lower bound on the key space, assume that the parameters are kept stable and only the four initial conditions are changed. This means that we employ at least four values that are represented in floating-point data type with mantissa precision of 53 bits (Kahan, 1996). Thus, the lower bound of the key space, for maps (9), (13) and (14), is $10^{4x16} = 10^{64} = (10^3)^{21.3} \gg (2^{10})^{21.3} = 2^{213}$. The key space can be increased by introducing different values of the nonlinear parameters for the simulated maps. So for example, by assuming in the map (14), based on the bifurcation diagram, that the parameter j varies from 2 to 6 where the system is chaotic for almost all values of j, then the key space is raised to about $4\bullet10^{15}\bullet10^{4x16}=2^2\bullet10^{79}= 2^2\bullet(10^3)^{26.3} \approx 2^{265}$. Of course this is just an indication, since the parameter span of j for which the system is chaotic is even wider, which means that the upper bound for the key space is larger. Moreover, the two maps to be compared can be generated with different parameter values. So all-around, even with a minimum set of parameters, the system based on the proposed generator can resist the brute force attack, since its key space is more than 2^{100} (Li et al., 2018).

Table 1: NIST test results for the scaled sine-sine map (9),

NIST Statistical test	P-value	Pass rate
Frequency	0.383827	95/100
Block frequency, M = 1001	0.983453	98/100
Runs	0.657933	99/100
Longest Run of Ones, M = 10000	0.383827	100/100
Binary Matrix Rank, M = 32, Q = 32	0.366918	97/100
Spectral	0.262249	100/100
Non-overlapping Template Matching, M = 10000, pattern 110	0.494392	98/100
Overlapping Template Matching, M = 10000, pattern 110	0.055361	100/100
Universal Statistical Test, L = 7	0.759756	100/100
Linear Complexity, M = 1000	0.699313	100/100
Serial Test 1	0.739918	97/100
Serial Test 2	0.595549	99/100
Approximate Entropy, M = 10	0.897763	99/100
Cumulative Sums (forward)	0.090936	97/100
Cumulative Sums (backward)	0.699313	98/100
Random Excursions, state -1	0.897763	99/100
Random Excursions Variant, state -1	0.003996	99/100

Overall, it is seen that using the proposed 2d maps, simple and efficient random bit generators can easily be constructed with desired randomness properties. Of course, using more advanced techniques can lead to even more secure designs with a higher key space, which can be addressed in the future.

CONCLUSION AND FUTURE DIRECTIONS

In this chapter, various chaotic two dimensional discrete time maps with chaotic behavior were introduced. These are derived from previously introduced and well-studied maps like the Hénon map, the Lozi map, the Tinkerbell map and the recently introduced Sine-sine map. The new maps are derived from adding nonlinear terms on these systems, or by multiplying any nonlinear term in the equations by a scaling factor. Thus, new chaotic maps are proposed, taking different shapes for different parameter values and different initial conditions. As shown from all the simulations, simple modifications on the already known chaotic maps can lead to new mappings with very interesting results. Furthermore, a choice of the proposed maps was applied to the problem of pseudo-random bit generation, yielding positive results. Subjects for future research would be the problem of adaptive control and synchronization of these maps. In addition, the construction of novel discrete hyperchaotic maps from the ones introduced here is a subject of interest. Finally, the problem of image encryption in combination with the bit generation is a problem of study. For all the simulations above, a 16-digit accuracy was used in Matlab.

Table 2: NIST test results for the squared sine-sine map (13).

NIST Statistical test	P-value	Pass rate
Frequency	0.924076	99/100
Block frequency, M = 1001	0.129620	98/100
Runs	0.739918	100/100
Longest Run of Ones, M = 10000	0.304126	98/100
Binary Matrix Rank, M = 32, Q = 32	0.719747	98/100
Spectral	0.224821	99/100
Non-overlapping Template Matching, M = 10000, pattern 110	0.437274	99/100
Overlapping Template Matching, M = 10000, pattern 110	0.401199	98/100
Universal Statistical Test, L = 7	0.122325	98/100
Linear Complexity, M = 1000	0.867692	98/100
Serial Test 1	0.595549	99/100
Serial Test 2	0.739918	100/100
Approximate Entropy, M = 10	0.003712	98/100
Cumulative Sums (forward)	0.595549	99/100
Cumulative Sums (backward)	0.798139	99/100
Random Excursions, state -1	0.699313	99/100
Random Excursions Variant, state -1	0.883171	98/100

Table 3: NIST test results for the cubic sine-sine map (14).

NIST Statistical test	P-value	Pass rate
Frequency	0.964295	100/100
Block frequency, M = 1001	0.262249	100/100
Runs	0.437274	99/100
Longest Run of Ones, M = 10000	0.062821	99/100
Binary Matrix Rank, M = 32, Q = 32	0.000184	99/100
Spectral	0.759756	99/100
Non-overlapping Template Matching, M = 10000, pattern 110	0.115387	100/100
Overlapping Template Matching, M = 10000, pattern 110	0.779188	98/100
Universal Statistical Test, L = 7	0.935716	98/100
Linear Complexity, M = 1000	0.419021	97/100
Serial Test 1	0.350485	99/100
Serial Test 2	0.437274	99/100
Approximate Entropy, M = 10	0.759756	99/100
Cumulative Sums (forward)	0.983453	100/100
Cumulative Sums (backward)	0.798139	99/100
Random Excursions, state -1	0.657933	100/100
Random Excursions Variant, state -1	0.249284	98/100

ACKNOWLEDGMENT

This research is co-financed by Greece and the European Union (European Social Fund- ESF) through the Operational Programme «Human Resources Development, Education and Lifelong Learning» in the context of the project "Reinforcement of Postdoctoral Researchers - 2nd Cycle" (MIS-5033021), implemented by the State Scholarships Foundation (IKY).

This study was supported by the grant of the Russian Science Foundation (RSF), project 20-79-10334.

REFERENCES

Aboites, V., Wilson, M., del Bosque, L., & del Campestre, L. (2009). Tinkerbell chaos in a ring phase-conjugated resonator. *International Journal of Pure and Applied Mathematics*, *54*(3), 429–435.

Aguilar-Bustos, A. Y., & Cruz-Hernández, C. (2009). Synchronization of discrete-time hyperchaotic systems: An application in communications. *Chaos, Solitons, and Fractals*, *41*(3), 1301–1310. doi:10.1016/j.chaos.2008.05.012

Akhmet, M., & Fen, M. O. (2016). Chaos Generation in Continuous/Discrete-Time Models. In Replication of Chaos in Neural Networks, Economics and Physics, (pp. 183–264). Springer Berlin Heidelberg.

Alain, K. S. T., Romanic, K., Azar, A. T., Vaidyanathan, S., Bertrand, F. H., & Adele, N. M. (2018). *Dynamics Analysis and Synchronization in Relay Coupled Fractional Order Colpitts Oscillators. Advances in System Dynamics and Control*. IGI-Global. doi:10.4018/978-1-5225-4077-9.ch011

Alain, K. S. T., Romanic, K., & Bertrand, F. H. (2017). Dynamics and Improved Robust Adaptive Control Strategy for the Finite Time Synchronization of Uncertain Nonlinear Systems. *International Journal of System Dynamics Applications*, *6*(4), 34–62. doi:10.4018/IJSDA.2017100103

Alawida, M., Samsudin, A., & Teh, J. S. (2020). Enhanced digital chaotic maps based on bit reversal with applications in random bit generators. *Information Sciences*, *512*, 1155–1169. doi:10.1016/j.ins.2019.10.055

Alligood, K. T., Sauer, T. D., & Yorke, J. A. (1996). *Chaos: An Introduction to Dynamical Systems*. Springer-Verlag. doi:10.1007/b97589

Alpar, O. (2015). Three winged lateen shaped chaotic attractor. *Nonlinear Dynamics*, *82*(1-2), 435–449. doi:10.100711071-015-2166-2

Azar, A. T., Volos, C., Gerodimos, N. A., Tombras, G. S., Pham, V. T., Radwan, A. G., Vaidyanathan, S., Ouannas, A., & Munoz-Pacheco, J. M. (2017). A novel chaotic system without equilibrium: Dynamics, synchronization and circuit realization. Complexity. doi:10.1155/2017/7871467

Aziz-Alaoui, M., Robert, C., & Grebogi, C. (2001). Dynamics of a Hénon–Lozi-type map. *Chaos, Solitons, and Fractals*, *12*(12), 2323–2341. doi:10.1016/S0960-0779(00)00192-2

Balibrea-Iniesta, F., Lopesino, C., Wiggins, S., & Mancho, A. M. (2015). Chaotic dynamics in nonautonomous maps: Application to the nonautonomous Hénon map. *International Journal of Bifurcation and Chaos in Applied Sciences and Engineering*, *25*(12), 1550172. doi:10.1142/S0218127415501722

Benedicks, M., & Carleson, L. (1991). The dynamics of the Hénon map. *Annals of Mathematics, 133*(1), 73–169. doi:10.2307/2944326

Bi, C., Zhang, Q., Xiang, Y., & Wang, J. (2013). Nonlinear dynamics of two-dimensional sinusoidal discrete map. In *Communications, Circuits and Systems (ICCCAS), 2013 International Conference on,* (vol. 2, pp. 438–441). IEEE. 10.1109/ICCCAS.2013.6765377

Biham, O., & Wenzel, W. (1989). Characterization of unstable periodic orbits in chaotic attractors and repellers. *Physical Review Letters, 63*(8), 819–822. doi:10.1103/PhysRevLett.63.819 PMID:10041193

Boyacı, O., & Tantuğ, A. C. (2017, August). A random number generation method based on discrete time chaotic maps. In *2017 IEEE 60th International Midwest Symposium on Circuits and Systems (MWSCAS)* (pp. 1212-1215). IEEE. 10.1109/MWSCAS.2017.8053147

Cvitanović, P., Gunaratne, G. H., & Procaccia, I. (1988). Topological and metric properties of Hénon-type strange attractors. *Physical Review A, 38*(3), 1503–1520. doi:10.1103/PhysRevA.38.1503 PMID:9900529

Das, S., Goswami, D., Chatterjee, S., & Mukherjee, S. (2014). Stability and chaos analysis of a novel swarm dynamics with applications to multi-agent systems. *Engineering Applications of Artificial Intelligence, 30,* 189–198. doi:10.1016/j.engappai.2013.12.014

Davidchack, R. L., Lai, Y.-C., Klebanoff, A., & Bollt, E. M. (2001). Towards complete detection of unstable periodic orbits in chaotic systems. *Physics Letters. [Part A], 287*(1), 99–104. doi:10.1016/S0375-9601(01)00463-7

dos Santos Coelho, L., & Mariani, V. C. (2012). Firefly algorithm approach based on chaotic Tinkerbell map applied to multivariable PID controller tuning. *Computers & Mathematics with Applications (Oxford, England), 64*(8), 2371–2382. doi:10.1016/j.camwa.2012.05.007

El-Ghazaly, G. (2012). Adaptive synchronization of unknown chaotic systems using Mamdani fuzzy approach. *International Journal of System Dynamics Applications, 1*(3), 122–138. doi:10.4018/ijsda.2012070104

Elhadj, Z. (2013). *Lozi Mappings: Theory and Applications.* CRC Press. doi:10.1201/b15363

Farmer, J. D., Ott, E., & Yorke, J. A. (1983). The dimension of chaotic attractors. In *The Theory of Chaotic Attractors* (pp. 142–169). Springer. doi:10.1007/978-0-387-21830-4_11

Filali, R. L., Benrejeb, M., & Borne, P. (2014). On observer-based secure communication design using discrete-time hyperchaotic systems. *Communications in Nonlinear Science and Numerical Simulation, 19*(5), 1424–1432. doi:10.1016/j.cnsns.2013.09.005

Galias, Z. (2016). Are numerical studies of long term dynamics conclusive: The case of the Hénon map. *Journal of Physics: Conference Series, 692,* 012001. doi:10.1088/1742-6596/692/1/012001

Gaspard, P. (1999). Microscopic chaos and chemical reactions. *Physica A, 263*(1), 315–328. doi:10.1016/S0378-4371(98)00504-4

Goldsztejn, A., Hayes, W., & Collins, P. (2011). Tinkerbell is chaotic. *SIAM Journal on Applied Dynamical Systems, 10*(4), 1480–1501. doi:10.1137/100819011

Grassberger, P., & Procaccia, I. (1983). Characterization of strange attractors. *Physical Review Letters*, *50*(5), 346–349. doi:10.1103/PhysRevLett.50.346

Grassberger, P., & Procaccia, I. (2004). Measuring the strangeness of strange attractors. In *The Theory of Chaotic Attractors* (pp. 170–189). Springer. doi:10.1007/978-0-387-21830-4_12

Guckenheimer, J., & Holmes, P. J. (2013). *Nonlinear oscillations, dynamical systems, and bifurcations of vector fields* (Vol. 42). Springer Science & Business Media.

Guegan, D. (2009). Chaos in economics and finance. *Annual Reviews in Control*, *33*(1), 89–93. doi:10.1016/j.arcontrol.2009.01.002

Hamza, R. (2017). A novel pseudo random sequence generator for image-cryptographic applications. *Journal of Information Security and Applications*, *35*, 119–127. doi:10.1016/j.jisa.2017.06.005

Heagy, J. (1992). A physical interpretation of the Hénon map. *Physica D. Nonlinear Phenomena*, *57*(3-4), 436–446. doi:10.1016/0167-2789(92)90012-C

Hénon, M. (1976). A two-dimensional mapping with a strange attractor. *Communications in Mathematical Physics*, *50*(1), 69–76. doi:10.1007/BF01608556

Huang, X., Liu, L., Li, X., Yu, M., & Wu, Z. (2019). A new pseudorandom bit generator based on mixing three-dimensional Chen chaotic system with a chaotic tactics. *Complexity*, *2019*, 2019. doi:10.1155/2019/6567198

Irfan, M., Ali, A., Khan, M. A., Ehatisham-ul-Haq, M., Mehmood Shah, S. N., Saboor, A., & Ahmad, W. (2020). Pseudorandom Number Generator (PRNG) Design Using Hyper-Chaotic Modified Robust Logistic Map (HC-MRLM). *Electronics (Basel)*, *9*(1), 104. doi:10.3390/electronics9010104

Kahan, W. (1996). IEEE standard 754 for binary floating-point arithmetic. *Lecture Notes on the Status of IEEE*, *754*, 11.

Kengne, J., Chedjou, J. C., Kenne, G., & Kyamakya, K. (2012). Dynamical properties and chaos synchronization of improved Colpitts oscillators. *Communications in Nonlinear Science and Numerical Simulation*, *17*(7), 2914–2923. doi:10.1016/j.cnsns.2011.10.038

Khan, A., Singh, S., & Azar, A. T. (2020a). Synchronization between a Novel Integer-Order Hyperchaotic System and a Fractional-Order Hyperchaotic System Using Tracking Control. In *The International Conference on Advanced Machine Learning Technologies and Applications (AMLTA2019). AMLTA 2019. Advances in Intelligent Systems and Computing*, (vol 921). Springer. 10.1007/978-3-030-14118-9_38

Khan, A., Singh, S., & Azar, A. T. (2020b). Combination-Combination Anti-Synchronization of Four Fractional Order Identical Hyperchaotic Systems. In *The International Conference on Advanced Machine Learning Technologies and Applications (AMLTA2019). AMLTA 2019. Advances in Intelligent Systems and Computing*, (vol 921). Springer. 10.1007/978-3-030-14118-9_41

Khettab, K., Bensafia, Y., Bourouba, B., & Azar, A. T. (2018). Enhanced Fractional Order Indirect Fuzzy Adaptive Synchronization of Uncertain Fractional Chaotic Systems Based on the Variable Structure Control: Robust H? Design Approach. In Mathematical Techniques of Fractional Order Systems, Advances in Nonlinear Dynamics and Chaos (ANDC) series, (pp. 559-595). Elsevier.

Kyriazis, M. (1991). Applications of chaos theory to the molecular biology of aging. *Experimental Gerontology*, *26*(6), 569–572. doi:10.1016/0531-5565(91)90074-V PMID:1800131

Li, C., Liao, X., & Wong, K. W. (2005). Lag synchronization of hyperchaos with application to secure communications. *Chaos, Solitons, and Fractals*, *23*(1), 183–193. doi:10.1016/j.chaos.2004.04.025

Li, C., Lin, D., Feng, B., Lü, J., & Hao, F. (2018). Cryptanalysis of a chaotic image encryption algorithm based on information entropy. *IEEE Access: Practical Innovations, Open Solutions*, *6*, 75834–75842. doi:10.1109/ACCESS.2018.2883690

Li, N., Pan, W., Yan, L., Luo, B., & Zou, X. (2014). Enhanced chaos synchronization and communication in cascade-coupled semiconductor ring lasers. *Communications in Nonlinear Science and Numerical Simulation*, *19*(6), 1874–1883. doi:10.1016/j.cnsns.2013.09.036

Lichtenberg, A. J., & Lieberman, M. A. (2013). *Regular and stochastic motion* (Vol. 38). Springer Science & Business Media.

Liu, L., Miao, S., Hu, H., & Deng, Y. (2016). Pseudorandom bit generator based on non-stationary logistic maps. *IET Information Security*, *10*(2), 87–94. doi:10.1049/iet-ifs.2014.0192

Longge, Z., & Xiangjie, L. (2013). The synchronization between two discrete-time chaotic systems using active robust model predictive control. *Nonlinear Dynamics*, *74*(4), 905–910. doi:10.100711071-013-1009-2

Lopesino, C., Balibrea-Iniesta, F., Wiggins, S., & Mancho, A. M. (2015). The Chaotic Saddle in the Lozi Map, Autonomous and Nonautonomous Versions. *International Journal of Bifurcation and Chaos in Applied Sciences and Engineering*, *25*(13), 1550184. doi:10.1142/S0218127415501849

Lorenz, E. N. (1963). Deterministic nonperiodic flow. *Journal of the Atmospheric Sciences*, *20*(2), 130–141. doi:10.1175/1520-0469(1963)020<0130:DNF>2.0.CO;2

Lorenz, E. N. (2008). Compound windows of the Hénon-map. *Physica D. Nonlinear Phenomena*, *237*(13), 1689–1704. doi:10.1016/j.physd.2007.11.014

Lozi, R. (1978). Un attracteur étrange (?) du type attracteur de Hénon. *Journal de Physique. Colloque*, *39*(C5), C5–C9. doi:10.1051/jphyscol:1978505

Merah, L., Ali-Pacha, A., Hadj-Said, N., Mecheri, B., & Dellassi, M. (2017). FPGA hardware co-simulation of new chaos-based stream cipher based on Lozi map. *International. Journal of Engineering Technology*, *9*(5), 420–425.

Meranza-Castillón, M. O., Murillo-Escobar, M. A., López-Gutiérrez, R. M., & Cruz-Hernández, C. (2019). Pseudorandom number generator based on enhanced Hénon map and its implementation. *AEÜ. International Journal of Electronics and Communications*, *107*, 239–251. doi:10.1016/j.aeue.2019.05.028

Michelitsch, M., & Rössler, O. (1989). A new feature in Hénon's map. *Computers & Graphics*, *13*(2), 263–265. doi:10.1016/0097-8493(89)90070-8

Misiurewicz, M. (1980). Strange attractors for the lozi mappings. *Annals of the New York Academy of Sciences*, *357*(1), 348–358. doi:10.1111/j.1749-6632.1980.tb29702.x

Moysis, L., & Azar, A. T. (2017). New discrete time 2d chaotic maps. *International Journal of System Dynamics Applications*, *6*(1), 77–104. doi:10.4018/IJSDA.2017010105

Murali, K., & Lakshmanan, M. (1998). Secure communication using a compound signal from generalized synchronizable chaotic systems. *Physics Letters. [Part A]*, *241*(6), 303–310. doi:10.1016/S0375-9601(98)00159-5

Murillo-Escobar, M. A., Cruz-Hernández, C., Cardoza-Avendaño, L., & Méndez-Ramírez, R. (2017). A novel pseudorandom number generator based on pseudorandomly enhanced logistic map. *Nonlinear Dynamics*, *87*(1), 407–425. doi:10.100711071-016-3051-3

Nepomuceno, E. G., Nardo, L. G., Arias-Garcia, J., Butusov, D. N., & Tutueva, A. (2019). Image encryption based on the pseudo-orbits from 1D chaotic map. *Chaos (Woodbury, N.Y.)*, *29*(6), 061101. doi:10.1063/1.5099261 PMID:31266342

Nosoohi, I., & Parvizian, J. (2013). Effects of Discount Scenarios on Chaotic Behavior of Inventory Level Under Price-Dependent Demand. *International Journal of System Dynamics Applications*, *2*(3), 58–72. doi:10.4018/ijsda.2013070104

Nusse, H. E., & Yorke, J. A. (2012). *Dynamics: numerical explorations: accompanying computer program dynamics* (Vol. 101). Springer.

Ott, E. (1981). Strange attractors and chaotic motions of dynamical systems. In *The Theory of Chaotic Attractors* (pp. 103–119). Springer. doi:10.1007/978-0-387-21830-4_9

Ouannas, A., Azar, A. T., & Abu-Saris, R. (2017a). A new type of hybrid synchronization between arbitrary hyperchaotic maps. *International Journal of Machine Learning and Cybernetics*, *8*(6), 1887–1894. doi:10.100713042-016-0566-3

Ouannas, A., Azar, A. T., & Vaidyanathan, S. (2017b). New hybrid synchronisation schemes based on coexistence of various types of synchronisation between master-slave hyperchaotic systems. *International Journal of Computer Applications in Technology*, *55*(2), 112–120. doi:10.1504/IJCAT.2017.082868

Ouannas, A., Azar, A. T., & Vaidyanathan, S. (2017c). A Robust Method for New Fractional Hybrid Chaos Synchronization. *Mathematical Methods in the Applied Sciences*, *40*(5), 1804–1812. doi:10.1002/mma.4099

Ouannas, A., Azar, A. T., & Vaidyanathan, S. (2017d). A new fractional hybrid chaos synchronization. Int. J. Modelling. *Identification and Control*, *27*(4), 314–322. doi:10.1504/IJMIC.2017.084719

Ouannas, A., Azar, A. T., & Vaidyanathan, S. (2017e). On A Simple Approach for Q-S Synchronization of Chaotic Dynamical Systems in Continuous-Time. Int. J. *Computing Science and Mathematics*, *8*(1), 20–27.

Ouannas, A., Azar, A. T., Ziar, T., & Vaidyanathan, S. (2017f). Fractional Inverse Generalized Chaos Synchronization Between Different Dimensional Systems. *Studies in Computational Intelligence*, *688*, 525–551. doi:10.1007/978-3-319-50249-6_18

Ouannas, A., Azar, A. T., Ziar, T., & Vaidyanathan, S. (2017g). A New Method To Synchronize Fractional Chaotic Systems With Different Dimensions. *Studies in Computational Intelligence*, *688*, 581–611. doi:10.1007/978-3-319-50249-6_20

Ouannas, A., Grassi, G., & Azar, A. T. (2020a). A New Generalized Synchronization Scheme to Control Fractional Chaotic Systems with Non-identical Dimensions and Different Orders. In *The International Conference on Advanced Machine Learning Technologies and Applications (AMLTA2019). AMLTA 2019. Advances in Intelligent Systems and Computing*, (vol 921). Springer. 10.1007/978-3-030-14118-9_42

Ouannas, A., Grassi, G., & Azar, A. T. (2020b). Fractional-Order Control Scheme for Q-S Chaos Synchronization. In *The International Conference on Advanced Machine Learning Technologies and Applications (AMLTA2019). AMLTA 2019. Advances in Intelligent Systems and Computing*, (vol 921). Springer. 10.1007/978-3-030-14118-9_44

Ouannas, A., Grassi, G., Azar, A. T., Khennaouia, A. A., & Pham, V. T. (2020c). Chaotic Control in Fractional-Order Discrete-Time Systems. In *AISI 2019. The International Conference on Advanced Intelligent Systems and Informatics AISI 2019. Advances in Intelligent Systems and Computing*, (vol 1058, pp. 207-217). Springer. 10.1007/978-3-030-31129-2_20

Ouannas, A., Grassi, G., Azar, A. T., Khennaouia, A. A., & Pham, V. T. (2020d). Synchronization of Fractional-Order Discrete-Time Chaotic Systems. In *The International Conference on Advanced Intelligent Systems and Informatics AISI 2019. Advances in Intelligent Systems and Computing*, (vol 1058, pp. 218-228). Springer.

Ouannas, A., Grassi, G., Azar, A. T., & Singh, S. (2019). New Control Schemes for Fractional Chaos Synchronization. In *Proceedings of the International Conference on Advanced Intelligent Systems and Informatics 2018. AISI 2018. Advances in Intelligent Systems and Computing*, (vol 845). Springer. 10.1007/978-3-319-99010-1_5

Parlitz, U. (2016). Estimating Lyapunov Exponents from Time Series. In *Chaos Detection and Predictability* (pp. 1–34). Springer., doi:10.1007/978-3-030-31129-2_21.

Patidar, V., Sud, K. K., & Pareek, N. K. (2009). A pseudo random bit generator based on chaotic logistic map and its statistical testing. *Informatica (Vilnius)*, *33*(4).

Pham, V. T., Gokul, P. M., Kapitaniak, T., Volos, C., & Azar, A. T. (2018) Dynamics, Synchronization and Fractional Order Form of a Chaotic System With Infinite Equilibria. In Mathematical Techniques of Fractional Order Systems, Advances in Nonlinear Dynamics and Chaos (ANDC) series, (pp. 475-502). Elsevier.

Pham, V. T., Vaidyanathan, S., Volos, C. K., Azar, A. T., Hoang, T. M., & Yem, V. V. (2017). A Three-Dimensional No-Equilibrium Chaotic System: Analysis, Synchronization and Its Fractional Order Form. Studies in Computational Intelligence, 688, 449-470.

Qu, Z. (2011). Chaos in the genesis and maintenance of cardiac arrhythmias. *Progress in Biophysics and Molecular Biology*, *105*(3), 247–257. doi:10.1016/j.pbiomolbio.2010.11.001 PMID:21078337

Rafikov, M., & Balthazar, J. M. (2008). On control and synchronization in chaotic and hyperchaotic systems via linear feedback control. *Communications in Nonlinear Science and Numerical Simulation, 13*(7), 1246–1255. doi:10.1016/j.cnsns.2006.12.011

Rezk, A. A., Madian, A. H., Radwan, A. G., & Soliman, A. M. (2020). Multiplierless chaotic Pseudo random number generators. *AEÜ. International Journal of Electronics and Communications, 113*, 152947. doi:10.1016/j.aeue.2019.152947

Richter, H. (2002). The generalized Hénon maps: Examples for higher-dimensional chaos. *International Journal of Bifurcation and Chaos in Applied Sciences and Engineering, 12*(06), 1371–1384. doi:10.1142/S0218127402005121

Rössler, O. E. (1979). An equation for hyperchaos. *Physics Letters. [Part A], 71*(2), 155–157. doi:10.1016/0375-9601(79)90150-6

Rukhin, A., Soto, J., Nechvatal, J., Smid, M., & Barker, E. (2001). *A statistical test suite for random and pseudorandom number generators for cryptographic applications*. Booz-Allen and Hamilton Inc.

Sharma, A., Patidar, V., Purohit, G., & Sud, K. K. (2012). Effects on the bifurcation and chaos in forced Duffing oscillator due to nonlinear damping. *Communications in Nonlinear Science and Numerical Simulation, 17*(6), 2254–2269. doi:10.1016/j.cnsns.2011.10.032

Singh, S., Azar, A. T., Bhat, M. A., Vaidyanathan, S., & Ouannas, A. (2018b). *Active Control for Multi-Switching Combination Synchronization of Non-Identical Chaotic Systems. Advances in System Dynamics and Control*. IGI-Global. doi:10.4018/978-1-5225-4077-9.ch005

Singh, S., Azar, A. T., Ouannas, A., Zhu, Q., Zhang, W., & Na, J. (2017) Sliding Mode Control Technique for Multi-switching Synchronization of Chaotic Systems. *Proceedings of 9th International Conference on Modelling, Identification and Control (ICMIC 2017)*, 880-885. 10.1109/ICMIC.2017.8321579

Singh, S., Azar, A. T., Vaidyanathan, S., Ouannas, A., & Bhat, M. A. (2018a). Multiswitching Synchronization of Commensurate Fractional Order Hyperchaotic Systems Via Active Control. In Mathematical Techniques of Fractional Order Systems, Advances in Nonlinear Dynamics and Chaos (ANDC) series, (pp. 319-345). Elsevier. doi:10.1016/B978-0-12-813592-1.00011-8

Singh, S., Azar, A. T., & Zhu, Q. (2018c). Multi-switching Master-Slave Synchronization of Nonidentical Chaotic Systems. In Q. Zhu, J. Na, & X. Wu (Eds.), *Innovative Techniques and Applications of Modelling, Identification and Control. Lecture Notes in Electrical Engineering* (Vol. 467, pp. 321–330). Springer. doi:10.1007/978-981-10-7212-3_20

Sprott, J., & Xiong, A. (2015). Classifying and quantifying basins of attraction. *Chaos (Woodbury, N.Y.), 25*(8), 083101. doi:10.1063/1.4927643 PMID:26328552

Stoyanov, B., & Kordov, K. (2015). Novel secure pseudo-random number generation scheme based on two Tinkerbell maps. *Advanced Studies in Theoretical Physics, 9*(9), 411–421. doi:10.12988/astp.2015.5342

Strogatz, S. H. (2014). *Nonlinear dynamics and chaos: with applications to physics, biology, chemistry, and engineering*. Westview press.

Tanase, R. (2016). Complex Hénon maps and discrete groups. *Advances in Mathematics*, *295*, 53–89. doi:10.1016/j.aim.2016.02.034

Todorova, M., Stoyanov, B., Szczypiorski, K., & Kordov, K. (2018). *SHAH: hash function based on irregularly decimated chaotic map.* arXiv preprint arXiv:1808.01956

Tutueva, A. V., Nepomuceno, E. G., Karimov, A. I., Andreev, V. S., & Butusov, D. N. (2020). Adaptive chaotic maps and their application to pseudo-random numbers generation. *Chaos, Solitons, and Fractals*, *133*, 109615. doi:10.1016/j.chaos.2020.109615

Vaidyanathan, S., & Azar, A. T. (2015b). Anti-synchronization of Identical Chaotic Systems Using Sliding Mode Control and an Application to Vaidyanathan–Madhavan Chaotic Systems. In *Advances and Applications in Sliding Mode Control systems* (pp. 527–547). Springer International Publishing. doi:10.1007/978-3-319-11173-5_19

Vaidyanathan, S., & Azar, A. T. (2015c). Anti-Synchronization of Identical Chaotic Systems using Sliding Mode Control and an Application to Vaidyanathan-Madhavan Chaotic Systems. Studies in Computational Intelligence, 576, 527-547. doi:10.1007/978-3-319-11173-5_19

Vaidyanathan, S., & Azar, A. T. (2015d). Hybrid Synchronization of Identical Chaotic Systems using Sliding Mode Control and an Application to Vaidyanathan Chaotic Systems. Studies in Computational Intelligence, 576, 549-569. doi:10.1007/978-3-319-11173-5_20

Vaidyanathan, S., & Azar, A. T. (2016a). Generalized Projective Synchronization of a Novel Hyperchaotic Four-Wing System via Adaptive Control Method. Advances in Chaos Theory and Intelligent Control. *Studies in Fuzziness and Soft Computing*, *337*, 275–296. doi:10.1007/978-3-319-30340-6_12

Vaidyanathan, S., & Azar, A. T. (2016b). Qualitative Study and Adaptive Control of a Novel 4-D Hyperchaotic System with Three Quadratic Nonlinearities. *Studies in Fuzziness and Soft Computing*, *337*, 179–202. doi:10.1007/978-3-319-30340-6_8

Vaidyanathan, S., & Azar, A. T. (2016c). Dynamic Analysis, Adaptive Feedback Control and Synchronization of an Eight-Term 3-D Novel Chaotic System with Three Quadratic Nonlinearities. *Studies in Fuzziness and Soft Computing*, *337*, 155–178. doi:10.1007/978-3-319-30340-6_7

Vaidyanathan, S., Azar, A. T., Akgul, A., Lien, C. H., Kacar, S., & Cavusoglu, U. (2019). A memristor-based system with hidden hyperchaotic attractors, its circuit design, synchronisation via integral sliding mode control and voice encryption. *Int. J. Automation and Control*, *13*(6), 644–667. doi:10.1504/IJAAC.2019.102665

Vaidyanathan, S., Azar, A. T., Rajagopal, K., & Alexander, P. (2015a). Design and SPICE implementation of a 12-term novel hyperchaotic system and its synchronization via active control Int. J. Modelling *Identification and Control*, *23*(3), 267–277. doi:10.1504/IJMIC.2015.069936

Vaidyanathan, S., Sampath, S., & Azar, A. T. (2015e). Global chaos synchronisation of identical chaotic systems via novel sliding mode control method and its application to Zhu system. *Int. J. Modelling Identification and Control*, *23*(1), 92–100. doi:10.1504/IJMIC.2015.067495

Valandar, M. Y., Barani, M. J., & Ayubi, P. (2020). A blind and robust color images watermarking method based on block transform and secured by modified 3-dimensional Hénon map. *Soft Computing*, *24*(2), 771–794. doi:10.100700500-019-04524-z

Volos, C., Kyprianidis, I., Stouboulos, I., & Vaidyanathan, S. (2016). Design of a chaotic random bit generator using a Duffing-van der Pol system. *International Journal of System Dynamics Applications*, *5*(3), 94–111. doi:10.4018/IJSDA.2016070105

Volos, C. K., Pham, V. T., Azar, A. T., Stouboulos, I. N., & Kyprianidis, I. M. (2018). Synchronization Phenomena in Coupled Dynamical Systems with Hidden Attractors. In V. T. Pham, S. Vaidyanathan, C. Volos, & T. Kapitaniak (Eds.), *Nonlinear Dynamical Systems with Self-Excited and Hidden Attractors. Studies in Systems, Decision and Control* (Vol. 133, pp. 375–401). Springer. doi:10.1007/978-3-319-71243-7_17

Wang, L., & Cheng, H. (2019). Pseudo-Random Number Generator Based on Logistic Chaotic System. *Entropy (Basel, Switzerland)*, *21*(10), 960. doi:10.3390/e21100960

Wang, Y., Liu, Z., Ma, J., & He, H. (2016). A pseudorandom number generator based on piecewise logistic map. *Nonlinear Dynamics*, *83*(4), 2373–2391. doi:10.100711071-015-2488-0

Wang, Z., Volos, C., Kingni, S. T., Azar, A. T., & Pham, V. T. (2017). Four-wing attractors in a novel chaotic system with hyperbolic sine nonlinearity. *Optik (Stuttgart)*, *131*, 1071–1078. doi:10.1016/j.ijleo.2016.12.016

Witte, C. L., & Witte, M. H. (1991). Chaos and predicting varix hemorrhage. *Medical Hypotheses*, *36*(4), 312–317. doi:10.1016/0306-9877(91)90002-G PMID:1809849

Wolf, A., Swift, J. B., Swinney, H. L., & Vastano, J. A. (1985). Determining Lyapunov exponents from a time series. *Physica D. Nonlinear Phenomena*, *16*(3), 285–317. doi:10.1016/0167-2789(85)90011-9

Yuan, G., Zhang, X., & Wang, Z. (2014). Generation and synchronization of feedback-induced chaos in semiconductor ring lasers by injection-locking. *Optik (Stuttgart)*, *125*(8), 1950–1953. doi:10.1016/j.ijleo.2013.11.007

Yuan, S., Jiang, T., & Jing, Z. (2011). Bifurcation and chaos in the Tinkerbell map. *International Journal of Bifurcation and Chaos in Applied Sciences and Engineering*, *21*(11), 3137–3156. doi:10.1142/S0218127411030581

Zaher, A. A., & Abu-Rezq, A. (2011). On the design of chaos-based secure communication systems. *Communications in Nonlinear Science and Numerical Simulation*, *16*(9), 3721–3737. doi:10.1016/j.cnsns.2010.12.032

Zhang, Q., Guo, L., & Wei, X. (2013). A novel image fusion encryption algorithm based on DNA sequence operation and hyper-chaotic system. *Optik (Stuttgart)*, *124*(18), 3596–3600. doi:10.1016/j.ijleo.2012.11.018

Zou, Y. L., Zhu, J., Chen, G., & Luo, X. S. (2005). Synchronization of hyperchaotic oscillators via single unidirectional chaotic-coupling. *Chaos, Solitons, and Fractals*, *25*(5), 1245–1253. doi:10.1016/j.chaos.2004.11.064

APPENDIX 1. MATLAB CODE

In this appendix we provide the matlab code for all the maps presented here.

```
%% A Collection of nonlinear 2d maps
% For each map different parameter values can be choses, as shown in the
% figures
format long
%% Henon map
clear
x(1)=0; y(1)=0;
a=1.4; b=0.3;
for i=2:100000
    x(i)=1-a*(x(i-1)^2)+y(i-1);
    y(i)=b*x(i-1);
end
plot(x,y,'.','MarkerSize',4)
title('Henon Map')

%% Lozi map
clear
x(1)=0; y(1)=0;
a=1.7; b=0.5;
for i=2:100000
    x(i)=1-a*abs(x(i-1))+ b*y(i-1);
    y(i)=b*x(i-1);
end
plot(x,y,'.','MarkerSize',4)
title('Lozi Map')

%% Henon-Lozi like maps
%% Map (3)
clear
x(1)=0; y(1)=0;
a=0.45; b=0.6;
for i=2:100000
    x(i)=1-a*(x(i-1)^2)+y(i-1)^2;
    y(i)=b*x(i-1);
end
plot(x,y,'.','MarkerSize',4)

%% Map (4)
clear
```

```
a=1.5; b=0.7;
x(1)=0; y(1)=0;
for i=2:100000
    x(i)=1-a*abs(x(i-1))+y(i-1)^2;
    y(i)=b*x(i-1);
end
plot(x,y,'.','MarkerSize',4)

%% Map (5)
clear
a=0.4; b=0.9;
x(1)=.0; y(1)=0.;
for i=2:1000000
    x(i)=1-a*(x(i-1)+y(i-1))^3;
    y(i)=b*abs((x(i-1)));
end
plot(x,y,'.','MarkerSize',4)

%% Map (6)
clear
x(1)=0.; y(1)=0.;
a=.68; b=-1.1;
for i=2:100000
    x(i)=1-a*(x(i-1))^6+y(i-1)^5;
    y(i)=b*(x(i-1));
end
plot(x,y,'.','MarkerSize',4)

%% Sine-Sine map (7)
clear
x(1)=1; y(1)=1;
for i=2:100000
    x(i)=sin(x(i-1))-sin(2*y(i-1));
    y(i)=x(i-1);
end
plot(x,y,'.', 'MarkerSize',4)

%% Abs Sine Sine maps
%% sine-sine map with factors (8)
clear
x(1)=1; y(1)=1;
a=.1; b=.8; c=1.5;
for i=2:400000
```

```
    x(i)=a*sin(x(i-1))-sin(b*y(i-1));
    y(i)=c*x(i-1);
end
plot(x,y,'.','MarkerSize',4)

%% sine-sine map with factors (9)
clear
x(1)=1; y(1)=1;
a=1.9; b=2.5; c=.9;
for i=2:40000
    x(i)=a*sin(x(i-1))-c*sin(b*y(i-1));
    y(i)=x(i-1);
end
plot(x,y,'.','MarkerSize',4)

%% Abs sine-sine map (10)
clear
x(1)=1; y(1)=1;
for i=2:400000
    x(i)=abs(sin(x(i-1))-sin(2*y(i-1)));
    y(i)=x(i-1);
end
plot(x,y,'.', 'MarkerSize',4)

%% Abs sine-sine map (11)
clear
x(1)=1; y(1)=1;
for i=2:400000
    x(i)=abs(sin(x(i-1)))-abs(sin(2*y(i-1)));
    y(i)=x(i-1);
end
plot(x,y,'.', 'MarkerSize',1)

%% Abs sine-sine map (12)
clear
x(1)=1; y(1)=1;
for i=2:400000
    x(i)=abs(sin(x(i-1)))-abs(sin(2*y(i-1)));
    y(i)=abs(x(i-1));
end
plot(x,y,'.', 'MarkerSize',4)

%% Square Sine-sine map (13)
```

```
clear
x(1)=1; y(1)=1;
j=.6;
for i=2:400000
    x(i)=j*sin(x(i-1))^2-sin(2*y(i-1));
    y(i)=x(i-1);
end
plot(x,y,'.','MarkerSize',4)

%% Cubic Sine-sine map (14)
clear
x(1)=0.5; y(1)=-1;
j=0.65;
for i=2:300000
    x(i)=j*sin(x(i-1))^3-sin(2*y(i-1));
    y(i)=x(i-1);
end
plot(x,y,'.','MarkerSize',1)

%% Higher order Sine-cos map (15)
clear
x(1)=.5; y(1)=1.;
j=1.; k=.7; l=.6;
for i=2:250000
    x(i)=j*sin(x(i-1))^3+k*cos(2*y(i-1))^3+l*cos(x(i-1)+y(i-1))^3;
    y(i)=x(i-1);
end
plot(x,y,'.','MarkerSize',4)

%% Tinkerbell map (16)
clear
x(1)=-0.1; y(1)=-0.1;
a=0.9; b=-0.6; c=2; d=0.5;
for i=2:100000
x(i)=x(i-1)^2-y(i-1)^2+a*x(i-1)+b*y(i-1);
y(i)=2*x(i-1)*y(i-1)+c*x(i-1)+d*y(i-1);
end
plot(x,y,'.','MarkerSize',1)
title('Tinkerbell Map')

%% Tinkerbell like maps
%% Map (17)
clear
```

```
x(1)=.5; y(1)=-1.;
a=0.3; b=0.3; c=2; d=0.9;
for i=2:100000
    x(i)=x(i-1)^2-y(i-1)^2+a*x(i-1)+b*y(i-1)^3;
    y(i)=2*x(i-1)*y(i-1)+c*x(i-1)+d*y(i-1);
end
plot(x,y,'.','MarkerSize',4)

%% Map (18)
clear
x(1)=.3; y(1)=-.5;
a=0.9; b=-0.6; c=2; d=0.5; g=1.25;
for i=2:100000
    x(i)=x(i-1)*y(i-1)+a*x(i-1)^3+b*y(i-1);
    y(i)=2*x(i-1)*y(i-1)+c*x(i-1)+d*y(i-1)+g*x(i-1)^3;
end
plot(x,y,'.','MarkerSize',4)

%% Map (19)
clear
x(1)=.3; y(1)=-.5;
a=0.9; b=-0.6; c=2; d=0.5;
for i=2:100000
    x(i)=x(i-1)*y(i-1)+a*x(i-1)^3+b*y(i-1);
    y(i)=6*x(i-1)^2*y(i-1)+c*x(i-1)+d*y(i-1)+1.25*x(i-1)^3;
end
plot(x,y,'.','MarkerSize',4)

%% For the bit generation
clear
format long
% set initial conditions for both maps
x(1)=1; y(1)=1; z(1)=.1; w(1)=.1;
a=6; b=2.3; c=0.8;
bit(1)=0; %initialize bit sequence
bitlength=100*10^16% set bitstream length
for i=2: bitlength
    %first map
        x(i)=a*sin(x(i-1))-c*sin(b*y(i-1));
        y(i)=x(i-1);
    % second map
        z(i)=a*sin(z(i-1))-c*sin(b*w(i-1));
        w(i)=z(i-1);
```

```
    if mod(10^11*x(i),1)>= mod(10^11*z(i),1)
        bit(i)=1;
    else
        bit(i)=0;
    end
end

% write bits into a file
bit=bit';
fid = fopen('bitfile.txt','w');
fprintf(fid,'%1d\n',bit)
fclose(fid)
```

Chapter 23
Antimonotonicity, Crisis, and Route to Chaos in a Tumor Growth Model

Dionysios Sourailidis
Laboratory of Nonlinear Systems – Circuits & Complexity, Physics Department, Aristotle University of Thessaloniki, Greece

Christos Volos
Laboratory of Nonlinear Systems – Circuits & Complexity, Physics Department, Aristotle University of Thessaloniki, Greece

Lazaros Moysis
https://orcid.org/0000-0002-5652-2532
Laboratory of Nonlinear Systems – Circuits & Complexity, Physics Department, Aristotle University of Thessaloniki, Greece

Ioannis Stouboulos
Laboratory of Nonlinear Systems – Circuits & Complexity, Physics Department, Aristotle University of Thessaloniki, Greece

ABSTRACT

In this chapter, a new model of a tumor growth is dynamically investigated. The model is presented in a form of a system of three ordinary differential equations, which describe the avascular, vascular, and metastasis tumor growth, respectively. For the investigation of system's dynamics and especially of the population of the immune cells in system's behavior, some of the most well-known tools from nonlinear theory, such as the phase portrait, the Poincaré map, the bifurcation diagram the Kaplan-Yorke dimension, and the Lyapunov exponents have been used. Interesting phenomena related with chaos theory, such as a period-doubling route to chaos, crisis phenomena, and antimonotonicity, have been revealed for the first time in this model.

DOI: 10.4018/978-1-7998-5788-4.ch023

1. INTRODUCTION

Cancer is a group of diseases, defined by an abnormal, uncontrolled proliferation of cells in a part of the body (Tiwari, 2012). It is the second leading cause of death globally, accounting for an estimated 9.6 million deaths, or one in six deaths, in 2018, according to World Health Organization (WHO, 2020). In spite of the advances of molecular biology and genomics, humanity's understanding of cancerous cell growth is still vague. As a result, it poses a daunting task to detect it early with the goal of taking appropriate measures to hinder its growth and reduce the victims' suffering, let alone find a cure. Developed countries are heavily investing into cancer research and many scientific disciplines are involved, including physics and mathematics, and this is where this chapter will focus.

In many cases, mathematical models have vastly contributed to comprehending carcinogenesis and cancer progression (Adam & Bellomo, 2012; Deisboeck *et al.*, 2011; Dominic *et al.*, 2005; Frank, 2007; Majhi, 2018; Nikulin, 2007; Tan, 2020; Voiculetz *et al.*, 1991). Consequently, that renders the need for formulating newer and more accurate models depicting cancer growth in all its stages more compelling. It is important to highlight the fact that cancer has three distinct stages: *avascular*, *vascular* and *metastatic* (Hogea *et al.*, 2005). During the avascular phase, the tumor obtains nutrients and "feeds" itself via diffusion processes alone, with nutrients already existing in the environment (Breward *et al.*, 2002; Byrne, & Chaplain, 1998; Chaplain, 1996; Izquierdo-Kulich *et al.*, 2013; Knighton *et al.*, 1977; Ward & King, 1997). In the vascular phase, when the tumor grows faster through angiogenesis (the process by which new blood vessels form), malignant tumor cells secrete chemicals that can diffuse into the surrounding healthy tissues and stimulate the growth of new capillary blood vessels (Ribatti *et al.*, 1999; Huang *et al.*, 2009; Keogan *et al.*, 1997; Verstovsek *et al.*, 2002; Yao *et al.*, 2008). The newly born blood vessels penetrate into the tumor mass providing it with nutrients and resulting in rapid growth of the tumor. Tumor growth occurs basically via the following mechanism. When tumor is supplied with a sufficient amount of nutrient, malignant cells are divided (cellular mitosis). So, when the density of malignant cells in a specific volume becomes too high, the cells are compressed by their neighbors, so they tend to move to less compressed areas, where they are allowed to continue the division process, and this process is repeated. In the third phase, metastasis, malignant cell migrate through the blood vessels and/or the lymphatic system towards other parts of the body, giving rise to secondary tumors (Chaffer & Weinberg, 2011; Gupta & Massagué, 2006; Nguyen & Massagué, 2007; Pani *et al.*, 2010; Poste & Fidler, 1980). This process is the primary cause for the host death.

Due to the fact of the existence of these distinct phases, researchers often attempt to answer specific questions for each of those stages separately. In many attempts these questions have been attempted to be answered by using dynamical system analysis. The reason of this is because growth of malignant tumors can be described as ordinary differential equation (ODE) models including two main populations: effector cells and tumor cells. Especially, dynamical systems with chaotic behavior have been used widely for this purpose (Ahmed, 1993; Brinkley, 2001; Dirks, 2008; Duarte *et al.*, 2013; Duesberg, 2007; Saleem & Agrawal, 2012). Chaotic dynamical systems have the feature of the sensitive dependence to initial conditions, which is a key component of chaos theory (Strogatz, 2015). Furthermore, dynamical systems are systems in which a function describes the time dependence of a point in a geometrical space (Verhulst, 1996), and they can be divided into discrete and continuous systems. A discrete dynamical system is a dynamical system, which state evolves over state space in discrete time steps according to a fixed rule (Moysis & Azar, 2017a). A continuous dynamical system is a dynamical system, which state evolves over state space continuously over according to a fixed rule.

The idea of chaos originated in 1961, when Edward Lorenz was working on simulating weather patterns (Lorenz, 2000). He found that even a seemingly minuscule difference in the initial conditions would alter the results in the long run completely. As a result, it was showed that even detailed atmospheric modeling is not, as a rule, capable of making accurate long-term weather predictions. Since then, the framework of chaos theory has been applied in every field of sciences, like physics, biology, engineering, chemistry, medicine, finance etc (Degn, 1986; Field & Gyorgyi, 1993; Gholizad *et al.*, 2017; Nicolay, 2014; Strogatz, 2015).

Therefore, as every system in nature, tumor evolution, too, can be described with the use of dynamical systems. The proliferation and death rate of malignant cells, healthy cells and T-cells, as well as the surroundings of the tumor are described by equations. Treating the cancer's evolution as a dynamical system allows one to exploit all the intricacies of dynamical system theory and reach profound conclusions. Another reason to consider this approach is because aspects of control theory (Alain *et al.*, 2017; Djouima *et al.*, 2018; Moysis *et al.*, 2017b) can be applied to the system and the tumor evolution can be mitigated. That groundbreaking idea was first introduced in 1980 by Carol Newton (Newton, 1980) and Arthur Albert *et al.* (Albert *et al.*, 1980) and shed new light in the area of cancer modeling, paving the way for numerous scientists to further effectively model and mitigate cancer growth. As our understanding of oncology evolves along with the evolution of technology, more precise methods are discovered.

In the present work, we aim to analyze further the dynamical behavior of a mathematical model, which describes tumor growth and exhibits chaotic behavior. It was first proposed by Llanos-Perez *et al.* (Llanos-Perez *et al.*, 2016). Their work is based on a chemical network, which is used to introduce a mechanism for avascular, vascular and metastasis tumor growth. The reader could find more details for this model in the paper of Llanos-Perez *et al.* (Llanos-Perez *et al.*, 2016) and we will expand on it in the next section.

Llanos-Perez *et al.* paved the way for many researchers to build upon their work. To begin with, Betancourt-Mar *et al.* (Betancourt-Mar *et al.*, 2017) extended it on existing thermodynamics formalism for the metabolic rate of human cancer cells. Montero *et al.* (Montero *et al.*, 2018) worked on a biological approach of this problem. Guerra *et al.* (Guerra *et al.*, 2018) generalized the aforementioned work with the inclusion of the epithelial-mesenchymal transition, which is a biologic process that allows a polarized epithelial cell, which normally interacts with basement membrane via its basal surface, to undergo multiple biochemical changes that enable it to assume a mesenchymal cell phenotype that includes enhanced migratory capacity, invasiveness, and increased resistance to apoptosis (Kalluri *et al.*, 2009). Also, a paper that was based on Llanos-Perez *et al.* work was by Jaime *et al.* (Jaime *et al.*, 2018) that greatly enhanced the model in order to mitigate malignant cell growth relative to healthy cell growth. Their analysis focused on the stages of the vascular tumor growth and metastasis. This work investigates the evolution of tumors subject to periodic perturbations.

This chapter is organized as follows. In the next Section the mathematical model of tumor growth, which has been used, is presented in more details. The dynamical behavior of the tumor growth system, by using well-known tools from nonlinear theory, such as the phase portrait, the Poincaré map, the bifurcation diagram the Kaplan-Yorke dimension and the Lyapunov exponents, is investigated thoroughly in Section 3. Finally the conclusion remarks and some thoughts for future work are presented in the last Section.

2. THE CHEMICAL NETWORK MODEL OF THE TUMOR GROWTH

As mentioned, in 2016 Llanos-Pérez *et al.* (Llanos-Pérez *et al.* 2016) presented a novel model of tumor growth, which describes the development of cancer through the three basic stages, avascular, vascular and metastasis, on the basis of experimental evidences. In that work a chemical network model, in which a classic chemical kinetics method that applies the law of mass action has been adopted. By using the aforementioned model a system of three ordinary differential equations (1), which describes the avascular, vascular and metastasis tumor growth has been produced.

$$\begin{cases} \dot{x} = x(2N - x) - Hxz \\ \dot{y} = y(4 - 0.14y) + 0.5x^2 - Iy - 0.5Hyz + 0.001z^2 \\ \dot{z} = -Iz + 0.07y^2 + 0.5Hyz - 0.002z^2 \end{cases} \tag{1}$$

In this system the variables x, y, z, represent the proliferating tumor cells in an avascular, vascular and metastasis phases, respectively. Also, parameter N represents the population of normal cells, H the population of the host cells and I the population of immune cells, which are the T lymphocytes, as well as the natural killer cells. In this chapter the numerical values of the constants are kept the same as in work of Llanos-Pérez *et al.* (Llanos-Pérez *et al.* 2016), which have been chosen empirically trying to have a greater generality and simplicity as possible, based on the adopted chemical network model.

Llanos-Pérez *et al.* in their paper have considered the parameters N and H as constants ($N = 5$, $H = 3$), because these cell groups are much more numerous than cancer cells, as well as for practical reasons due to the fact that their number does not change, while I has been considered as the control parameter, because the population of immune cells against these cancer cells may increase or decrease. The same strategy has been adopted in this work, however various cases with different values of the parameters N and H have been examined, in order to reveal the plethora of dynamical phenomena that the proposed system presents. Furthermore, in comparison to the work of Llanos-Pérez *et al.* (Llanos-Pérez *et al.*, 2016) this chapter has introduced system's dynamical analysis mainly through the classic from the nonlinear theory bifurcation diagram. This diagram is very useful in the direction of study of the dynamics of a nonlinear system revealing interesting phenomena concerning chaos theory.

3. DYNAMICAL ANALYSIS OF THE TUMOR GROWTH MODEL

In this section, the influence of the proposed tumor growth model's (1) parameter I to its dynamical behavior, by keeping the value of $N = 5$ and for various values of H, with initial conditions (x_0, y_0 z_0) = (1, 0, 0), is performed. For the investigation of system's dynamical behavior, system (1) is integrated numerically using the classical fourth-order Runge–Kutta integration algorithm. For each set of parameters used in this work, the time step is always $\Delta t = 0.001$ and the calculations are performed using variables and parameters in extended precision mode. For each parameter settings, the system is integrated for a sufficiently long time and the transient is discarded. Furthermore, Lyapunov exponents of system (1), for the chosen set of parameters and initial conditions, are calculated each time by using the Wolf *et al.* algorithm (Wolf *et al.*, 1985).

To study the type of scenario giving rise to chaos by considering the population of immune cells, namely the parameter I in system (1), as the main bifurcation control parameter, the bifurcation diagram in Fig. 1(a) is obtained, while the other parameters remain fixed as $N = 5$ and $H = 3$. As mentioned before, these values have been chosen initially due to the fact that they have been used in the paper of Llanos-Péırez *et al.* The bifurcation diagram is obtained by plotting the variable z when the trajectory cuts the plane $x = 0$ with $dx/dt < 0$, as the control parameter I is increased in tiny steps in the range of $0 \leq I \leq 0.8$. Furthermore, the corresponding spectrum of the three largest Lyapunov exponents is shown in Fig. 1(b). It can be seen that the bifurcation diagram well coincides with the spectrum of the Lyapunov exponents.

In more details, as it is known from the theory, the Lyaponov Exponents (LE_i) determine the notion of predictability of a system. A positive Lyaponov exponent in the range of $\hat{I\Pi}$[0.143, 0.506] in Fig. 1(b) is an indication that system (1) is chaotic. This chaotic behavior is also confirmed by the phase portraits in Fig. 3, for $I = 0.337$. Moreover, the system displays a periodic state when the maximum LE is equal to zero, i.e. for $I < 0.143$ and $I > 0.506$ and for small windows inside the chaotic region. As shown in Fig. 1(b), the results from the study of the diagram of Lyapunov exponents regarding system's (1) behavior are consistent with the results produced from the bifurcation diagram of Fig. 1(a).

In applied mathematics, the Kaplan–Yorke conjecture concerns the dimension of an attractor, using Lyapunov exponents (Kaplan & Yorke, 1979; Frederickson *et al.*, 1983). By arranging the Lyapunov exponents in order from largest to smallest LE_1 ³ LE_2 ³ ··· ³ LE_n, and by letting j the index for which $\sum_{i=1}^{j} LE_i \geq 0$ and $\sum_{i=1}^{j+1} LE_i < 0$, the conjecture, which is called Lyapunov dimension of the attractor, is defined by

$$D_{KY} = j + \frac{\sum_{i}^{j} LE_i}{\left|LE_{j+1}\right|} \tag{3}$$

In Fig. 2 the Lyapunov dimension D_{KY} of system (1) versus the parameter I is depicted. From this diagram the maximum value of the Lyapunov dimension ($D_{KY} = 2.042$) of the system (1) for $I = 0.337$ is found.

Another important feature of system (1) is the mechanism through which this system is driven to chaos. According to the bifurcation diagram of Fig. 1(b) system (1) starting for period-1 state, for low values of I, is driven to chaos through the well-known period-doubling route. In Fig. 4 the phase portraits in x–y plane, for increasing values of parameter I, as the system follows this route, is depicted.

For $I = 0.247$ the system suddenly exits from chaos and enters to a period-3 state, which is another important feature of chaos, as depicted in the phase portrait of Fig. 5. The occurrence of sudden qualitative changes of chaotic dynamics as a parameter varies is known from 1983. That year Grebogi *et al.* reported that such changes may result from the collision of an unstable periodic orbit and a coexisting chaotic attractor (Grebogi *et al.*, 1983). These phenomena were called crises, which have been observed in many dynamical systems since then (He, 2003; Naseradinmousavi & Nataraj, 2012; Kengne *et al.*, 2016; Stavrinides *et al.*, 2009).

Figure 1. (a) Bifurcation diagram of system (1) for increasing values of I and (b) the corresponding graph of the three largest Lyapunov exponents plotted in the range of $0 \leq I \leq 0.8$, for $N = 5$ and $H = 3$.

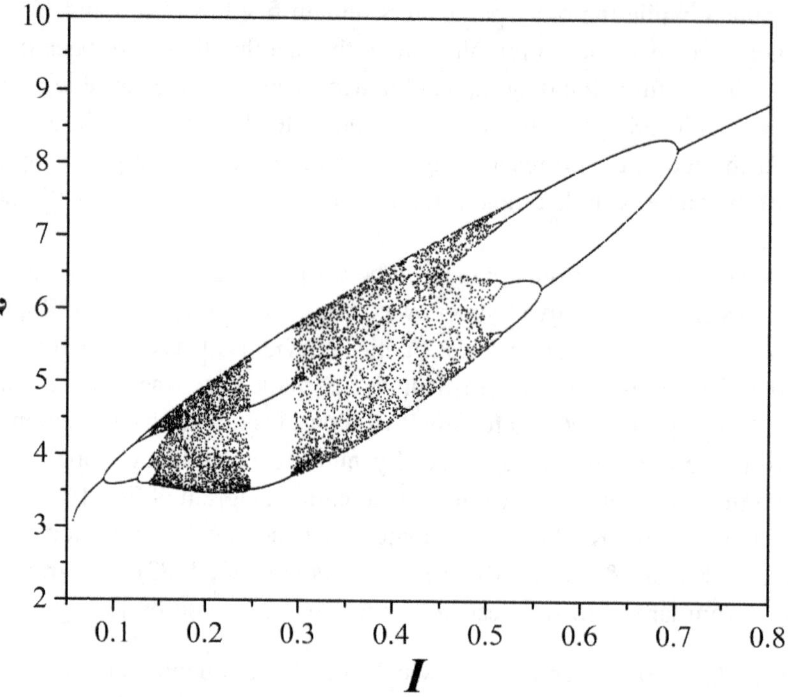

Figure 2. The diagram of Lyapunov dimension D_{KY} of system (1) versus the parameter I, for $N = 5$ and $H = 3$, in the ranges of $0 \leq I \leq 0.8$ and $2 \leq D_{KY} \leq 2.045$.

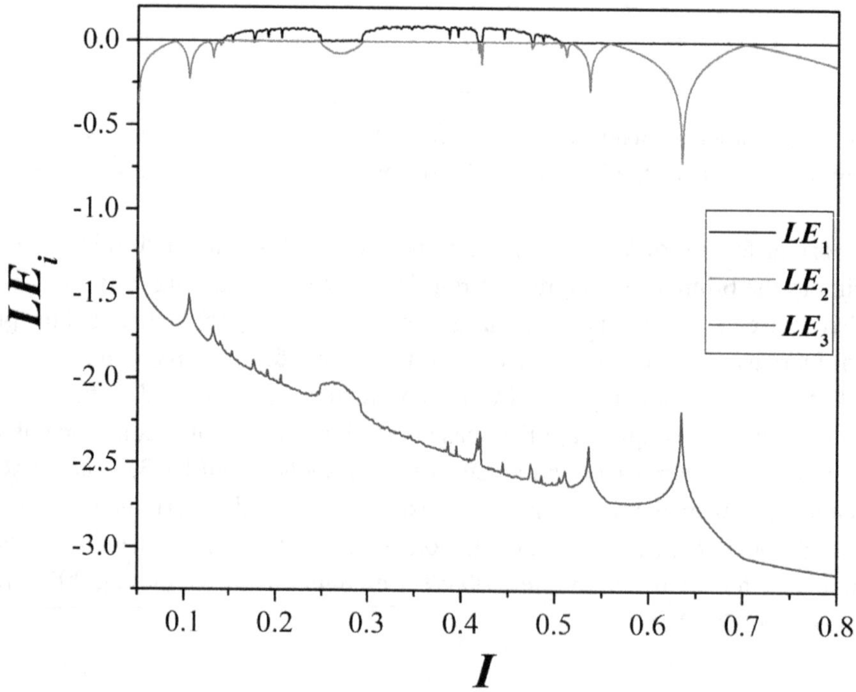

Figure 3. Chaotic attractors of system (1) for N = 5, H = 3 and I = 0.337, in (a) x–y plane, (b) x–z plane and (c) y–z plane.

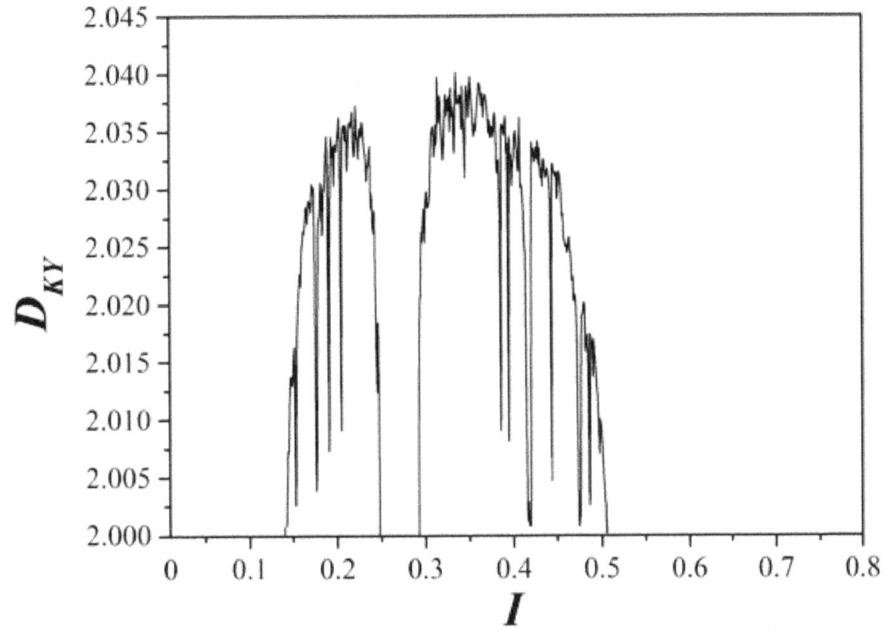

Figure 4. System's (1) periodic attractors in x–y plane, for N = 5, H = 3, and (a) I = 0.075 (period-1), (b) I = 0.110 (period-2), (c) I = 0.135 (period-4), (d) I = 0.140 (period-8).

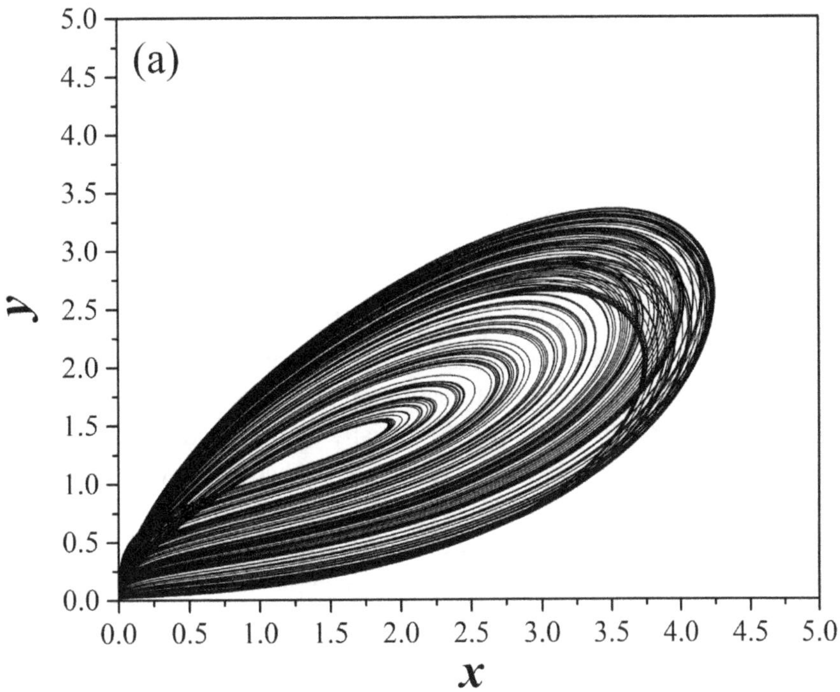

Also, with a more clear view on the bifurcation diagram of Fig. 1(a), a second interesting phenomenon concerning system's (1) dynamical behavior can be observed. This is the antimonotonicity, which has been introduced by Dawson *et al.* (Dawson *et al.*, 1992), which is a fundamental phenomenon in bifurcations for a large class of nonlinear dissipative systems (Kocarev *et al.*, 1993; Kyprianidis *et al.*, 2000). In this phenomenon periodic orbits are not only created but also destroyed, when one increases the control parameter monotonically (smoothly) in any neighborhood of a homoclinic tangency value-inevitable reversals of period-doubling cascades. So, antimonotonicity is due to tangential intersections between the stable and unstable manifolds of a system.

According to this phenomenon, the system enters to chaos with the well-known period-doubling route (period-1 → period-2 → … → chaos) and exits from the chaos following the reverse period-doubling route (chaos →… → period-2 → period-1). This route forms a "*chaotic bubble*" as it is known (Fig. 1(a)). As a result, a shape of chaotic bubble is formed in the bifurcation diagram. This phenomenon is relevant because it describes a complex scenario of how a nonlinear system creates or destroys unstable periodic orbits by parameter alterations. At the parameter value where the Lyapunov exponent is maximal, the chaotic attractor has embedded on it the largest amount of unstable periodic orbits that the dynamics can create.

In Fig. 6 the bifurcation diagrams of the variable z versus the parameter I, for $N = 5$ and various values of H, are presented. In these diagrams the route to the forming of the chaotic bubble is revealed. As the parameter H increased, starting from the value of $H = 2$, in which the system is always in period-1 steady state (Fig. 6(a)), consecutive doubling of period occurs forming the primary bubble (Fig. 6(b)), as well as bubbles in which period-4 (Fig. 6(c)) and period-8 (Fig. 6(d)) regions has been created. In this way, by increasing further the value of H, the system appears the chaotic bubble for $H = 3$ (Fig. 1(a)).

4. CONCLUSION

In this chapter the recently new model of tumor growth, which has been proposed by Llanos-Pérez *et al.* in 2016, was investigated from the dynamics perspective. As presented in a previous section, this model is a simple third-order dynamical system, which describes the avascular, vascular and metastasis tumor growth. Furthermore, as far as the authors know, it is the first published system that connects these tumor stages. So, it is very important to study the effect of each system's parameter, namely the population of normal cells N, the population of the host cells H and the population of immune cells I. However, this work was concentrated to the study of system's dynamics by using the population of immune cells as a control parameter. For this reason, some of the most well-known tools from nonlinear theory, such as the phase portrait, the Poincaré map, the bifurcation diagram, the Kaplan-Yorke dimension and the Lyapunov exponents were used. These tools and especially the bifurcation diagrams revealed, for the first time for this model, interesting phenomena related with chaos theory, such as a period-doubling route to chaos, crisis phenomena, and antimonotonicity. As a future work a more thoroughly investigation of the aforementioned model's dynamics by considering the population of normal cells N and the population of the host cells H as control parameters, has been planned.

Figure 5. System's (1) period-3 attractor in x–y plane, for N = 5, H = 3, and I = 0.27.

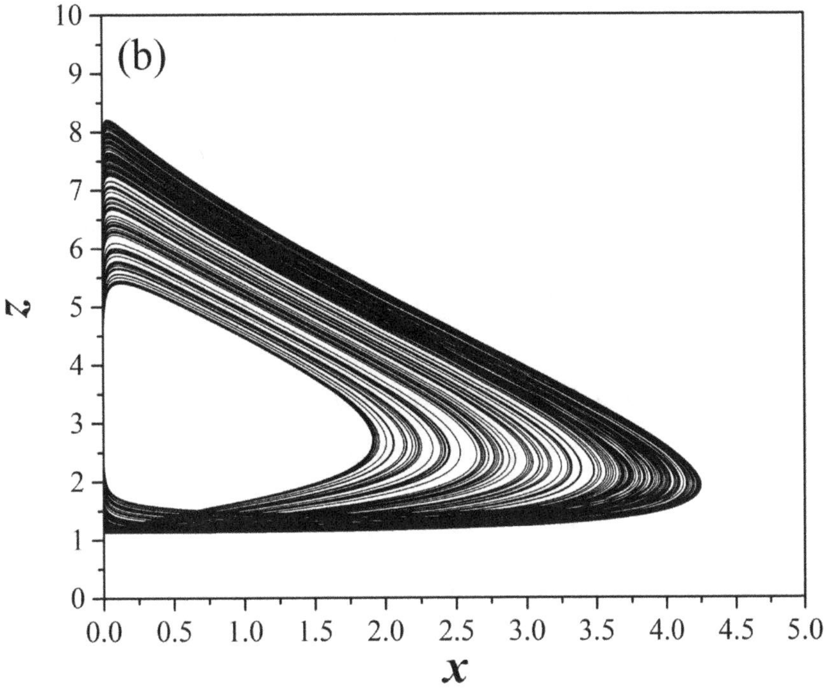

Figure 6. Bifurcation diagrams of system (1) for increasing values of I, for N = 5 and (a) H = 2, (b) H = 2.5, (c) H = 2.55, (d) H = 2.65.

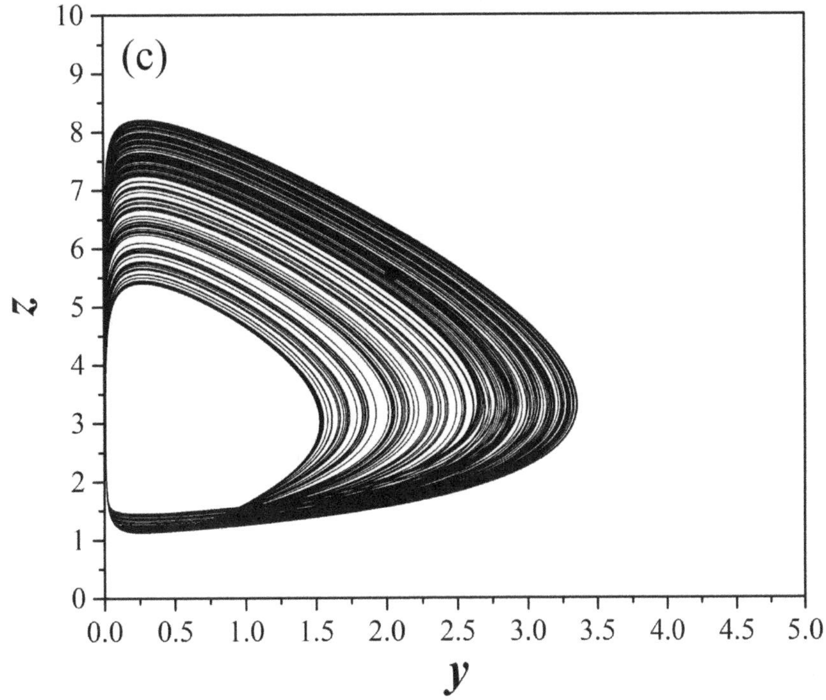

REFERENCES

Adam, J. A., & Bellomo, N. (2012). *A survey of models for tumor-immune system dynamics*. Springer Science & Business Media.

Ahmed, E. (1993). Fractals and chaos in cancer models. *International Journal of Theoretical Physics*, *32*(2), 353–355. doi:10.1007/BF00673722

Alain, K. S. T., Romanic, K., & Bertrand, F. H. (2017). Dynamics and Improved Robust Adaptive Control Strategy for the Finite Time Synchronization of Uncertain Nonlinear Systems. *International Journal of System Dynamics Applications*, *6*(4), 34–62. doi:10.4018/IJSDA.2017100103

Albert, A., Freedman, M., & Perelson, A. S. (1980). Tumor and the immune system: The effects of a tumor growth modulator. *Mathematical Biosciences*, *50*(1-2), 25–58. doi:10.1016/0025-5564(80)90120-0

Betancourt-Mar, J. A., Llanos-Pérez, J. A., Cocho, G., Mansilla, R., Martin, R. R., Montero, S., & Nieto-Villar, J. M. (2017). Phase transitions in tumor growth: IV relationship between metabolic rate and fractal dimension of human tumor. *Physica A*, *473*, 344–351. doi:10.1016/j.physa.2016.12.089

Breward, C. J. W., Byrne, H. M., & Lewis, C. E. (2002). The role of cell-cell interactions in a two-phase model for avascular tumour growth. *Journal of Mathematical Biology*, *45*(2), 125–152. doi:10.1007002850200149 PMID:12181602

Brinkley, B. R. (2001). Managing the centrosome numbers game: From chaos to stability in cancer cell division. *Trends in Cell Biology*, *11*(1), 18–21. doi:10.1016/S0962-8924(00)01872-9 PMID:11146294

Byrne, H. M., & Chaplain, M. A. J. (1998). Necrosis and apoptosis: Distinct cell loss mechanisms in a mathematical model of avascular tumour growth. *Computational and Mathematical Methods in Medicine*, *1*(3), 223–235.

Cattani, C., & Ciancio, A. (2008). Qualitative analysis of second-order models of tumor–immune system competition. *Mathematical and Computer Modelling*, *47*(11-12), 1339–1355. doi:10.1016/j.mcm.2007.07.005

Chaffer, C. L., & Weinberg, R. A. (2011). A perspective on cancer cell metastasis. science. *JSTOR*, *331*(6024), 1559–1564. PMID:21436443

Chaplain, M. A. J. (1996). Avascular growth, angiogenesis and vascular growth in solid tumours: The mathematical modelling of the stages of tumour development. *Mathematical and Computer Modelling*, *23*(6), 47–87. doi:10.1016/0895-7177(96)00019-2

Dawson, S. P., Grebogi, C., Yorke, J. A., Kan, I., & Kocak, H. (1992). Antimonotonicity: Inevitable reversals of period-doubling cascades. *Physics Letters. [Part A]*, *163*(3), 249–254. doi:10.1016/0375-9601(92)90442-O

Degn, H., Holden, A. V., & Olsen, L. F. (1986). *Chaos in Biological Systems*. Springer-Verlag.

Deisboeck, T. S., Wang, Z., Macklin, P., & Cristini, V. (2011). Multiscale cancer modeling. *Annual Review of Biomedical Engineering*, *13*(1), 127–155. doi:10.1146/annurev-bioeng-071910-124729 PMID:21529163

Dirks, P. B. (2008). Brain tumor stem cells: Bringing order to the chaos of brain cancer. *Journal of Clinical Oncology, 26*(17), 2916–2924. doi:10.1200/JCO.2008.17.6792 PMID:18539973

Djouima, M., Azar, A. T., Drid, S., & Mehdi, D. (2018). Higher Order Sliding Mode Control for Blood Glucose Regulation of Type 1 Diabetic Patients. *International Journal of System Dynamics Applications, 7*(1), 65–84. doi:10.4018/IJSDA.2018010104

Dominic, W., & Natalia, K. (2005). *Computational biology of cancer: lecture notes and mathematical modeling*. World Scientific.

Duarte, J., Januário, C., Rodrigues, C., & Sardanyes, J. (2013). Topological Complexity and Predictability in the Dynamics of a Tumor Growth Model with Shilnikov's Chaos. *International Journal of Bifurcation and Chaos in Applied Sciences and Engineering, 23*(07), 1350124. doi:10.1142/S0218127413501241

Duesberg, P. (2007). Chromosomal chaos and cancer. *Scientific American, 296*(5), 52–59. doi:10.1038 cientificamerican0507-52 PMID:17500414

Field, R. J., & Gyorgyi, L. (1993). *Chaos in Chemistry and Biochemistry*. World Scientific. doi:10.1142/1706

Frank, S. A. (2007). *Dynamics of cancer: incidence, inheritance, and evolution* (Vol. 1). Princeton University Press. doi:10.1515/9780691186863

Frederickson, P., Kaplan, J. L., Yorke, E. D., & Yorke, J. A. (1983). The Liapunov dimension of strange attractors. *Journal of Differential Equations, 49*(2), 185–207. doi:10.1016/0022-0396(83)90011-6

Gholizad, A., Ahmadi, L., Hassannayebi, E., Memarpour, M., & Shakibayifar, M. (2017). A system dynamics model for the analysis of the deregulation in electricity market. *International Journal of System Dynamics Applications, 6*(2), 1–30. doi:10.4018/IJSDA.2017040101

Grebogi, C., Ott, E., & Yorke, J. A. (1983). Crises, sudden changes in chaotic attractors and transient chaos. *Physica D. Nonlinear Phenomena, 7*(1-3), 181–200. doi:10.1016/0167-2789(83)90126-4

Guerra, A., Rodriguez, D. J., Montero, S., Betancourt-Mar, J. A., Martin, R. R., Silva, E., Bizzarri, M., Cocho, G., Mansilla, R., & Nieto-Villar, J. M. (2018). Phase transitions in tumor growth VI: Epithelial–Mesenchymal transition. *Physica A, 499*, 208–215. doi:10.1016/j.physa.2018.01.040

Gupta, G. P., & Massagué, J. (2006). Cancer metastasis: Building a framework. *Cell, 127*(4), 679–695. doi:10.1016/j.cell.2006.11.001 PMID:17110329

He, K. (2003). Critical phenomenon, crisis and transition to spatiotemporal chaos in plasmas. In *Advances in Space Environment Research* (pp. 475–494). Springer. doi:10.1007/978-94-007-1069-6_47

Hogea, C. S., Murray, B. T., & Sethian, J. A. (2005). Computational modeling of solid tumor evolution via a general Cartesian mesh/level set method. *Fluid Dynamics & Materials Processing, 1-2*, 1–37.

Huang, Y., Song, N., Ding, Y., Yuan, S., Li, X., Cai, H., Shi, H., & Luo, Y. (2009). Pulmonary vascular destabilization in the premetastatic phase facilitates lung metastasis. *Cancer Research, 69*(19), 7529–7537. doi:10.1158/0008-5472.CAN-08-4382 PMID:19773447

Izquierdo-Kulich, E., Rebelo, I., Tejera, E., & Nieto-Villar, J. M. (2013). Phase transition in tumor growth: I avascular development. *Physica A, 392*(24), 6616–6623. doi:10.1016/j.physa.2013.08.010

Jaime, J. C., Mesa-Álvarez, M. D., Martin, R. R., Betancourt-Mar, J. A., Cocho, G., Mansilla, R., & Nieto-Villar, J. M. (2019). Chronotherapy of cancer: Periodic perturbations in vascular growth and metastasis. *Biological Rhythm Research, 50*(4), 495–504. doi:10.1080/09291016.2018.1465698

Kalluri, R., & Weinberg, R. A. (2009). The basics of epithelial-mesenchymal transition. *The Journal of Clinical Investigation, 119*(6), 1420–1428. doi:10.1172/JCI39104 PMID:19487818

Kaplan, J. L., & Yorke, J. A. (1979). Chaotic behavior of multidimensional difference equations. In *Functional differential equations and approximation of fixed points* (pp. 204–227). Springer. doi:10.1007/BFb0064319

Kengne, J., Tabekoueng, Z. N., & Fotsin, H. B. (2016). Coexistence of multiple attractors and crisis route to chaos in autonomous third order Duffing–Holmes type chaotic oscillators. *Communications in Nonlinear Science and Numerical Simulation, 36*, 29–44. doi:10.1016/j.cnsns.2015.11.009

Keogan, M. T., McDermott, V. G., Paulson, E. K., Sheafor, D. H., Frederick, M. G., de Long, D. M., & Nelson, R. C. (1997). Pancreatic malignancy: Effect of dual-phase helical CT in tumor detection and vascular opacification. *Radiology, 205*(2), 513–518. doi:10.1148/radiology.205.2.9356637 PMID:9356637

Knighton, D., Ausprunk, D., Tapper, D., & Folkman, J. (1977). Avascular and vascular phases of tumour growth in the chick embryo. *British Journal of Cancer, 35*(3), 347–356. doi:10.1038/bjc.1977.49 PMID:856240

Kocarev, L., Halle, K. S., Eckert, K., & Chua, L. O. (1993). Experimental observation of antimonotonicity in Chua's circuit. *International Journal of Bifurcation and Chaos in Applied Sciences and Engineering, 3*(04), 1051–1055. doi:10.1142/S0218127493000878

Kyprianidis, I. M., Stouboulos, I. N., Haralabidis, P., & Bountis, T. (2000). Antimonotonicity and chaotic dynamics in a fourth-order autonomous nonlinear electric circuit. *International Journal of Bifurcation and Chaos in Applied Sciences and Engineering, 10*(08), 1903–1915. doi:10.1142/S0218127400001171

Llanos-Pérez, J. A., Betancourt-Mar, J. A., Cocho, G., Mansilla, R., & Nieto-Villar, J. M. (2016, November). Nieto-Villar José Manuel. Phase transitions in tumor growth: III vascular and metastasis behavior. *Physica A, 462*, 560–568. doi:10.1016/j.physa.2016.06.086

Lorenz, E. (2000). The butterfly effect. World Scientific.

Majhi, S. K. (2018). An efficient feed foreword network model with sine cosine algorithm for breast cancer classification. *International Journal of System Dynamics Applications, 7*(2), 1–14. doi:10.4018/IJSDA.2018040101

Montero, S., Martin, R., Mansilla, R., Cocho, G., & Nieto-Villar, J. M. (2018). Parameters estimation in phase-space landscape reconstruction of cell fate: A systems biology approach. In *Systems Biology* (pp. 125–170). Humana Press. doi:10.1007/978-1-4939-7456-6_8

Moysis, L., & Azar, A. T. (2017a). New discrete time 2d chaotic maps. *International Journal of System Dynamics Applications, 6*(1), 77–104. doi:10.4018/IJSDA.2017010105

Moysis, L., Azar, A. T., Kafetzis, I., Tsiaousis, M., & Charalampidis, N. (2017b). Introduction to Control Systems Design Using Matlab. *International Journal of System Dynamics Applications*, *6*(3), 130–170. doi:10.4018/IJSDA.2017070107

Naseradinmousavi, P., & Nataraj, C. (2012). Transient chaos and crisis phenomena in butterfly valves driven by solenoid actuators. *Communications in Nonlinear Science and Numerical Simulation*, *17*(11), 4336–4345. doi:10.1016/j.cnsns.2012.01.034

Newton, C. M. (1980). Biomathematics in oncology: Modeling of cellular systems. *Annual Review of Biophysics and Bioengineering*, *9*(1), 541–579. doi:10.1146/annurev.bb.09.060180.002545 PMID:6994595

Nguyen, D. X., & Massagué, J. (2007). Genetic determinants of cancer metastasis. *Nature Reviews. Genetics*, *8*(5), 341–352. doi:10.1038/nrg2101 PMID:17440531

Nicolay, D. (2014). *Asymptotic Chaos Expansions in Finance: Theory and Practice*. Springer-Verlag. doi:10.1007/978-1-4471-6506-4

Nikulin, M. S., Commenges, D., & Huber, C. (Eds.). (2006). *Probability, statistics and modelling in public health*. Springer. doi:10.1007/b136588

Pani, G., Galeotti, T., & Chiarugi, P. (2010). Metastasis: Cancer cell's escape from oxidative stress. *Cancer and Metastasis Reviews*, *29*(2), 351–378. doi:10.100710555-010-9225-4 PMID:20386957

Poste, G., & Fidler, I. J. (1980). The pathogenesis of cancer metastasis. *Nature*, *283*(5743), 139–146. doi:10.1038/283139a0 PMID:6985715

Ribatti, D., Vacca, A., & Dammacco, F. (1999). The role of the vascular phase in solid tumor growth: A historical review. *Neoplasia (New York, N.Y.)*, *1*(4), 293–302. doi:10.1038j.neo.7900038 PMID:10935483

Saleem, M., & Agrawal, T. (2012). Chaos in a tumor growth model with delayed responses of the immune system. *Journal of Applied Mathematics*, *2012*, 2012. doi:10.1155/2012/891095

Stavrinides, S. G., Miliou, A. N., Anagnostopoulos, A. N., Konstantakos, V., Magafas, L., & Laopoulos, T. (2009). Desynchronization crisis induced intermittency in a master–slave PLL configuration. *Chaos, Solitons, and Fractals*, *42*(1), 33–39. doi:10.1016/j.chaos.2008.10.017

Strogatz, S. (2015). *Nonlinear dynamics and chaos: with applications to physics, biology, chemistry, and engineering (studies in nonlinearity)*. Westview Press.

Tan, W. Y. (2020). *Stochastic Models for Carcinogenesis*. CRC Press. doi:10.1201/9780367810832

Tiwari, M. (2012). *Tumor Marker and Carcinogenesis*. River Publishers.

Verhulst, F. (1996). *Nonlinear Differential Equations and Dynamical Systems*. Springer-Verlag. doi:10.1007/978-3-642-61453-8

Verstovsek, S., Kantarjian, H., Manshouri, T., Cortes, J., Giles, F. J., Rogers, A., & Albitar, M. (2002). Prognostic significance of cellular vascular endothelial growth factor expression in chronic phase chronic myeloid leukemia. Blood. *The Journal of the American Society of Hematology*, *99*(6), 2265–2267. PMID:11877311

Voiculetz, N., Balaban, A. T., Niculescu-Duvaz, I., & Simon, Z. (1991). *Modeling of cancer genesis and prevention*. CRC press.

Ward, J. P., & King, J. R. (1997). Mathematical modelling of avascular-tumour growth. *Mathematical Medicine and Biology. The Journal of IMA / Islamic Medical Association of North America*, *14*(1), 39–69.

WHO. (2020). https://www.who.int/health-topics/cancer#tab=tab_1

Wolf, S. A., Swift, J. B., Swinney, H. L., & Vastano, J. A. (1985). Determining Lyapunov exponents from a time series. *Physica D. Nonlinear Phenomena*, *16*(3), 285–317. doi:10.1016/0167-2789(85)90011-9

Yao, J. C., Phan, A., Hoff, P. M., Chen, H. X., Charnsangavej, C., Yeung, S. C. J., Hess, K., Ng, C., Abbruzzese, J. L., & Ajani, J. A. (2008). Targeting vascular endothelial growth factor in advanced carcinoid tumor: A random assignment phase II study of depot octreotide with bevacizumab and pegylated interferon alfa-2b. *Journal of Clinical Oncology*, *26*(8), 1316–1323. doi:10.1200/JCO.2007.13.6374 PMID:18323556

Compilation of References

Aassve, A., Engelhardt, H., Francavilla, F., Kedir, A., Kim, J., Mealli, F., … Prskawetz, A. (2005). *Poverty and Fertility in Less Developed Countries: A Comparative Analysis* (ISER Working Paper Series No. 2005–13). ISER Working Paper 2005-13. Retrieved from http://hdl.handle.net/10419/92124

Abdel-Hamid, T. K. (1989). The dynamics of software project staffing: A system dynamics based simulation approach. *IEEE Transactions on Software Engineering*, *15*(2), 109–119. doi:10.1109/32.21738

Abdel, M. S. E. D. A., Elsamahy, M., Hassan, M. A. M., & Bendary, F. M. (2017). Enhancement of Turbo-Generators Phase Backup Protection Using Adaptive Neuro Fuzzy Inference System. *International Journal of System Dynamics Applications*, *6*(1), 58–76. doi:10.4018/IJSDA.2017010104

Abdel, M. S. E.-D. A., El Samahy, M., Hassan, M. A. M., & El Bendary, F. (2016). Applications of ANFIS in Loss of excitation faults detection in hydro-generators. *International Journal of System Dynamics Applications*, *5*(2), 63–79.

Abdel, M. S. E.-D. A., Hassan, M. A. M., & Bendary, F. M. (2018). Loss of Excitation Protection of Medium Voltage Hydro Generators Using Adaptive Neuro Fuzzy Inference System. In *Advances in System Dynamics and Control* (pp. 253–285). IGI Global.

Abdelsalam, H. A., & El-Tagy, A. O. (2012). A Simulation Model for Managing Marketing Multi-Channel Conflict. *International Journal of System Dynamics Applications*, *1*(4), 59–76. doi:10.4018/ijsda.2012100103

Abdelwahab, O. M., & Shaaban, M. F. (2019). PV and EV charger allocation with V2G capabilities. *Proceedings - 2019 IEEE 13th International Conference on Compatibility, Power Electronics and Power Engineering, CPE-POWERENG. IEEE, 2019*, 1–5. 10.1109/CPE.2019.8862370

Abdollahi, H. (2020). An Adaptive Neuro-Based Fuzzy Inference System (ANFIS) for the Prediction of Option Price: The Case of the Australian Option Market. *International Journal of Applied Metaheuristic Computing*, *11*(2), 99–117. doi:10.4018/IJAMC.2020040105

Abdollahi, H., Ebrahimi, S. B., & Farmani, A. (2016). The inter-relationship between profitability and growth in the emerging Iranian market. *International Journal of Industrial Engineering & Production Research*, *27*(3), 209–217.

Abdollahi, H., & Talooni, M. (2016). Coordinating Contracts for Two-Echelon Supply Chain with Price and Promotional Sensitive Demand. *Modern Applied Science*, *10*(7), 132–140. doi:10.5539/mas.v10n7p132

Abdullah, M., & Iqbal, S. (2020). The Dawn of Paternal HR: An Exploratory Study Using System Dynamics Approach. In Occupational Stress: Breakthroughs in Research and Practice (pp. 217-237). IGI Global.

Abdullah, F. (2006). Measuring service quality in higher education: Three instruments compared. *International Journal of Research & Method in Education*, *29*(1), 7189. doi:10.1080/01406720500537445

Abeygunawardhana, P. K. W., & Murakami, T. (2008). Stability improvement of two wheel driven mobile manipulator using nonlinear PD controller. *IEEE Transactions on Industry Applications, 128*(10), 1149–1156. doi:10.1541/ieejias.128.1149

Aboamer, M. A., Azar, A. T., Wahba, K., Schumann, A., & Bär, K. J. (2019). Nonlinear single-input single-output model-based estimation of cardiac output for normal and depressed cases. *Neural Computing & Applications, 31*(7), 2955–2978. doi:10.100700521-017-3245-8

Aboamer, M. M., Azar, A. T., Mohamed, A. S. A., Bär, K. J., Berger, B. S., & Wahba, K. (2014b). Nonlinear Features of Heart Rate Variability in Paranoid Schizophrenic. *Neural Computing & Applications, 25*(7-8), 1535–1555. doi:10.100700521-014-1621-1

Aboamer, M. M., Azar, A. T., Wahba, K., & Mohamed, A. S. A. (2014a). Linear model-based estimation of blood pressure and cardiac output for Normal and Paranoid cases. *Neural Computing & Applications, 25*(6), 1223–1240. doi:10.100700521-014-1566-4

Aboites, V., Wilson, M., del Bosque, L., & del Campestre, L. (2009). Tinkerbell chaos in a ring phase-conjugated resonator. *International Journal of Pure and Applied Mathematics, 54*(3), 429–435.

Abreu, F. B., & Carapuça, R. (1994). Object-oriented software engineering: Measuring and controlling the development process. *Proceedings of the 4th International Conference on Software Quality, 186*, 1–8.

Abu-Seada, H. F., Mansor, W. M., Bendary, F. M., Emery, A. A., & Moustafa Hassan, M. A. (2013). Application of Particle Swarm Optimization in Design of PID Controller for AVR System. *International Journal of System Dynamics Applications, 3*(2), 1–17. doi:10.4018/ijsda.2013070101

Ackoff, R. L. (1967). Management Misinformation Systems. *Management Science, 14*(4), B147–B156. doi:10.1287/mnsc.14.4.B147

Ackoff, R. L. (1994). Systems thinking and thinking systems. *System Dynamics Review, 10*(2–3), 175–188. doi:10.1002dr.4260100206

Acuña, L. G., Ríos, D. R., Arboleda, C. P., & Ponzón, E. G. (2018). Cooperation model in the electricity energy market using bi-level optimisation and Shapley value. *Operations Research Perspectives, 5*, 161–168.

Adak, M. F. (2018), Software Defect Detection by using Data Mining based Fuzzy Logic. *2018 Sixth International Conference on Digital Information, Networking, and Wireless Communications (DINWC)*, 65-69. 10.1109/DINWC.2018.8356997

Adam, J. A., & Bellomo, N. (2012). *A survey of models for tumor-immune system dynamics*. Springer Science & Business Media.

Afshar, H.R. (2013). An introduction to the economy of theater in Iran. *Journal of Dramatic Art, 18*(2), 13-20. (in Persian)

Afshariasl, I. (2013). Economy in Iran Theater. [in Persian]. *Journal of Namayesh, 167*, 62–65.

Aguilar-Bustos, A. Y., & Cruz-Hernández, C. (2009). Synchronization of discrete-time hyperchaotic systems: An application in communications. *Chaos, Solitons, and Fractals, 41*(3), 1301–1310. doi:10.1016/j.chaos.2008.05.012

Ahlburg, D. A. (1996). Population Growth and Poverty. In *The Impact of Population Growth on Well-being in Developing Countries* (pp. 219–258). Springer., doi:10.1007/978-3-662-03239-8_7

Ahmad, S., Mat Tahar, R., Muhammad-Sukki, F., Munir, A. B., & Abdul Rahim, R. (2016). Application of system dynamics approach in electricity sector modelling: A review. *Renewable & Sustainable Energy Reviews, 56*, 29–37. doi:10.1016/j.rser.2015.11.034

Ahmad, S., Siddique, N. H., & Tokhi, M. O. (2011). A Modular fuzzy control approach for two-wheeled wheelchair. *Journal of Intelligent & Robotic Systems*, *64*(3), 401–426. doi:10.100710846-011-9541-0

Ahmed, E. (1993). Fractals and chaos in cancer models. *International Journal of Theoretical Physics*, *32*(2), 353–355. doi:10.1007/BF00673722

Ahn, J. Y., Han, K. S., Oh, S. Y., & Lee, C. D. (2011). An application of interval-valued intuitionistic fuzzy sets for medical diagnosis of headache. *International Journal of Innovative Computing, Information, & Control*, *7*, 2755–2762.

Ajeil, F. H., Ibraheem, I. K., Azar, A. T., & Humaidi, A. J. (2020). Grid-Based Mobile Robot Path Planning Using Aging-Based Ant Colony Optimization Algorithm in Static and Dynamic Environments. *Sensors (Basel)*, *20*(7), 1880. doi:10.339020071880 PMID:32231091

Akhmet, M., & Fen, M. O. (2016). Chaos Generation in Continuous/Discrete-Time Models. In Replication of Chaos in Neural Networks, Economics and Physics, (pp. 183–264). Springer Berlin Heidelberg.

Akkaya, M. (2020). Startup Valuation: Theories, Models, and Future. In Valuation Challenges and Solutions in Contemporary Businesses (pp. 137–156). IGI Global.

Alain, K. S. T., Azar, A. T., Kengne, R., & Bertrand, F. H. (2020). Stability analysis and robust synchronisation of fractional-order modified Colpitts oscillators. *Int. J. Automation and Control*, *14*(1), 52–79. doi:10.1504/IJAAC.2020.103806

Alain, K. S. T., Azar, A. T., Kengne, R., & Bertrand, F. H. (2020). Stability analysis and robust synchronisation of fractional-order modified Colpitts oscillators. *International Journal of Automation and Control*, *14*(1), 52–79.

Alain, K. S. T., Romanic, K., Azar, A. T., Vaidyanathan, S., Bertrand, F. H., & Adele, N. M. (2018). *Dynamics Analysis and Synchronization in Relay Coupled Fractional Order Colpitts Oscillators. Advances in System Dynamics and Control.* IGI-Global. doi:10.4018/978-1-5225-4077-9.ch011

Alain, K. S. T., Romanic, K., & Bertrand, F. H. (2017). Dynamics and Improved Robust Adaptive Control Strategy for the Finite Time Synchronization of Uncertain Nonlinear Systems. *International Journal of System Dynamics Applications*, *6*(4), 34–62. doi:10.4018/IJSDA.2017100103

Alawida, M., Samsudin, A., & Teh, J. S. (2020). Enhanced digital chaotic maps based on bit reversal with applications in random bit generators. *Information Sciences*, *512*, 1155–1169. doi:10.1016/j.ins.2019.10.055

Albach, P. G. (1993). The dilemma of change in Indian higher education. *Higher Education*, *26*(1), 3–20. doi:10.1007/BF01575104

Albert, A., Freedman, M., & Perelson, A. S. (1980). Tumor and the immune system: The effects of a tumor growth modulator. *Mathematical Biosciences*, *50*(1-2), 25–58. doi:10.1016/0025-5564(80)90120-0

Alcantud, J. C. R., Varela, G., Santos-Buitrago, B., Santos-Garcia, G., & Jimenez, M. F. (2019). Analysis of survival for lung cancer resections cases with fuzzy and soft set theory in surgical decision making. *PLoS One*, *14*(6), e0218283. doi:10.1371/journal.pone.0218283 PMID:31216304

Al-Dahan, Z. T., Bachache, N. K., & Bachache, L. N. (2016). Design and implementation of fall detection system using MPU 6050 Arduino. In *Proceedings of 14th International Conference on Inclusive Smart Cities and Digital Health*, (pp. 180-187). 10.1007/978-3-319-39601-9_16

Ali, R. M., & Deif, A. M. (2016). Assessing leanness level with demand dynamics in a multi-stage production system. *Journal of Manufacturing Technology Management*, *27*(5), 614–639. doi:10.1108/JMTM-08-2015-0064

Al-Kadeem, R., Backar, S., Eldardiry, M., & Haddad, H. (2017). Review on using system dynamics in designing work systems of project organizations: Product development process case study. *International Journal of System Dynamics Applications*, *6*(2), 52–70. doi:10.4018/IJSDA.2017040103

Alkadeem, R., Backar, S., Haddad, H., & Eldardiry, M. (2017). NPD projects simulation model incorporating managerial aspects using system dynamics approach. *International Journal of System Dynamics Applications*, *6*(3), 38–58. doi:10.4018/IJSDA.2017070103

Allen, C., Metternicht, G., & Wiedmann, T. (2017). An Iterative Framework for National Scenario Modelling for the Sustainable Development Goals (SDGs). *Sustainable Development*, *25*(5), 372–385. doi:10.1002d.1662

Allen, C., Metternicht, G., & Wiedmann, T. (2019). Prioritising SDG targets: Assessing baselines, gaps and interlinkages. *Sustainability Science*, *14*(2), 421–438. doi:10.100711625-018-0596-8

Alligood, K. T., Sauer, T. D., & Yorke, J. A. (1996). *Chaos: An Introduction to Dynamical Systems*. Springer-Verlag. doi:10.1007/b97589

Almeshal, M., Tokhi, M. O., & Goher, K. O. (2013). Stabilization of a new configurable Two-wheeled machine using a PD-PID and a hybrid FL control strategies: A Comparative study. *International Scholarly and Scientific Research & Innovation*, *6*(10), 61–66.

Alpar, O. (2015). Three winged lateen shaped chaotic attractor. *Nonlinear Dynamics*, *82*(1-2), 435–449. doi:10.100711071-015-2166-2

Alsaeedi, A., & Khan, M. Z. (2019). Software Defect Prediction Using Supervised Machine Learning and Ensemble Techniques: A Comparative Study. *Journal of Software Engineering and Applications*, *12*(5), 85–100. doi:10.4236/jsea.2019.125007

Al-Sawalha, A. (2009). Chaos anti-synchronization of two non-identical chaotic systems with known or fully unknown parameters. *Chaos, Solitons, and Fractals*, *42*(3), 1926–1932. doi:10.1016/j.chaos.2009.03.100

Amadei, B. (2015). *A Systems Approach to Modeling Community Development Projects* (M. Noori, Ed.). Momentum Press Engineering.

Amadei, B. (2019). Engineering for peace and diplomacy. *Sustainability (Switzerland)*, *11*(20). Advance online publication. doi:10.3390u11205646

Amadei, B., & Zelinka, D. (2018). *Integrated Modeling of the Water-Energy-Land-Food Nexus: Importance of Context, Scale, Boundary, and Community Capacity*. In *24th International Sustainability Development Research Society Conference*. Messina, Italy.

Amara, K., Malek, A., Bakir, T., Fekik, A., Azar, A. T., Almustafa, K. M., Bourennane, E., & Hocine, D. (2019). Adaptive neuro-fuzzy inference system based maximum power point tracking for stand-alone photovoltaic system. Int. J. Modelling. *Identification and Control*, *33*(4), 311–321. doi:10.1504/IJMIC.2019.107480

Ammar, H.H., Azar, A.T., Shalaby, R., & Mahmoud, M. I. (2019). Metaheuristic Optimization of Fractional Order Incremental Conductance (FO-INC) Maximum Power Point Tracking (MPPT). *Complexity*. doi:10.1155/2019/7687891

Ammari, H. M. (2009). *Challenges and opportunities of connected K-covered wireless sensor networks: From sensor deployment to data gathering*. Springer Science & Business Media.

Amos, Rogriguez, & Wafula (2018). A Systems Dynamics Model for Mobile Industry Governance in the Context of the Kenyan Vision 2030. *International Journal of System Dynamics Applications, 70*(1), 65–94.

Andersen, D. F. (1994). Scripts for group model building. *System Dynamics Review, 13*(2), 102–129.

Andersen, D. F., Richardson, G. P., & Vennix, J. A. M. (2007). Group model Building: Problem structuring, Policy Simulation and Decision Support. *The Journal of the Operational Research Society, 58*(5), 691–694. doi:10.1057/palgrave.jors.2602339

Anderson, M. B., & Woodrow, P. J. (1991). Reducing vulnerability to drought and famine: Developmental approaches to relief. *Disasters, 15*(1), 43-54.

Anderson, K. E. (2020). Getting acquainted with social networks and apps: It is time to talk about TikTok. *Library Hi Tech News, 37*(4), 7–12. doi:10.1108/LHTN-01-2020-0001

Anderson, M. B., & Woodrow, P. J. (1989). *Rising from the Ashes: Development Strategies in Times of Disasters*. Westview.

An, Q., & Sun, L. (2008). On-line parameter identification for vector controlled PMSM drives using adaptive algorithm. *2008 IEEE Vehicle Power and Propulsion Conference, VPPC 2011*, 59(11), 4169-4178. 10.1109/VPPC.2008.4677634

Ansah, P. J. (2017). Causal Analyses of Public Debt and Structural Adjustment. *International Journal of System Dynamics Applications, 6*(1), 21–37. doi:10.4018/IJSDA.2017010102

Anter, A. M., Azar, A. T., & Fouad, K. M. (2020). Intelligent Hybrid Approach for Feature Selection. In *The International Conference on Advanced Machine Learning Technologies and Applications (AMLTA2019). AMLTA 2019. Advances in Intelligent Systems and Computing*, (vol. 921). Springer. 10.1007/978-3-030-14118-9_8

Anter, A. M., Abu ElSoud, M., Azar, A. T., & Hassanien, A. E. (2014). A Hybrid Approach to Diagnosis of Hepatic Tumors in Computed Tomography Images. *International Journal of Rough Sets and Data Analysis, 1*(2), 31–48. doi:10.4018/ijrsda.2014070103

Anter, A. M., Hassanien, A. E., Abu ElSoud, M., & Azar, A. T. (2015). Automatic Liver Parenchyma Segmentation System from Abdominal CT Scans using Hybrid Techniques. *International Journal of Biomedical Engineering and Technology, 17*(2), 148–168. doi:10.1504/IJBET.2015.068052

Anwar, M. A. (2010). Role of information management in the preservation of indigenous knowledge. *Pakistan Journal of Information Management and Libraries, 11*(1).

Arcade, J., Godet, M., Meunier, F., & Roubelat, F. (1999). *Structural analysis with the MICMAC method & Actor's strategy with MACTOR method.* Futures Research Methodology, American Council for the United Nations University: The Millennium Project.

Archana, D., Kalyani, K., & Prasad, B. S. (2012). Efficiency optimization control of induction motor using Fuzzy logic. *International Journal of Soft Computing and Engineering, 2*(3), 327–332.

Arend, R. J., & Bromiley, P. (2009). Assessing the dynamic capabilities view: Spare change, everyone? *Strategic Organization, 7*(1), 75–90. doi:10.1177/1476127008100132

Arias-Gaviria, J., Carvajal-Quintero, S. X., & Arango-Aramburo, S. (2019). Understanding dynamics and policy for renewable energy diffusion in Colombia. *Renewable Energy, 139*, 1111–1119. doi:10.1016/j.renene.2019.02.138

Armenia, S., Pompei, A., Castaño Barreto, A. C., Atzori, A. S., & Fonseca, J. M. (2019). The Rural-Urban Food Systems' Links with the Agenda 2030: From FAO Guidelines on Food Supply and Distribution Systems to a Dairy Sector Application in the Area of Bogota. *Systems, 7*(3), 45. doi:10.3390ystems7030045

Arndt, H. (2007). Using System Dynamics based Learning Environments to Enhance System Thinking. *Conference Proceedings, the 2007 International Conference of the System Dynamics Society and 50th Anniversary Celebration.*

Asad, A. H., Azar, A. T., & Hassanien, A. E. (2013). Ant Colony-based System for Retinal Blood Vessels Segmentation. *Proceedings of Seventh International Conference on Bio-Inspired Computing: Theories and Applications (BIC-TA 2012) Advances in Intelligent Systems and Computing, 201*, 441-452. 10.1007/978-81-322-1038-2_37

Asad, A. H., Azar, A. T., & Hassaanien, A. E. (2012). Integrated Features based on Gray-level and Hu moment-invariants with Ant Colony System for Retinal Blood Vessels Segmentation. *International Journal of Systems Biology and Biomedical Technologies, 1*(4), 60–73. doi:10.4018/ijsbbt.2012100105

Asad, A. H., Azar, A. T., & Hassanien, A. E. (2014a). A New Heuristic Function of Ant Colony System for Retinal Vessel Segmentation. *International Journal of Rough Sets and Data Analysis, 1*(2), 15–30. doi:10.4018/ijrsda.2014070102

Asad, A. H., Azar, A. T., & Hassanien, A. E. (2014b). A Comparative Study on Feature Selection for Retinal Vessel Segmentation Using Ant Colony System. *Recent Advances in Intelligent Informatics Advances in Intelligent Systems and Computing, 235*, 1–11. doi:10.1007/978-3-319-01778-5_1

Asongu, Nwachukwu, & Aziz. (2018) Conditional Determinants of Mobile Phones Penetration and Mobile Banking in Sub-Saharan Africa. *Journal of the Knowledge Economy, 9*(1), 81-110.

Atanassov K., (1999). *Intuitionistic Fuzzy Sets: Theory and Applications*. Wyrzburg: Physica-Verlag.

Atanassov, K. (1986). Intuitionistic fuzzy sets. *Fuzzy Sets and Systems, 20*(1), 87–96. doi:10.1016/S0165-0114(86)80034-3

Atanassov, K. (2005). Answer to D. Dubois, S. Gottwald, P. Hajek, J. Kacprzyk and H. Prade's paper "Terminological difficulties in fuzzy set theory—The case of Intuitionistic Fuzzy Sets." *Fuzzy Sets and Systems, 156*(3), 496–499. doi:10.1016/j.fss.2005.06.003

Atanassov, K., & Gargov, G. (1989). Interval valued intuitionistic fuzzy sets. *Fuzzy Sets and Systems, 31*(3), 343–349. doi:10.1016/0165-0114(89)90205-4

Atig, M., Bouheraoua, M., & Fekik, A. (2018). An Experimental Investigation of Heating in Induction Motor under Open Phase Fault. *Iranian Journal of Electrical and Computer Engineering, 8*(3), 1288.

Auvigne, C., Germano, P., Civet, Y., & Perriard, Y. (2014). Design considerations for a contactless battery charger. *2014 16th European Conference on Power Electronics and Applications, EPE-ECCE Europe 2014*. 10.1109/EPE.2014.6910704

Auxilia, M., Raja, K., & Kannan, K. (2020). Cloud-Based Access Control Framework for Effective Role Provisioning in Business Application. *International Journal of System Dynamics Applications, 9*(1), 63–80. doi:10.4018/IJSDA.2020010104

Avdonina, E. D., & Ibragimov, N. H. (2013). Conservation laws and exact solutions for nonlinear diffusion in anisotropic media. *Communications in Nonlinear Science and Numerical Simulation, 18*(10), 2595–2603. doi:10.1016/j.cnsns.2013.02.009

Azar, A. T., El-Said, S. A., Balas, V. E., & Olariu, T. (2013) Linguistic Hedges Fuzzy Feature Selection For Erythemato-Squamous Diseases. In Soft Computing Applications, Advances in Intelligent Systems and Computing (AISC). Springer-Verlag. Doi:10.1007/978-3-642-33941-7_43

Azar, A.T. (2012). System dynamics as a useful technique for complex systems. *International Journal of Industrial and System Engineering, 10*(4), 377-440.

Azar, A.T., & Hassanien, A.E. (2014). Dimensionality Reduction of Medical Big Data Using Neural-Fuzzy Classifier. *Soft Computing, 19*(4), 1115-1127. DOI doi:10.100700500-014-1327-4

Azar, A.T., (2012). System dynamics as a useful technique for complex systems. *International Journal of Industrial and Systems Engineering, 10*(4), 377-410.

Azar, A.T., Volos, C., Gerodimos, N.A., Tombras, G.S., Pham, V.T., Radwan, A.G., Vaidyanathan, S., Ouannas, A., & Munoz-Pacheco, J.M. (2017a). *A novel chaotic system without equilibrium: Dynamics, synchronization and circuit realization*. . doi:10.1155/2017/7871467

Azar, A. T. (2011). Neuro-Fuzzy System for Cardiac Signals Classification. International Journal of Modelling. *Identification and Control, 13*(1/2), 108–116. doi:10.1504/IJMIC.2011.040495

Azar, A. T. (2012). System dynamics as a useful technique for complex systems. *International Journal of Industrial and Systems Engineering, 10*(4), 377–410. doi:10.1504/IJISE.2012.046298

Azar, A. T. (2013a). Neuro-Fuzzy Applications in Dialysis Systems. In A. T. Azar (Ed.), *Biofeedback Systems and Soft Computing Techniques of Dialysis* (Vol. 405, pp. 1223–1274)., doi:10.1007/978-3-642-27558-6_10

Azar, A. T. (2013b). Multi-Adaptive Neural-Fuzzy System as a Novel Predictor of in-vivo Blood Side Dialyzer Urea Clearance. *International Journal of Computer Applications in Technology, 46*(2), 77–90. doi:10.1504/IJCAT.2013.052290

Azar, A. T. (2014). Neuro-fuzzy feature selection approach based on linguistic hedges for medical diagnosis. *International Journal of Modelling Identification and Control, 22*(3), 195–206. doi:10.1504/IJMIC.2014.065338

Azar, A. T., El-Said, S. A., & Hassanien, A. E. (2013). Fuzzy and Hard Clustering Analysis For Thyroid Disease. *Computer Methods and Programs in Biomedicine, 111*(1), 1–16. doi:10.1016/j.cmpb.2013.01.002 PMID:23357404

Azar, A. T., Hassanien, A. E., & Kim, T. H. (2012) Expert System Based On Neural-Fuzzy Rules for Thyroid Diseases Diagnosis. *International Conference on Bio-Science and Bio-Technology (BSBT 2012)*. 10.1007/978-3-642-35521-9_13

Azar, A. T., & Moysis, L. (2017). Introduction to Control System Design using Matlab. *International Journal of System Dynamics Applications, 6*(3), 130–170. doi:10.4018/IJSDA.2017070107

Azar, A. T., & Moysis, L. (2017). New Discrete Time 2D Chotic Maps. *International Journal of System Dynamics Applications, 6*(1), 77–104. doi:10.4018/IJSDA.2017010105

Azar, A. T., & Serrano, F. E. (2016). Stabilization of Mechanical Systems with Backlash by PI Loop Shaping. *International Journal of System Dynamics Applications, 5*(3), 20–47. doi:10.4018/IJSDA.2016070102

Azar, A. T., Serrano, F. E., Flores, M. A., Vaidyanathan, S., & Zhu, Q. (2020). Adaptive neural-fuzzy and backstepping controller for port-Hamiltonian systems. *International Journal of Computer Applications in Technology, 62*(1), 1–12.

Azar, A. T., Serrano, F. E., Rossell, J. M., Vaidyanathan, S., & Zhu, Q. (2020). Adaptive self-recurrent wavelet neural network and sliding mode controller/observer for a slider crank mechanism. *International Journal of Computer Applications in Technology, 63*(4), 273–285. doi:10.1504/IJCAT.2020.110404

Azar, A. T., & Vaidyanathan, S. (2015a). *Handbook of Research on Advanced Intelligent Control Engineering and Automation*. IGI Global. doi:10.4018/978-1-4666-7248-2

Azar, A. T., & Vaidyanathan, S. (2015b). *Computational Intelligence applications in Modeling and Control, Studies in Computational Intelligence* (Vol. 575). Springer-Verlag.

Azar, A. T., & Vaidyanathan, S. (2015c). *Chaos Modeling and Control Systems Design, Studies in Computational Intelligence* (Vol. 581). Springer-Verlag.

Azar, A. T., Vaidyanathan, S., & Ouannas, A. (2017). *Fractional Order Control and Synchronization of Chaotic Systems, Studies in Computational Intelligence* (Vol. 688). Springer-Verlag., doi:10.1007/978-3-319-50249-6

Azar, A. T., & Valentina, E. B. (2012). Statistical methods and Artificial neural network techniques in electromyography. *International Journal of System Dynamics Applications, 1*(1), 39–47. doi:10.4018/ijsda.2012010103

Azhari, F., Abdi, F., & Abbas Shojaie, A. (2014). Using System Dynamics Method to Manage Construction and Demolition Waste. *International Journal of System Dynamics Applications*, *3*(4), 65–86. doi:10.4018/ijsda.2014100104

Aziz, A. S. A., Azar, A. T., Hassanien, A. E., & Hanafy, S. E. (2012). Continuous Features Discretizaion for Anomaly Intrusion Detectors Generation. *The 17th Online World Conference on Soft Computing in Industrial Applications (WSC17).*

Aziz, A. S. A., Hassanien, A. E., Azar, A. T., & Hanafy, S. E. (2013). Genetic Algorithm with Different Feature Selection Techniques for Anomaly Detectors Generation. *2013 Federated Conference on Computer Science and Information Systems (FedCSIS).*

Aziz, A. S. A., Hassanien, A. E., Azar, A. T., & Hanafy, S. E. (2013). Machine learning techniques for anomalies detection and classification. *Advances in Security of Information and Communication Networks Communications in Computer and Information Science*, *381*, 219–229.

Aziz-Alaoui, M., Robert, C., & Grebogi, C. (2001). Dynamics of a Hénon–Lozi-type map. *Chaos, Solitons, and Fractals*, *12*(12), 2323–2341. doi:10.1016/S0960-0779(00)00192-2

Baccour, N., Koubaa, A., Ben Jamaa, M., Youssef, H., Zuniga, M., & Alves, M. (2009). A comparative simulation study of link quality estimators in wireless sensor networks. *2009 IEEE International Symposium on Modeling, Analysis & Simulation of Computer and Telecommunication Systems*. 10.1109/MASCOT.2009.5366798

Baines, T. S., & Harrison, D. K. (1999). An opportunity for system dynamics in manufacturing system modelling. *Production Planning and Control*, *10*(6), 542–552. doi:10.1080/095372899232830

Bajracharya, A. (2018). A System Dynamics Approach in Understanding the Mode Choice. In Intelligent Transportation and Planning: Breakthroughs in Research and Practice (pp. 1066–1086). IGI Global. doi:10.4018/978-1-5225-5210-9.ch048

Bajracharya, A. (2014). A Qualitative Systems Thinking Approach in Understanding the Implementation of Innovation on Construction Projects. *International Journal of System Dynamics Applications*, *3*(4), 87–108. doi:10.4018/ijsda.2014100105

Bajracharya, A. (2016). Public transportation and Private car A system dynamic approach in understanding the mode choice. *International Journal of System Dynamics Applications*, *5*(2), 1–18. doi:10.4018/IJSDA.2016040101

Bala, B. K., Arshad, F. M., & Noh, K. M. (2017). *System Dynamics: Modelling and Simulation*. Springer. doi:10.1007/SpringerReference_7284

Balcik, B., Beamon, B., & Smilowitz, K. (2008). Last Mile Distribution in Humanitarian Relief. *Journal of Intelligent Transport Systems*, *12*(2), 51–63. doi:10.1080/15472450802023329

Baldassarri, D. (2015). Cooperative networks: Altruism, group solidarity, reciprocity, and sanctioning in ugandan producer organisations. *American Journal of Sociology*, *121*(2), 355–395. doi:10.1086/682418 PMID:26594712

Baldin, A., & Bille, T. (2018). Modelling preference heterogeneity for theatre tickets: A discrete choice modelling approach on Royal Danish Theatre booking data. *Applied Economics*, *50*(5), 545–558. doi:10.1080/00036846.2017.1335386

Baldin, A., Bille, T., Ellero, A., & Favaretto, D. (2018). Revenue and attendance simultaneous optimization in performing arts organizations. *Journal of Cultural Economics*, *42*(4), 677–700. doi:10.100710824-018-9323-7

Bale, C. S. E., Varga, L., & Foxon, T. J. (2015). Energy and complexity: New ways forward. *Applied Energy*, *138*, 150–159. doi:10.1016/j.apenergy.2014.10.057

Balibrea-Iniesta, F., Lopesino, C., Wiggins, S., & Mancho, A. M. (2015). Chaotic dynamics in nonautonomous maps: Application to the nonautonomous Hénon map. *International Journal of Bifurcation and Chaos in Applied Sciences and Engineering*, *25*(12), 1550172. doi:10.1142/S0218127415501722

Balochian, S., & Rajaee, N. (2018). Fractional-order optimal control of fractional-order linear vibration systems with time delay. *International Journal of System Dynamics Applications*, *7*(3), 72–93. doi:10.4018/IJSDA.2018070104

Bank, W. (2017). *GNI per capita, PPP*. Retrieved July 12, 2017, from https://data.worldbank.org/indicator/NY.GNP.PCAP.PP.CD

Banu, P. K. N., Inbarani, H. H., Azar, A. T., Hala, S., Own, H. S., & Hassanien, A. E. (2014). Rough Set Based Feature Selection for Egyptian Neonatal Jaundice. In *Advanced Machine Learning Technologies and Applications: Second International Conference, AMLTA 2014, Cairo, Egypt, November 28-30, 2014. Proceedings, Communications in Computer and Information Science*, (*Vol. 488*). Springer-Verlag GmbH Berlin/Heidelberg. 10.1007/978-3-319-13461-1_35

Banu, P. K. N., Azar, A. T., & Inbarani, H. H. (2017). Fuzzy firefly clustering for tumor and cancer analysis. *International Journal of Modelling Identification and Control*, *27*(2), 92–103. doi:10.1504/IJMIC.2017.082941

Bao, X., Li, Q., Wu, T., Deng, Y., Hu, M., & Zeng, R. (2020). WDM-based bidirectional chaotic communication for semiconductor lasers system with time delay concealment. *Optics Communications*, *472*, 472. doi:10.1016/j.optcom.2020.125868

Baporikar, N. (2016). Infrastructure development as a catalyst for social economic advancement. *International Journal of System Dynamics Applications*, *5*(4), 31–42. doi:10.4018/IJSDA.2016010103

Baporikar, N. (2016). Infrastructure Development as a Catalyst for Social-Economic Advancement. *International Journal of System Dynamics Applications*, *5*(4), 101–113. doi:10.4018/IJSDA.2016100106

Barakat, M. H., Azar, A. T., & Ammar, H. H. (2020). Agricultural Service Mobile Robot Modeling and Control Using Artificial Fuzzy Logic and Machine Vision. In *The International Conference on Advanced Machine Learning Technologies and Applications (AMLTA2019). AMLTA 2019. Advances in Intelligent Systems and Computing*, (vol. 921, pp. 453-465). Springer. 10.1007/978-3-030-14118-9_46

Barbier, E. B. & Burgess, J. C. (2017). The sustainable development goals and the systems approach to sustainability. *Economics, 11*.

Barlas, Y. (1989). Multiple tests for validations of system dynamics type of simulation models. *European Journal of Operational Research*, *42*(1), 59–87. doi:10.1016/0377-2217(89)90059-3

Barlas, Y. (1996). Formal aspects of model validity and validation in system dynamics. *System Dynamics Review*, *12*(3), 183–210. doi:10.1002/(SICI)1099-1727(199623)12:3<183::AID-SDR103>3.0.CO;2-4

Barlas, Y., & Diker, G. (2000, September). A Dynamic Simulation Game (UNIGAME) for Strategic University Management. *Simulation & Gaming*, *31*(3), 331–358. doi:10.1177/104687810003100302

Barlas, Y., & Diker, G. D. (1996). Decision Support for Strategic University Management: A Dynamic Interactive Game. *Proceedings of the 14th International System Dynamics Conference*.

Barnett, M. L., Jermier, J. M., & Lafferty, B. A. (2006). Corporate Reputation: The Definitional Landscape. *Corporate Reputation Review*, *9*(1), 26–38. doi:10.1057/palgrave.crr.1550012

Barorikar, N. (2020). Logistics Effectiveness Through Systems Thinking. *International Journal of System Dynamics Applications*, *9*(2), 16. doi:10.4018/IJSDA.2020040104

Barton, J., & Haslett, T. (2007). Analysis, synthesis, systems thinking and the scientific method: Rediscovering the importance of open systems. *Systems Research and Behavioral Science*, *24*(2), 143–155. doi:10.1002res.816

Bar-Yam, Y. (1997). Dynamics of complex systems. Addison-Wesley.

Bashir, K., Li, T., & Yohannese, C. W. (2019). An Empirical Study for Enhanced Software Defect Prediction Using a Learning-Based Framework. *International Journal of Computational Intelligence Systems*, *12*(1), 282–298. doi:10.2991/ijcis.2018.125905638

Bateman, N. (2005). Sustainability: The elusive element of process improvement. *International Journal of Operations & Production Management*, *25*(3), 261–276. doi:10.1108/01443570510581862

Bature, A. A., Buyamin, S., Ahmad, M. N., & Muhammad, M. (2014). A comparison of controllers for balancing two wheeled inverted pendulum robot. *International Journal of Mechanical & Mechatronics Engineering*, *14*(3), 62–68.

Baumol, W. J., & Bowen, W. G. (1993). *Performing arts-the economic dilemma: a study of problems common to theater, opera, music and dance*. The Twentieth Century Fund.

Beauregard, R., & Corriveau, P. (2007). User Experience Quality: A Conceptual Framework for Goal Setting and Measurement. In V. G. Duffy (Ed.), *Digital Human Modeling* (pp. 325–332). Springer Berlin Heidelberg. doi:10.1007/978-3-540-73321-8_38

Becerra, M., González, E. C., Herrera, M. M., & Romero, O. R. (2016). Collaborative Planning Capacities in Distribution Centers. In L. Zhang, X. Song, & Y. Wu (Eds.), *Theory, Methodology, Tools and Applications for Modeling and Simulation of Complex Systems* (pp. 622–632). Springer Singapore. doi:10.1007/978-981-10-2663-8_64

Becker, H. S. (1984). *Art worlds*. University of California Press.

Beckert, J., & Rössel, J. (2013). The price of art: Uncertainty and reputation in the art field. *European Societies*, *15*(2), 178–195. doi:10.1080/14616696.2013.767923

Belarbi, M. (2006). *Modélisation et Simulation d'un Système de Pompage Photovoltaïque (Master's)*. University of Oran.

Belkacem & Naceri. (2005). Speed Sensorless DTC of Induction Motor Based on an improved Adaptive Flux Observer. *IEEE International Conference on Industrial Technology,* 1192-1197.

Bell, G., Cooper, M., Kennedy, M., & Warwick, J. (2000). The Development of the Holon Planning and Costing Framework for Higher Education Management. *Proceedings of the 18th System Dynamics Conference.*

Ben Smida, M., Sakly, A., Vaidyanathan, S., & Azar, A. T. (2018). *Control-Based Maximum Power Point Tracking for a Grid-Connected Hybrid Renewable Energy System Optimized by Particle Swarm Optimization. Advances in System Dynamics and Control*. IGI-Global. doi:10.4018/978-1-5225-4077-9.ch003

Benedicks, M., & Carleson, L. (1991). The dynamics of the Hénon map. *Annals of Mathematics*, *133*(1), 73–169. doi:10.2307/2944326

Benedict, O. (2016). A System dynamics model for subsistence farmers food security Resilience in sub Saharan Africa. *International Journal of System Dynamics Applications*, *5*(1), 17–30. doi:10.4018/IJSDA.2016010102

Beni, G., & Wang, J. (1993). Swarm intelligence in cellular robotic systems. *Robots and Biological Systems: Towards a New Bionics,* 703-712. doi:10.1007/978-3-642-58069-7_38

Benson, T., & Marlin, S. (2017). *Systems Thinking Overview*. Retrieved from http://watersfoundation.org/systems-thinking/overview/

Bentahar, J., Khosravifar, B., Serhani, M. A., & Alishahi, M. (2012). On the analysis of reputation for agent-based web services. *Expert Systems with Applications*, *39*(16), 12438–12450.

Beregov, R. Y., & Melkikh, A. V. (2015). De-synchronization and chaos in two inductively coupled Van der Pol autogenerators. *Chaos, Solitons, and Fractals*, *73*, 17–28. doi:10.1016/j.chaos.2014.12.017

Berne, M., Vialle, P., & Whalley, J. (2019). An analysis of the disruptive impact of the entry of Free Mobile into the French mobile telecommunications market. *Telecommunications Policy*, *43*(3), 262–277. Advance online publication. doi:10.1016/j.telpol.2018.07.007

Bessen, J. E. (2011). *A Generation of Software Patents* (SSRN Scholarly Paper ID 1868979). Social Science Research Network. https://papers.ssrn.com/abstract=1868979

Betancourt-Mar, J. A., Llanos-Pérez, J. A., Cocho, G., Mansilla, R., Martin, R. R., Montero, S., & Nieto-Villar, J. M. (2017). Phase transitions in tumor growth: IV relationship between metabolic rate and fractal dimension of human tumor. *Physica A*, *473*, 344–351. doi:10.1016/j.physa.2016.12.089

Bhushnan. S., (2012). System Dynamics Integrative Modeling and Simulation for Mobile Telephony Innovation Diffusion: A Study of Emerging Indian Telecom Market. *International Journal of System Dynamics Applications, 12*(1), 84-121.

Bi, C., Zhang, Q., Xiang, Y., & Wang, J. (2013). Nonlinear dynamics of two-dimensional sinusoidal discrete map. In *Communications, Circuits and Systems (ICCCAS), 2013 International Conference on*, (vol. 2, pp. 438–441). IEEE. 10.1109/ICCCAS.2013.6765377

Biham, O., & Wenzel, W. (1989). Characterization of unstable periodic orbits in chaotic attractors and repellers. *Physical Review Letters*, *63*(8), 819–822. doi:10.1103/PhysRevLett.63.819 PMID:10041193

Bishnu, P. S., & Bhattacherjee, V. (2012). Software Fault Prediction Using Quad Tree-Based K-Means Clustering Algorithm. *IEEE Transactions on Knowledge and Data Engineering*, *24*(6), 1146–1150. doi:10.1109/TKDE.2011.163

Bi, Z., Kan, T., Mi, C. C., Zhang, Y., Zhao, Z., & Keoleian, G. A. (2016). A review of wireless power transfer for electric vehicles: Prospects to enhance sustainable mobility. *Applied Energy*, *179*, 413–425. doi:10.1016/j.apenergy.2016.07.003

Blank, S., & Euchner, J. (2018). The genesis and future of Lean Startup: An interview with Steve Blank. *Research Technology Management*, *61*(5), 15–21. doi:10.1080/08956308.2018.1495963

Blaschke, F. (1972). The principle of field orientation as applied to the new transvector closed-loop system for rotating-field machines. *Siemens Review, 34*(3), 217-220.

Boadway, R., & Tremblay, J.-F. (2003). *Public Economics and Startup Entrepreneurs* (SSRN Scholarly Paper ID 385781). Social Science Research Network. https://papers.ssrn.com/abstract=385781

Bollen, K. A. (1987). Total, direct, and indirect effects in structural equation models. *Sociological Methodology*, *17*, 37–69. doi:10.2307/271028

Borgatti, S. P., Everett, M. G., & Freeman, L. (2002). *UCINET 6 for Windows: Software for Social Network Analysis*. Analytic Technologies.

Borgatti, S. P., Everett, M. G., & Johnson, J. C. (2013). *Analyzing Social Networks*. SAGE Publications.

Boudinar, A. H., Bendiabdellah, A., & Benouzza, N. (2019). Diagnosis of the Broken Rotor Bars Faults by Root-MUSIC Method. In *Advanced Condition Monitoring and Fault Diagnosis of Electric Machines* (pp. 59–88). IGI Global.

Boulkroune, A., Bouzeriba, A. B., Azar, A.T. (2016a). Fuzzy Adaptive Synchronization of Uncertain Fractional-order Chaotic Systems. In Advances in Chaos Theory and Intelligent Control, Studies in Fuzziness and Soft Computing (Vol. 337). Springer-Verlag.

Bounechba, H., Bouzid, A., Snani, H., & Lashab, A. (2016). Real-time simulation of MPPT algorithms for PV energy system. *Electrical Power and Energy Systems, Elsevier.*, *3*(8), 201–212. doi:10.1016/j.ijepes.2016.03.041

Bouzaida, S., & Sakly, A. (2018). Adaptive neuro-fuzzy sliding mode controller. *International Journal of System Dynamics Applications*, *7*(2), 34–54. doi:10.4018/IJSDA.2018040103

Boyacı, O., & Tantuğ, A. C. (2017, August). A random number generation method based on discrete time chaotic maps. In *2017 IEEE 60th International Midwest Symposium on Circuits and Systems (MWSCAS)* (pp. 1212-1215). IEEE. 10.1109/MWSCAS.2017.8053147

Braun, W. (2002). The System Archetypes. *System, 2002*, 27. https://www.albany.edu/faculty/gpr/PAD724/724WebArticles/sys_archetypes.pdf

Breiger, R. L. (2005). Culture and classification in markets: An introduction. *Poetics*, *33*(3-4), 157–162. doi:10.1016/j.poetic.2005.10.003

Breiman, L. (2001). Random Forests. *Machine Learning*, *45*(1), 5–32. doi:10.1023/A:1010933404324

Breward, C. J. W., Byrne, H. M., & Lewis, C. E. (2002). The role of cell-cell interactions in a two-phase model for avascular tumour growth. *Journal of Mathematical Biology*, *45*(2), 125–152. doi:10.1007002850200149 PMID:12181602

Brinkley, B. R. (2001). Managing the centrosome numbers game: From chaos to stability in cancer cell division. *Trends in Cell Biology*, *11*(1), 18–21. doi:10.1016/S0962-8924(00)01872-9 PMID:11146294

BrockJ. M. (2018) Inequality of Opportunity and Beliefs About Success and Failure, SSRN. doi:10.2139srn.3178010

Buis, A. (2020). *Milankovitch (Orbital) Cycles and Their Role in Earth's Climate – Climate Change: Vital Signs of the Planet.* Retrieved May 4, 2020, from https://climate.nasa.gov/news/2948/milankovitch-orbital-cycles-and-their-role-in-earths-climate/

Bustince, H., & Burillo, P. (1996). Vague sets are intuitionistic fuzzy sets. *Fuzzy Sets and Systems*, *79*(3), 403–405. doi:10.1016/0165-0114(95)00154-9

Byrne, H. M., & Chaplain, M. A. J. (1998). Necrosis and apoptosis: Distinct cell loss mechanisms in a mathematical model of avascular tumour growth. *Computational and Mathematical Methods in Medicine*, *1*(3), 223–235.

Byungkook, Y., & Ham, W. (1998). Adaptive fuzzy sliding mode control of nonlinear systems. *IEEE Transactions on Fuzzy Systems*, *6*(2), 315–321. doi:10.1109/91.669032

Cabrera, V. E., Hagevoort, R., Solís, D., Kirksey, R., & Diemer, J. A. (2008). The New Mexico dairy industry: An economic engine. *Journal of Dairy Science*, *91*, 2144–2150. doi:10.3168/jds.2007-0791 PMID:18420646

Cahyani, N. D. W., Martini, B., Choo, K.-K. R., & Al-Azhar, A. M. N. (2017). Forensic data acquisition from cloud-of-things devices: Windows Smartphones as a case study. *Concurrency and Computation*, *29*(14), e3855. doi:10.1002/cpe.3855

Cai, H., & Xu, L. (2015). Modeling and Control for Cage Rotor Dual Mechanical Port Electric Machine - Part II: Independent Control of Two Rotors. *IEEE Transactions on Energy Conversion*, *30*(3), 966–973. doi:10.1109/TEC.2015.2400971

Camocardi, P., Battaiotto, P., & Mantz, R. (2008). Wind generator with double stator induction machine. Control strategy for a water pumping application. In *University Power Engineering Conference, UPEC.* September. *43rd International* (pp. 1-5). IEEE.

Capra, F., & Luisi, P. L. (2016). *The Systems View of Life: A Unifying Vision.* Cambridge University Press.

Cárdenas, L., Zapata, M., Franco, C. J., & Dyner, I. (2017). Assessing the combined effect of the diffusion of solar rooftop generation, energy conservation and efficient appliances in households. *Journal of Cleaner Production*, *162*, 491–503. doi:10.1016/j.jclepro.2017.06.068

Carmel, E. (1994). Time-to-completion in software package startups. *Proceedings of the Twenty-Seventh Hawaii International Conference on System Sciences*, 4, 498–507. 10.1109/HICSS.1994.323468

Carmel, E., & Sawyer, S. (1998). Packaged software development teams: What makes them different? *Information Technology & People*, *11*(1), 7–19. doi:10.1108/09593849810204503

Carnoy, M. (2000). *Globalization and Educational Reforms in Globalization and Education*. Rowman and Littlefield Publishers.

Casadei, D., Profumo, F., Serra, G., & Tani, A. (2002). FOC and DTC: Two viable schemes for induction motors torque control. *IEEE Transactions on Power Electronics*, *17*(5), 779–787. doi:10.1109/TPEL.2002.802183

Casado, E. (2005). Expanding Data Mining Power with System Dynamics. Encyclopedia of Information Science and Technology, 1155–1161. doi:10.4018/978-1-59140-553-5.ch204

Casey, R., & Toyli, J. (2012). Mobile voice diffusion and service competition: A system dynamic analysis of regulatory policy. *Telecommunications Policy*, *36*(3), 162–174. doi:10.1016/j.telpol.2011.07.002

Castaneda, M., Franco, C. J., & Dyner, I. (2017). Evaluating the effect of technology transformation on the electricity utility industry. *Renewable & Sustainable Energy Reviews*, *80*(65), 341–351. doi:10.1016/j.rser.2017.05.179

Castaneda, M., Jimenez, M., Zapata, S., Franco, C. J., & Dyner, I. (2017). Myths and facts of the utility death spiral. *Energy Policy*, *110*(65), 105–116. doi:10.1016/j.enpol.2017.07.063

Castaneda, M., Zapata, S., Cherni, J., Aristizabal, A. J., & Dyner, I. (2020). The long-term effects of cautious feed-in tariff reductions on photovoltaic generation in the UK residential sector. *Renewable Energy*, *155*, 1432–1443. doi:10.1016/j.renene.2020.04.051

Castiglione, C., Infante, D., & Zieba, M. (2018). Technical efficiency in the Italian performing arts companies. *Small Business Economics*, *51*(3), 609–638. doi:10.100711187-017-9931-1

Cattani, C., & Ciancio, A. (2008). Qualitative analysis of second-order models of tumor–immune system competition. *Mathematical and Computer Modelling*, *47*(11-12), 1339–1355. doi:10.1016/j.mcm.2007.07.005

Central Bank of Iran. (2014). *The classification of individual consumption by purpose reports (publication No. HBS 1394)*. Central Bank of the Islamic Republic of Iran. (in Persian)

Chady, Mendes, & Razek. (2002). Decoupled Direct Control for PWM Inverter-Fed Induction Motor Drives. *IEEE Transactions on Industry Applications*, *38*(5).

Chaffer, C. L., & Weinberg, R. A. (2011). A perspective on cancer cell metastasis. science. *JSTOR*, *331*(6024), 1559–1564. PMID:21436443

Chakravarty, S. (2007). Determinants of mobile penetration in Asia-a panel study. *Indian Journal of Economics and Business.*, *6*(2), 147–165.

Challagulla, V. U. B., Bastani, F. B., Yen, I-L., & Paul, R. A. (2005). Empirical Assessment of Machine Learning based Software Defect Prediction Techniques. *Proceedings of the 10th IEEE International Workshop on Object-Oriented Real-Time Dependable Systems*, 263-270. 10.1109/WORDS.2005.32

Chambers, R. (1983). *Rural Development: Putting the Last First*. Longman Inc.

Chantar, H., Mafarja, M., Alsawalqah, H., Heidari, A. A., Aljarah, I., & Faris, H. (2019). Feature selection using binary grey wolf optimizer with elite-based crossover for Arabic text classification. *Neural Computing & Applications*, 1–20. doi:10.100700521-019-04368-6

Chaplain, M. A. J. (1996). Avascular growth, angiogenesis and vascular growth in solid tumours: The mathematical modelling of the stages of tumour development. *Mathematical and Computer Modelling*, *23*(6), 47–87. doi:10.1016/0895-7177(96)00019-2

Chapman, L., & Plewes, S. (2014). A UX Maturity Model: Effective Introduction of UX into Organizations. In A. Marcus (Ed.), *Design, User Experience, and Usability. User Experience Design Practice* (pp. 12–22). Springer International Publishing. doi:10.1007/978-3-319-07638-6_2

Charmz, K. (2014). *Constructing Grounded Theory* (J. Seaman, Ed.; 2nd ed.). SAGE Publications.

Chau, K. T., Chan, C. C., & Liu, C. (2008). Overview of permanent-magnet brushless drives for electric and hybrid electric vehicles. *IEEE Transactions on Industrial Electronics*, *55*(6), 2246–2257. doi:10.1109/TIE.2008.918403

Checkland, P., & Poulter, J. (2006). *Learning for action: a short definitive account of soft systems methodology and its use for practitioner, teachers, and students*. Wiley. Retrieved from https://library.wur.nl/WebQuery/clc/1928849

Checkland, P., & Poulter, J. (2006). *Learning for action: a short definitive account of soft systems methodology and its use for practitioner, teachers, and students*. Wiley.

Chen, C., & Liao, C. (2010). Model-based object tracking in wireless sensor networks. *Wireless Networks*, *17*(2), 549–565. doi:10.100711276-010-0296-5

Cheng, Y. C. (2003). Quality assurance in education: Internal, interface, and future. *Quality Assurance in Education*, *11*(4), 202–213. doi:10.1108/09684880310501386

Chengying, M., Xinxin, Y., & Jifu, C. (2012). Swarm Intelligence-Based Test Data Generation for Structural Testing. *2012 IEEE/ACIS 11th International Conference on Computer and Information Science*, 623-628. 10.1109/ICIS.2012.103

Chen, H. H. (2009). Chaos control and global synchronization of Liu chaotic systems using linear Balanced feedback control. *Chaos, Solitons, and Fractals*, *40*(1), 466–473. doi:10.1016/j.chaos.2007.07.098

Chen, H. H., Lee, A. H. I., & Chen, S. (2014). Strategic policy to select suitable intermediaries for innovation to promote PV solar energy industry in China. *Applied Energy*, *115*, 429–437. doi:10.1016/j.apenergy.2013.10.034

Chen, S. H., Sia, C. H., & Shiue, M. C. (2020). On mathematical analysis of synchronization to bidirectionally coupled Kuramoto oscillators. *Nonlinear Analysis Real World Applications*, *56*, 103169. doi:10.1016/j.nonrwa.2020.103169

Chen, T. Y. (2007). A note on distances between intuitionistic fuzzy sets and/or interval-valued fuzzy sets based on the Hausdorff metric. *Fuzzy Sets and Systems*, *158*(22), 2523–2525. doi:10.1016/j.fss.2007.04.024

Chen, Y. Q., Fink, O., & Sansavini, G. (2017). Combined fault location and classification for power transmission lines fault diagnosis with integrated feature extraction. *IEEE Transactions on Industrial Electronics*, *65*(1), 561–569.

Chen, Y., Lin, Y., & Sun, T. (2013). A prediction scheme for object tracking in grid wireless sensor networks. *2013 Seventh International Conference on Innovative Mobile and Internet Services in Ubiquitous Computing*. 10.1109/IMIS.2013.67

Chetia, B., & Das, P. K. (2010). An application of Interval Valued Fuzzy Soft Set in Medical Diagnosis. *Int. J. Contempt. Math. Science.*, *5*(38), 1887–1894.

Chikh, A., & Chandra, A. (2009, October). Optimization and control of a photovoltaic powered water pumping system. In Electrical Power & Energy Conference (EPEC), 2009 IEEE (pp. 1-6). IEEE.

Choi, H. M., Mun, G. S., & Ahn, J. Y. (2012). A medical diagnosis based on interval valued fuzzy set. *Biomedical Engineering: Applications, Basis and Communications*, *24*(4), 349.

Choudhary, G. R., Kumar, S., Kumar, K., Mishra, A., & Catal, C. (2018). Empirical analysis of change metrics for software fault prediction. *Computers & Electrical Engineering*, *67*, 15–24. doi:10.1016/j.compeleceng.2018.02.043

Choudhury, A. (2019). Vibration Monitoring of Rotating Electrical Machines: Vibration Monitoring. In Advanced Condition Monitoring and Fault Diagnosis of Electric Machines (pp. 163-188). IGI Global.

Chou, H. G., Chuang, C. F., Wang, W. J., & Lin, J. C. (2013). A Fuzzy-Model-Based Chaotic Synchronization and Its Implementation on a Secure Communication System. *IEEE Trans. Signal Processing Society*, *8*, 2177–2185.

Christopher, M., & Gattorna, J. (2005). Supply chain cost management and value-based pricing, *34*, 115–121. doi:10.1016/j.indmarman.2004.07.016

Chua, L. O. (1993). Article. *IEICE Transactions on Fundamentals of Electronics, Communications and Computer Science*, *E76-A*(5), 704–734.

Chua, L. O., & Lin, G.-N. (1990). Canonical realization of Chua's circuit family. *IEEE Transactions on Circuits and Systems*, *37*(7), 885–902. doi:10.1109/31.55064

Cockburn, I. M., & MacGarvie, M. J. (2009). Patents, Thickets and the Financing of Early-Stage Firms: Evidence from the Software Industry. *Journal of Economics & Management Strategy*, *18*(3), 729–773. doi:10.1111/j.1530-9134.2009.00228.x

Collste, D., Pedercini, M., & Cornell, S. E. (2017). Policy coherence to achieve the SDGs: Using integrated simulation models to assess effective policies. *Sustainability Science*, *12*(6), 921–931. doi:10.100711625-017-0457-x PMID:30147764

COLREGS. (2013). Convention on the International Regulations for Preventing Collisions at Sea. International Maritime Organization, 8(7), 14-20.

Communication Authority of Kenya. (2019). *Second Quarter Sector Statistics Report for the Financial Year 2019/2020.* Author.

Cooper, R. G., & Kleinschmidt, E. J. (1987a). Success factors in product innovation. *Industrial Marketing Management*, *16*(3), 215–223. doi:10.1016/0019-8501(87)90029-0

Cooper, R. G., & Kleinschmidt, E. J. (1987b). New products: What separates winners from losers? *Journal of Product Innovation Management*, *4*(3), 169–184. doi:10.1111/1540-5885.430169

Coopman, A., Osborn, D., Ullah, F., Auckland, E., & Long, G. (2016). *Seeing the whole: Implementing the SDGs in an Integrated and Coherent Way*. Academic Press.

Coopman, A., Osborn, D., Ullah, F., Auckland, E., & Long, G. (2016). *Seeing the whole: Implementing the SDGs in an Integrated and Coherent Way*. Retrieved from https://sf.stakeholderforum.org/index.php/our-publications-sp-1224407103/reports-in-our-publications/625-seeing-the-whole-implementing-the-sdgs-in-the-an-integrated-and-coherent-way

Cosenz, F., & Noto, G. (2016). Applying System Dynamics Modelling to Strategic Management: A Literature Review. *Systems Research and Behavioral Science*, *33*(6), 703–741. doi:10.1002res.2386

Cosenz, F., Rodrigues, V. P., & Rosati, F. (2019). Dynamic business modeling for sustainability: Exploring a system dynamics perspective to develop sustainable business models. *Business Strategy and the Environment*, (April), 1–14. doi:10.1002/bse.2395

Costanza, R., Daly, L., Fioramonti, L., Giovannini, E., Kubiszewski, I., Fogh, L., … Wilkinson, R. (2016). The UN Sustainable Development Goals and the Dynamics of Well-being. *Solutions*, *7*(1), 20–22. Retrieved from https://www.thesolutionsjournal.com/article/the-un-sustainable-development-goals-and-the-dynamics- of-well-being/%0AThe

Covert, A. (2013, September 3). *Microsoft and Nokia: A marriage of necessity.* CNNMoney. https://money.cnn.com/2013/09/03/technology/mobile/microsoft-nokia-smartphones/index.html

Coyle, G. (2009). Qualitative and quantitative modelling in system dynamics. *SYSTEM DYNAMICS-Volume, II*, 33.

Coyle, R. G. (1996). *System dynamics modelling: a practical approach.* Chapman & Hall. doi:10.1007/978-1-4899-2935-8

Cozzi, G., & Spinesi, L. (2006). Intellectual appropriability, product differentiation, and growth. *Macroeconomic Dynamics, 10*(01), 39–55. doi:10.1017/S1365100506050115

Cropley, D. H. (2003). A case of Compulsory Teaching Accreditation of Engineering Faculty, Guest Editorial. *IEEE Transactions on Education, 46*(4), 24–32. doi:10.1109/TE.2003.819952

Crowne, M. (2002). Why software product startups fail and what to do about it. Evolution of software product development in startup companies. Engineering Management Conference, 2002. IEMC '02. 2002 IEEE International, 1, 338–343. 10.1109/IEMC.2002.1038454

Cruz, P. P., & Rivas, J. R. (2000). Induction motor space vector control using adaptive reference model direct and indirect methods. In *Industrial Electronics, Proceedings of the IEEE International Symposium on* (Vol. 1, pp. 300-305). IEEE.

Cui, S., Yuan, Y., & Wang, T. (2008). Research on switched reluctance double-rotor motor used for hybrid electric vehicle. *2008 International Conference on Electrical Machines and Systems.*

Curcio, K., Navarro, T., Malucelli, A., & Reinehr, S. (2018). Requirements engineering: A systematic mapping study in agile software development. *Journal of Systems and Software, 139*, 32–50. doi:10.1016/j.jss.2018.01.036

Curran, K., Morrison, S., & Mc Cauley, S. (2012). Google+ vs Facebook: The Comparison. *Telkomnika, 10*(2), 379. doi:10.12928/telkomnika.v10i2.814

Cvitanović, P., Gunaratne, G. H., & Procaccia, I. (1988). Topological and metric properties of Hénon-type strange attractors. *Physical Review A, 38*(3), 1503–1520. doi:10.1103/PhysRevA.38.1503 PMID:9900529

Czibula, I.-G., Czibula, G., Marian, Z., & Ionescu, V.-S. (2016). A Novel Approach Using Fuzzy Self-Organizing Maps for Detecting Software Faults. *Studies in Informatics and Control, 25*(2), 207–216. doi:10.24846/v25i2y201608

D'Ambros, M., Lanza, M., & Robbes, R. (2012). Evaluating defect prediction approaches a benchmark and an extensive comparison. *Empirical Software Engineering, 17*(4), 531–577. doi:10.100710664-011-9173-9

D'Ausilio, A. (2012). Arduino: A low cost multipurpose lab equipment. *Behavior Research Methods, 44*(2), 305–313. doi:10.375813428-011-0163-z PMID:22037977

Dahane, A., & Berrached, N. (2019). *Wireless sensor networks: A survey.* Wireless and Sensor Networks., doi:10.1201/9781351190756-1

Daher, B., Hannibal, B., Portney, K. E., & Mohtar, R. H. (2019). Towards creating an environment of cooperation between water, energy, and food stakeholders in San Antonio. *The Science of the Total Environment, 651*, 2913–2926. doi:10.1016/j.scitotenv.2018.09.395 PMID:30463143

Daher, B., Mohtar, R. H., Pistikopoulos, E. N., Portney, K. E., Kaiser, R., & Saad, W. (2018). Developing socio-techno-econo-political (STEP) solutions for addressing resource nexus hotspots. *Sustainability, 2018*(10), 512. doi:10.3390u10020512

Damasio, A-R. (1994). *Descartes' error.* Putman's Sons.

Damasio, A.-R., Damasio, H., & Christen, Y. (Eds.). (1996). *Neurology of decision making.* Springer. doi:10.1007/978-3-642-79928-0

Das, M., & Ghosh, S. K. (2020). Advanced Bayesian Network Models with Fuzzy Extension. In Enhanced Bayesian Network Models for Spatial Time Series Prediction (pp. 101-113). Springer.

Das, S., Goswami, D., Chatterjee, S., & Mukherjee, S. (2014). Stability and chaos analysis of a novel swarm dynamics with applications to multi-agent systems. *Engineering Applications of Artificial Intelligence*, *30*, 189–198. doi:10.1016/j.engappai.2013.12.014

Davidchack, R. L., Lai, Y.-C., Klebanoff, A., & Bollt, E. M. (2001). Towards complete detection of unstable periodic orbits in chaotic systems. *Physics Letters. [Part A]*, *287*(1), 99–104. doi:10.1016/S0375-9601(01)00463-7

Davis, A. M., & Zweig, A. S. (2005). The Rise and Fall of a Software Startup. *Journal of Information Technology Case and Application Research*, *7*(2), 31–48. doi:10.1080/15228053.2005.10856064

Davis, J. P., & Bingham, C. B. (2007). Developing Theory Through Simulation Methods. *Academy of Management Review*, *32*(650), 480–499.

Davvaz, B., & Sadrabadi, E. H. (2016). An application of intuitionistic fuzzy sets in medicine. *International Journal of Biomathematics*, *9*(3), 16500371–15. doi:10.1142/S1793524516500376

Dawson, S. P., Grebogi, C., Yorke, J. A., Kan, I., & Kocak, H. (1992). Antimonotonicity: Inevitable reversals of period-doubling cascades. *Physics Letters. [Part A]*, *163*(3), 249–254. doi:10.1016/0375-9601(92)90442-O

de Grandbois, Y. (2016). The Story of Service Science. In *Service Science and the Information Professional* (pp. 27–46). Chandos Publishing. doi:10.1016/B978-1-84334-649-4.00002-0

de Santiago, J., Bernhoff, H., Ekergård, B., Eriksson, S., Ferhatovic, S., Waters, R., & Leijon, M. (2012). Electrical Motor Drivelines in Commercial All-Electric Vehicles: A Review. *IEEE Transactions on Vehicular Technology*, *61*(2), 475–484. Advance online publication. doi:10.1109/TVT.2011.2177873

de Wilton, A. (2011). Patent Value: A Business Perspective for Technology Startups. *Technology Innovation Management Review*, *1*(3), 5–11. doi:10.22215/timreview/501

Deepa, G., Praba, B., Manimaran, A., Chandrasekaran, V. M., & Rajakumar, K. (2018). Medical diagnosis using intuitionistic fuzzy set in terms shortest distance measure. *Research Journal of Pharmacy and Technology*, *11*(3), 949–952. doi:10.5958/0974-360X.2018.00176.2

Degn, H., Holden, A. V., & Olsen, L. F. (1986). *Chaos in Biological Systems*. Springer-Verlag.

Deif, A. M. (2010). Computer Simulation to Manage Lean Manufacturing Systems. *2nd International Conference on Computer Engineering and Technology (ICCET)*, 677–681. 10.1109/ICCET.2010.5485976

Deif, A. M. (2012). Dynamic analysis of a lean cell under uncertainty. *International Journal of Production Research*, *50*(4), 1127–1139. doi:10.1080/00207543.2011.556154

Deif, A. M., & Elmaraghy, H. (2014). Cost performance dynamics in lean production leveling. *Journal of Manufacturing Systems*, *33*(4), 613–623. doi:10.1016/j.jmsy.2014.05.010

Deisboeck, T. S., Wang, Z., Macklin, P., & Cristini, V. (2011). Multiscale cancer modeling. *Annual Review of Biomedical Engineering*, *13*(1), 127–155. doi:10.1146/annurev-bioeng-071910-124729 PMID:21529163

Demars, C. E. (2002). *Modeling student Outcomes in a General Education Course with Hierarchial linear models*. Paper presented at 42nd AIR Forum, Ontario, Canada.

Deng, T., Xia, G. Q., Cao, L. P., Chen, J. G., Lin, X. D., & Wu, Z. M. (2009). Bidirectional chaos synchronization and communication in semiconductor lasers with optoelectronic feedback. *Optics Communications*, *282*(11), 2243–2249. doi:10.1016/j.optcom.2009.02.040

De, S. K., Biswas, R., & Roy, A. R. (2001). An application of intuitionistic fuzzy sets in medical diagnosis. *Fuzzy Sets and Systems*, *117*(2), 209–213. doi:10.1016/S0165-0114(98)00235-8

Dieter, D. G., & Lenze, J. S. (2018). Google+ Case Study: Create a Social Network or Risk Everything. *Performance Improvement*, *57*(10), 26–36. doi:10.1002/pfi.21826

Dirks, P. B. (2008). Brain tumor stem cells: Bringing order to the chaos of brain cancer. *Journal of Clinical Oncology*, *26*(17), 2916–2924. doi:10.1200/JCO.2008.17.6792 PMID:18539973

Dominic, W., & Natalia, K. (2005). *Computational biology of cancer: lecture notes and mathematical modeling*. World Scientific.

dos Santos Coelho, L., & Mariani, V. C. (2012). Firefly algorithm approach based on chaotic Tinkerbell map applied to multivariable PID controller tuning. *Computers & Mathematics with Applications (Oxford, England)*, *64*(8), 2371–2382. doi:10.1016/j.camwa.2012.05.007

Dove, L., & Sethumadhavan, A. (2012). Breaking Down Barriers: The Interdependence of Research and Design. *Interaction*, *19*(6), 76–80. doi:10.1145/2377783.2377799

Duarte, J., Januário, C., Rodrigues, C., & Sardanyes, J. (2013). Topological Complexity and Predictability in the Dynamics of a Tumor Growth Model with Shilnikov's Chaos. *International Journal of Bifurcation and Chaos in Applied Sciences and Engineering*, *23*(07), 1350124. doi:10.1142/S0218127413501241

Dubois, S., Gottwald, S., Hajek, P., Kacprzyk, J., & Prade, H. (2005). Terminological difficulties in fuzzy set theory—The case of "Intuitionistic Fuzzy Sets. *Fuzzy Sets and Systems*, *156*(3), 485–491. doi:10.1016/j.fss.2005.06.001

Duchesneau, D. A., & Gartner, W. B. (1990). A profile of new venture success and failure in an emerging industry. *Journal of Business Venturing*, *5*(5), 297–312. doi:10.1016/0883-9026(90)90007-G

Duesberg, P. (2007). Chromosomal chaos and cancer. *Scientific American*, *296*(5), 52–59. doi:10.1038cientificamerican0507-52 PMID:17500414

Durant, W. (1954). *The story of civilization*. Simon and Schuster.

Dutta, P., & Doley, D. (2019). Fuzzy decision making for medical diagnosis using arithmetic of generalised parabolic fuzzy numbers. *Granular Computing*, 1-12.

Dutta, P., & Ali, T. (2019, March). Decision Making for Medical Diagnosis Through Credibility Theory. In *International Conference on Information Technology and Applied Mathematics* (pp. 713-724). Springer.

Dutta, P., & Doley, D. (2020). Medical diagnosis under uncertain environment through bipolar-valued fuzzy sets. In *Computer vision and machine intelligence in medical image analysis* (pp. 127–135). Springer. doi:10.1007/978-981-13-8798-2_13

Dutta, P., & Limboo, B. (2017). Bell-shaped fuzzy soft sets and their application in medical diagnosis. *Fuzzy Information and Engineering*, *9*(1), 67–91. doi:10.1016/j.fiae.2017.03.004

Dyner, I., & Franco, C. (2004). *Consumer´s Bounded Rationality: The case of competitive Energy Markets*. Medellin, Colombia: John Wiley & Sons Ltda. InterScience.

Dyner, I. (2000). Energy modelling platforms for policy and strategy support. *The Journal of the Operational Research Society*, *51*(2), 136–144.

Dyner, I., & Larsen, E. R. (2001). From planning to strategy in the electricity industry. *Energy Policy*, *29*(13), 1145–1154. doi:10.1016/S0301-4215(01)00040-4

Eadicicco, L. (2015, April 27). *Why Google+ failed, according to Google insiders.* Business Insider Australia. https://www.businessinsider.com.au/what-happened-to-google-plus-2015-4

Eason, K. (2001). Changing perspectives on the organizational consequences of information technology. *Behaviour & Information Technology*, *20*(5), 323–328. doi:10.1080/01449290110083585

Easterly, W. (2007). *The White Man's Burden: Why the West's Efforts to Aid the Rest Have Done So Much Ill and So Little Good.* Oxford University Press.

Ebrahim, AbdelHadi, & Mahmoud. (2016). Optimal Design of MPPT Controllers for Grid Connected Photovoltaic Array System. *International Journal of Emerging Electric Power Systems, 4*(17), 511–517.

Ebrahim, Elyan, Wadie, & Abd-Allah. (2017). Optimal design of RC snubber circuit for mitigating transient overvoltage on VCB via hybrid FFT/Wavelet Genetic approach. *Electric Power Systems Research, 3*(4), 451-461.

Edghill, J., & Towill, D. (1989). The use of system dynamics in manufacturing systems engineering. *Transactions of the Institute of Measurement and Control*, *11*(4), 208–216. doi:10.1177/014233128901100406

Eichhorn, R., Stefan, J. L., & Hanggi, P. (2002). Simple polynomial classes of chaotic jerky dynamics. *Chaos, Solitons, and Fractals*, *13*(1), 1–15. doi:10.1016/S0960-0779(00)00237-X

Elbaset & Hassan. (2015). Design and Implantation of 1oo kW Roof Top Grid Connected PV System. *3rd International Conference on Energy Systems and Technologies*, 16–19.

ElBedwehy, M. N., Ghoneim, M. E., Hassanien, A. E., & Azar, A. T. (2014). A Computational Knowledge Representation Model for Cognitive Computers. *Neural Computing & Applications*, *25*(7-8), 1517–1534. doi:10.100700521-014-1614-0

El-Din, Aziz. Elsamahy, Hassan, & Bendary. (2017). Enhancement of Turbo-Generators Phase Backup Protection Using Adaptive Neuro-Fuzzy Inference System. *International Journal of System Dynamics Applications*, *4*(6), 58–76.

El-Fouly, F. & Ramadan, R. (2020). Optimal and Meta-Heuristic Algorithms for Object Tracking in Multi-Sink Wireless Sensor Networks. *WAS Science Nature Journal*, (1), 20-38.

El-Ghazaly, G. (2012). Adaptive synchronization of unknown chaotic systems using Mamdani fuzzy approach. *International Journal of System Dynamics Applications*, *1*(3), 122–138. doi:10.4018/ijsda.2012070104

Elhadj, Z. (2013). *Lozi Mappings: Theory and Applications*. CRC Press. doi:10.1201/b15363

Elharakany, R. A., Moscardini, A., Khalifa, N. M., & Elghany, M. M. A. (2018). Modelling the Effect on Quality of Information and Communications Technology (ICT) facilities in Higher Education: Case Study—Egyptian Universities. *International Journal of System Dynamics Applications*, *7*(3), 1–30. doi:10.4018/IJSDA.2018070101

Elizabeth, S., & Sujatha, L. (2013). Application of Fuzzy Membership Matrix in Medical Diagnosis and Decision Making. *Applied Mathematical Sciences*, *7*(127), 6297–6307. doi:10.12988/ams.2013.38485

Elizabeth, S., & Sujatha, L. (2014). Medical Diagnosis Based on Interval Valued Fuzzy Number Matrices. *Ann. of Pure Appl. Math.*, *7*, 91–96.

Elkholy, H. A., Azar, A. T., Magd, A., Marzouk, H., & Ammar, H. H. (2020a). Classifying Upper Limb Activities Using Deep Neural Networks. In *Proceedings of the International Conference on Artificial Intelligence and Computer Vision (AICV2020). AICV 2020. Advances in Intelligent Systems and Computing,* (vol. 1153, pp. 268-282). Springer. 10.1007/978-3-030-44289-7_26

Elkholy, H. A., Azar, A. T., Shahin, A. S., Elsharkawy, O. I., & Ammar, H. H. (2020b) Path Planning of a Self Driving Vehicle Using Artificial Intelligence Techniques and Machine Vision. In *Proceedings of the International Conference on Artificial Intelligence and Computer Vision (AICV2020). AICV 2020. Advances in Intelligent Systems and Computing,* (vol. 1153, pp. 532-542). Springer. 10.1007/978-3-030-44289-7_50

Ellefsen, A. L., Æsøy, V., Ushakov, S., & Zhang, H. J. I. T. o. R. (2019). *A comprehensive survey of prognostics and health management based on deep learning for autonomous ships*. Academic Press.

Elshazly, H. I., Elkorany, A. M., Hassanien, A. E., & Azar, A. T. (2013b). Ensemble Classifiers for Biomedical Data: Performance Evaluation. *IEEE 8th International Conference on Computer Engineering & Systems (ICCES),* 184-189. DOI: 10.1109/ICCES.2013.6707198

Elshazly, H. I., Azar, A. T., Elkorany, A. M., & Hassanien, A. E. (2013a). Hybrid System based on Rough Sets and Genetic Algorithms for Medical Data Classifications. *International Journal of Fuzzy System Applications*, *3*(4), 31–46. doi:10.4018/ijfsa.2013100103

Emary, E., Zawbaa, H., Hassanien, A. E., Schaefer, G., & Azar, A. T. (2014a). Retinal Vessel Segmentation based on Possibilistic Fuzzy c-means Clustering Optimised with Cuckoo Search. *IEEE 2014 International Joint Conference on Neural Networks (IJCNN 2014).* 10.1109/IJCNN.2014.6889932

Emary, E., Zawbaa, H., Hassanien, A. E., Schaefer, G., & Azar, A. T. (2014b). Retinal Blood Vessel Segmentation using Bee Colony Optimization and Pattern Search. *IEEE 2014 International Joint Conference on Neural Networks (IJCNN 2014).*

Emary, E., Zawbaa, H. M., & Hassanien, A. E. (2016a). Binary ant lion approaches for feature selection. *Neurocomputing*, *213*, 54–65. doi:10.1016/j.neucom.2016.03.101

Emary, E., Zawbaa, H. M., & Hassanien, A. E. (2016b). Binary grey wolf optimization approaches for feature selection. *Neurocomputing*, *172*, 371–381. doi:10.1016/j.neucom.2015.06.083

Enev, S. (2007, June). Input-output decoupling control of induction motors with rotor resistance and load torque identification. In *IEEE Mediterranean Conference on Control & Automation 2007. MED'07* (pp. 1-5). IEEE.

Engelbrecht. (2018). *Computational intelligence: An introduction* (3rd ed.). Wiley-Blackwell.

Engineering, C., & Sciences, N. (2018). Optimal Design of High-Performance Interior Pm. *The International Journal of Energy & Engineering Sciences*, *3*(2), 26–35.

Eng, T. J. E., & Sci, C. (2011). Sliding mode controller gain adaptation and chattering reduction techniques for DSP-based PM DC motor. *Turkish Journal of Electrical Engineering and Computer Sciences*, *19*(4), 531–549.

Enshaei, N., Hammad, A., & Naderkhani, F. (2020). A Comprehensive Review on Advanced Maintenance Strategies for Smart Railways. In *Handbook of Research on Interdisciplinary Approaches to Decision Making for Sustainable Supply Chains* (pp. 433–457). IGI Global.

Eras, J. J. C., Morejón, M. B., Gutiérrez, A. S., García, A. P., Ulloa, M. C., Martínez, F. J. R., & Rueda-Bayona, J. G. (2019). A look to the electricity generation from non-conventional renewable energy sources in Colombia. *International Journal of Energy Economics and Policy*, *9*(1), 15–25. doi:10.32479/ijeep.7108

Erik, P., & Logtens, T. (2016). Model based Exploration of Societal aging in the Netherlands. *International Journal of System Dynamics Applications*, *4*(1), 57–95.

Eriksen, S. D. (1995). TQM and the transformation from elite to a mass system of higher education in the UK. *Quality Assurance in Education*, *3*(4), 14–29. doi:10.1108/09684889510146795

Esmaeili, N., Alfi, A., & Khosravi, H. (2017). Balancing and trajectory tracking of two-wheeled mobile robot using backstepping sliding mode control: Design and experiments. *Journal of Intelligent & Robotic Systems*, *87*(3-4), 601–613. doi:10.100710846-017-0486-9

Eswari, T., & Vanitha, V. (2013). A novel rule based intrusion detection framework for wireless sensor networks. *2013 International Conference on Information Communication and Embedded Systems (ICICES)*. 10.1109/ICICES.2013.6508172

Faktor, S. (2014, May 1). *What Killed Google+ And What Can Save It*. Forbes. https://www.forbes.com/sites/stevefaktor/2014/05/01/what-killed-google-and-what-will-save-it/

Faris, H., Mafarja, M. M., Heidari, A. A., Aljarah, I., Ala'M, A. Z., Mirjalili, S., & Fujita, H. (2018). An efficient binary salp swarm algorithm with crossover scheme for feature selection problems. *Knowledge-Based Systems*, *154*, 43–67. doi:10.1016/j.knosys.2018.05.009

Farmer, J. D., Ott, E., & Yorke, J. A. (1983). The dimension of chaotic attractors. In *The Theory of Chaotic Attractors* (pp. 142–169). Springer. doi:10.1007/978-0-387-21830-4_11

Fath, B. D. (2017). Systems ecology, energy networks, and a path to sustainability. *International Journal of Design & Nature and Ecodynamics*, *12*(1), 1–15. doi:10.2495/DNE-V12-N1-1-15

Fatma, L., Ghabi, J., & Dhouibi, H. (2020). Applying Interval Fuzzy Petri Net to Failure Analysis. *International Journal of Service Science, Management, Engineering, and Technology*, *11*(1), 14–30.

Fayad, M. E., Laitinen, M., & Ward, R. P. (2000). Thinking Objectively: Software Engineering in the Small. *Communications of the ACM*, *43*(3), 115–118. doi:10.1145/330534.330555

Fazioli, R., & Filippini, M. (1997). Cost structure and product mix of local public theatres. *Journal of Cultural Economics*, *21*(1), 77–86. doi:10.1023/A:1007341613767

Featherston, C. R., & Doolan, M. (2013). Using System Dynamics to Inform Scenario Planning : A Case Study. In *31 st International Conference of the System Dynamics Society*. System Dynamics Society. Retrieved from https://www.systemdynamics.org/conferences/2013/proceed/papers/P1193.pdf

Feder, T., & Katz-Gerro, T. (2012). Who benefits from public funding of the performing arts? Comparing the art provision and the hegemony–distinction approaches. *Poetics*, *40*(4), 359–381. doi:10.1016/j.poetic.2012.05.004

Fekik, A., Denoun, H., Hamida, M. L., Azar, A. T., Atig, M., & Zhu, Q. M. (2018b) Neural Network Based Switching State Selection for Direct Power Control of Three Phase PWM-Rectifier. In *2018 10th International Conference on Modelling, Identification and Control (ICMIC)*. 10.1109/ICMIC.2018.8529997

Fekik, A., Denoun, H., Azar, A. T., Zaouia, M., Benyahia, N., Hamida, M. L., Benamrouche, N., & Vaidyanathan, S. (2018a). *Artificial Neural Network for PWM Rectifier Direct Power Control and DC Voltage Control. Advances in System Dynamics and Control*. IGI-Global. doi:10.4018/978-1-5225-4077-9.ch010

Feldman, D. P. (2019). *Chaos and Dynamical Systems*. Princeton University Press.

Feld, W. M. (2001). *Lean manufacturing: tools, techniques, and how to use them*. CRC Press.

Fernández-Blanco, V., Rodríguez-Álvarez, A., & Wiśniewska, A. (2019). Measuring technical efficiency and marginal costs in the performing arts: The case of the municipal theatres of Warsaw. *Journal of Cultural Economics*, *43*(1), 97–119. doi:10.100710824-018-9330-8

Field, R. J., & Gyorgyi, L. (1993). *Chaos in Chemistry and Biochemistry*. World Scientific. doi:10.1142/1706

Filali, R. L., Benrejeb, M., & Borne, P. (2014). On observer-based secure communication design using discrete-time hyperchaotic systems. *Communications in Nonlinear Science and Numerical Simulation*, *19*(5), 1424–1432. doi:10.1016/j.cnsns.2013.09.005

Finstad, K., Xu, W., Kapoor, S., Canakapalli, S., & Gladding, J. (2009). FEATURE: Bridging the Gaps Between Enterprise Software and End Users. *Interaction*, *16*(2), 10–14. doi:10.1145/1487632.1487635

Flah, A. (2017). HEV recharge battery algorithm using a fuzzy controller. *International Conference on Green Energy Conversion Systems (GECS).*

Flah, A. (2018). Internal Fuzzy Hybrid Charger System for a Hybrid Electrical Vehicle. *ASME. Journal of Energy Resources Technology*, *140*(1), 12003–12008. doi:10.1115/1.4037352

Flavier, J. M. (1995). The regional program for the promotion of indigenous knowledge in Asia. Academic Press.

Fletcher, M. A., & Zuber-Skerritt, O. (2008). Professional Development through Action Research: Case Examples in South African Higher Education. *Systemic Practice and Action Research*, *21*(1), 73–96. doi:10.100711213-007-9083-3

Ford, A. (1997). System Dynamics and the Electric Power Industry. *System Dynamics Review*, *13*(1), 57–85. doi:10.1002/(SICI)1099-1727(199721)13:1<57:AID-SDR117>3.0.CO;2-B

Ford, A. (1999). Cycles in competitive electricity markets: A simulation study of the western United States. *Energy Policy*, *27*(11), 637–658. doi:10.1016/S0301-4215(99)00050-6

Ford, A. (2010). *Modeling the environment* (2nd ed.). Island Press.

Ford, A. (2010). *Modeling the Environment* (2nd ed.). Island Press.

Ford, D. N., & Sterman, J. D. (1998). Dynamic modeling of product development processes. *System Dynamics Review*, *14*(1), 31–68.

Forenbacher, I., Husnjak, S., Cvitić, I., & Jovović, I. (2019). Determinants of mobile phone ownership in Nigeria. *Telecommunications Policy*, *43*(7), 101812. Advance online publication. doi:10.1016/j.telpol.2019.03.001

Forgionne, G. A., Gupta, J. N. D., & Mora, M. (2003). *Decision Making Support Systems: Achievements, Challenges and Opportunities*. Available at https://www.igi-global.com/chapter/decision-making-supportsystems/8080?camid=4v1

Forrester, J. W. (1998). *Designing the Future*. Retrieved April 15, 2016, from http://sysdyn.clexchange.org/people/jay-forrester.html

Forrester, J. W. (1961). *Industrial dynamics*. M.I.T. Press.

Forrester, J. W. (1971). Counterintuitive behavior of social systems. *Technological Forecasting and Social Change*, *3*, 1–22. doi:10.1016/S0040-1625(71)80001-X

Forrester, J. W. (1994). Policies, Decisions, and Information Sources for Modelling. In J. D. W. Morecroft & J. D. Sterman (Eds.), *Modelling for Learning Organisations*. Productivity Press.

Forrester, J. W. (1994). System Dynamics, System Thinking, and Soft OR. *System Dynamics Review*, *10*(2-3), 245–256. doi:10.1002dr.4260100211

Forrester, J. W., Mass, N. J., & Ryan, C. J. (1976). The system dynamics national model: Understanding socio-economic behavior and policy alternatives. *Technological Forecasting and Social Change*, *9*(1-2), 51–68. doi:10.1016/0040-1625(76)90044-5

Forrester, J. W., & Senge, P. M. (1980). Tests for building confidence in system dynamics models. In A. A. Legasto, J. W. Forrester, & J. M. Lyneis (Eds.), *System dynamics* (pp. 209–228). TIMS Studies in the Management Science.

Foster, R. O. (1972). Education in the City D-2144. In MIT System Dynamic Group Literature Collection. System Dynamics Society.

Frances, C. (1995). *Using System Dynamics Technology to Improve Planning for Human Resource Development*. Presentation at the Congress of Political Economists, International 6th Annual Conference, Seoul, Korea.

Frances, C. (1995). *Using System Dynamics Technology to Improve Planning For Human Resource Development*. Presentation at the Congress of Political Economists, International 6th Annual Conference, Seoul, Korea.

Frances, C. (2000), Using System Dynamics as a Tool for Decision Making in Higher Education Management- U.S. Experience. Academic Press.

Frances, C. (2000). *Selected papers presented at an international seminar on Using System Dynamics as a Tool for Decision Making in Higher Education Management held in June 1999 at the Royal Society, London and South Bank University, under the auspices of the Society for Research into Higher Education*. South Bank University Technical Report SBU-CISM-12-00.

Frances, C., Van Alstyne, M., Ashton, A., & Hochstettler, T. (1994). Using System Dynamics Technology to Improve Planning Higher Education: Results in Arizona and Houston, Texas. *Proceedings of the 13th International System Dynamics Conference*, 444-453.

Franco, C. J., Castaneda, M., & Dyner, I. (2015). Simulating the new British Electricity-Market Reform. *European Journal of Operational Research*, *245*(1), 273–285. doi:10.1016/j.ejor.2015.02.040

Frank, S. A. (2007). *Dynamics of cancer: incidence, inheritance, and evolution* (Vol. 1). Princeton University Press. doi:10.1515/9780691186863

Frederickson, P., Kaplan, J. L., Yorke, E. D., & Yorke, J. A. (1983). The Liapunov dimension of strange attractors. *Journal of Differential Equations*, *49*(2), 185–207. doi:10.1016/0022-0396(83)90011-6

Friedman, S., & Raphael, M. (2007). Individual Learning Style and Systems Tool Preferences. *Conference Proceedings, The 2007 International Conference of the System Dynamics Society and 50th Anniversary Celebration.*

Gafi, E. G., & Javadian, N. (2018). A System Dynamics Model for Studying the Policies of Improvement of Chicken Industry Supply Chain. *International Journal of System Dynamics Applications*, *7*(4), 20–37. doi:10.4018/IJSDA.2018100102

Galbraith, P. L. (1982). *Forecasting Futures for Higher Education in Australia: An Application of Dynamic Modelling* (Ph.D. Thesis). The University of Queensland.

Galbraith, P.L. (1998a). When Strategic Plans are not Enough: Challenges in University Management. *System Dynamics: An International Journal of Policy Modelling*, 55-84.

Galbraith, P. L. (1998b). System Dynamics and University Management. *System Dynamics Review*, *10*(2), 69–84. doi:10.1002/(SICI)1099-1727(199821)14:1<69::AID-SDR139>3.0.CO;2-T

Galbraith, P. L., & Carss, B. W. (1989). Strategies for Institutional Resource Allocation: Insights from a Dynamic Model. *Higher Education Policy*, *2*(2), 31–38. doi:10.1057/hep.1989.23

Galias, Z. (2016). Are numerical studies of long term dynamics conclusive: The case of the Hénon map. *Journal of Physics: Conference Series*, *692*, 012001. doi:10.1088/1742-6596/692/1/012001

Galli. (2020). The Application of Systems Engineering to Project Management A Review of Their Relationship. *International Journal of System Dynamics Applications, 9*(1), 81-95.

Galli, B. J. (2019). Thoughts of Using Economic Decision-Making to Systems Engineering and Systems Thinking: An Exploratory Study. *International Journal of System Dynamics Applications*, *8*(3), 1–14.

Galli, B. J. (2020). Economic-Decision-Making in New Product Development: A Review of the Relationship. *International Journal of Applied Management Sciences and Engineering*, *7*(1), 1–27.

Galli, B. J. (2020). The Application of Systems Engineering to Project Management: A Review of Their Relationship. *International Journal of System Dynamics Applications*, *9*(1), 81–106. doi:10.4018/IJSDA.2020010105

Gangsar, P., & Tiwari, R. (2019). Online Diagnostics of Mechanical and Electrical Faults in Induction Motor Using Multiclass Support Vector Machine Algorithms Based on Frequency Domain Vibration and Current Signals. *ASCE-ASME Journal of Risk and Uncertainty in Engineering Systems, Part B. Mechanical Engineering*, *5*(3).

Gan, P. Y., & Li, Z. D. (2015). Quantitative study on long term global solar photovoltaic market. *Renewable & Sustainable Energy Reviews*, *46*, 88–99. doi:10.1016/j.rser.2015.02.041

Gardner, J. S. (2002). Natural Hazards Risk in the Kullu District, Himachal Pradesh, India. *Geographical Review, 92*(2), 282-306.

Garwood, T. L., Hughes, B. R., Oates, M. R., O'Connor, D., & Hughes, R. (2018). A review of energy simulation tools for the manufacturing sector. *Renewable and Sustainable Energy Reviews, 81*(August), 895–911. doi:10.1016/j.rser.2017.08.063

Gary, M. S., Kunc, M., Morecroftc, J., & Scott, F. R. (2008). System dynamics and strategy. *System Dynamics Review*, *24*(4), 407–429. doi:10.1002dr.402

Gaspard, P. (1999). Microscopic chaos and chemical reactions. *Physica A*, *263*(1), 315–328. doi:10.1016/S0378-4371(98)00504-4

Gatignon, H., & Xuereb, J.-M. (1997). Strategic Orientation of the Firm and New Product Performance. *JMR, Journal of Marketing Research*, *34*(1), 77–90. doi:10.2307/3152066

Geethanjali, M., & Ramadoss, H. (2019). Fault Diagnosis of Induction Motors Using Motor Current Signature Analysis: A Review. In Advanced Condition Monitoring and Fault Diagnosis of Electric Machines (pp. 1-37). IGI Global.

Ge, F. L., & Fan, Y. (2013). A system dynamics model of coordinated development of central and provincial economy and oil enterprises. *Energy Policy*, *60*, 41–51. doi:10.1016/j.enpol.2013.05.033

Gehrke, C., Walker, C., & Walker, E. (1996). Some Comments on Interval-Valued Fuzzy Sets. *International Journal of Intelligent Systems*, *11*(10), 751–759. doi:10.1002/(SICI)1098-111X(199610)11:10<751::AID-INT3>3.0.CO;2-Y

Georgantzas, N. C., & Katsamakas, E. G. (2008). Information systems research with system dynamics. *System Dynamics Review*, *24*(3), 247–264. doi:10.1002dr.420

Georgiadis, P., & Michaloudis, C. (2012). Real-time production planning and control system for job-shop manufacturing: A system dynamics analysis. *European Journal of Operational Research*, *216*(1), 94–104. doi:10.1016/j.ejor.2011.07.022

Ghabi, J., Rhif, A., & Vaidyanathan, S. (2018). Discrete Time Sliding Mode Control Scheme for Nonlinear Systems With Bounded Uncertainties. *International Journal of System Dynamics Applications*, *7*(2), 15–33. doi:10.4018/IJSDA.2018040102

Ghabim, Rhif, & Vaidyanathan. (2018). Discrete Time Sliding Mode Control Scheme for Nonlinear Systems With Bounded Uncertainties. *International Journal of System Dynamics Applications, 7*(2), 15-23.

Gharbi, Gharsellaoui, & Khalgui. (2017). Real-Time Reconfigurations of Embedded Control Systems. *International Journal of System Dynamics Applications*, *4*(6), 71–93.

Gholizad, A., Ahmadi, L., Hassannayebi, E., Memarpour, M., & Shakibayifar, M. (2017). A system dynamics model for the analysis of the deregulation in electricity market. *International Journal of System Dynamics Applications*, *6*(2), 1–30. doi:10.4018/IJSDA.2017040101

Giachetti, C., & Marchi, G. (2017). Successive changes in leadership in the worldwide mobile phone industry: The role of windows of opportunity and firms' competitive action. *Research Policy*, *46*(2), 352–364.

Giardino, C., Bajwa, S. S., Wang, X., & Abrahamsson, P. (2015). Key Challenges in Early-Stage Software Startups. In C. Lassenius, T. Dingsøyr, & M. Paasivaara (Eds.), *Agile Processes in Software Engineering and Extreme Programming* (pp. 52–63). Springer International Publishing.

Giardino, C., Wang, X., & Abrahamsson, P. (2014). Why Early-Stage Software Startups Fail: A Behavioral Framework. In C. Lassenius & K. Smolander (Eds.), *Software Business. Towards Continuous Value Delivery* (pp. 27–41). Springer International Publishing., doi:10.1007/978-3-319-08738-2_3

Giove, S., Azar, A. T., & Nordio, M. (2013). Fuzzy Logic Control for Dialysis Application. In A. T. Azar (Ed.), *Biofeedback Systems and Soft Computing Techniques of Dialysis. Springer-Verlag GmbH Berlin/Heidelberg*. doi:10.1007/978-3-642-27558-6_9

Glenn, J. C. (2004). *Introduction to the Futures Research Methods Series*. Futures Research Methodology - Version 2.0. Retrieved from http://mp.cim3.net/file/project/mp-sofi-sd/reference/01-Introduction.PDF

Global Goals. (2016). i. Retrieved from http://www.globalgoals.org

Godinho Filho, M., & Barco, C. F. (2015). A framework for choosing among different lean-based improvement programs. *International Journal of Advanced Manufacturing Technology*, *81*(1–4), 183–197. doi:10.100700170-015-7181-4

Godinho Filho, M., & Uzsoy, R. (2011). The effect of shop floor continuous improvement programs on the lot size-cycle time relationship in a multi-product single-machine environment. *International Journal of Advanced Manufacturing Technology*, *52*(5–8), 669–681. doi:10.100700170-010-2770-8

Goel, D.R., & Biswal, A. (1998). TQM in Education, Reality, Desirability and Feasibility. *University News, 31*(44), 10-13.

Goher, K. M., Tokhi, M. O., & Siddique, N. H. (2010). Dynamic modeling and control of a two wheeled robotic vehicle with a virtual payload. *Journal of Engineering and Applied Sciences (Asian Research Publishing Network)*, *6*(3), 7–41.

Goldsztejn, A., Hayes, W., & Collins, P. (2011). Tinkerbell is chaotic. *SIAM Journal on Applied Dynamical Systems*, *10*(4), 1480–1501. doi:10.1137/100819011

Gómez, C. R., Arango-Aramburo, S., & Larsen, E. R. (2017). Construction of a Chilean energy matrix portraying energy source substitution: A system dynamics approach. *Journal of Cleaner Production*, *162*, 903–913. doi:10.1016/j.jclepro.2017.06.111

Gonçalves, P., Hines, J., & Sterman, J. (2005). The impact of endogenous demand on push-pull production system. *System Dynamics Review*, *21*(3), 187–216. doi:10.1002dr.318

Gondra, I. (2008). Applying machine learning to software fault-proneness prediction. *Journal of Systems and Software*, *81*(2), 186–195. doi:10.1016/j.jss.2007.05.035

Gong, L., Jiang, S., & Jiang, L. (2019). Tackling Class Imbalance Problem in Software Defect Prediction Through Cluster-Based Over-Sampling with Filtering. *IEEE Access: Practical Innovations, Open Solutions*, *7*, 145725–145737. doi:10.1109/ACCESS.2019.2945858

Gonzalez, R., Cuevas, R., Motamedi, R., Rejaie, R., & Cuevas, A. (2013). Google+ or Google-? Dissecting the evolution of the new OSN in its first year. *Proceedings of the 22nd International Conference on World Wide Web*, 483–494.

Goodwin, N. C. (1987). Functionality and Usability. *Communications of the ACM*, *30*(3), 229–233. doi:10.1145/214748.214758

Gordon, T. J. (2014). Cross-Impact Analysis. In *Introduction to the futures methods research series. Futures research methodology, V3.0*. The Millennium Project.

Gordon, T. J., & Hayward, H. (1968). Initial experiments with the cross impact matrix method of forecasting. *Futures*, *1*(2), 100–116. doi:10.1016/S0016-3287(68)80003-5

Gornitzka, A., & Maasen, P. (2000). The Economy, Higher Education, and European integration: An Introduction. *Higher Education Policy*, *13*(3), 217–223. doi:10.1016/S0952-8733(00)00008-8

Gorripotu, T. S., Samalla, H., Jagan Mohana Rao, C., Azar, A. T., & Pelusi, D. (2019). TLBO Algorithm Optimized Fractional-Order PID Controller for AGC of Interconnected Power System. In J. Nayak, A. Abraham, B. Krishna, G. Chandra Sekhar, & A. Das (Eds.), *Soft Computing in Data Analytics. Advances in Intelligent Systems and Computing* (Vol. 758). Springer. doi:10.1007/978-981-13-0514-6_80

Grabowski, F., & Strzałka, D. (2008). Simple, complicated and complex systems - The brief introduction. *2008 Conference on Human System Interaction, HSI 2008*, 570–573. 10.1109/HSI.2008.4581503

Graedel, T. E., & Allenby, B. R. (2010). *Industrial Ecology and Sustainability Engineering*. Pearson.

Graham, S. J. H., & Sichelman, T. (2008). Why Do Start-Ups Patent? *Berkeley Technology Law Journal*, *23*(3), 1063–1097.

Grandzol, J. R. (2005). *Improving the Faculty Selection Process in Higher Education: A Case for the Analytic Hierarchy Process* (Vol. 6). Association for Institutional Research.

Grassberger, P., & Procaccia, I. (1983). Characterization of strange attractors. *Physical Review Letters*, *50*(5), 346–349. doi:10.1103/PhysRevLett.50.346

Grassberger, P., & Procaccia, I. (2004). Measuring the strangeness of strange attractors. In *The Theory of Chaotic Attractors* (pp. 170–189). Springer. doi:10.1007/978-0-387-21830-4_12

Grassi, G., Ouannas, A., Azar, A. T., Radwan, A. G., Volos, C., Pham, V. T., Ziar, T., Madian, A., Kyprianidis, I. M., & Stouboulos, I. N. (2017). *Chaos Synchronisation Of Continuous Systems Via Scalar Signal*. The 6th International Conference on Modern Circuits and Systems Technologies (MOCAST), Thessaloniki Greece. 10.1109/MOCAST.2017.7937629

Grebogi, C., Ott, E., & Yorke, J. A. (1983). Crises, sudden changes in chaotic attractors and transient chaos. *Physica D. Nonlinear Phenomena*, *7*(1-3), 181–200. doi:10.1016/0167-2789(83)90126-4

Green, D. (1994). *What is Quality in Higher Education*. St. Edmundsbury Press.

Grishma, B. R., & Anjali, C. (2015). Software root cause prediction using clustering techniques: A review. *2015 Global Conference on Communication Technologies (GCCT)*. 10.1109/GCCT.2015.7342714

Grisolía, J. M., & Willis, K. G. (2012). A latent class model of theatre demand. *Journal of Cultural Economics*, *36*(2), 113–139. doi:10.100710824-012-9158-6

Groen, A. J., Wakkee, I. A. M., & De Weerd-Nederhof, P. C. (2008). Managing Tensions in a High-tech Start-up: An Innovation Journey in Social System Perspective. *International Small Business Journal*, *26*(1), 57–81. doi:10.1177/0266242607084659

Grzegorzewski, P. (2004). Distances between intuitionistic fuzzy sets and/or interval-valued fuzzy sets based on the Hausdorff metric. *Fuzzy Sets and Systems*, *148*(2), 319–328. doi:10.1016/j.fss.2003.08.005

Grzegorzewski, P., & Mrówka, E. (2005). Some notes on (Atanassov's) intuitionistic fuzzy sets. *Fuzzy Sets and Systems*, *156*(3), 492–495. doi:10.1016/j.fss.2005.06.002

Guan, X., Zhai, Q., & Lai, F. (2002). New lagrangian relaxation based algorithm for resource scheduling with homogeneous Subproblems. *Journal of Optimization Theory and Applications*, *113*(1), 65–82. doi:10.1023/A:1014805213554

Guckenheimer, J., & Holmes, P. J. (2013). *Nonlinear oscillations, dynamical systems, and bifurcations of vector fields* (Vol. 42). Springer Science & Business Media.

Guedes, J. J., Castoldi, M. F., Goedtel, A., Agulhari, C. M., & Sanches, D. S. (2019). Differential evolution applied to line-connected induction motors stator fault identification. *Soft Computing*, *23*(21), 11217–11226.

Guegan, D. (2009). Chaos in economics and finance. *Annual Reviews in Control*, *33*(1), 89–93. doi:10.1016/j.arcontrol.2009.01.002

Guerra, A., Rodriguez, D. J., Montero, S., Betancourt-Mar, J. A., Martin, R. R., Silva, E., Bizzarri, M., Cocho, G., Mansilla, R., & Nieto-Villar, J. M. (2018). Phase transitions in tumor growth VI: Epithelial–Mesenchymal transition. *Physica A*, *499*, 208–215. doi:10.1016/j.physa.2018.01.040

Guffey, R. V. (2015). Globalization, Cross Border Education, and Student Migration: Determining Student Institutional. *Choice Factors*. Available at https://www.igi-global.com/chapter/globalization-cross-border-education-andstudent-migration/133730?camid=4v1

Guma, I. P., Rwashana, A. S., & Oyo, B. (2018). Food Security Indicators for Subsistence Farmers Sustainability: A System Dynamics Approach. *International Journal of System Dynamics Applications*, *7*(1), 45–64. doi:10.4018/IJSDA.2018010103

Gunes, M., Sorges, U., & Bouazizi, I. (2002). ARA-the ant-colony based routing algorithm for MANETs. *Proceedings. International Conference on Parallel Processing Workshop*. 10.1109/ICPPW.2002.1039715

Guo, X., & Guo, X. (2015). China's photovoltaic power development under policy incentives: A system dynamics analysis. *Energy*, *93*, 589–598. doi:10.1016/j.energy.2015.09.049

Gupta, G. P., & Massagué, J. (2006). Cancer metastasis: Building a framework. *Cell*, *127*(4), 679–695. doi:10.1016/j.cell.2006.11.001 PMID:17110329

Gupta, V., Narayanamurthy, G., & Acharya, P. (2018). Can lean lead to green? Assessment of radial tyre manufacturing processes using system dynamics modelling. *Computers & Operations Research*, *89*, 284–306. doi:10.1016/j.cor.2017.03.015

Gurumurthy, G., & Das, D. K. (2020). Terminal sliding mode disturbance observer based adaptive super twisting sliding mode controller design for a class of nonlinear systems. *European Journal of Control*. Advance online publication. doi:10.1016/j.ejcon.2020.05.004

Haddad, M., & Otayek, R. (2018). Addressing the challenge of lean manufacturing sustainment with system dynamics modeling: a case study on apparel manufacturing in a developing country. *2nd European Conference on Industrial Engineering and Operations Management*, 442–450. http://www.ieomsociety.org/paris2018/papers/89.pdf

Haddad, M., & Otayek, R. (2019). Assessing the Sustainment of a Lean Implementation Using System Dynamics Modeling : A Case Study of Apparel Manufacturing in Lebanon. *International Journal of System Dynamics Applications*, *8*(4), 14–29. doi:10.4018/IJSDA.2019100102

Halbower, M. C. (1993). The First Three Hours D-4230-3. In MIT System Dynamics Group Literature Collection. System Dynamics Society.

Hamidi, A., Weber, L., & Nasiri, A. (2013). EV charging station integrating renewable energy and second-life battery. *Proceedings of 2013 International Conference on Renewable Energy Research and Applications, ICRERA 2013*, 1217–1221. 10.1109/ICRERA.2013.6749937

Hamrawi, H. (2011). *Type-2 fuzzy α-cuts* (Ph. D. Thesis). De Montfort University.

Hamrawi, H., Coupland, S., & John, R. (2010). A novel α-cut representation for type-2 fuzzy sets in fuzzy systems. *IEEE International Conference*.

Hamza, R. (2017). A novel pseudo random sequence generator for image-cryptographic applications. *Journal of Information Security and Applications*, *35*, 119–127. doi:10.1016/j.jisa.2017.06.005

Hannola, L., Friman, J., & Niemimuukko, J. (2013). Application of agile methods in the innovation process. *International Journal of Business Innovation and Research*, *7*(1), 84–98. doi:10.1504/IJBIR.2013.050557

Han, X., Li, Z., Dahhou, B., Cabassud, M., & He, M. (2020). Nonlinear Observer Based Fault Diagnosis for an Innovative Intensified Heat-Exchanger/Reactor. *Proceedings of the 11th International Conference on Modelling, Identification and Control (ICMIC2019)*.

Harrington, P. (2012). *Machine Learning in Action*. Manning.

Harvey, L., & Green, D. (1993). Defining Quality. *Assessment & Evaluation in Higher Education*, *18*(1), 9–34. doi:10.1080/0260293930180102

Hasan, H., & Ilias, A. (2008). Service quality and student satisfaction: A case study at private higher education institutions. *International Business Research*, *3*(3), 10–21.

Hashemi, Z., & Rahideh, A. (2019). Rotor Electrical Fault Detection of Wind Turbine Induction Generators Using an Unscented Kalman Filter. *Iranian Journal of Science and Technology, Transactions of Electrical Engineering*, 1-10.

Haslett, T. (2012). Simulating Hamlet: Questions, Cautions, and Critique. *International Journal of System Dynamics Applications*, *1*(4), 77–87. doi:10.4018/ijsda.2012100104

Hassanien, A. E., & Azar, A. T. (2015). Brain Computer Interfaces: Current Trends and Applications. In Intelligent Systems Reference Library, (Vol. 74). Springer-Verlag GmbH Berlin/Heidelberg.

Hassanien, A. E., Azar, A. T., Gaber, T., Bhatnagar, R., & Tolba, M. F. (2019a). The International Conference on Advanced Machine Learning Technologies and Applications (AMLTA2019). In *Advances in Intelligent Systems and Computing*, (Vol. 921). Springer.

Hassanien, A. E., Azar, A. T., Gaber, T., Oliva, D., & Tolba, M. F. (2020). Proceedings of the International Conference on Artificial Intelligence and Computer Vision (AICV2020). In *Advances in Intelligent Systems and Computing*, (Vol. 1153). Springer.

Hassanien, A. E., Azar, A. T., Snasel, V., Kacprzyk, J., & Abawajy, J. H. (2015). Big Data in Complex Systems: Challenges and Opportunities. Springer-Verlag GmbH Berlin/Heidelberg.

Hassanien, A. E., Tolba, M. F., Shaalan, K., & Azar, A. T. (2019b). Proceedings of the International Conference on Advanced Intelligent Systems and Informatics 2018. In *Advances in Intelligent Systems and Computing*, (Vol. 845). Springer.

Hassanien, A. E., Tolba, M., & Azar, A. T. (2014a). Advanced Machine Learning Technologies and Applications. In *Second International Conference, AMLTA 2014, Cairo, Egypt, November 28-30, 2014. Proceedings, Communications in Computer and Information Science*, (*Vol. 488*). Springer-Verlag GmbH Berlin/Heidelberg. 10.1007/978-3-319-13461-1

Hassanien, A.E., Moftah, H.M., Azar, A.T., & Shoman, M. (2014b). MRI Breast cancer diagnosis hybrid approach using adaptive Ant-based segmentation and Multilayer Perceptron neural networks classifier. *Applied Soft Computing, 14*(Part A), 62-71.

Hassenzahl, M., & Tractinsky, N. (2006). User experience—A research agenda. *Behaviour & Information Technology, 25*(2), 91–97. doi:10.1080/01449290500330331

Hayes, A. (2019). *Game Theory Definition*. Retrieved May 4, 2020, from https://www.investopedia.com/terms/g/gametheory.asp

Heagy, J. (1992). A physical interpretation of the Hénon map. *Physica D. Nonlinear Phenomena, 57*(3-4), 436–446. doi:10.1016/0167-2789(92)90012-C

He, F., Xie, G., & Luo, J. (2020). Electrical bearing failures in electric vehicles. *Friction, 8*(1), 4–28.

Heilbrun, J., & Gray, C. M. (2001). *The economics of art and culture*. Cambridge University Press. doi:10.1017/CBO9780511754135

Heisler, Y. (2015, August 3). Microsoft engineer exposes the inside story: Why Windows Phone was doomed from the start. *BGR*. https://bgr.com/2015/08/03/windows-phone-failure-microsoft-engineer/

He, K. (2003). Critical phenomenon, crisis and transition to spatiotemporal chaos in plasmas. In *Advances in Space Environment Research* (pp. 475–494). Springer. doi:10.1007/978-94-007-1069-6_47

Hellmann, T., & Puri, M. (2000). The Interaction between Product Market and Financing Strategy: The Role of Venture Capital. *Review of Financial Studies, 13*(4), 959–984. doi:10.1093/rfs/13.4.959

Helm, S., & Klode, C. (2011). Challenges in measuring corporate reputation. In *Reputation management* (pp. 99–110). Springer.

Hendrickson, C., Matthews, H. S., Matthews, D. H., Joshi, S., Maclean, H. L., Lave, L. B., … Bergerson, J. A. (2006). Environmental Life Cycle Assessment of Goods and Services: An Input-Output Approach. Taylor and Francis.

Hénon, M. (1976). A two-dimensional mapping with a strange attractor. *Communications in Mathematical Physics, 50*(1), 69–76. doi:10.1007/BF01608556

Herrera, M. M., Carvajal-Prieto, L. A., Uriona-Maldonado, M., & Ojeda, F. (2019a). Modeling the Customer Value Generation in the Industry's Supply Chain. *International Journal of System Dynamics Applications, 8*(4), 1–13. doi:10.4018/ijsda.2019100101

Herrera, M. M., Cosenz, F., & Dyner, I. (2019b). How to support energy policy coordination? Findings from the Brazilian wind industry. *The Electricity Journal, 32*(8), 106636. doi:10.1016/j.tej.2019.106636

Herrera, M. M., Uriona, M., & Dyner, I. (2020). Dynamics performance of the wind-power supply chain with transmission capacity constraints. *Iranian Journal of Electrical and Computer Engineering, 10*(2), 1142–1148. doi:10.11591/ijece.v10i1.pp1142-1148

Herrick, J. (2018, March 30). *We know exactly why Windows Phone failed.* TechnoBuffalo. https://www.technobuffalo.com/windows-phone-failure-reason-terry-myerson

Hewitt, K. (Ed.). (1983). *Interpretations of Calamity from the Viewpoint of Human Ecology, The Risks and Hazards Series: 1.* Allen and Unwin.

Hill, F. M. (1995). Managing service quality in higher education: The role of the student as primary consumer. *Quality Assurance in Education, 3*(3), 10–21. doi:10.1108/09684889510093497

Hilmola, O.-P., Helo, P., & Ojala, L. (2003). The value of product development lead time in software startup. *System Dynamics Review, 19*(1), 75–82. doi:10.1002dr.255

Hirpara & Sharma. (2020). An analysis of a Wind Turbine-Generator System In The Presence of Stochasticity and Fokker-Planck Equations. *International Journal of System Dynamics Applications, 9*(1), 18–43.

Hogea, C. S., Murray, B. T., & Sethian, J. A. (2005). Computational modeling of solid tumor evolution via a general Cartesian mesh/level set method. *Fluid Dynamics & Materials Processing, 1-2*, 1–37.

Hokkanen, L., Kuusinen, K., & Väänänen, K. (2016). Minimum Viable User EXperience: A Framework for Supporting Product Design in Startups. Agile Processes, in Software Engineering, and Extreme Programming, 66–78. doi:10.1007/978-3-319-33515-5_6

Holling, C. S. (2001). Understanding the Complexity of Economic, Ecological, and Social Systems. *Ecosystems (New York, N.Y.), 4*(5), 390–405. doi:10.100710021-001-0101-5

Hopfensitz, A., Miquel-Florensa, J., & Associates. (2015). How forced displacement flows affect public good contributions: The social consequences of conflict in Colombia. Toulouse School of Economics, 1–22.

Hopper, M., & Stave, K. (2008). Assessing the Effectiveness of Systems Thinking Interventions in the Classroom. *Conference Proceedings, The 2008 International Conference of the System Dynamics Society.*

Hopper, M., & Stave, K. A. (2008). Assessing the Effectiveness of Systems Thinking Interventions in the Classroom. *The 26th International Conference of the System Dynamics Society*, 1-26.

Horne, H., & Pierce, A. (1996). *A Practical Guide to Staff Development and Appraisal in Schools.* Kogan Pages.

Hosseini-nasab, H., Dehghani, M., & Hosseini-nasab, A. (2013). Analysis of Technology Effectiveness of Lean. *Industrial Engineering (American Institute of Industrial Engineers), 2013*, 1–10.

Hossein, S. (2018). The Stability of a Class of Fractional Order Switching System with Time-Delay Actuator. *International Journal of System Dynamics Applications, 7*(1), 85–96. doi:10.4018/IJSDA.2018010105

Hou, Z., Yang, Y., Liu, L., & Wang, Y. (2019). Terminal sliding mode control based impact time and angle constrained guidance. *Aerospace Science and Technology, 93*, 1270–9638. doi:10.1016/j.ast.2019.04.050

Ho, W., Dey, K. P., & Higson, H. E. (2006). Multiple criteria decision- making techniques in higher education. *International Journal of Educational Management, 20*(5), 319–337. doi:10.1108/09513540610676403

Ho, W., Higson, H. E., & Dey, P. K. (2007). An integrated multiple criteria decision making approach for resource allocation in higher education. *International Journal of Innovation and Learning*, *4*(5), 471–486. doi:10.1504/IJIL.2007.012958

Huangfu, Y., Ma, R., & Miraoui, A. (2012). Avoidance high-frequency chattering second-order sliding mode controller design: Buck converter in wind power system. *International Journal of Antennas and Propagation*, *2012*, 176830. doi:10.1155/2012/176830

Huang, H. L., & Guo, Y. (2019). An Improved Correlation Coefficient of Intuitionistic Fuzzy Sets. *Journal of Intelligent Systems*, *28*(2), 231–243. doi:10.1515/jisys-2017-0094

Huang, X., Liu, L., Li, X., Yu, M., & Wu, Z. (2019). A new pseudorandom bit generator based on mixing three-dimensional Chen chaotic system with a chaotic tactics. *Complexity*, *2019*, 2019. doi:10.1155/2019/6567198

Huang, Y., Song, N., Ding, Y., Yuan, S., Li, X., Cai, H., Shi, H., & Luo, Y. (2009). Pulmonary vascular destabilization in the premetastatic phase facilitates lung metastasis. *Cancer Research*, *69*(19), 7529–7537. doi:10.1158/0008-5472.CAN-08-4382 PMID:19773447

Hurni, H., Giger, M., Liniger, H., Mekdaschi Studer, R., Messerli, P., Portner, B., Schwilch, G., Wolfgramm, B., & Breu, T. (2015). Soils, agriculture and food security: The interplay between ecosystem functioning and human well-being. *Current Opinion in Environmental Sustainability*, *15*, 25–34. doi:10.1016/j.cosust.2015.07.009

Hussein, H. A. T., Ammar, M. E., & Hassan, M. A. M. (2017). Three Phase Induction Motor's Stator Turns Fault Analysis Based on Artificial Intelligence. *International Journal of System Dynamics Applications*, *6*(3), 1–19. doi:10.4018/IJSDA.2017070101

Hussien, A. G., Hassanien, A. E., Houssein, E. H., Amin, M., & Azar, A. T. (2020). New binary whale optimization algorithm for discrete optimization problems. *Engineering Optimization*, *52*(6), 945–959. doi:10.1080/0305215X.2019.1624740

Hwang, J. J., Kuo, J. K., Wu, W., Chang, W. R., Lin, C. H., & Wang, S. E. (2013). Lifecycle performance assessment of fuel cell/battery electric vehicles. *International Journal of Hydrogen Energy*, *38*(8), 3433–3446. doi:10.1016/j.ijhydene.2012.12.148

Ibanez-Lopez, A. S., Martinez-Val, J. M., & Moratilla-Soria, B. Y. (2017). A dynamic simulation model for assessing the overall impact of incentive policies on power system reliability, costs and environment. *Energy Policy*, *102*(December), 170–188. doi:10.1016/j.enpol.2016.12.026

Ibragimov, N. K. (1985). Transformation groups applied to mathematical physics. Dordrecht: Kammogne ST, Ahmad TA, Kengne R, Fotsin HB. (2020). Stability analysis and robust synchronization of fractional or- der modified colpitts oscillators. *International Journal of Automation and Control*, *14*(1), 52–79.

Ibraheem, G. A. R., Azar, A. T., Ibraheem, I. K., & Humaidi, A. J. (2020). A Novel Design of a Neural Network based Fractional PID Controller for Mobile Robots Using Hybridized Fruit Fly and Particle Swarm Optimization, Complexity, vol. 2020. *Article ID*, *3067024*, 1–18. doi:10.1155/2020/3067024

Ibrahim, H. A., Azar, A. T., Ibrahim, Z. F., & Ammar, H. H. (2020). A Hybrid Deep Learning Based Autonomous Vehicle Navigation and Obstacles Avoidance. In *Proceedings of the International Conference on Artificial Intelligence and Computer Vision (AICV2020)*. AICV 2020. Springer. 10.1007/978-3-030-44289-7_28

ICSU & ISSC. (2015). *Review of Targets for the Sustainable Development Goals: The Science Perspective*. Paris, France: International Council for Science. Retrieved from http://www.icsu.org/publications/reports-and-reviews/review-of-targets-for-the-sustainable-development-goals-the-science-perspective-2015/SDG-Report.pdf

IDEAM. (2017). *Promedios mensuales de irradiación global media recibida en superficie para las principales ciudades del país (Wh/m2 por día)*. Retrieved from http://atlas.ideam.gov.co/basefiles/Anexo-Promedios-mensuales-de-Irradiacion-Global-Media.pdf

Ilyas, M., & Mahgoub, I. (2004). *Handbook of sensor networks: Compact wireless and wired sensing systems*. CRC Press.

Inbarani, H. H., Azar, A. T., & Jothi, G. (2014). Supervised hybrid feature selection based on PSO and rough sets for medical diagnosis. *Computer Methods and Programs in Biomedicine*, *113*(1), 175–185. doi:10.1016/j.cmpb.2013.10.007 PMID:24210167

Inbarani, H. H., Azar, A. T., & Jothi, G. (2020). Leukemia Image Segmentation using a Hybrid Histogram based Soft Covering Rough K-Means Clustering Algorithm. *Electronics (Basel)*, *9*(1), 188. doi:10.3390/electronics9010188

Inbarani, H. H., Bagyamathi, M., & Azar, A. T. (2015b). A novel hybrid feature selection method based on rough set and improved harmony search. *Neural Computing & Applications*, *26*(8), 1859–1880. doi:10.100700521-015-1840-0

Inbarani, H. H., Banu, P. K. N., & Azar, A. T. (2014). Feature selection using swarm-based relative reduct technique for fetal heart rate. *Neural Computing & Applications*, *25*(3-4), 793–806. doi:10.100700521-014-1552-x

Inbarani, H. H., Kumar, S. S., Azar, A. T., & Hassanien, A. E. (2015a). Hybrid TRS-PSO Clustering Approach for Web2.0 Social Tagging System. *International Journal of Rough Sets and Data Analysis*, *2*(1), 22–37. doi:10.4018/ijrsda.2015010102

Inbarani, H. H., Kumar, S. U., Azar, A. T., & Hassanien, A. E. (2018). Hybrid Rough-Bijective Soft Set Classification system. *Neural Computing & Applications*, *29*(8), 67–78. doi:10.100700521-016-2711-z

Inelmen, E. (2006). Philosophy of Education. *The Proceedings of the Twenty-First World Congress of Philosophy*, *4*, 57-64. http://inelmen.boun.edu.tr

Irfan, M., Ali, A., Khan, M. A., Ehatisham-ul-Haq, M., Mehmood Shah, S. N., Saboor, A., & Ahmad, W. (2020). Pseudorandom Number Generator (PRNG) Design Using Hyper-Chaotic Modified Robust Logistic Map (HC-MRLM). *Electronics (Basel)*, *9*(1), 104. doi:10.3390/electronics9010104

Ismail, M. M. (2012). Applications of ANFIS and fuzzy algorithms for improvement of the DTC performance for the three phase saturated model of induction motor. *International Journal of System Dynamics Applications*, *1*(3), 54–83. doi:10.4018/ijsda.2012070102

Ismail, M. M., Fattah, H. A. A., & Bahgat, A. (2012). Adaptive output feedback voltage-based control of magnetically-saturated induction motors. *International Journal of System Dynamics Applications*, *1*(3), 1–53. doi:10.4018/ijsda.2012070101

Izquierdo-Kulich, E., Rebelo, I., Tejera, E., & Nieto-Villar, J. M. (2013). Phase transition in tumor growth: I avascular development. *Physica A*, *392*(24), 6616–6623. doi:10.1016/j.physa.2013.08.010

Jagdeesh, R. (2001). Improvement of quality of higher education in engineering sciences with an emphasis on International aspects. *The Indian Journal of Technical Education, 24*(2), 50-55.

Jahns, T. M., Han, S., El-Refaie, A. M., Baek, J., Aydin, M., Guven, M. K., & Soong, W. L. (2006). Design and Experimental Verification of a 50 kW Interior Permanent Magnet Synchronous Machine. In *Conference Record of the 2006 IEEE Industry Applications Conference Forty-First IAS Annual Meeting* (Vol. 4, pp. 1941–1948). 10.1109/IAS.2006.256801

Jaime, J. C., Mesa-Álvarez, M. D., Martin, R. R., Betancourt-Mar, J. A., Cocho, G., Mansilla, R., & Nieto-Villar, J. M. (2019). Chronotherapy of cancer: Periodic perturbations in vascular growth and metastasis. *Biological Rhythm Research*, *50*(4), 495–504. doi:10.1080/09291016.2018.1465698

Jain, G., & Umashankar, S. (2016, April). Modelling and simulation of solar photovoltaic fed induction motor for water pumping application using perturb and observer MPPT algorithm. In *Energy Efficient Technologies for Sustainability (ICEETS), 2016 International Conference on* (pp. 250-254). IEEE.

Jain, V., & Raj, T. (2018). An adaptive neuro-fuzzy inference system for makespan estimation of flexible manufacturing system assembly shop: A case study. *International Journal of System Assurance Engineering and Management*, *9*(6), 1302–1314.

Jalalipor, B. (2009). Theater and a concern named funding: a comparative study between theater in Iran and France. *Journal of Art and Architecture, 79*, 70-97. (in Persian)

Jangholi, M., & Sedaghati, A. (2016). Design and Optimization of MPPT Routing in Solar Car Cells with the Maximum Power Point Tracking. *International Journal of Advanced Research in Electrical. Electronics and Instrumentation Engineering*, (5), 4.

Jaziri, S., & Jemli, K. (2013, May). Optimization of a photovoltaic powered water pumping system. In *Control, Decision and Information Technologies (CoDIT), International Conference on* (pp. 422-428). IEEE.

Jensen, R. T. (2007). The Digital Provide: Information (Technology), Market Performance and Welfare in the South Indian Fisheries Sector. *The Quarterly Journal of Economics*, *122*(3), 879–924. doi:10.1162/qjec.122.3.879

Jia, C., & Chan, W. K. (2013). Which compiler optimization options should I use for detecting data races in multithreaded programs. *2013 8th International Workshop on Automation of Software Test (AST),* 53-56. 10.1109/IWAST.2013.6595791

Jia, L. (2018). A Hybrid Feature Selection Method for Software Defect Prediction. *IOP Conf. Ser.: Mater. Sci. Eng., ACMME, 394*(3). doi:10.1088/1757-899X/394/3/032035

Jia, R., & Ding, R. (2002). *System Dynamics Analysis of Feedback Dynamic Complexity*. Higher Education Press.

Jimenez, M., Franco, C. J., & Dyner, I. (2016). Diffusion of renewable energy technologies: The need for policy in Colombia. *Energy*, *111*, 818–829. doi:10.1016/j.energy.2016.06.051

Ji, Y., Liu, T., & Min, L. (2008). Generalized chaos synchronization theorems for bidirectional differential equations and discrete systems with applications. *Physics Letters. [Part A]*, *372*(20), 3645–3652. doi:10.1016/j.physleta.2008.02.022

Johnson, K., Olson, E. A., & Manandhar, S. (1982). Environmental Knowledge and Response to Natural Hazards in Mountainous Nepal. *Mountain Research and Development*, *2*(2), 175–188. doi:10.2307/3672962

Joshi, S., & Talange, D. (2016). Fault tolerant control of an auv using periodic output feedback with multi model approach. *International Journal of System Dynamics Applications*, *5*(2), 41–62.

Jothi, G., Inbarani, H. H., & Azar, A. T. (2013). Hybrid Tolerance-PSO Based Supervised Feature Selection For Digital Mammogram Images. *International Journal of Fuzzy System Applications*, *3*(4), 15–30. doi:10.4018/ijfsa.2013100102

Jothi, G., Inbarani, H. H., Azar, A. T., & Devi, K. R. (2019). Rough set theory with Jaya optimization for acute lymphoblastic leukemia classification. *Neural Computing & Applications*, *31*(9), 5175–5194. doi:10.100700521-018-3359-7

Jothi, G., Inbarani, H. H., Azar, A. T., Koubaa, A., Kamal, N. A., & Fouad, K. M. (2020). Improved Dominance Soft Set based Decision Rules with Pruning for Leukemia Image Classification. *Electronics (Basel)*, *9*(5), 794. doi:10.3390/electronics9050794

Jothi, G., Inbarani, H. H., Azar, A. T., & Polat, K. (2017). Tolerance rough set firefly-based quick reduct. *Neural Computing & Applications*, *28*(10), 2995–3008. doi:10.100700521-016-2514-2

Jureczko, M., & Madeyski, L. (2010). Towards identifying software project clusters with regard to defect prediction. In *Proceedings of the 6th International Conference on Predictive Models in Software Engineering, ser. PROMISE '10*. New York, NY: ACM. Available: https: //doi.acm.org/10.1145/1868328.1868342

Kacprzyk, J. (1997). *Multistage Fuzzy Control*. Wiley.

Kaen, E., & Algarni, A. (2019). Feature Selection Approach for Improving the Accuracy of Software Bug Prediction. *JKAU: Comp. IT. Sci.*, *8*(1), 35–44.

Kahan, W. (1996). IEEE standard 754 for binary floating-point arithmetic. *Lecture Notes on the Status of IEEE, 754*, 11.

Kalluri, R., & Weinberg, R. A. (2009). The basics of epithelial-mesenchymal transition. *The Journal of Clinical Investigation, 119*(6), 1420–1428. doi:10.1172/JCI39104 PMID:19487818

Kamal, N. A., Azar, A. T., Elbasuony, G. S., Almustafa, K. A., & Almakhles, D. (2020). PSO-based Adaptive Perturb and Observe MPPT Technique for Photovoltaic Systems. In *The International Conference on Advanced Intelligent Systems and Informatics AISI 2019*. Springer.

Kammogne, S.T., Fotsin, H.B., Kounchou, M. & Louodop, P., (2013) A robust observer design for passivity-based synchronization of uncertain modified Colpitts oscillators and circuit simulation. *Asian Journal of Science and Technology, 5*, 29-41.

Kammogne, A. S. T., Kountchou, M. N., Kengne, R., Azar, A. T., Fotsin, H. B., & Ouagni, S. T. M. (2020). Polynomial Robust Observer Implementation based-passive Synchronization of Nonlinear Fractional-Order Systems with Structural Disturbances. *Frontiers of Information Technology & Electronic Engineering.*, *21*(9), 1369–1386. doi:10.1631/FITEE.1900430

Kammogne, S. T., Azar, A. T., Fotsin, H. B., & Kengne, R. (2019). Robust observer-based synchronization of chaotic oscillators with structural perturbations and input nonlinearity. *International Journal of Automation and Control, 13*(4), 387–412. doi:10.1504/IJAAC.2019.100467

Kammogne, S. T., & Fotsin, H. B. (2011). Synchronization of modified Colpitts oscillators with structural perturbations. *Physica Scripta, 83*(6), 65011–65018. doi:10.1088/0031-8949/83/06/065011

Kammogne, S. T., & Fotsin, H. B. (2015). Robust adaptive exponential synchronization of two different stochastic perturbed chaotic systems with structural perturbations. *International Journal of Engineering Mathematics.*, *2015*, 1–10. doi:10.1155/2015/535317

Kammogne, S. T., Kengne, R., & Fotsin, H. (2017). Dynamics and improve robust adaptive control strategy for the finite time synchronization of uncertain nonlinear system. *International Journal of System Dynamics Applications*, *6*(4), 32–62.

Kaplan, J. L., & Yorke, J. A. (1979). Chaotic behavior of multidimensional difference equations. In *Functional differential equations and approximation of fixed points* (pp. 204–227). Springer. doi:10.1007/BFb0064319

Karampuri, R., Jain, S., & Somasekhar, V. T. (2014, December). A single-stage solar PV power fed Open-End Winding Induction Motor pump drive with MPPT. In *Power Electronics, Drives and Energy Systems (PEDES), 2014 IEEE International Conference on* (pp. 1-6). IEEE.

Karner, M., Glawar, R., Sihn, W., & Matyas, K. (2019). An industry-oriented approach for machine condition-based production scheduling. *Procedia CIRP*, *81*, 938–943.

Kashfi, P., Nilsson, A., & Feldt, R. (2016). *Integrating User eXperience Practices into Software Development Processes: Implications of Subjectivity and Emergent Nature of UX*. ArXiv Preprint ArXiv:1605.03783.

Kate, K. (2002). Science and the Convention on Biological Diversity. *Science*, *295*(5564), 2371–2372. doi:10.1126cience.1070725 PMID:11923511

Kemell, K.-K., Wang, X., Nguyen-Duc, A., Grendus, J., Tuunanen, T., & Abrahamsson, P. (2020). Startup Metrics That Tech Entrepreneurs Need to Know. In *Fundamentals of Software Startups* (pp. 111–127). Springer. doi:10.1007/978-3-030-35983-6_7

Kengne, J., Chedjou, J. C., Kenne, G., & Kyamakya, K. (2012). Dynamical properties and chaos synchronization of improved Colpitts oscillators. *Communications in Nonlinear Science and Numerical Simulation*, *17*(7), 2914–2923. doi:10.1016/j.cnsns.2011.10.038

Kengne, J., Tabekoueng, Z. N., & Fotsin, H. B. (2016). Coexistence of multiple attractors and crisis route to chaos in autonomous third order Duffing–Holmes type chaotic oscillators. *Communications in Nonlinear Science and Numerical Simulation*, *36*, 29–44. doi:10.1016/j.cnsns.2015.11.009

Kennedy, M. (2008). A Taxonomy of System Dynamics Pedagogic Techniques. *Conference Proceedings, The 2008 International Conference of the System Dynamics Society.*

Kennedy, M. (1998a). A Pilot System Dynamics Model to Capture and Monitor Quality Issues in Higher Education Institutions Experiences Gained. *Proceedings of the 16th System Dynamics Conference*, 1-7.

Kennedy, M. (1998b). Some Issues in System Dynamics Model Building to Support Quality Monitoring in Higher Education. *Proceedings of the 16th System Dynamics Conference*, 1-8.

Kennedy, M. (2002). An extended Taxonomy of System Dynamics Models of Higher Education. *Proceedings of the 20th System Dynamics Conference*, 1-19.

Kennedy, M., & Clare, C. (1999). Some Issues in Building System Dynamics Model for Improving the Resource Management Process in Higher Education. *Proceedings of the 17th International System Dynamics Conference.*

Keogan, M. T., McDermott, V. G., Paulson, E. K., Sheafor, D. H., Frederick, M. G., de Long, D. M., & Nelson, R. C. (1997). Pancreatic malignancy: Effect of dual-phase helical CT in tumor detection and vascular opacification. *Radiology*, *205*(2), 513–518. doi:10.1148/radiology.205.2.9356637 PMID:9356637

Khaligh, A., Member, S., Li, Z., & Member, S. (2010). Battery, Ultracapacitor, Fuel Cell, and Hybrid Energy Storage Systems for Electric, Hybrid Electric, Fuel Cell, and Plug-In Hybrid Electric Vehicles : State of the Art. *IEEE Transactions on Vehicular Technology*, *59*(6), 2806–2814. doi:10.1109/TVT.2010.2047877

Khan, A., Singh, S., & Azar, A. T. (2020a). Combination-Combination Anti-Synchronization of Four Fractional Order Identical Hyperchaotic Systems. In *The International Conference on Advanced Machine Learning Technologies and Applications (AMLTA2019). AMLTA 2019. Advances in Intelligent Systems and Computing*, (vol 921, pp. 406-414). Springer. 10.1007/978-3-030-14118-9_41

Khan, A., Singh, S., & Azar, A. T. (2020b). Synchronization between a Novel Integer-Order Hyperchaotic System and a Fractional-Order Hyperchaotic System Using Tracking Control. In *The International Conference on Advanced Machine Learning Technologies and Applications (AMLTA2019). AMLTA 2019. Advances in Intelligent Systems and Computing*, (vol 921, pp. 382-391). Springer. 10.1007/978-3-030-14118-9_38

Kharola, A., & Patil, P. (2017). Position and tilt control of two-wheeled robot (TWR): A neuro-fuzzy approach. *International Journal of System Dynamics Applications*, *6*(4), 1–17. doi:10.4018/IJSDA.2017100102

Khataie, A. H., & Bulgak, A. A. (2013). A cost of quality decision support model for lean manufacturing : Activity-based costing application. *International Journal of Quality & Reliability Management, 30*(7), 751–764. doi:10.1108/IJQRM-Jan-2011-0016

Khennaoui, A. A., Ouannas, A., Grassi, G., & Azar, A. T. (2020). Dynamic Analysis of a Fractional Map with Hidden Attractor. In *Proceedings of the International Conference on Artificial Intelligence and Computer Vision (AICV2020). AICV 2020. Advances in Intelligent Systems and Computing*, (vol 1153, pp. 731-739). Springer. 10.1007/978-3-030-44289-7_68

Khennaoui, A. A., Ouannas, A., Bendoukha, S., Grassi, G., Lozi, R. P., & Pham, V. T. (2019). On fractional–order discrete–time systems: Chaos, stabilization and synchronization. *Chaos, Solitons, and Fractals, 119*, 150–162. doi:10.1016/j.chaos.2018.12.019

Khettab, K., Bensafia, Y., Bourouba, B., & Azar, A. T. (2018). Enhanced Fractional Order Indirect Fuzzy Adaptive Synchronization of Uncertain Fractional Chaotic Systems Based on the Variable Structure Control: Robust H? Design Approach. In Mathematical Techniques of Fractional Order Systems, Advances in Nonlinear Dynamics and Chaos (ANDC) series, (pp. 559-595). Elsevier.

Kim, K., Park, S., Park, H., & Ham, Y. H. (2008). Reliable and real-time data dissemination in wireless sensor networks. *MILCOM 2008 - 2008 IEEE Military Communications Conference*. doi:10.1109/MILCOM.2008.4753643

Kimmerer, R. W. (2002). Weaving traditional ecological knowledge into biological education: A call to action. *Bioscience, 52*(5), 432–438. doi:10.1641/0006-3568(2002)052[0432:WTEKIB]2.0.CO;2

Kim, S., & Kwon, S. J. (2015). Dynamic modeling of a two-wheeled inverted pendulum balancing mobilr robot. *International Journal of Control, Automation, and Systems, 13*(4), 926–933. doi:10.100712555-014-0564-8

Kishimoto, J., Goto, M., & Inoue, K. (2017). Do acquisitions by electric utility companies create value? Evidence from deregulated markets. *Energy Policy, 105*(February), 212–224. doi:10.1016/j.enpol.2017.02.032

Klein, L. (2013). *UX for Lean Startups: Faster, Smarter User Experience Research and Design*. O'Reilly Media, Inc.

Kljajic, M., Borštnar, M. K., Škraba, A., & Kofjac, D. (2012). System Approach to MIS and DSS and its Modeling within SD. Regional Development: Concepts, Methodologies, Tools, and Applications, 315–335. doi:10.4018/978-1-4666-0882-5.ch208

Klotins, E., Unterkalmsteiner, M., & Gorschek, T. (2015). Software Engineering Knowledge Areas in Startup Companies: A Mapping Study. In J. M. Fernandes, R. J. Machado, & K. Wnuk (Eds.), *Software Business* (pp. 245–257). Springer International Publishing. doi:10.1007/978-3-319-19593-3_22

Knighton, D., Ausprunk, D., Tapper, D., & Folkman, J. (1977). Avascular and vascular phases of tumour growth in the chick embryo. *British Journal of Cancer, 35*(3), 347–356. doi:10.1038/bjc.1977.49 PMID:856240

Kocarev, L., Halle, K. S., Eckert, K., & Chua, L. O. (1993). Experimental observation of antimonotonicity in Chua's circuit. *International Journal of Bifurcation and Chaos in Applied Sciences and Engineering, 3*(04), 1051–1055. doi:10.1142/S0218127493000878

Kohm, K., Boersma, P. D., Meffe, G. K., & Noss, R. (2000). Putting the science into practice and the practice into science. *Conservation Biology, 14*(3), 593–594. doi:10.1046/j.1523-1739.2000.00000-e25.x

Kollock, P., Blumstein, P., & Schwartz, P. (1994). The judgment of equity in intimate relationships. *Social Psychology Quarterly*, 340–351.

Kotler, P., Keller, K. L., Manceau, D., & Hémonnet-Goujot, A. (2015). *Marketing management* (Vol. 14). Prentice Hall Englewood Cliffs.

Koubaa, A., Ammar, A., Alahdab, M., Kanhouch, A., & Azar, A.T. (2020). DeepBrain: Experimental Evaluation of Cloud-Based Computation Offloading and Edge Computing in the Internet-of-Drones for Deep Learning Applications. *Sensors 2020, 20*(18), 1-25. doi:10.339020185240

Krakauer, D. C. (Ed.). (2019). *World Hidden in Plain Sight: the Evolving Idea of Complexity at the Sante Fe Institute*. SFI Press.

Krause, P. C., Wasynczuk, O., Sudhoff, S. D., & Pekarek, S. (2002). *Analysis of electric machinery and drive systems* (Vol. 2). Wiley Online Library.

Krause, P., Wasynczuk, O., Sudhoff, S. D., & Pekarek, S. D. (2013). *Analysis of Electric Machinery and Drive Systems* (Vol. 75). John Wiley & Sons.

Kroeze, R. C., & Krein, P. T. (2008). Electrical battery model for use in dynamic electric vehicle simulations. *IEEE Power Electronics Specialists Conference, 2008. PESC 2008*, 1336–1342. 10.1109/PESC.2008.4592119

Kubota & Matsuse. (1993). DSP based speed adaptive flux observer of induction motor. *IEEE Transactions on Industry Applications*, *29*(2), 344–348.

Kubota, Matsuse, & Nakano. (1990). New adaptive flux observer of induction motor for wide speed range motor drives. *Industrial Electronics Society, 1990. IECON APOS, 90. 16th Annual Conference of IEEE, 2,* 921 – 926.

Kudryashov, N. A. (2005). Simplest equation method to look for exact solutions of nonlinear differential equations. *Chaos, Solitons, and Fractals*, *24*, 1217–1231.

Kudryashov, N. A., & Loguinova, N. B. (2008). Extended simplest equation method for nonlinear differential equations. *Applied Mathematics and Computation*, *205*(1), 396–40. doi:10.1016/j.amc.2008.08.019

Kumar Sahu, B. (2015). A study on global solar PV energy developments and policies with special focus on the top ten solar PV power producing countries. *Renewable & Sustainable Energy Reviews*, *43*, 621–634. doi:10.1016/j.rser.2014.11.058

Kumar, A., Shankar, R., & Momaya, K. (2015). The Bass Diffusion Model does not explain diffusion. *Proceedings of the 33rd International Conference of the System Dynamics Society*. Retrieved from https://www.systemdynamics.org/conferences/2015/papers/P1233.pdf

Kumar, J., Azar, A. T., Kumar, V., & Rana, K. P. S. (2018) Design of Fractional Order Fuzzy Sliding Mode Controller for Nonlinear Complex Systems. In Mathematical Techniques of Fractional Order Systems, Advances in Nonlinear Dynamics and Chaos (ANDC) series, (pp. 249-282). Elsevier. doi:10.1016/B978-0-12-813592-1.00009-X

Kumar, K. V., Ramya, K., & Irfan, M. (2019). Advanced Fault Diagnosis Monitoring Scheme in Asynchronous Motor Using Soft Computing Method. In Advanced Condition Monitoring and Fault Diagnosis of Electric Machines (pp. 89-97). IGI Global.

Kumar, S., Inbarani, H.H., Azar, A.T., Own, H.S., & Balas, V.E. (2014). Optimistic Multi-Granulation Rough set based Classification for Neonatal Jaundice Diagnosis. *Soft Computing Applications, 356*, 307-317. DOI doi:10.1007/978-3-319-18296-4_26

Kumar, L., Hota, C., Tummalapalli, S., & Murthy Neti, L. B. (2020). Selection of Significant Metrics for Improving the Performance of Change-Proneness Modules. In *Automated Software Engineering: A Deep Learning-Based Approach. Learning and Analytics in Intelligent Systems* (Vol. 8). Springer. doi:10.1007/978-3-030-38006-9_1

Kumar, S. S., Inbarani, H. H., Azar, A. T., & Hassanien, A. E. (2015a). Rough Set Based Meta-Heuristic Clustering Approach for Social E-Learning Systems. *International Journal of Intelligent Engineering Informatics*, *3*(1), 23–41. doi:10.1504/IJIEI.2015.069098

Kumar, S. S., Inbarani, H. H., Azar, A. T., & Polat, K. (2017). Covering-based rough set classification system. *Neural Computing & Applications*, *28*(10), 2879–2888. doi:10.100700521-016-2412-7

Kumar, S. U., Azar, A. T., Inbarani, H. H., Liyaskar, O. J., & Almustafa, K. M. (2019). Weighted Rough set Theory for Fetal Heart Rate Classification. *Int. J Sociotechnology and Knowledge Development*, *11*(4), 1–19. doi:10.4018/IJSKD.2019100101

Kumar, S. U., Inbarani, H. H., & Azar, A. T. (2015b). Hybrid Bijective soft set - Neural network for ECG arrhythmia classification. *International Journal of Hybrid Intelligent Systems*, *12*(2), 103–118. doi:10.3233/HIS-150209

Kumar, V., & Reinartz, W. (2016). Creating Enduring Customer Value. *Journal of Marketing*, *80*(6), 36–68. doi:10.1509/jm.15.0414

Kupiainen, E., Mäntylä, M. V., & Itkonen, J. (2015). Using metrics in Agile and Lean Software Development–A systematic literature review of industrial studies. *Information and Software Technology*, *62*, 143–163. doi:10.1016/j.infsof.2015.02.005

Kyprianidis, I. M., Stouboulos, I. N., Haralabidis, P., & Bountis, T. (2000). Antimonotonicity and chaotic dynamics in a fourth-order autonomous nonlinear electric circuit. *International Journal of Bifurcation and Chaos in Applied Sciences and Engineering*, *10*(08), 1903–1915. doi:10.1142/S0218127400001171

Kyriazis, M. (1991). Applications of chaos theory to the molecular biology of aging. *Experimental Gerontology*, *26*(6), 569–572. doi:10.1016/0531-5565(91)90074-V PMID:1800131

Labedz, C., & Gray, J. (2013). Accounting for Lean Implementation in Government Enterprise: Intended and Unintended Consequences. *International Journal of System Dynamics Applications*, *2*(1), 14–26. doi:10.4018/ijsda.2013010102

Lagrosen, S., Seyyed-Hashemi, R., & Leitner, M. (2004). Examination of the dimensions of quality in higher Education. *Quality Assurance in Education*, *12*(2), 61–69. doi:10.1108/09684880410536431

Lahoti, G., Shah, U., & Kadam, P. (2016). Robust control of boost converter for flexible operation in PV based systems. *Energy Efficient Technologies for Sustainability (ICEETS), International Conference on,* 345-349.

Lakshmanan, M., & Senthilkumar, D. V. (2010). *Dynamics of Nonlinear Time-Delay Systems*. Springer.

Lamamra, K., Azar, A. T., & Ben Salah, C. (2017). *Chaotic system modelling using a neural network with optimized structure. Studies in Computational Intelligence* (Vol. 688). Springer-Verlag.

Lamba, T., Kavita, & Mishra, A. K. (2019). Optimal Machine learning Model for Software Defect Prediction. *I. J. Intelligent Systems and Applications*, *2*(2), 36–48. doi:10.5815/ijisa.2019.02.05

Lambert, C. (2012, February). The future of theater in a digital era, is the play still the thing? *Harvard Magazine*. Retrieved from http://www.harvardmag.com/pdf/2012/01-pdfs/0112-34.pdf

Larsen, M. A. D., & Drews, M. (2019). Water use in electricity generation for water-energy nexus analyses: The European case. *The Science of the Total Environment*, *651*, 2044–2058. doi:10.1016/j.scitotenv.2018.10.045 PMID:30321726

Last, A. K., & Wetzel, H. (2010). The efficiency of German public theaters: A stochastic frontier analysis approach. *Journal of Cultural Economics*, *34*(2), 89–110. doi:10.100710824-009-9111-5

Lázaro, J. G. M., & Pinilla, C. B. (2020). Detection and Classification of Wear Fault in Axial Piston Pumps: Using ANNs and Pressure Signals. In Pattern Recognition Applications in Engineering (pp. 286-316). IGI Global.

Le Blanc, D. (2015). Towards Integration at Last? The Sustainable Development Goals as a Network of Targets. *Sustainable Development*, *23*(3), 176–187. doi:10.1002d.1582

Lee, C. S., Jeong, J. B., Lee, B. H., & Hur, J. (2011). Study on 1.5 kW battery chargers for neighborhood electric vehicles. *2011 IEEE Vehicle Power and Propulsion Conference. VPPC, 1–4*. Advance online publication. doi:10.1109/VPPC.2011.6043129

Lee, H., Kim, E., Kang, H.-J., & Park, M. (2001). A new sliding-mode control with fuzzy boundary layer. *Fuzzy Sets and Systems, 120*(1), 135–143. doi:10.1016/S0165-0114(99)00072-X

Lee, H., & Utkin, V. I. (2007). Chattering suppression methods in sliding mode control systems. *Annual Reviews in Control, 31*(2), 179–188. doi:10.1016/j.arcontrol.2007.08.001

Lee, S., Cherry, J., Safoutin, M., McDonald, J., & Olechiw, M. (2018). Modeling and Validation of 48V Mild Hybrid Lithium-Ion Battery Pack. *SAE International Journal of Alternative Powertrains, 7*(3), 273–288. doi:10.4271/2018-01-0433

Lei, Y., Yang, B., Jiang, X., Jia, F., Li, N., & Nandi, A. K. (2020). Applications of machine learning to machine fault diagnosis: A review and roadmap. *Mechanical Systems and Signal Processing, 138*, 106587.

Lena, J. C., & Lindemann, D. J. (2014). Who is an artist? New data for an old question. *Poetics, 43*, 70–85. doi:10.1016/j.poetic.2014.01.001

Leroi-Werelds, S., Streukens, S., Brady, M. K., & Swinnen, G. (2014). Assessing the value of commonly used methods for measuring customer value: A multi-setting empirical study. *Journal of the Academy of Marketing Science, 42*(4), 430–451. doi:10.100711747-013-0363-4

Lessmann, S., Baesens, B., Mues, C., & Pietsch, S. (2008). Benchmarking classification models for software defect prediction: A proposed framework and novel findings. *IEEE Transactions on Software Engineering, 34*(4), 485–496. doi:10.1109/TSE.2008.35

Levy, S. (2011, June 28). *Inside Google+—How the Search Giant Plans to Go Social | WIRED*. https://www.wired.com/2011/06/inside-google-plus-social/

Lewis, L. S., & Altbach, P. G. (1996). Faculty Versus Administration: A Universal Problem. *Higher Education Policy, 9*(3), 255–258. doi:10.1016/S0952-8733(96)00016-5

Liang, H., Liu, Y., Sheng, G., & Jiang, X. (2019). Fault-cause identification method based on adaptive deep belief network and time–frequency characteristics of travelling wave. *IET Generation, Transmission & Distribution, 13*(5), 724–732.

Li, C. (2009). Distances between interval-valued fuzzy sets. *Proc. Society Annl. Conf. (NAFIPS2009)*. DOI:10.1109/NAFIPS.2009.5156438

Li, C., Liao, X., & Wong, K. W. (2005). Lag synchronization of hyperchaos with application to secure communications. *Chaos, Solitons, and Fractals, 23*(1), 183–193. doi:10.1016/j.chaos.2004.04.025

Li, C., Lin, D., Feng, B., Lü, J., & Hao, F. (2018). Cryptanalysis of a chaotic image encryption algorithm based on information entropy. *IEEE Access: Practical Innovations, Open Solutions, 6*, 75834–75842. doi:10.1109/ACCESS.2018.2883690

Lichtenberg, A. J., & Lieberman, M. A. (2013). *Regular and stochastic motion* (Vol. 38). Springer Science & Business Media.

Lichtenthaler, U. (2020). Agile innovation: The complementarity of design thinking and lean startup. *International Journal of Service Science, Management, Engineering, and Technology, 11*(1), 157–167. doi:10.4018/IJSSMET.2020010110

Li, N., Pan, W., Yan, L., Luo, B., & Zou, X. (2014). Enhanced chaos synchronization and communication in cascade-coupled semiconductor ring lasers. *Communications in Nonlinear Science and Numerical Simulation, 19*(6), 1874–1883. doi:10.1016/j.cnsns.2013.09.036

Lindsay, A. W. (1992). Concepts of Quality in Higher Education. *Journal of Tertiary Education Administration, 14*(2), 153–163. doi:10.1080/1036970920140203

Lin, F. Y., & Lee, C. (2010). An efficient Lagrangean relaxation-based object tracking algorithm in wireless sensor networks. *Sensors (Basel), 10*(9), 8101–8118. doi:10.3390100908101 PMID:22163643

Lingden, P., Alsadoon, A., Prasad, P.W.C., Alsadoon, O. H., Ali, R. S. & Nguyen. (2019). A novel modified undersampling (MUS) technique for software defect prediction. *Computational Intelligence*, 1–18. Doi:10.1111/coin.12229

Lin, H. X., Choong, Y.-Y., & Salvendy, G. (1997). A proposed index of usability: A method for comparing the relative usability of different software systems. *Behaviour & Information Technology, 16*(4–5), 267–277. doi:10.1080/014492997119833

Linkenback-Fuchs, A. (2002). *Cultural Strategies for Risk Adjustments in Mountains*. Asia High Summit 2002, Kathmandu, Nepal.

Lin, T. C., & Kuo, C. H. (2012). Adaptive Hybrid Intelligent Tracking Control for Uncertain Fractional Order Chaotic Systems. *International Journal of System Dynamics Applications, 1*(1), 1–16. doi:10.4018/ijsda.2012010101

Liu, D., Chen, X., & Peng, D. (2018a). Cosine similarity measure between hybrid intuitionistic fuzzy sets and its application in medical diagnosis. *Computational and Mathematical Methods in Medicine, 2018*, 2018. doi:10.1155/2018/3146873 PMID:30416537

Liu, D., Liu, G., & Liu, Z. (2018b). Some Similarity Measures of Neutrosophic Sets Based on the Euclidean Distance and Their Application in Medical Diagnosis. *Computational and Mathematical Methods in Medicine*. doi:10.1155/2018/7325938 PMID:30622626

Liu, J., Hull, V., Godfray, H. C. J., Tilman, D., Gleick, P., Hoff, H., Pahl-Wostl, C., Xu, Z., Chung, M. G., Sun, J., & Li, S. (2018). Nexus approaches to global sustainable development. *Nature Sustainability, 1*(9), 466–476. doi:10.103841893-018-0135-8

Liu, L., Ma, D., Azar, A. T., & Zhu, Q. (2020). Neural computing Enhanced Parameter Estimation for Multi-Input and Multi-Output Total Nonlinear Dynamic Models. *Entropy (Basel, Switzerland), 22*(5), 510. doi:10.3390/e22050510

Liu, L., Miao, S., Hu, H., & Deng, Y. (2016). Pseudorandom bit generator based on non-stationary logistic maps. *IET Information Security, 10*(2), 87–94. doi:10.1049/iet-ifs.2014.0192

Liu, L., Zhang, X., & Ma, H. (2010). Optimal node selection for target localization in wireless camera sensor networks. *IEEE Transactions on Vehicular Technology, 59*(7), 3562–3576. doi:10.1109/TVT.2009.2031454

Liu, T., Liu, Y., Cui, X., Xu, G., & Qian, D. (2012). MOLTS: Mobile object localization and tracking system based on wireless sensor networks. *2012 IEEE Seventh International Conference on Networking, Architecture, and Storage*. 10.1109/NAS.2012.34

Li, Z., Chen, D., Zhu, J., & Liu, Y. (2015). Nonlinear dynamics of fractional order Duffing system. *Chaos, Solitons, and Fractals, 81*, 111–116. doi:10.1016/j.chaos.2015.09.012

Llanos-Pérez, J. A., Betancourt-Mar, J. A., Cocho, G., Mansilla, R., & Nieto-Villar, J. M. (2016, November). Nieto-Villar José Manuel. Phase transitions in tumor growth: III vascular and metastasis behavior. *Physica A, 462*, 560–568. doi:10.1016/j.physa.2016.06.086

Loganathan, C., & Pushpalatha, V. (2018). An Application of Interval Valued Fuzzy Matrices in Agricultural Field. *Int. J. Math. And Appl., 6*(3), 387–397.

Lomas, L. (2004). Embedding quality: The challenges for higher education. *Quality Assurance in Education*, *12*(4), 157–165. doi:10.1108/09684880410561604

Longge, Z., & Xiangjie, L. (2013). The synchronization between two discrete-time chaotic systems using active robust model predictive control. *Nonlinear Dynamics*, *74*(4), 905–910. doi:10.100711071-013-1009-2

Lopesino, C., Balibrea-Iniesta, F., Wiggins, S., & Mancho, A. M. (2015). The Chaotic Saddle in the Lozi Map, Autonomous and Nonautonomous Versions. *International Journal of Bifurcation and Chaos in Applied Sciences and Engineering*, *25*(13), 1550184. doi:10.1142/S0218127415501849

Lorenz, E. (2000). The butterfly effect. World Scientific.

Lorenz, E. N. (1963). Deterministic nonperiodic flow. *Journal of the Atmospheric Sciences*, *20*(2), 130–141. doi:10.1175/1520-0469(1963)020<0130:DNF>2.0.CO;2

Lorenz, E. N. (2008). Compound windows of the Hénon-map. *Physica D. Nonlinear Phenomena*, *237*(13), 1689–1704. doi:10.1016/j.physd.2007.11.014

Lozi, R. (1978). Un attracteur étrange (?) du type attracteur de Hénon. *Journal de Physique. Colloque*, *39*(C5), C5–C9. doi:10.1051/jphyscol:1978505

Lucke, M., Stief, A., Chioua, M., Ottewill, J. R., & Thornhill, N. F. (2020). Fault detection and identification combining process measurements and statistical alarms. *Control Engineering Practice*, *94*, 104195.

Luo, M., & Liang, J. (2018). A Novel Similarity measure for interval-valued intuitionistic fuzzy sets and its applications. *Symmetry*, *10*(10), 441. doi:10.3390ym10100441

Lv, Y. (2012). Routing in wireless sensor networks using swarm intelligence. *Proceedings of the 10th World Congress on Intelligent Control and Automation*. 10.1109/WCICA.2012.6357964

Lv, Y. (2014). An energy efficient routing based on swarm intelligence for wireless sensor networks. *Journal of Software*, *9*(10). Advance online publication. doi:10.4304/jsw.9.10.2758-2763

Maani, K. (2017). *Multi-Stakeholder Decision Making for Complex Problems: A Systems Thinking Approach With Cases*. World Scientific. doi:10.1142/9294

Mackintosh, M., Jarvis, R., & Henry, E. (1994, November). On Managing Hybrids: Some Dilemmas in Higher Education Management. *Financial Management & Accountability*, *10*(4), 339–353. doi:10.1111/j.1468-0408.1994.tb00156.x

MacMillan, I. C., Zemann, L., & Subbanarasimha, P. N. (1987). Criteria distinguishing successful from unsuccessful ventures in the venture screening process. *Journal of Business Venturing*, *2*(2), 123–137. doi:10.1016/0883-9026(87)90003-6

Mahboubi, H., Momeni, A., Aghdam, A. G., Sayrafian-Pour, K., & Marbukh, V. (2012). An efficient target monitoring scheme with controlled node mobility for sensor networks. *IEEE Transactions on Control Systems Technology*, *20*(6), 1522–1532. doi:10.1109/TCST.2011.2167151

Maheshwari, S., & Srivastava, A. (2016). Study on divergence measures for intuitionistic fuzzy sets and its application in medical diagnosis. *Journal of Applied Analysis and Computation*, *6*(3), 772–789. doi:10.11948/2016050

Mahmood, A. (2019). Gazing Lean through the lens of System of Systems Dynamics: A case of weaving mill. *Journal of Engineered Fibers and Fabrics*, *14*, 1–16. doi:10.1177/1558925019870705

Mahmoud, M. I., El-Araby, S. M., Shehata, S. A., AbouZaid, R. M. F., & El-Samie, F. A. (2015). Design and Implementation of a Fast General Purpose Fuzzy Processor. In Research Methods: Concepts, Methodologies, Tools, and Applications (pp. 543-558). IGI Global. doi:10.4018/978-1-4666-7456-1.ch025

Mahmoudi, C., Flah, A., & Sbita, L. (2017). Prototype design of a compact plug in solar Electric Vehicle. *International Conference on Green Energy Conversion Systems (GECS)*, 1-4. http://batteryuniversity.com/learn/article/electric_vehicle

Ma, J., Song, X., Jin, W., & Wang, C. (2015). Autapse-induced synchronization in a coupled neuronal network. *Chaos, Solitons, and Fractals*, *80*, 31–38. doi:10.1016/j.chaos.2015.02.005

Majhi, S. K. (2018). An efficient feed foreword network model with sine cosine algorithm for breast cancer classification. *International Journal of System Dynamics Applications*, *7*(2), 1–14. doi:10.4018/IJSDA.2018040101

Maji, S. (2003). Quality Assurance in Technical Education, Possibilities through E-Learning in Distance Education. *The Indian Journal of Technical Education*, *26*(2), 19–25.

Makariye, N. (2017). Towards shortest path computation using Dijkstra algorithm. *2017 International Conference on IoT and Application (ICIOT).* 10.1109/ICIOTA.2017.8073641

Malhotra, R. (2016). An empirical framework for defect prediction using machine learning techniques with Android software. *Appl. Soft Comput.*, *49*(C), 1034–1050.

Maltz, E. N., Murphy, K. E., & Hand, M. L. (2007). Decision support for university enrollment management: Implementation and experience. *Decision Support Systems*, *44*(1), 106123. doi:10.1016/j.dss.2007.03.008

Mandl, C. E. (2019). *Managing Complexity in Social Systems*. Springer., doi:10.1007/978-3-030-01645-6

Mani, R.S., & Devi, U. (2009). Quality Assurance in Education through Quality Circles. *Global and Indian Context. Online Journal*, *7*(4), 1–17.

Manjula, C., & Florence, L. (2018). Deep neural network-based hybrid approach for software defect prediction using software metrics. *Cluster Computing*, *22*(S4), 9847–9863. doi:10.100710586-018-1696-z

Mann, R. J. (2006). *Do Patents Facilitate Financing in the Software Industry?* Social Science Research Network. https://papers.ssrn.com/abstract=510103

Mann, R. J., & Sager, T. W. (2007). Patents, venture capital, and software start-ups. *Research Policy*, *36*(2), 193–208. doi:10.1016/j.respol.2006.10.002

Mariapresenti, L., & Arockiarani, I. (2018). A New Approach Of Interval Valued Vague Set In Plants. *International Journal of Pure and Applied Mathematics*, *120*(7), 325–333.

Marino, R., Peresada, S., & Valigi, P. (1993). Adaptive input-output linearizing control of induction motors. *IEEE Transactions on Automatic Control*, *38*(2), 208–221. doi:10.1109/9.250510

Martínez-Guerra, R., Corona-Fortunio, D. M. G., & Mata-Machuca, J. L. (2013). Synchronization of chaotic Liouvillian systems: An application to Chua's oscillator. *Applied Mathematics and Computation*, *219*(23), 10934–10944. doi:10.1016/j.amc.2013.04.056

Mata-Machuca, J. L., Martínez-Guerra, R., & Aguilar-López, R. (2011). Chaotic systems synchronization via high order observer design. *Journal of Applied Research and Technology*, *9*(1), 57–68. doi:10.22201/icat.16656423.2011.9.01.448

Mata-Machuca, J., Martinez-Guerra, R., & Aguilar-Lopez, R. (2010). An exponential polynomial observer for synchronization of chaotic systems. *Communication and Nonlinear Scieciences Numerical Simulation*, *15*(12), 4114–4130. doi:10.1016/j.cnsns.2010.01.040

Matos, S., & Silvestre, B. S. (2013). Managing stakeholder relations when developing sustainable business models: The case of the Brazilian energy sector. *Journal of Cleaner Production*, *45*, 61–73. doi:10.1016/j.jclepro.2012.04.023

May, B. (2012). Applying Lean Startup: An Experience Report – Lean amp;amp; Lean UX by a UX Veteran: Lessons Learned in Creating amp;amp; Launching a Complex Consumer App. *Agile Conference (AGILE)*, 2012, 141–147. 10.1109/Agile.2012.18

Mazzarol, T. (1998). Critical success factors for international education marketing. *International Journal of Educational Management*, *12*(4), 163–175. doi:10.1108/09513549810220623

McCune, R. R., & Madey, G. R. (2014). Control of artificial swarms with DDDAS. *Procedia Computer Science*, *29*, 1171–1181. doi:10.1016/j.procs.2014.05.105

Meadows, D. (1999). *Leverage Points: Places to Intervene in a System*. The Sustainability Institute. Retrieved from http://donellameadows.org/archives/leverage-points-places-to-intervene-in-a-system/

Meadows, D. (2008). *Thinking in Systems: A Primer* (D. Wright, Ed.). Chelsea Green Publishing.

Mearns, M. A., Du Toit, A. S. A., & Mukuka, G. (2006). Conservation of indigenous knowledge at cultural villages: An exploratory study. *Mousaion*, *24*(1), 23–50.

Meenakshi, A. R., & Kaliraja, M. (2011). An application of interval valued fuzzy matrices in medical diagnosis. *International Journal of Mathematical Analysis.*, *5*(36), 1791–1802.

Meghni, B., Dib, D., & Azar, A. T. (2017). A Second-order sliding mode and fuzzy logic control to Optimal Energy Management in PMSG Wind Turbine with Battery Storage. *Neural Computing & Applications*, *28*(6), 1417–1434. doi:10.100700521-015-2161-z

Mekki, H, Boukhetala, D., & Azar, A.T. (2015). Sliding Modes for Fault Tolerant Control. *Studies in Computational Intelligence, 576*, 407-433. DOI doi:10.1007/978-3-319-11173-5_15

Melkou, L., & Hamerlain, L. (2014). Classical sliding and generalized variable structure controls for a manipulator robot arm with pneumatic artificial muscles. *International Journal of System Dynamics Applications*, *3*(1), 1–24. doi:10.4018/ijsda.2014010103

Meng, Z., Sun, C., An, Y., & Chen, R. (2011, August). Nonlinear decoupling control of induction motor based on parameter adaptive identification. In *Intelligent Human-Machine Systems and Cybernetics (IHMSC), 2011 International Conference on* (Vol. 2, pp. 98-102). IEEE. 10.1109/IHMSC.2011.94

Menger, P. M. (1999). Artistic labor markets and careers. *Annual Review of Sociology*, *25*(1), 541–574. doi:10.1146/annurev.soc.25.1.541

Merah, L., Ali-Pacha, A., Hadj-Said, N., Mecheri, B., & Dellassi, M. (2017). FPGA hardware co-simulation of new chaos-based stream cipher based on Lozi map. *International. Journal of Engineering Technology*, *9*(5), 420–425.

Meranza-Castillón, M. O., Murillo-Escobar, M. A., López-Gutiérrez, R. M., & Cruz-Hernández, C. (2019). Pseudorandom number generator based on enhanced Hénon map and its implementation. *AEÜ. International Journal of Electronics and Communications*, *107*, 239–251. doi:10.1016/j.aeue.2019.05.028

Merriam-Webster. (2020). *Nexus*. Author.

Meziane, S. (2009). *Commande Adaptative et Prédictive de la Machine Asynchrone* (Doctoral dissertation). Constantine University.

Miao, Y., Hwang, I., Liu, M., & Wang, F. (2019). Adaptive fast nonsingular terminal sliding mode control for attitude tracking of flexible spacecraft with rotating appendage. *Aerospace Science and Technology*, *93*, 1270–9638. doi:10.1016/j.ast.2019.105312

Michelitsch, M., & Rössler, O. (1989). A new feature in Hénon's map. *Computers & Graphics, 13*(2), 263–265. doi:10.1016/0097-8493(89)90070-8

Mikyeong, P., & Hong, E. (2014). Software Fault Prediction Model using Clustering Algorithms Determining the Number of Clusters Automatically. *International Journal of Software Engineering and Its Applications, 8*, 199–204.

Milić, S. D. (2020). Fuzzy-Decision Algorithms for Cyber Security Analysis of Advanced SCADA and Remote Monitoring Systems. In Cyber Security of Industrial Control Systems in the Future Internet Environment (pp. 131-155). IGI Global.

Milinski, M. (2016). Reputation, a universal currency for human social interactions. *Philosophical Transactions of the Royal Society B: Biological Sciences, 371*(1687). doi:10.1098/rstb.2015.0100

Millennium Institute. (n.d.). *iSDG - Integrated Sustainable Development Goals Model.* Retrieved May 6, 2020, from https://www.millennium-institute.org/isdg

Millonas, M. (1993). Swarms, Phase Transitions, and Collective Intelligence. Addison Wesley. arXiv:adap-org/9306002

Mindil, A., Rodrigues, A., McCormick, D., & Mwangi, R. (2014, January-March). ICT Powered Strategic Flexibility System Dynamic Model: A Pillar for Economic Development in Micro and Small Enterprises. *International Journal of System Dynamics Applications, 3*(1), 90–110. doi:10.4018/ijsda.2014010105

Ministry of Electricity & Energy New & Renewable Energy Authority (NREA). (n.d.). *Annual Report 2019/020.* Author.

Minten, B., Koru, B., & Stifel, D. (2013). The last mile(s) in modern input distribution: Pricing, profitability, and adoption. *Agricultural Economics, 44*(6), 629–646. doi:10.1111/agec.12078

Mirembe, N. M. (2001). *Theatre and the market in Uganda* (Doctoral dissertation). Available from Makerere university Institutional repository: http://hdl.handle.net/10570/1671

Mirjalili, S., Mirjalili, S. M., & Lewis, A. (2014). Grey wolf optimizer. *Advances in Engineering Software, 69*, 46–61. doi:10.1016/j.advengsoft.2013.12.007

Mishra, V., Das, S., Jafari, H., & Ong, S. H. (2016). Study of fractional order Van der Pol equation. *Journal of King Saud University - Science, 28*(1), 55-60.

Mishra, D., & Mishra, A. (2009). Software process improvement in SMEs: A comparative view. *Computer Science and Information Systems, 6*(1), 111–140. doi:10.2298/CSIS0901111M

Misiurewicz, M. (1980). Strange attractors for the lozi mappings. *Annals of the New York Academy of Sciences, 357*(1), 348–358. doi:10.1111/j.1749-6632.1980.tb29702.x

Misra, S. C., Kumar, V., & Kumar, U. (2009). Identifying some important success factors in adopting agile software development practices. *Journal of Systems and Software, 82*(11), 1869–1890. doi:10.1016/j.jss.2009.05.052

Mitchell, M. (2011). *Complexity: A Guided Tour.* Oxford University Press.

Mittal, S., Azar, A. T., & Kamal, N. A. (2021). Nonlinear Fractional Order System Synchronization via Combination-Combination Multi-switching. In *Proceedings of the International Conference on Advanced Intelligent Systems and Informatics 2020. AISI 2020. Advances in Intelligent Systems and Computing*, (vol 1261). Springer. 10.1007/978-3-030-58669-0_75

Mjahed, S., Bouzaachane, K., Azar, A. T., Hadaj, S. E., & Raghay, S. (2020). Hybridization of Fuzzy and Hard Semi-Supervised Clustering Algorithms Tuned with Ant Lion Optimizer Applied to Higgs Boson Search. *CMES-Computer Modeling in Engineering & Sciences, 125*(2), 459–494. doi:10.32604/cmes.2020.010791

Mnasser, A., Bounani, F., Manar, T. E., & Ksouri, M. (2014). Neural networls predictive controller using an adaptive control rate. *International Journal of System Dynamics Applications*, *3*(3), 1–21. doi:10.4018/ijsda.2014070106

Moaddy, K., Hashim, I., & Momani, S. (2011). Non-standard finite difference schemes for solving fractional-order Rössler chaotic and hyperchaotic systems. *Computers & Mathematics with Applications (Oxford, England)*, *62*(3), 1068–1074. doi:10.1016/j.camwa.2011.03.059

Mobayen, S., Baleanu D., & Tchier, F. (2016). Second-order fast terminal sliding mode control design based on LMI for a class of non-linear uncertain systems and its application to chaotic systems. *Journal of Vibration and Control*. . doi:10.1177/1077546315623887

Mobus, G. E., & Kalton, M. C. (2015). *Principles of Systems Science*. Springer. doi:10.1007/978-1-4939-1920-8

Moffatt, I., & Hanley, N. (2001). Modeling sustainable development: Systems dynamic and input-ouput approaches. *Environmental Modelling & Software*, *16*(6), 545–557. doi:10.1016/S1364-8152(01)00024-X

Mohamed & Ibrahim. (2012). Optimal Sizing and Economic Analysis of Stand-Alone Photovoltaic System. *The 2012 World Congress on Power and Energy Engineering (WCPEE'12)*, *4*(5), 188-195.

Mohamed, Ibrahim, Youssef, & Rakha. (2014). Optimal Sizing and Economic Analysis of Different Configurations of Photovoltaic Systems. *International Review of Electrical Engineering*, *4*(9), 111-121.

Mohamed, N. A., Azar, A. T., Abbas, N. E., Ezzeldin, M. A., & Ammar, H. H. (2020) Experimental Kinematic Modeling of 6-DOF Serial Manipulator Using Hybrid Deep Learning. In *Proceedings of the International Conference on Artificial Intelligence and Computer Vision (AICV2020). AICV 2020*. Springer.

Mohamed, N., Aymen, F., Ben Hamed, M., & Lassaad, S. (2019). Analysis of battery-EV state of charge for a dynamic wireless charging system. *Energy Storage, 2020*, *2*(2). doi:10.1002/est2.117

Mohamed, R. G., Ebrahim, M. A., & Bendary, F. M. (2017). *Transient Stability Enhancement for 20 MW PV Power Plant via Incremental Conductance Controller*. doi:10.4018/IJSDA.2017100106

Mohamed, N., Aymen, F., & Mouna, B. H. (2018). Wireless Charging System for a Mobile Hybrid Electric Vehicle. *2018 International Symposium on Advanced Electrical and Communication Technologies (ISAECT)*, 1–5. 10.1109/ISAECT.2018.8618829

Mohamed, R. G., Ebrahim, M. A., Bendary, F. M., & Osman, S. A. A. (2017). Transient Stability Enhancement for 20 MW PV power plant via incremental conductance controller. *International Journal of System Dynamics Applications*, *6*(4), 102–123.

Mohamed, R. G., Ebrahim, M. A., Bendary, F. M., & Osman, S. A. A. (2017). Transient Stability Enhancement for 20 MW PV Power Plant via Incremental Conductance Controller. *International Journal of System Dynamics Applications*, *6*(4), 102–123.

Mohrehkesh, S., & Nadeem, T. (2011). Toward a wireless charging for battery electric vehicles at traffic intersections. *IEEE Conference on Intelligent Transportation Systems, Proceedings, ITSC*, 113-118. 10.1109/ITSC.2011.6083137

Mohtar, R. H., & Daher, B. (2012). Water, Energy, and Food: The Ultimate Nexus. In *Encyclopedia of Agricultural, Food, and Biological Engineering* (2nd ed., pp. 1–15). Taylor & Francis.

Mohtar, R. H., & Daher, B. (2016). Water-Energy-Food Nexus Framework for facilitating multi-stakeholder dialogue. *Water International*, *41*(5), 655–661. doi:10.1080/02508060.2016.1149759

Mokhtabad, S. M. (2013). A cultural economic model for privatizing theater in Iran. *Journal of Art and Architecture, 167*, 55-57. (in Persian)

Mollona, E. (2008). Strategic Modelling and Business Dynamics. A Feedback Systems Approach. *The Journal of Management and Governance*, (12), 213–217. doi:10.100710997-008-9050-0

Montero, S., Martin, R., Mansilla, R., Cocho, G., & Nieto-Villar, J. M. (2018). Parameters estimation in phase-space landscape reconstruction of cell fate: A systems biology approach. In *Systems Biology* (pp. 125–170). Humana Press. doi:10.1007/978-1-4939-7456-6_8

Montoya, D. G., Carlos, A. R. P., & Giral, R. (2016). Maximum power point tracking of photovoltaic systems based on the sliding mode control of the module admittance. *Electric Power Systems Research, Elsevier, 136*, 125–134. doi:10.1016/j.epsr.2016.02.001

Morcillo, J., Angulo, F., & Franco, C. J. (2020). Analysing the Hydroelectricity Variability on Power Markets from a System Dynamics and Dynamic Systems Perspective : Seasonality and. *Energies, 13*(2381), 1–25. doi:10.3390/en13092381

Morcillo, J., Franco, C., & Angulo, F. (2017). Delays in electricity market models. *Energy Strategy Reviews, 16*, 24–32. doi:10.1016/j.esr.2017.02.004

Morecroft, J. D., & Sterman, J.D. (Eds.). (1992). Modelling for Learning. European Journal of Operations Research, 59.

Morecroft, J. D. W. (1986). The dynamics of a fledgling high-technology growth market: Understanding and managing growth cycles. *System Dynamics Review, 2*(1), 36–61. doi:10.1002dr.4260020104

Moulay, E., & Perruquetti, W. (2006). Finite time stability and stabilization of a class of continuous systems. *Journal of Mathematical Analysis and Applications, 323*(2), 1430–1443. doi:10.1016/j.jmaa.2005.11.046

Mouly, C. M. C. (2003). Quality in Technical Education, A Critical analysis of the Governing Factors. *The Indian Journal of Technical Education, 26*(2), 50–54.

Mounir, D., Azar, A. T., Drid, S., & Mehdi, D. (2018). Higher Order Sliding Mode Control for Blood Glucose Regulation of Type 1 Diabetic Patients. *International Journal of System Dynamics Applications, 7*(1), 65–84. doi:10.4018/IJSDA.2018010104

Mousa, M. E., Ebrahim, M. A., & Hassan, M. A. M. (2015). Stabilizing and Swinging-Up the Inverted Pendulum Using PI and PID Controllers Based on Reduced Linear Quadratic Regulator Tuned by PSO. *International Journal of System Dynamics Applications, 4*(4), 52–69. doi:10.4018/IJSDA.2015100104

Mousa, M. E., Ebrahim, M. A., & Hassan, M. M. (2017). Optimal Fractional Order Proportional Integral Differential Controller for Inverted Pendulum with Reduced Order Linear Quadratic Regulator. In *Fractional Order Control and Synchronization of Chaotic Systems* (pp. 225–252). Springer International Publishing. doi:10.1007/978-3-319-50249-6_8

Mousavi Agah, S. M., & Abbasi, A. (2012). The impact of charging plug-in hybrid electric vehicles on residential distribution transformers. *2012 2nd Iranian Conference on Smart Grids, ICSG 2012, 25*(1), 371–380.

Moustafa, S., ElNainay, M. Y., El Makky, N., & Abougabal, M. S. (2018). Software bug prediction using weighted majority voting techniques. *Alexandria Engineering Journal, 57*(4), 2763–2774. doi:10.1016/j.aej.2018.01.003

Moyo, D. (2009). Dead Aid: Why Aid Is Not Working and How There Is Another Way For Africa. New York: Academic Press.

Muchande, S., Kadam, A., Unni, K., & Thale, S. (2013). Design and Implementation of a Direct Torque Control Space Vector Modulated Three Phase Induction Motor Drive. In *Advances in Computing, Communication, and Control* (pp. 659–672). Springer. doi:10.1007/978-3-642-36321-4_61

Müller, A., Janetschek, H., & Weigelt, J. (2015). Towards a governance heuristic for sustainable development. *Current Opinion in Environmental Sustainability*, *15*, 49–56. doi:10.1016/j.cosust.2015.08.007

Munoz-Pacheco, J. M., Volos, C., Campos-Canton, E., Azar, A. T., Pham, V. T., & Radwan, A. G. (2017). New Trends on Modeling, Design, and Control of Chaotic Systems. *Mathematical Problems in Engineering*, *2017*, 3737681. Advance online publication. doi:10.1155/2017/3737681

Murali, K., & Lakshmanan, M. (1998). Secure communication using a compound signal from generalized synchronizable chaotic systems. *Physics Letters. [Part A]*, *241*(6), 303–310. doi:10.1016/S0375-9601(98)00159-5

Murillo-Escobar, M. A., Cruz-Hernández, C., Cardoza-Avendaño, L., & Méndez-Ramírez, R. (2017). A novel pseudorandom number generator based on pseudorandomly enhanced logistic map. *Nonlinear Dynamics*, *87*(1), 407–425. doi:10.100711071-016-3051-3

Nabavi, Branch, Balochian, & Branch. (2018). The Stability of a Class of Fractional Order Switching System with Time-Delay Actuator. *International Journal of System Dynamics Applications, 7*(1), 85-99.

Naglaa, K. (2014). Load Frequency Control in Power System via Improving PID Controller Based on Particle Swarm Optimization and ANFIS Techniques. *International Journal of System Dynamics Applications*, *3*(3), 1–24. doi:10.4018/ijsda.2014070101

Najm, A. A., Ibraheem, I. K., Azar, A. T., & Humaidi, A. J. (2020). Genetic Optimization-Based Consensus Control of Multi-Agent 6-DoF UAV System. *Sensors (Basel)*, *20*(12), 3576. doi:10.339020123576 PMID:32599862

Namujju, D. L., Yücel, G., Pruyt, E., & Okou, R. (2015). A Simulation-Based Analysis of Electricity Access in Uganda. *International Journal of System Dynamics Applications*, *4*(1), 1–30.

Naoui, M. (2019). Inductive charger efficiency under internal and external parameters variation for an electric vehicle in motion. *International Journal of Powertrains*, *8*(4), 343–358. doi:10.1504/IJPT.2019.104674

Naseradinmousavi, P., & Nataraj, C. (2012). Transient chaos and crisis phenomena in butterfly valves driven by solenoid actuators. *Communications in Nonlinear Science and Numerical Simulation*, *17*(11), 4336–4345. doi:10.1016/j.cnsns.2012.01.034

Nasution, F. B., Bazin, N. E., Rosalyn, R., & Hasanuddin, H. (2018). Public Policymaking Framework Based on System Dynamics and Big Data. *International Journal of System Dynamics Applications*, *7*(4), 38–53. doi:10.4018/IJSDA.2018100103

Natrajan, R. (2000). The Role of Accreditation in Promoting Quality Assurance of Technical Education. *International Journal of Engineering Education*, *16*(2), 88–96.

Natrajan, R. (2003). The Role of Social and Societal Responsibility as a Core value of the University and Corporate Sectors. *The Indian Journal of Technical Education*, *26*(1), 11–21.

Nepomuceno, E. G., Nardo, L. G., Arias-Garcia, J., Butusov, D. N., & Tutueva, A. (2019). Image encryption based on the pseudo-orbits from 1D chaotic map. *Chaos (Woodbury, N.Y.)*, *29*(6), 061101. doi:10.1063/1.5099261 PMID:31266342

New York Times. (1946, May 25). Atomic Education Urged by Einstein: Scientist in Plea for $200,000 to Promote New Type of Essential Thinking. *New York Times*, p. 11. Retrieved from https://timesmachine.nytimes.com/timesmachine/1946/05/25/100998236.html?pageNumber=11

Newman, M. E. J. (2018). *Networks: An Introduction* (2nd ed.). Oxford University Press. doi:10.1093/oso/9780198805090.001.0001

Newton, C. M. (1980). Biomathematics in oncology: Modeling of cellular systems. *Annual Review of Biophysics and Bioengineering*, *9*(1), 541–579. doi:10.1146/annurev.bb.09.060180.002545 PMID:6994595

Nguyen, S., & Park, A. (2020). A Comparison of Machine Learning Algorithms of Big Data for Time Series Forecasting Using Python. In Open Source Software for Statistical Analysis of Big Data: Emerging Research and Opportunities (pp. 197-218). IGI Global.

Nguyen, B. H., Xue, B., & Zhang, M. (2020). A survey on swarm intelligence approaches to feature selection in data mining. *Swarm and Evolutionary Computation*, *54*, 100663. doi:10.1016/j.swevo.2020.100663

Nguyen, D. X., & Massagué, J. (2007). Genetic determinants of cancer metastasis. *Nature Reviews. Genetics*, *8*(5), 341–352. doi:10.1038/nrg2101 PMID:17440531

Nguyen-Duc, A., Seppänen, P., & Abrahamsson, P. (2015). Hunter-gatherer Cycle: A Conceptual Model of the Evolution of Software Startups. *Proceedings of the 2015 International Conference on Software and System Process*, 199–203. 10.1145/2785592.2795368

Nicolay, D. (2014). *Asymptotic Chaos Expansions in Finance: Theory and Practice*. Springer-Verlag. doi:10.1007/978-1-4471-6506-4

Nightingale, D., & Srinivasan, J. (2011). *Beyond the lean revolution: achieving successful and sustainable enterprise transformation*. AMACOM.

Nikabadi, M. S., & Hajihoseinali, A. (2018). E-Waste Recycling System in Closed Loop Supply Chain. *International Journal of System Dynamics Applications*, *7*(2), 55–80. doi:10.4018/IJSDA.2018040104

Nikulin, M. S., Commenges, D., & Huber, C. (Eds.). (2006). *Probability, statistics and modelling in public health*. Springer. doi:10.1007/b136588

Niles, R. (1991). Pre- College Education Project UROP Evaluation D-4229. In MIT System Dynamics Group Literature Collection. System Dynamics Society.

Nilsson, M. (2017). *Important interactions among the Sustainable Development Goals under review at the High-Level Political Forum 2017* (No. 2017–6). Retrieved from www.sei-international.org

Nilsson, M. (2017). *Important interactions among the Sustainable Development Goals*. High-Level Political Forum 2017 (No. 2017-06). Retrieved from www.sei-international.org

Nilsson, M., Griggs, D., Visbeck, M., Ringler, C., Mcollum, D., Mollier, L., … Unger, S. (2017). A Guide to SDG Interactions. In Science to Implementation. Paris, France: International Council for Science. doi:10.24948/2017.01

Nilsson, M., Griggs, D., & Visback, M. (2016). Map the interactions between Sustainable Development Goals. *Nature*, *534*(15), 320–322. doi:10.1038/534320a PMID:27306173

Nilsson, M., Lucas, P., & Yoshida, T. (2013). Towards an integrated framework for SDGs: Ultimate and enabling goals for the Case of energy. *Sustainability*, *5*(10), 4124–4151. doi:10.3390u5104124

Njajra, Z., Flah, A., Elhak Chariag, D., & Sbita, L. (2017). A PMDRM decoupling vector control strategy for EVT drive in hybrid vehicles. *International Conference on Green Energy Conversion Systems (GECS)*. 10.1109/GECS.2017.8066227

Nodenof, T., Latorcade, P., & Sallaberry, C. (2004). Model based Engineering of Learning for Adaptive Web Based Educational Systems. *Proceedings for the 13th International World Wide Web Conference*. Available from http://Oportal.acm.org.lispac.1sbu.ac.uk/

Noel, M., & Schankerman, M. (2013). Strategic Patenting and Software Innovation. *The Journal of Industrial Economics*, *61*(3), 481–520. doi:10.1111/joie.12024

North Carolina Climate Office. (2019). *Global Patterns*. Retrieved May 5, 2020, from https://climate.ncsu.edu/climate/patterns

Nosoohi, I., & Parvizian, J. (2013). Effects of Discount Scenarios on Chaotic Behavior of Inventory Level Under Price-Dependent Demand. *International Journal of System Dynamics Applications*, *2*(3), 58–72. doi:10.4018/ijsda.2013070104

Novic, A. (2011). Variable structure systems with sliding modes in motion control—A survey. *IEEE Transactions on Industrial Informatics*, *7*(2), 212–223. doi:10.1109/TII.2011.2123907

Nurmuliani, N., Zowghi, D., & Powell, S. (2004). Analysis of requirements volatility during software development life cycle. *2004 Australian Software Engineering Conference. Proceedings.*, 28–37. 10.1109/ASWEC.2004.1290455

Nusse, H. E., & Yorke, J. A. (2012). *Dynamics: numerical explorations: accompanying computer program dynamics* (Vol. 101). Springer.

O'Neil, K. M. (2008). Bringing art to market: The diversity of pricing styles in a local art market. *Poetics*, *36*(1), 94–113. doi:10.1016/j.poetic.2007.12.002

Ochoa, C., Dyner, I., & Franco, C. J. (2013). Simulating power integration in Latin America to assess challenges, opportunities, and threats. *Energy Policy*, *61*, 267–273. doi:10.1016/j.enpol.2013.07.029

Ochoa, C., & van Ackere, A. (2014). Does size matter? Simulating electricity market coupling between Colombia and Ecuador. *Renewable & Sustainable Energy Reviews*, *50*, 1108–1124. doi:10.1016/j.rser.2015.05.054

Oh, H., & Kim, K. (2017). Customer satisfaction, service quality, and customer value: Years 2000-2015. *International Journal of Contemporary Hospitality Management*, *29*(1), 2–29. doi:10.1108/IJCHM-10-2015-0594

Olariu, S., & Zomaya, A. Y. (2013). *Handbook of Bioinspired algorithms and application* (3rd ed.). Chapman and Hall/CRC.

Oliva, R. (2003). Model calibration as a testing strategy for system dynamics models. *European Journal of Operational Research*, *151*(3), 552–568. doi:10.1016/S0377-2217(02)00622-7

Omamo, A. O., Rodrigues, A. J., & Muliaro, W. J. (2020). A System Dynamics Model of Technology and Society: In the Context of a Developing Nation. *International Journal of System Dynamics Applications*, *9*(2), 42–63. doi:10.4018/IJSDA.2020040103

Omamo, A. O., Rodriguez, A. J., & Muliaro, J. W. (2018). A Systems Dynamics Model for Mobile Industry Governance in the Context of the Kenyan Vision 2030. *International Journal of System Dynamics Applications*, *7*(2), 81–100. doi:10.4018/IJSDA.2018040105

Omogbai, O., & Salonitis, K. (2016). A lean assessment tool based on systems dynamics. *26th CIRP Design Conference*, *50*, 106–111. 10.1016/j.procir.2016.04.169

Omogbai, O., & Salonitis, K. (2017). The Implementation of 5S Lean Tool Using System Dynamics Approach. *Procedia CIRP*, *60*, 380–385. doi:10.1016/j.procir.2017.01.057

Onggo, S. (2012). Adult social care workforce analysis in England: A system dynamics approach. *International Journal of System Dynamics Applications, 1*(4), 1–20. doi:10.4018/ijsda.2012100101

Orojloo, H., Moghadam, R. A., & Haghighat, A. T. (2012). Energy and path aware ant colony optimization based routing algorithm for wireless sensor networks. *Communications in Computer and Information Science, 269*, 182–191. doi:10.1007/978-3-642-29219-4_22

Ortega, R., Pyrkin, A., Bobtsov, A., Efimov, D., & Aranovskiy, S. (2020). A Globally Convergent Adaptive Indirect Field-Oriented Torque Controller for Induction Motors. *Asian Journal of Control, 22*(1), 11–24.

Osterwalder, A., & Pigneur, Y. (2010). *Business model generation: A handbook for visionaries, game changers, and challengers*. John Wiley & Sons.

Ott, E. (1981). Strange attractors and chaotic motions of dynamical systems. In *The Theory of Chaotic Attractors* (pp. 103–119). Springer. doi:10.1007/978-0-387-21830-4_9

Ouannas, A., Azar, A. T., & Radwan, A. G. (2016). On Inverse Problem of Generalized Synchronization Between Different Dimensional Integer-Order and Fractional-Order Chaotic Systems. In *The 28th International Conference on Microelectronics*. IEEE. 10.1109/ICM.2016.7847942

Ouannas, A., Azar, A. T., Ziar, T., & Radwan, A. G. (2017c). Study on coexistence of different types of synchronization between different dimensional fractional chaotic systems. In Fractional Order Control and Synchronization of Chaotic Systems, Studies in Computational Intelligence (Vol. 688, pp. 637-6690). Springer Verlag.

Ouannas, A., Grassi, G., & Azar, A. T. (2020b). A New Generalized Synchronization Scheme to Control Fractional Chaotic Systems with Non-identical Dimensions and Different Orders. In *The International Conference on Advanced Machine Learning Technologies and Applications (AMLTA2019). AMLTA 2019. Advances in Intelligent Systems and Computing*, (vol 921, pp. 415-424). Springer. 10.1007/978-3-030-14118-9_42

Ouannas, A., Grassi, G., & Azar, A. T. (2020c). Fractional-Order Control Scheme for Q-S Chaos Synchronization. In *The International Conference on Advanced Machine Learning Technologies and Applications (AMLTA2019). AMLTA 2019. Advances in Intelligent Systems and Computing*, (vol 921, pp. 434-441). Springer. 10.1007/978-3-030-14118-9_44

Ouannas, A., Grassi, G., Azar, A. T., & Gasri, A. (2019b). A New Control Scheme for Hybrid Chaos Synchronization. In *Proceedings of the International Conference on Advanced Intelligent Systems and Informatics 2018. AISI 2018. Advances in Intelligent Systems and Computing*, (vol 845). Springer. 10.1007/978-3-319-99010-1_10

Ouannas, A., Grassi, G., Azar, A. T., & Singh, S. (2019a). New Control Schemes for Fractional Chaos Synchronization. In *Proceedings of the International Conference on Advanced Intelligent Systems and Informatics 2018. AISI 2018. Advances in Intelligent Systems and Computing*, (vol 845). Springer. 10.1007/978-3-319-99010-1_5

Ouannas, A., Grassi, G., Azar, A. T., Khennaouia, A. A., & Pham, V. T. (2020a). Chaotic Control in Fractional-Order Discrete-Time Systems. In *AISI 2019. The International Conference on Advanced Intelligent Systems and Informatics AISI 2019. Advances in Intelligent Systems and Computing*, (vol 1058, pp. 207-217). Springer. 10.1007/978-3-030-31129-2_20

Ouannas, A., Grassi, G., Azar, A. T., Khennaouia, A. A., & Pham, V. T. (2020d). Synchronization of Fractional-Order Discrete-Time Chaotic Systems. In *The International Conference on Advanced Intelligent Systems and Informatics AISI 2019. Advances in Intelligent Systems and Computing*, (vol 1058, pp. 218-228). Springer. 10.1007/978-3-030-31129-2_21

Ouannas, A., Grassi, G., Azar, A. T., Radwan, A. G., Volos, C., Pham, V. T., Ziar, T., Kyprianidis, I. M., & Stouboulos, I. N. (2017g). Dead-Beat Synchronization Control in Discrete-Time Chaotic Systems. *The 6th International Conference on Modern Circuits and Systems Technologies (MOCAST)*. 10.1109/MOCAST.2017.7937628

Ouannas, A., Azar, A. T., & Abu-Saris, R. (2016a). A new type of hybrid synchronization between arbitrary hyperchaotic maps. *International Journal of Machine Learning and Cybernetics*. Advance online publication. doi:10.100713042-016-0566-3

Ouannas, A., Azar, A. T., & Vaidyanathan, S. (2017a). New hybrid synchronisation schemes based on coexistence of various types of synchronisation between master-slave hyperchaotic systems. *International Journal of Computer Applications in Technology*, *55*(2), 112–120. doi:10.1504/IJCAT.2017.082868

Ouannas, A., Azar, A. T., & Vaidyanathan, S. (2017c). A Robust Method for New Fractional Hybrid Chaos Synchronization. *Mathematical Methods in the Applied Sciences*, *40*(5), 1804–1812. doi:10.1002/mma.4099

Ouannas, A., Azar, A. T., & Vaidyanathan, S. (2017d). A new fractional hybrid chaos synchronization. Int. J. Modelling. *Identification and Control*, *27*(4), 314–322. doi:10.1504/IJMIC.2017.084719

Ouannas, A., Azar, A. T., & Vaidyanathan, S. (2017g). On A Simple Approach for Q-S Synchronization of Chaotic Dynamical Systems in Continuous-Time. Int. J. *Computing Science and Mathematics*, *8*(1), 20–27.

Ouannas, A., Azar, A. T., Vaidyanathan, S., & Radwan, A. G. (2016b). New Control Schemes of Synchronization for Incommensurate and Commensurate Fractional Order Chaotic Systems. *Proceedings of the 4th International Conference on Control Engineering & Information Technology CEIT '16.*

Ouannas, A., Azar, A. T., & Ziar, T. (2019c). Control of Continuous-Time Chaotic (Hyperchaotic) Systems: F-M Synchronization. *Int. J. Automation and Control*, *13*(2), 226–242. doi:10.1504/IJAAC.2019.098215

Ouannas, A., Azar, A. T., & Ziar, T. (2020f). On Inverse Full State Hybrid Function Projective Synchronization for Continuous-time Chaotic Dynamical Systems with Arbitrary Dimensions. *Differential Equations and Dynamical Systems*, *28*(4), 1045–1058. doi:10.100712591-017-0362-x

Ouannas, A., Azar, A. T., Ziar, T., & Radwan, A. G. (2017e). *A Study On Coexistence of Different Types of Synchronization Between Different dimensional Fractional Chaotic Systems. Studies in Computational Intelligence* (Vol. 688). Springer-Verlag.

Ouannas, A., Azar, A. T., Ziar, T., & Radwan, A. G. (2017f). *Generalized Synchronization of Different Dimensional Integer-order and Fractional Order Chaotic Systems. Studies in Computational Intelligence* (Vol. 688). Springer-Verlag.

Ouannas, A., Azar, A. T., Ziar, T., & Vaidyanathan, S. (2017a). Fractional inverse generalized chaos synchronization between different dimensional systems. *Fractional Order Control and Synchronization of Chaotic Systems Studies in Computational Intelligence*, *688*, 525–551.

Ouannas, A., Azar, A. T., Ziar, T., & Vaidyanathan, S. (2017b). A new method to synchronize fractional chaotic systems with different dimensions. *Fractional Order Control and Synchronization of Chaotic Systems Studies in Computational Intelligence*, *688*, 581–611.

Ouannas, A., Azar, A. T., Ziar, T., & Vaidyanathan, S. (2017b). *On New Fractional Inverse Matrix Projective Synchronization Schemes. Studies in Computational Intelligence* (Vol. 688). Springer-Verlag.

Ouannas, A., Azar, A. T., Ziar, T., & Vaidyanathan, S. (2017c). *Fractional Inverse Generalized Chaos Synchronization Between Different Dimensional Systems. Studies in Computational Intelligence* (Vol. 688). Springer-Verlag.

Ouannas, A., Azar, A. T., Ziar, T., & Vaidyanathan, S. (2017d). *A New Method To Synchronize Fractional Chaotic Systems With Different Dimensions. Studies in Computational Intelligence* (Vol. 688). Springer-Verlag.

Ouannas, A., Azar, A. T., Ziar, T., & Vaidyanathan, S. (2017f). Fractional Inverse Generalized Chaos Synchronization Between Different Dimensional Systems. *Studies in Computational Intelligence*, *688*, 525–551. doi:10.1007/978-3-319-50249-6_18

Ouannas, A., Azar, A. T., Ziar, T., & Vaidyanathan, S. (2017g). A New Method To Synchronize Fractional Chaotic Systems With Different Dimensions. *Studies in Computational Intelligence*, *688*, 581–611. doi:10.1007/978-3-319-50249-6_20

Ouannas, A., Grassi, G., Azar, A. T., & Khennaoui, A. A. (2020e). Synchronization Control in Fractional Discrete-Time Systems with Chaotic Hidden Attractors. In A. Hassanien, R. Bhatnagar, & A. Darwish (Eds.), *Advanced Machine Learning Technologies and Applications. AMLTA 2020. Advances in Intelligent Systems and Computing* (Vol. 1141, pp. 661–669). Springer.

Ouannas, A., Grassi, G., Azar, A. T., Khennaouia, A. A., & Pham, V. T. (2020d). Synchronization of Fractional-Order Discrete-Time Chaotic Systems. In *The International Conference on Advanced Intelligent Systems and Informatics AISI 2019. Advances in Intelligent Systems and Computing*, (vol 1058, pp. 218-228). Springer.

Owlia, M. S., & Aspinwall, E. M. (1996). Quality in higher education – a survey. Total Quality Management, 7(2).

Owlia, M. S., & Aspinwall, E. M. (1998). A framework for measuring quality in engineering education. Total Quality Management, 7(2).

Own, M. (2009). Switching between type-2 fuzzy sets and intuitionistic fuzzy sets: An application in medical diagnosis. *Applied Intelligence*, *31*(1), 283–291. doi:10.100710489-008-0126-y

Oyo, B. (2008). A system Dynamics Tool for Higher Education Funding and Quality Policy Analysis. *Proceedings of System Dynamics Conference.*

Pai, R. R., Hebbar, S., Kamath, V., & Kamath, G. (2013). Improvement of Process Productivity through Just-in-Time. *Research Journal of Management Sciences*, *2*(12), 1–6.

Palomba, F., Zanoni, M., Fontana, F. A., Lucia, A. D., & Oliveto, R. (2019). Toward a smell-aware bug prediction model. *IEEE Transactions on Software Engineering*, *45*(2), 194–218. doi:10.1109/TSE.2017.2770122

Panda, M. (2019). Software defect prediction using hybrid distribution base balance instance selection and radial basis function classifier. *International Journal of System Dynamics Applications*, *8*(3), 53–75. doi:10.4018/IJSDA.2019070103

Panda, M., & Patra, M. R. (2009). Evaluating machine learning algorithms for detecting network intrusions. *International Journal of Recent Trends in Engineering*, *1*(1), 472–477.

Pandey, B., Rhmann, W., Ansari, G. & Pandey, D. K. (2019). Software fault prediction based on change metrics using hybrid algorithms: An empirical study. *Journal of King Saud University – Computer and Information Sciences.* doi:10.1016/j.jksuci.2019.03.006

Pandey, D. N. (1998). *Ethno-forestry: Local Knowledge for Sustainable Forestry and Livelihood Security*. Himanshu/AFN.

Pandey, S., Dwivedi, B., & Tripathi, A. (2016). Performance analysis of SMC controlled PV fed boost converter. *7th India International conference on In Power Electronics (IICPE),* 1-4.

Pandey, S., & Srivastava, S. (2014). Data Driven Enterprise UX: A Case Study of Enterprise Management Systems. In S. Yamamoto (Ed.), *Human Interface and the Management of Information. Information and Knowledge in Applications and Services* (pp. 205–216). Springer International Publishing., doi:10.1007/978-3-319-07863-2_21

Pani, G., Galeotti, T., & Chiarugi, P. (2010). Metastasis: Cancer cell's escape from oxidative stress. *Cancer and Metastasis Reviews*, *29*(2), 351–378. doi:10.100710555-010-9225-4 PMID:20386957

Pant, A. S. (2013). *National Workshop on Quality Assurance and Accreditation*. Academic Press.

Papachristos, G. (2019). System dynamics modelling and simulation for socio-technical transitions research. *Environmental Innovation and Societal Transitions, 31*(September), 248–261. doi:10.1016/j.eist.2018.10.001

Papakostas, A., Hatzimichailidis, A. G., & Kaburlasos, V. G. (2013). Distance and similarity measures between intuitionistic fuzzy sets: A comparative analysis from a pattern recognition point of view. *Pattern Recognition Letters, 34*(14), 1609–1622. doi:10.1016/j.patrec.2013.05.015

Park, H., Lim, K. M., Park, J. S., & Kwun, Y. C. (2008). Distances between interval-valued intuitionistic fuzzy sets. *Journal of Physics: Conference Series, 96*, 012089. doi:10.1088/1742-6596/96/1/012089

Parlitz, U. (2016). Estimating Lyapunov Exponents from Time Series. In *Chaos Detection and Predictability* (pp. 1–34). Springer., doi:10.1007/978-3-030-31129-2_21.

Parry, M. E., & Song, X. M. (1994). Identifying new product successes in China. *Journal of Product Innovation Management, 11*(1), 15–30. doi:10.1111/1540-5885.1110015

Paternoster, N., Giardino, C., Unterkalmsteiner, M., Gorschek, T., & Abrahamsson, P. (2014). Software development in startup companies: A systematic mapping study. *Information and Software Technology, 56*(10), 1200–1218. doi:10.1016/j.infsof.2014.04.014

Patidar, V., Sud, K. K., & Pareek, N. K. (2009). A pseudo random bit generator based on chaotic logistic map and its statistical testing. *Informatica (Vilnius), 33*(4).

Pham, V. T., Gokul, P. M., Kapitaniak, T., Volos, C., & Azar, A. T. (2018) Dynamics, Synchronization and Fractional Order Form of a Chaotic System With Infinite Equilibria. In Mathematical Techniques of Fractional Order Systems, Advances in Nonlinear Dynamics and Chaos (ANDC) series, (pp. 475-502). Elsevier.

Pham, V. T., Gokul, P. M., Kapitaniak, T., Volos, C., & Azar, A. T. (2018). Dynamics, Synchronization and Fractional Order Form of a Chaotic System With Infinite Equilibria. In Mathematical Techniques of Fractional Order Systems, Advances in Nonlinear Dynamics and Chaos (ANDC) series, (pp. 475-502). Elsevier.

Pham, V. T., Vaidyanathan, S., Volos, C. K., Azar, A. T., Hoang, T. M., & Yem, V. V. (2017). A Three-Dimensional No-Equilibrium Chaotic System: Analysis, Synchronization and Its Fractional Order Form. Studies in Computational Intelligence, 688, 449-470.

Pham, V. T., Vaidyanathan, S., Volos, C. K., Azar, A. T., Hoang, T. M., & Yem, V. V. (2017a). *A Three-Dimensional No-Equilibrium Chaotic System: Analysis, Synchronization and Its Fractional Order Form. Studies in Computational Intelligence* (Vol. 688). Springer-Verlag.

Pham, V. T., Volos, C. K., Vaidyanathan, S., & Azar, A. T. (2017b). Dynamics, Synchronization and Fractional Order Form of a Chaotic System without Equilibrium. In C. K. Volos (Ed.), *Nonlinear Systems: Design, Applications and Analysis*. Nova Science Publishers.

Pham, V. T., Volos, C., & Vaidyanathan, S. (2017). A Chaotic Time-Delay System With Saturation Nonlinearity. *International Journal of System Dynamics Applications*, *6*(3), 111–129.

Pham, V. T., Volos, C., & Vaidyanathan, S. (2017). A chaotic time-delay system with saturation nonlinearity. *International Journal of System Dynamics Applications*, *6*(3), 1–19. doi:10.4018/IJSDA.2017070106

Pikovsky, A., Rosenblum, M., & Kurths, J. (2001). *Synchronization: A universal concept in nonlinear sciences*. Cambridge University Press.

Pilla, R., Azar, A. T., & Gorripotu, T. S. (2019). Impact of Flexible AC Transmission System Devices on Automatic Generation Control with a Metaheuristic Based Fuzzy PID Controller. *Energies*, *12*(21), 4193. doi:10.3390/en12214193

Pilla, R., Botcha, N., Gorripotu, T. S., & Azar, A. T. (2020). Fuzzy PID Controller for Automatic Generation Control of Interconnected Power System Tuned by Glow-Worm Swarm Optimization. In J. Nayak, V. Balas, M. Favorskaya, B. Choudhury, S. Rao, & B. Naik (Eds.), *Applications of Robotics in Industry Using Advanced Mechanisms. ARIAM 2019. Learning and Analytics in Intelligent Systems* (Vol. 5, pp. 140–149). Springer. doi:10.1007/978-3-030-30271-9_14

Pišek, P., Štumberger, B., Marčič, T., & Virtič, P. (2013). Design analysis and experimental validation of a double rotor synchronous PM machine used for HEV. *IEEE Transactions on Magnetics*, *49*(1), 152–155. doi:10.1109/TMAG.2012.2220338

Poles, R. (2013). System Dynamics modelling of a production and inventory system for remanufacturing to evaluate system improvement strategies. *International Journal of Production Economics*, *144*(1), 189–199. doi:10.1016/j.ijpe.2013.02.003

Pomerol, J.-C. (2003). *From Human Decision Making to DMSS Architecture*. www.igi-global.com/chapter/human-decision-making-dmssarchitecture/

Pomerol, J.-C. (1997, May 16). Artificial intelligence and human decision making. *European Journal of Operational Research*, *99*(1), 3–25. doi:10.1016/S0377-2217(96)00378-5

Poppendieck, M., & Cusumano, M. A. (2012). Lean Software Development: A Tutorial. *IEEE Software*, *29*(5), 26–32. doi:10.1109/MS.2012.107

Poste, G., & Fidler, I. J. (1980). The pathogenesis of cancer metastasis. *Nature*, *283*(5743), 139–146. doi:10.1038/283139a0 PMID:6985715

Potash, P., & Heinbokel, J. (2005). Unleashing the Revolutionary Implications of a System Dynamics Education. *The 23rd International Conference of the System Dynamics Society*.

Pottie, G. J., & Kaiser, W. J. (2000). Wireless integrated network sensors. *Communications of the ACM*, *43*(5), 51–58. doi:10.1145/332833.332838

Pourpanah, F., Wang, R., Lim, C. P., Wang, X., Seera, M., & Tan, C. J. (2019). An improved fuzzy ARTMAP and Q-learning agent model for pattern classification. *Neurocomputing*, *359*, 139–152.

Prasad, K. (2009). Justification of world class maintence systems using analytic hieracy constant sum method. Journal of Quality in maintence. *Engineering (London)*, *15*(1), 47–77.

Premalatha, H. M., & Srikrishna, C. V. (2018). Software Fault Prediction and Classification using Cost based Random Forest in Spiral Life Cycle Model. *International Journal of Intelligent Engineering and Systems*, *11*(2), 1–8. doi:10.22266/ijies2018.0430.02

Project Strobe: Protecting your data, improving our third-party APIs, and sunsetting consumer Google+. (2018, October 8). Google. https://blog.google/technology/safety-security/project-strobe

Pruyt, E. (2017). Integrating Systems Modelling and Data Science: The Joint Future of Simulation and 'Big Data' Science. Artificial Intelligence: Concepts, Methodologies, Tools, and Applications, 822–840. doi:10.4018/978-1-5225-1759-7.ch033

Pruyt, E. (2016, January-March). Integrating Systems Modelling and Data Science: The Joint Future of Simulation and 'Big Data' Science. *International Journal of System Dynamics Applications*, *5*(1), 1–16. Advance online publication. doi:10.4018/IJSDA.2016010101

Pruyt, E., & Logtens, T. (2015). Model-based exploration of societal aging in the Netherlands. *International Journal of System Dynamics Applications*, *4*(1), 57–95. doi:10.4018/IJSDA.2015010103

Qudrat-Ullah, H. (2009). System Dynamics Based Technology for Decision Support. Encyclopedia of Information Science and Technology, Second Edition, 3647–3650. doi:10.4018/978-1-60566-026-4.ch581

Qudrat-Ullah, H., & Seong, B. S. (2010). How to do structural validity of a system dynamics type simulation model: The case of an energy policy model. *Energy Policy*, *38*(5), 2216–2224. doi:10.1016/j.enpol.2009.12.009

Quiceno, G., Álvarez, C., Ávila, R., Fernández, Ó., Franco, C. J., Kunc, M., & Dyner, I. (2019). Scenario analysis for strategy design: A case study of the Colombian electricity industry. *Energy Strategy Reviews, 23*(December), 57–68. doi:10.1016/j.esr.2018.12.009

Qu, Z. (2011). Chaos in the genesis and maintenance of cardiac arrhythmias. *Progress in Biophysics and Molecular Biology*, *105*(3), 247–257. doi:10.1016/j.pbiomolbio.2010.11.001 PMID:21078337

Rabiee, H., Ataei, M., & Ekramian, M. (2019). Continuous nonsingular terminal sliding mode control based on adaptive sliding mode disturbance observer for uncertain nonlinear systems. *Automatica, 109.*

Radomes, A. A., & Arango, S. (2015). Renewable energy technology diffusion: An analysis of photovoltaic-system support schemes in Medellín, Colombia. *Journal of Cleaner Production*, *92*, 152–161. doi:10.1016/j.jclepro.2014.12.090

Radzicki, M. J. (2009). System dynamics and its contribution to economics and economic modeling. In *Complex Systems in Finance and Econometrics* (pp. 727–737). Springer. doi:10.1007/978-1-4419-7701-4_39

Rafikov, M., & Balthazar, J. M. (2008). On control and synchronization in chaotic and hyperchaotic systems via linear feedback control. *Communications in Nonlinear Science and Numerical Simulation*, *13*(7), 1246–1255. doi:10.1016/j.cnsns.2006.12.011

Rahimi, A., Kumar, K. D., & Alighanbari, H. (2020). Fault detection and isolation of control moment gyros for satellite attitude control subsystem. *Mechanical Systems and Signal Processing*, *135*, 106419.

Rahmandad, H., & Weiss, D. M. (2009). Dynamics of concurrent software development. *System Dynamics Review*, *25*(3), 224–249. doi:10.1002dr.425

Raie, A., & Rashtchi, V. J. E. E. (2002). *Using a genetic algorithm for detection and magnitude determination of turn faults in an induction motor*. Academic Press.

Raisbeck, I. (1994). Royal mail: developing a total quality organization. In D. Green (Ed.), *What is Quality in Higher Education?* SRHE and Open University Press.

Rajkumar, G., & Alagarsamy, K. (2013). The Most Common Factors for The Failure of Software development Project. International Journal of Computer. *Science and Applications*, *1*(11), 74–77.

Rajput, J.S. & Walia, K. (1997). Quality in Higher Education. *The Journal of Higher Education*, *20*(4), 553–567.

Ramos, A., Gago, A., Labandeira, X., & Linares, P. (2015). The role of information for energy efficiency in the residential sector. *Energy Economics*, *52*, S17–S29.

Rana, Z. A., Mian, M. A., & Shamail, S. (2015). Improving Recall of software defect prediction models using association mining. *Knowledge-Based Systems*, *90*, 1–13. doi:10.1016/j.knosys.2015.10.009

Rand, D. G., & Nowak, M. A. (2013). Human cooperation. *Trends in Cognitive Sciences*, *17*(8), 413–425. doi:10.1016/j.tics.2013.06.003 PMID:23856025

Ranger, S. (2015, July 9). *Microsoft and Windows Phone: What went wrong, and where can they go from here?* ZDNet. https://www.zdnet.com/article/microsoft-and-windows-phone-what-went-wrong-and-where-can-they-go-from-here/

Rashed, M., MacConnell, P. F., & Stronach, A. F. (2006). Nonlinear adaptive state-feedback speed control of a voltage-fed induction motor with varying parameters. *IEEE Transactions on Industry Applications*, *42*(3), 723–732. doi:10.1109/TIA.2006.872951

Raumer, T. V., Dion, J. M., & Dugard, L. (1993). Adaptive non-linear speed control of induction motors. *International Journal of Adaptive Control and Signal Processing*, *7*(5), 435–455. doi:10.1002/acs.4480070511

Redondo, J. M., Olivar, G., Ibarra-Vega, D., & Dyner, I. (2018). Modeling for the regional integration of electricity markets. *Energy for Sustainable Development*, *43*, 100–113. doi:10.1016/j.esd.2017.12.003

Rego, E. E., & de Oliveira Ribeiro, C. (2018). Successful Brazilian experience for promoting wind energy generation. *The Electricity Journal*, *31*(2), 13–17. doi:10.1016/j.tej.2018.02.003

Reidel. Olver, P. (1993). *Applications of Lie groups to differential equations*. Springer. doi:10.1007/978-1-4612-4350-2

Ren, J -H & Liu, F. (2019). Predicting Software Defects Using Self-Organizing Data Mining. *IEEE Access*, *7*(2019), 122796 – 122810. DOI-. doi:10.1109/ACCESS.2019.2927489

Repenning, N. P., & Sterman, J. D. (2002). Capability traps and self-confirming attribution errors in the dynamics of process improvement. *Administrative Science Quarterly*, *47*(2), 265–295. doi:10.2307/3094806

Rezk, A. A., Madian, A. H., Radwan, A. G., & Soliman, A. M. (2020). Multiplierless chaotic Pseudo random number generators. *AEÜ. International Journal of Electronics and Communications*, *113*, 152947. doi:10.1016/j.aeue.2019.152947

Rhmann, W. (2018). Cross-project defect prediction using hybrid search-based algorithms. *Int. J. Inf. Tecnol.* doi:10.100741870-018-0244-7

Ribatti, D., Vacca, A., & Dammacco, F. (1999). The role of the vascular phase in solid tumor growth: A historical review. *Neoplasia (New York, N.Y.)*, *1*(4), 293–302. doi:10.1038j.neo.7900038 PMID:10935483

Richardson, G. P., & Andersen, D. F. (2010). Systems Thinking, Mapping, and Modeling in Group Decision and Negotiation. In C. Eden & D. N. Kilgour (Eds.), *Handbook for Group Decision and Negotiation* (pp. 313–324). Springer. doi:10.1007/978-90-481-9097-3_19

Richmond, B. (1994). Systems thinking/system dynamics: let's get on with it. In *1994 International Systems Dynamics Conference*, (vol. 10, pp. 135–157). 10.1002dr.4260100204

Richter, H. (2002). The generalized Hénon maps: Examples for higher-dimensional chaos. *International Journal of Bifurcation and Chaos in Applied Sciences and Engineering*, *12*(06), 1371–1384. doi:10.1142/S0218127402005121

Rickles, D., Hawe, P., & Shiell, A. (2007). A simple guide to chaos and complexity. *Journal of Epidemiology and Community Health*, *61*(11), 933–937. doi:10.1136/jech.2006.054254 PMID:17933949

Ries, E. (2011). *The Lean Startup: How Today's Entrepreneurs Use Continuous Innovation to Create Radically Successful Businesses*. Crown Publishing Group. https://books.google.com.au/books?id=tvfyz-4JILwC

Ringler, C., Bhaduri, A., & Lawford, R. (2013). The nexus across water, energy, land and food (WELF): Potential for improved resource use efficiency? *Current Opinion in Environmental Sustainability*, *5*(6), 617–624. doi:10.1016/j.cosust.2013.11.002

Rintamäki, T., & Kirves, K. (2017). From perceptions to propositions: Profiling customer value across retail contexts. *Journal of Retailing and Consumer Services*, *37*(October), 159–167. doi:10.1016/j.jretconser.2016.07.016

Ritchie-Dunham, J. L., & Rabbino, H. T. (2001). *Managing from Clarity: Identifying, Aligning, and Leveraging Strategic Resources*. John Wiley.

Rittel, H., & Webber, M. L. (1973). *Dilemmas in a General Theory of Planning. Policy Sciences 4*. Elsevier Science.

Rizzo, I., & Throsby, D. (2006). Cultural heritage: economic analysis and public policy. Handbook of the Economics of Art and Culture, 1, 983-1016.

Roberts, N. H. (Ed.). (1976). System Dynamics Curriculum Development Project for Elementary and Secondary Education D-2435. In MIT System Dynamics Group Literature Collection. System Dynamics Society.

Roberts, N. H. (Ed.). (1978). System Simulation of Student Performance in the Elementary Classroom, In Management Applications of System Dynamics. Productivity Press.

Roberts, N. H. (Ed.). (1978). System Simulation of Student Performance in the Elementary Classroom. In Management Applications of System Dynamics. Productivity Press.

Robertson, D. P., & Hull, R. B. (2001). Beyond biology: Toward a more public ecology for conservation. *Conservation Biology*, *15*(4), 970–979. doi:10.1046/j.1523-1739.2001.015004970.x

Robson, C. (2002). Real world research: A Resource for Social Scientists and Practitioner-Researchers (2nd ed.). Blackwell.

Rocca, P., Benedetti, M., Donelli, M., Franceschini, D., & Massa, A. (2009). Evolutionary optimization as applied to inverse scattering problems. *Inverse Problems*, *25*(12), 123003. doi:10.1088/0266-5611/25/12/123003

Rodden, K., Hutchinson, H., & Fu, X. (2010). Measuring the User Experience on a Large Scale: User-centered Metrics for Web Applications. *Proceedings of the SIGCHI Conference on Human Factors in Computing Systems*, 2395–2398. 10.1145/1753326.1753687

Rodriguez, A., De Leon, J., & Fridman, L. (2005). Synchronization in reduced-order of chaotic systems via control approaches based on high-order sliding-mode observer. *Chaos Soliton Fractals, 42*, 3219–3233. doi:10.1007/978-3-319-50249-6_22

Roller, L. H., & Waverman, L. (2001). Telecommunications infrastructure and economic growth:A simultaneous approach. *The American Economic Review*, *91*(4), 909–923. doi:10.1257/aer.91.4.909

Rössler, O. E. (1979). An equation for hyperchaos. *Physics Letters. [Part A]*, *71*(2), 155–157. doi:10.1016/0375-9601(79)90150-6

Rosu, S. G. (2016). A Dynamic Wireless Charging System for Electric Vehicles Based on DC / AC Converters with SiC MOSFET-IGBT Switches and Resonant Gate-Drive. IEEE.

Rother, M., & Shook, J. (2009). *Learning to see: value stream mapping to add value and eliminate muda*. Lean Enterprise Institute.

Rothwell, R. (1994). Towards the fifth-generation innovation process. *International Marketing Review*, *11*(1), 7–31. doi:10.1108/02651339410057491

Roy, P., Goswami, S., Chakraborty, S., Azar, A. T., & Dey, N. (2014). Image Segmentation Using Rough Set Theory: A Review. *International Journal of Rough Sets and Data Analysis*, *1*(2), 62–74. doi:10.4018/ijrsda.2014070105

Ruan, X., Cai, J., & Chan, J. (2008). Learning to control two-wheeled self-balancing robot using reinforcement learning rules and fuzzy neural networks. In *Proceedings of 4th IEEE International Conference on Natural Computation*, (pp. 395-398). 10.1109/ICNC.2008.361

Ruan, X., Chan, J., Cai, J., & Dai, L. (2009). Balancing control of the two-wheeled upstanding robot using adaptive fuzzy control method. In *Proceedings of IEEE International Conference on Intelligent Computing and Intelligent Systems*, (pp 95-98). 10.1109/ICICISYS.2009.5358152

Rubino, D. (2014, February 23). *Windows Phone is losing some hardware differentiation as Microsoft woos Android manufacturers*. Windows Central. https://www.windowscentral.com/windows-phone-losing-hardware-differentiation

Rukhin, A., Soto, J., Nechvatal, J., Smid, M., & Barker, E. (2001). *A statistical test suite for random and pseudorandom number generators for cryptographic applications*. Booz-Allen and Hamilton Inc.

Runge, D. (1997). A Note on Teaching System Dynamics D-2653. In MIT System Dynamic Group Literature Collection. System Dynamics Society.

Ruth, B., Patrizio, C., Dugard, L., Allgöwer, F., Kleimenov, A., & Scherer, C. (2008). Trends in Theory of Control System Design. *Proceedings of the 17th World Congress, the International Federation of Automatic Control*.

Saarinen, T. F. (1969). *Perception of the Drought Hazard on the Great Plains*. University of Chicago Press.

Sachs, J., Schmidt-Traub, G., Kroll, C., Durand-Delacre, D., & Teksoz, K. (2016). *SDG Index & Dashboards - Global Report*. Bertelsmann Stiftung and Sustainable Development Solutions Network.

Sadeghi, G. (1997). The economy of Iran Theater. *Journal of Art, 34*, 53-83. (in Persian)

Sadeghi, G. (2011). This implies the death of theater. *Journal of Art and Architecture, 144*, 42-43. (in Persian)

Saeed, K. (1997). *System Dynamics as a Technology for New Liberal Education*, Worcester Polytechnic Institute, Report No.3.

Saeed, K. (1997). *System Dynamics as a Technology for New Liberal Education*. Worcester Polytechnic Institute, Report No. 3.

Saeed, K. (1990). Bringing Practicum to Learning: A System Dynamics Modelling Approach. *Higher Education Policy, 3*(2), 44–49. doi:10.1057/hep.1990.30

Saeed, K. (1993). Bringing experimental learning to the social sciences: A simulation laboratory on economic development. *System Dynamics Review, 9*(2), 153–164. doi:10.1002dr.4260090205

Saeed, K. (1996). The Dynamics of Collegial Systems in the Developing Countries. *Higher Education Policy, 9*(2), 75–86. doi:10.1016/S0952-8733(96)90040-9

Saeed, K. (1998). Maintaining professional competence in innovation organizations. *Human Systems Management, 17*, 69–87.

Salazar, P., & Martinez, M. M. (2007). Achievements teaching Systems Thinking and Systems Dynamics to Graduate students through e-learning. *Conference Proceedings, The 2007 International Conference of the System Dynamics Society and 50th Anniversary Celebrations*.

Saldana, J. (2016). *The Coding Manual for Qualitative Researchers* (J. Seaman, Ed.; 3rd ed.). SAGE Publications.

Saleem, M., & Agrawal, T. (2012). Chaos in a tumor growth model with delayed responses of the immune system. *Journal of Applied Mathematics, 2012*, 2012. doi:10.1155/2012/891095

Salem Khalifa, A. (2004). *Customer value: a review of recent literature and an integrative configuration. Management Decision* (Vol. 42). doi:10.1108/00251740410538497

Salima, M., Loubna, A., & Riad, T. (2018). A Global Stability of Linearizing Control of Induction Motor for PV Water Pumping Application. *International Journal of System Dynamics Applications, 7*(3), 31–56.

Sallam, O. K., Azar, A. T., Guaily, A., & Ammar, H. H. (2020). Tuning of PID controller using Particle Swarm Optimization for cross flow heat Exchanger based on CFD System Identification. In *The International Conference on Advanced Intelligent Systems and Informatics AISI 2019*. Springer. 10.1007/978-3-030-31129-2_28

Sambuc, R. (1975). *Fonctions f-floues aplication 'l' aide au diagnostic en pathologie thyroidienne* (Ph.D. Thesis). University of Marseille.

Samuel, E., & Balamurugan, M. (2012a). Intuitionistic fuzzy set with rank correlation technique in medical diagnosis, In *Proceedings of the International Conference on Mathematics in Engineering & Business Management*. Stella Maris College.

Samuel, E., & Balamurugan, M. (2012b). Intuitionistic fuzzy set in medical diagnosis using ranking function. *Surveys in Mathematics and Mathematical Science*, *2*(1), 23–34.

Samuel, E., & Balamurugan, M. (2013). IFS with N-Parameters In Medical Diagnosis. *International Journal of Pure and Applied Mathematics*, *84*(3), 185–192. doi:10.12732/ijpam.v84i3.5

Sanchez, E. (1979). Medical Diagnosis and Composite Fuzzy Relations. In Advances in Fuzzy Set Theory and Applications. Amsterdam: North-Holland.

Sánchez, M. A. (2014). Modeling for System´s Understanding. Formal Languages for Computer Simulation: Transdisciplinary Models and Applications, 38–61. doi:10.4018/978-1-4666-4369-7.ch002

Sanchez, E. (1976). Resolution of Composite Fuzzy Relation Equations. *Information and Control*, *30*(1), 38–48. doi:10.1016/S0019-9958(76)90446-0

Sang, Y., Sun, W., Duan, F., & Zhao, J. (2019). Bidirectional synchronization control for an electrohydraulic servo loading system. *Mechatronics*, *62*, 62. doi:10.1016/j.mechatronics.2019.102254

Santoro, A., Mancini, E., & Azar, A. T. (2013). Biofeedback Systems and Their Application in the Hemodialysis Therapy. In A. T. Azar (Ed.), *Biofeedback Systems and Soft Computing Techniques of Dialysis,* (Vol. 405, pp. 1081–1107). doi:10.1007/978-3-642-27558-6_6

Santos, Santos, Silva, & Ferreira. (2020). A System Dynamics Model for Sales and Operations Planning: An Integrated Analysis for the Lime Industry. *International Journal of System Dynamics Applications, 9*(1), 1-15.

Sarafian, G., & Kaplan, B. Z. (1995). Article. *IEEE Trans. Circ. Systems I*, *42*(6), 376–376. doi:10.1109/81.390276

Sarkar, J., & Bhattacharyya, S. (2012). Application of graphene and graphene-based materials in clean energy-related devices Minghui. *Archives of Thermodynamics*, *33*(4), 23–40. doi:10.2478/v10173-012-0026-8

Satsangi, A. (2011). Disaster Management. Academic Press.

Saufi, M. S. R. B. M., Ahmad, Z. A. B., Leong, M. S., & Lim, M. H. (2020). Gearbox fault diagnosis using a deep learning model with limited data sample. *IEEE Transactions on Industrial Informatics.*

Savov, V. (2012, January 26). *Microsoft paid Nokia $250 million to use Windows Phone in Q4 2011*. The Verge. https://www.theverge.com/2012/1/26/2736132/microsoft-nokia-250-million-platform-support-payment

Sawada. (2015). Energy Storage Management in Grid Connected Solar Photovoltaic System. *Journal of Engineering Research and Applications, 5*, 1-5.

Sayed, A. S., Azar, A. T., Ibrahim, Z. F., Ibrahim, H. A., Mohamed, N. A., & Ammar, H. H. (2020). Deep Learning Based Kinematic Modeling of 3-RRR Parallel Manipulator. In *Proceedings of the International Conference on Artificial Intelligence and Computer Vision (AICV2020). AICV 2020.* Springer.

Sayed, G. I., Hassanien, A. E., & Azar, A. T. (2019). Feature selection via a novel chaotic crow search algorithm. *Neural Computing & Applications*, *31*(1), 171–188. doi:10.100700521-017-2988-6

Sbita, L., & Flah, A. (2013). High speed PMSM control without speed encoder based on Q-MRAS observer. *CEIT-13EIT-13*, 174–178.

Schilderman, T. (2004). Adapting traditional shelter for disaster mitigation and reconstruction: experiences with community-based approaches. *Journal of Building Research and Information, 32*(5), 414-426.

Schmidt-Traub, G., Karoubi, E. de la M., & Espey, J. (2015). *Indicators and a Monitoring Framework for the Sustainable Development Goals: Launching a Data Revolution*. Sustainable Development Solutions Network.

Scholl, H. J. (2004). Action Research and System Dynamics: Can they benefit from each other.Paper presented at the 2004 Hawaiian Internal Conference on System Sciences,Waikoloa,Hawaii, 1-11. 10.1109/HICSS.2004.1265243

Schonfeld, E. (2011, July 11). *Google+ Added $20 Billion To Google's Market Cap | TechCrunch*. https://techcrunch.com/2011/07/10/google-plus-20-billion-market-cap/

Schot, S. H. (1978). The time rate of change of acceleration. *American Journal of Physics*, *46*(11), 1090–1094. doi:10.1119/1.11504

Schoumaker, B. (2004). Poverty and fertility in sub-Saharan Africa: evidence from 25 countries. In Intergovernmental Panel on Climate Change (Ed.), Population Association of America Meeting (pp. 1–30). Boston: Cambridge University Press. 10.1017/CBO9781107415324.004

Schwaber, K., & Beedle, M. (2001). *Agile Software Development with Scrum* (1st ed.). Prentice Hall PTR.

Schwaninger, M., & Grosser, S. (2008). System Dynamics as Model- Based Theory Building. *Systems Research and Behavioral Science*, *25*(4), 447–465. doi:10.1002res.914

Scott, J. (2000). *Social Network Analysis: A Handbook*. SAGE Publications.

Segan, S. (2014, February 27). *OEM: Windows Phone License Fee Plummeting*. PCMag Australia. https://au.pcmag.com/mobile-apps/4192/oem-windows-phone-license-fee-plummeting

Segura, M. G., Oleghe, O., & Salonitis, K. (2019). Analysis of lean manufacturing strategy using system dynamics modelling of a business model. *International Journal of Lean Six Sigma*. Advance online publication. doi:10.1108/ijlss-05-2017-0042

Sehgal, K., Jain, A., Nagrath, P., & Kumar, A. (2019). Recent advances in networks and data security survey on various mobile operating systems. *International Conference on Innovative Computing and Communications*, 181–190. 10.1007/978-981-13-2324-9_18

Sehl, K. (2020, March 2). *Everything Brands Need to Know About TikTok in 2020*. Hootsuite Social Media Management. https://blog.hootsuite.com/what-is-tiktok/

Senge, P. M. (1988). New System Dynamics Learning Tools for Management Education and Training D-3999. In MIT System Dynamics Group Literature Collection. System Dynamics Society.

Senge, P. (1990). *The Fifth Discipline*. The Art and Practice of the Learning Organisation, Currency Doubleday.

Senge, P. M., Kleiner, A., Robert, C., Ross, R. B., & Smith, B. J. (1994). *The Fifth Discipline Fieldbook: Strategies and Tools for Building a Learning Organization*. Marsh.

Serdar Asan, S., & Asan, U. (2007). Qualitative cross-impact analysis with time consideration. *Technological Forecasting and Social Change*, *74*(5), 627–644. doi:10.1016/j.techfore.2006.05.011

Shaffer, W. A. (1976). An Initial Concept for Organizing System Dynamics Curriculum D-2438. In MIT System Dynamics Group Literature Collection. System Dynamics Society.

Shafiei, M., & Hakaki, A. (2018). A Dynamic Model of Effective Factors on Open Innovation in Manufacturing Small and Medium Sized Companies. *International Journal of System Dynamics Applications*, *7*(1), 1–26. doi:10.4018/IJSDA.2018010101

Shahin, M., Saied, E., Hassan, M. M., & Bendary, F. (2014). Voltage swell mitigation using flexible AC transmission systems based on evolutionary computing methods. *International Journal of System Dynamics Applications*, *3*(3), 73–95. doi:10.4018/ijsda.2014070104

Shahverdiev, E. M. (2019). Effect of parameter mismatches on chaos synchronization between Josephson junctions coupled in series and driven by a central junction. *Physica C: Superconductivity and its Applications, 31*(1), 230-235.

Shahverdiev, E. M., Bayramov, P. A., & Qocayeva, M. V. (2013). Inverse chaos synchronization between bidirectionally coupled variable multiple time delay systems. *Optik (Stuttgart)*, *124*(18), 3427–3429. doi:10.1016/j.ijleo.2012.10.055

Shahverdiev, E. M., & Shore, K. A. (2009). Synchronization of chaos in unidirectionally and bidirectionally coupled multiple time delay laser diodes with electro-optical feedback. *Optics Communications*, *282*(2), 310–316. doi:10.1016/j.optcom.2008.10.009

Shaik, A. S., & Rodrigues, L. L. R. (2018). Plant layout optimisation with implementation of technical cleanliness in an automotive industry: A system dynamics approach. *Int. J. Technology. Policy and Management*, *18*(3), 201–221. doi:10.1504/IJTPM.2018.093850

Shanbhag, N., & Pardede, E. (2019). A Metrics Framework for Product Development in Software Startups. *Journal of Enterprising Culture*, *27*(03), 283–307. doi:10.1142/S0218495819500110

Shanbhag, N., & Pardede, E. (2019). The Dynamics of Product Development in Software Startups: The Case for System Dynamics. *International Journal of System Dynamics Applications*, *8*(2), 51–77. doi:10.4018/IJSDA.2019040104

Sharma, A., Verma, P., Mathew, L., & Chatterji, S. (2018). *Using Motor Current Analysis for Broken Rotor Bar Fault Detection in Rotary Machines*. Paper presented at the 2018 3rd International Conference on Communication and Electronics Systems (ICCES).

Sharma, A., Patidar, V., Purohit, G., & Sud, K. K. (2012). Effects on the bifurcation and chaos in forced Duffing oscillator due to nonlinear damping. *Communications in Nonlinear Science and Numerical Simulation*, *17*(6), 2254–2269. doi:10.1016/j.cnsns.2011.10.032

Sharvani, S., Cauvery, C., & Rangaswamy, T. (2009). Different types of swarm intelligence algorithm for routing. *2009 International Conference on Advances in Recent Technologies in Communication and Computing*. 10.1109/ARTCom.2009.157

Shattock, M. (1997, July). The New Managerial Implications of the New Priorities. *Higher Education Management*, *9*(2), 27–34.

Shattock, M. (1999, March). The Impact of the Dearing Report on UK Higher Education. *Higher Education Management, 11*(1), 7–17.

Shepperd, M., Bowes, D., & Hall, T. (2014). Researcher bias: The use of machine learning in software defect prediction. *IEEE Transactions on Software Engineering*, *40*(6), 603–616. doi:10.1109/TSE.2014.2322358

Shereen, A., Ramadan, R. A., & Ghali, N. I. (2016). Classification of EEG Signals for Motor Imagery based on Mutual Information and Adaptive Neuro Fuzzy Inference System. *International Journal of System Dynamics Applications*, *5*(4), 64–82. doi:10.4018/IJSDA.2016100104

Shetty, D. K., Abakari, G., Rodrigues, L. L. R., Oommen Mathew, A., & Motlagh, F. G. (2019). To develop Lean Systems in Electronic Automotive Parts Manufacturing Industry: A System Dynamics Modelling Approach. *2019 International Conference on Automation, Computational and Technology Management (ICACTM)*, 277–283. 10.1109/ICACTM.2019.8776803

Shi, Y. (2018). Unified swarm intelligence algorithms. *Critical Developments and Applications of Swarm Intelligence*, 1-26. doi:10.4018/978-1-5225-5134-8.ch001

Shi, H., Liu, J., Wu, Y., Zhang, K., Zhang, L., & Xue, P. (2016). Fault diagnosis of nonlinear and large-scale processes using novel modified kernel Fisher discriminant analysis approach. *International Journal of Systems Science*, *47*(5), 1095–1109.

Shi, P., Xia, Y. Q., Liu, G. P., & Rees, D. (2006). On designing of sliding-mode control for stochastic jump systems. *IEEE Transactions on Automatic Control*, *51*(1), 97–103. doi:10.1109/TAC.2005.861716

Shi, S. L., Li, J. X., & Fang, Y. M. (2018). Extended-state-observer-based chattering free sliding mode control for nonlinear systems with mismatched disturbance. *IEEE Access: Practical Innovations, Open Solutions*, *6*, 22952–22957. doi:10.1109/ACCESS.2018.2828868

Shtessel, Y. B., Shkolnikov, I. A., & Brown, M. D. J. (2003). An asymptotic second-order smooth sliding mode control. *Asian Journal of Control*, *5*(4), 498–504. doi:10.1111/j.1934-6093.2003.tb00167.x

Shukla, M. K., Sharma, B. B., & Azar, A. T. (2018). Control and Synchronization of a Fractional Order Hyperchaotic System via Backstepping and Active Backstepping Approach. In Mathematical Techniques of Fractional Order Systems, Advances in Nonlinear Dynamics and Chaos (ANDC) series, (pp. 597-624). Elsevier. doi:10.1016/B978-0-12-813592-1.00019-2

Siddique, N., & Adeli, H. (2013). *Computational intelligence: synergies of fuzzy logic, neural networks and evolutionary computing*. John Wiley & Sons.

Silva, A. R., Pimenta, F. M., Assireu, A. T., & Spyrides, M. H. C. (2016). Complementarity of Brazils hydro and offshore wind power. *Renewable & Sustainable Energy Reviews*, *56*, 413–427. doi:10.1016/j.rser.2015.11.045

Singh, S., & Azar, A. T. (2020a). Multi-switching Combination Synchronization of Fractional Order Chaotic Systems. In *Proceedings of the International Conference on Artificial Intelligence and Computer Vision (AICV2020). AICV 2020. Advances in Intelligent Systems and Computing*, (vol 1153, pp. 655-664). Springer. 10.1007/978-3-030-44289-7_61

Singh, S., & Azar, A. T. (2020b). Controlling Chaotic System via Optimal Control. In *The International Conference on Advanced Intelligent Systems and Informatics AISI 2019. Advances in Intelligent Systems and Computing*, (vol 1058, pp. 277-287). Springer. 10.1007/978-3-030-31129-2_26

Singh, S., Azar, A. T., Vaidyanathan, S., Ouannas, A., & Bhat, M. A. (2018a). Multiswitching Synchronization of Commensurate Fractional Order Hyperchaotic Systems Via Active Control. In Mathematical Techniques of Fractional Order Systems, Advances in Nonlinear Dynamics and Chaos (ANDC) series, (pp. 319-345). Elsevier. doi:10.1016/B978-0-12-813592-1.00011-8

Singh, K. V., Bansal, H. O., & Singh, D. (2019). A comprehensive review on hybrid electric vehicles: Architectures and components. *Journal of Modern Transportation*, *27*(2), 77–107. doi:10.100740534-019-0184-3

Singh, S., Azar, A. T., Bhat, M. A., Vaidyanathan, S., & Ouannas, A. (2018b). *Active Control for Multi-Switching Combination Synchronization of Non-Identical Chaotic Systems. Advances in System Dynamics and Control*. IGI-Global. doi:10.4018/978-1-5225-4077-9.ch005

Singh, S., Azar, A. T., Ouannas, A., Zhu, Q., Zhang, W., & Na, J. (2017). Sliding Mode Control Technique for Multi-switching Synchronization of Chaotic Systems. *Proceedings of 9th International Conference on Modelling, Identification and Control (ICMIC 2017)*, 880-885. 10.1109/ICMIC.2017.8321579

Singh, S., Azar, A. T., & Zhu, Q. (2018c). Multi-switching Master-Slave Synchronization of Nonidentical Chaotic Systems. In Q. Zhu, J. Na, & X. Wu (Eds.), *Innovative Techniques and Applications of Modelling, Identification and Control. Lecture Notes in Electrical Engineering* (Vol. 467, pp. 321–330). Springer. doi:10.1007/978-981-10-7212-3_20

Slotine, J. E. (1991). *Applied nonlinear control*. Prentice-Hall.

Soliman, M., Azar, A. T., Saleh, M. A., & Ammar, H. H. (2020) Path Planning Control for 3-Omni Fighting Robot Using PID and Fuzzy Logic Controller. In *The International Conference on Advanced Machine Learning Technologies and Applications (AMLTA2019). AMLTA 2019*. Springer.

Song, L. Z., Benedetto, C. D., & Song, M. (2010). Competitive Advantages in the First Product of New Ventures. *IEEE Transactions on Engineering Management*, *57*(1), 88–102. doi:10.1109/TEM.2009.2013836

Soni, S., & Chorasia, B. K. (2017). Policy Planning in Higher Technical Education: A System Dynamic Approach. *International Journal of System Dynamics Applications*, *6*(3), 87–110. doi:10.4018/IJSDA.2017070105

Spichkova, M., & Hamilton, M. (2016). Dynamic Decision Making System for Public Transport Routes. *International Journal of System Dynamics Applications, 5*(3), 47-70. doi:10.4018/IJSDA.2016070103

Sprott, J. C. (2000). Simple chaotic systems and circuits. *American Journal of Physics*, *68*(8), 758–763. doi:10.1119/1.19538

Sprott, J. C. (2010). *Elegant chaos: algebraically simple flow*. Singapore: World Scientific Publishing.

Sprott, J., & Xiong, A. (2015). Classifying and quantifying basins of attraction. *Chaos (Woodbury, N.Y.)*, *25*(8), 083101. doi:10.1063/1.4927643 PMID:26328552

Sreedhar, V. (2006). Plug-in hybrid electric vehicles with full performance. *2006 IEEE Conference on Electric and Hybrid Vehicles, ICEHV*, 1-2. 10.1109/ICEHV.2006.352291

Srinivasan, S., & LD, D. B. (2020). A Neuro-Fuzzy Approach to Detect Rumors in Online Social Networks. *International Journal of Web Services Research, 17*(1), 64–82.

Srinivas, Lakshmi, & Ramesh. (2016). Simulation of Incremental Conductance MPPT Algorithm for PV Systems using Lab View. *International Journal of Innovative Research in Electrical, Electronics. Instrumentation and Control Engineering*, *1*(4), 34–38.

Srivastava, A. (2019). Medical diagnosis using intuitionistic fuzzy sets. In. AIP Conference Proceedings: Vol. 2061. *No. 1* (p. 020029). AIP Publishing LLC.

Stadnicka, D., & Litwin, P. (2019). Value stream mapping and system dynamics integration for manufacturing line modelling and analysis. *International Journal of Production Economics*, *208*, 400–411. doi:10.1016/j.ijpe.2018.12.011

Stavrinides, S. G., Miliou, A. N., Anagnostopoulos, A. N., Konstantakos, V., Magafas, L., & Laopoulos, T. (2009). De-synchronization crisis induced intermittency in a master–slave PLL configuration. *Chaos, Solitons, and Fractals*, *42*(1), 33–39. doi:10.1016/j.chaos.2008.10.017

Stephan, R. M., Mohtar, R. H., Daher, B., Irujo, A. E., Hillers, A., Ganter, J. C., Karlberg, L., Martin, L., Nairizi, S., Rodriguez, D. J., & Sarni, W. (2018). Water–energy–food nexus: A platform for implementing the Sustainable Development Goals. *Water International*, *43*(3), 472–479. Advance online publication. doi:10.1080/02508060.2018.1446581

Sterman, J. D. (1984). Appropriate summary statistics for evaluating the historical fit of system dynamics models. *Dynamica*. Retrieved from https://www.systemdynamics.org/conferences/1983/proceed/plenary/sterm203.pdf

Sterman, J., & Sterman, J. D. (2000). *Business Dynamics: Systems Thinking and Modeling for a Complex World with CD-ROM* (HAR/CDR edition). McGraw-Hill Education.

Sterman, J.D. (1992). *Teaching Takes Off OR/MS Today*. Academic Press.

Sterman, J.D. (2000). *Business Dynamics. Systems Thinking and Modeling for a complex World.* Irwin McGraw-Hill.

Sterman, J. D. (1994). Learning in and about complex systems. *System Dynamics Review*, *10*(2–3), 291–330. doi:10.1002dr.4260100214

Sterman, J. D. (2000). *Business dynamics: systems thinking and modeling for a complex world.* Irwin/McGraw-Hill.

Sterman, J. D. (2000). *Business Dynamics: Systems Thinking and Modeling for a Complex World.* McGraw-Hill Higher Education.

Sterman, J. D. (2001). System Dynamics Modeling: Tools for Learning in a Complex World. *California Management Review*, *43*(4), 8–25. doi:10.2307/41166098

Sterman, J. D. (2002). All models are wrong: Reflections on becoming a system scientist. *System Dynamics Review*, *18*(4), 501531. doi:10.1002dr.261

Sterman, J. D. (2006). Learning from evidence in a complex world. *American Journal of Public Health*, *96*(3), 505–514. doi:10.2105/AJPH.2005.066043 PMID:16449579

Stoyanov, B., & Kordov, K. (2015). Novel secure pseudo-random number generation scheme based on two Tinkerbell maps. *Advanced Studies in Theoretical Physics*, *9*(9), 411–421. doi:10.12988/astp.2015.5342

Strogatz, S. (2015). *Nonlinear dynamics and chaos: with applications to physics, biology, chemistry, and engineering (studies in nonlinearity)*. Westview Press.

Strogatz, S. H. (2014). *Nonlinear dynamics and chaos: with applications to physics, biology, chemistry, and engineering*. Westview press.

Stuntz, L., & Lyneis, D. (2007). System Dynamics in K-12 Education: Lessons Learned. *Conference Proceedings, The 2007 International Conference of the System Dynamics Society and 50th Anniversary Celebration.*

Sudimac, B., Ugrinović, A., & Jurčević, M. (2020). The application of photovoltaic systems in sacred buildings for the purpose of electric power production: The case study of the Cathedral of St. Michael the Archangel in Belgrade. *Sustainability (Switzerland)*, *12*(4), 1408. doi:10.3390u12041408

Sureshchandar, G. S., & Rajendra, C. (2001). A holistic model for total quality service. *International Journal of Service Industry Management*, *12*(4), 378–412. doi:10.1108/09564230110405299

Sutton, S. M. (2000). The role of process in software start-up. *IEEE Software*, *17*(4), 33–39. doi:10.1109/52.854066

Swakman, V., Molleman, L., Ule, A., & Egas, M. (2016). Reputation-based cooperation: Empirical evidence for behavioral strategies. *Evolution and Human Behavior*, *37*(3), 230–235.

Sweeney, L., & Sterman, J. (2007). Thinking about systems: student and teacher conceptions of natural and social systems. System Dynamics Review. Accepted July.

Swinerd, C., & McNaught, K. R. (2014). Simulating the diffusion of technological innovation with an integrated hybrid agent-based system dynamics model. *Journal of Simulation*, *8*(3), 231–240. doi:10.1057/jos.2014.2

Sylwester, K., & Roberts, G. (2013). Reputation-based partner choice is an effective alternative to indirect reciprocity in solving social dilemmas. *Evolution and Human Behavior*, *34*(3), 201–206.

System Dynamics Society. (2016). *Introduction to System Dynamics.* Retrieved July 4, 2017, from https://www.systemdynamics.org/what-is-s/

Szmidt, E. (2014). *Distances and Similarities in Intuitionistic Fuzzy Sets.* Springer. doi:10.1007/978-3-319-01640-5

Szmidt, E., & Janusz, K. (2011). Intuitionistic fuzzy sets two and three term representations in the context of a hausdorff distance. *ACTA Univ. Matthiae BelII Ser. Math*, *19*, 53–62.

Szmidt, E., & Kacprzyk, J. (1997). On measuring distances between intuitionistic fuzzy sets. *Notes on IFS*, *3*(4), 1–13.

Szmidt, E., & Kacprzyk, J. (2000). Distances between intuitionistic fuzzy sets. *Fuzzy Sets and Systems*, *114*(3), 505–518. doi:10.1016/S0165-0114(98)00244-9

Szmidt, E., & Kacprzyk, J. (2001). Intuitionistic fuzzy sets in some medical applications. *Lecture Notes in Computer Science*, *2206*, 148–151. doi:10.1007/3-540-45493-4_19

Szmidt, E., & Kacprzyk, J. (2006). Distances between intuitionistic fuzzy sets: straightforward approaches may not work. *3rd International IEEE Conference Intelligent Systems IS06*, 716–721. 10.1109/IS.2006.348507

Szmidt, E., & Kacprzyk, J. (2008). Dilemmas with distances between intuitionistic fuzzy sets: Straightforward approaches may not work. *Studies in Computational Intelligence*, *109*, 415–430. doi:10.1007/978-3-540-77623-9_24

Taha, H. A., Sakr, A. H., & Yacout, S. (2019). *Aircraft Engine Remaining Useful Life Prediction Framework for Industry 4.0.* Paper presented at the 4th North America conference on Industrial Engineering and Operations Management, Toronto, Canada.

Tahre, M. A., & Abdeljawad, M. (2013). A new modular strategy for action sequence automation using neural networks and hidden markov models. *International Journal of System Dynamics Applications*, *2*(3), 1–18. doi:10.4018/ijsda.2013070102

Talukdar, P., & Dutta, P. (2018). Disease diagnosis using an advanced distance measure for intuitionistic fuzzy sets. *Int. Res. J. Microbiol*, *7*(2), 29-42.

Tambouratzis, T., Giannatsis, J., Kyriazis, A., & Siotropos, P. (2020). Applying the Computational Intelligence Paradigm to Nuclear Power Plant Operation: A Review (1990-2015). *International Journal of Energy Optimization and Engineering*, *9*(1), 27–109.

Tanase, R. (2016). Complex Hénon maps and discrete groups. *Advances in Mathematics*, *295*, 53–89. doi:10.1016/j.aim.2016.02.034

Tang, R., Fong, S., Yang, X.-S., & Deb, S. (2012). Wolf search algorithm with ephemeral memory. In *IEEE Seventh International Conference on Digital Information Management (ICDIM 2012)* (pp.165-172). IEEE.

Tan, W. Y. (2020). *Stochastic Models for Carcinogenesis*. CRC Press. doi:10.1201/9780367810832

Taradeh, M., Mafarja, M., Heidari, A. A., Faris, H., Aljarah, I., Mirjalili, S., & Fujita, H. (2019). An evolutionary gravitational search-based feature selection. *Inf Sci*, *497*, 219–239. doi:10.1016/j.ins.2019.05.038

Tarai, A., Poria, S., & Chatterjee, P. (2009). Synchronization of generalised linearly bidirectionally coupled unified chaotic system. *Chaos, Solitons, and Fractals*, *40*(2), 885–892. doi:10.1016/j.chaos.2007.08.039

Tayal, D. (2017). Leveraging innovation for electricity utilities. *The Electricity Journal*, *30*(3), 23–29. doi:10.1016/j.tej.2017.02.008

Tcvetkov, R., Szmidt, E., & Kacprzyk, J. (2009). On Some Issues Related to the Distances between the Atanassov Intuitionistic Fuzzy Sets. *Cybernetics and Information Technologies*, *9*(2), 54–61.

Thakker, R., Eveleigh, T., Holzer, T., & Sarkani, S. (2013). A System Dynamics Approach to Quantitatively Analyze the Effects of Mobile Broadband Ecosystem's Variables on Demands and Allocation of Wireless Spectrum for the Cellular Industry. *International Journal of System Dynamics Applications*, *2*(3), 73–93. doi:10.4018/ijsda.2013070105

Thompson, K., & Reimann, P. (2007). Do school learn more about the environment from a system dynamics model by themselves or with a partner. *Conference Proceedings, The 2007 International Conference of the System Dynamics Society and 50th Anniversary Celebration.*

Thounthong, P., Raël, S., & Davat, B. (2007). Control strategy of fuel cell and supercapacitors association for a distributed generation system. *IEEE Transactions on Industrial Electronics*, *54*(6), 3225–3233. doi:10.1109/TIE.2007.896477

Thwink.org. (2014). *The Three Pillars of Sustainability*. Retrieved from http://www.thwink.org/sustain/glossary/Sustainability.htm

Tidd, J., & Bessant, J. R. (2018). *Managing innovation: Integrating technological, market and organizational change*. John Wiley & Sons.

Tiwari, M. (2012). *Tumor Marker and Carcinogenesis*. River Publishers.

Tizhoosh, H. R. (2005). Image thresholding using type II fuzzy sets. *Pattern Recognition*, *38*(12), 2363–2372. doi:10.1016/j.patcog.2005.02.014

Tizhoosh, H. R. (2008). Interval-valued versus intuitionistic fuzzy sets: Isomorphism versus semantics. *Pattern Recognition*, *41*(5), 1812–1813. doi:10.1016/j.patcog.2007.11.003

Todorova, M., Stoyanov, B., Szczypiorski, K., & Kordov, K. (2018). *SHAH: hash function based on irregularly decimated chaotic map*. arXiv preprint arXiv:1808.01956

Todorov, V., & Marinova, D. (2011). Modelling sustainability. *Mathematics and Computers in Simulation*, *81*(7), 1397–1408. doi:10.1016/j.matcom.2010.05.022

Torres, J. P., Kunc, M., & O'Brien, F. (2017). Supporting strategy using system dynamics. *European Journal of Operational Research*, *260*(3), 1081–1094. doi:10.1016/j.ejor.2017.01.018

Tosi, Z., & Yoshimi, J. (2016). Simbrain 3.0: A flexible, visually-oriented neural network simulator. *Neural Networks*, *83*, 1–10. doi:10.1016/j.neunet.2016.07.005 PMID:27541049

Toufouti, R., Meziane, S., & Benalla, H. (2009). New direct torque neuro-fuzzy control based SVM for dual two level inverter-fed induction motor. *Journal of Control Engineering and Applied Informatics*, *11*(2), 3–13.

Trading Economics Website. (2015). *Iran Economic forecasts*. http://www.tradingeconomics.com/iran/inflation-cpi/forecast/

Tremblay, D., & Dekkiche. (2007). A Generic Battery Model for the Dynamic Simulation of Hybrid Electric Vehicles. *IEEE Vehicle Power and Propulsion Conference,* 284-289.

Tremblay, O., & Dessaint, L. (2009). *Experimental Validation of a Battery Dynamic Model for EV Applications*. Academic Press.

Tripathi, N., Oivo, M., Liukkunen, K., & Markkula, J. (2019). Startup ecosystem effect on minimum viable product development in software startups. *Information and Software Technology*, *114*, 77–91. doi:10.1016/j.infsof.2019.06.008

Tsai, C. C., Lin, S. C., & Lin, B. C. (2010). Intelligent adaptive motion control using fuzzy basis function networks for self-balancing two-wheeled transporters. In *Proceedings of IEEE International Conference on Fuzzy Systems*, (pp. 1-6). 10.1109/FUZZY.2010.5584070

Tulinayo, P. F., Hoppenbrouwers, S. J. B. A., & Proper, H. A. E. (2008). Integrating System Dynamics with Object- Role Modeling. *IFIP International Federation for Information Processing*.

Tutelea, L. N., Boldea, I., & Deaconu, S. I. (2012). Optimal design of dual rotor single stator PMSM drive for automobiles. *2012 IEEE International Electric Vehicle Conference, IEVC 2012*. 10.1109/IEVC.2012.6183224

Tutueva, A. V., Nepomuceno, E. G., Karimov, A. I., Andreev, V. S., & Butusov, D. N. (2020). Adaptive chaotic maps and their application to pseudo-random numbers generation. *Chaos, Solitons, and Fractals*, *133*, 109615. doi:10.1016/j.chaos.2020.109615

Tvedt, J. D., & Gollofello, J. S. (1995). Evaluating the effectiveness of process improvements on software development cycle time via system dynamics modelling. *Computer Software and Applications Conference, 1995. COMPSAC 95. Proceedings., Nineteenth Annual International*, 318–325. 10.1109/CMPSAC.1995.524796

Umashanker, V., & Dutta, K. (2007). Balanced scorecards in managing higher education institutions: An Indian perspective. *International Journal of Educational Management*, *21*(1), 54–67. doi:10.1108/09513540710716821

UN Water. (2016). *Integrated Monitoring Guide for SDG 6: Targets and global indicators*. UN.

United Nations Development Programme. (2017). *Human Development Index*. Retrieved July 12, 2017, from http://hdr.undp.org/en/indicators/137506

United Nations Economic and Social Council. (2015). *A Nexus Approach For The SDGs*. Retrieved from https://www.un.org/ecosoc/sites/www.un.org.ecosoc/files/files/en/2016doc/interlinkages-sdgs.pdf

United Nations General Assembly. (2015). *Transforming our world: the 2030 Agenda for Sustainable Development* (Vol. 16301). United Nations General Assembly. doi:10.100713398-014-0173-7.2

United Nations General Assembly. (2015). Transforming our world: the 2030 Agenda for Sustainable Development, Pub. L. No. 70/1, 16301 General Assembly Seventieth Session 35. New York.

United Nations. (2016). *Sustainable Development Goals Report*. Author.

Unluturk, A., & Aydogdu, O. (2017). Adaptive control of two-wheeled mobile balance robot capable to adapt different surfaces using a novel artificial neural network based real-time switching dynamic controller. *International Journal of Advanced Robotic Systems*, *14*(2), 1–9. doi:10.1177/1729881417700893

Un-Noor, F., Padmanaban, S., Mihet-Popa, L., Mollah, M. N., & Hossain, E. (2017). A comprehensive study of key electric vehicle (EV) components, technologies, challenges, impacts, and future direction of development. *Energies*, *10*(8), 1217. doi:10.3390/en10081217

Unterkalmsteiner, M., Abrahamsson, P., Wang, X., Nguyen-Duc, A., Shah, S., Bajwa, S. S., Baltes, G. H., Conboy, K., Cullina, E., & Dennehy, D. (2016). Software startups–a research agenda. *E-Informatica Software Engineering Journal*, *10*(1).

UPME & BID. (2015). *Integración de las energías renovables no convencionales en Colombia. Ministerio de Minas y Energía*. Retrieved from http://www.upme.gov.co/Estudios/2015/Integracion_Energias_Renovables/INTEGRACION_ENERGIAS_RENOVANLES_WEB.pdf

UPME. (2016). *Boletin estadístico de Minas y energía 2012 – 2016*. Retrieved from http://www1.upme.gov.co/PromocionSector/SeccionesInteres/Documents/Boletines/Boletin_Estadistico_2012_2016.pdf

US Office of Management and Budget. (2017). *North American Industry Classification System*. NAICS.

Utkarsh, G., & Gadgil, M., & Rao, P.R.S. (1999). Intellectual property rights on biological resources: Benefiting from biodiversity and people's knowledge. *Geographical Review*, *92*(2), 282–306.

Vadakkan, J. J., & Raghavanpilla, V. (2015). An energy efficient multi-target tracking in wireless sensor networks based on polygon tracking method. *International Journal of Emerging Trends in Science and Technology*. doi:10.18535/ijetst/v2i8.03

Vahdatzad, M. A., & Mojtahedzadeh, M. T. (2000). Some Issues in the Strategic Management in a fast growing academic Institution: The Case of University of Yadz. *Proceedings of the 18th International Conference of the System Dynamics Society*.

Vaidyanathan, S., & Azar, A.T. (2015a). Anti-Synchronization of Identical Chaotic Systems using Sliding Mode Control and an Application to Vaidyanathan-Madhavan Chaotic Systems. *Studies in Computational Intelligence, 576*, 527-547. DOI . doi:10.1007/978-3-319-11173-5_19

Vaidyanathan, S., & Azar, A.T. (2015b). Hybrid Synchronization of Identical Chaotic Systems using Sliding Mode Control and an Application to Vaidyanathan Chaotic Systems. *Studies in Computational Intelligence, 576*, 549-569. DOI . doi:10.1007/978-3-319-11173-5_20

Vaidyanathan, S., & Azar, A.T. (2015c). Analysis, Control and Synchronization of a Nine-Term 3-D Novel Chaotic System. *Studies in Computational Intelligence, 581*, 3-17. DOI . 63. doi:10.1007/978-3-319-13132-0_1

Vaidyanathan, S., & Azar, A.T. (2015d). Analysis and Control of a 4-D Novel Hyperchaotic System. *Studies in Computational Intelligence, 581*, 19-38. DOI . doi:10.1007/978-3-319-13132-0_2

Vaidyanathan, S., Idowu, B.A., & Azar, A.T. (2015c). Backstepping Controller Design for the Global Chaos Synchronization of Sprott's Jerk Systems. *Studies in Computational Intelligence, 581*, 39-58. DOI . doi:10.1007/978-3-319-13132-0_3

Vaidyanathan, S., & Azar, A. T. (2016a). *Dynamic Analysis, Adaptive Feedback Control and Synchronization of an Eight-Term 3-D Novel Chaotic System with Three Quadratic Nonlinearities*. Springer-Verlag. doi:10.1007/978-3-319-30340-6_7

Vaidyanathan, S., & Azar, A. T. (2016a). *Dynamic Analysis, Adaptive Feedback Control and Synchronization of an Eight-Term 3-D Novel Chaotic System with Three Quadratic Nonlinearities. Studies in Fuzziness and Soft Computing* (Vol. 337). Springer-Verlag.

Vaidyanathan, S., & Azar, A. T. (2016b). *Qualitative Study and Adaptive Control of a Novel 4-D Hyperchaotic System with Three Quadratic Nonlinearities*. Springer-Verlag. doi:10.1007/978-3-319-30340-6_8

Vaidyanathan, S., & Azar, A. T. (2016b). *Qualitative Study and Adaptive Control of a Novel 4-D Hyperchaotic System with Three Quadratic Nonlinearities. Studies in Fuzziness and Soft Computing* (Vol. 337). Springer-Verlag.

Vaidyanathan, S., & Azar, A. T. (2016c). *A Novel 4-D Four-Wing Chaotic System with Four Quadratic Nonlinearities and its Synchronization via Adaptive Control Method. Advances in Chaos Theory and Intelligent Control. Studies in Fuzziness and Soft Computing* (Vol. 337). Springer-Verlag.

Vaidyanathan, S., & Azar, A. T. (2016d). *Adaptive Control and Synchronization of Halvorsen Circulant Chaotic Systems. Advances in Chaos Theory and Intelligent Control. Studies in Fuzziness and Soft Computing* (Vol. 337). Springer-Verlag.

Vaidyanathan, S., & Azar, A. T. (2016e). *Adaptive Backstepping Control and Synchronization of a Novel 3-D Jerk System with an Exponential Nonlinearity. Advances in Chaos Theory and Intelligent Control. Studies in Fuzziness and Soft Computing* (Vol. 337). Springer-Verlag.

Vaidyanathan, S., & Azar, A. T. (2016f). *Adaptive Control and Synchronization of Halvorsen Circulant Chaotic Systems. Advances in Chaos Theory and Intelligent Control.* Springer-Verlag.

Vaidyanathan, S., & Azar, A. T. (2016f). *Generalized Projective Synchronization of a Novel Hyperchaotic Four-Wing System via Adaptive Control Method. Advances in Chaos Theory and Intelligent Control. Studies in Fuzziness and Soft Computing* (Vol. 337). Springer-Verlag.

Vaidyanathan, S., & Azar, A. T. (2016g). *Generalized Projective Synchronization of a Novel Hyperchaotic Four-Wing System via Adaptive Control Method.* Springer-Verlag. doi:10.1007/978-3-319-30340-6_12

Vaidyanathan, S., & Azar, A. T. (2016g). Takagi-Sugeno Fuzzy Logic Controller for Liu-Chen Four-Scroll Chaotic System. *Int. J. of Intelligent Engineering Informatics, 4*(2), 135–150. doi:10.1504/IJIEI.2016.076699

Vaidyanathan, S., Azar, A. T., Akgul, A., Lien, C. H., Kacar, S., & Cavusoglu, U. (2019). A memristor-based system with hidden hyperchaotic attractors, its circuit design, synchronisation via integral sliding mode control and voice encryption. *Int. J. Automation and Control, 13*(6), 644–667. doi:10.1504/IJAAC.2019.102665

Vaidyanathan, S., Azar, A. T., & Boulkroune, A. (2018d). A novel 4-D hyperchaotic system with two quadratic nonlinearities and its adaptive synchronization. *Int. J. Automation and Control, 12*(1), 5–26. doi:10.1504/IJAAC.2018.088612

Vaidyanathan, S., Azar, A. T., & Ouannas, A. (2017a). An Eight-Term 3-D Novel Chaotic System with Three Quadratic Nonlinearities, its Adaptive Feedback Control and Synchronization. *Fractional Order Control and Synchronization of Chaotic Systems Studies in Computational Intelligence, 688*, 719–746.

Vaidyanathan, S., Azar, A. T., & Ouannas, A. (2017a). *An Eight-Term 3-D Novel Chaotic System with Three Quadratic Nonlinearities, its Adaptive Feedback Control and Synchronization. Studies in Computational Intelligence* (Vol. 688). Springer-Verlag.

Vaidyanathan, S., Azar, A. T., & Ouannas, A. (2017b). Hyperchaos and Adaptive Control of a Novel Hyperchaotic System with Two Quadratic Nonlinearities. *Fractional Order Control and Synchronization of Chaotic Systems Studies in Computational Intelligence, 688*, 773–803.

Vaidyanathan, S., Azar, A. T., & Ouannas, A. (2017c). *Hyperchaos and Adaptive Control of a Novel Hyperchaotic System with Two Quadratic Nonlinearities. Studies in Computational Intelligence* (Vol. 688). Springer-Verlag.

Vaidyanathan, S., Azar, A. T., Rajagopal, K., & Alexander, P. (2015b). Design and SPICE implementation of a 12-term novel hyperchaotic system and its synchronization via active control. *Int. J. Modelling Identification and Control, 23*(3), 267–277. doi:10.1504/IJMIC.2015.069936

Vaidyanathan, S., Azar, A. T., Rajagopal, K., Sambas, A., Kacar, S., & Cavusoglu, U. (2018c). A new hyperchaotic temperature fluctuations model, its circuit simulation, FPGA implementation and an application to image encryption. *Int. J. Simulation and Process Modelling, 13*(3), 281–296. doi:10.1504/IJSPM.2018.093113

Vaidyanathan, S., Azar, A. T., Sambas, A., Singh, S., Alain, K. S. T., & Serrano, F. E. (2018a). *A Novel Hyperchaotic System With Adaptive Control, Synchronization, and Circuit Simulation. Advances in System Dynamics and Control.* IGI-Global. doi:10.4018/978-1-5225-4077-9.ch013

Vaidyanathan, S., Jafari, S., Pham, V. T., Azar, A. T. E., & Alsaadi, F. (2018b). A 4-D chaotic hyperjerk system with a hidden attractor, adaptive backstepping control and circuit design. *Archives of Control Sciences.*, *28*(2), 239–254.

Vaidyanathan, S., Sampath, S., & Azar, A. T. (2015a). Global chaos synchronisation of identical chaotic systems via novel sliding mode control method and its application to Zhu system. *Int. J. Modelling Identification and Control*, *23*(1), 92–100. doi:10.1504/IJMIC.2015.067495

Vaidyanathan, S., Volos, C., Pham, V.-T., Madhavan, K., & Idowu, B. A. (2014). Adaptive backstepping control, synchronization and circuit simulation of a 3-D novel jerk chaotic system with two hyperbolic sinusoidal nonlinearities. *Archives of Control Sciences*, *24*(3), 375–403. doi:10.2478/acsc-2014-0022

Vaidyanathan, S., Zhu, Q., & Azar, A. T. (2017b). *Adaptive Control of a Novel Nonlinear Double Convection Chaotic System. Studies in Computational Intelligence* (Vol. 688). Springer-Verlag.

Valandar, M. Y., Barani, M. J., & Ayubi, P. (2020). A blind and robust color images watermarking method based on block transform and secured by modified 3-dimensional Hénon map. *Soft Computing*, *24*(2), 771–794. doi:10.100700500-019-04524-z

van Gelderen, M., Thurik, R., & Bosma, N. (2005). Success and Risk Factors in the Pre-Startup Phase. *Small Business Economics*, *24*(4), 365–380. doi:10.100711187-004-6994-6

Velthuis, O. (2013). *Talking prices: Symbolic meanings of prices on the market for contemporary art*. Princeton University Press.

Venkataram, P., & Giridharan, A. (2009). Quality Assurance and Assessment in Technical Education System: A web Based Approach. *International Conference on Engineering Education-ICEE 2007*, 1-6.

Vennix, J. A. (1999). Group model building: Tackling messy problems. *System Dynamics Review*, *15*(4), 379–401. doi:10.1002/(SICI)1099-1727(199924)15:4<379::AID-SDR179>3.0.CO;2-E

Vennix, J. A. M. (1996). *Group Model Building: Facilitating Team Learning Using System Dynamics. Quality Progress* (Vol. 30). Wiley. Retrieved from https://www.amazon.com/Group-Model-Building-Facilitating-Learning/dp/0471953555

Ventana Systems Inc. (2013). *Vensim® User Manual*. Ventana System, Inc. http://www.vensim.com

Verghese, A. (2007). OD Interventions for Sustainable Excellence in Higher Education. *Conference on Global Competition and Competitiveness of Indian Corporate*, 39-51.

Verhulst, F. (1996). *Nonlinear Differential Equations and Dynamical Systems*. Springer-Verlag. doi:10.1007/978-3-642-61453-8

Vermeiren, L., Dequidt, A., Guerra, T. M., Rago-Tirmant, H., & Parent, M. (2011). Modeling, control and experimental verification on a two-wheeled vehicle with free inclination: An urban transportation system. *Control Engineering Practice*, *19*(7), 744–756. doi:10.1016/j.conengprac.2011.04.002

Verstovsek, S., Kantarjian, H., Manshouri, T., Cortes, J., Giles, F. J., Rogers, A., & Albitar, M. (2002). Prognostic significance of cellular vascular endothelial growth factor expression in chronic phase chronic myeloid leukemia. Blood. *The Journal of the American Society of Hematology*, *99*(6), 2265–2267. PMID:11877311

Villas-Boas, A. (2017, October 10). *Microsoft's Windows Phone platform is dead*. Business Insider Australia. https://www.businessinsider.com.au/microsofts-windows-phone-platform-dead-2017-10

Vishal, K., & Agrawal, S. K. (2017). On the dynamics, existence of chaos, control and synchronization of a novel complex chaotic system. *Zhongguo Wuli Xuekan*, *55*(2), 519–532. doi:10.1016/j.cjph.2016.11.012

Vladimirova, K., & Le Blanc, D. (2016). Exploring Links Between Education and Sustainable Development Goals Through the Lens of UN Flagship Reports. *Sustainable Development, 24*(4), 254–271. doi:10.1002d.1626

Voiculetz, N., Balaban, A. T., Niculescu-Duvaz, I., & Simon, Z. (1991). *Modeling of cancer genesis and prevention.* CRC press.

Vokshi, M., Youssef, D., & Toçi, V. (2019). *Mobile phone penetration and its impact on inequality in the Western Balkan countries. Zagreb International Review of Economics & Business, 22*(2), 111–131. doi:10.2478/zireb-2019-0023

Volos, C. K., Kyprianidis, I. M., & Stouboulos, I. N. (2011). Various synchronization phenomena in bidirectionally coupled double scroll circuits. *Communications in Nonlinear Science and Numerical Simulation, 16*(8), 3356–3366. doi:10.1016/j.cnsns.2010.11.015

Volos, C. K., Pham, V. T., Azar, A. T., Stouboulos, I. N., & Kyprianidis, I. M. (2018). Synchronization Phenomena in Coupled Dynamical Systems with Hidden Attractors. In V. T. Pham, S. Vaidyanathan, C. Volos, & T. Kapitaniak (Eds.), *Nonlinear Dynamical Systems with Self-Excited and Hidden Attractors. Studies in Systems, Decision and Control* (Vol. 133, pp. 375–401). Springer. doi:10.1007/978-3-319-71243-7_17

Volos, C., Akgul, A., Pham, V.-T., Stouboulos, I., & Kyprianidis, I. (2017). A simple chaotic circuit with a hyperbolic sine function and its use in a sound encryption scheme. *Nonlinear Dynamics, 89*(2), 1047–1106. doi:10.100711071-017-3499-9

Volos, C., Kyprianidis, I., Stouboulos, I., & Vaidyanathan, S. (2016). Design of a chaotic random bit generator using a Duffing-van der Pol system. *International Journal of System Dynamics Applications, 5*(3), 94–111. doi:10.4018/IJSDA.2016070105

Von Stechow, C., Minx, J. C., Riahi, K., Jewell, J., McCollum, D. L., Callaghan, M. W., Bertram, C., Luderer, G., & Baiocchi, G. (2016). 2°C and SDGs: United they stand, divided they fall? *Environmental Research Letters, 11*(3), 1–15. doi:10.1088/1748-9326/11/3/034022

Vrat, P. (2003). Quality Assurance in Technical education, Recent trends and challenges ahead. *The Indian Journal of Technical Education, 27*(2), 12–14.

Wade, S., Dunnigan, M. W., & Williams, B. W. (1997). Modeling and simulation of induction machine vector control with rotor resistance identification. *IEEE Transactions on Power Electronics, 12*(3), 495–506. doi:10.1109/63.575677

Wai, R. J., & Lin, K. M. (2005). Robust decoupled control of direct field-oriented induction motor drive. *IEEE Transactions on Industrial Electronics, 52*(3), 837–854. doi:10.1109/TIE.2005.847585

Wang, Y., Hong, H., Guo, J., Wang, X., & Shang, W. (2020). Configuration error function design and application to fixed-time geometric terminal sliding-mode control on SE(3). *Acta Astronautica, 174.*

Wang, L., & Cheng, H. (2019). Pseudo-Random Number Generator Based on Logistic Chaotic System. *Entropy (Basel, Switzerland), 21*(10), 960. doi:10.3390/e21100960

Wang, S., Cao, T., & Chen, B. (2017). Urban energy–water nexus based on modified input–output analysis. *Applied Energy, 196*(February), 208–217. doi:10.1016/j.apenergy.2017.02.011

Wang, W. Q., & Xin, X. L. (2005). Distance measure between intuitionistic fuzzy sets. *Pattern Recognition Letters, 26*(13), 2063–2069. doi:10.1016/j.patrec.2005.03.018

Wang, Y., Liu, Z., Ma, J., & He, H. (2016). A pseudorandom number generator based on piecewise logistic map. *Nonlinear Dynamics, 83*(4), 2373–2391. doi:10.100711071-015-2488-0

Wang, Z., Volos, C., Kingni, S. T., Azar, A. T., & Pham, V. T. (2017). Four-wing attractors in a novel chaotic system with hyperbolic sine nonlinearity. *Optik (Stuttgart)*, *131*, 1071–1078. doi:10.1016/j.ijleo.2016.12.016

Ward, J. P., & King, J. R. (1997). Mathematical modelling of avascular-tumour growth. *Mathematical Medicine and Biology. The Journal of IMA / Islamic Medical Association of North America*, *14*(1), 39–69.

Warren, D. M. (1991). *Using Indigenous Knowledge in Agricultural Development*. World Bank Discussion Paper No.127. Washington, DC: The World Bank.

Warren, K. (2005). Improving strategic management with the fundamental principles of system dynamics. *System Dynamics Review*, *21*(4), 329–350. doi:10.1002dr.325

Wasserman, S., & Faust, F. (1994). *Social Network Analysis: Methods and Applications*. Cambridge University Press. doi:10.1017/CBO9780511815478

Watanabe, K., Tang, J., Nakamura, M., Koga, S., & Fukuda, T. (1993). Mobile robot control using fuzzy gaussian neural networks. In *Proceedings of IEEE/RSJ International Conference on Intelligent Robots and Systems (IROS)*, (pp. 919-925). 10.1109/IROS.1993.583253

Watson, D. (2000, January). Managing in Higher Education: The 'Wicked Issues'. *Higher Education Quarterly*, *54*(1), 5–21. doi:10.1111/1468-2273.00142

Watty, K. (2003). When will Academics Learn about Quality. *Quality in Higher Education*, *10*(1), 17–25.

Wei, L. (2013). Hybrid Vehicle System Modeling. In *Introduction to Hybrid Vehicle System Modeling and Control*. Academic Press.

Weigelt, J., Janetschek, H., Müller, A., & Töpfer, K. (2015). Editorial overview: Environmental change issues: Soils in the nexus. *Current Opinion in Environmental Sustainability*, *15*, v–viii. doi:10.1016/j.cosust.2015.09.006

Weimer-Jehle, W. (2006). Cross-impact balances: A system-theoretical approach to cross-impact analysis. *Technological Forecasting and Social Change*, *73*(4), 334–361. doi:10.1016/j.techfore.2005.06.005

Weimer-Jehle, W. (2008). Cross-impact balances: Applying pair interaction systems and multi-value Kauffman nets to multidisciplinary systems analysis. *Physica A*, *387*(14), 3689–3700. doi:10.1016/j.physa.2008.02.006

Weir, A. (2012, January 19). *ZTE reveals cost of Windows Phone OS licensing*. Neowin. https://www.neowin.net/news/zte-reveals-cost-of-windows-phone-os-licensing

Weitz, N., Nilsson, M., & Davis, M. (2014). A Nexus Approach to the Post-2015 Agenda: Formulating Integrated Water, Energy, and Food SDGs. *SAIS Review (Paul H. Nitze School of Advanced International Studies)*, *34*(2), 37–50. doi:10.1353ais.2014.0022

Wembe, E. T., & Yamapi, R. (2009). Linearly and nonlinearly bidirectionally coupled synchronization of hyperchaotic systems. *Chaos, Solitons, and Fractals*, *14*(4), 1439–1453.

Wen, Y. H., Lin, Y. S., & Leu, Y. G. (2013). Design and implementation of the balance of two-wheeled robots. In *Proceedings of IEEE International Conference on Advanced Robotics and Intelligent Systems (ARIS)*, (pp. 130-134). 10.1109/ARIS.2013.6573547

Wheat, D. (2010). What Can System Dynamics Learn From the Public Policy Implementation Literature? *Systems Research and Behavioral Science*, *27*(3), 425–442. doi:10.1002res.1039 PMID:32327920

WHO. (2020). https://www.who.int/health-topics/cancer#tab=tab_1

Williams, D. (2002). An Application of System Dynamics to Requirements Engineering Process Modelling (Unpublished PhD). South Bank University.

Williams, D. (2003). *Challenges of System Dynamics to Deliver Requirements Engineering Projects: faster, better and cheaper*. Paper presented at the 2003 System Dynamics Conference, New York., NY

Williams, R., & Van Dyke, N. (2007). Measuring the international standing of universities with an application to Australian universities. *Higher Education*, *53*(6), 819841. doi:10.100710734-005-7516-4

Witte, C. L., & Witte, M. H. (1991). Chaos and predicting varix hemorrhage. *Medical Hypotheses*, *36*(4), 312–317. doi:10.1016/0306-9877(91)90002-G PMID:1809849

Wolf, A., Swift, J. B., Swinney, H. L., & Vastano, J. A. (1985). Determining Lyapunov exponents from a time series. *Physica D. Nonlinear Phenomena*, *16*(3), 285–317. doi:10.1016/0167-2789(85)90011-9

Wolstenholme, E. F. (1999). Qualitative vs Quantitative Modelling: The Evolving Balance. *The Journal of the Operational Research Society*, *50*(4), 422–428. doi:10.1057/palgrave.jors.2600700

Womack, J. P., & Jones, D. T. (2010). *Lean Thinking: Banish Waste and Create Wealth in Your Corporation*. Simon and Schuster.

Womack, J., & Jones, D. (2011). *Seeing the Whole Value Stream* (2nd ed.). Lean Enterprise Institute.

Wu, J., & Zhang, W. (2011). Design of fuzzy logic controller for two-wheeled self-balancing robot. In *Proceedings of 6th International Forum on Strategic Technology (IFOST)*, (pp. 1266-1270). 10.1109/IFOST.2011.6021250

Wu, J., Balliet, D., & Van Lange, P. A. M. (2016). Reputation, Gossip, and Human Cooperation. *Social and Personality Psychology Compass*, *10*(6), 350–364. doi:10.1111pc3.12255

Wyburn, J., & Roach, P. A. (2013). A System Dynamics Model of the American Collectable Comic Book Market. *International Journal of System Dynamics Applications*, *2*(1), 37–58. doi:10.4018/ijsda.2013010103

Xiao, R., Yang, X., Zhu, W., & Cao, J. (2009). Modeling and simulation of dual-rotor permanent-magnet synchronous motor based on MATLAB. *Proceedings - The 12th International Conference on Electrical Machines and Systems, ICEMS 2009*, 1–4. 10.1109/ICEMS.2009.5382660

Xiaojian, X., Saleh, M., Haipeng, R., & Sajad, J. (2017). Robust finite-time synchronization of a class of chaotic systems via adaptive global sliding mode control. *Journal of Vibration and Control*, *0*(0), 1–13.

Xu, B., & Wang, H. (2010). A reliability transmission routing metric algorithm for wireless sensor network. *2010 International Conference on E-Health Networking Digital Ecosystems and Technologies (EDT)*. 10.1109/EDT.2010.5496534

Xu, J., Wang, F., Xie, S., Xu, J., & Feng, J. (2004). A new control method for permanent magnet synchronous machines with observer. *PESC Record - IEEE Annual Power Electronics Specialists Conference, IEEE,* 1404-1408. 10.1109/PESC.2004.1355628

Xu, C., Peak, D., & Prybutok, V. (2015). A customer value, satisfaction, and loyalty perspective of mobile application recommendations. *Decision Support Systems*, *79*(September), 171–183. doi:10.1016/j.dss.2015.08.008

Xu, Z., Khoshgoftaar, T. M., & Allen, E. B. (2000). Prediction of software faults using fuzzy nonlinear regression modelling. In *Fifth IEEE international symposium on high assurance systems engineering*, (pp. 281–290). IEEE.

Yamakawa, P., Gareth, H. R., Jose, M. S., & Nikolai, A. (2013). The diffusion of mobile telephones: An empirical analysis for Peru. *Telecommunications Policy*, *37*(6-7), 594–606. doi:10.1016/j.telpol.2012.12.010

Yamany, W., Emary, E., & Hassanien, A. E. (2014). Wolf search algorithm for attribute reduction in classification. *2014 IEEE Symposium on Computational Intelligence and Data Mining (CIDM)*, 351–358. 10.1109/CIDM.2014.7008689

Yang, Y., & Yan, Y. (2016). Attitude regulation for unmanned quadrotors using adaptive fuzzy gain-scheduling sliding mode control. *Aerospace Science and Technology*, *54*, 208–217. doi:10.1016/j.ast.2016.04.005

Yao, J. C., Phan, A., Hoff, P. M., Chen, H. X., Charnsangavej, C., Yeung, S. C. J., Hess, K., Ng, C., Abbruzzese, J. L., & Ajani, J. A. (2008). Targeting vascular endothelial growth factor in advanced carcinoid tumor: A random assignment phase II study of depot octreotide with bevacizumab and pegylated interferon alfa-2b. *Journal of Clinical Oncology*, *26*(8), 1316–1323. doi:10.1200/JCO.2007.13.6374 PMID:18323556

Yao, J. F. F., & Yao, J. S. (2001). Fuzzy decision making for medical diagnosis based on fuzzy number and compositional rule of inference. *Fuzzy Sets and Systems*, *120*(2), 351–366. doi:10.1016/S0165-0114(99)00071-8

Yasuda, G. (2016). Design and Implementation of Distributed Autonomous Cordinators for cooperative multi-robot systems. *International Journal of System Dynamics Applications*, *5*(4), 1–15. doi:10.4018/IJSDA.2016100101

Yatimi & Aroudam. (2018). Standalone Photovoltaic System with Maximum Power Point Tracking: Modeling and Simulation. *International Journal of System Dynamics Applications, 7*(3), 94-117.

Yatimi, H., & Aroudam, E. (2018). Standalone Photovoltaic System with Maximum Power Point Tracking: Modeling and Simulation. *International Journal of System Dynamics Applications*, *7*(3), 94–111. doi:10.4018/IJSDA.2018070105

Yilmaz, M., & Krein, P. T. (2013). Review of battery charger topologies, charging power levels, and infrastructure for plug-in electric and hybrid vehicles. *IEEE Transactions on Power Electronics*, *28*(5), 2151-2169. doi:10.1109/TPEL.2012.2212917

Yoo, W., Ji, D., & Won, S. (2010). Synchronization of two different non-autonomous chaotic systems using fuzzy disturbances observer. *Physics Letters. [Part A]*, *374*(11-12), 1354–1361. doi:10.1016/j.physleta.2010.01.023

Yuan, G., Zhang, X., & Wang, Z. (2014). Generation and synchronization of feedback-induced chaos in semiconductor ring lasers by injection-locking. *Optik (Stuttgart)*, *125*(8), 1950–1953. doi:10.1016/j.ijleo.2013.11.007

Yuan, S., Jiang, T., & Jing, Z. (2011). Bifurcation and chaos in the Tinkerbell map. *International Journal of Bifurcation and Chaos in Applied Sciences and Engineering*, *21*(11), 3137–3156. doi:10.1142/S0218127411030581

Yuan, Z., Xu, Z., & Guo, L. (2012). Generalized synchronization of two bidirectionally coupled discrete dynamical systems. *Communications in Nonlinear Science and Numerical Simulation*, *17*(2), 992–1002. doi:10.1016/j.cnsns.2011.07.014

Yujing, W., Yongkui, L., & Peidong, G. (2015). Modelling Construction Project Management Based on System Dynamics. *Metallurgical and Mining Industry*, *9*, 1056 – 1060. Available at: https://www.metaljournal.com.ua/assets/Journal/english-edition/MMI_2015_9/0166Yujing-Wang.pdf

Zadeh, L. A. (1965). Fuzzy sets. *Information and Control*, *8*(3), 338–356. doi:10.1016/S0019-9958(65)90241-X

Zaher, A. A., & Abu-Rezq, A. (2011). On the design of chaos-based secure communication systems. *Communications in Nonlinear Science and Numerical Simulation*, *16*(9), 3721–3737. doi:10.1016/j.cnsns.2010.12.032

Zamalloa, M. Z., & Krishnamachari, B. (2007). An analysis of unreliability and asymmetry in low-power wireless links. *ACM Transactions on Sensor Networks*, *3*(2), 7. doi:10.1145/1240226.1240227

Zandi, M., Payman, A., Martin, J., Pierfederici, S., Davat, B., & Meibody-Tabar, F. (2011). Energy Management of a Fuel Cell / Supercapacitor / Battery Power Source for Electric Vehicular Applications. *IEEE Transactions on Vehicular Technology, 2010, 60*(2), 433-443. doi:10.1109/TVT.2010.2091433

Zapata, S., Castaneda, M., Franco, C. J., & Dyner, I. (2019). Clean and secure power supply: A system dynamics based appraisal. *Energy Policy*, *131*(April), 9–21. doi:10.1016/j.enpol.2019.04.028

Zelinka, D., & Amadei, B. (2017). A Methodology to Model the Integrated Nature of the Sustainable Development Goals: Importance for Engineering Education. *American Society for Engineering Education Conference.*

Zelinka, D. (2019). Amadei, Bernard., Systems Approach for Modeling Interactions Among the Sustainable Development Goals Part 1: Cross-Impact Network Analysis. *International Journal of System Dynamics Applications*, *8*(1).

Zelinka, D. (2019). *Systems Approaches to Transdisciplinary Sustainable Development Issues (No. 13810159).* Proquest. ProQuest.

Zelinka, D., & Amadei, B. (2018a). A new insecure-secure diffusion archetype to model complex anthropocentric systems. *36th International Conference of the System Dynamics Society.*

Zelinka, D., & Amadei, B. (2018b). A Systems Approach to Modeling Corruption: Bridging the Disciplinary Divide Between the Social and Technical Sciences. In *24th International Sustainability Development Research Society Conference.* Messina, Italy: International Sustainability Development Research Society.

Zelinka, D., & Amadei, B. (2019). Systems Approach for Modeling Interactions Among the Sustainable Development Goals Part 1: Cross-Impact Network Analysis. *International Journal of System Dynamics Applications*, *8*(1), 23–40. doi:10.4018/IJSDA.2019010102

Zelinka, D., & Amadei, B. (2019a). A Systems Approach for Modeling Interactions Among the Sustainable Development Goals Part 1: Cross-Impact Network Analysis. *International Journal of System Dynamics Applications*, *8*(1), 41–59. doi:10.4018/IJSDA.2019010103

Zeng, W. Y., & Guo, P. (2008). Normalized distance, similarity measure, inclusion measure and entropy of interval-valued fuzzy sets and their relationship. *Inform. Sci.*, *178*(5), 1334–1342. doi:10.1016/j.ins.2007.10.007

Zhang, R., Tao, J., & Zhou, H. (2019). Fuzzy Optimal Energy Management for Fuel Cell and Supercapacitor Systems Using Neural Network Based Driving Pattern Recognition. *IEEE Transactions on Fuzzy Systems, 27*(1), 45-57. doi:10.1109/TFUZZ.2018.2856086

Zhang, Q., Guo, L., & Wei, X. (2013). A novel image fusion encryption algorithm based on DNA sequence operation and hyper-chaotic system. *Optik (Stuttgart)*, *124*(18), 3596–3600. doi:10.1016/j.ijleo.2012.11.018

Zhou, D., Gao, F., Ravey, A., Al-Durra, A., & Simões, M. G. (2017). Online energy management strategy of fuel cell hybrid electric vehicles based on time series prediction. *2017 IEEE Transportation Electrification Conference and Expo (ITEC)*, 113-118. 10.1109/ITEC.2017.7993256

Zhou, F., Yang, S., Fujita, H., Chen, D., & Wen, C. (2020). Deep learning fault diagnosis method based on global optimization GAN for unbalanced data. *Knowledge-Based Systems*, *187*, 104837.

Zhou, Q., Mo, H., & Deng, Y. (2020). A new divergence measure of pythagorean fuzzy sets based on belief function and its application in medical diagnosis. *Mathematics*, *8*(1), 142. doi:10.3390/math8010142

Zhou, X., & Moinuddin, M. (2017). *Sustainable Development Goals Interlinkages and Network Analysis: A practical tool for SDG integration and policy coherence. The Institute for Global Environmental Strategies.* IGES. doi:10.1128/MCB.05460-11

Zhu, T., & Yi, P. (2012). Reliable and energy-efficient networking protocol design in wireless sensor networks. *Intelligent Sensor Networks*, 411-426. doi:10.1201/b14300-22

Zhu, J., Xu, B., Wang, X., Feng, H., & Xu, X. (2013). The research of sensorless vector control for permanent magnet linear synchronous motor. *Journal of Computers*, *8*(5), 1184–1191. doi:10.4304/jcp.8.5.1184-1191

Zhu, Q., & Azar, A. T. (2015). *Complex system modelling and control through intelligent soft computations* (Vol. 319). Springer.

Zhu, Q., & Azar, A. T. (2015). *Complex system modelling and control through intelligent soft computations. Studies in Fuzziness and Soft Computing* (Vol. 319). Springer-Verlag.

Zou, Y. L., Zhu, J., Chen, G., & Luo, X. S. (2005). Synchronization of hyperchaotic oscillators via single unidirectional chaotic-coupling. *Chaos, Solitons, and Fractals*, *25*(5), 1245–1253. doi:10.1016/j.chaos.2004.11.064

Zuidervaart, L. (2010). *Art in public: politics, economics, and a democratic culture*. Cambridge University Press. doi:10.1017/CBO9780511760341

About the Contributors

Ahmad Taher Azar received the M.Sc. degree in 2006, and Ph.D. degree in 2009 from the Faculty of Engineering, Cairo University, Egypt, and he got his post-doctoral from the USA in 2015. He is a Research Professor at Prince Sultan University, Riyadh, Kingdom Saudi Arabia. He is also a professor at the Faculty of Computers and Artificial intelligence, Benha University, Egypt. Prof. Azar is the Editor in Chief of the International Journal of System Dynamics Applications (IJSDA) and International Journal of Service Science, Management, Engineering, and Technology (IJSSMET) published by IGI Global, USA. He is also the Editor in Chief of International Journal of Intelligent Engineering Informatics (IJIEI), Inderscience Publishers, Olney, UK. Prof. Azar has worked as associate editor of IEEE Trans. Neural Networks and Learning Systems from 2013 to 2017 and Associate Editor of ISA Transactions, Elsevier from 2018 to 2020. He is currently associate editor of IEEE systems journal and Human-centric Computing and Information Sciences, Springer. Dr. Ahmad Azar has worked in Control Theory & Applications, Robotics, Process Control, Artificial Intelligence, and machine learning. He has authored/co-authored over 350 research publications in peer-reviewed reputed journals, book chapters, and conference proceedings. He is an editor of many books in Fuzzy logic systems, modeling techniques, control systems, Robotics, computational intelligence, Chaos Modeling, and Machine learning published by Springer and Elsevier. Dr. Ahmad Azar is closely associated with several international journals as a reviewer. He serves as an international program committee member in many international and peer-reviewed conferences. Dr. Ahmad Azar is a Senior Member of IEEE since Dec. 2013 due to his significant contributions to the profession. Dr. Ahmad Azar is the recipient of several awards, including the Benha University Prize for Scientific Excellence (2015, 2016, 2017, and 2018) and the Highest Citation Award from Benha University (2015, 2016, 2017, and 2018). In June 2018, Prof. Azar was awarded the Egyptian State Encouragement Award in Engineering Sciences, the Academy of Scientific Research and Technology of Egypt, 2017. In July 2018, he was selected as an Energy and Electricity Research council member, Academy of Scientific Research, Ministry of Higher Education. In Aug. 2018, he was selected as a senior Member of the International Rough Set Society (IRSS). In February 2020, Prof. Azar has been awarded the Distinguished Egyptian Order of the first class from the Egyptian President. In October 2020, Prof. Azar has been awarded Abdul Hameed Shoman Arab Researchers Award in Machine learning and Big data analytics. In October 2020, Prof. Azar has been selected as a Distinguished researcher at Prince Sultan University, Riyadh, Saudi Arabia. In November 2020, Prof. Azar has been selected from Stanford university among the top 2% scientists in the world in Artificial Intelligence. These statistics were published by Stanford University in PLOS journal and based on SCOPUS Database. Prof. Ahmad Azar is the Chair of IEEE Computational Intelligence Society (CIS) Egypt Chapter, Vice chair of IEEE Computational Intelligence Society Interdisciplinary Emergent Technologies Task Force, vice-Chair Research Activities of IEEE Robotics and Automation Society Egypt Chapter, Committee member of IEEE CIS Task Force on Fuzzy Logic in Medical Sciences Also, he is the Vice-president (North) of System dynamics Africa Regional Chapter and an Academic Member of IEEE Systems, Man, and Cybernetics Society Technical Committee on Computational Collective Intelligence.

Nashwa Kamal is the managing editor of some international journals like International Journal of System Dynamics Applications (IJSDA), International Journal of Service Science, Management, Engineering, and Technology (IJSSMET) and International Journal of Sociotechnology and Knowledge Development (IJSKD) published by IGI Global, USA. Also, she is the managing editor of International Journal of Intelligent Engineering Informatics (IJIEI), Inderscience Publishers, Olney, UK. Dr. Nashwa Ahmad Kamal has worked in the areas of Control Theory & Applications, Process Control, Nonlinear control, Robotics and Renewable Energy and has authored/coauthored some research publications in peer-reviewed reputed journals book chapters and conference proceedings. She is a member of international group of control systems and International Society of Automation (ISA), USA.

* * *

Hooman Abdollahi is a Ph.D. candidate in Finance at UiT The Arctic University of Norway. His principal research interests are finance, data science, and system dynamics modelling. He has actively collaborated with researchers in several other disciplines of finance and business, evidenced by publications in leading journals such as Energy, Applied Energy, etc. He also acts as a reviewer for various journals such as Applied Soft Computing Journal, Journal of Global Information Management, Soft Computing, Clean Technologies and Environmental Policy, Transport Policy, etc.

Kammogne Soup Tewa Alain received the BSc degree in Physics from the University of Dschang, Cameroon, in 2005, the MSc in electronics from the same university in 2008 and the PhD degree in electronic and control theory in 2014. In 2012, he was awarded the DIPES II from the High Teacher College of Bambili, Cameroon. He is currently a Lecturer at the University of Dschang and National Polytechnic of Bamenda. His research interests are in the areas of nonlinear systems, focusing on chaos control, robust control, noises analysis and their potential applications to secure communication, biological systems and power electronics.

Mohammed E. Ammar received the B.Sc. and M.Sc. degrees in electrical engineering from Cairo University, Cairo, Egypt, in 2000 and 2003, respectively, and the Ph.D. degree in electrical engineering from the University of British Columbia, Vancouver, Canada, in 2009.,He was a Postdoctoral Research Fellow at the University of British Columbia and a Research Associate at the University of Waterloo, Canada. From 2010–2012, he worked as a Lead Engineer for Invensys Operations Management. He is currently an Assistant Professor at Cairo University, Cairo, Egypt.,His research interests include control applications in power systems, distributed generation, renewable energy integration, and smart grids.

Flah Aymen was born in Tunisia in Gabès. Dr Flah his Ph.D in electrical engineering from National school of engineering of Gabès in Tunisia. He received MS.c. in automatic and intelligent techniques from the same school. He was a supervisor on a lot of scientific projects since 2009. He has extensive experience in motor control applications and intelligent control methods. Since 2012, Electrical vehicle, as power management and control, and renewable energy are his research fields.

Denis Butusov is a Head of Youth Research Institute and an Associate Professor in St. Petersburg Electrotechnical University "LETI". His research interests include nonlinear dynamics, geometrical integration, signal processing, ODE solvers, chaos theory and applications.

Lina Carvajal-Prieto is currently a fulltime Professor of Marketing in the Faculty of Business at Universidad Piloto de Colombia, Bogotá, Colombia. She earned Bachelor of Basic Education with Major in Humanities from Universidad Pedagógica Nacional, B.S. in Marketing Engineering from Universidad Piloto de Colombia, master's in administration from Universidad Nacional de Colombia. She has published journal papers. Her research interests include marketing, image, identity, satisfaction and education.

Basant Chourasia is working as a professor in Department of Industrial and Production Engineering. Achieved PhD from IIT Delhi.

Bassel Daher, Ph.D., is an Assistant Research Scientist at the Texas A&M Energy Institute, where he leads its Convergence Research Incubator. Daher is also a Research Fellow at the Institute for Science, Technology & Public Policy (ISTPP) of The Bush School of Government and Public Service and an Adjunct Assistant Professor at the Department of Biological and Agricultural Engineering. Daher is passionate about building bridges between research disciplines and cross sectoral stakeholders with the goal of arriving to a future that is more sustainable, equitable, and resource secure for all. Daher's research focuses on developing tools to catalyze evidence-based multi-stakeholder dialogue around the trade-offs associated with technological, policy, and social interventions for addressing the interconnected water, energy, and food (WEF) security challenges. He is particularity interested in bridging natural and social sciences methodologies, with the goal of unlocking new potential to address these interconnected WEF challenges. Daher focuses on doing so guided by circular economy principles and in the context of implementing the Sustainable Development Goals (SDGs). Daher has co-authored over 30 journal articles, book chapters, and policy briefs related to the interconnected water-energy-food securities and nexus assessment tools. These include with Chatham House, Sustainability, Water International, Science of the Total Environment, Environmental Science and Policy, and for multiple United Nations reports. Daher is a frequent speaker and moderator at academic and non-academic events, including the International Conference for Sustainable Development, Stockholm World Water Week, World Water Forum, World Water Congress, American Geophysical Union, American Society of Agricultural and Biological Engineers, and Water Innovation Labs. Daher's professional experience includes diverse themes, employment and internships in Lebanon, Qatar, Poland, France, and the USA (Indiana and Texas). Daher's research excellence was recognized with the 2019 Biological & Agricultural Engineering Department's Excellence Award for Outstanding Research. In 2017, he was the 2017 recipient of the Montgomery Award, given by the Office of Graduate and Professional Studies to those student leaders making major contributions to academic opportunities and quality of life of their fellow graduate students at Texas A&M University. Bassel was Research Associate at Texas &M's Water-Energy-Food Nexus Research Group at the Department of Biological and Agricultural Engineering (2014-2020), and served as Project Coordinator at Texas A&M's Water-Energy-Food Nexus Initiative (2015-2018). He was Research Associate at Qatar Environment and Energy Research Institute (2012-2014). Daher holds a PhD in Water Management and Hydrologic Sciences from Texas A&M University (2019), an MSE in Biological and Agricultural Engineering from Purdue University (2012), and a Bachelor of Civil Engineering from the American University of Beirut (2010).

Palash Dutta received his M. Sc. Degree in Mathematics from Dibrugarh University, Dibrugarh, India, in 2006, and his M. Phil. and Ph. D. degree in Mathematics from the same University in 2009 and 2012 respectively. He is an Assistant Professor in the Department of Mathematics, Dibrugarh University, since 2013. His research interests are uncertainty modelling using possibilistic approaches and decision making under the situations of uncertainty.

Fatma Hanafy El-Fouly is currently working as assistant professor at the department of communication and computer engineering, higher Institute of Engineering, El-Shorouk Academy. She received her PhD degree in electronic engineering from Menoufia University, Menouf, Egypt in 2016. Her current research interests include wireless sensor networks and digital communication fields.

Marc Haddad is an Associate Professor in the Department of Industrial and Mechanical Engineering at the Lebanese American University. He earned B.E. and M.Sc. degrees in Aerospace Engineering and a M.Sc. in Transportation Systems Engineering from the Georgia Institute of Technology, and a PhD in Technology Management and Policy from the Engineering Systems Division at the Massachusetts Institute of Technology (MIT). His research interests include systems modeling and simulation for policy analysis with applications in aviation, transportation and energy systems, and applied research on the use of systems thinking and lean thinking approaches in industrial and business contexts. Dr. Haddad is also an entrepreneur with experience in technology startups, and he has over 10 years of professional experience as an engineer, consultant and project manager in the aerospace and transportation industries.

Milton M. Herrera holds a BSc degree in Production Engineering and MSc degree in Industrial Engineering at the Universidad Distrital Francisco José de Caldas. Also, he holds a PhD in Model based policy and public management at the Università Degli Studi di Palermo. He is an Assistant Professor in Production and Logistics and Researcher at the Economic Sciences Research Centre at the Universidad Militar Nueva Granada. His research interests are system dynamics modelling and the energy supply chains.

Fotsin Hilaire Bertrand is a Professor of physics and electronics at the University of Dschang, Cameroon, where he is the Deputy Head of the Condensed Matter, Electronics, and Signal Processing Laboratory. He was awarded the "Doctorat de Troisième Cycle" degree in physics (electronics) from the University of Yaoundé I (Cameroon) in 2000. In 2005, he was awarded the "Doctorat d'Etat" degree from the University of Yaoundé I, Cameroon. He served as an Assistant Lecturer at the Physics Department, Faculty of Science, University of Dschang, Cameroon, from 1996 to 2000. From 2000 to 2009 he worked as a Lecturer in the same Department. From 2004 to 2005 he was a Visiting Scientist at the Centre de Recherche en Automatique de Nancy (CRAN) in France. In 2009, he became an associate professor. In 2016, he became a full professor of Physics. His research interests include nonlinear dynamics, focusing on chaos control and synchronization in electronic circuits. Professor Fotsin is the author and coauthor of more than 150 scientific articles and chapters in the area of nonlinear dynamics, chaos control and chaos synchronization, and renewable energies.

Nermine M. Khalifa got her PhD degree in Engineering & Information Science in (2010) from Middlesex University, London, UK. She became an associate professor in College of Management & Technology in (2014) and holds the post of Deputy Dean for postgraduates studies and scientific research since (2015) till present. She is specialized in E-business, E-commerce application and Supply

Chain Management. She had many publications in the field of E-commerce, E-Supply Chain, Enterprise Resource Planning, E-Customer Relationship Management, Performance measures, System dynamic & Simulation, RFID and Global Supply Chain and Software Engineering. These papers have been published in international conferences, well-known journals and book chapters. She contributed as a reviewer in international conferences, Journals and book chapters such as: IADIS Conferences, IBIMA Conferences, ICOSCM & ICSCMIS conferences, System Dynamics Conferences, International Journal of Supply Chain, Journal of Supply Chain, IGI publications.

Ashwani Kharola completed his PhD in Mechanical Engineering from Graphic Era University, Dehradun. He received B.Tech (with Honors) in Mechanical Engineering from Dehradun Institute of Technology, Dehradun in 2010 and M.Tech in CAD/CAM & Robotics from Graphic Era University, Dehradun in 2013. He obtained Silver Medal in M.Tech for (2011-13) batch. Currently he is working as Assistant Professor in Tulas Institute, Dehradun. Earlier he has worked as Senior Research Fellow (SRF) in Institute of Technology Management (ITM), One of premier training institute of Defence Research & Development Organisation (DRDO), Ministry of Defence, Govt. of India. He is also pursuing PhD in Mechanical Engineering from Graphic Era University (Deemed University), Dehradun. He has published many National/International papers in peer reviewed ISSN Journals and IEEE Conferences. His current areas of work includes Fuzzy logic reasoning, Adaptive Neuro-fuzzy inference system (ANFIS) control, Neural Networks, PID, Mathematical Modeling & Simulation.

Habib Kraiem received the Ph.D. degree in electrical engineering on October 2010 from the National Engineering School of Gabes, Tunisia. He joined the Electrical Department, Higher Institute of Industrial Systems Gabes, Tunisia. He is currently with the Department of Electrical Engineering, Faculty of Engineering, Northern Border University, Saudi Arabia. His current research interests include the areas of power electronics, machine drives, automatic control and renewable energy.

Atarsia Loubna in 2009 received the Engineer degree from the University of Mohamed CherifMessaadia Souk Ahras. Algeria. In 2014 received the M.S. degrees in electrotechnical, Option control of electrical machine. In 2015 inscription in doctor's degree, in Oum El Bouaghi University, Algeria Associate teacher in Souk Ahras University from October to 2015.

Rania Gamal Mohamed was born in Giza, Egypt, in 1985. Received her B.Sc. in 2008 from the Electrical Machines and Power System department, Shoubra faculty of engineering, Zagazig University-Benha Branch, Egypt. She received M.Sc. degree in 2013 from faculty of engineering, Cairo University, Egypt and the Ph.D. degree in electrical engineering from the - Al-Azhar University, Cairo, in 2017. From 2008 to 2011, she was a technical support engineer at Mitsubishi Company (Elevator & Escalator). From 2011 to 2013 she was a demonstrator at the Electrical Machines and Power System department, Thebes Academy, Cairo, Egypt. From 2014 to 2015 she was an assistant lecturer at Pyramids Higher Institute for Engineering and Technology, 6th October, Giza, Egypt. Currently, she is a lecture at Marg High Institute for Engineering and Modern Technology, Cairo, Egypt. Her research interests include modeling and control of interconnected power systems, Renewable Energy, Nonlinear control, and Power system optimization.

Mohamed Ahmed Ebrahim Mohamed received his B.Sc., M.Sc. and Ph.D. degrees in electrical engineering from Faculty of Engineering at Shoubra, Benha University, Cairo, Egypt, in 2004, 2009 and 2013 respectively. Mohamed took up the rank of Demonstrator, Assistant Lecturer and a Lecturer in Benha University in 2005, 2009 and 2013, respectively and associate professor since 2018. His research interests include analysis, design, and control of electric power systems, new and renewable energy applications. He published 63 scientific papers and 5 book chapters in international book series. He did several post-doctoral research missions with Federation of Research of Fuel Cells and the University of Technology of Belfort-Montbéliard, Belfort, France. He is the PI and coordinator of different Egyptian-French projects. He is a reviewer for several IEEE Transactions, IET and different Elsevier journals.

Mohamed Ahmed Moustafa Hassan is Professor (Control of Power Systems) in Electric Power Dept., Cairo University. Also He was awarded a DAAD Scholarship to study in Germany for his PhD program. Furthermore, He worked in many universities inside Egypt (Cairo University, AUC, Suez Canal University, Banha University, South Valley University, PHI, IAEMS), and outside Egypt (University of Calgary in CANADA, UAEU in UAE, TSI in UAE, AUST in UAE, and JEGC in KSA). He is offering many practical consultations for different sectors in Egypt like Water utilities (HCWW, and its affiliated companies, as well as October Market, Some Cities Stadium), especially in the field of Automation, Fuzzy Logic, SCADA, and DCS. He attended many international scientific conferences. He published more than 145 scientific articles in journals as well as in International Conferences inside Egypt and outside Egypt (in UK, Belgium, Canada, UAE, Cyprus, Russia, Italy). He is being supervising c.a 20 M.Sc. thesis at Cairo University, Egypt (26 Finished). He examined many M.Sc. theses (c.a. 60) at different Universities (Cairo Uni., Banha Uni., Helwan University, Al Azhar Uni.) in Egypt.

Lazaros Moysis received the B.Sc., M.Sc. and Ph.D. degrees in 2011, 2013 and 2017 from the Department of Mathematics, Aristotle University of Thessaloniki, Greece. He is currently a post-doctoral researcher at the Physics Department, Aristotle University of Thessaloniki, at the Laboratory of Nonlinear Systems - Circuits and Complexity. His research interests include the theory of control systems, descriptor systems and chaotic systems, and problems like controllability, observer design, synchronization and its application in secure communications, encryption and robotics.

Mohamed Naoui was born in Nefta, Tunis, in 1991. degrees in Electrical Engineering from the University of Gabès, Tunisia, in 2015. PhD student in electrical engineering at the research unit Proceeds, energies, environment and electrical systems, National Engineering School of Gabès.

Rami Otayek holds a B.E. in Industrial Engineering from the Lebanese American University. He has over 5 years of professional experience in implementing lean manufacturing in different industries. He has interests in applied research using system dynamics modeling, discrete event simulation and lean manufacturing principles.

Mrutyunjaya Panda holds a Ph.D degree in Computer Science from Berhampur University. He obtained his Master in Communication System Engineering from Sambalpur University, MBA in HRM from IGNOU, New Delhi, Bachelor in Electronics and Tele- Communication Engineering from Utkal University in 2002, 2009, 1997 respectively. He is having 21 years of teaching and research experience. He is presently working as Reader in Computer Science Department, Utkal University, Vani Vihar Bhubaneswar-4, Odisha, India. He is a member of MIR labs(USA), KES(Australia), IAENG(Hong Kong), ACEEE(I), IETE(I), CSI(I), ISTE(I). He has published about 75 papers in International and national journals and conferences. He has also published 8 book chapters to his credit. He has authored 2 text books and 4 edited books. He is a program committee member of various international conferences. He is acting as a member of various international journals. His active area of research includes Data Mining, Intrusion detection and prevention. Social networking, Mobile Communication, wireless sensor networks etc.

Eric Pardede received his PhD in Computer Science from La Trobe University, Melbourne, Australia in 2007. He is currently a Senior Lecturer in the Department of Computer Science and Information Technology at La Trobe University, Australia. He has wide range of teaching and research experience. His current research interests include higher education pedagogy, data analytics and entrepreneurship.

Rabie Ramadan is an associate professor in the fields of Internet of Things (IoT), Mobile Computing, and Computational Intelligence. He got his Ph.D. from Computer Engineering from Southern Methodist University (SMU), Dallas, Texas, USA. He has work experience for more than 27 years. He has industrial experience as an executive manager for a software development company, "Ibda" and "Smartec" companies and worked at Bank of America Securities, Dallas, TX, US, for a three of years. He is leading three research groups in the fields of IoT and Computational Intelligence and an author of 125 journal and conference articles in the field of IoT, Sensor Networks, Mobile Computing, and Computational Intelligence. He also served as a chair, co-chair, or TPC for several conferences, Also, he is an editor of WAS SCIENCE NATURE (WASSN) and associate editor of the International Journal of System Dynamics Applications (IJSDA). He has different roles in many journals including IEEE Transactions on Industrial Informatics and IEEE Access.

Toufouti Riad received his Engineer, Master Doctorate and Professor degrees in Electrical Drive from the University of Constantine Algeria, respectively, in 1999, 2003, 2008 and 2017. Since 2003, he is currently a Professor of Electrical Engineering in Department of Electrical Engineering at Souk Ahras University, Algeria. His research interest includes on Electric Drives and Hybrid Power System, and Smart Grid.

Meziane Salima was born in Algeria. She received her Engineer, and Master degrees respectively in 2000, 2003 in Electrical Drive. In 2009 she received her Doctorate degree in Electrical Engineering from University of Constantine Algeria. Recently, she is a professor and she is member in Electrotechnical and renewable energy at Souk Ahras University Algeria.

Lassaad Sbita was born in Hammam Lif, Tunisia, in 1962. He received the B.E. degree in electrical engineering from the University of Tunis, Tunisia, in 1985, and the D.E.A. and Thesis degrees in electrical engineering from the ENSET of Tunis, Tunisia, in 1987 and 1997, respectively. In 1988, he joined the Department of Electrical Engineering, ENIS, University of Sfax, as a Professor Assistant, and in 1991 with the Department of Electrical Engineering, ENIG, University of Gabes, became an Associate Professor in 1998, and a Professor in 2009. His current research interests include power electronics, electrical machines and drives, active filters, renewables energies, electrical grid and power quality. Dr. Sbita is a director of a research unit of PV, wind and geothermal systems, which ministry code is "url1es82". He was the General Chair of the IEEE International Conference on Green Energy Conversion Systems (GECS 2017).

Narendranath Shanbhag received his Masters degree in Information Systems Management from La Trobe University, Melbourne, Australia in 2013. He is currently pursuing the Ph.D. degree in Computer Science at La Trobe University, Melbourne, Australia. His research is in the area of entrepreneurship with a focus on software product startups.

Sanjay Soni is working as Assistant Professor in Industrial and Production Engineering at Jabalpur Engineering College. Working in the area of System Dynamics from last 10 years.

Dionysios Sourailidis is a PhD in Physics candidate at Aristotle University of Thessaloniki (Greece).

Hussein A. Taha is a Ph.D. student at Polytechnique Montréal University, Canada since 2019. He received his BSc and MSc from Cairo University, Electrical Power and machines engineering. He worked as an undergraduate and graduate professional engineer in SIEMENS AG, ASIC Industrial Automation, and SAMSUNG Electronics industrial organizations. He worked as a research assistant in multidisciplinary research projects to serve the industry and this research were published in journal and conference papers. His research interests are Artificial Intelligent control systems, conditional monitoring and fault diagnosis, and energy management.

Christos Volos's research interests are mainly but not exhaustively in area of applied complexity in systems as well as in chaotic electronics and their applications. Research regarding nonlinear dynamical systems with hidden attractors is of great interest. He has collaborations with several laboratories (Politechnika Łódzka-Poland, University of Catania, University of Aberdeen-UK, University of Wisconsin-USA, St. Petersburg State University-Russia, China University of Geosciences, City University of Hong Kong, Instituto Nacional de Astrofísica, Óptica y Electrónica-Mexico, Benemérita Universidad Autónoma de Puebla-Mexico, Sakarya University-Turkey, Cairo University-Egypt, Vel Tech University-India, Amirkabir University of Technology-Iran). Dr. Volos has been invited to give lectures in the areas of his expertise i.e. nonlinear systems and circuits with hidden attractors and their applications. He has been a co-author of a book related with systems with hidden attractors. Dr. Volos has authored or co-authored 132 papers in international scientific journals, 51 papers in proceedings of international conferences, 16papers in proceeding of national conferences and 38 book chapters. Also, he is a co-author in one book and editor or co-editor in 7 books. He has also written or co-written teaching notes for an undergraduate course. His published work has received more than 1850 citations while he enjoys an h-index = 24.

Dave Zelinka, Ph.D., is an expert in systems thinking, system dynamics modeling, and data science. He currently consults for various organizations in peace engineering, systems approaches to sustainability, industrial ecology, and many issues related to global development, issues for which he is passionate about. Additionally, his postdoctoral work is at the interface of data science, geospatial analysis, the global power sector, and climate change. He is also researching for his book on applied complexity science, which is an expansion of his PhD dissertation entitled, Systems Approaches to Transdisciplinary Sustainable Development Issues. Dave holds a doctorate in civil systems engineering and a second master's degree in global engineering from the Mortenson Center for Global Engineering in the Department of Civil, Environmental, and Architectural Engineering from the University of Colorado (CU-Boulder). He holds his first MS in environmental and sustainability engineering from the University of Colorado (UC-Denver) and a BS in environmental and ecological engineering from Purdue University in West Lafayette, Indiana. Dave's research involves applied complexity science and dynamical modeling. He specializes in the water-energy-land-food (WELF) nexus, the United Nations' Sustainable Development Goal (SDGs), and engineering approaches to analyzing complex interconnected issues in socio-political systems, particularly political corruption, conflict, and urbanization. His research has been published in the International Journal of System Dynamics Applications the International Conference of the System Dynamics Society, International Sustainable Development Research Society Conference, and the American Society for Engineering Education Conference. He also has two book chapters coming out as part of Modeling, Analysis, and Control of Complex Systems and one in Encyclopedia of the United Nations Sustainable Development Goals.

Index

F

G

H

I

K

L

M

N

O

P

Q

R

S

T

U

V

W

www.ingramcontent.com/pod-product-compliance
Lightning Source LLC
LaVergne TN
LVHW080617300926
849162LV00005B/1187
* 9 7 8 1 7 9 9 8 5 7 8 8 4 *